ELECTRON FLOW IN ORGANIC CHEMISTRY

ELECTRON FLOW IN ORGANIC CHEMISTRY

A Decision-Based Guide to Organic Mechanisms

Second Edition

Paul H. Scudder
New College of Florida

WILEY

A JOHN WILEY & SONS, INC., PUBLICATION

Published by John Wiley & Sons, Inc., Hoboken, New Jersey

Published simultaneously in Canada

Library of Congress Cataloging-in-Publication Data

Scudder, Paul H.
Electron flow in organic chemistry : a decision-based guide to organic mechanisms / Paul H. Scudder, New College of Florida. — Second edition.
pages cm
Includes index.
Summary: "The revised 2nd edition builds on and improves this legacy, continuing the rigorous mechanistic approach to organic chemistry. Each mechanistic process is divided into its basic units, the dozen common electron flow pathways that become the building blocks of all the common mechanistic processes" — Provided by publisher.
ISBN 978-0-470-63804-0 (pbk.)
1. Chemistry, Organic. 2. Chemical bonds. 3. Charge exchange. I. Title.
QD251.3.S38 2013
547'.128—dc23 2012025745

Printed in the United States of America.

10 9 8 7 6 5 4 3 2 1

Contents in Brief

Contents

Preface

TO THE STUDENT

Critical Thinking Approach

Organic chemistry courses have a well-deserved reputation for being highly memorization based. But it does not have to be so. An organic chemistry course is a great place to learn critical thinking. My students kept asking me, "Why didn't it do this instead of that?" Soon I was mapping out alternatives and getting them to decide the answers to those questions. My course had evolved from "know the answer" to "explain the answer," and also "predict what would happen here." As the course progressed, my students developed a good chemical intuition and felt they understood why reactions occurred. They could write reasonable mechanisms for unfamiliar reactions and predict what might happen for reactions they had never seen. How to learn organic chemistry by using this critical thinking approach is the essence of this book.

Dealing with Informational Overload

Ideally, college is where you learn to think, but there is often so much factual material to cover in an organic chemistry course that memorization can take over. In the face of the sheer mass of content to be learned, the development of the necessary skills critical to becoming a scientist—logic and analysis—can be lost. This "tyranny of content" in an ever-growing field means that the "why" of the field may get swamped under a flood of facts. The mystery of a good puzzle, the draw of the sciences, can also be lost.

You might approach organic chemistry with the idea that only memorization can get you through it; this may have been true many years ago, but it is not the case now. With memorization, if you have not seen it before, you are usually in trouble. Information learned through memorization is also the first to be forgotten, and the volumes of information required in organic chemistry seem to be lost particularly rapidly. This loss can even occur before the cumulative second semester final.

An Expert Systems Approach to Organic Chemistry

To "explain the answer," you need to know what the alternatives are, and why one of them succeeds and others fail. You need to "generate and select" alternatives, which is the essence of a good critical thinking process. The map of all alternatives from the start point can be represented as a tree, and is our "problem space." You need an efficient way to navigate this problem space to the correct answer. For that, you need a small set of essential principles, or "control knowledge," to guide the route selection decisions toward the correct answer. Good intuition arises from the automatic use of control knowledge to guide the decision process.

The impressive advantage that a decision-based approach to organic chemistry has over memorization is that it engages you in critical thinking, a skill everyone can benefit from improving. This approach allows for extrapolation into the unknown and provides room for the joy of discovery. If you are going to learn how to think in organic chemistry, you need to know what the alternative paths are and how to decide between them.

The development of decision-making algorithms for artificial intelligence systems has led to a new way of thinking about the decision process. Computers have to make use of decision trees and problem spaces, where all possible choices are examined and weighed and the best of the options selected. This same methodology can be applied to organic chemistry. You will have to learn about problem spaces, search trees, and methods to decide the best path. This text extracts the essence of the field: the conceptual tools, the general rules, the trends, the modes of analysis, and everything that one would use to construct an expert system. It explains and makes use of analysis tools more common to expert systems, but rare in undergraduate organic chemistry texts. If you can internalize this expert-system decision process, you will develop a chemical intuition and are well on your way to becoming an expert yourself.

Unique to This Text

This book organizes reactions by similar processes, as you would in an expert system. Reactants are grouped into generic groups that behave similarly. By being able to classify hundreds of different structural types into a small number of electron sources and sinks, you take control of the information overload and make it manageable. You will be able to make a good guess at how new reaction partners might behave.

All mechanisms are viewed as composed of simple elemental processes, the electron flow paths. Even the most complex reactions can be simplified into a sequence of basic electron flow paths. These elemental processes are limited in number and are repeated, again and again, making them easier to both learn and retain. In this way, a mere dozen electron flow paths can explain nearly all of the common reactions found in an undergraduate organic course. This decision-based book shows how to choose which of the dozen common electron flow paths are reasonable to use, and in what order. New reactions become puzzles to solve, not just another item to be memorized. Reactions are much easier to remember if you can understand how they work.

Motivation and Relevance

The most important question that you as a student have to answer is, "Why am I in this course?" Organic chemistry is a lot of work no matter what approach is used, and you will need to see the personal relevance in order to have the motivation to succeed. Premedical students need the ability to reason through complex problems; this is the essence of diagnosis. Biology majors need to have a good chemical intuition, so that they can understand the chemistry of life, what makes it work or malfunction. A good organic chemistry course will give you precisely these skills: good chemical intuition and the ability to approach and solve complex problems. Rote memorization will provide neither. The critical thinking skills and methods of analysis learned in a decision-based organic course are highly valuable and easily transportable to other areas.

TO THE INSTRUCTOR

Critical Thinking Approach

Critical thinking has become a major emphasis in undergraduate education. Science students respond well to being given puzzles to solve rather than content to memorize. Organic reaction mechanisms can provide the hook to interest students in analyzing and thinking like a scientist. "How does this reaction work?" We need to prepare our students to enter a world where content is easily accessible on the web, but critical analysis of all this content is not easy at all. We want our future scientists and physicians to be good at critical thinking, for the web will be at their fingertips to aid their recall. Adding this text as a critical thinking supplement to your organic chemistry course can make the course more important to students and should help them succeed.

This second edition provides students with something that they cannot get anywhere else: a chemical intuition based on learning and internalizing a cross-checked decision process. An important part of the scientific method (or diagnosis) is the ability to postulate a reasonable hypothesis, fitting the data. This text teaches students how to write reasonable reaction mechanisms, and assumes only a general chemistry background.

Unique Decision-Based Approach

To be able to teach students to make good decisions, we need to teach "control knowledge," which is the essence of a good intuition. These are checks of reasonability that include, among other things: stability trends, compatibility with the media pH, evaluation of energetics, and similarity to known processes. The second edition uses flowcharts and energy surfaces as problem space maps to help with illustration of these concepts, while continuing the rigorous mechanistic approach to organic chemistry.

Unique to this text is the concept of mechanisms being built from a limited number of elementary electron flow pathways, and the concept that learning to assemble these pathways in a reasonable manner is all that is necessary to master mechanisms in organic chemistry. The impressive advantage that a decision-based approach has over memorization is that it engages the student. The instructor can ask questions like, "Why did it go this way and not that way?" New reactions become puzzles to solve, not simply more items to memorize.

This text uses several concepts and tools not present in most undergraduate organic chemistry texts to aid in understanding the most difficult sections of the course. Hard-soft acid–base theory is used to guide decisions and to explain and predict the dual reactivity of many species. Energy diagrams and surfaces are presented so that students have a physical model to help with the more complex decisions.

An optional level of explanation is included that makes use of frontier molecular orbital theory to explain reactivity. A beginner who has difficulty with molecular orbital concepts can skip these sections without penalty.

Changes From the First Edition

Besides the usual clarifications and modifications necessary to bring the text up to date, the text has been expanded to reinforce a decision-based approach. There are more flowcharts, correlation matrices, and algorithms that illustrate decision processes. Energy surfaces, normally the domain of graduate texts, serve as concept maps and allow students to visualize alternatives. The text has been made more accessible to beginning students and meshes better with standard texts.

A new Chapter 3, "Proton Transfer and the Principles of Stability," has been added to thoroughly develop how structure determines reactivity using a reaction from general chemistry. Proton transfer mechanisms and product predictions are introduced, setting up the discussion of organic reactions.

A new Chapter 4, "Important Reaction Archetypes," was added so that the main mechanistic reaction types that form the core of an organic course are emphasized first. This chapter shows the problem space for each archetype and how reactant structure influences the favored route. Electron flow paths are introduced gradually with these reaction archetypes. The ΔpK_a rule is used for deciding reasonable reaction energetics.

The later chapters develop a general approach to all organic reactions by showing how to focus on the most reactive centers and choose the best route. This book provides tools for handling large amounts of information. It emphasizes the "why" of organic chemistry in order to help make sense of all the material. Common errors are now placed within the appropriate sections. A new Chapter 12, "Qualitative Molecular Orbital Theory and Pericyclic Reactions," collects most of the more difficult orbital control related topics into a final chapter. A larger collection of important tools is gathered together in the Appendix, including a new section on structure elucidation strategies.

More Biochemical Examples

Biochemical examples give added relevance for the biology majors and premedical students who make up a significant portion of undergraduate organic chemistry students. The elegance of biochemical processes in optimizing a low-energy route can be appreciated and understood by looking at mechanisms. These examples also provide a bridge if this text is to be used for review of organic chemistry before a biochemistry or enzymology course.

Online Aids

No matter what you hand out on the first day of class, your exams are your syllabus. Unfortunately, the students' universal test of importance of any material is, "Is this going to be on the exam?" Therefore, if you do not alter the way you test on the material, you have not significantly changed your course. In addition to the answers to the exercises, material is online at the Wiley instructor's website for this text to aid in implementing a decision-based approach to organic chemistry.

Applications

This textbook is designed to be flexible in its instructive role. It can be used in the major's sophomore undergraduate organic chemistry course as a short, highly mechanistic supplemental text. It can be used as the primary text in an advanced undergraduate or beginning graduate course in organic reaction mechanisms, or as a supplemental review text for graduate courses in physical organic chemistry, enzymatic reaction mechanisms, or biochemistry. This text is the product of over thirty years of teaching organic chemistry at New College, the Honors College of the State of Florida.

Acknowledgments

I would like to thank all that have helped to bring the first and second editions of this book to fruition, especially my father, Prof. Harvey I. Scudder, who helped me refine an algorithm-based teaching approach, and my Ph.D. mentor, Prof. Barry M. Trost. I am indebted to my students, who helped me work through the many versions of this text, to my colleagues at New College, and to the reviewers of this manuscript. I will maintain an errata list and encourage anyone to send me errors not on the list. I gratefully acknowledge the encouragement of my parents and my wife, son, and daughter, who inspired me to keep writing in the face of an ever-growing project. Finally, I would like to thank all those at John Wiley & Sons who made the publication of this book possible.

This book is dedicated to my students, who have taught me to question everything.

1

BONDING AND ELECTRON DISTRIBUTION

Electron Flow In Organic Chemistry: A Decision-Based Guide To Organic Mechanisms, Second Edition.
By Paul H. Scudder

1.1 THE DECISION-BASED APPROACH TO ORGANIC CHEMISTRY

As mentioned in the Preface to the Student, this decision-based approach to organic chemistry is modeled after the scientific method. A good hypothesis is just a reasonable guess. You will learn how to recognize alternatives and how to judge which alternative is most reasonable. This is the essence of critical thinking, a crucial skill for scientists, physicians, and life in general. You will develop a good intuition, for intuition can be considered just an internalized decision process. We will use the artificial intelligence concepts of problem spaces and tree searches to help you develop this intuition for organic chemistry.

1.1.1 Introduction to Problem Spaces

If you were planning a road trip across the US, you would need a map of the highways. It would allow you to see all routes from your starting city to your goal city. You would then choose the best route for what you wanted to see and the time you had for the trip. This is exactly the process you want to go through for understanding organic chemistry. We need a map and the ability to choose the best route. Our maps of problems are called problem spaces and are often shown as trees, with a decision to be made at each branch point.

Figure 1.1A illustrates a generic problem space and some of the approaches to working from the start at the top of the tree at point S down to the correct answer. If the correct route is from S to A to D to I, some students may attempt to memorize "S goes to I" without understanding the process involved. In order to have a greater understanding, instructors spend book and class time explaining a correct route to the answer. However, students may see the "lightning strike" to product as shown in Figure 1.1B but not understand the choices that were made along the way. Instructors may feel that, if shown enough times, students will be able to do it themselves. But we can't expect you to make good choices if you have not been taught how to make those decisions. When you work problems and depart from a reasonable route, you need to see your answer discussed to understand how better choices could have been made.

1.1.2 Introduction to Tree Searches

There are several ways to search a problem space tree. Computers often do a "breadth-first" search by considering every possible route and selecting the most efficient one. We might do this, if we planned our road trip with a map of all major roads. However, while working most science problems we are not provided with a complete problem space map, but rather must generate one on our own. We need to employ a different search strategy. New students often approach science problems "depth-first" by going with the first thing that occurs to them and hoping to get lucky (Fig. 1.1C). If they don't reach a correct answer, they back up to the last branch and try another route. However, the more complex the problem space is, the less successful this depth-first search process is. Organic chemistry can have complex highly branched problem spaces, so we again need a different tree search strategy. The tree search strategy that excels in complex problem spaces is a "best-first" search, and that is the focus of this text.

A "best-first" search process is shown in Figure 1D, where alternatives are generated and the best alternative is picked at each branch point. Mastering and internalizing this "generate and select" process is crucial to becoming a good scientist or physician. After all, diagnosis is just a physician's search of a complex problem space. An internalized best-first search becomes the essence of a good intuition. If you can take this "generate and select" process and apply it to other fields, it may be the most valuable thing you can take away from a course in organic chemistry.

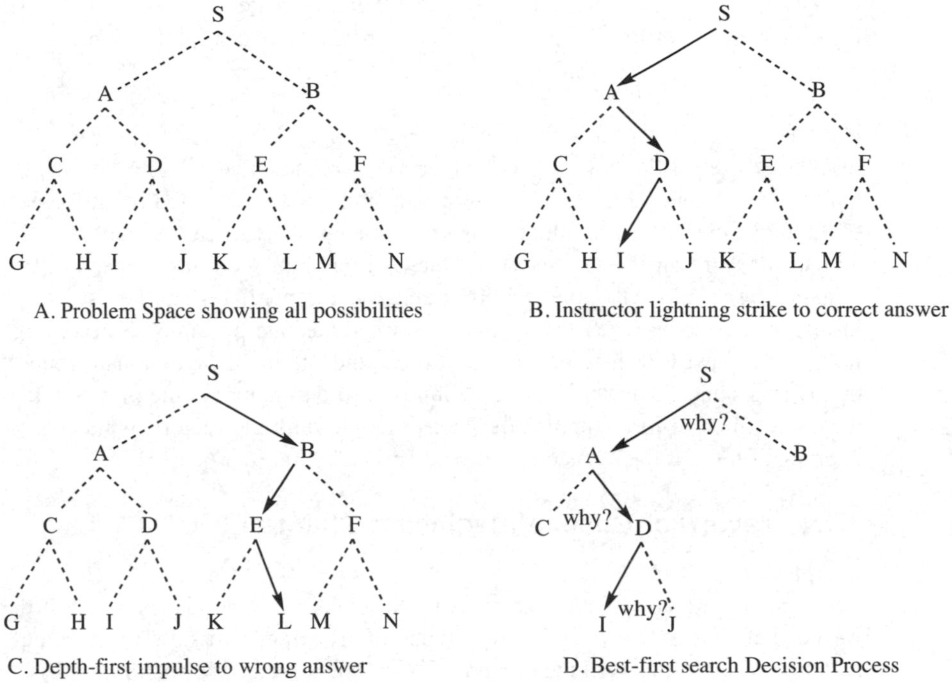
A. Problem Space showing all possibilities
B. Instructor lightning strike to correct answer
C. Depth-first impulse to wrong answer
D. Best-first search Decision Process

Figure 1.1 A generic problem space and some strategies for tree searches in the problem space.

1.1.3 Introduction to Control Knowledge

First we need to recognize when we are at a branch point and that a decision needs to be made. Then, to be able to make good decisions in a best-first search, we need to master and utilize "control knowledge." These are checks of reasonableness to help you select the best route. For our road trip, this may be the quality of the highway, the season of year, and how much time we have. For organic problems, control knowledge includes the stability of intermediates, reactivity trends of reactants, whether the medium is acidic or basic, evaluation of energetics, and similarity to known processes. We need to understand the principles of stability that will determine whether a route is of lower energy. This text will use flowcharts and energy surfaces as problem space maps to illustrate what decisions need to be made. The good news is that the control knowledge that applies to thousands of organic reactions is limited and relatively easy to understand.

An important part of control knowledge is to recognize the boundaries of the problem. When we say an answer is in "the ballpark," we are just testing for reasonable bounds. If a friend bragged he or she could run a mile in a minute, you would have good reason not to believe, considering the fastest runners barely break a four-minute mile. You have tested incoming data with control knowledge before deciding, the essence of critical thinking. Table 1.1 illustrates a few common values and upper bounds that are useful in organic chemistry. Each of the dozen electron flow path alternatives with which we build our explanations of more complex organic reactions has boundaries that need to be considered. These path limitations are included as each electron flow path is introduced and are summarized in Chapter 7.

Table 1.1 Common Values and Upper Bounds for Common Chemical Reactions

Measurement	Common for Organic	Other Examples	Upper Bound
Density	Often <1 g/cm^3	1.0 g/cm^3 (water)	22.61 g/cm^3 (osmium)
Concentration	<1 mol/L	55.6 mol/L (water)	118.9 mol/L (osmium)
Temperature	–78 to +220°C	25°C (room temp.)	1175°C (quartz melts)

Another important aspect of control knowledge is the use of the reactivity trends of reactants to select the hottest site for reaction in a molecule. This allows us to focus on only the most important part of the molecule and not be distracted by differences in parts of the molecule that do not matter, like the unreactive hydrocarbon skeleton. In this way you won't "slip on the grease" when the hydrocarbon section changes but the hot spot remains the same. Also, the stability trends of intermediates can be used to predict the lowest-energy route when two or more intermediates are possible. Since energy is often limited, the lowest-energy route is the fastest and often the predominant route. Stability trends of products determine the route in reversible systems, as the most stable product is the one formed. More complex decisions involve multiple factors, which contribute to a "tipping point" for the decision, as discussed in Chapter 9.

1.1.4 Preview of Goals of Beginning Chapters

Chapter 1

Since favorable organic reactions usually break weak bonds and make stronger ones, we need to understand the different types of bonding, and how to represent both the structure of molecules and the process of bond breakage and formation.

Chapter 2

To be able to judge the most favorable route, we must understand qualitatively both thermodynamics and kinetics. Therefore we need to understand the process of bond making and breaking, what makes bonds strong or weak, and how the energetics of a process makes some more favorable than others.

Chapter 3

To ease into the discussion of organic reactions, it is best to start with a familiar reaction like proton transfer and thoroughly understand how structure determines reactivity. If we can understand proton transfer and the energetics associated with various possible routes, we have gone a long way toward understanding organic reactions. Organic reactions make use of the same principles and often have proton transfer as a first step.

Chapter 4

Since just a few mechanistic reaction types form the core of an organic course, we strive to understand these important reaction archetypes. What does the problem space for each archetype look like, and how do the reactant structure and reaction conditions influence the most favored route?

Later Chapters

The later chapters develop a general approach to all organic reactions by showing how to focus on the most reactive centers and choose the best route. This book provides tools for handling large amounts of information. It will emphasize the grammar of organic chemistry that will help you make sense of all the material. Without this essential logic to guide you, organic chemistry can become a flood of disconnected reactions to be memorized that will challenge even those good at memorization.

In this book, reactions are organized by similar process. Reactants are gathered into generic groups that behave similarly. A reaction mechanism is our current best guess at all the steps for how the reaction actually occurs. Even the most complex reactions can be broken down into a sequence of basic electron flow steps, called electron flow pathways, which become the building blocks of all the common mechanistic processes. The purpose of this text is to teach the understanding and proper combination of these building blocks. Just a dozen electron flow paths can explain almost all of the common reactions found in an undergraduate organic chemistry course. The mechanistic problem space becomes the correct use of these pathways. The best advantage that a decision-based approach has over memorization is that it engages the mind. We can ask questions like, "Why did it do this and not that?" New reactions become puzzles to solve, not another thing to be memorized.

Our approach is to understand why and when a reaction will occur and to establish a command of organic chemistry based more upon understanding the basics than upon memorization. The most impressive result of organizing the material by mechanistic process is that you will develop a sound chemical intuition and will be capable of making good educated guesses. In addition, the how and why of organic chemistry is far more interesting than the memorization of its components. We will make use of trends and general rules as tools to aid us in understanding what has happened in an organic reaction. These tools have their exceptions and limitations, but they allow us to develop an overall "feel" for organic reactions. We can worry about the exceptions, if necessary, after the bulk of the material is mastered.

Molecular orbital interaction diagrams and frontier molecular orbital theory are used in an additional, more advanced level of explanation (indicated by a sidebar) that can be skipped without penalty. Also indicated are the more mathematical sections of thermodynamics and kinetics that provide support for the formulas and can be skipped.

1.1.5 The Principle of Electron Flow

A reaction will occur when there is an energetically accessible path by which electrons can flow from an electron source to an electron sink. A reaction is a flow of electron density from an electron-rich region to an electron-deficient region (the obvious exceptions to this are free radical reactions, treated separately in Chapter 11). To know which regions are electron rich or electron deficient, we must be able to predict the distribution of electron density over a molecule.

A *nucleophile* ("nucleus-loving," $Nu:^-$) is a *Lewis base* (electron pair donor) that has an available electron pair for bonding. Nucleophiles act as electron sources and can be negatively charged like hydroxide anion or neutral like water. An *electrophile* ("electron-loving," E^+) is a *Lewis acid* (electron pair acceptor) that can accept two electrons to form a bond. Electrophiles act as electron sinks and can be positively charged like hydronium cation or neutral like bromine.

Any bond formed is the combination of a nucleophile and an electrophile. The most probable product of a reaction results from the best electron source bonding with the best electron acceptor. The curved-arrow notation (Section 1.4) allows us to describe the flow of electrons from source to sink.

The concept of flow is very important. Just as water flows under the influence of gravity, electrons flow under the influence of charge: from electron-rich atoms to electron-deficient atoms. A poor electron source will not react with a poor electron sink within a useful length of time. One might consider this a case of no pull and no push, so no appreciable electron movement occurs. A poor electron sink requires combination with a good electron source for a reaction to occur.

We need to answer the questions: What properties distinguish a good electron source, a good electron acceptor, and a good pathway for electron flow? There are relatively few pathways through which the common electron sources and sinks react.

The use of generic electron sources and sinks and generic electron flow pathways makes the similarities and interrelationships of the major reactions in organic chemistry become obvious. The electron flow pathways become the building blocks of even complex organic reaction mechanisms, so all the mechanisms seem to "flow" from first principles.

Before we explore the problem space for a simple proton transfer reaction, we need to understand the basics of bonding and define a consistent nomenclature. In order to use the electron flow paths, you first need to be able to keep track of atoms and electrons—write Lewis structures correctly and easily.

1.2 IONIC AND COVALENT BONDING

Before trying to understand stable molecules, let's try to understand stable atomic electron configurations. Figure 1.2 shows the valence electrons of the atoms of elements common in organic chemistry. An atom might achieve the stable filled shell electron configuration of a noble gas (group VIIIA) in any of three ways: lose electrons, gain electrons, or share electrons.

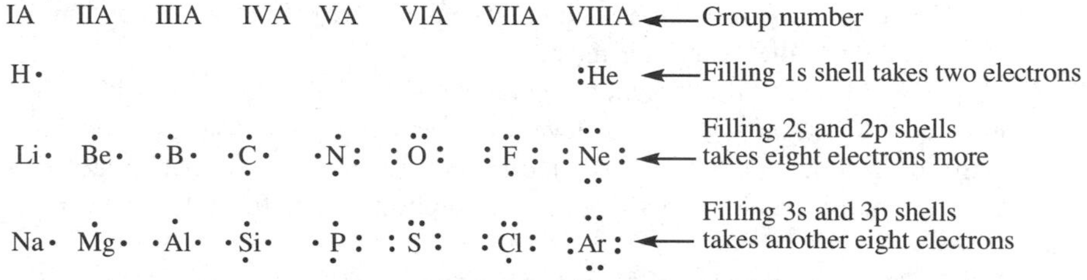
IA IIA IIIA IVA VA VIA VIIA VIIIA
Group number
H
He
Filling 1s shell takes two electrons
Li Be B C N O F Ne
Filling 2s and 2p shells takes eight electrons more
Na Mg Al Si P S Cl Ar
Filling 3s and 3p shells takes another eight electrons

Figure 1.2 The valence electrons of elements in the first three rows of the periodic table.

In the first row of the periodic table, the two valence electrons of helium fill the 1*s* shell, making it very stable. If lithium loses one of its three total electrons, it now has the helium duet and is stable as the Li^+ cation. The hydrogen atom needs two electrons to form the helium duet, and it can achieve that either by gaining an electron outright to form hydride anion or by sharing an electron with another atom. The first row is special in needing only an electron duet for stability; higher rows require an octet of valence electrons for stability. Because hydrogen can add one electron to form hydride, it is sometimes drawn in two places on the periodic table, in Group IA and in Group VIIA.

In the second row, neon has a complete valence shell with both the 2*s* and 2*p* orbitals filled (in addition to the filled 1*s* helium core). Sodium has one more electron than neon; losing that electron gives Na^+, which has neon's stable filled shell. If a fluorine atom gains one electron to become fluoride, it has the complete octet of neon and a filled shell.

To generalize, elements of Groups IA and IIA commonly lose their valence electrons to step back to the last noble gas electron configuration. For this reason it is normal to write all compounds of elements in Group IA and IIA as ionic. Halogens of Group VIIA can gain an electron to step forward to the next noble gas electron configuration of the halide ion. Groups IIIA through VIIA can gain stable valence octets by sharing electrons in covalent bonds.

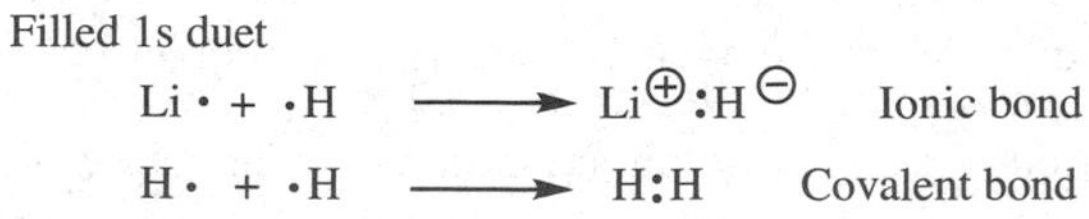
Filled 1s duet
Li + H
Li H
Ionic bond
H + H
H:H
Covalent bond

Filled 2s and 2p octet

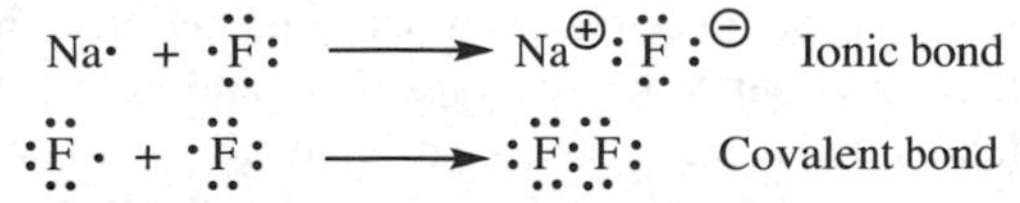
Ionic bond
Covalent bond

Ionic and covalent bonding are the two extremes of bonding. In a covalent bond, a pair of electrons is shared between two atoms. The shared pair of electrons is in a bonding orbital made from the overlap of an atomic orbital from each atom. For a strong covalent bond, these shared atomic orbitals must overlap well and must be similar in energy. In an ionic bond, that pair of electrons resides primarily on one of the two atoms, producing two oppositely charged ions that are attracted to each other. This attraction is expressed mathematically by Coulomb's law, which says the attractive force is directly proportional to the charge of each ion and inversely proportional to the square of the distance between them. A strong ionic bond has opposite charges very close to each other. Reactions tend to break weak bonds and make strong ones, so a reaction can be favored by the formation of strong ionic bonds or strong covalent bonds or both (the HSAB principle, Section 2.4).

Most bonds fall between these two extremes. Pauling defined *electronegativity* as the power of an atom in a molecule to attract electrons to itself. The electronegativity difference of two bonded atoms is an indication of how polarized, how ionic, the bond is. In homonuclear diatomic molecules like H_2 there is pure covalency, no polarization, for the electronegativity difference is zero. Common table salt, Na^+Cl^-, contains a good example of an ionic bond in which the electronegativity difference is large. Table 1.2 is a partial periodic table of the average electronegativities of commonly encountered elements. Elements more electronegative than carbon make up the upper right corner in the periodic table and are in boldface. Electronegativity is the result of increasing nuclear charge on the valence electrons. For example, the electrons in the 2*p* orbital get pulled closer to the nucleus in going from boron to fluorine and the nuclear charge increases. Likewise, Na^+ and F^- both have 10 electrons, but Na^+ is much smaller and its electrons are held tighter than fluoride due to sodium's greater nuclear charge, 11 protons vs. 9.

Table 1.2 Average Electronegativities of Selected Elements

H 2.20									
Li 0.98					B 2.04	C 2.55	**N** **3.04**	**O** **3.44**	**F** **3.98**
Na 0.93	Mg 1.31				Al 1.61	Si 1.90	P 2.19	**S** **2.58**	**Cl** **3.16**
K 0.82			Cu (I) 1.90	Zn 1.65				Se 2.55	**Br** **2.96**
				Cd 1.69					**I** **2.66**

The electronegativity of an atom will vary depending on what else is bonded to it. For example, the carbon atom of a CF_3 group is more electronegative than the carbon atom of a CH_3 group. In the CF_3 group, the three highly electronegative fluorines withdraw electron density from the CF_3 carbon atom, which in turn will withdraw more electron density from whatever it's bonded to.

Both the degree and the direction of the polarization of a bond can be predicted by the electronegativity difference. The bonding electron pair is more likely to be found around the more electronegative atom. Carbon can be either partially plus, δ+, or partially minus, δ–, depending on the electronegativity of the group bonded to it. The polarization of the bonding electrons will become very important in understanding how two species interact, since like charges repel and unlike charges attract. A partially plus carbon will attract a negative ion; a partly negative carbon will attract a positive one.

C–Cl
δ+ δ-
C–Li
δ- δ+

1.3 LEWIS STRUCTURES AND RESONANCE FORMS

Electron flow paths are written in the language of Lewis dot structures and curved arrows. Lewis dot structures are used to keep track of all electrons, and curved arrows are used to symbolize electron movement. You must be able to draw a proper Lewis structure complete with formal charges accurately and quickly. Your command of curved arrows must also be automatic. These two points cannot be overemphasized, since all explanations of reactions will be expressed in the language of Lewis structures and curved arrows. A Lewis structure contains the proper number of electrons, the correct distribution of those electrons over the atoms, and the correct formal charge. We will show all valence electrons; lone pairs are shown as darkened dots and bonds by lines.

An atom in a molecule is most stable if it can achieve the electronic configuration of the nearest noble gas, thus having a completely filled valence shell. Hydrogen with two electrons around it, a duet, achieves the configuration of helium. Second-row elements achieve the configuration of neon with an octet of valence electrons. Third-row elements achieve an octet but may also expand their valence shell; for example SF_6 is a stable molecule with six single bonds to sulfur (12 bonding electrons total).

1.3.1 Procedure for Drawing Lewis Structures

Use the periodic table to find the valence electrons contributed from each atom. Add an additional electron for a negative charge, or subtract one to account for a positive charge to get the **total number of valence electrons**. Then **draw single bonds between all connected atoms to establish a skeleton**, or preliminary structure. You need to know the pattern in which the atoms are connected. If you have to guess at the connectivity, the most symmetrical structure is often correct. Sulfuric acid, for example, has the one sulfur in the center surrounded by the four oxygens, two of which have attached hydrogens. Since hydrogen forms one covalent bond, it is always on the outside of the structure. **Place any additional bonds between adjacent atoms that both have incomplete octets to satisfy the following general bonding trends**. There may be more than one way to do this (see the next section). Do not exceed the octet for second-row elements. Higher-row elements like phosphorus or sulfur can exceed an octet.

General Bonding Trend	Example
One bond: H, F, Cl, Br, I	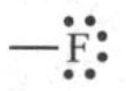
Two bonds: O, S	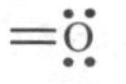
Three bonds: N, P	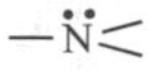
Four bonds: C, Si	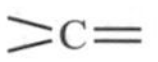

Count the number of bonds used in the structure and multiply by two electrons per bond to get the number of electrons shared in bonds. **Subtract the number of shared electrons from the number of valence electrons to get the number of unshared**

electrons (if any). Place those unshared electrons as lone pairs on atoms that still need them to complete octets.

Assign formal charges. Formal charge is a comparison of the number of electrons an atom "owns" in the Lewis structure with the number it would have if it were free. The atom is assigned only half of the electrons that it shares in a bond, but all of its unshared electrons.

Formal charge = free atom valence – (# of bonds + unshared # of electrons)

The sum of the formal charges must be equal to the total charge of the species. As a crosscheck, usually if an atom has more than the number of bonds listed in the general bonding trends, it will have a positive formal charge; if it has less, it will bear a negative formal charge. Formal charge is usually a good indicator of the electron polarization in a molecule and thus is helpful in identifying electron-rich and electron deficient regions.

Example: Methoxide ion, CH_3O^-

The total number of valence electrons is 14; we get 6 from O, 4 from C, 1 from each of 3 H's, and 1 for the minus charge. It took four bonds to connect the atoms, so the number of shared electrons is 8. We have used eight electrons, and there are six electrons remaining to be added as lone pairs to complete oxygen's octet.

Valence	14
- Shared	-8
Left over	6

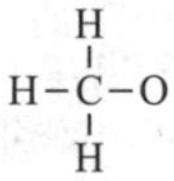

All that is left to do is assign formal charge. Oxygen started with six valence electrons, and in this structure has one bond to it and six unshared electrons, so –1 must be its formal charge. A check shows that the shells are correct for all the atoms, all the valence electrons have been used, and that the sum of the formal charges equals the total charge. The final structure is:

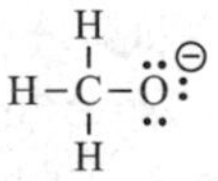

Example: Acetaldehyde, CH_3CHO

The total number of valence electrons is 18; we get 6 from the O, 4 from each of 2 Cs, 1 from each of 4 Hs. The six-bond skeleton shares 12 electrons.

Valence	18
- Shared	-12
Left over	6

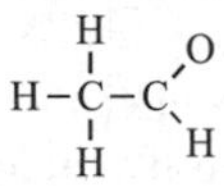

Another bond must be made because two adjacent atoms, C and O, have less than expected from the general bonding trends. This bond should go between those two atoms giving the structure:

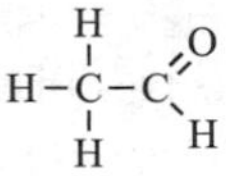

Seven bonds used 14 electrons; the 4 unshared valence electrons are added as lone pairs to complete the octet of oxygen. Formal charges are all zero; the final structure is:

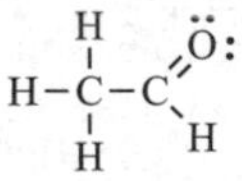

Crosscheck: Valid Lewis structures:

- **have correct number of valence electrons,**
- **have correct formal charge (sums to total charge),**
- **do not exceed octets for second-row elements.**

Common Errors

Throughout the text the incorrect example will appear on the left with the corrected version to the right of it, followed by an explanation on why the left example is incorrect.

Lewis Structure Errors

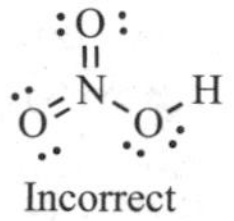
Incorrect

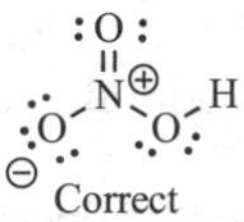
Correct

It is impossible to flow electrons correctly if you cannot keep track of them accurately. The structure on the left has exceeded the octet for nitrogen. Since nitrogen is a second-row element, it cannot expand its valence shell.

1.3.2 Resonance Hybrids

Often, one Lewis structure is not sufficient to describe the electron distribution in a molecule. In many cases, the use of a *resonance hybrid* is necessary. All resonance forms are valid Lewis structures. In resonance forms, **only the electrons move and not the atomic nuclei**. Each resonance form does not have a separate existence but is a part of a hybrid whole. The use of a double-headed arrow, ↔, between the forms reinforces the notion of a hybrid representation of a single structure. The forms are not in equilibrium with each other (equilibrium is shown by two opposing arrows: ⇆).

Only one hybrid exists, not individual equilibrating resonance forms. We are trying to describe a molecule's rather diffuse electron cloud with lines and dots; it is not surprising that often one Lewis structure is insufficient. The acetate ion, CH_3COO^-, requires two resonance forms to describe it. Both structures are of equal importance; the carbon–oxygen bond lengths are equal. Generally, the more resonance forms of similar energy an ion has (implied by similar structure), the more stable it is.

Frequently, the hybrid cannot be represented by equally weighted resonance forms. The hybrid may be more like one form than the other, and thus the resonance forms are denoted as major and minor. For formaldehyde, $H_2C{=}O$, the resonance forms represent the uneven electron distribution of a polarized multiple bond; the charge-separated minor resonance form places a negative charge on the more electronegative atom. Oxygen is winning the electronegativity tug-of-war for the bonding electrons, and mixing in this minor resonance form acknowledges that. The form on the far right does not contribute because it is contrary to the polarization of the bond. A good general rule is to avoid having an incomplete octet on a highly electronegative atom, like nitrogen or oxygen.

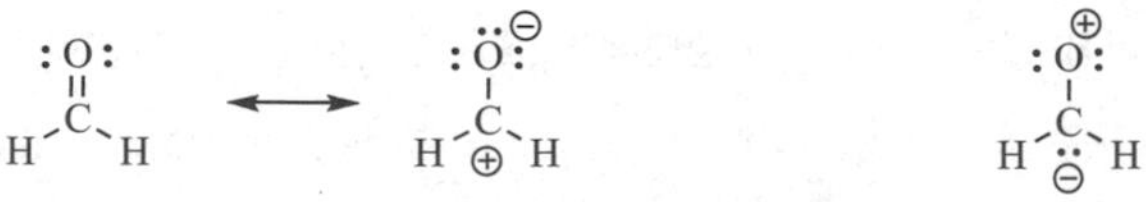

Major Minor Insignificant (incomplete octet on O)

Resonance forms for acetamide, CH_3CONH_2, reveal the polarization of the amide group. The partially negative oxygen, not the partially positive nitrogen, will be the reactive site when an amide is used as an electron source.

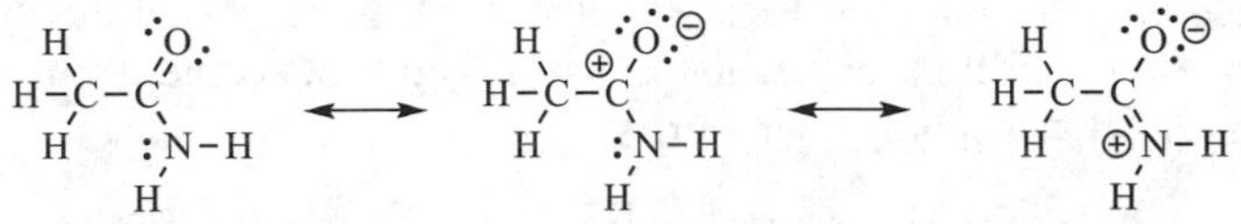

Major Minor Minor

The major resonance form will have the most covalent bonds and complete shells and the least amount of charge separation. It will seem to be the most stable of the possible resonance forms. Place any negative charges on electronegative atoms. Resonance structures with an incomplete shell are usually minor. Resonance forms having an electronegative atom with an incomplete octet are insignificant.

Occasionally "no-bond" resonance forms are needed to describe the electron distribution in a molecule, and are usually associated with very acidic hydrogens. The polarized bond in hydrochloric acid can be represented by using resonance structures to show the partial ionic nature of the bond. The atoms do not move, only the electrons.

Major Minor

Crosscheck: The major resonance contributor will have:

- **the most covalent bonds,**
- **the least amount of charge separation,**
- **the most complete octets and duets possible,**
- **any negative charges on the most electronegative atoms.**

1.4 CURVED-ARROW NOTATION

A full-headed curved arrow indicates the movement of two electrons from the tail of the arrow to the head. A half-headed curved arrow indicates the shift of one electron likewise. The two ways that a bond can break are *heterolytic* (two electrons) or *homolytic* (one electron). Homolytic processes are unusual and will be treated separately in Chapter 11.

Heterolytic cleavage Homolytic cleavage

Arrows indicate a movement or flow of electrons that **must come from a site of electron density**, either a lone pair or a bond, and move to a site that can accept additional electron density.

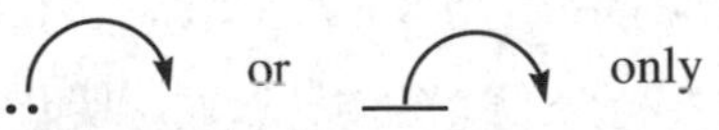
or
only

If an arrow comes from a bond, that bond is broken. If an arrow comes from a lone pair, the lone pair is removed and a new bond is formed at the head of the arrow. If the head of the arrow points between two atoms, it forms a new bond between them. If it points to an atom, it forms a new lone pair on that atom.

A source of confusion for beginning students is that for intermolecular bond-forming reactions, some authors will point the arrow between the two atoms, whereas others will point it directly at the second atom. We will try a compromise—when an arrow goes between two molecules, the head of the arrow is drawn close to the appropriate atom on the second molecule. The following are slightly different ways to show the formation of a bond between Nu:$^-$ and E$^+$ to give the Nu–E bond; we will use the arrow notation on the right. A useful generalization is that an arrow that comes from a lone pair will always form a bond, not another lone pair.

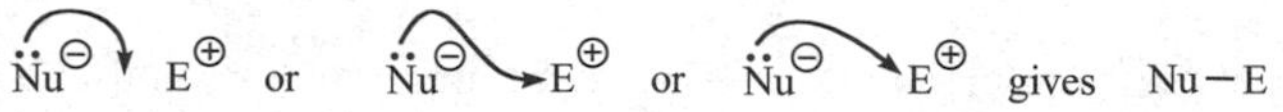
Nu
E
or
Nu
E
or
Nu
E
gives
Nu−E

The bond or lone pair from which the first arrow in an electron flow originates is called the *electron source*. The head of the last arrow in an electron flow points to the *electron sink*. **Arrows will always point away from negative charges and toward positive charges.** Sometimes it is useful to use arrows to interconvert resonance structures, but those arrows do not really indicate electron flow.

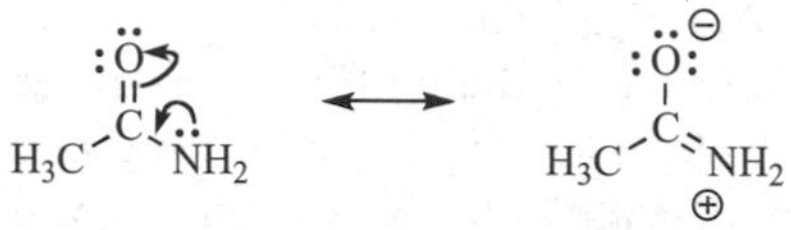

Some specific examples will help illustrate the correct use of arrows in reactions:

The first arrow on the left comes from the lone pair on the electron-rich hydroxide anion and makes a bond between the hydroxide oxygen and the hydrogen. The second arrow breaks the O–H bond and makes a new lone pair on oxygen. Note that with correct electron bookkeeping **the charge on one side of the transformation arrow will be the same as on the other side** (in this case one minus charge). Charge is conserved. If the electron movement signified by the curved arrows is correct, the products will also be valid Lewis structures.

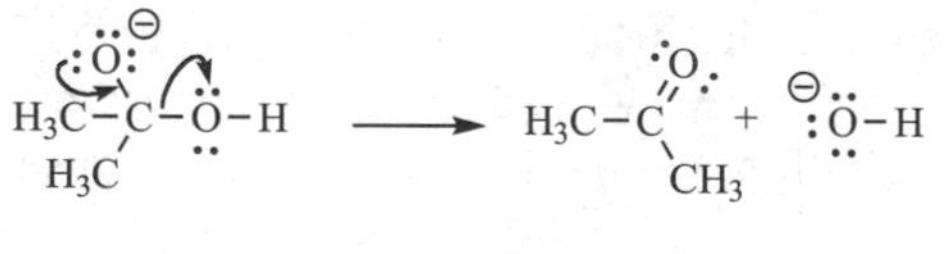

In this second example, the flow comes from the lone pair on the negative oxygen and forms a double bond. The flow continues by breaking the carbon–oxygen bond to form a new lone pair on the oxygen of the hydroxide ion. The electron source in this example changes from negative to neutral because the flow removes electrons from it; the sink becomes negative in accepting the electron flow.

In this last example, the flow starts with the electron-rich sulfur anion and forms a carbon–sulfur bond with the CH_2 group. The pi bond breaks and forms a new pi bond. The flow finishes by breaking the carbon–oxygen pi bond and forming a new lone pair on the electronegative oxygen. Every time an arrow forms a bond to an atom that already has a complete octet, another bond to that atom must break, so the octet is not exceeded.

Exercise: Cover the right side of the previous reactions and draw the product.

A good way to see whether you have mastered arrows and the concept of electron flow is to provide the arrows given the reactants and products. In these one-step mechanism problems, you must decide which bonds were made and broken and in which direction the electron flow went. Here is an example:

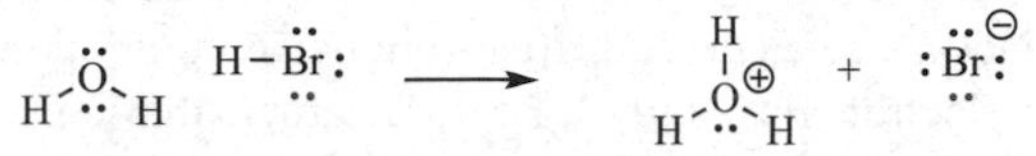

Note the bonding changes: Oxygen has one less lone pair and has formed a new bond to the hydrogen from HBr; the H–Br bond is broken, and bromine now has another lone pair. Now look at the charges: Oxygen is now positive, and bromine is now negative. Electron flow must have come from oxygen (the source) and ended up on bromine (the sink) to account for the change in charge. Only one set of arrows could be correct: The first arrow must come from the oxygen lone pair and form a new O–H bond; the second arrow must break the H–Br bond and form a new lone pair on bromine.

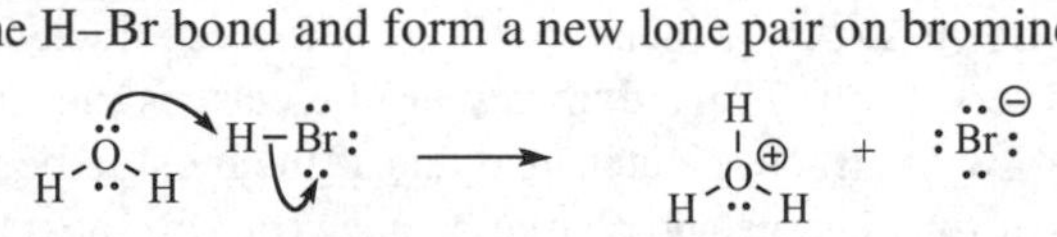

A more complex example might be useful.

Oxygen again has lost a lone pair and formed a bond to hydrogen. A carbon–carbon double bond has formed, the carbon–chlorine bond is broken, and a new lone pair is on chlorine. The minus charge on oxygen in the reactants is now on chlorine in the products; the flow must have come from oxygen (the source) to chlorine (the sink). Again only one set of arrows could be correct: The first arrow must come from the lone pair on the negative oxygen and form an O–H bond; the second arrow must break the C–H bond and form a double bond; the third arrow must break the C–Cl bond and form a lone pair on chlorine.

The flow of negative charge tells you the direction of electron flow in this last example.

The lone pair on nitrogen is gone and a new carbon–nitrogen bond has formed; the carbon–iodine bond is broken and iodine now has an extra lone pair. The charges indicate that electron density has been drained away from nitrogen (which has become positive in the product) and deposited on iodine to give the negative iodide anion. Only one set of arrows could be correct: The first arrow starts from the nitrogen lone pair and forms the N–C bond; the second arrow breaks the C–I bond and forms a new lone pair on iodine.

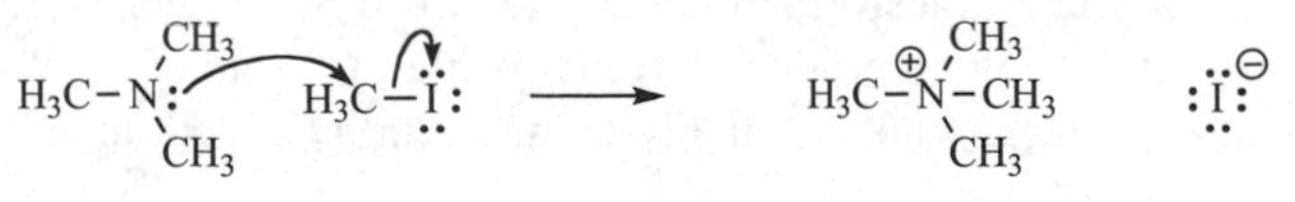

Exercise: Cover the answers in these previous examples and draw the arrows.

Precautions: It is very important that you pay strict attention to the Lewis structures and arrow positions. Lack of care can lead to some rather absurd structures and proposals. **Arrows always point in the same direction as the electron flow, never against it.** Never use a curved arrow to indicate the motion of atoms; curved arrows are reserved for electron flow only. Be forewarned that some texts may combine several steps on one structure to avoid redrawing a structure; others may show a partial set of arrows and expect you to fill in the rest mentally. In your study and practice always draw out each electron flow step completely, for errors that would otherwise be easy to find may become difficult to locate if several steps are jumbled together.

Good Arrow Pushing Habits

- Draw Lewis structures near any arrows (to keep track of atoms and electrons).
- Start every arrow from a **drawn** pair of electrons (not from a minus formal charge).
- Check Lewis structure octets and duets (the most important crosscheck).
- Check formal charge and charge balance (or your next step goes astray).
- Use the **known** electron flow paths (introduced over the next few chapters).

Common Errors

Ligand-Rich Versus Electron-Deficient Errors

Groups like NH_4^+ and H_3O^+ are cationic but not electron deficient because both nitrogen and oxygen have complete octets. For example, the nitrogen in NH_4^+ has four groups bound to it, completing its octet and giving it a formal charge of +1.

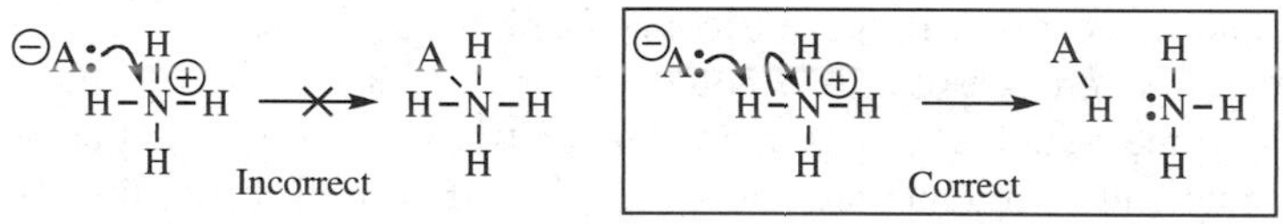
Incorrect
Correct

Any attack by an anion on the nitrogen would form a fifth bond and exceed its octet. Second-row elements, like C, N, and O, cannot exceed their octet. The other resonance forms of the ammonium ion indicate that nitrogen is **winning** the electronegativity tug-of-war. The hydrogen atoms in the ammonium ion are partially positive, and therefore acidic.

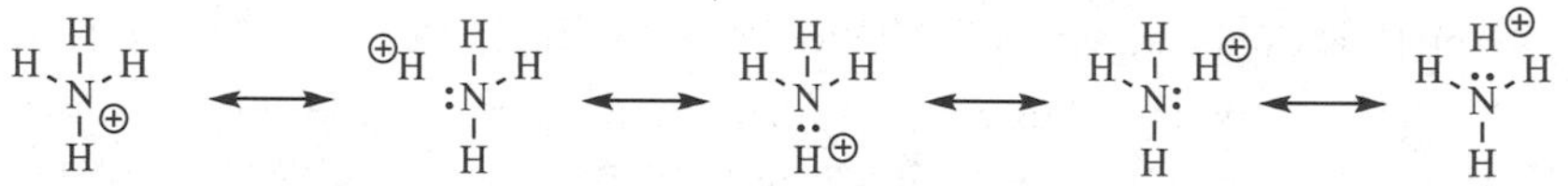

Similar errors are also seen with the even more acidic hydronium ion, H_3O^+, where again oxygen cannot exceed its octet and is much more electronegative than hydrogen.

Off-the-Path Errors

Until you have the principles of mechanistic organic chemistry thoroughly mastered, it is best to restrict your mechanistic proposals to simple combinations of the electron flow pathways, shown in Chapter 7. You may see a shortcut that with several arrows would allow you to transform the lines and dots of the Lewis structure of the reactant into the lines and dots of the product, but that is not the point of it. What you are trying to do with arrows is guess what is actually going on in the reaction, and for that you should use processes that are actually known to exist. The pathways are a very powerful mechanistic vocabulary, and there are very few mechanistic processes that cannot be expressed as a simple combination of them.

Charge Errors

Typical organic solvents have difficulty stabilizing adjacent like charges. Therefore, avoid forming adjacent like charges unless the reaction solvent is water. **Very few organic intermediates have multiple charges.** Charge must be conserved; the total charge on both sides of the reaction arrow must be the same. Charge is part of your electron count and must balance through each step. Since arrows move electrons, the charges on your final structure must be consistent with the electron movement indicated by your arrows.

Becomes more positive
or less negative
Becomes more negative
or less positive

Electron Flow Backwards Error

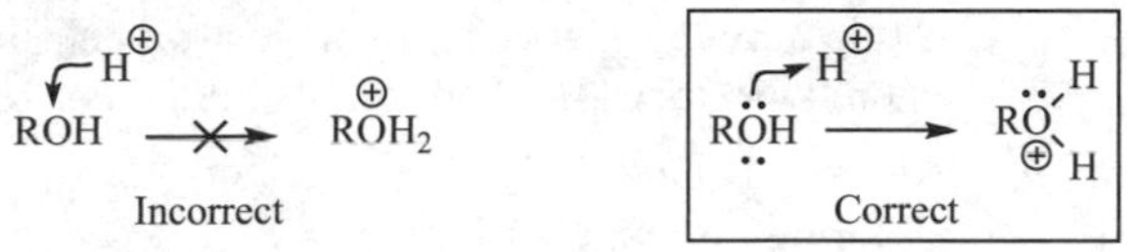
Incorrect
Correct

A proton has no electrons at all, so an arrow cannot start from it. This arrow error is seen in some biochemistry texts to mean "put the proton there." Always use arrows only for electron movement, nothing else. **Arrows will always head toward positive centers and away from negative ones.**

Electron Flow Continuity Errors

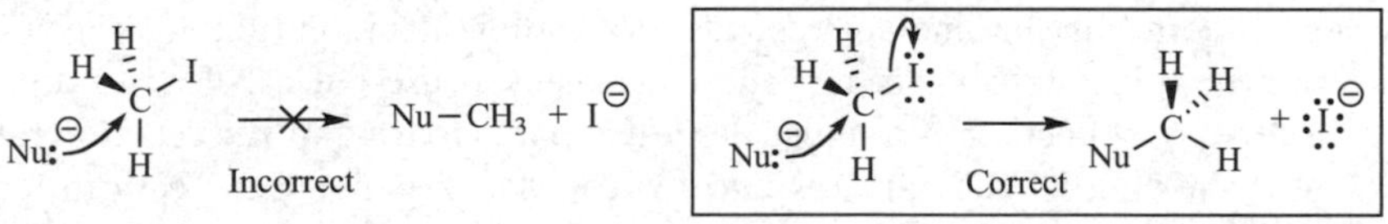
Incorrect
Correct

To be fair, some authors just draw the first arrow of an electron flow and expect the reader to supply the rest; it is best to see the entire flow drawn out.

Curved Arrows That Bounce or Miss Their Target

Beginners are often not careful in their drawing of arrows and may start an arrow from the wrong bond or put the head in between or on the wrong atoms. Since it is easy to spot, we won't give an example here. Always draw out all the electron pairs on any group that you are going to draw arrows near. Start the first arrow from a bond or electron pair, not from the formal charge.

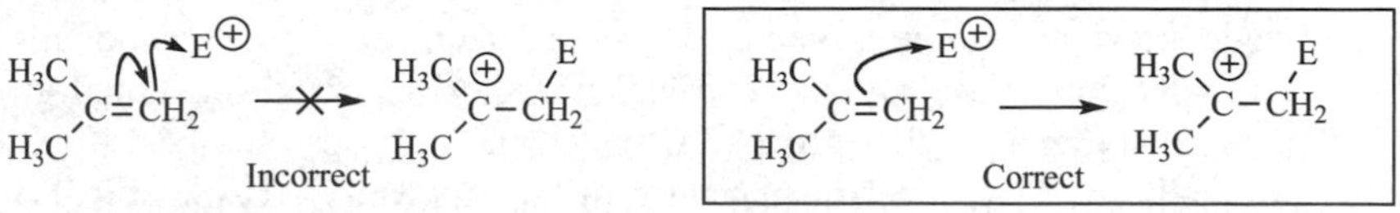
Incorrect
Correct

This "electron bounce" uses two arrows when only one is required. The second arrow is not starting from a drawn electron pair. **All** arrows must start from a **drawn** electron pair. It is much harder to detect errors when arrows are not used carefully.

1.5 NOMENCLATURE AND ABBREVIATIONS

1.5.1 Common Abbreviations

A complete table of abbreviations used in this text is in the Appendix, but a few are so common they need to be learned early.

b	Brønsted base, accepts a proton from a Brønsted acid
δ+	A partial positive charge
δ–	A partial negative charge
E	Electrophile or Lewis acid, an electron pair acceptor
erg	Electron-releasing group
ewg	Electron-withdrawing group
Et	Ethyl, CH_3CH_2
HA	Brønsted acid, donates a proton to a Brønsted base
L	Leaving group, departs with its bonding electron pair
M	Metal
MO	Molecular orbital
Nu	Nucleophile or Lewis base, an electron pair donor
Ph	Phenyl group, C_6H_5, a monosubstituted benzene
R	Any alkyl chain
X	Chlorine, bromine, or iodine
Y, Z	Heteroatoms, commonly oxygen, nitrogen, or sulfur
......	Partially broken bond (or weak complexation)

1.5.2 Line Structure

A line structure or skeletal formula is a very easy way to represent organic structures. Each corner in the line corresponds to a carbon atom, and the hydrogen atoms are not drawn. Line structure is fast and convenient to use but can lead to difficult to locate errors. **Any portion of a molecule that is participating in the reaction must be drawn out, showing all the carbons and hydrogens.** It is much too easy to forget about the hydrogen atoms omitted in line structure. The parts of the molecule not participating in the reaction may be shown in line structure or abbreviated. Draw out functional groups (see Table 1.3) and show all H's on heteroatoms.

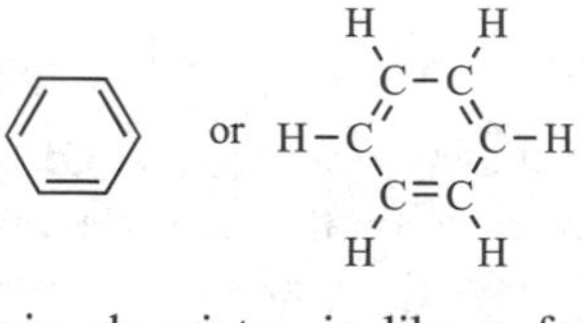

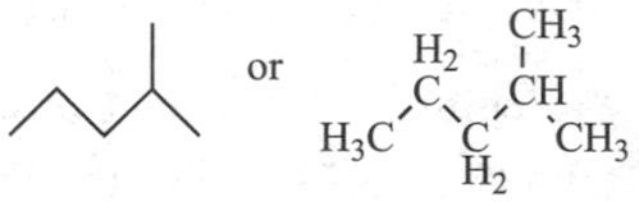

Organic chemistry is like a foreign language: It is cumulative and requires that vocabulary be learned in addition to grammar. Vocabulary is best learned as you need it, but there is so much to learn that a head start is helpful. You must be able to count in organic, know the common functional groups, and name simple compounds. Compounds are usually numbered from the end with the most oxidized functional group down the longest chain, giving other groups off the chain the lowest numbers if there is a choice. For detail, see Notes on Nomenclature in the Appendix. **Use Table 1.3 to learn the Lewis structures of the common functional groups, so you won't have to draw them from scratch every time they are needed.** A much larger functional group glossary is in the Appendix. In tables, lines off of a carbon are used to denote a bond to R or H.

Table 1.3 Lewis Structures of Common Functional Groups

Name	Functional Group	Example	Typed Version
Acyl halide	$-C(=O)-X$	$H_3C-C(=O)-Cl$	CH_3COCl
Alcohol	$>C-O-H$	H_3C-O-H	CH_3OH
Aldehyde	$-C(=O)-H$	$H_3C-C(=O)-H$	CH_3CHO
Alkane	$>C-C<$	H_3C-CH_3	CH_3CH_3
Alkene	$>C=C<$	$H_2C=CH_2$	CH_2CH_2
Alkyl halide	$>C-X$	H_3C-Br	CH_3Br
Alkyne	$-C\equiv C-$	$HC\equiv CH$	HCCH
Amide	$-C(=O)-N<$	$H_3C-C(=O)-NH_2$	CH_3CONH_2
Amine	$>C-N<$	H_3C-NH_2	CH_3NH_2
Carboxylic acid	$-C(=O)-O-H$	$H_3C-C(=O)-O-H$	CH_3COOH
Ester	$-C(=O)-O-C<$	$H_3C-C(=O)-O-CH_3$	$CH_3CO_2CH_3$
Ether	$>C-O-C<$	$H_3C-O-CH_3$	CH_3OCH_3
Ketone	$>C-C(=O)-C<$	$H_3C-C(=O)-CH_3$	CH_3COCH_3
Nitrile	$>C-C\equiv N$:	$H_3C-C\equiv N$:	CH_3CN

Exercise: Cover all but the far left side of the page and draw Lewis structures for all functional groups. Cover the left side and name all functional groups from the Lewis structures. Flash cards are useful in learning vocabulary like this.

1.5.3 List of the First Ten Alkanes

1	Methane	CH_4
2	Ethane	CH_3CH_3
3	Propane	$CH_3CH_2CH_3$
4	Butane	$CH_3CH_2CH_2CH_3$
5	Pentane	$CH_3CH_2CH_2CH_2CH_3$
6	Hexane	$CH_3CH_2CH_2CH_2CH_2CH_3$
7	Heptane	$CH_3CH_2CH_2CH_2CH_2CH_2CH_3$
8	Octane	$CH_3CH_2CH_2CH_2CH_2CH_2CH_2CH_3$
9	Nonane	$CH_3CH_2CH_2CH_2CH_2CH_2CH_2CH_2CH_3$
10	Decane	$CH_3CH_2CH_2CH_2CH_2CH_2CH_2CH_2CH_2CH_3$

1.6 AN ORBITAL VIEW OF BONDING

(A Supplementary, More Advanced Explanation)

1.6.1 Electrons as Waves

At the turn of the century in 1900, the behaviors of electromagnetic waves like light and particles like an electron were clearly separate in classical physics. In 1905, Einstein explained the photoelectric effect by saying light sometimes behaved like a particle. In 1924, De Broglie predicted that a particle like an electron could behave like a wave. It was soon confirmed that a beam of electrons could be diffracted similar to light. Diffraction was considered a property of waves, and so the wave-particle duality was firmly established. The idea that an electron has predictable wavelike properties became the foundation of quantum mechanics and modern atomic theory. A negative electron is constrained in space to be near the positive nucleus by the attraction of opposite charges. Whenever we constrain a wave to a limited space, we get standing waves. Our next step is to understand standing waves. We will start with a wave confined in one-dimensional space, or as quantum mechanics would call it, a particle in a one-dimensional box. From one-dimensional standing waves, we will move to two- and then three-dimensional standing waves. These standing waves are described by mathematical equations called wave functions. Schrödinger's wave functions describe these three-dimensional standing waves, the specific quantized orbitals that an electron may occupy in an atom.

1.6.2 Standing Waves in One and Two Dimensions

Since the electron has a wavelike behavior, the electrons in a molecule will have properties similar to a wave confined to a limited space. The stable configurations will be standing waves. We can view atomic and molecular orbitals like standing waves on a guitar string, because they too are waves confined to a limited space. Since the ends of the guitar string are fixed, only the waves that have zero amplitude at the ends will work. The lowest frequency, longest wavelength, that will fit is called the fundamental (Fig. 1.3), or first harmonic. The fundamental has one half-wavelength between the ends. The next one to fit is two half-wavelengths, the second harmonic. Three half-wavelengths gives the third harmonic, and four half-wavelengths gives the fourth harmonic.

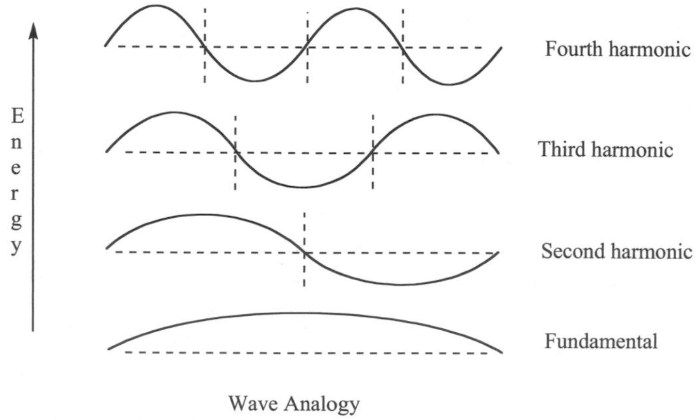
Energy
Fourth harmonic
Third harmonic
Second harmonic
Fundamental
Wave Analogy

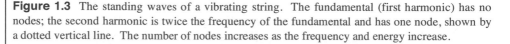

Figure 1.3 The standing waves of a vibrating string. The fundamental (first harmonic) has no nodes; the second harmonic is twice the frequency of the fundamental and has one node, shown by a dotted vertical line. The number of nodes increases as the frequency and energy increase.

As we fit more half-wavelengths into this limited space, the actual wavelength, λ, gets shorter, and the frequency, *v*, gets higher. The frequency is inversely proportional to the wavelength, $\nu = c/\lambda$ (where *c* is the speed of light), and frequency is also directly proportional to the energy, $E = h\nu$ (where *h* is Planck's constant). The fundamental is of lowest energy, and each higher harmonic gets higher in energy. The nodes, indicated by vertical dotted lines in Figure 1.3, are the regions where the wave goes through zero, the horizontal dotted line. The nodes divide the standing wave into half-wavelengths. Note that the nodes are symmetrical with the center of the system. Also note the simple fact that if we have a smaller box (shorter string), the fundamental wavelength is shorter, and thus of higher energy and frequency. The pushing of a guitar string down at a fret does exactly this to change the frequency.

The standing waves in Figure 1.3 are sine waves. The value of the sine wave has a positive mathematical sign when it is above the dashed horizontal zero line (peak), and it has a negative mathematical sign when it is below the zero line (trough). With orbitals, discussed in the next section, most texts just use shading to indicate mathematical sign, to avoid confusion between it and charge.

Electrons confined to an atom or molecule behave similarly; as the energy of the orbital increases, the number of nodes increases, like the higher harmonics on a vibrating string. Electrons are confined in three-dimensional space, so a quick look at standing waves in two dimensions will help the transition to standing waves in three dimensions. Instead of a guitar string, now visualize the head of a drum (Fig. 1.4). The fundamental again has no nodes, so the entire drumhead moves up and down. The second harmonic has one side of the drumhead moving down as the other side moves up; the single node is a line crossing the exactly in the middle of the drum. The third harmonic has two nodal lines, perpendicular to each other. This means the third harmonic has the diagonal quadrants going up as the opposite ones are going down.

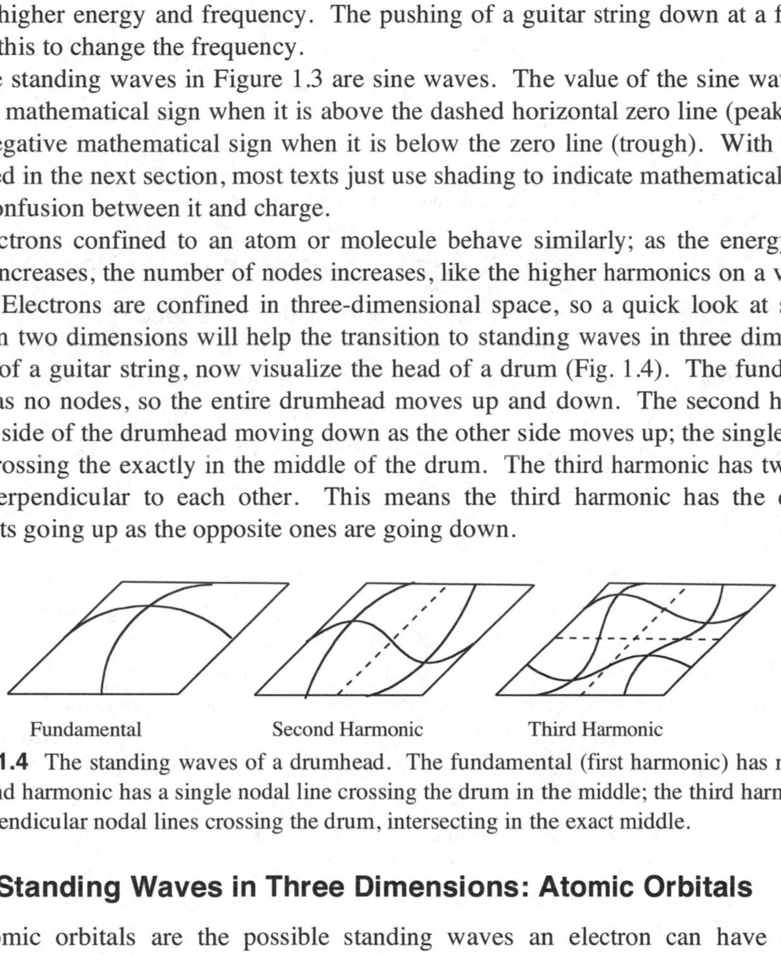
Fundamental
Second Harmonic
Third Harmonic

Figure 1.4 The standing waves of a drumhead. The fundamental (first harmonic) has no nodes; the second harmonic has a single nodal line crossing the drum in the middle; the third harmonic has two perpendicular nodal lines crossing the drum, intersecting in the exact middle.

1.6.3 Standing Waves in Three Dimensions: Atomic Orbitals

Atomic orbitals are the possible standing waves an electron can have in three dimensions (Fig. 1.5). The lowest-energy orbital, 1*s*, has zero nodes. The orbitals of the second electron shell all have one node. We use shading to indicate where the wave function has a different mathematical sign, like the peak and trough of our standing wave. The 2*s* orbital's node is spherical, so the orbital looks like a ball inside a shell. Each 2*p* orbital has a single nodal plane through the nucleus. The orbitals of the third electron shell all have two nodes. The 3*s* has two nodal spheres, so that it looks like a ball in a shell inside another shell. The 3*p* has a nodal surface inside each lobe in addition to a plane going through the nucleus. Four of the 3*d* orbitals have two perpendicular nodal planes, each intersecting at the nucleus. The fifth, the $3dz^2$, has two nodal cones intersecting at the nucleus.

We actually try to make orbital drawings like Figure 1.5 show two things: phase and probability density. The wave function itself has mathematical signs that are important when we go to overlap it with another wave function. We need to be able to show if the two orbital wave functions overlap in phase (peak with peak) or out of phase (peak with trough). The square of the wave function is the probability density of finding the electron (the odds of finding it in a particular region of space) and does not have any negative areas. But the probability density is a diffuse cloud, making drawing such a fuzzball a problem. A compromise is to draw the surface that would contain about 90% of the electron density. So we take the shape of the probability density (often stylized) and shade it to show the sign of the original wavefunction, so that the drawing shows both phase and probability density, the familiar orbitals from general chemistry.

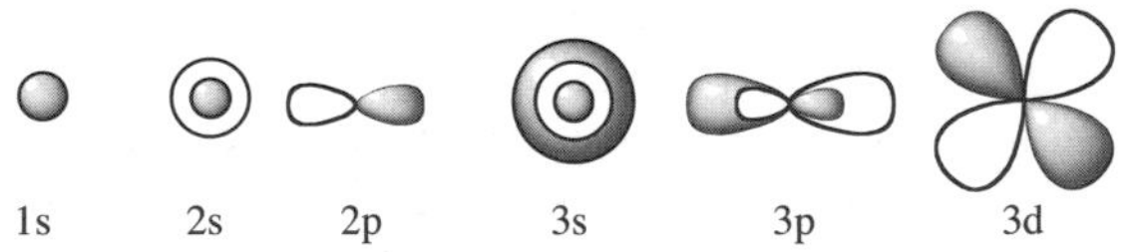
1s
2s
2p
3s
3p
3d

Figure 1.5 Representative cross-sectional drawings of atomic orbitals, which are standing waves in three dimensions. Shading indicates mathematical sign.

1.6.4 Mixing Atomic Orbitals into Molecular Orbitals

Molecular orbitals involve the combination of atomic orbitals on different atoms. Figure 1.6 shows how two 1*s* orbitals can be combined by addition and subtraction of the 1*s* orbitals to give the molecular orbitals for the hydrogen molecule, H_2. To subtract the 1*s* orbital, we mix it in with its mathematical sign (shading) reversed. The addition combines the atomic orbital waves in phase (peak with peak), and the subtraction combines the waves out of phase (peak with trough). The number of orbitals is conserved. **If we combine two atomic orbitals, we must get two molecular orbitals.** A covalent bond is the sharing of an electron pair between two nuclei. The *bonding molecular orbital* has no nodes (the highest probability of finding the electrons is between the two nuclei) and is lower in energy. The bonding orbital is lower in energy simply because the "box" has gotten bigger, so the fundamental has a longer wavelength. The *antibonding molecular orbital* has a node between the two nuclei (the probability goes through zero there) and is the higher-energy molecular orbital. Orbitals are filled from lowest energy up. Since the hydrogen molecule has two electrons, one from each atom, the bonding molecular orbital is filled and the antibonding orbital is empty. For a mnemonic to remember the two hydrogen molecular orbitals, think of a black-eyed cow.

Since the antibonding molecular orbital is raised more than the bonding is lowered, if we try to fill both molecular orbitals (with 4 electrons), the overall energy is increased compared to the isolated atoms. This is why helium does not form a diatomic molecule. With both the bonding and antibonding orbitals filled with two electrons, there is no net bond and the helium atoms drift apart. With three electrons, the antibonding orbital has just one electron in it, so the bond order of $He_2{}^+$ is 1/2, resulting in a very weak bond.

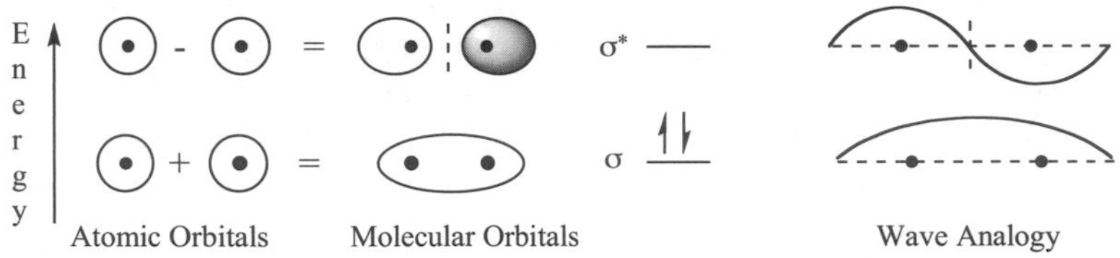
Energy
-
=
σ*
+
=
σ
Atomic Orbitals
Molecular Orbitals
Wave Analogy

Figure 1.6 The bonding and antibonding molecular orbitals of hydrogen. The dots in the drawing approximate the nuclei positions. The standing wave extends beyond the nuclei because the orbitals do.

1.7 THE SHAPES OF MOLECULES

1.7.1 Valence Shell Electron Pair Repulsion (VSEPR) Theory

Once you can draw a good Lewis structure of a molecule, you can use valence shell electron pair repulsion theory to get its shape. VSEPR theory arises from the fact that rcpulsion bctwccn pairs of bonding and lone pair electrons determine the shape of the molecule. For the purpose of VSEPR, a double or triple bond is considered as a group, and a lone pair acts as if it is larger than a bonding pair. Figure 1.7 shows the shapes of molecules as a function of how many groups crowd around the center atom. The solid wedge in Figure 1.7 symbolizes that the bond is coming out of the paper, and the dashed wedge symbolizes that the bond is going back into the paper. You can demonstrate this VSEPR effect by tying together balloons of the same size, each corresponding to a group of electrons. Only third-row and higher atoms (not carbon) can bond five or six groups.

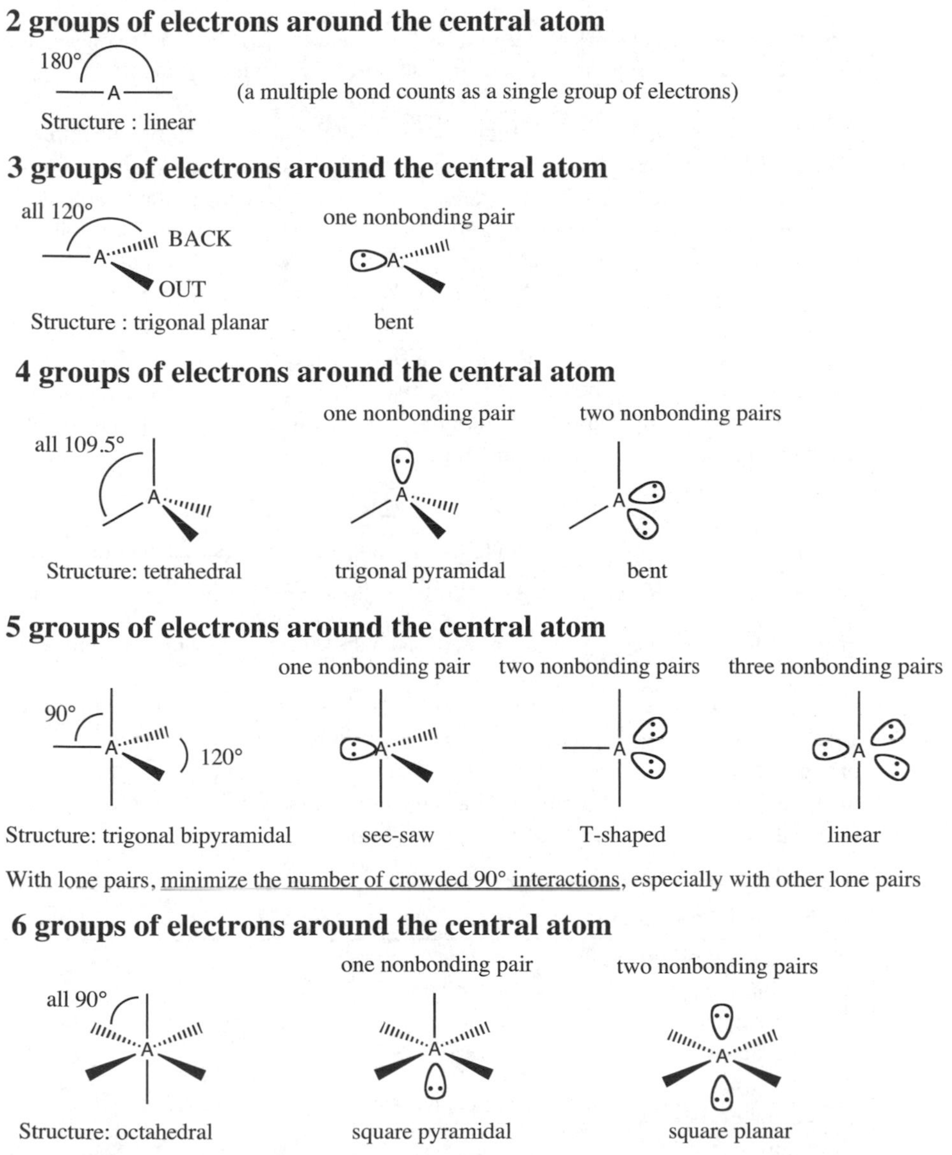
2 groups of electrons around the central atom
180°
A
(a multiple bond counts as a single group of electrons)
Structure : linear
3 groups of electrons around the central atom
all 120°
BACK
OUT
A
Structure : trigonal planar
one nonbonding pair
bent
4 groups of electrons around the central atom
all 109.5°
Structure: tetrahedral
one nonbonding pair
trigonal pyramidal
two nonbonding pairs
bent
5 groups of electrons around the central atom
90°
120°
Structure: trigonal bipyramidal
one nonbonding pair
see-saw
two nonbonding pairs
T-shaped
three nonbonding pairs
linear
With lone pairs, minimize the number of crowded 90° interactions, especially with other lone pairs
6 groups of electrons around the central atom
all 90°
Structure: octahedral
one nonbonding pair
square pyramidal
two nonbonding pairs
square planar

Figure 1.7 VSEPR-predicted shapes of molecules.

1.7.2 Hybridization

To review, an atomic orbital holds two electrons and is described by a mathematical expression called a *wave function*. From wave functions, we can get the energies of the orbitals and the probability distribution in space of any electrons occupying each orbital. A three-dimensional probability distribution of electron density is difficult to display graphically. Figure 1.8 shows several ways of representing a $2p$ orbital. We arbitrarily shade one lobe of the $2p$ orbital to show that its mathematical sign is different from the other lobe. There is a planar node in the center as the math expression goes through zero.

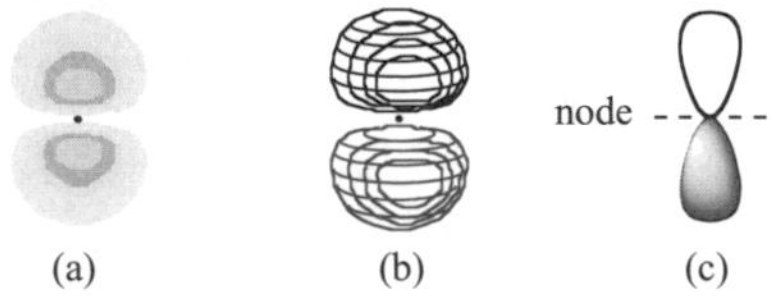

node
(a)
(b)
(c)

Figure 1.8 Representations of a $2p$ orbital. The nucleus is indicated by a heavy dot. (*a*) Dot density cross section. (*b*) Three-dimensional surface. (*c*) Simplified version.

It is convenient to have orbitals for bonding that point along the directions found from VSEPR. Toward this end, we can mathematically mix atomic orbitals on the same atom to get hybrid orbitals that point in the right directions, and are useful for the description of bonding with other atoms. Figure 1.9 shows how a $2s$ orbital can be added to and subtracted from a $2p$ orbital to give the two *sp* hybrid orbitals. To subtract, we change the sign (shading) of the $2s$ orbital and then add it to the $2p$ orbital. The two orbital wave functions reinforce where the mathematical signs are the same (the lobe gets larger) and cancel out where the sign is different (the lobe gets smaller). The number of orbitals is conserved: When we combine two atomic orbitals, we get two hybrid orbitals. We now have two hybrid orbitals pointing 180° from each other able to be used to bond to two other atoms in a linear arrangement.

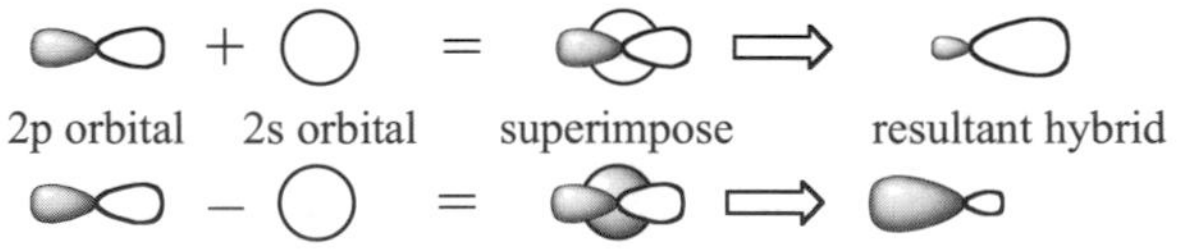

2p orbital
2s orbital
superimpose
resultant hybrid

Figure 1.9 The mixing of a $2p$ and a $2s$ orbital on the same atom giving two *sp* hybrids.

Carbon has one $2s$ and three $2p$ orbitals that can be used for hybridization (Table 1.4). Combining the $2s$ and one of the $2p$ orbitals to get two *sp* hybrids leaves two $2p$ orbitals remaining. Because the $2s$ orbital is lower in energy and closer to the nucleus than a $2p$ orbital, **hybrid orbitals that contain a higher %*s* character will form bonds that will be shorter, stronger, and lower in energy.**

Table 1.4 Summary of the Properties of Hybrid Orbitals

Hybridization of Carbon	*sp*	sp^2	sp^3
Number of hybrid orbitals	2	3	4
Interorbital angle	180°	120°	109.5°
%*s* character	50	33	25
%*p* character	50	67	75
Orientation	Linear	Trigonal	Tetrahedral
Electronegativity of C	3.29	2.75	2.48
Remaining *p* orbitals	2	1	0

To determine the hybridization of a carbon atom, just count the atoms it is bonded to: An sp^3 carbon bonds to four other atoms, an sp^2 carbon bonds to three, and an *sp* carbon bonds to only two. Use the Lewis structure for this rather than the line structure because the latter does not show all the hydrogens.

1.7.3 Single Bonds

We can overlap two sp^3 orbitals to produce a sigma bond. The in-phase combination overlaps lobes of the same mathematical sign and is the *bonding* orbital. The higher-energy out-of-phase combination is called the *antibonding* orbital (denoted by *). Sigma (σ) bond orbitals are cylindrical along the axis of the bond. **Strong bonds are due to good overlap of the bonding orbitals**.

To show that an sp^3 carbon is tetrahedral, organic molecules are usually drawn in three dimensions. The solid wedge in Figure 1.10 symbolizes the bond is coming out of the paper, and the dashed wedge symbolizes the bond is going back into the paper. The tetrahedral shape of sp^3 carbon has profound biochemical implications. *Stereochemistry* is the description of how identically connected atoms of a molecule are arranged in space (discussed in Section 4.2.6). The stereochemistry of a pharmaceutical can make all the difference between potent useful biological activity and undesired effects.

C
C
Sigma bonding
C
C
Sigma antibonding

Figure 1.10 The bonding and antibonding orbitals of a carbon–carbon single bond are made from the combination of two sp^3 hybrid orbitals.

Conformational isomers differ only in rotations about single bonds. The eclipsed and staggered conformational isomers of ethane, CH_3CH_3, are shown in Figure 1.11. Although single bonds are drawn as if they were "frozen in space," rotation usually has no significant barrier at room temperature (free rotation). Ethane's rotational barrier is tiny, only 3 kcal/mol (13 kJ/mol). One kilocalorie/mol is equal to 4.184 kilojoules/mol.

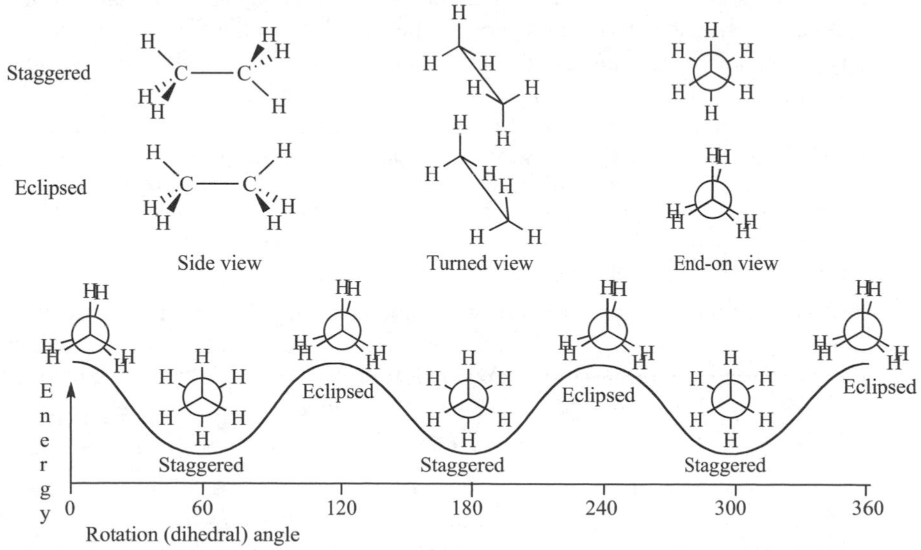
Staggered
Eclipsed
Side view
Turned view
End-on view
Energy
Eclipsed
Staggered
0
60
120
180
240
300
360
Rotation (dihedral) angle

Figure 1.11 Views of two conformational isomers of ethane, staggered and eclipsed. The end-on view is called a Newman projection. Below them is a plot of the energy as the C–C bond rotates.

The staggered conformation is the lowest energy; the eclipsed conformation is highest since the C–H bonds are in the process of passing each other and their electron clouds repel. Rotation about the center C–C bond in butane requires passing two methyl groups with a slightly larger barrier of 4.5 kcal/mol (19 kJ/mol). The staggered form when the methyl groups are opposite (*anti*) is about 0.8 kcal/mol (3.4 kJ/mol) lower in energy than the staggered form, where the methyl groups are 60° to each other (*gauche*). A significant rotation barrier may arise if two very large groups would bump into one another when rotation about the single bond occurs. The best way to learn what organic compounds actually look like is by working with molecular models.

1.7.4 Double Bonds

Pi bonds are made by two orbitals interacting side by side in the same plane, and are relatively reactive due to less effective overlap. The in-phase and out-of-phase combination of two 2*p* orbitals yield the pi molecular orbitals for ethene, CH_2=CH_2, shown in Figure 1.12. Double bonds result from a sigma bond and a pi bond between the two bonding atoms. Since that sigma bond lies in the nodal plane of the pi bond, the sigma and pi bonds of a double bond are considered independent.

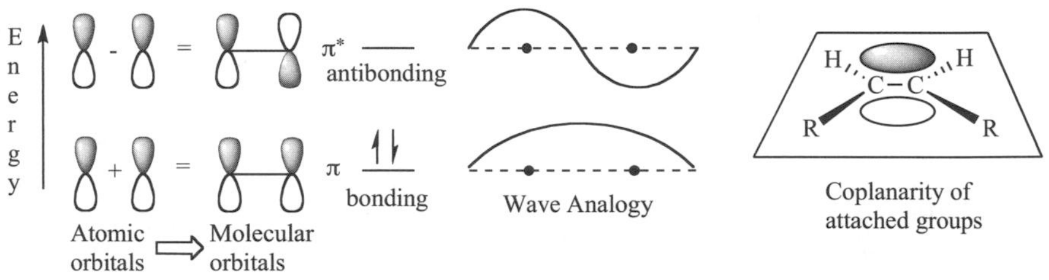
Energy
-
=
π*
antibonding
+
=
π
bonding
Atomic orbitals
Molecular orbitals
Wave Analogy
H
H
C–C
R
R
Coplanarity of attached groups

Figure 1.12 The bonding and antibonding orbitals of a carbon–carbon pi bond are made from the combination of two 2*p* orbitals by subtraction and addition. Atoms adjacent to the pi bond lie in a plane, as shown on the right. The R groups in this example are *cis* to each other.

The requirement that the *p* orbitals overlap causes a very large barrier to rotation about the double bond, about 63 kcal/mol (264 kJ/mol). *Cis* (two groups on the same side) and *trans* (two groups on opposite sides) double-bond isomers do not interconvert at any reasonable temperatures. The *trans* isomer tends to be slightly more stable than the *cis* isomer because the groups may bump into one another when they are *cis*. More alkyl substitution on the double bond (replacing an H on the double bond by an R) makes the alkene slightly more stable. For example, an equilibrium mixture of butenes is found to contain 3% 1-butene, 23% *cis*-2-butene (R equals CH_3 above), and 74% *trans*-2-butene.

1.7.5 Triple Bonds

A triple bond is composed of one sigma bond and two pi bonds. Since the two pi bonds are perpendicular to each other, they are treated as two separate, noninteracting pi bonds. **Perpendicular orbitals do not interact.** The triple bond is linear, and atoms bonded to it lie in a straight line (Fig. 1.13). Organic chemists tend to draw skinny *p* orbitals (see Fig. 1.8); the triple bond is overall cylindrically symmetrical.

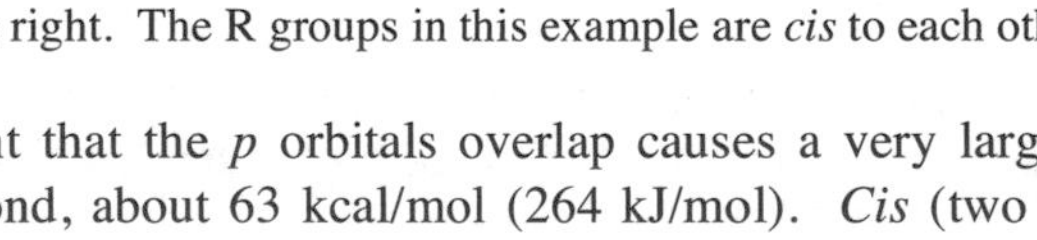
H
C
C
H
H
C
C
H

Figure 1.13 The pi orbitals of a triple bond. Cyclooctyne is shown on right.

The average bond strength of a triple bond is 200 kcal/mol (837 kJ/mol); therefore each pi bond is worth about 59 kcal/mol (247 kJ/mol). If placed in a ring, the triple bond can be bent from colinearity and become weaker and much more reactive. Cyclooctyne is the smallest ring that contains a triple bond and is still stable enough to be isolated.

1.7.6 Cumulenes

Cumulenes are compounds with two adjacent perpendicular double bonds (Figure 1.14). Each double bond can be considered separately. The central carbon is *sp* hybridized, and the two CH_2s lie in perpendicular planes. In heterocumulenes, heteroatoms have replaced one or more of the carbons, as in carbon dioxide, O=C=O.

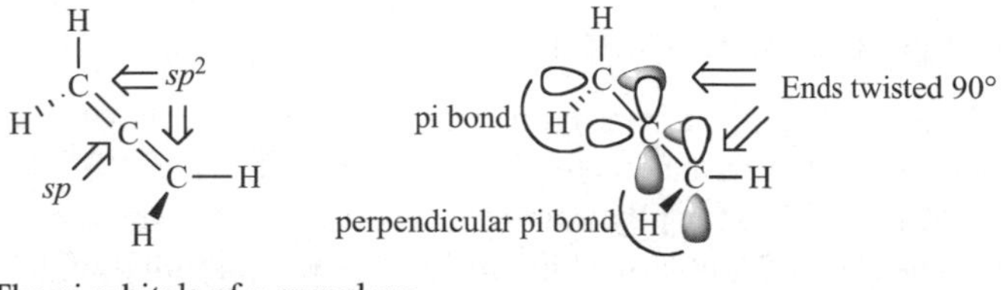
sp2
sp
pi bond
Ends twisted 90°
perpendicular pi bond

Figure 1.14 The pi orbitals of a cumulene.

1.8 MOLECULAR REPULSIONS, ATTRACTIONS, AND HYDROGEN BONDING

1.8.1 Nonbonded Repulsions

All groups on a molecule take up space, and this occupied space is called *steric bulk*. Interactions that attempt to force two groups to occupy the same space are very unfavorable. The like-charged electron clouds on the two groups repel each other, and as they are forced together the extra energy required is called *steric hindrance*, or nonbonded repulsion. If the path that the nucleophile must follow to satisfy orbital overlap requirements has too much steric hindrance, the reaction will not occur. The nucleophile will just bounce off the group that is in the way. Bulky nucleophiles need very open sites for attack, and hindered sites can be attacked only by small nucleophiles. The steric interactions between site and nucleophile play an important part in the competition between an anion acting as a nucleophile or acting as a base.

How close is too close? If the distance between two groups is less than the sum of their van der Waals radii, then strong repulsion begins. For most groups this distance is between 3 and 4 Å, which is about twice the normal C–C single bond length of 1.54 Å. Models are very useful for comparing the relative sizes of groups. Shown below are some common groups ordered by increasing size: hydrogen, methyl, ethyl, isopropyl, phenyl, and *tert*-butyl.

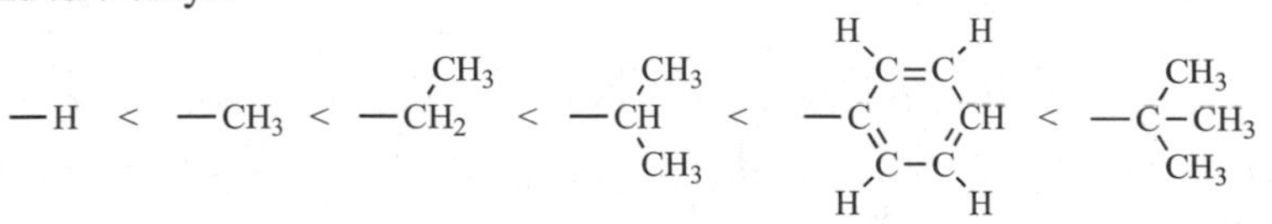

Smallest Largest

1.8.2 Dipole Attractions and Repulsions

Unlike charges attract, and like charges repel each other. The uneven distribution of electrons in a molecule can result in a pair of partial charges, represented by δ+ and δ− or by +→, a dipole moment arrow with the plus end indicated. Polar covalent bonds create bond dipoles. Another common dipole arises from a lone pair of electrons in a hybrid orbital. If the center of positive charge (from the nuclei positions) is different from the center of negative charge (from the electron cloud) then the molecule will have an overall dipole. An example is water, which has both bond dipoles and lone pair dipoles.

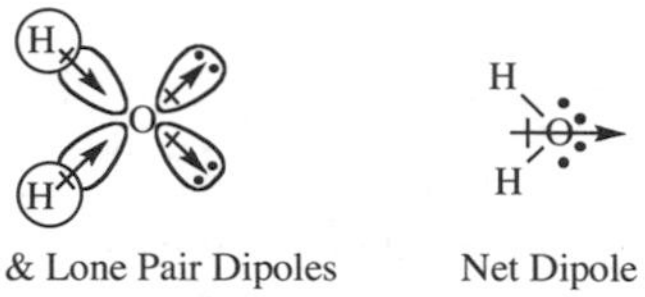
Bond & Lone Pair Dipoles
Net Dipole

Figure 1.15 The net dipole is consists of components of bond dipoles and lone pair dipoles.

1.8.3 Hydrogen Bonding

Hydrogen bonding is the overlap of a lone pair orbital of a heteroatom with an H atom bonded to another heteroatom, commonly O, N, S, and F. With rare exceptions, the C–H bond is not significantly polarized, and does not hydrogen bond. The attraction of the permanent δ+ of a polarized hydrogen–heteroatom bond with the δ– of a heteroatom lone pair provides an ionic component. Although the hydrogen bond strength averages about 5 kcal/mol (21 kJ/mol), its importance should not be underestimated. Hydrogen bonding causes water to be a liquid rather than a gas at room temperature. Hydrogen bonds with a linear arrangement of all three atoms are most common. Often a very weak bond like a hydrogen bond or a complexation is indicated by a dotted line, ·······, which is also used to symbolize a partially broken or made bond. Don't start electron flow arrows from such a line; always use Lewis structures near arrows to keep track of electrons.

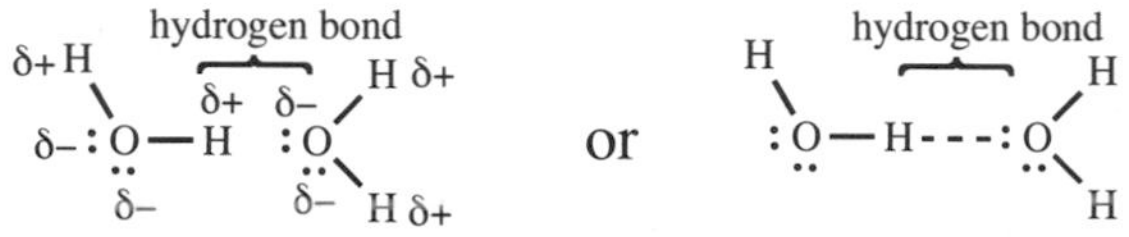
hydrogen bond
or
hydrogen bond

1.8.4 Cation Pi-Complexes

Pi-complexes, also called donor–acceptor complexes, are often a weak association of an electron-rich molecule with an electron-poor species. The donor is commonly the electron cloud of a pi bond or aromatic ring; the acceptor can be a metal ion, a halogen, or another organic compound. In the absence of solvent, as can occur in an enzyme cavity, the cation-pi interaction can be stronger than hydrogen bonding. The cation snuggles into the face of the aromatic pi cloud (see aromaticity, Sections 1.9.3 and 12.3).

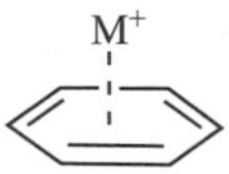
M^+

The gas phase attraction between benzene and various cations decreases with increasing cation size (Li^+ at 38 kcal/mol, Na^+ at 27 kcal/mol, and K^+ at 18 kcal/mol). Figure 1.16 shows the cation-pi interaction between an alkene and silver ion.

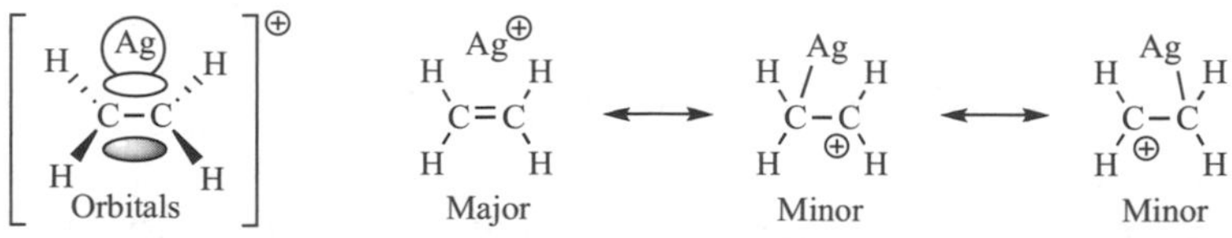
Orbitals
Major
Minor
Minor

Figure 1.16 A silver ion pi-complex. Resonance forms describe the orbital overlap shown at left.

1.9 CONJUGATION, VINYLOGY, AND AROMATICITY

1.9.1 Conjugation

Two pi systems connected by a single bond behave as one *conjugated* system because the *p* orbitals of the two pi systems are close enough to have good overlap and additional pi bonding. Figure 1.17 shows how a double bond and a carbonyl group can be placed in a molecule to form both a conjugated system and an unconjugated system. In the conjugated system, the partial plus of the carbonyl is delocalized into the second pi bond. The additional pi bonding makes a conjugated system more stable than the corresponding unconjugated system by about 3.5 kcal/mol (14.6 kJ/mol); any equilibrium between the two favors the conjugated system.

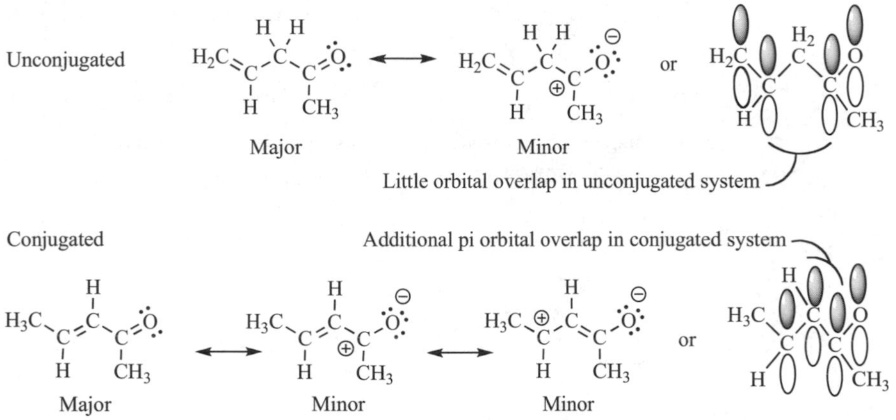
Unconjugated
Major
Minor
or
Little orbital overlap in unconjugated system
Conjugated
Additional pi orbital overlap in conjugated system
Major
Minor
Minor
or

Figure 1.17 The resonance forms and pi overlap of an unconjugated system (top) and a conjugated system (bottom) of a carbon–carbon pi bond and a carbon–oxygen pi bond.

Another example of conjugation is the nitrogen lone pair joined by resonance with a carbonyl group shown in Figure 1.18 for amides. The nitrogen lone pair is not in an sp^3 orbital but in a *p* orbital that overlaps the *p* orbitals of the pi bond to make a three-*p* orbital pi system. Any three-*p* orbital pi system is called an *allylic system*. The conjugated system will have properties different from its individual parts because it now acts as a hybrid and not as two independent functional groups. The third resonance form requires the *p* orbital of the nitrogen lone pair to line up with the *p* orbital on carbon to get the proper overlap for the double bond. This required alignment gives amides a significant rotational barrier of 15 to 20 kcal/mol.

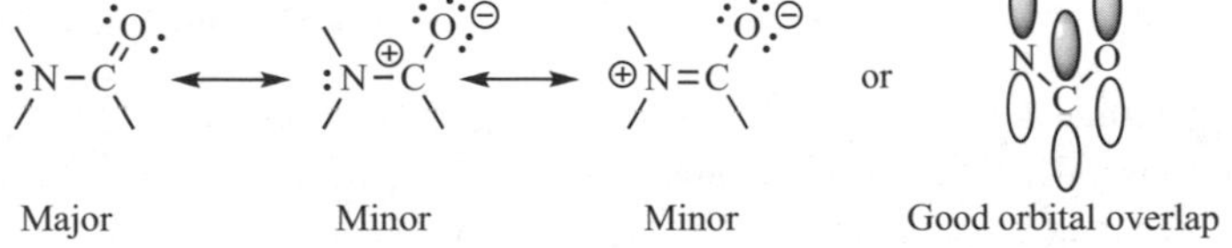
or
Major
Minor
Minor
Good orbital overlap

Figure 1.18 Conjugation of a nitrogen lone pair with a carbonyl to give an amide group.

Still another example is the carboxylic acid group in which a hydroxyl is conjugated with a carbonyl group. The resonance forms of the carboxylic acid explain the increased acidity of the hydrogen (major form is on the left, with more minor forms to the right):

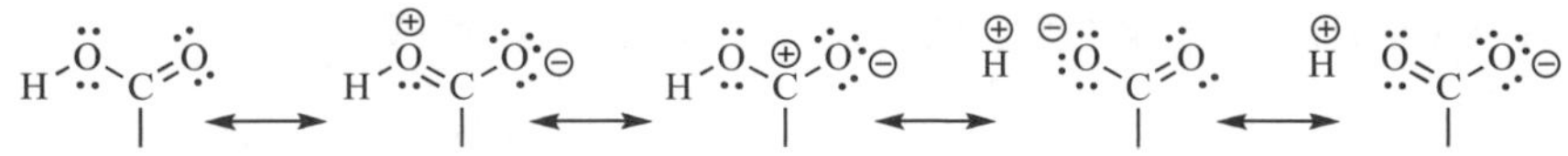

1.9.2 Vinylogy

Vinylogy is the extension of the properties of a system by the insertion of a carbon–carbon double bond. The conjugated system increases in length, but the properties remain approximately the same. A good example of this is the comparison of a carboxylic acid and a vinylogous carboxylic acid (Fig. 1.19). The hydroxyl group is still conjugated with the carbonyl through the intervening pi bond.

Carboxylic acid Vinylogous carboxylic acid

Figure 1.19 The insertion of a carbon–carbon double bond between the OH and the C=O of the carboxylic acid creates the vinylogous carboxylic acid on the right.

Similar resonance forms in a vinylogous carboxylic acid also account for an acidic hydrogen.

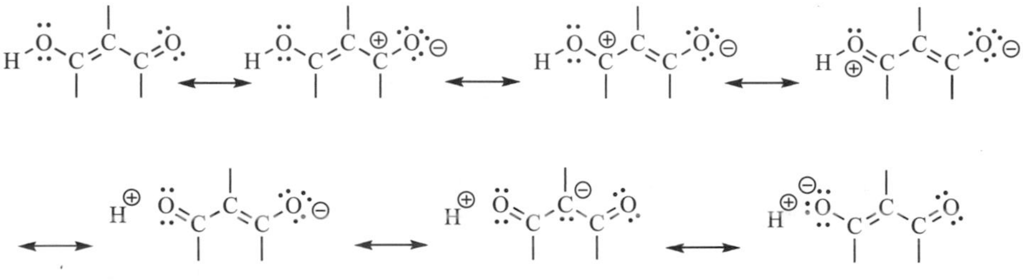

The following structures compare several other vinylogous systems with their respective parent systems.

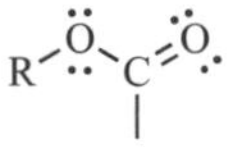

Ester

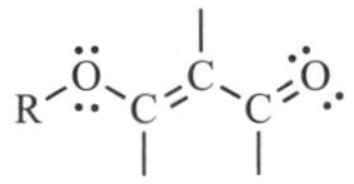

Vinylogous ester

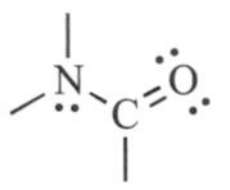

Amide

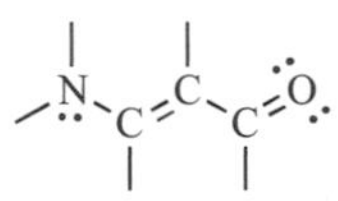

Vinylogous amide

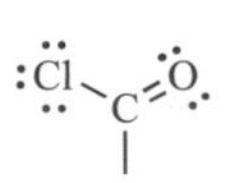

Acyl chloride

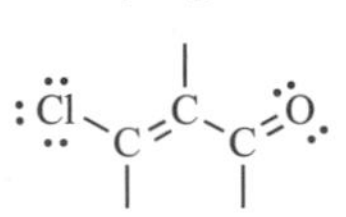

Vinylogous acyl chloride

1.9.3 Aromaticity

Benzene, C_6H_6, is much more stable and less reactive to electrophiles than simple alkenes, is described by two neutral resonance structures of equal importance, and is the prime example of an aromatic compound:

Any cyclic conjugated compound will be especially stable when the ring contains **$4n+2$ pi electrons**, where n is any integer (*Hückel's rule*). This extra stability, "resonance stabilization," of the 2, 6, 10, 14, 18, ... cyclic electron systems is termed *aromaticity* and results from the pattern of molecular orbital energies of cyclic conjugated systems (Fig. 1.20). Although the derivation of the pattern of these molecular orbitals is beyond the scope of this text, a qualitative explanation of them is given in Chapter 12. As with atomic orbital energy levels that are filled in each noble gas, filled molecular energy levels are more stable than partially filled levels. It takes two electrons to fill the bottom level, then four more to fill the next level, and four more for the next, giving the 2, 6, 10 pattern of stability.

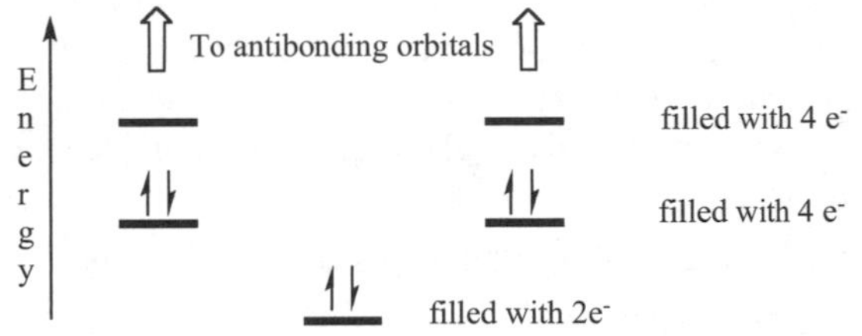
Energy
To antibonding orbitals
filled with 4 e⁻
filled with 4 e⁻
filled with 2e⁻

Figure 1.20 The general pattern of molecular orbital energy levels for cyclic conjugated systems. Six electrons are shown, filling the second MO level (stable). Four electrons would leave that MO half-filled (unstable).

The resonance stabilization of benzene is commonly estimated to be about 36 kcal/mol (151 kJ/mol), but this is not an experimentally verifiable number. The $4n$ pi electron systems (4, 8, 12, 16, ...), with half-filled energy levels, not only are less stable than the $4n+2$ systems but are destabilized relative to their open-chain analog and are called *antiaromatic*.

To summarize, **aromatic stabilization occurs in rings that have an unbroken conjugated loop of *p* orbitals (a *p* orbital on each atom in the loop with no big twists between) and $4n+2$ electrons in the loop.**

If the molecule has an unbroken loop of *p* orbitals, you will be able to draw at least two resonance structures that shift the double bonds around the loop, as we did for benzene. If any atom in the ring has four groups bound to it, then there can be no *p* orbital on that atom; the pi loop is broken, and the ring cannot be aromatic.

To determine how many electrons occupy the loop, count each double bond as contributing two electrons. A lone pair on an atom in the loop is counted as contributing two electrons if there is no double bond to that atom. (If an atom is doubly bonded and contains a lone pair, the lone pair cannot be counted since it must be in an orbital that is perpendicular to the loop.) If an atom has two lone pairs, just one is counted (only one can align with the loop). Count a triple bond as contributing only two electrons (one of the two pi bonds must be perpendicular to the loop).

Example problem

Which of the following compounds are stabilized by aromaticity?

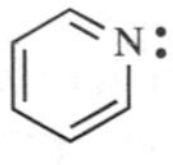
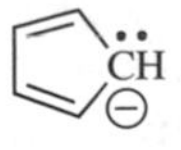
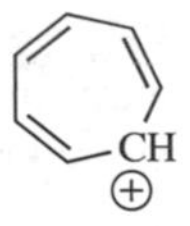
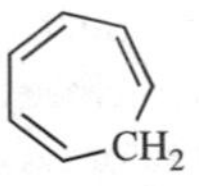

Answer: The first three on the left are aromatic. The far left compound has six electrons in the loop because the nitrogen lone pair is perpendicular to the loop, and so is not counted. The anion and the cation both have six electrons in the loop and a p orbital on every atom. Neither of the two compounds on the right has a complete loop of p orbitals because of the CH_2 group(s) breaking the pi conjugation.

1.10 SUMMARY

In this chapter we have explored the structure of organic compounds. This is important since structure determines reactivity. We have seen that weak bonds are a source of reactivity. Strong bonds are made by good overlap of similar-sized orbitals (same row on periodic table). Bends or twists that decrease orbital overlap weaken bonds. Lewis structures and resonance forms along with electron flow arrows allow us to keep track of electrons and explain the changes that occur in reactions. VSEPR will help us predict the shape of molecules. Next we must review how bonds are made and broken, and what makes reactions favorable. Critical concepts and skills from this chapter are:

Best-First Search

- Generate all reasonable alternatives and select the best one.

Valid Lewis Structures:

- have correct number of valence electrons,
- have correct formal charges that sum to the total charge,
- do not exceed octets for second-row elements.

The Major Resonance Contributor Will Have:

- the most covalent bonds,
- the least amount of charge separation,
- the most complete octets and duets as possible,
- any negative charges on the most electronegative atoms.

Good Arrow Pushing Habits

- Draw full Lewis structures near all arrows (so you can keep track of electrons).
- Start every arrow from a **drawn** pair of electrons (not from a minus formal charge).
- Check Lewis structure octets and duets (the most important crosscheck).
- Check formal charge and charge balance (or your next step goes astray).
- Use the **known** electron flow paths (introduced over the next few chapters).

VSEPR

- Groups of electrons, bonds and nonbonding pairs, repel each other and arrange themselves in space to minimize this repulsion. Lone pairs are given more space if possible (minimize 90° lone pair interactions).

Conjugation, Vinylogy, and Aromaticity

- Conjugation is the favorable overlap of adjacent p orbitals or pi systems to form a single larger pi system.
- Vinylogy is the extension of the properties of a pi system by the insertion of a pi bond in conjugation.
- Aromatic stabilization is found for cyclic conjugated rings with $4n + 2$ pi electrons.

Essential Skills

- Lewis structures—be able to rapidly draw any compound's major resonance form with the correct formal charges.
- Arrows—understand and be able to supply the arrows for converting between resonance forms, and for one-step reactions given the reactants and products.

ADDITIONAL EXERCISES

1.1 Draw Lewis structures with resonance for the following neutral compounds.

O–O–O H_2C–N–N O–C–O H–O–N–O H–O–NO_2

1.2 Draw Lewis structures with resonance for the following charged species. Decide the major resonance form, if any.

$H_2CNH_2^+$ OCN^- $HOCO_2^-$ H_2COH^+ $H_2CCHCH_2^+$

1.3. What is the polarization of the indicated bond?

Br–Br HO–Cl H_2B–H H_2C=O I–Cl

1.4 Circle the electrophiles and underline the nucleophiles in the following group.

BF_3 H^+ Ne NH_3 $^-C{\equiv}N$

1.5 Draw the Lewis structure(s) that would be the product of the arrows.

(a)
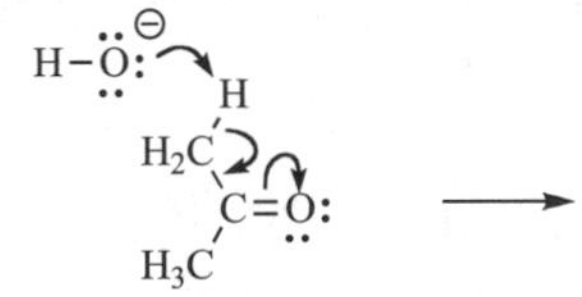

(b)
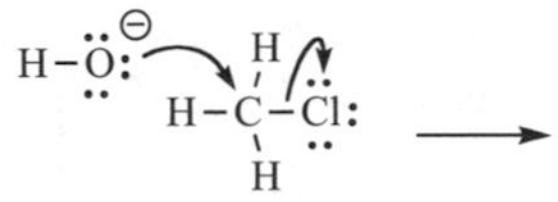

(c)
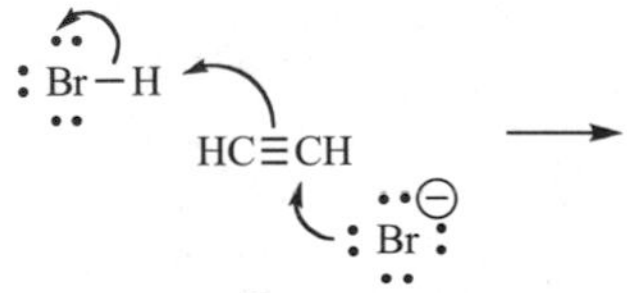

(d)
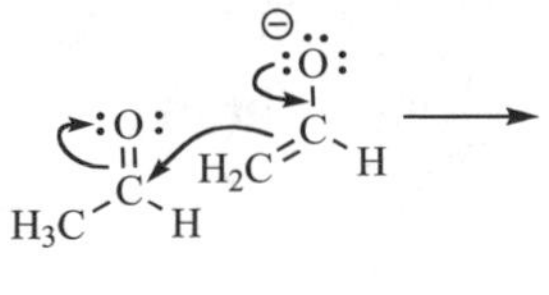

(e)
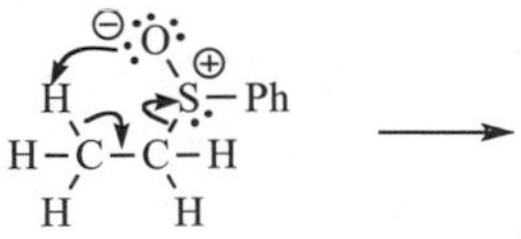

1.6 Give the curved arrows necessary for the following reactions.

(a)

(b)
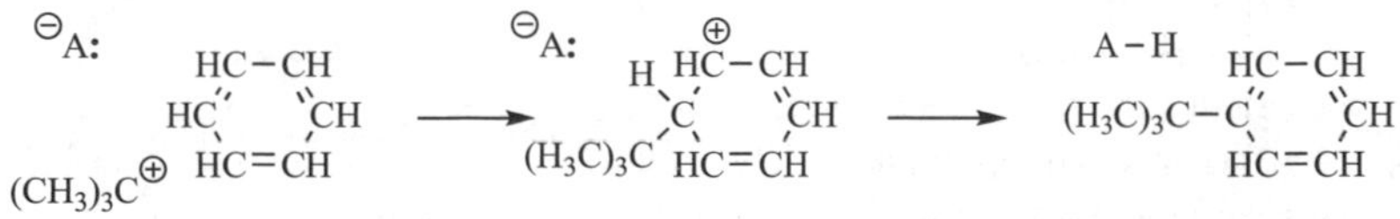

(c)

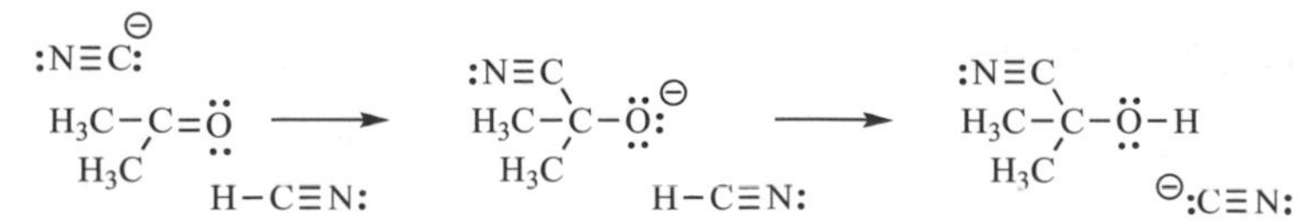

(d)

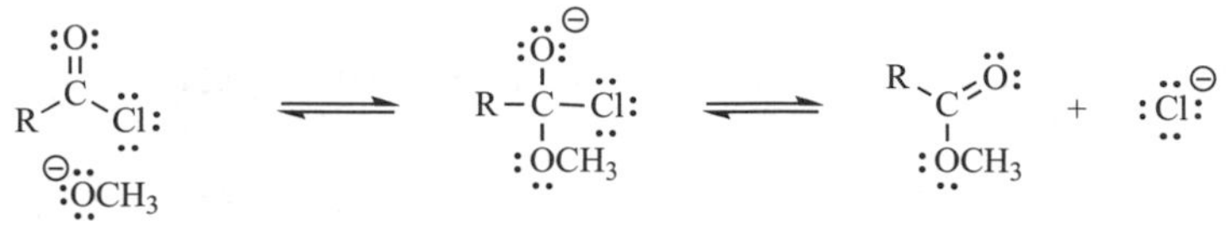

(e)

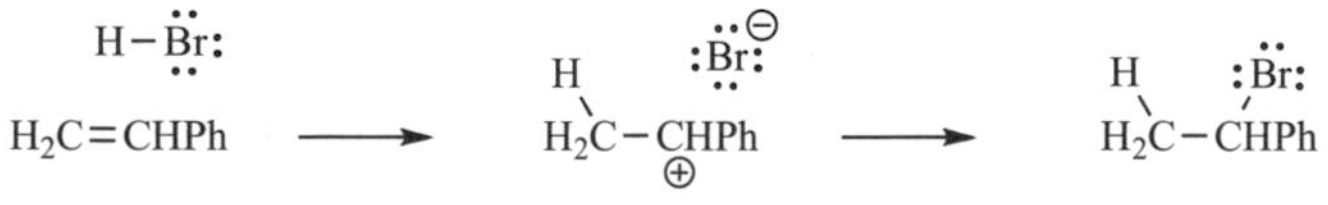

1.7 Draw full Lewis structures for the following line structures.

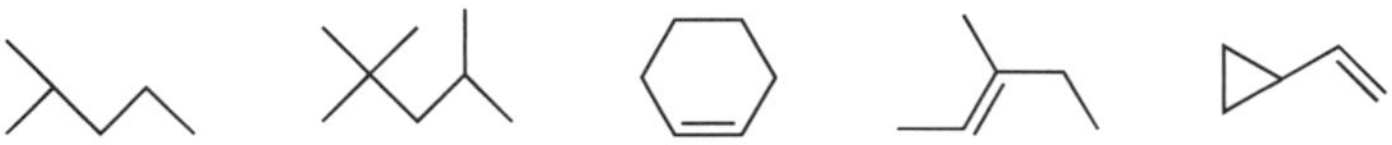

1.8 In the following structures circle any carbon atom that bears a significant partial positive charge. (Hint: look for electronegative atoms.)

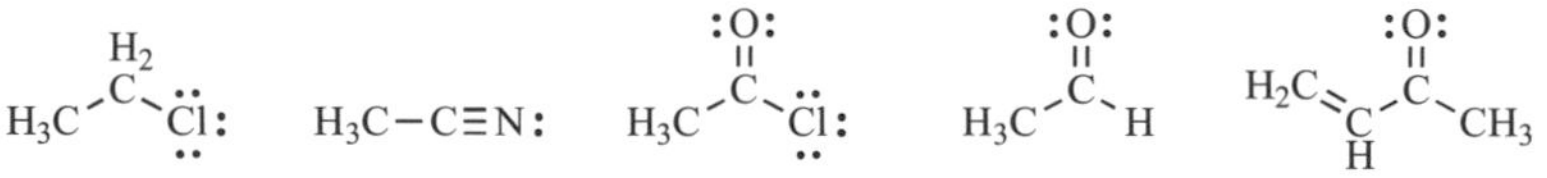

1.9 In the following structures circle any carbon atom that bears a negative or significant partial negative charge. (Hint: draw some resonance forms.)

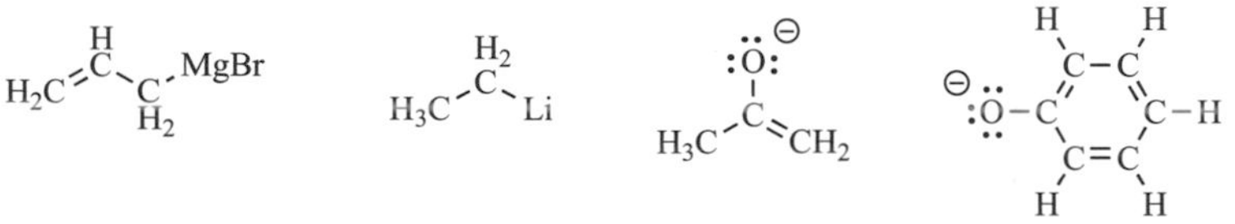

1.10 Give the hybridization of the carbons in these structures.

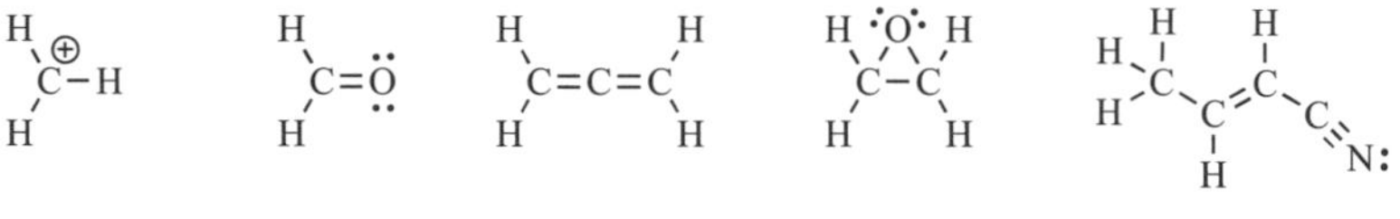

1.11 For each of the resonance pairs below, determine which is the major contributor.

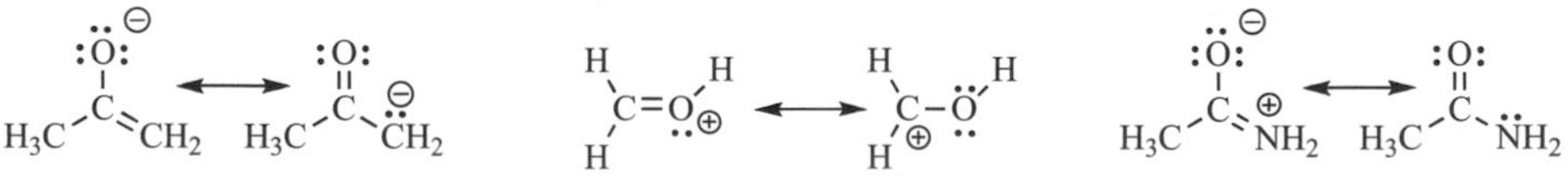

1.12 Draw the arrows to interconvert the resonance forms in problem 1.11

1.13 Draw the pi overlap for:

(a) an amide
(b) carbon dioxide
(c) an ester
(d) a vinylogous amide
(e) a nitrile

1.14 Circle the aromatic compounds in the following list.

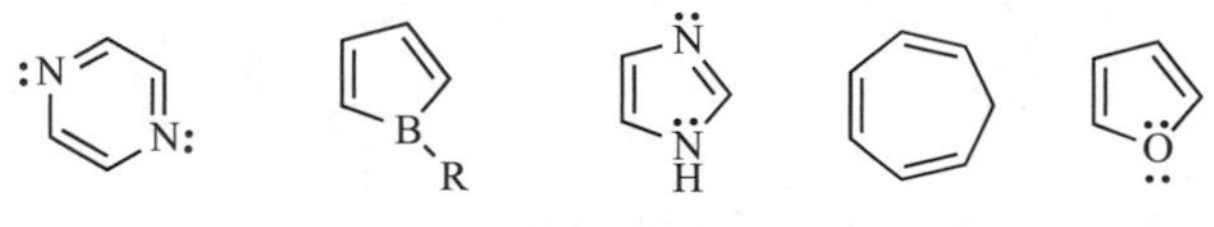

1.15 Circle the conjugated systems in the following list.

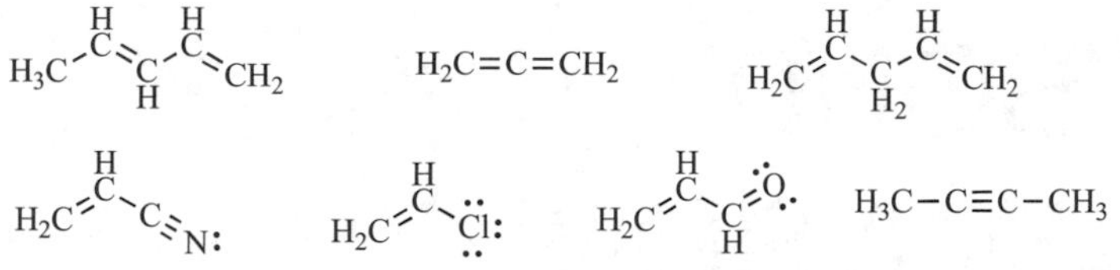

1.16 First draw a good Lewis structure, then decide the structure of the following compounds, using VSEPR. For connectivity on these, put the unique atom in the center.

SF_4 CO_2 BF_3 H_2O NH_3

1.17 First draw a good Lewis structure, then decide the structure of the following ions, using VSEPR. For connectivity on these, put the unique atom in the center.

NO_2^- NO_2^+ NO_3^- SO_4^{-2} PO_4^{-3}

1.18 Circle the molecules that have a dipole moment.

NO_2^- NO_2^+ NO_3^- NH_4^+ $AlCl_4^-$

1.19 Rank the following on the size of the atom or ion (use #1 for largest).

Ne Na^+ F^- Mg^{2+} Al^{3+}

1.20 Circle the polar covalent and underline the mostly ionic compounds below.

Br_2 LiBr HCl KBr H_2O

1.21 Draw full Lewis structures for the following line structures of biomolecules.

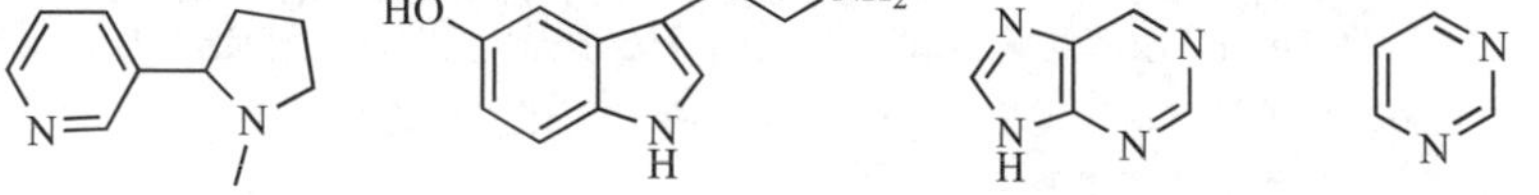

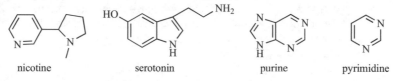

nicotine serotonin purine pyrimidine

1.22 Decide which of the following anions is the most stable, the least stable.

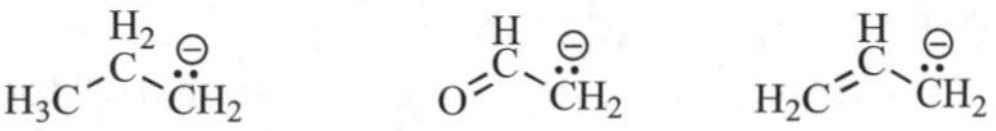

1.23 Decide which of the following cations is the most stable, the least stable.

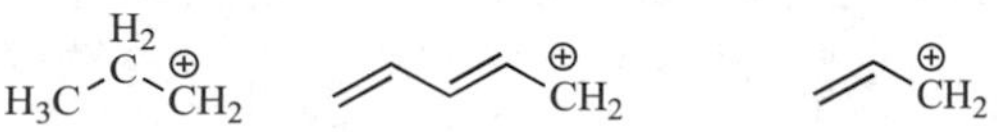

1.24 Decide whether ends of this compound are flat or twisted: $H_2C{=}C{=}C{=}CH_2$

1.25 Decide whether the nitrogen atom in an amide is tetrahedral or trigonal planar.

2

THE PROCESS OF BOND FORMATION

Electron Flow In Organic Chemistry: A Decision-Based Guide To Organic Mechanisms, Second Edition.
By Paul H. Scudder

2.8 ENZYMATIC CATALYSIS—LESSONS FROM BIOCHEMISTRY

Catalysts; Boltzmann Distribution of Energies; Collision Frequency; Separate Binding and Chemical Steps; New Lower-Energy Routes; Enzyme Active Sites; Enzymes Bind the Transition State Best; Selectivity

2.9 SUMMARY

A favorable reaction will occur whenever the reactant can overcome the energy barrier to reaction and produce a product that is lower in energy. To determine when a reaction will occur, we need to understand the basics of energetics.

2.1 ENERGETICS CONTROL KNOWLEDGE

Energetically favorable reactions usually break weak bonds and form strong ones. To be able to judge the most favorable route we must understand qualitatively both thermodynamics, which determines the position of equilibrium, and kinetics, which determines the rate of reaction. The products of reversible reactions are controlled by thermodynamics; products of irreversible reactions are controlled by kinetics. Therefore, we need a quick way to tell if a reaction is expected to be reversible. Energetics control knowledge allows us to decide whether a reaction is too energetically uphill to occur (within bounds). If the reverse reaction is too uphill, then the process is considered irreversible. We need to understand the process of bond making and breaking and how the energy of the process makes some reactions more favorable than others, in order to predict reaction products.

2.2 ORBITAL OVERLAP IN COVALENT BOND FORMATION

Sigma Bonding

Strong bonds require good overlap of the bonding orbitals. **The formation of sigma bonds requires approximate colinearity of the reacting orbitals** because that produces the best overlap, thus the strongest bond. A weak bond is a site of reactivity. The average bond strength for a C–C single bond is 83 kcal/mol (347 kJ/mol). Deformed sigma bonds can be made, as illustrated by three-membered ring formation, but at a cost. Rings containing three and four atoms are strained because the orbitals can no longer be directed along a line between the atoms, and their overlap and bond strength decreases (see Fig. 2.1). This "*ring strain* makes small rings easy to break. Strain energy destabilizes a three-membered ring by 27 kcal/mol (113 kJ/mol). Rings of five, six, and seven atoms are relatively strain-free.

Orbitals of different sizes overlap poorly and therefore form weak bonds. In general, the overlap with a carbon-based orbital decreases as one goes down a column in the periodic table. For example, the strengths of a $2p$, $3p$, $4p$, and $5p$ halogen to carbon bond are, respectively, C–F, 116 kcal/mol (485 kJ/mol); C–Cl, 81 kcal/mol (339 kJ/mol); C–Br, 68 kcal/mol (285 kJ/mol); and C–I, 51 kcal/mol (213 kJ/mol). Also, bonds between atoms that both contain lone pairs of electrons (for example, bromine) are weak because of lone pair–lone pair repulsion.

Figure 2.1 The larger six-membered ring on the left does not distort the sigma bond; the three-membered ring on the right does.

Pi Bonding

The formation of pi bonds requires approximate coplanarity of the reacting orbitals. The average bond strength of a double bond is 146 kcal/mol (611 kJ/mol); after subtraction of the sigma bond strength, the pi bond is worth about 63 kcal/mol (264 kJ/mol). For best overlap, the two *p* orbitals of the pi bond must lie in the same plane. The overlap of a pi bond is greatly diminished if one of the ends is twisted so that the two *p* orbitals are no longer coplanar. The greater the amount of this twist, the less stable the pi bond. At a 90° twist angle the pi bond no longer exists, for the two *p* orbitals are perpendicular to each other and no longer interact. Double bonds with as little as a 30° twist are reactive and difficult to make. A *trans* double bond in a ring is twisted as the ring size gets smaller (Fig. 2.2).

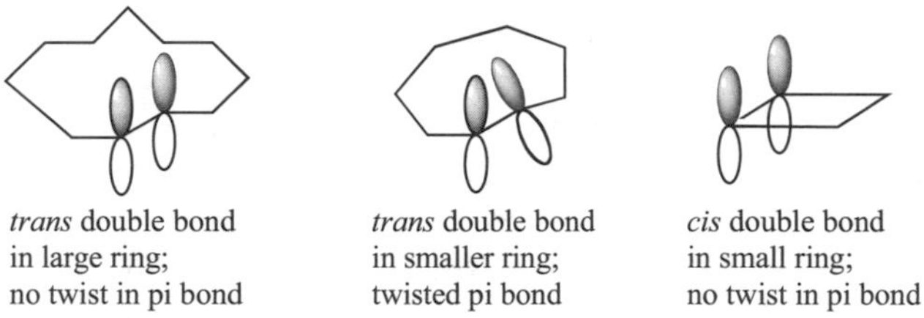
trans double bond
in large ring;
no twist in pi bond
trans double bond
in smaller ring;
twisted pi bond
cis double bond
in small ring;
no twist in pi bond

Figure 2.2 The distortion of a pi bond with varying ring size.

The smallest ring that contains a *trans* double bond and is still stable enough to be isolated at room temperature is *trans*-cyclooctene, which is about 11 kcal/mol (46 kJ/mol) more strained than *cis*-cyclooctene. Cyclobutene, with an untwisted *cis* double bond, is stable well above room temperature. Pi bonds between orbitals of different sizes are very weak because of poor overlap; a good example is the 2*p*–3*p* carbon–sulfur double bond (Fig. 2.3).

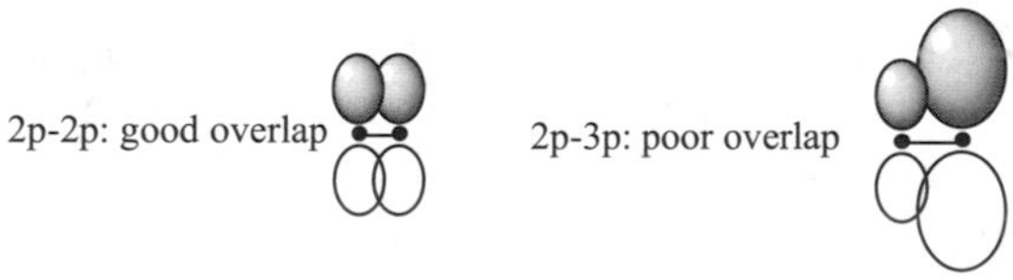
2p-2p: good overlap
2p-3p: poor overlap

Figure 2.3 The better pi overlap in a 2*p*–2*p* versus a 2*p*–3*p* pi bond.

The orbital alignment requirements have some slack; an error of 10° off the proper angle appears to have little effect. Pi-bond-forming reactions that involve deformations from coplanarity of up to 30° can occur in rare cases. Pi overlap falls off with the cosine of the twist angle. The extreme, a 90° twist, would have no pi bonding at all, a loss of about 63 kcal/mol (264 kJ/mol) bonding energy stabilization because perpendicular orbitals do not interact.

Electron flow must occur through overlapping orbitals. Deformed sigma and pi bonds are less stable, and so the paths forming them are necessarily of higher energy; this is important since we will be attempting to predict the lowest-energy path on the energy surface. If the orbital overlap is poor, the path is less likely to occur. Good orbital overlap must occur in the product (thus strong bonds) and also all along the path of electron flow to the product. Reactions that occur in two or more steps have the opportunity between steps to rotate groups into orbital alignment for the next step. Reactions that occur in one step (concerted reactions) do not have this option; their pathways may be of higher energy because of rigid orientation requirements in the transition state. It is relatively easy to forget the three-dimensional nature of organic compounds when they are written flat on a page. Whenever you have the slightest suspicion that the orbital alignment may be poor, build the molecular model.

Stereoelectronic effects occur when the position of an orbital affects the course of a reaction. These orbital-position effects are a direct consequence of the need for the best orbital overlap in forming a new bond. Enzymes use the orbital alignment requirements of a reaction to achieve selectivity. Depending on the enzyme, the reactant will be held in such a way that the appropriate bonds are rotated into the alignment required for the desired reaction. In the following example, the aromatic ring conjugated to the C=N pi bond forms the pi system of the electron sink. The N–C single bond can rotate to bring either the C–R, C–H, or C–COO⁻ bond into alignment with the *p* orbitals of the pi system of the electron sink. One enzyme holds the molecule in the active site such that only the C–COO⁻ bond, indicated in bold, aligns with the pi system of the electron sink and is selectively broken. With minor deviations, the entire molecule lies in the plane of the paper except the bond in bold, which comes up out of the plane. In this conformation, only the bond in bold is properly aligned to form the N=C pi bond in the product.

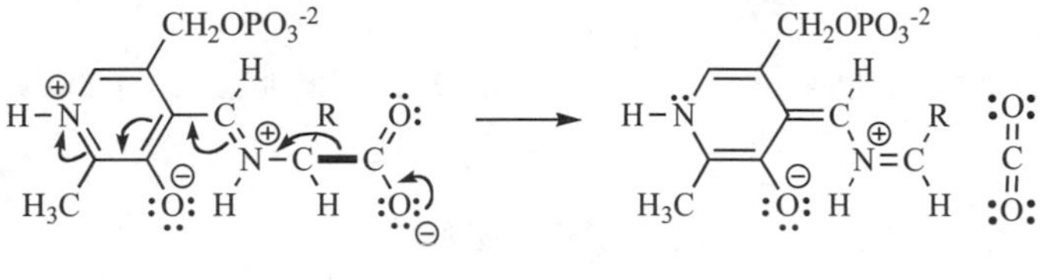

Microscopic Reversibility

The most basic rule of energetics is **the principle of microscopic reversibility: The forward and reverse reactions follow the same lowest-energy route but in opposite directions. There is only one energetically best pathway.** If A to B is the lowest-energy path forward, then B to C to A cannot be the lowest-energy path back. Remember to look in both directions when deciding on the lowest-energy path. The lowest-energy path for addition of a nucleophile is reversed to get the best path for kicking it back out.

Often the orbital alignment in one direction is clearer than in the opposite direction, so microscopic reversibility can help. For example, when a nucleophile attacks a three-*p* orbital (allylic) system such as an ester on the central carbon atom, it must approach the central *p* orbital along its axis, as shown in Figure 2.4.

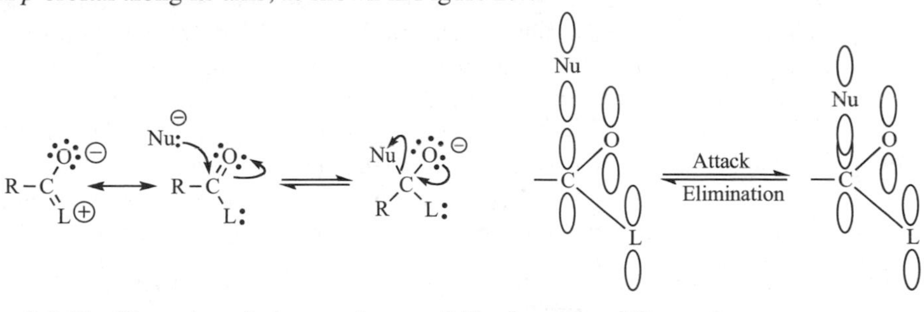
Attack
Elimination

Figure 2.4 The illustration of microscopic reversibility for nucleophilic attack.

The remaining *p* orbitals in the product are parallel to the newly formed bond. It is less clear in the reverse reaction, which follows the same path but in the opposite direction, that the lone pair orbitals of O and L must be aligned parallel to each other and to the breaking bond at the transition state. The orbitals are then lined up so that they can easily become the allylic pi system of the ester. If one of the lone pairs were not lined up, the allylic system could not be established at the transition state, and that transition state would be much higher in energy. Allylic stabilization is about 14 to 25 kcal/mol.

Another example is the attack by an electrophile on the carboxylate anion, an allylic system bearing more than one lone pair (Fig. 2.5). Since the allylic system stabilizes both starting material and product, the lowest-energy route would maintain that allylic system throughout the reaction path. In general, **the transition state of lowest energy maximizes the extent of bonding.** The loss of the electrophile (a proton, for example) would create a new lone pair in the plane of the carboxylate, perpendicular to the allylic pi system. The reverse reaction must be electrophilic attack on a lone pair that is not part of the allylic pi system. Because of the principle of microscopic reversibility, we use either the forward or reverse reaction to decide the best route.

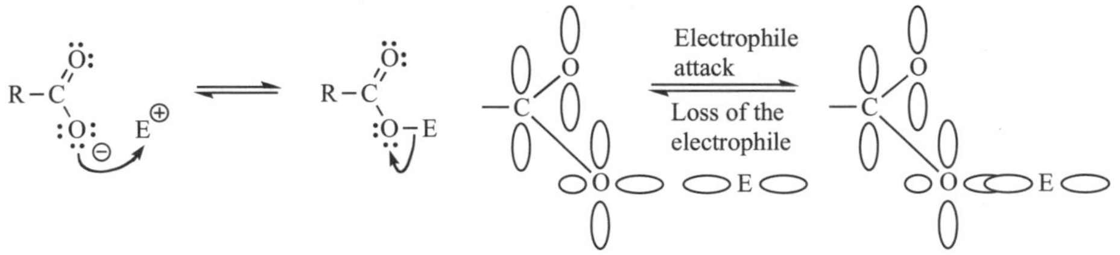

Electrophile attack
Loss of the electrophile

Figure 2.5 The illustration of microscopic reversibility for electrophilic attack

2.3 ORBITAL INTERACTION DIAGRAMS

(A Supplementary, More Advanced Explanation)

An understanding of interaction diagrams is not absolutely necessary for using the principle of electron flow to predict organic reaction products. However, it is useful for understanding reactivity trends and the stability of reactive intermediates. This section relies on the principles discussed in Section 1.6, An Orbital View of Bonding.

One way to describe the mixing of orbitals between two atoms is to use an orbital interaction diagram. The orbitals to be interacted are drawn on each side of the diagram, and their interaction is shown in the center of the diagram. When two orbitals interact, the result is two new orbitals. Shown on the left side of Figure 2.6 is the interaction of two hydrogen 1*s* orbitals containing one electron each to form a bonding orbital into which both electrons are placed. The two electrons in the bonding orbital have a lower energy than they had before the atomic orbitals interacted to form the bonding and antibonding molecular orbitals. The electrons are stabilized; energy is released, and a sigma bond is formed between the two hydrogen atoms. Shown on the right side of Figure 2.6 is the interaction of two helium 1*s* orbitals containing two electrons each. Now both the bonding and antibonding orbitals must be filled. The two electrons in the antibonding orbital are destabilized more than the two electrons in the bonding orbital are stabilized, resulting in net destabilization and no bond between the helium atoms. The two helium atoms drift apart; there is no bond to hold them together.

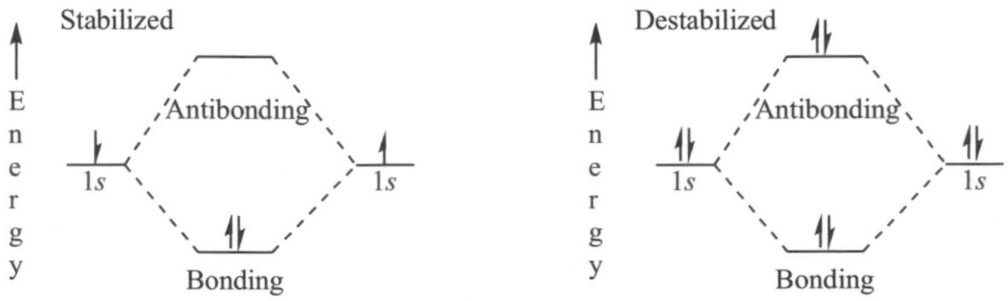

Stabilized
Energy
Antibonding
1s
1s
Bonding
Destabilized
Energy
Antibonding
1s
1s
Bonding

Figure 2.6 The left side shows the orbital interaction diagram for two hydrogen atoms and the right side shows the interaction of two helium atoms (see molecular orbitals drawn in Fig. 1.6).

The better the overlap between two orbitals, the greater their interaction. If the orbitals are oriented such that their antibonding overlap exactly cancels their bonding overlap, then there is no interaction (Fig. 2.7). Alternatively, if the orbitals are too far apart to overlap, there is no interaction.

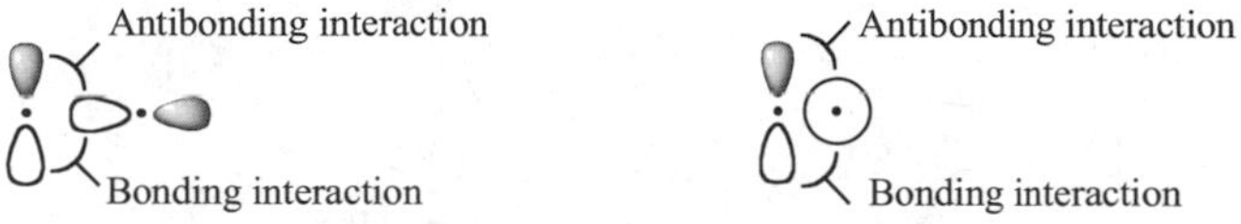
Antibonding interaction
Bonding interaction
Antibonding interaction
Bonding interaction

Figure 2.7 Orbitals that don't interact because their bonding and antibonding overlap cancels.

The closer in energy the two interacting orbitals are, the greater their interaction. A greater interaction results in both a larger rising of the antibonding orbital and a larger lowering of the bonding molecular orbitals formed. When a full orbital interacts with an empty orbital, a large stabilization of the bonding electron pair is possible. Figure 2.8 shows a good example of the decrease in stabilization of the bonding molecular orbital as the energy difference between the interacting orbitals increases.

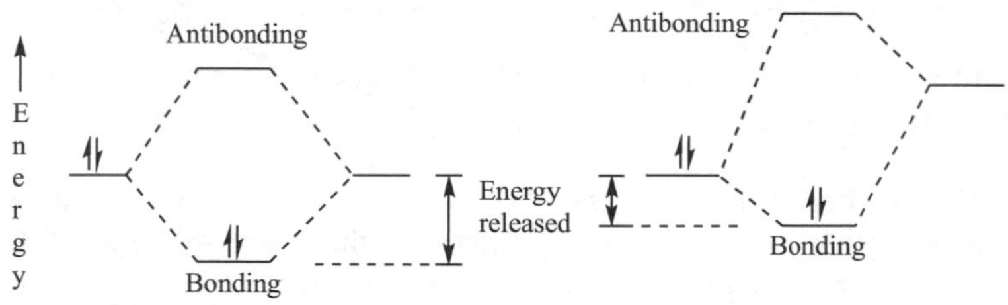
Energy
Antibonding
Bonding
Energy released
Antibonding
Bonding

Figure 2.8 The energy released upon interaction of a full with an empty orbital. On the left side the two orbitals are of identical energy and the interaction is large. On the right is the small interaction of two orbitals of greatly differing energies.

Electronegativity Effects

Electronegativity affects orbital energies and electron distribution. Because oxygen has a higher nuclear charge than carbon, its $2p$ orbital is of lower energy than the carbon $2p$ orbital. As illustrated in Figure 2.9, the lower-energy oxygen $2p$ orbital tends to pull down the energy of both the π bonding and π^* antibonding molecular orbitals. The lowered π^* orbital energy will have an important influence on reactivity.

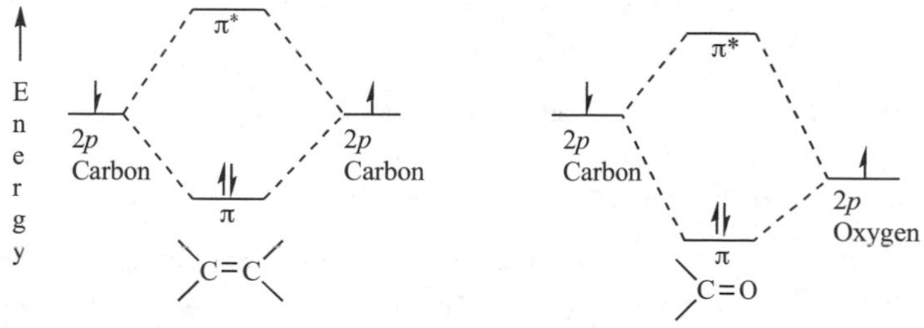
Energy
π*
2p
Carbon
π
2p
Carbon
C=C
π*
2p
Carbon
π
2p
Oxygen
C=O

Figure 2.9 The effect of increasing electronegativity on the interaction of two $2p$ orbitals lowers the energy of both π and π^*.

When two orbitals of differing energy interact, the molecular orbital contains a greater percentage of the atomic orbital closest to it in energy. This is easily demonstrated (Fig. 2.10) by examining the two extremes. As we have seen for a carbon–carbon pi bond, when the two orbitals are of equal energy each contributes equally to the molecular orbital. The other extreme is when the orbitals are so far apart in energy that they do not interact: The higher orbital would contain only the higher atomic orbital. An intermediate case in which the energies of the two orbitals differ but the orbitals still interact is shown in the center of Figure 2.10.

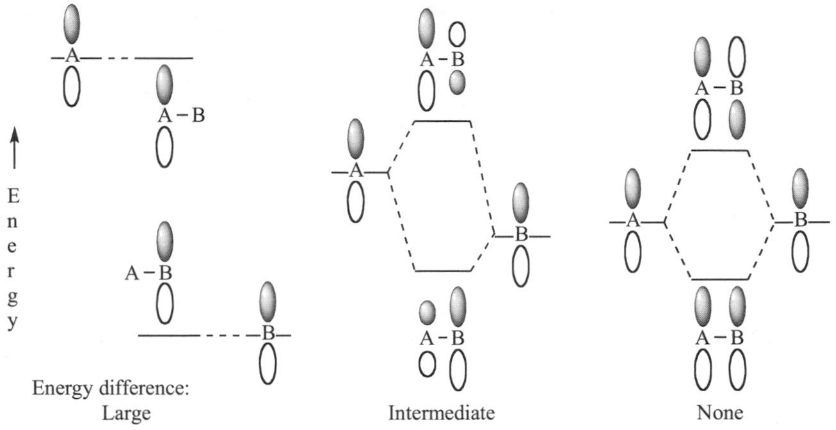
A
A-B
A-B
B
E
n
e
r
g
y
Energy difference:
Large
A-B
A
B
A-B
Intermediate
A-B
A
B
A-B
None

Figure 2.10 The interaction of orbitals of differing energy. The orbital with the larger contribution is drawn larger.

In the π molecular orbitals of a *carbonyl*, the carbon–oxygen double bond, the lower-energy oxygen 2*p* orbital contributes more to the π than to the π* molecular orbital, and the higher-energy carbon 2*p* orbital contributes more to the π* than to the π molecular orbital (Fig. 2.11). Because of the distortion of the bonding molecular orbital, the electrons in the carbonyl pi bond have a greater probability of being found around the oxygen. The valence bond resonance forms are just another way of describing this electron distribution in the molecular orbital.

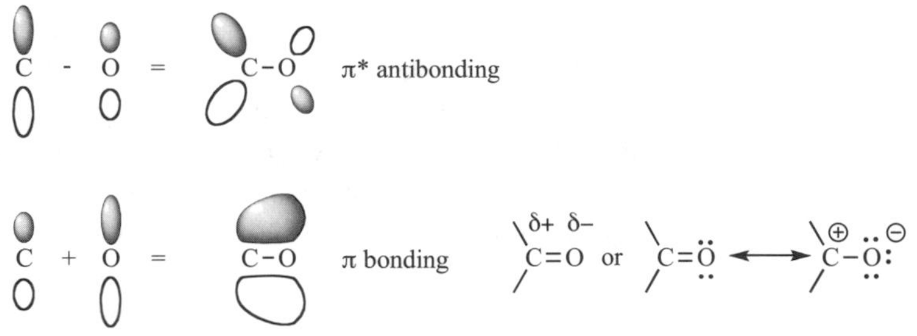
C - O = C-O π* antibonding
C + O = C-O π bonding
δ+ δ−
C=O or C=O
C−O

Figure 2.11 The pi molecular orbitals of a carbonyl group.

Nucleophile–Electrophile Interactions

Nucleophile–electrophile interactions involve the overlap and interaction of the least stabilized filled orbital (highest occupied molecular orbital, HOMO) of the nucleophile with the lowest unoccupied molecular orbital (LUMO) of the electrophile to yield the new bonding orbital and a new antibonding orbital. **The closer in energy the two interacting orbitals are, the greater their interaction. The better the overlap is between the two orbitals, the greater the interaction.** Just as we used interaction diagrams earlier to explain bonding stabilization, we can use them now to predict the bonding stabilization that occurs in a transition state when two different molecules react to form a new bond. The full orbital of the nucleophile interacts with the empty orbital of the electrophile to form a new bonding orbital and also a new antibonding orbital. The electrons of the nucleophile flow into the new bonding orbital and fill it. The nucleophile and electrophile approach each other so as to maximize the overlap of the HOMO with the LUMO, resulting in greater interaction and greater stabilization of the transition state. A generalized example is shown in Figure 2.12.

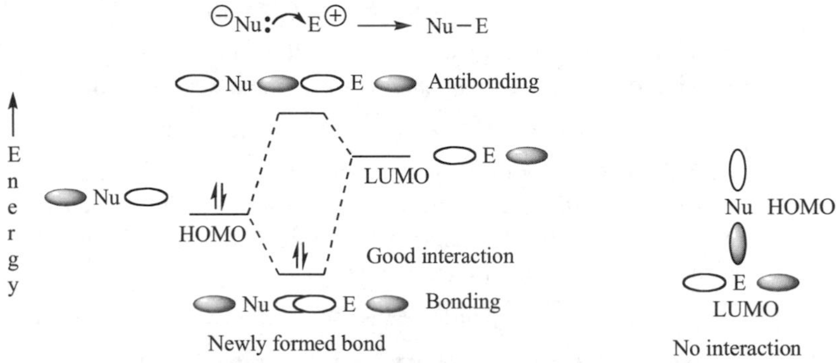
Nu: E ⊕ → Nu−E
Nu E Antibonding
Energy
Nu
LUMO
E
HOMO
Good interaction
Nu E Bonding
Newly formed bond
Nu HOMO
E
LUMO
No interaction

Figure 2.12 The interaction of the HOMO of the nucleophile and the LUMO of the electrophile to produce a sigma bond. The orbital arrangement on the right of the figure gives no interaction because the orbitals have equal positive and negative overlap (recall Fig. 2.7).

Although resonance structures generally give enough information about the partial pluses and minuses to explain a molecule's reactivity (on the basis of opposite charges attracting each other), we occasionally need to refer to HOMO–LUMO interactions for additional help. Few explanations in this text rely just on HOMO–LUMO arguments.

2.4 POLARIZABILITY AND HARD AND SOFT ACID–BASE THEORY

We have seen that strong covalent bonds are made by good overlap of two orbitals that are close in energy. Strong ionic bonds, on the other hand, are made when two oppositely charged ions get close together. Since favorable reactions tend to form the strongest bonds, either ionic or covalent, we can predict some reactions by checking whether the reactants or products have the stronger bonds. A good ionic bond partner has a small radius and a high charge and holds its electrons tightly and is said to be **hard**. A good covalent bond partner is the opposite, **soft**, able to share its valence electrons with valence orbitals that can be distorted toward the partner.

Hardness or *softness* is a property of a Lewis acid or base that is independent of its strength. An important aspect of electron availability is **polarizability, the ease of distortion of the valence electron shell of an atom by an adjacent charge**. For example, the valence electrons of iodide are shielded from the nuclear charge by all of the core electrons and thus are capable of being distorted toward a partially positive reactive site much more easily than the valence electrons of fluoride. **Polarizability increases going down a column in the periodic table.** For our purposes, it is more important to be able to compare the relative hardness of acids and bases; several trends are helpful.

Hard base—It is most often a small negatively charged ion of a strongly electronegative element with **low polarizability** that is difficult to oxidize. More electronegative elements hold their valence electrons tighter and are harder:

$$\text{harder } F^- > HO^- > H_2N^- > H_3C^- \text{ softer}$$

Soft base—It is often a large, neutral species, or one with a diffuse charge of a weakly electronegative element with **high polarizability** that is easy to oxidize. **More resonance forms of an ion indicate a more diffuse charge.** The less charged or more diffuse the charge (delocalization softens), the softer the base is. The more polarizable the base, the softer it is:

harder $F^- > Cl^- > Br^- > I^-$ softer

Hard acid—It is most often a small positively charged ion of an element with **low polarizability**. The more charge it has, the harder the acid is:

harder $Al^{3+} > Mg^{2+} > Na^+$ softer (but still reasonably hard)

Soft acid—It is often a large, neutral, or diffusely charged species or element with **high polarizability**. Again, the hard-to-soft trends for acids are more important than absolute values. The less charged or more diffuse the charge, the softer the acid is. The more polarizable acid is softer:

harder $Mg^{2+} > Cu^{2+} > Cd^{2+} > Hg^{2+}$ softer

The HSAB principle: Hard bases favor binding with hard acids; soft bases favor binding with soft acids.

The soft–soft interaction is a covalent bond, which is favored by good overlap of orbitals that are relatively close in energy. The hard–hard interaction is an ionic bond, which is favored by highly charged species that have small radii so that they can get close together to form a strong ionic bond. The soft–hard interaction is relatively weak. The HSAB principle simply expresses the tendency of reactions to form strong bonds, either ionic or covalent. **When a pair of molecules collides, two attractive forces lead to reaction: the hard–hard attraction (opposite charges attracting each other) and the soft–soft attraction (the interaction of filled orbitals with empty orbitals).**

An organometallic can react with a metal salt to produce a new organometallic and a new salt in the process called *transmetallation*. As expected from the HSAB principle, this equilibrium favors the formation of the more covalent organometallic from the softer pair and the more ionic salt from the harder pair. Transmetallation allows the conversion of a reactive organometallic into a more covalent, less reactive organometallic:

$$2\,RMgCl \quad + \quad CdCl_2 \rightarrow \quad R_2Cd \quad + \quad 2\,MgCl_2$$

R soft–Mg hard + Cd soft–Cl hard → R soft–Cd soft + Mg hard–Cl hard

Hard and Soft from a HOMO–LUMO Perspective
(A Supplementary, More Advanced Explanation)

Hard and soft can be explained in terms of the HOMO and LUMO energies and their separation (Fig. 2.13). Soft species have high-energy HOMOs and a small energy separation between the HOMO and LUMO. The soft–soft interaction, a covalent bond, is favored by a good interaction between the HOMO of the base and the LUMO of the acid. The closer in energy these two orbitals are, the larger the interaction, and the more stable the covalent bond formed is.

Hard species have stable low-energy HOMOs and a large energy separation between the HOMO and LUMO. This large energy separation between the HOMO and LUMO means that there is little covalent interaction. The hard–hard interaction is the electrostatic attraction of opposite charges.

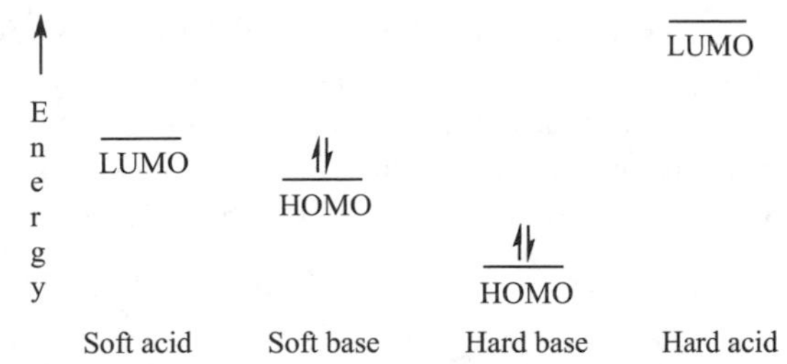
LUMO
E
n
e
r
g
y
LUMO
HOMO
HOMO
Soft acid
Soft base
Hard base
Hard acid

Figure 2.13 Representative HOMO and LUMO levels for hard and soft acids and bases.

The hard–soft interaction has two problems: The more diffuse charge (if any) of the soft species makes a poor ionic bond. The large energy gap between the LUMO of one and the HOMO of the other significantly decreases their interaction, resulting in a weak covalent bond (Fig. 2.14).

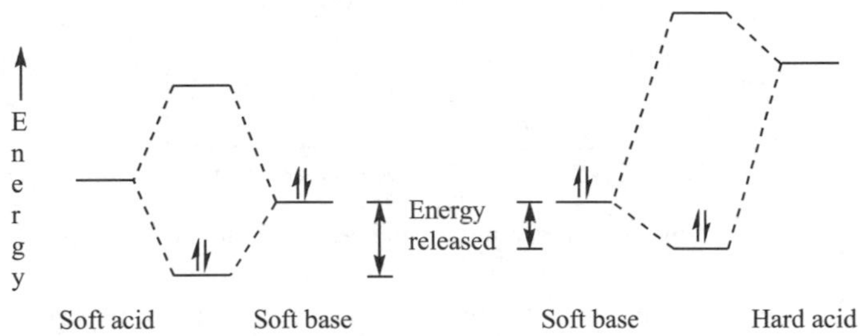
E
n
e
r
g
y
Energy
released
Soft acid
Soft base
Soft base
Hard acid

Figure 2.14 The soft–soft versus the soft–hard interaction.

2.5 THERMODYNAMICS, POSITION OF EQUILIBRIUM

This text will use energy surfaces as problem space maps, and also will use the related energy diagrams to explain why a particular reaction may be favorable and how factors influence that favorability. We need to understand why and if a process is energetically downhill to predict whether it has a chance of happening.

An energy diagram is a plot of energy versus the reaction coordinate. The reaction coordinate is a measure of the degree of a molecule's progress toward complete reaction. The energy diagram is actually a slice along the lowest-energy path on an energy surface. Although the actual energy surface for all but the simplest of reactions is very complex, we can learn much from a simple three-dimensional surface. The vertical axis is energy, and the two horizontal axes will each be the distance between atoms undergoing bond breaking or bond making.

For example, in a simple reaction, $Nu:^- + Y{-}L \rightarrow Nu{-}Y + L:^-$, a nucleophile collides with the Y end of Y–L and knocks off the leaving group L. One horizontal axis will be the distance between Nu and Y; the other horizontal axis will be the distance between Y and L. As the reaction progresses, the distance between Nu and Y decreases while that between Y and L increases. The reaction would go through a point of highest energy, a *transition state*, symbolized by ‡, in which Y is partially bonded to both Nu and L. When the reaction is over, the leaving group (or nucleofuge), L, has departed with its bonding pair of electrons.

For simplicity, we can start with two dimensions by looking down at the surface from above. The reaction's simple energy surface as viewed from the top is shown in Figure 2.15. The dark line from reactants, R, to products, P, is the path of lowest energy that the reacting species might follow. As the reactants proceed along this diagonal lowest-energy path, the Y–L bond begins to break (any movement on the surface to the right stretches the Y–L bond), and the Nu–Y bond begins to form (any movement on the surface in the downward direction brings Nu and Y closer together). The midpoint of the diagonal corresponds to a point where the Y–L bond is half-broken and the Nu–Y bond is half-formed.

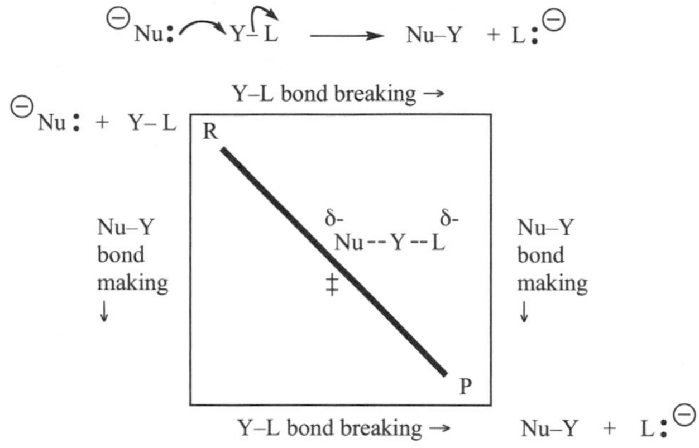
Nu: Y–L ⟶ Nu–Y + L:
Y–L bond breaking →
Nu: + Y–L
R
Nu–Y
bond
making
↓
δ- δ-
Nu--Y--L
‡
Nu–Y
bond
making
↓
P
Y–L bond breaking →
Nu–Y + L:

Figure 2.15 A simplified energy surface viewed from the top.

The third dimension, energy, becomes apparent in Figure 2.16 as we view the surface from off to the side. The lowest-energy path, again shown by the dark line, starts back at R and goes up through the "mountain pass," the transition state, and then down to point P in the front corner of the surface (the path is shaded when it is behind the surface). It is downhill to the products from the transition state.

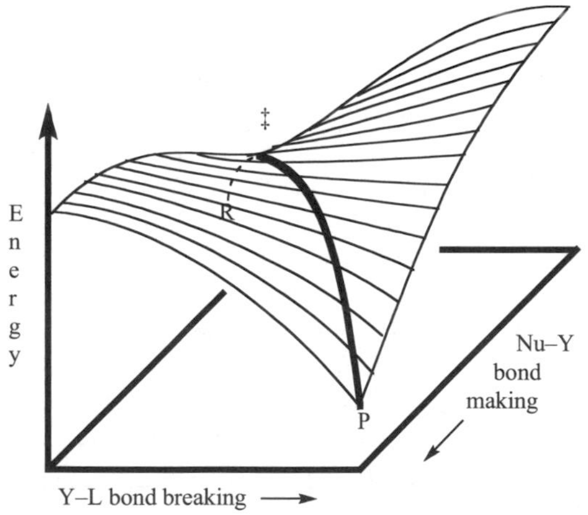
‡
E
n
e
r
g
y
R
P
Nu–Y
bond
making
Y–L bond breaking ⟶

Figure 2.16 A simplified energy surface viewed from the side.

The slice through the surface from point R to P along the lowest-energy path is the energy diagram (Fig. 2.17). For this particular simple surface, the reaction coordinate was the diagonal in the top view of the surface. Energy diagrams are easier to visualize than the lowest-energy path displayed on a more complex surface; there the path may weave around, fall into, and climb out of the energy minima of several intermediate compounds along the way to product.

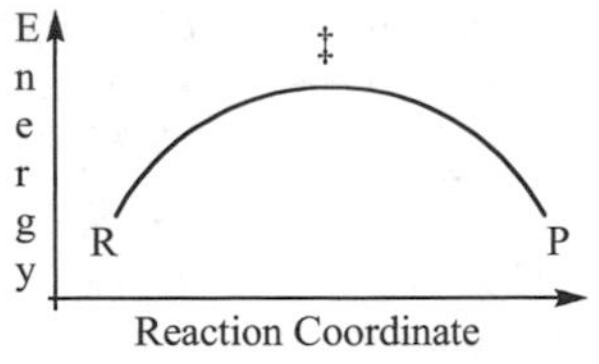

E
n
e
r
g
y
‡
R
P
Reaction Coordinate

Figure 2.17 The lowest-energy slice of the simplified energy surface.

The example in Figure 2.17 with the slight modification that P is more stable than R would produce an energy diagram as shown in Figure 2.18. Energy is expressed as G, free energy, the energy free to do work. The standard free-energy difference between the reactants and the products is $\Delta G°$ and is related to the equilibrium constant, K_{eq}, by the formulas given below:

$$\boldsymbol{\Delta G° = -RT \ln K_{eq}} = -2.303\ RT \log K_{eq}$$

or $$\log K_{eq} = -\Delta G°/2.303RT$$

or $$K_{eq} = 10^{-\Delta G°/2.303RT}$$

where $R = 1.99 \times 10^{-3}$ kcal/mol-K (8.33×10^{-3} kJ/mol-K) and T = temperature in K.

At a room temperature of 25°C, T = 298 K, so $2.303RT$ = 1.36 kcal/mol (5.70 kJ/mol). If $\Delta G°$ is in kcal/mol, $K_{eq} = 10^{-\Delta G°/1.36}$ or, conversely, $\Delta G° = -1.36\log K_{eq}$. **At room temperature, every 1.36 kcal/mol (5.73 kJ/mol) change in $\Delta G°$ changes the equilibrium constant K_{eq} by a factor of 10.** Thus, we have a quick way to interconvert $\Delta G°$ in kcal/mol to K_{eq} at room temperature: Divide $\Delta G°$ by −1.36 to obtain the exponent of K_{eq}; conversely, multiply the exponent of K_{eq} by −1.36 to get $\Delta G°$ in kcal/mol. For kJ/mol just use −5.73 instead.

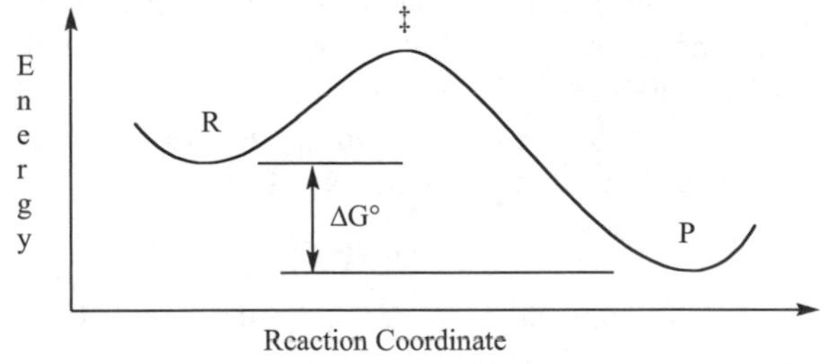

E
n
e
r
g
y
‡
R
ΔG°
P
Reaction Coordinate

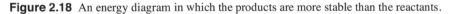

Figure 2.18 An energy diagram in which the products are more stable than the reactants.

As the standard free-energy difference, $\Delta G°$, gets more negative, K_{eq} gets larger, and the more the lower-energy compound P will predominate in the equilibrium mixture. If $\Delta G°$ is a positive value, then the product P is uphill in energy from the reactants. Table 2.1 gives representative values for 25°C, with P becoming more stable than R.

Example problem

The K_{eq} of a proton transfer reaction at 25°C is 10,000. What is $\Delta G°$ in kcal/mol?

Answer: We can use the formula $\Delta G° = -1.36\log K_{eq}$. Since a K_{eq} of 10,000 is also expressed as 10^4; log K_{eq} is 4. This gives $\Delta G° = -1.36(4) = -5.44$ kcal/mol. A K_{eq} greater than 1 favors products, and so does a negative $\Delta G°$; therefore the negative sign cross-checks.

Table 2.1 $\Delta G°$ and K_{eq} Values for 25°C (room temperature)

$\Delta G°$ kcal/mol	K_{eq}	Reactant	Product	$\Delta G°$ (kJ/mol)
+5.44	0.0001	99.99	0.01	+22.79
+4.08	0.001	99.9	0.1	+17.09
+2.72	0.01	99	1	+11.39
+1.36	0.1	91	9	+5.70
+1.0	0.18	85	15	+4.18
+0.5	0.43	70	30	+2.09
0	1	50	50	0
−0.5	2.33	30	70	−2.09
−1.0	5.41	15	85	−4.18
−1.36	10	9	91	−5.70
−2.72	100	1	99	−11.39
−4.08	1000	0.1	99.9	−17.09
−5.44	10,000	0.01	99.99	−22.79
−9.52	10^7	Essentially complete		−39.90

Exercise: Cover the value of K_{eq} in Table 2.1 and give the approximate K_{eq} from the $\Delta G°$; then cover the $\Delta G°$ and give the approximate $\Delta G°$ from the K_{eq}.

A reaction can have a positive $\Delta G°$ and still proceed. For example, if the reaction had an unfavorable equilibrium constant with $\Delta G°$ as +2.72 kcal/mol (+11.39 kJ/mol), the equilibrium mixture would contain a 99:1 R to P mixture. If we started the reaction with 100 mol of pure R, the reaction would be spontaneous and produce 1 mol of P to give the equilibrium mixture. By Le Châtelier's principle, a system at equilibrium will respond to any change by shifting to restore the equilibrium mixture. If a subsequent reaction removed the small amount of P that was formed, then more P would be produced to restore the equilibrium mixture. That following reaction would act as a *driving force* to overcome an unfavorable R to P equilibrium.

In order to understand what contributes to the standard free energy change of a reaction, it can be separated into its components:

$$\Delta G° = \Delta H° - T\Delta S°$$

The standard *enthalpy* change, **$\Delta H°$, is a measure of the heat absorbed or evolved in a reaction** and bears a negative sign if heat is given off (exothermic reactions). All exothermic reactions break weak bonds and make strong ones. Conversely, if heat is absorbed, the reaction is endothermic and $\Delta H°$ is positive. A bond-strength table is included in the Appendix. The heat of reaction can be calculated from the difference in the heats of formation of the reactants and products, but often those heats of formation are not known. A method to get an approximate heat of reaction is to consider just the bonds being formed and broken. Since bond breaking requires heat and bond making releases it, the heat of reaction can be approximated as:

$$\Delta H° = \Delta H_{\text{(bonds broken)}} - \Delta H_{\text{(bonds made)}}$$

Another use for the bond-strength table is to predict which of two products has most likely a lower heat of formation and is the preferred product of the reaction. The product that overall has the strongest bonds is preferred. Only the bonds that differ in each structure need be considered. We can tell which of two possible products is more stable by just finding out which way is exothermic for product A equilibrating with product B.

The standard *entropy* change, **$\Delta S°$, is a measure of the change in the randomness within the system**. If there is more disorder after the reaction, for example, one molecule breaking in two (with no complicating factors such as solvation), $\Delta S°$ will be positive. The entropy term arises from a combination of translational, rotational, and vibrational motions. It is estimated that the entropy loss of two molecules combining into one at 25°C is about 13 kcal/mol (56 kJ/mol). In many organic reactions the entropy term is small and does not dominate over the enthalpy term, so if the reaction is exothermic ($\Delta H°$ is negative) then it is most often also exergonic ($\Delta G°$ is negative). The calculated heat of reaction can be used as a guide to predicting the overall energetics of the process.

Example problem

Calculate $\Delta H°$ for $H_2C{=}CH_2$ + HBr → $H_3C{-}CH_2Br$

Answer: Bonds broken = C=C 146 kcal/mol (611 kJ/mol) + H–Br 87 kcal/mol (364 kJ/mol) = 233 kcal/mol (975 kJ/mol). Bonds made = C–C 83 kcal/mol (347 kJ/mol) + C–H 99 kcal/mol (414 kJ/mol) + C–Br 68 kcal/mol (285 kJ/mol) = 250 kcal/mol (1046 kJ/mol). Since $\Delta H°$ = bonds broken − bonds made = 233 − 250 = −17 kcal/mol (−71 kJ/mol). Because the sign of $\Delta H°$ is negative, the reaction is exothermic; it gives off heat. And even though it has joined two molecules into one, costing disorder, the overall reaction is still energetically favorable. It is important to notice that we broke the C=C double bond, at 146 kcal/mol (611 kJ/mol), and replaced it by making a C–C single bond, at 83 kcal/mol (347 kJ/mol).

2.6 KINETICS, RATE OF REACTION

We now know how to tell whether a reaction is energetically favorable, but a reaction may be downhill in energy and still not proceed at a reasonable rate. **The rate of reaction is controlled by the barrier height to reaction, $\Delta G^‡$, the free energy of activation**. No reaction occurs if the reactants do not have enough energy to overcome the reaction barrier. If sufficient energy is added, then the reaction proceeds. Paper can sit in air without burning for quite a long time at room temperature, but if the temperature is raised high enough it will burst into flames. In Figure 2.19, note that the $\Delta G^‡$ values for the forward and reverse reactions are different. Since B starts out at a lower energy it must climb more of an energy hill than A to get to the same transition state, denoted by the symbol ‡.

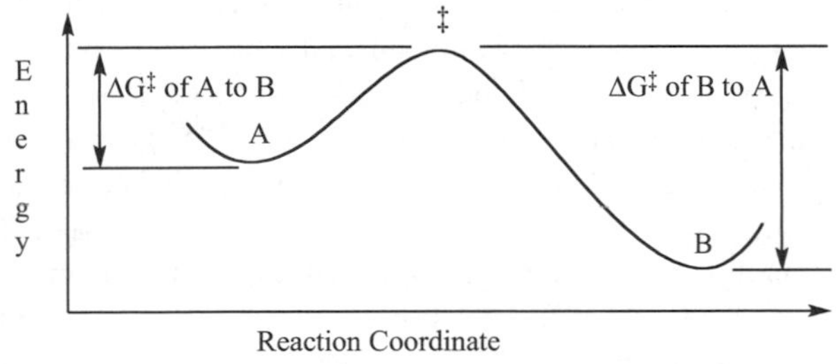
‡
Energy
ΔG‡ of A to B
A
ΔG‡ of B to A
B
Reaction Coordinate

Figure 2.19 The free energy of activation, $\Delta G^‡$ for the forward and reverse reactions.

If the barrier for A forward to B is small, and for B reversing to A is very large, then the reaction is an irreversible *unimolecular* process, A → B, and the rate of B formation is denoted by k_f[A] where k_f is the forward rate constant times [A] the molar concentration of compound A. The left side of Figure 2.20 shows this irreversible unimolecular reaction and illustrates the reaction half-life. The half-life of a reaction is the amount of time it will take for half of the reactant to react. After two half-lives just 1/4 of the reactant is left, and after three half-lives just 1/8 of it is left. A reaction is 97% complete after five half-lives.

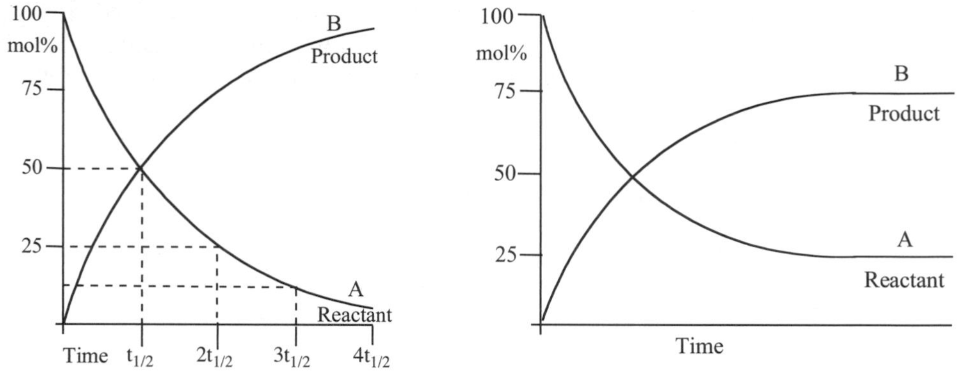
100
mol%
75
50
25
B
Product
A
Reactant
Time $t_{1/2}$ $2t_{1/2}$ $3t_{1/2}$ $4t_{1/2}$
100
mol%
75
50
25
B
Product
A
Reactant
Time

Plot of relative concentrations for A $\rightarrow$ B   Plot of relative concentrations for A $\rightleftharpoons$ B

Figure 2.20 The relative concentrations versus time for an irreversible and a reversible reaction.

An example reversible reaction is shown on the right side of Figure 2.20. If the reaction is reversible, the overall rate is the reverse rate, k_r[B], subtracted from the forward rate, k_f[A], or just k_f[A] – k_r[B]. The initial rate (when [B] = 0) is just k_f[A], but as the concentration of B grows, so does the extent of reverse reaction. At equilibrium, the A and B concentrations do not change, and the overall rate is zero, making k_f[A] = k_r[B], which rearranges to [B]/[A] = k_f/k_r = K_{eq}. In Figure 2.20, the K_{eq} is 0.75/0.25 = 3.

The rate of an irreversible *bimolecular* (two molecule) reaction in which A collides with B to produce C is expressed as a rate constant, *k*, times the concentrations.

$$A + B \rightarrow C \qquad \text{Rate} = k[A][B]$$

Concentrations for reactions vary, but the average organic reaction is run at 1 M or less. Reactive intermediates often have large rate constants, which can more than make up for their low concentrations. The largest value that the bimolecular rate constant can have is 10^{10} L/mol-s, which corresponds to a reaction upon every collision. This limit, the rate at which the two molecules collide, is called the *diffusion-controlled limit*.

Using the formula in the supplementary section that follows, we can calculate the room temperature unimolecular rate constant *k* for various values of $\Delta G^{\ddagger}$, shown in Table 2.2. A unimolecular reaction with a $\Delta G^{\ddagger}$ of 15 kcal/mol (63 kJ/mol) would be 97% complete in 0.05 seconds. If the barrier were 25 kcal/mol (105 kJ/mol), then 97% completion would take almost 2 weeks. The number of molecules in the reaction mixture at room temperature that have enough energy to traverse a 20 kcal/mol (84 kJ/mol) barrier is small; the average energy of a molecule at room temperature is only 1 kcal/mol (4.2 kJ/mol). However, if the reaction barrier is lower, many more molecules will have sufficient energy to react.

At room temperature, dropping the $\Delta G^{\ddagger}$ by 1.36 kcal/mol (5.70 kJ/mol) increases the rate tenfold. A common landmark is that **the $\Delta G^{\ddagger}$ of a reaction that proceeds at a reasonable rate at room temperature is about 20 kcal/mol (84 kJ/mol).** Our landmark value for a reasonable reaction rate at room temperature is highly dependent on $\Delta G^{\ddagger}$, the molecularity of the reaction, the reactant concentrations, and on how long the experimenter considers a reasonable time to wait. This value is still useful as a reference point. An approximate general rule for the temperature dependence of a reaction is that **the rate doubles for every 10°C increase in temperature**.

Table 2.2 $\Delta G^{\ddagger}$, Rate Constants, Half-Lives, and Completion Times of a Unimolecular Reaction at Room Temperature, 298 K (25°C)

$\Delta G^{\ddagger}$ in kcal/mol	k_{298} in s^{-1}	Half-Life	97% Complete
15 (63 kJ/mol)	64.1	0.01 sec	0.05 sec
16 (67 kJ/mol)	11.9	0.06 sec	0.3 sec
17 (71 kJ/mol)	2.2	0.3 sec	1.6 sec
18 (75 kJ/mol)	4.1×10^{-1}	1.7 sec	8.5 sec
19 (79 kJ/mol)	7.5×10^{-2}	9.2 sec	45.9 sec
20 (84 kJ/mol)	1.4×10^{-2}	49.6 sec	4.1 min
21 (88 kJ/mol)	2.6×10^{-3}	4.5 min	22.3 min
22 (92 kJ/mol)	4.8×10^{-4}	24 min	2 hr
23 (96 kJ/mol)	8.9×10^{-5}	130 min	10.8 hr
24 (100 kJ/mol)	1.6×10^{-5}	11.7 hr	2.4 days
25 (105 kJ/mol)	3.0×10^{-6}	63.2 hr	13.2 days

If the reaction goes by a stepwise mechanism, it may have an energy diagram similar to that shown in Figure 2.21, in which intermediates lie in minima, energy valleys, and transition states are always at maxima, the top of the energy hills. The highest energy hill, $\ddagger_2$, is a major factor in determining the rate of reaction along with the concentration of intermediate C, which may not be known. The *lifetime* of intermediate C is defined as the reciprocal of its rate constant for unimolecular decomposition. As the height of the lowest barrier to decomposition of C drops, the lifetime of C gets shorter. When the lifetime of the intermediate gets less than 10^{-12} s (only long enough for several molecular vibrations), then A + B → D is considered concerted, occurring in one step. Still, the various bond-forming and bond-breaking processes need not be exactly synchronous. If the first step in a multistep process is the slowest, this "bottleneck" step will be the *rate-determining step* and not any of the faster steps that follow in the overall process.

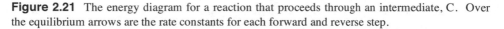

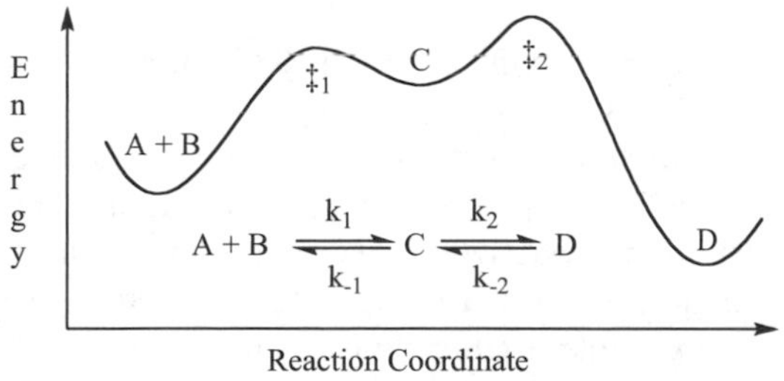
E
n
e
r
g
y
A + B
‡1
C
‡2
D
A + B
k1
k-1
C
k2
k-2
D
Reaction Coordinate

Figure 2.21 The energy diagram for a reaction that proceeds through an intermediate, C. Over the equilibrium arrows are the rate constants for each forward and reverse step.

A transition state contains partially broken and formed bonds, but the degree of this varies and depends on how close to the starting materials or products the point of transition is. The *Hammond postulate*, illustrated by Figure 2.22, states that the transition states of exothermic reactions resemble starting materials in energy and geometry whereas transition states of endothermic reactions resemble products. Transition states close to starting materials, A → B, have rather little bond breaking or making. Transition states close to products, C → D, have almost completed bond breaking and making.

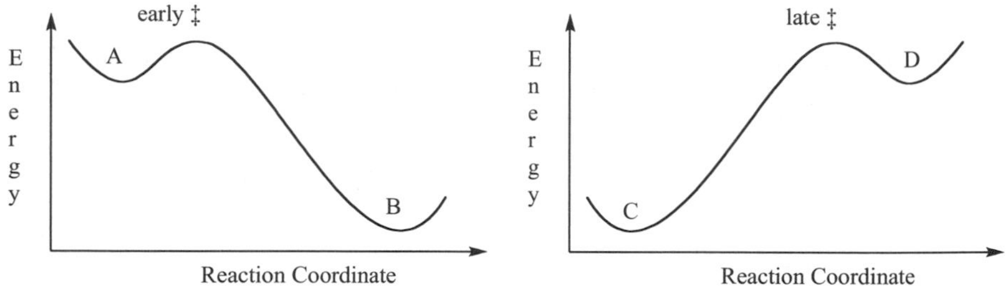
early ‡
A
B
Energy
Reaction Coordinate
late ‡
D
C
Energy
Reaction Coordinate

Figure 2.22 The transition state shifts position as the overall energy of the process changes.

Exercise

Hold a strip of paper in your hand so that both ends are at the same level and bow it up as in Figure 2.17. Notice that the high point, our transition state, is in the exact center. Now lower the end of the strip held in your right hand and notice that the transition point moves toward your left hand. Now raise your right hand higher than your left hand and notice how the transition state shifts and compares with that shown in Figure 2.22.

As shown in Figure 2.23, the stability of the product has a great influence on the energy of the transition states of endothermic reactions (D → E and F) but has very little influence on the energy of the transition states of exothermic reactions (A → B and C).

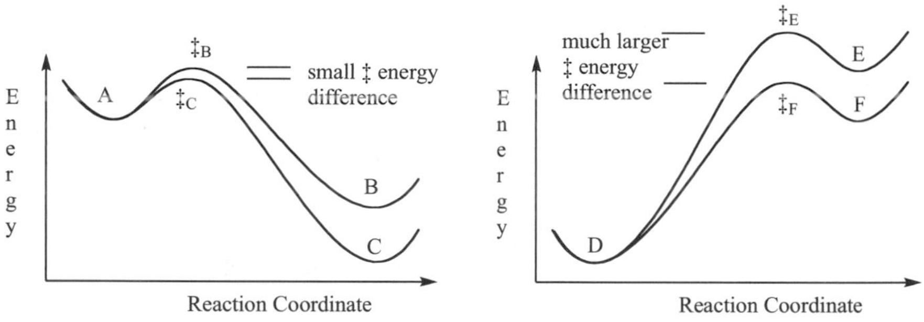
‡B
‡C
small ‡ energy difference
A
B
C
Energy
Reaction Coordinate
‡E
much larger ‡ energy difference
E
‡F
F
D
Energy
Reaction Coordinate

Figure 2.23 The degree to which the product stability influences the transition state.

Therefore, with few exceptions, the more exothermic a reaction is, the less the product stability affects the energy of the transition state and the less selective the reaction is. **The more reactive a species is, the less selective it is.** A general rule of reactivity is, **the more stable a compound is, the less reactive it is**. The converse is also true: The less stable a compound is, the more reactive it is. Basically, a highly reactive, unstable compound might react with almost anything, whereas a less reactive, more stable compound is more limited in what it can react with. It would react at the best site for reaction in preference to other, less favorable sites.

Some reactions can produce two different products, and it is necessary to determine whether the reaction is reversible to be able to predict the product. For example, on the left side of Figure 2.24, if the reaction is completely reversible, equilibrium is established in which the major species in the reaction pot is G, the most stable, lowest-energy species. A reaction that produces the more stable product is under *thermodynamic control*. The equilibrium product ratio is determined by the difference in free energy of the two products, $\Delta\Delta G°$. If the reaction is not reversible, shown on the right side of Figure 2.24, then the product that is formed faster, F, will be the major product. When the relative rates of reaction (barrier heights), $\Delta\Delta G^{\ddagger}$, determine the product ratio, the reaction is under *kinetic control*. As the temperature drops and the molecules have less energy, the reaction takes the lower-barrier pathway.

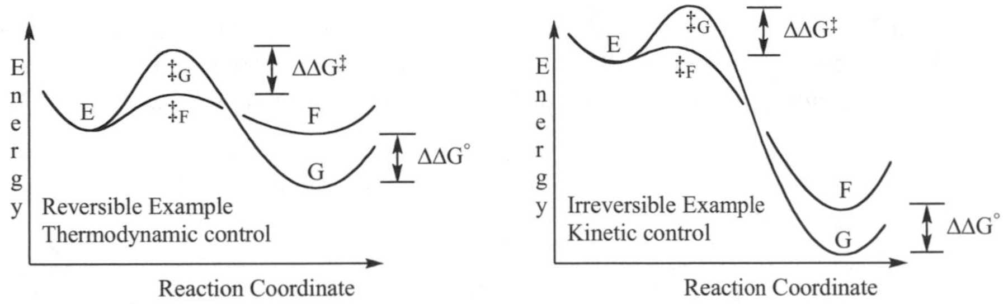

Energy
E
‡G
‡F
F
G
ΔΔG‡
ΔΔG°
Reversible Example
Thermodynamic control
Reaction Coordinate
Irreversible Example
Kinetic control

Figure 2.24 The energy diagrams for two competing reactions. Reaction E to F has the lower barrier to reaction, whereas reaction E to G produces the more stable product. As the reaction becomes more exothermic on the right, it becomes less reversible as the return barriers increase.

When the relative rates of reaction control the product ratio, how much does a small difference in $\Delta G^{\ddagger}$ make? At room temperature, a 1.36 kcal/mol (5.70 kJ/mol) difference in barrier heights, $\Delta\Delta G^{\ddagger}$, means that F is formed about ten times as fast as G. This would mean that the product mixture would contain ten times as much F, the product of traversing the lower barrier, as it would G. In Figure 2.24, if the barrier height for G were less than that for F, then G would be both the kinetic **and** thermodynamic product.

The Basics of Barriers

What causes a reaction barrier? Just like $\Delta G°$, the $\Delta G^{\ddagger}$ can be broken into two terms:

$$\Delta G^{\ddagger} = \Delta H^{\ddagger} - T\Delta S^{\ddagger}$$

The $\Delta H^{\ddagger}$ term arises because bond breaking must usually precede bond making to a slight degree. The old bond must be partially broken before the new one can begin to form. Since it costs energy to break a bond and we have not yet gained any energy from bond formation, extra energy in the form of $\Delta H^{\ddagger}$ is required to start the reaction. Any reaction that requires the complete breakage of a bond before any bond formation can begin will have a larger $\Delta H^{\ddagger}$ than a reaction whose bond breaking and making occur at almost the same time. The larger the $\Delta H^{\ddagger}$ is, the larger the barrier is.

The $\Delta S^{\ddagger}$ term is a measure of the degree of disorder in the transition state. If the transition state for the reaction requires that the reacting molecules be aligned relative to each other in a specific organization, then disorder is sacrificed and the value of $\Delta S^{\ddagger}$ is a negative number. A negative $\Delta S^{\ddagger}$ makes the $-T\Delta S^{\ddagger}$ term a positive number, which adds to the $\Delta H^{\ddagger}$ raising the barrier by increasing the value of $\Delta G^{\ddagger}$. The reaction would proceed slowly because few collisions have the proper alignment. As the temperature increases and molecular motion gets more violent, the alignment problem increases.

There are many factors that will increase the barrier height. **As an intermediate increases in energy, the transition states leading up to it must also increase in energy.** Therefore, the stability of any intermediates will directly affect the barrier height. As shown in Figure 2.25, anything that tends to stabilize the intermediate **relative to the reactant** lowers the barrier, and conversely, anything that stabilizes the reactant **relative to the intermediate** raises the barrier. As an example, if the reactant has additional bonding that is lost in going to the transition state, the barrier is raised. If the transition state has additional bonding that is not present in the reactant, the barrier is lowered. We will spend the next few chapters discussing the stability of intermediates.

Several generalizations on barriers are useful: Formation of charge from neutral reactants tends to raise the barrier, whereas neutralization of charge lowers the barrier. If reactant groups have trouble achieving the proper alignment for reaction, the barrier is raised. If the site for reaction is not very accessible, the barrier is raised. If the intermediate and the transition state leading up to it are better stabilized by solvation than the reactants, the barrier is lowered by this solvent stabilization (see Section 2.7).

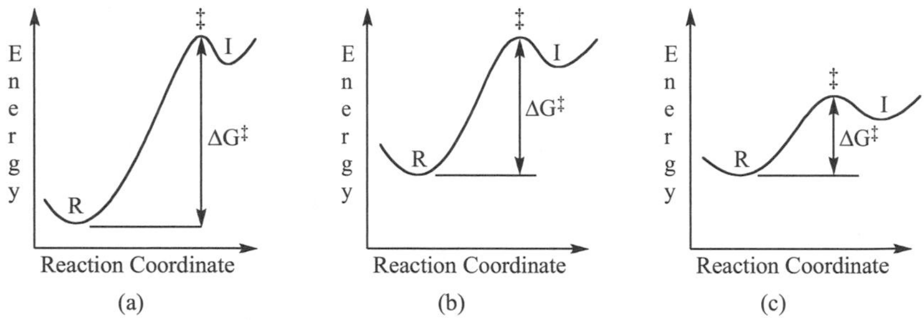
Energy
‡
I
ΔG‡
R
Reaction Coordinate
(a)
Energy
‡
I
ΔG‡
R
Reaction Coordinate
(b)
Energy
‡
I
ΔG‡
R
Reaction Coordinate
(c)

Figure 2.25 Use diagram *b* for reference. In *a* compared to *b* the reactant R is stabilized relative to the intermediate I; therefore the barrier increases. In *c* compared to *b* the intermediate is stabilized relative to reactant R; the barrier decreases.

Mathematical Perspective on Kinetics
(A Supplementary, More Advanced Explanation)

The rate constant, k, is related to the barrier height, $\Delta G^{\ddagger}$, by the Eyring equation:

$$k = (\kappa T/h)e^{-\Delta G^{\ddagger}/RT}$$

where

κ = Boltzmann's constant = 1.38×10^{-23} J/K

T = temperature in K

h = Planck's constant = 6.63×10^{-34} J-s

e = 2.71828

R = gas constant = 1.99×10^{-3} kcal/mol-K (8.33×10^{-3} kJ/mol-K)

Using the above equation, we can calculate the rate constant k for various values of $\Delta G^{\ddagger}$, shown in Table 2.2. The half-life of a unimolecular reaction is equal to $(\ln 2)/k$ or $0.693/k$, and a reaction is 97% complete after five half-lives. Table 2.2 uses the formula $\ln(c°/c) = kt$ to relate the rate constant, k, and the original concentration, $c°$, to the concentration, c, at time t seconds for a unimolecular reaction. Reactant concentrations are very important in a bimolecular reaction. If reactant B in a bimolecular reaction is in large excess, then its concentration will not change significantly over the course of the reaction and can be considered a constant, making the reaction *pseudo-first order*. We can then use the above formula to calculate the concentration of A at time t by substituting the pseudo-first-order rate constant, $k' = k[B]$.

We can estimate the half-life and 97% completion times for a bimolecular reaction in which both reactants are at the same concentration by using the k from Table 2.2 and the formula $(1/c) - (1/c°) = kt$. For example, a reaction with a 20 kcal/mol (84 kJ/mol) barrier and both reactants at 1 M has a 97% completion time of 38.5 min at 25°C (298K).

Example problem

If the concentration of both reactants of a bimolecular reaction were both at 0.1 M, how long would it take to be 97% complete at 25°C if $\Delta G^{\ddagger}$ is 20 kcal/mol (84 kJ/mol)?

Answer: 97% complete means 3% starting material remains, so c is 0.003 M if $c°$ was 0.1 M. Substitution of these values and a k of 0.014 from Table 2.2 into the formula $(1/c) - (1/c°) = kt$ gives the time as 23,095 seconds or 6.4 hours.

Any proposed reaction mechanism must fit the experimentally derived rate expression. The simplified rate expression for the reaction below, from Figure 2.21, is obtained if two assumptions are made: The second step is irreversible (k_{-2} is negligible), and the second step is slow compared to the first (k_2 is much smaller than k_1 or k_{-1}).

$$\mathrm{A + B} \underset{k_{-1}}{\overset{k_1}{\rightleftharpoons}} \mathrm{C} \underset{k_{-2}}{\overset{k_2}{\rightleftharpoons}} \mathrm{D}$$

The reaction rate for the formation of D is k_2[C], where [C] is determined by the equilibrium constant, k_1/k_{-1} = [C]/[A][B]. Solving for [C] gives [C] = k_1[A][B]/k_{-1}. Substitution into the rate expression gives

$$\text{rate} = \frac{d\,[\mathrm{D}]}{d\,t} = k_2[\mathrm{C}] = \frac{k_2 k_1[\mathrm{A}][\mathrm{B}]}{k_{-1}}$$

The rate expression without any restrictions on the rate constants can be obtained by assuming that intermediate C does not build up, the *steady-state approximation*. The rate of formation of C, k_1[A][B] + k_{-2}[D], equals its rate of destruction, k_{-1}[C] + k_2[C]. Solving for [C] gives

$$[\mathrm{C}] = \frac{k_1[\mathrm{A}][\mathrm{B}] + k_{-2}[\mathrm{D}]}{k_{-1} + k_2}$$

Substitution for [C] into the rate expression $d\,[\mathrm{D}]/dt = k_2[\mathrm{C}] - k_{-2}[\mathrm{D}]$ gives

$$\frac{d\,[\mathrm{D}]}{d\,t} = \frac{k_2(k_1[\mathrm{A}][\mathrm{B}] + k_{-2}[\mathrm{D}])}{k_{-1} + k_2} - k_{-2}[\mathrm{D}]$$

This simplifies to a rate expression with a forward and a reverse term just like our simple equilibrium example.

$$\text{rate} = \frac{d\,[\mathrm{D}]}{d\,t} = \frac{k_1 k_2[\mathrm{A}][\mathrm{B}]}{k_{-1} + k_2} - \frac{k_{-1} k_{-2}[\mathrm{D}]}{k_{-1} + k_2}$$

If C → D is irreversible, $k_{-2} = 0$, the second term drops out, giving

$$\text{rate} = \frac{d\,[\mathrm{D}]}{d\,t} = \frac{k_1 k_2[\mathrm{A}][\mathrm{B}]}{k_{-1} + k_2}$$

If the second step is slow compared to the first, $k_{-1} >> k_2$, the expression becomes the same as our example with the two assumptions.

2.7 SOLVENT STABILIZATION OF IONS

Rates of reactions that produce a charged intermediate from a neutral reactant can be accelerated, as in Figure 2.25c, by running the reaction in a polar solvent. Since many nucleophiles, electrophiles, and reactive intermediates are charged, it is important to understand how charged species are stabilized. Nature abhors an isolated, localized charge. There are two ways, intramolecular and intermolecular, that charges are spread out and thereby stabilized. Intramolecular stabilization of charge is discussed in the next chapter. Intermolecular stabilization of charge is the essence of solvation.

Charge can be partially stabilized by polar solvent molecules. The clustering of the solvent dipoles of the opposite charge around the charged molecule becomes, in effect, an intermolecular sharing of the charge (Fig. 2.26). Water can solvate cations and anions well; it has lone pairs with which to stabilize cations and polarized O–H bonds that can hydrogen-bond to anions. A proton in water is commonly represented in organic mechanisms as the hydronium ion, H_3O^+, or sometimes as a proton bonded to four other water molecules as $H_9O_4^+$, but the aggregate is certainly much larger.

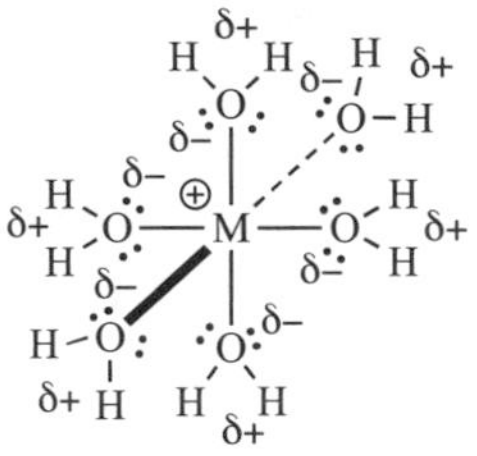

Figure 2.26 The solvation of an inorganic cation in water.

The arrangement of common solvents in the following list is from most to least polar. Solvents are also classified as either *protic* (solvents with acidic hydrogens) or *aprotic* (no acidic hydrogens). The most polar protic solvent, water, has two polarized O–H bonds and two lone pairs. Polar protic solvents can stabilize both cations and anions. Hexane, a nonpolar aprotic solvent, has no polarized bonds (C–H bonds are not polarized) and no lone pairs. **The more alkyl "grease" a solvent has, the less polar it is**. Remember that "**like dissolves like**." Ionic compounds such as salts dissolve in polar solvents and not in nonpolar ones. Nonpolar compounds dissolve in nonpolar solvents and not in polar ones. There are "highly polar aprotic" solvents, such as dimethyl sulfoxide (DMSO) and hexamethylphosphoramide (HMPA) that can stabilize cations much better than anions. The negative dipole end of these solvents is easily accessible and therefore can nestle in close to a cation, solvating it well. However, their positive dipole end is highly hindered and can't get very close to an anion, therefore solvating it poorly. Anions are less stabilized by these solvents, and thus are more reactive. Bases are much more basic in these solvents.

DMSO HMPA

Polar protic

water, acetic acid (carboxylic acids), methanol, ethanol, isopropanol (alcohols)

Polar aprotic

dimethyl sulfoxide (DMSO), dimethylformamide (DMF), acetonitrile, ethyl acetate, acetone (ketones), hexamethylphosphoramide (HMPA), pyridine (amines)

Nonpolar

dichloromethane, chloroform, tetrahydrofuran (THF), diethyl ether (ethers), carbon tetrachloride, benzene, hexane

Solvation is a hard–hard interaction, and therefore the smaller and harder the ion, the more energy is released when it is solvated. The more energy released, the more stabilized and tightly solvated the ion is. Tightly solvated ions will be less useful in reactions because to react they must break out of this stabilizing solvation. The tightness of solvation will have a major effect in determining reactivity. The importance of solvation is often underestimated in chemical reactions.

Solvent polarity can restrict the possibilities for reaction paths. Whenever charged species are present in a reaction, the reaction barrier is highly dependent on the polarity of the solvent (Fig. 2.27). Solvent stabilization of the reactants more than the products decreases reactivity. Solvent stabilization of the products more than the reactants increases reactivity. In addition to stabilizing ions, polar protic solvents can also allow proton transfer and equilibration between the various ionic species in solution.

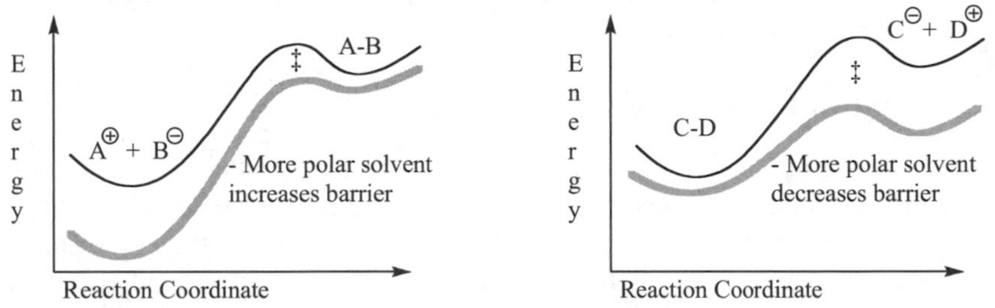
Energy
A-B
‡
A⊕ + B⊖
More polar solvent increases barrier
Reaction Coordinate
Energy
C⊖+ D⊕
‡
C-D
More polar solvent decreases barrier
Reaction Coordinate

Figure 2.27 The effect of a more polar solvent on the barrier to reaction depends on which the polar solvent stabilizes more: the reactants (left) or the products (right).

Example problem

Which of the following solvents, ethanol, hexane, diethyl ether, would dissolve the most KOH? The least?

Answer: KOH is a salt and therefore very polar. It will dissolve the most in a polar solvent, the least in a nonpolar. The most polar solvent of the three is ethanol. Hexane is nonpolar; diethyl ether is slightly polar.

2.8 ENZYMATIC CATALYSIS—LESSONS FROM BIOCHEMISTRY

A *catalyst* is a substance that speeds up a reaction and can be recovered unchanged. The most common catalysts for organic reactions are acids or bases. An acid can protonate a reactant, giving it a full positive charge and making it a better electron sink. A base can remove a proton, making the reactant anionic and a better electron source. Even though another step has been added (protonation or deprotonation), with a better electron source or sink the overall reaction proceeds faster. The reaction barrier, $\Delta G^{\ddagger}$, is lowered by catalysis, and therefore more molecules now have enough energy to react.

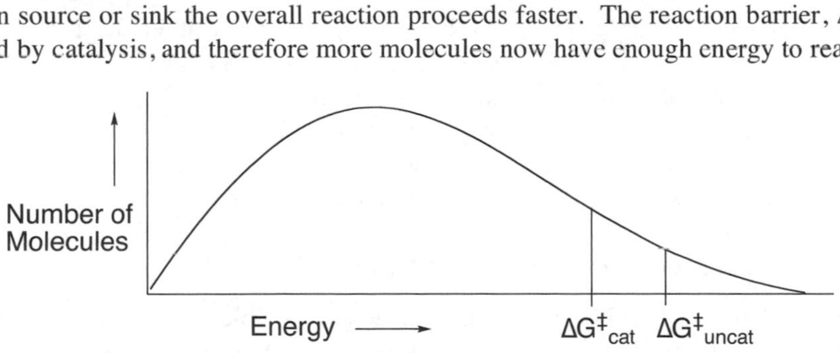
Number of Molecules
Energy
ΔG‡cat
ΔG‡uncat

Figure 2.28 Boltzmann energy distribution and the $\Delta G^{\ddagger}$ for catalyzed and uncatalyzed processes. More molecules have the energy to cross the lower $\Delta G^{\ddagger}{}_{cat}$ barrier than the higher $\Delta G^{\ddagger}{}_{uncat}$ barrier.

Figure 2.28 shows the distribution of energies near room temperature. The area to the right of the barrier value is the proportion of molecules with sufficient energy to react. As the $\Delta G^{\ddagger}$ is lowered by a catalyst, many more molecules have sufficient energy to cross the lowered barrier. Recall that the $\Delta G°$, on the other hand, is determined by the identity of the reactants and products and so is not affected by a catalyst. Just the rate of reaction is changed by a catalyst, not the position of equilibrium.

To really understand the basics of catalysis, we can try to understand how some of the best catalysts work. Enzymes are biochemical catalysts with exceptional abilities to speed up reactions. There are enzymes in your body that are highly selective and work in water and at mild temperatures, and some are so fast that they can produce their products as fast as the reactants can diffuse to them.

Collision Frequency

Molecules have to collide to react. Most reactions are run in solution, so as a solution gets more concentrated, collisions increase and so does the rate. But there is an upper limit, often determined by solubility, on how concentrated the solution can get. A concentration landmark is that pure water is 55.5 mol/liter; pure benzene, C_6H_6, is 11.1 mol/liter. Often chemists must run reactions at 1 mol/liter or less. In addition, not all collisions are effective because usually certain parts of the reactant molecules must hit to react; otherwise they just bounce off each other. This orientation requirement is part of the $\Delta S^‡$ portion of the barrier, $\Delta G^‡ = \Delta H^‡ - T\Delta S^‡$. If the reactive partners are part of the same molecule (intramolecular reactions), they collide as the molecule twists and turns much more frequently than if they were separate. In fact, intramolecular reactions can get effective concentrations of well over a million mol/liter!

Enzyme Active Sites

It is believed that proximity and orientation are critical to the way enzymes accelerate reactions. The key is to increase the number of effective collisions between reactive partners. Not only are the reactive partners undergoing a huge number of collisions within the active site of the enzyme, but the active site is also orienting them correctly. The slow step for some enzymes is just getting the reactants into the active site, the diffusion-controlled limit.

An enzyme active site often has a unique shape; only molecules that have the right shape will fit. This allows the enzyme to select for the correct reactants among all those swimming in the cell soup. Even better, the correct reactant usually has to fit in the active site in the proper orientation to react. This first step takes care of the $\Delta S^‡$ component of the $\Delta G^‡$ barrier by binding the reactant in the proper orientation for reaction. What is left of the barrier is the $\Delta H^‡$ component, which can be lowered by stabilization of the transition state(s) or intermediate(s).

The enzyme active site can also twist the reactant into a more reactive conformation, and can provide the perfect microenvironment for reaction: proper solvation, nearby acid or base catalysts, needed electrophiles and nucleophiles.

Enzymes Bind the Transition State Best

Outside of the enzyme we are limited to two molecule collisions, because the odds of a simultaneous collision of three independent molecules in the proper orientation for reaction are very slim. Three molecule collisions are considered a mechanistic possibility only when two of the three reacting molecules are loosely associated via hydrogen bonding or pi-complexation. However, within the enzyme's active site, there can be several groups snuggled around our reactant to contribute as needed to lower the reaction barrier by stabilizing the transition state.

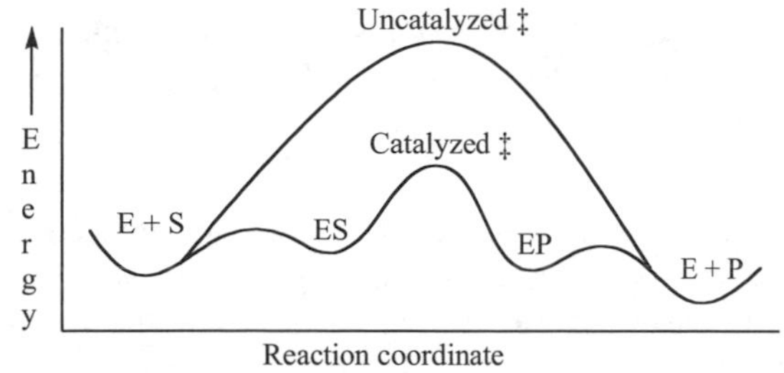
Uncatalyzed ‡
Catalyzed ‡
Energy
E + S
ES
EP
E + P
Reaction coordinate

Figure 2.29 The enzyme E binds the substrate S in the active site to form an enzyme-substrate complex ES. The chemical step occurs to give the enzyme-product complex EP that dissociates into product P and free enzyme E, ready for another cycle.

Figure 2.29 illustrates a simple energy diagram for enzymatic catalysis. For the enzyme-catalyzed chemical step, the transition state fits best in the active site, and so is stabilized the most. The active site can contain the needed acid and/or base catalysts in just the right position. Groups inside the active site often stabilize any reaction intermediates by electrostatics or hydrogen bonding, thus bringing down the barriers leading to those intermediates. Amazingly, the reaction barriers in an enzyme-catalyzed process can be brought so low that their reaction rates are incredibly fast, approaching 10^{20} times faster than the uncatalyzed reaction in water. Later sections will discuss lessons we can learn from the chemical mechanisms of enzymatic catalysis.

2.9 SUMMARY

- Strong bonds require good overlap of the bonding orbitals. The formation of sigma bonds requires approximate colinearity of the reacting orbitals. The formation of pi bonds requires approximate coplanarity of the reacting orbitals.

- The closer in energy the two interacting orbitals are, the greater their interaction. The better the overlap is between the two orbitals, the greater the interaction. Greater orbital interaction between two atoms means a stronger bond between them.

- Principle of microscopic reversibility: The forward and reverse reactions follow the same lowest-energy route but in opposite directions. There is only one energetically best pathway. The transition state of lowest energy maximizes the extent of bonding.

- The HSAB principle: Hard bases favor binding with hard acids; soft bases favor binding with soft acids. Softness is characterized by high polarizability and often little charge; hardness is the reverse. The soft–soft interaction is a covalent bond, which is favored by good overlap of orbitals that are close in energy. The hard–hard interaction is an ionic bond, which is favored by highly charged species with small radii so that they can get close together. When molecules collide, two attractive forces lead to reaction: the hard–hard attraction (opposite charges attracting each other) and the soft–soft attraction (the interaction of filled orbitals with empty orbitals). Solvation is a hard–hard interaction, so the smaller and harder the ion, the more energy is released when it is solvated.

- At room temperature, every drop of 1.36 kcal/mol in $\Delta G°$ increases the equilibrium constant by a factor of 10. Likewise, dropping the $\Delta G^{\ddagger}$ by 1.36 kcal/mol increases the reaction rate tenfold. The $\Delta G^{\ddagger}$ of a reaction that goes at a reasonable rate at room temperature is about 20 kcal/mol.

- Free energy is related to enthalpy and entropy by the equation $\Delta G° = \Delta H° - T\Delta S°$

The standard *enthalpy* change, $\Delta H°$, is a measure of the heat absorbed or evolved in a reaction and is approximated as $\Delta H° = \Delta H_{(\text{bonds broken})} - \Delta H_{(\text{bonds made})}$, whereas $\Delta S°$, the standard entropy change, is a measure of the change in the randomness within the system.

• A *catalyst* is a substance that speeds up a reaction and can be recovered unchanged. An enzyme catalyst increases reaction rates by the dramatic amplification of the number of effective collisions between the reactants within the enzyme's active site. Enzymes increase the rate of the chemical step by using binding to stabilize the transition state more than the reactants or products.

• The more reactive a species is, the less selective it is.

• The more stable a compound is, the less reactive it is.

• As an intermediate increases in energy, the transition states leading up to it must also increase in energy. Anything that stabilizes the intermediate drops the barrier to it.

ADDITIONAL EXERCISES

2.1 There are two ways that a compound can be stable: one due to $\Delta G°$ and the other due to $\Delta G^‡$. Draw energy diagrams to explain.

2.2 Use HSAB theory to predict the direction of the following reaction:

$$2\ CH_3Li + CuCl \rightleftharpoons (CH_3)_2Cu^- Li^+ + LiCl$$

2.3 Using this energy diagram:

(a) What is the slow step?

(b) What is the $\Delta G°$ of C → D?

(c) What is the $\Delta G^‡$ of A → C?

(d) What is the $\Delta G^‡$ of D → C?

(e) Can C be isolated at 25°C?

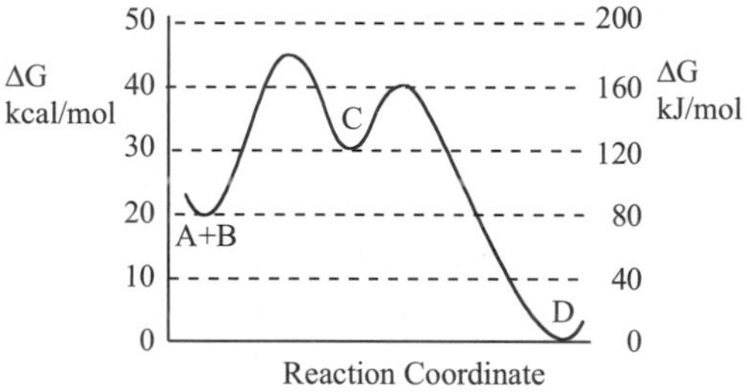
ΔG
kcal/mol
50
40
30
20
10
0
200
160
120
80
40
0
ΔG
kJ/mol
C
A+B
D
Reaction Coordinate

2.4 In problem 2.3, if the reactants A and B were charged and the intermediate C was not, would the barrier to reaction increase or decrease in a more polar solvent?

2.5 Calculate the $\Delta H°$ and decide whether this reaction is exothermic or endothermic.

$$CH_4 + F_2 \longrightarrow CH_3F + HF$$

2.6 Using this energy diagram, what is:

(a) the kinetic product?

(b) the thermodynamic product?

(c) the $\Delta G^‡$ for A → B?

(d) the $\Delta G^‡$ for A → C?

(e) the $\Delta G°$ for A → C?

(f) the most likely product at 0°C?

(g) the equilibrium ratio C/B at 25°C?

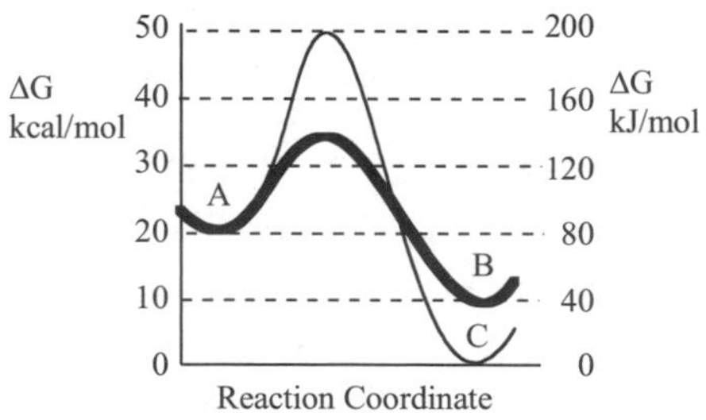
ΔG
kcal/mol
50
40
30
20
10
0
200
160
120
80
40
0
ΔG
kJ/mol
A
B
C
Reaction Coordinate

2.7 At room temperature, a reversible reaction equilibrates A into B with an 85:15 ratio respectively. What is the free-energy difference between A and B?

2.8 Rank all of the following:

(a) Use #1 to indicate most soluble in water.

CH_3CH_2OH $CH_3CH_2OCH_2CH_3$ $CH_3CH_2CH_2OH$ $CH_3CH_2CH_2CH_2CH_3$

(b) Use #1 to indicate the most polar solvent

$CH_3CH_2CH_2CH_2CH_3$ EtOEt H_2O CH_3OH $CH_3CH_2CH_2OH$

(c) Use #1 to indicate the softest Lewis base

F^- CH_3^- OH^- NH_2^-

(d) Use #1 to indicate the weakest H–X bond

HBr HCl HI HF

2.9 Explain why possibility B and not possibility A occurs even though they are technically the same process. Models may help.

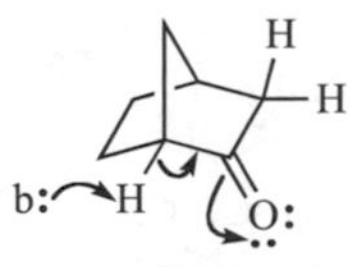

Possibility A

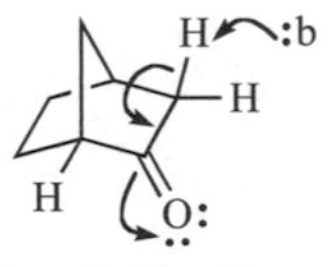

Possibility B

2.10 Use a ΔH calculation to tell which of the two possible products is the more stable.

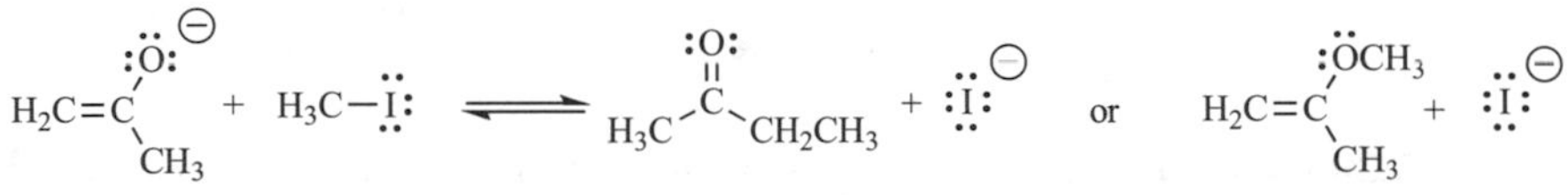

2.11 Using the bond-strength table in the Appendix, calculate the ΔH of reaction for each of the following processes. Would an equilibrium favor reactants or products?

(a)

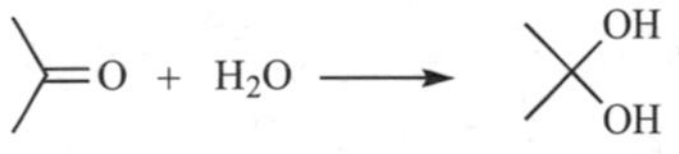

(b)

2.12 In each molecule which of the two boldface carbon atoms is softer?

$I\text{–}\mathbf{C}H_2CH_2CH_2\mathbf{C}H_2\text{–}OSO_2Ph$ $H_2\mathbf{C}{=}CH\text{–}\mathbf{C}HO$

2.13 A reaction equilibrates C to D with a $\Delta G°$ of –1 kcal/mol (–4.2 kJ/mol). Assuming that the reaction remains reversible, and assuming that $\Delta G°$ does not change significantly, what would be the D:C ratio if this reaction were run at –50°C? At 25°C? At 100°C?

2.14 For the reaction below, the equilibrium at 25°C favors the reactant but at 160°C favors the products. In this case, $\Delta G°$ does vary with significantly with temperature, but assume $\Delta H°$ and $\Delta S°$ do not. Explain how the favored species can vary with temperature.

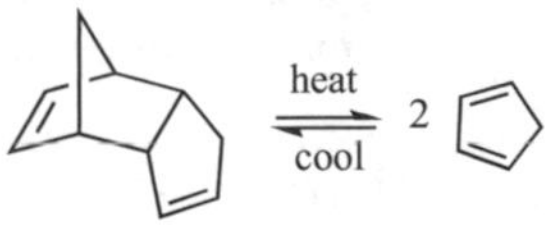
heat
cool
2

2.15 Calculate the ΔH of reaction for the reverse reaction in problem 2.14 and decide whether it is exothermic or endothermic.

2.16 The relative rates of two related unimolecular reactions is 10,000:1 under identical conditions at 25°C. What is the difference in barrier heights?

2.17 For the following allylic species, would attack by an electrophile at carbon occur in the plane of the allylic system (the plane in which the atoms of O–C=C lie) or perpendicular to it? Would attack by an electrophile at oxygen occur in the plane of the allylic system or perpendicular to it?

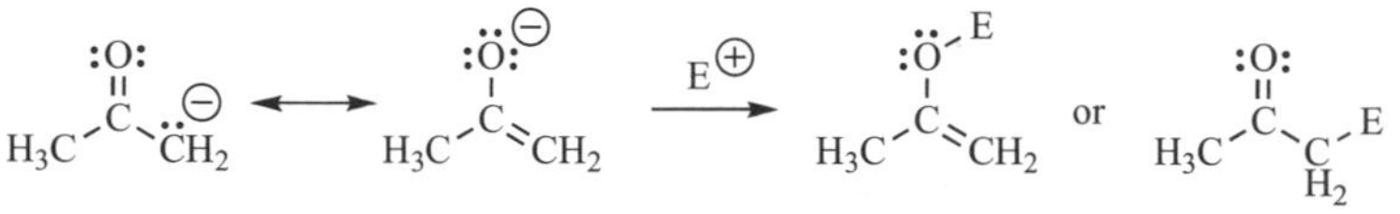

2.18 Calculate the $\Delta H°$ and decide whether this reaction is exothermic or endothermic.

$$H_2C{=}CH_2 + Cl_2 \overset{?}{\rightleftharpoons} ClCH_2CH_2Cl$$

2.19 Draw an energy diagram for the exothermic reaction below (assume ΔH dominates ΔG as it does often). Compound I is a reactive intermediate.

$$A + B \xrightarrow{\text{slow}} I \xrightarrow{\text{fast}} C$$

2.20 Using the energy diagram below for two similar reactions (one trace in bold), decide what determines the preferred product E or F from A + B. Note that intermediate C on route to product F is more stable than intermediate D on route to product E. The barriers from each intermediate to the products differ also. Assume neither route is reversible.

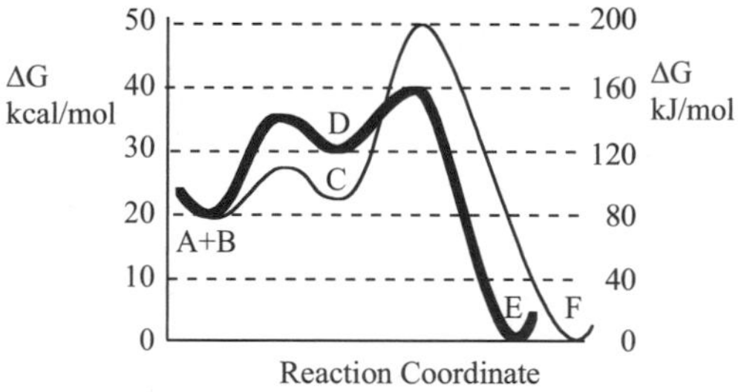
ΔG
kcal/mol
50
40
30
20
10
0
200
160
120
80
40
0
ΔG
kJ/mol
A+B
C
D
E
F
Reaction Coordinate

2.21 The stability of simple alkyl-substituted carbocations, R^+, is shown below.

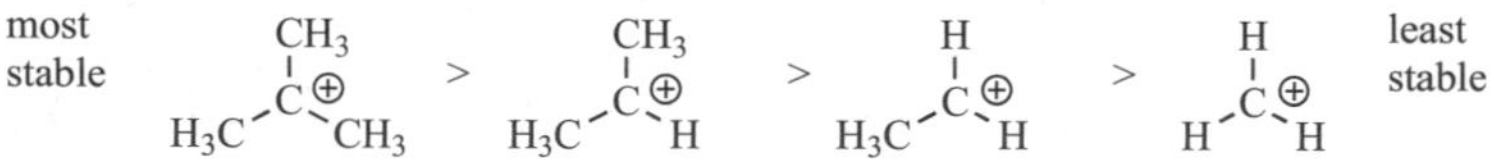
most
stable
least
stable

On a single energy diagram, draw a series of four individual traces for the ionization reaction of R–Br below that reflect these carbocation stabilities as the structure of R varies. Assume all reactions are endothermic. Arbitrarily assign R–Br to the same energy level in all traces, so that the comparison of carbocation stabilities is emphasized.

3

PROTON TRANSFER AND THE PRINCIPLES OF STABILITY

Electron Flow In Organic Chemistry: A Decision-Based Guide To Organic Mechanisms, Second Edition.
By Paul H. Scudder

3.1 INTRODUCTION TO PROTON TRANSFER

Most of the principles used to explain the acidity of compounds are also used to explain the reactivity and stability of organic compounds in general. Also, proton transfer reactions are often the first step in most organic reactions. For this reason, if we can thoroughly comprehend proton transfer, then we have a solid basis for understanding most organic reactions.

3.1.1 Acid and Base Generic Groups and the Proton Transfer Path

A Brønsted acid is a proton donor. A hydrogen atom attached to an electronegative atom is very often quite acidic because the electronegative atom can stabilize the lone pair created when the proton is removed. A Brønsted base accepts the proton to form the conjugate acid. Most anions and lone pairs can serve as Brønsted bases. The conjugate base and the conjugate acid are the base and acid for the reverse reaction. Initially, the base forms a hydrogen bond to the acid; the proton is transferred, and then the hydrogen-bonded conjugate acid/base pair dissociates. This text will symbolize a Brønsted acid as HA and base as b in a proton transfer (p.t.) reaction and use the **p.t. electron flow path. The tail of the first arrow starts from the electron pair on the base, and the arrowhead is near the proton removed. The second arrow starts from the H–A bond, breaks it, and deposits the bonding electron pair on the conjugate base:**

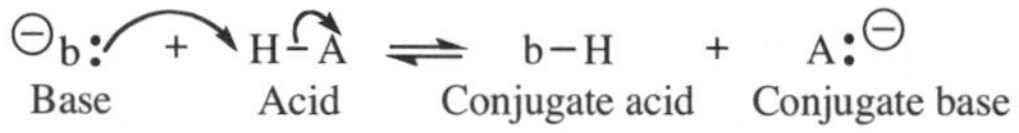
b
H
A
b−H
A
Base
Acid
Conjugate acid
Conjugate base

Figure 3.1 gives examples of four common charge types of proton transfer reactions. The base can be neutral or negative, and the acid can be neutral or positive. It is unusual for the base to be positive or the acid to be negative, because the proton transfer reaction would produce either a dication or dianion, respectively, and those tend to be less stable.

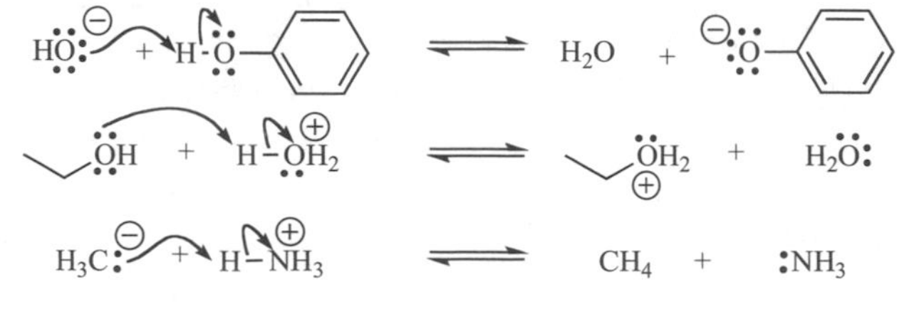

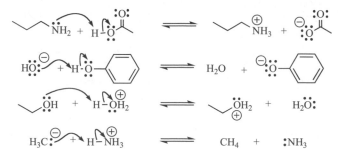

Figure 3.1 Four common charge types of proton transfer reactions (from top): neutral base + neutral acid, anionic base + neutral acid, neutral base + cationic acid, anionic base + cationic acid.

In general, **strong acids have weak, stable conjugate bases**. The more stable the conjugate base is, the stronger the acid. Similarly, the more stable the conjugate acid is, the stronger the base. **Proton transfer reactions favor neutralization, so they prefer to head toward the weaker acid and weaker base.** Some common acids and bases used in organic reactions are listed in the next section. We will first see how the strength of acids and bases are ranked, and then use that data to understand trends in acidity and basicity.

Example problem

Draw the acid, conjugate acid, and electron flow arrows for the reaction below.

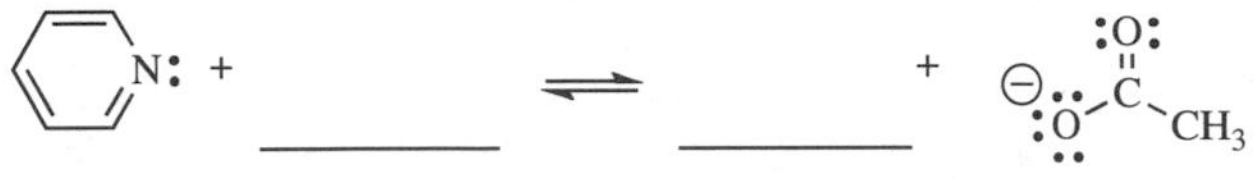

Answer: The missing acid is the conjugate base with a proton added on its most basic atom. The missing conjugate acid is the base with a proton added on its most basic atom. Note that charge has to balance, so the conjugate acid had to be positively charged.

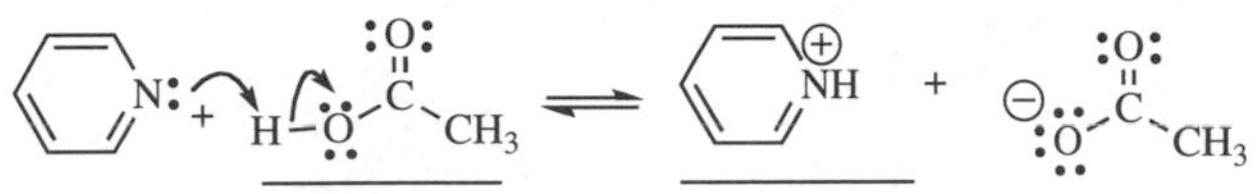

3.2 RANKING OF ACIDS AND BASES, THE pK_a CHART

3.2.1 The pK_a Chart and pH Relationships

$$\text{H–A} \rightleftharpoons \text{H}^+ + \text{A}^- \qquad\qquad K_a = \frac{[\text{H}^+][\text{A}^-]}{[\text{HA}]}$$

$$\text{pH} = -\log[\text{H}^+] \qquad [\text{H}^+] = 10^{-\text{pH}} \qquad \text{p}K_a = -\log K_a \qquad K_a = 10^{-\text{p}K_a}$$

The dissociation constant, K_a, of the acid HA is shown above. The larger the value of the K_a, the more ionized the acid. Since the range of values is so huge, we make use of a log scale similar to that of pH. The pK_a of an acid is the negative log of its K_a. A low pK_a corresponds to a large value for K_a, the dissociation constant. The pK_a chart (see Appendix) serves as a good ranking of acids. **Stronger acids have lower pK_a values. An acid with a negative pK_a is a strong acid**; a pK_a of 0 gives a K_a of 1.

Because the pK_a chart is a composite of many experimental observations under widely different conditions, it has the greatest extrapolation errors at its extremes and is most accurate in the center, between –2 and 16. While we will write for simplicity HA ionizing to give $H^+ + A^-$, it is always true that the proton is transferred to a base. For our pK_a chart (in the Appendix) this base is assumed to be water.

When [HA] = [A^-], these terms cancel out of the K_a expression, leaving K_a = [H^+]; therefore the pK_a of the acid matches the solution pH when [HA] = [A^-]. At this pH the acid HA is half-ionized. As the pH becomes more basic, [A^-] concentration increases and [HA] decreases, as Figure 3.2 illustrates. For each pH unit away from the pK_a the ratio changes by a factor of 10. The following buffer equation gives the relationship between pH, pK_a, and the ratio of [A^-]/[HA]. Buffers are always made near the pK_a of the weak acid. We can use this buffer equation or Figure 3.2 to see that if we wanted a buffer of pH 4 from an acid with a pK_a of 5, we would need a 1:10 ratio of $[A^-]/[HA]$ because log(1/10) = −1.

$$\mathrm{pH} = \mathrm{p}K_\mathrm{a} + \log \frac{[\mathrm{A}^-]}{[\mathrm{HA}]}$$

Figure 3.2 has important biochemical implications. At physiological pH of 7, which is two pH units more basic than its pK_a, the carboxylic acid symbolized by H–A in the figure would be 99% the conjugate base. Both of the amino acids aspartate and glutamate have similar carboxylic acids on their side chain. The overall charge of a protein will depend on pH and pK_as of the groups attached to it.

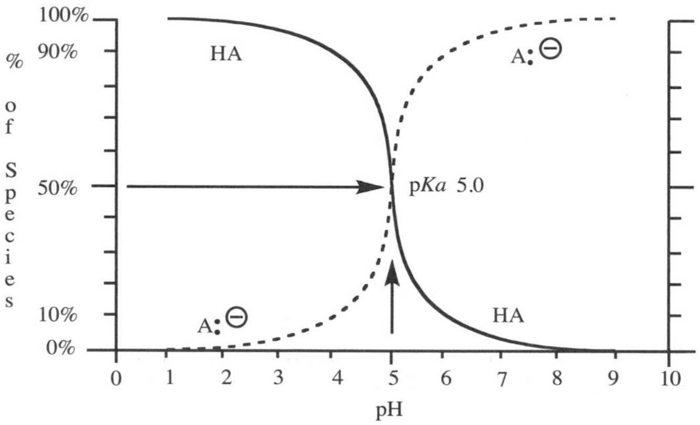
100%
90%
50%
10%
0%
% of Species
HA
A:⊖
pKa 5.0
A:⊖
HA
0
1
2
3
4
5
6
7
8
9
10
pH

Figure 3.2 Species concentration as a function of pH for an acid of pK_a = 5. The acid and conjugate base are equal concentrations at the pH = pK_a. At one pH unit away from the pK_a the species are in a 10 to 1 ratio, so 91:9. Two pH units away from the pK_a will give a ratio of 99:1.

Acids can have more than one acidic proton, and thus have more than one pK_a. The first (lowest number) pK_a is for the most acidic H. Phosphoric acid, H_3PO_4, with three acidic hydrogens, is shown in Figure 3.3.

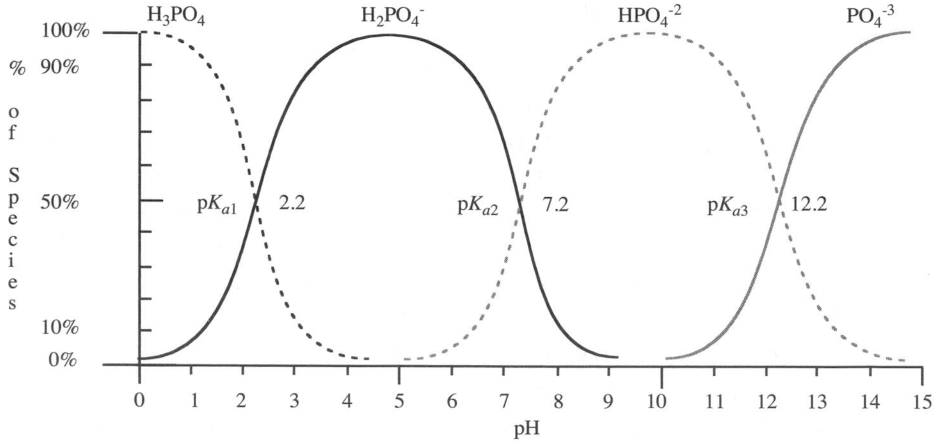
H₃PO₄
H₂PO₄⁻
HPO₄⁻²
PO₄⁻³
100%
90%
50%
10%
0%
% of Species
pKa1 2.2
pKa2 7.2
pKa3 12.2
0
1
2
3
4
5
6
7
8
9
10
11
12
13
14
15
pH

Figure 3.3 Species concentration as a function of pH for H_3PO_4, with pK_{a1} = 2.2, pK_{a2} = 7.2, pK_{a3} = 12.2. Note how each acid and conjugate base concentration varies with pH and that the concentrations of each coupled acid/conjugate base pair cross (become equal) at their related pK_a.

Example problem

Using Figure 3.3, which compounds and in what relative concentration would you have to use to make a phosphate buffer with a pH of 7.2? How about with a pH of 8.2?

Answer: pH 7.2 is at pK_{a2}; equal concentrations of $H_2PO_4^-$ and HPO_4^{2-} are needed. At pH of 8.2, one unit more basic, 10 times more conjugate base HPO_4^{2-} than conjugate acid $H_2PO_4^-$ is needed to get the correct pH. **For any mixture of an acid and its conjugate base, the pH will be within a few pH units of the pK_a of the acid used.**

3.2.2 Media pH Crosscheck

The old phrase "a snowball's chance in hell" illustrates the fact that the environment can play a critical part in the probability of existence of any species. Since $K_w = [H^+][OH^-] = 10^{-14}$, it is easy to calculate one concentration given the other. For example, in pH 1 water, the $[H^+]$ is 10^{-1} M; therefore the concentration of hydroxide ion is 10^{-13} M. The probability that a reactive species will encounter hydroxide in a medium this acidic is very, very low. Likewise it would be highly unlikely to find hydronium in very basic solutions. These principles generalize into several important control knowledge rules:

Acidic media contain powerful electrophiles and rather weak nucleophiles, whereas basic media contain excellent nucleophiles and weak electrophiles. Similarly, no medium can possibly be both strongly acidic and strongly basic.

You could prepare two different solutions, one with an excellent electrophile and the other with an excellent nucleophile, and pour them together. However, as in tossing a snowball into a blast furnace, you must be prepared for the process to be rapid and probably violent.

Forgetting the media restrictions is one of the most common mistakes of students learning to do mechanisms. Keeping Figures 3.2 and 3.3 in mind, you can easily decide whether the medium is acidic or basic by using the pK_a of the reagents. For example, if the solution contains EtOH/EtONa, and knowing that EtONa is ethanol's conjugate base, means that the approximate pH of the solution will be close to the ethanol's pK_a of 16, or very basic. We don't need to be exact, just in the ballpark. Often, all we need to know for the mechanism is whether the solution is acidic or basic.

3.2.3 Common Acids and Their pK_a Values

The next list gives the common acids used in organic reactions in rank order by pK_a.

	Strong acid	
pK_a =	−15	CF_3SO_3H, trifluoromethanesulfonic (triflic) acid
	≈ −9	H_2SO_4, sulfuric acid pK_{a1}
	−9	HBr, hydrobromic acid
	−7	HCl, hydrochloric acid
	−6.5	$HOSO_2C_7H_7$, toluenesulfonic acid
	−1.74	H_3O^+, hydronium ion
	−1.4	HNO_3, nitric acid
	2.2	H_3PO_4, phosphoric acid pK_{a1}
	4.8	CH_3COOH, acetic acid
	9.2	NH_4Cl, ammonium chloride
	Weak acid	

The strongest acid that can exist in a solvent is the protonated solvent. When mixed in water, all the acids above hydronium on the chart will just protonate water to form hydronium. This is called the leveling effect.

Example problem

Which of the following, HNO_3, HF, HCl, is the strongest acid? The weakest acid?

Answer: The acid with the lowest pK_a is the strongest. The strongest acid is HCl at $pK_a = -7$; the next is HNO_3 at -1.4; the weakest is HF at 3.2.

3.2.4 Common Bases and Their pK_{abH} Values

The pK_a chart also serves as a good ranking of bases. For the reaction $b^- + H^+ \rightleftharpoons$ b–H, the proton acceptor is the base b^- and b–H is the conjugate acid. **The basicity of the base increases as the pK_{abH} increases** (pK_{abH} is the pK_a of the conjugate acid). **Strong bases have weak conjugate acids.** Poor solvation of an anionic base will increase its basicity (Section 3.3.8). The following list gives in rank order by pK_{abH} the common bases used in organic reactions. Similar to what was seen in Section 3.2.3, the strongest possible base in a solvent is the deprotonated solvent (the leveling effect).

	Strong base	
pK_{abH} =	≈ 50	CH_3Li, *n*-BuLi
	≈ 36	Et_2NLi, $NaNH_2$, NaH, KH
	≈ 26	$[(CH_3)_3Si]_2NLi$
	≈ 19	*t*-BuOK
	≈ 16	EtONa, CH_3ONa, NaOH
	≈ 10	Na_2CO_3, Et_3N
	≈ 5	CH_3COONa, pyridine
	Weak base	

Example problem

Which of EtO^-, NH_3, NH_2^- is the strongest base? Which is the weakest?

Answer: The strongest base has the highest pK_{abH}. This means that NH_2^- at pK_{abH} 35 is the strongest base. The next strongest is EtO^-, at pK_{abH} 16. The weakest base is NH_3, with a pK_{abH} of 9.2. Remember to look up the conjugate acid: Add a proton to the base, then find it on the pK_a chart. For NH_3 as a base, look up $^+NH_4$ at 9.2 and not NH_3.

3.3 STRUCTURAL FACTORS THAT INFLUENCE ACID STRENGTH

The many factors contributing to acidity can be divided into intrinsic (related to just the compound itself) or extrinsic (related to the environment into which the compound is placed). For the following discussion we will use the example of a neutral acid that produces an anionic conjugate base upon deprotonation. Figure 3.4 shows how stabilization or destabilization of the acid or conjugate base affects the overall energy diagrams to give an unfavorable or a favorable proton transfer.

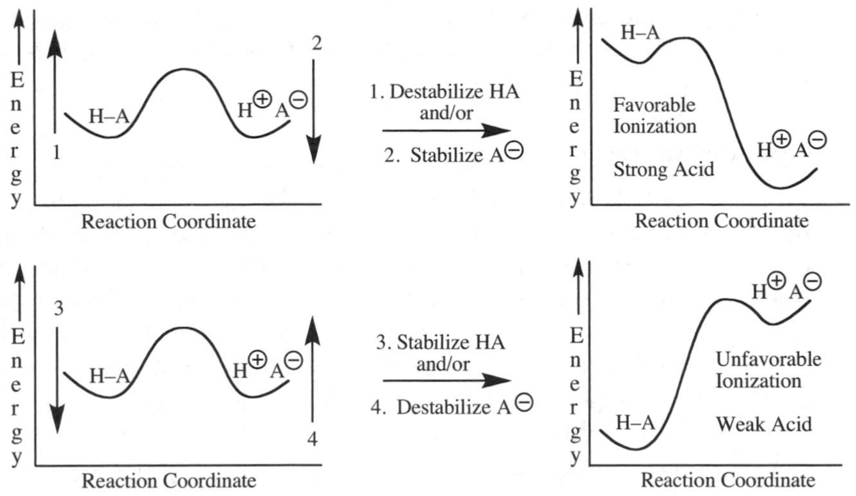
Energy
H–A
H⊕ A⊖
1
2
1. Destabilize HA
and/or
2. Stabilize A⊖
Reaction Coordinate
Favorable
Ionization
Strong Acid
3
4
3. Stabilize HA
and/or
4. Destabilize A⊖
Unfavorable
Ionization
Weak Acid

Figure 3.4 Energy diagrams for unfavorable and favorable proton transfer reactions.

3.3.1 Electronegativity

A hydrogen atom attached to a more electronegative atom will be more acidic. Electronegativity increases as we go to the right on the periodic table because the nuclear charge is increasing. The anionic conjugate base is more stable if the lone pair formed from deprotonation is on a more electronegative atom. The increasing nuclear charge as one goes to the right within a row of the periodic table means that the lone pair on the conjugate base resides in a more stable, lower-energy orbital.

H–CH_3 pK_a = 48	H–NH_2 pK_a = 35	H–OH pK_a = 15.7	H–F pK_a = 3.2
least acidic			most acidic

3.3.2 Strength of the Bond to H

If the H–A bond is stronger, it will be more difficult to break, and the acid will be weaker. The overlap to H decreases as we go down a column in the periodic table, so the H–A bond gets weaker. The acidity increases going down the column from fluorine to iodine: HF < HCl < HBr < HI. The overlap with the hydrogen's 1*s* orbital is worst with the 5*p* orbital of iodine, gets a little better with the 4*p* of bromine, better still with the 3*p* of chlorine, and is best with the 2*p* of fluorine.

3.3.3 Resonance Effects

Resonance effects allow us to delocalize the nonbonding electron pair formed from loss of a proton. An example of increasing acidity as more resonance stabilization is added is CH_3CN pK_a = 25, $CH_2(CN)_2$ pK_a = 11.2, $CH(CN)_3$ pK_a = –5. Resonance effects are dramatic and dominant. Tricyanomethane at pK_a = –5 is more acidic than nitric acid pK_a = –1.4, and is one of the strongest carbon acids. Since the cyano group delocalizes the anionic pair of electrons into its pi system, it is called a pi acceptor. The best electron-withdrawing groups (Section 3.5.1) are pi acceptors that can place the anion on an electronegative atom, like in the resonance form shown below. The stabilization by the cyano group also includes inductive and field effects, discussed next.

H₂C–C≡N:
H₂C=C=N:

With neutral oxygen acids, the more the conjugate base can distribute the negative charge, the stronger the acid. For example, HOCl at pK_a = 7.49, is a much weaker acid than $HClO_4$ at pK_a = −10, which delocalizes the resulting anion over four oxygen atoms.

3.3.4 Inductive/Field Effects

Inductive effects are electron withdrawal or donation through the sigma bond framework of the molecule. Field effects are through-space electrostatic attractions and repulsions. Since it is hard to separate the through-bond from through-space effects, inductive and field effects are grouped together. A good relative ranking of these inductive/field effects for various groups, G, is the acidity of a series of carboxylic acids, G–CH_2–COOH. Most groups, being more electronegative than carbon, tend to increase acidity. For example, if G is the very electronegative trimethylammonium group, then $(CH_3)_3N^+$–CH_2–COOH, is more acidic than acetic acid (G = H) by three pK_a units, a factor of 1000. Inductive/field effects fall off rapidly with distance; the effect of the group in G–CH_2–CH_2–COOH is only 40% of that for G–CH_2–COOH. Inductive/field effects are approximately additive. For example, the acidity of Cl_2CH–COOH is about halfway between that of ClCH_2–COOH and Cl_3C–COOH.

The resonance and inductive/field effects can conflict, and resonance effects usually dominate. For example, an oxygen atom has lone pair of electrons it can donate into a pi system by resonance (pi donor). At the same time the oxygen atom is electronegative, so is withdrawing by an inductive/field effect. Section 4.2.5 will cover the ranking of electron-releasing groups (erg). In the three phenols shown below, stabilization of the anionic conjugate base by a good electron-withdrawing group will increase the acidity of phenol. The lowest pK_a (most acidic) is 7.2, the nitrophenol. The unsubstituted phenol at pK_a 10.0 is next. Least acidic at pK_a 10.2 is the methoxyphenol. The inductive effect of the electronegative methoxy oxygen has fallen off with distance, whereas the resonance donation from the methoxy lone pair destabilizes the conjugate base and decreases the acidity of the phenol.

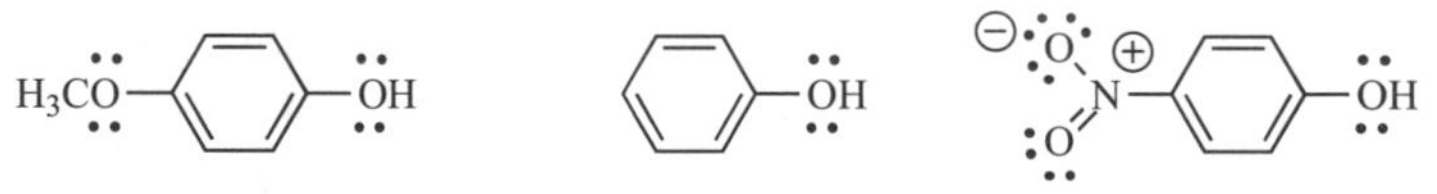
H3CO
OH
OH
O
N
O
OH

3.3.5 Hydrogen Bonding

Internal hydrogen bonding to the conjugate base will definitely increase acidity. Fumaric acid below has two acidic hydrogens, pK_{a1} = 3.02 and pK_{a2} = 4.38. The second carboxylic acid can inductively withdraw and lower the first pK_a but is held too far away to hydrogen bond. On the other hand, maleic acid's two acidic hydrogens differ much more in acidity, pK_{a1} = 1.92 and pK_{a2} = 6.23. For the first deprotonation, the other carboxylic acid in maleic acid can easily hydrogen bond to the conjugate base and stabilize the anion. The second deprotonation of maleic acid is more difficult than that of fumaric acid because of the destabilizing field/inductive effect of two like charges being held closer together.

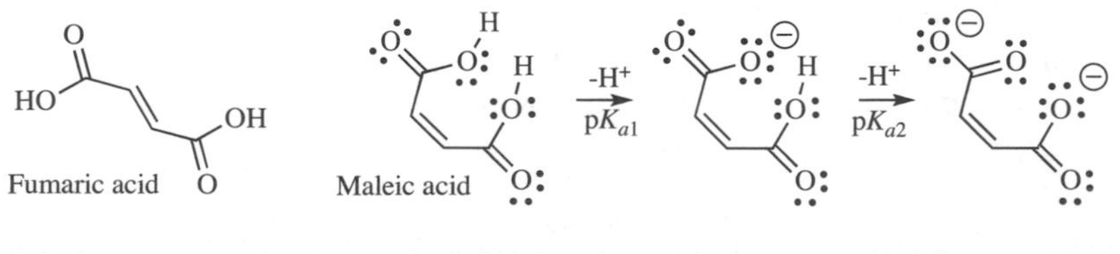
O
HO
OH
O
Fumaric acid
Maleic acid
-H+
pKa1
-H+
pKa2

3.3.6 Aromaticity

Cyclopentadiene is unusually acidic for a carbon acid, with a pK_a of 16. The conjugate base has six electrons in an unbroken loop of *p* orbitals and so is stabilized by aromaticity (Section 1.8). Aromaticity stabilizes this ion by more than 25 pK_a units.

3.3.7 Charge

Positively charged acids are more acidic than their corresponding neutrals, as illustrated by the lower pK_a of the cations of the following compounds.

Cation	pK_a	Neutral	pK_a
NH_4^+	9.2	NH_3	35
H_3O^+	−1.74	H_2O	15.7
More Acidic		Less Acidic	

3.3.8 Extrinsic Factors—Solvation and Ion Pairing

Besides resonance and inductive/field effects, solvation can affect acidity. Protic solvents like water or alcohols stabilize the conjugate base by hydrogen bonding. In fact, our common pK_a chart is referenced to water and would have very different numbers if done in a polar aprotic solvent like dimethyl sulfoxide (DMSO) that can't hydrogen bond. Anions are much more basic in DMSO [for example, hydroxide (pK_{abH} 15.7) has a pK_{abH} in DMSO of 31.4]. Bulky alkyl groups can sterically hinder solvation and will decrease acidity; for example, acidity decreases with CH_3OH > $(CH_3)_2CHOH$ > $(CH_3)_3COH$. Polar solvents also favor dissociation by stabilizing charged species by clustering the oppositely charged ends of the solvent dipoles around the charged species as shown in Figure 2.26. An acid will dissociate less in a nonpolar solvent.

Enzyme active sites can be very different from bulk water (often quite nonpolar), so the pK_as of groups within can differ by several orders of magnitude. Electrostatic stabilization or destabilization of charge by neighboring charged groups also alters the pK_as of acidic and basic groups.

Ion pairing (two close opposite charges) is a very significant source of stabilization via ionic bonding. Biochemical systems often make use of 2+ ions for catalysis. A water molecule that is complexed to a cation like Zn^{2+} is significantly more acidic than bulk water, because the conjugate base, hydroxide, is stabilized by ion pairing with the 2+ ion. The enzyme carbonic anhydrase does exactly this to create a hydroxide nucleophile at neutral pH to rapidly convert carbon dioxide to bicarbonate. Electrostatic catalysis is the acceleration of a reaction by ion pairing with the electric field of an opposite charge.

3.3.9 Super Acid Systems

Although trifluoromethanesulfonic acid, CF_3SO_3H, and fluorosulfonic acid, FSO_3H, are extremely strong acids, both with an estimated pK_a of about −15, even stronger acid systems exist. The combination of a strong acid and a Lewis acid to complex the conjugate base can produce an acid with an estimated pK_a less than −20. For example, the combination of FSO_3H with the Lewis acid SbF_5 is a super acid (sometimes called magic acid) that is a strong enough acid system to protonate an alkane. The strongest super acid system to date is HF and SbF_5, which has the very nonnucleophilic conjugate base SbF_6^- and has been used to study carbocations (themselves strong acids) under stable ion conditions.

3.4 STRUCTURAL FACTORS THAT INFLUENCE BASE STRENGTH

Since strong bases have weak conjugate acids, the information in the previous section also applies to base strength, only in reverse. Anything that stabilizes the conjugate acid lowers its acidity, and thus makes the base stronger. The bottom of Figure 3.4 shows that stabilization of HA or destabilization of A^- makes the acid HA weaker, thus making the base A^- stronger. Therefore, as was shown in Section 3.2.4, the higher the pK_{abH} the stronger the base.

3.4.1 Anions Are More Basic than the Corresponding Neutral Species

Anions are more electron rich and so are more basic as illustrated by the higher pK_{abH} of the anions compared to their neutrals in the following compounds.

Anion	pK_{abH}	Neutral	pK_{abH}
NH_2^-	35	NH_3	9.2
$PhNH_2^-$	27	$PhNH_3$	4.6
OH^-	15.7	H_2O	−1.74
EtO^-	16	EtOH	−2.4
More Basic		Less Basic	

3.4.2 Lone Pairs with Less *s* Character are More Basic

Keeping within the same atom, mixing more *s* orbital into the lone pair hybridization stabilizes the lone pair because a 2*s* orbital is lower in energy than a 2*p* orbital. The more stable lone pair is less basic and harder to protonate illustrated by the lower pK_{abH} of the following compounds. Sections 3.4.4 to 3.4.6 are exceptions to this and to the next trend.

sp^3 Lone Pair	pK_{abH}	sp^2 Lone Pair	pK_{abH}	*sp* Lone Pair	pK_{abH}
	10.6		5.2	$H_3C{-}C{\equiv}N$:	−10
Most Basic				Least Basic	

3.4.3 Lone Pairs on Less Electronegative Atoms are More Basic

Lone pairs on electronegative atoms are held tighter in lower-energy orbitals because of the increased nuclear charge. These stabilized lone pairs are less available for protonation, as illustrated by the lower pK_{abH} of the following compounds.

Less Electronegative	pK_{abH}		pK_{abH}	More Electronegative	pK_{abH}
CH_3^-	48	NH_2^-	35	OH^-	15.7
Most Basic				Least Basic	

3.4.4 Conjugated Systems Preserve Conjugation If At All Possible

Lone pairs conjugated with pi systems are less basic than lone pairs not involved with conjugation. Conjugation provides extra bonding, and that stabilizes the electrons in the conjugated system, making them less basic. The following examples show the stabilization coming from conjugation dominating. Amides protonate on the oxygen lone pairs not involved in the allylic conjugation of the amide even though oxygen is more electronegative than nitrogen.

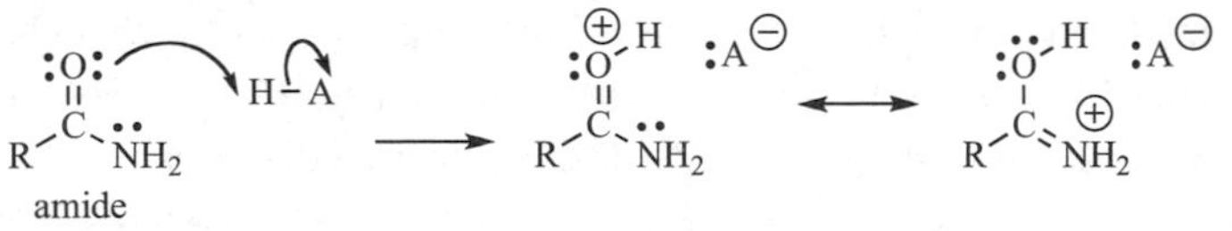

The base DMAP below protonates on the sp^2 nitrogen that is perpendicular to the pi system and not the lone pair involved in conjugation with the pyridine ring. Likewise, imidazole protonates on the lone pair not involved in the aromaticity of the ring system.

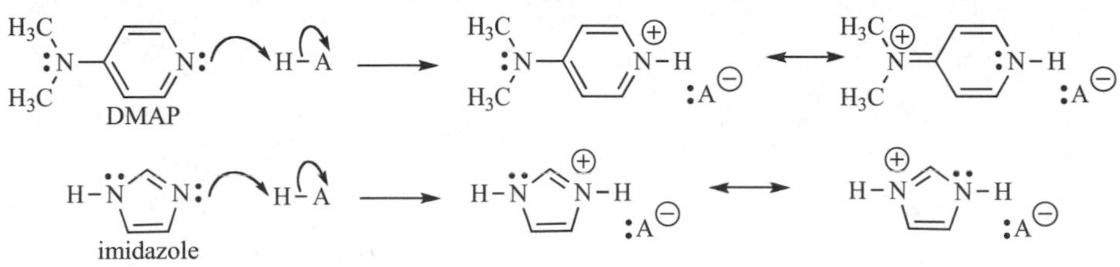

3.4.5 Inductive/Field Effects Affect the Stability of Conjugate Acids

Protonation of lone pairs adjacent to the partial positive charge of a carbonyl is less favored than protonation of the carbonyl itself. Protonation of the ester carbonyl produces a delocalized cation, whereas protonation of the ester's ether oxygen is destabilized by having a positive charge next to the partial positive carbonyl carbon.

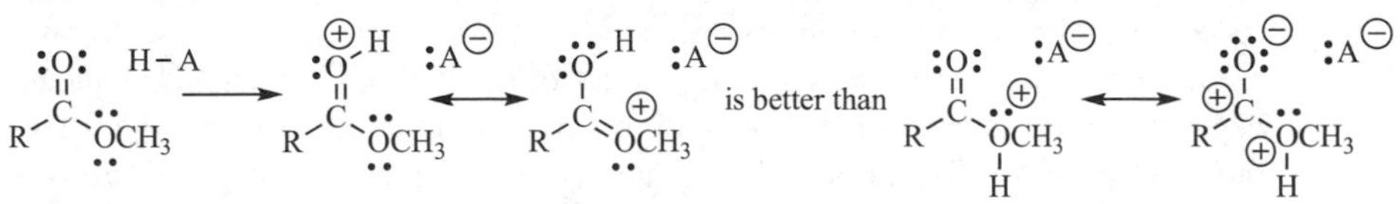

3.4.6 Resonance Effects Affect the Stability of Conjugate Acids

Arginine is an amino acid containing a guanidinium group that is positively charged at physiological pH. The resonance-stabilized cationic conjugate acid of this side chain group has a pK_{abH} of 13.2. At that very basic pH, it would be half protonated, half neutral. Arginine in a protein's active site can be a positive charge to attract a negatively charged substrate.

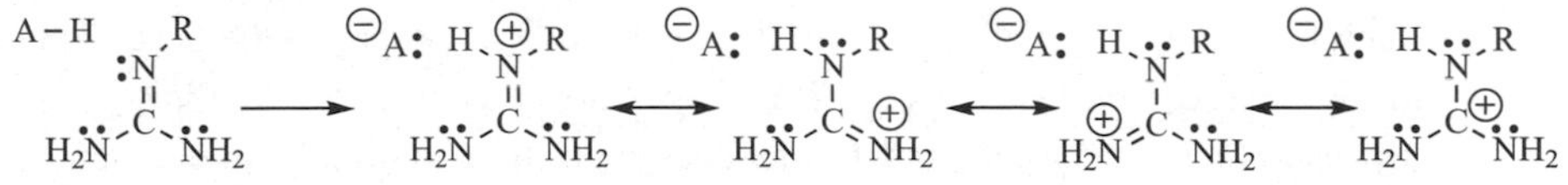

3.5 CARBON ACIDS AND RANKING OF ELECTRON-WITHDRAWING GROUPS

Carbon acids can serve as good examples for the principles of acidity. Often organic reactions in basic media start with deprotonation of an acidic C–H, so we will need to understand carbon acids in more detail. The major factor governing the acidity of the carbon–hydrogen bond is the stability of the resultant carbanion (conjugate base). A useful generalization is: **the more acidic the carbon acid (lower pK_a), the more stable the carbanion.** A pK_a chart then becomes a handy reference for carbanion stability.

3.5.1 Ranking of Electron-Withdrawing Groups

The major factor in the acidity of carbon acids is resonance delocalization of the conjugate base by electron-withdrawing groups. The pK_a chart becomes a guide (good enough for our purposes) for the approximate ranking of the common electron-withdrawing groups. The acidity of CH_3–ewg increases as groups better at withdrawing electrons more effectively stabilize the anionic conjugate base, $^-CH_2$–ewg. If the pK_as of CH_3–ewg's are compared, **the more acidic (lower pK_a) the CH_3–ewg is, the better the ewg group is at withdrawing electrons.** The following is a list of electron-withdrawing groups ranked by the pK_a of CH_3–ewg.

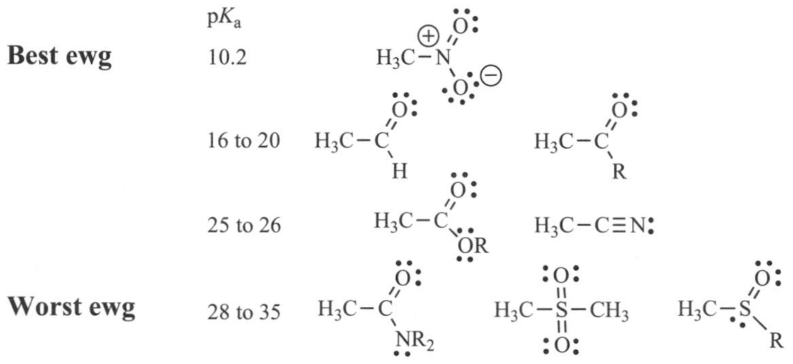
pKa
Best ewg
10.2
16 to 20
25 to 26
Worst ewg
28 to 35

The best electron-withdrawing groups can stabilize the anion by resonance; for this reason they are called pi acceptors. They delocalize the anion and distribute the charge onto more electronegative atoms. The two most common electron-withdrawing groups are the nitrile and the carbonyl. The ability of the carbonyl to withdraw electron density depends on the nature of the group bound on the other side of it. If the group attached to the other side of the carbonyl also has a lone pair donating in electron density, the carbonyl becomes a poorer electron-withdrawing group. For this reason, ketones are better electron-withdrawing groups than esters, which are better than amides.

Example problem

Which of the following, CN, Ph, NO_2, is the best electron-withdrawing group? Which is the poorest?

Answer: The best electron-withdrawing group has the lowest pK_a for CH_3–ewg. Therefore CH_3NO_2, with a pK_a of 10.2, makes the nitro group the best. The next is the nitrile, CH_3CN, with a pK_a of 25. The worst electron-withdrawing group is phenyl because $PhCH_3$ has a pK_a of 40. The phenyl group is not really an electron-withdrawing group; its field/inductive effect is mildly withdrawing. It stabilizes the charge simply by delocalization, something it does almost equally well for either anions or cations.

Groups that are not pi acceptors stabilize an adjacent anion less effectively, and their CH_3–ewg pK_as are not yet available. Very electronegative groups, such as trimethylammonium cation, $^+N(CH_3)_3$, and trifluoromethyl, CF_3, can withdraw electron density through field/inductive effects. The halogens are resonance electron donors, but their inductive effect is withdrawing. The field/inductive effect of an electron-withdrawing group (resonance effects blocked) is occasionally needed and can be ranked by the effect on the pK_a of the carboxylic acid group, ewg–CH_2COOH (Section 3.3.4).

3.5.2 Carbanion Stability Trends

Hybridization has a dramatic effect on carbanion stability. Table 3.2 illustrates the strength, pK_a, electronegativity, and hybridization trends for some simple C–H bonds. Since the 2*s* orbital is closer to the nucleus and lower in energy than the 2*p* orbital, an anion residing in a hybrid orbital with greater %*s* (therefore a more electronegative orbital) is more stable. For carbon acids, the electronegativity of the carbon is more important than the C-H bond strength.

Resonance likewise has a major effect. Conjugation with double bonds or aromatic rings stabilizes anions by distributing the charge over more atoms. An electron-withdrawing group (ewg) directly bonded to the carbon bearing the negative charge can stabilize the anion best. The better the electron withdrawal is, the greater the stabilization. The stabilization is greatest by resonance (through pi bonds) to a pi acceptor. The more electron-withdrawing groups on the carbanion, the more stable it will be. However, additional groups do not have as much effect as the first group. The attachment of one carbonyl drops the pK_a of methane from 48 to 19.2; the attachment of a second carbonyl group drops the pK_a less, from 19.2 to 9.0.

Table 3.2 Hybridization of the C–H bond and Acidity

Decreasing stability of carbanion formed →	HC≡CH	$H_2C{=}CH_2$	$H_3C{-}CH_3$
Hybridization of carbon	*sp*	sp^2	sp^3
%*s* character of carbon	50	33	25
pK_a of C–H bond	25	44	50
Electronegativity of C	3.29	2.75	2.48
Strength of C–H bond (kcal/mol)	131.9	110.7	101.1

Other factors contribute, but often to a much lesser degree. Inductive and field effects from electronegative atoms can also stabilize the carbanion somewhat. Hydrogen bonding is very rare for carbanions. Aromaticity greatly stabilizes carbanions, but the only common example is cyclopentadienyl anion $C_5H_5^-$.

Solvation of carbanions is complicated by their basicity. Less basic carbanions (pK_{abH} ≤ 25) can be made in protic media like alcohols that can hydrogen bond. But more basic carbanions will just deprotonate the protic solvent, and be quenched. Therefore nonprotic solvents such as diethyl ether are commonly used as solvents for these very basic carbanions.

Example problem

Which of the following, $^-$C≡CR, $^-CH_2Ph$, $^-$CH(C≡N)$_2$, is the most stable carbanion? Which is the least?

Answer: The most stable carbanion will be the weakest base and have the lowest pK_{abH}. This means that $^-$CH(C≡N)$_2$, with a pK_{abH} of 11.2, is the most stable. The next in stability is $^{-}C{\equiv}CR$, with a pK_{abH} of 25. The least stable carbanion is $^{-}CH_2Ph$, with a pK_{abH} of 40.

3.5.3 Estimation of the pK_a for Related Compounds

The pK_a for compounds not on the pK_a chart must be estimated using closely related compounds. Since the hydrogens of a simple alkyl chain are not acidic, we need to focus only with the hydrogens on electronegative heteroatoms or hydrogens immediately adjacent to functional groups. We are looking for anything that can intrinsically stabilize the conjugate base. Hydrogens on electronegative atoms are reasonably acidic (Section 3.31). Hydrogens on atoms adjacent to the δ+ end of a polarized multiple bond (C=Y or C≡Y) are acidic because the conjugate base can delocalize the electron pair formed (Section 3.3.3).

Identify the functional groups in the compound, and then find the pK_a for hydrogens on or adjacent to those functional groups. In this way the pK_a for $(CH_3)_2C{=}O$ on the chart must do for all simple ketones. Likewise the pK_a for ethanol will have to do for 1-butanol. For compounds of similar structure, the error from this estimation process is usually only one to two pK_a units. Get as close as you can to the structural piece that will stabilize the conjugate base. Do not confuse the stabilization by a ketone with that of an ester or amide, simply because all three have a carbonyl group.

Example problem

What is the pK_a of $CH_3CH_2CH_2CH_2CH_2C{\equiv}N$?

Answer: The functional group is a nitrile, R−C≡N. The only nitrile on the pK_a chart is $CH_3C{\equiv}N$ at a pK_a of 25. Therefore in $CH_3CH_2CH_2CH_2CH_2C{\equiv}N$ the CH_2 next to the nitrile will have a pK_a close to 25. The other CH_2s are not acidic because the conjugate base would not have any stabilization.

3.5.3 Finding the Most Acidic Hydrogen

Usually it is the most acidic hydrogen that is deprotonated in a proton transfer reaction. However, finding the most acidic hydrogen of a compound is not always easy, especially if the compound has more than one functional group. A hydrogen on a carbon situated between two polarized multiple bonds, for example, $N{\equiv}C{-}CH_2{-}C{\equiv}N$, would be even more acidic because the conjugate base is very delocalized. In conclusion, check all hydrogens on electronegative atoms and those adjacent to functional groups; watch for any H whose loss would produce a highly delocalized system. Compare representative pK_as for each to find the most acidic H.

Example problem

What is the most acidic hydrogen of the following compound?

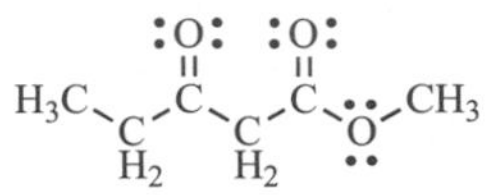

H3C
C
H2
C
O
C
H2
C
O
O
CH3

Answer: The functional groups are a ketone and an ester. Both CH_2s are adjacent to the δ+ end of a polarized multiple bond and are acidic. The pK_as of each set of hydrogens are given below.

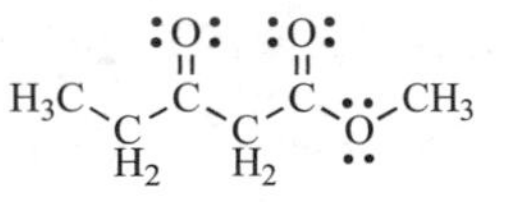
50 19.2 10.7 ≈ 50

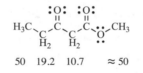

Neither methyl group is acidic, for both lack any sort of stabilization of their conjugate base. The CH_2 between the two carbonyls is the most acidic; it has the lowest pK_a because the anion is delocalized.

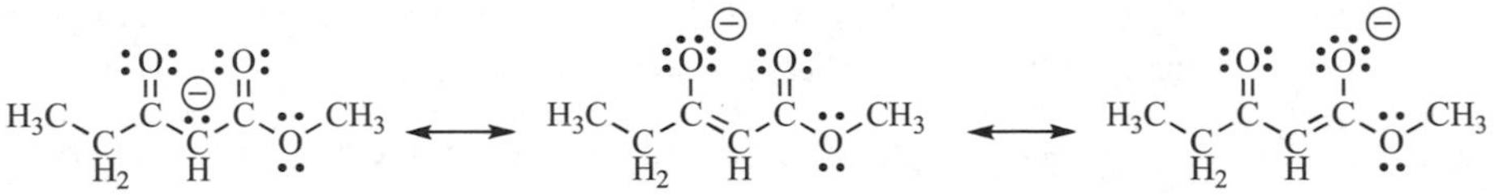

The previous examples illustrated the estimation of the pK_a for compounds that were structurally close to ones on the pK_a chart. It is necessary to recognize general structural features that affect acidity so that crude estimations of acidity can be made for compounds that do not have close relatives on the pK_a chart. Structural features that affect acidity are many, and examples can be found on the pK_a chart.

3.5.4 Carbanion Stabilization from a HOMO–LUMO Perspective
(A Supplementary, More Advanced Explanation)

The best electron withdrawal occurs when the energy of the pi acceptor's LUMO is close to that of the full 2*p* orbital of the carbanion. For example, compare how two groups, the C=C and the C=O, stabilize an adjacent carbanion. The pK_a of methane is about 48; if we attach a C=C, which can stabilize the anion by delocalization, the pK_a drops to 43. If we instead attach a C=O, which can stabilize by delocalization and is a more electronegative group, the pK_a drops all the way to 19.2, indicating a much greater stabilization of the anion. As illustrated by Figure 3.5, the LUMO of the carbonyl is lower in energy than that of a simple carbon–carbon double bond (recall Section 2.5) and therefore interacts more with the full 2*p* orbital of the carbanion.

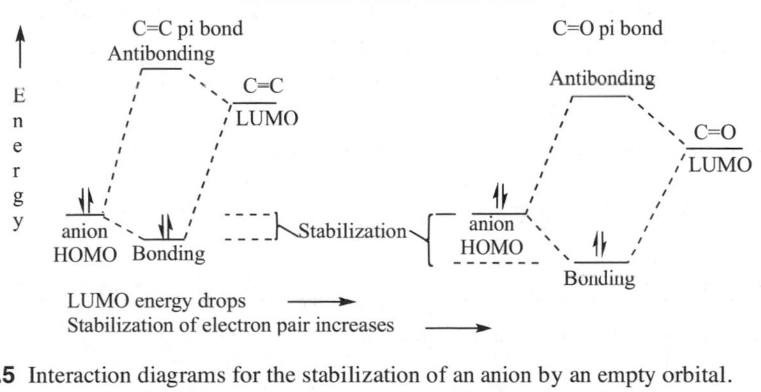
Interaction of a full P orbital with a –
C=C pi bond
Antibonding
C=C
LUMO
Energy
anion
HOMO
Bonding
Stabilization
C=O pi bond
Antibonding
C=O
LUMO
anion
HOMO
Bonding
LUMO energy drops
Stabilization of electron pair increases

Figure 3.5 Interaction diagrams for the stabilization of an anion by an empty orbital.

The lower in energy the LUMO of the electron-withdrawing group (ewg) is, the better the group will be at withdrawing electrons. The lower-energy LUMO can interact better with filled orbitals. The better electron-withdrawing groups contain a low-lying empty *p* orbital or π* orbital that can form a strong pi bond to the full *p* orbital of the anion.

3.6 CALCULATION OF K_{eq} FOR PROTON TRANSFER

Calculation of the equilibrium constant for the proton transfer reaction is easy. **Proton transfers go toward the formation of the weaker base.** Let's use the reaction of hydroxide with hydrochloric acid as an example. The reaction goes from the stronger base, hydroxide, pK_{abH} 15.7, to the weaker base, chloride, pK_{abH} –7.

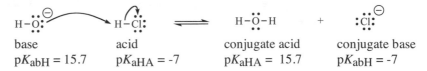

base	acid	conjugate acid	conjugate base
pK_{abH} = 15.7	pK_{aHA} = -7	pK_{aHA} = 15.7	pK_{abH} = -7

Because the reaction favors products, we expect the K_{eq} to be greater than 1. We can estimate the K_{eq} by knowing that each pK_a unit gives a factor of 10 in K_{eq}. So if the pK_{abH} in a proton transfer drops 2 pK_a units then the K_{eq} is 10^{+2}. In the above reaction the pK_{abH} dropped 22.7 pK_a units [15.7 – (–7)] so the $K_{eq} = 10^{+22.7}$.

Or using just the reactants, the $K_{eq} = 10^{(pK_{abH} - pK_{aHA})}$. The derivation of this quick method is straightforward. Any proton transfer reaction can be considered to be the sum of two K_a equilibria, one written in the forward direction and the second written in the reverse. The equilibrium constant for the reverse reaction is the reciprocal of K_a.

K_a forward	$\text{H–A} \rightleftharpoons \text{H}^+ + \text{A}^-$
K_a reverse	$\text{H}^+ + \text{b}^- \rightleftharpoons \text{b–H}$
Sum (the H^+ on both sides cancels):	$\text{H–A} + \text{b}^- \rightleftharpoons \text{A}^- + \text{b} - \text{H}$

When equilibria are added the equilibrium constants are multiplied.

$$K_{a\ forward} = K_{a\text{HA}} = \frac{[\text{H}^+][\text{A}^-]}{[\text{HA}]} \qquad K_{a\ reverse} = \frac{1}{K_{a\text{bH}}} = \frac{[\text{bH}]}{[\text{H}^+][\text{b}^-]}$$

$$K_{eq} = (K_{a\text{HA}})(\frac{1}{K_{a\text{bH}}}) = \frac{[\text{H}^+][\text{A}^-]}{[\text{HA}]}\frac{[\text{bH}]}{[\text{H}^+][\text{b}^-]} = \frac{[\text{A}^-][\text{bH}]}{[\text{HA}][\text{b}^-]}$$

$$\log K_{eq} = \log\frac{K_{a\text{HA}}}{K_{a\text{bH}}} = \log K_{a\text{HA}} - \log K_{a\text{bH}} = (-\log K_{a\text{bH}}) - (-\log K_{a\text{HA}})$$

$$\log K_{eq} = pK_{a\text{bH}} - pK_{a\text{HA}} \qquad K_{eq} = 10^{\{pK_{a\text{bH}} - pK_{a\text{HA}}\}}$$

Subtract the pK_a of the acid from the pK_{abH} of the base to get the exponent of K_{eq}.

How unfavorable can a proton transfer be and still be useful in a reaction?

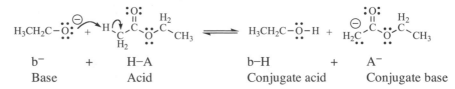

b^-	+	H–A	b–H	+	A^-
Base		Acid	Conjugate acid		Conjugate base

In this example, since the pK_a of bH is 16 and that of HA is almost 26, the K_{eq} would be very small, $10^{(16-26)}$ or 10^{-10}, indicating a strong preference for b^- and HA. If b^- and HA were initially equimolar, $[\text{b}^-] = [\text{HA}]$, then it follows from the stoichiometry of the reaction that $[\text{bH}] = [\text{A}^-]$. Substitution into the K_{eq} expression produces:

$$K_{eq} = \frac{[A^-][bH]}{[HA][b^-]} = \frac{[A^-]^2}{[b^-]^2} \text{ or } = \frac{[A^-]}{[b^-]} = \sqrt{10^{-10}} = 10^{-5} \text{ or } \frac{1}{100{,}000} = \frac{[A^-]}{[b^-]}$$

Therefore the concentration of b^- at equilibrium is 100,000 times greater than A^-. On occasion a K_{eq} as unfavorable as 10^{-10} can still be useful. Products arising from A^- will be significant only if b^- produces a less favored product that easily reverts back to reactants, or if A^- is much more reactive than b^-. Consumption of A^- will drive the equilibrium to replenish it.

If the proton transfer K_{eq} is greater than 10^{+10}, it can for all practical purposes be considered irreversible. An irreversible proton transfer is often the last step in a reaction and the driving force for the entire process. Each pK_a unit converts to a 1.36 kcal/mol (5.73 kJ/mol) contribution to the $\Delta G°$ of the reaction at room temperature.

Proton transfer is the first step in many important reactions. The ability to predict which protons will be pulled off in basic media or which sites will be protonated in acidic media is very important. The reaction solvent often behaves as the actual acid or base for proton transfer because the strongest acid that can occur in any medium is the protonated solvent, and the strongest base is the deprotonated solvent. Thus, the strongest acid in water is H_3O^+ and the strongest base is OH^-.

The speed of proton transfer is usually very fast, often at the diffusion-controlled limit. Exceptions occur when the acidic H is very hindered, and when a carbanion deprotonates a C–H bond, which can be so slow that other reactions can easily compete.

Proton Transfer K_{eq} Crosscheck and the ΔpK_a Rule

An unfavorable K_{eq} smaller than 10^{-10} is most likely too small to be useful. All favorable K_{eq} values are useful, and thus have no limit. The check of the proton transfer K_{eq} will be very important in mechanisms to decide if a proton transfer is reasonable. This lower limit principle generalizes into a very important piece of control knowledge: **the ΔpK_a rule: Avoid intermediates that are more than 10 pK_a units *uphill* from the reactants (either 10 pK_a units more basic or 10 pK_a units more acidic).** Reactions usually head toward neutralization, forming weaker acids and weaker bases, not stronger.

Example problem

Calculate the K_{eq} for proton transfer for the following reaction. Does it favor reactants or products?

$$EtO^- + PhOH \rightleftharpoons EtOH + PhO^-$$

Answer: The base in the forward reaction is EtO^- at a pK_{abH} = 16; the acid is PhOH at pK_a = 10. The K_{eq} equals $10^{(pK_{abH} - pK_{aHA})}$ or $10^{(16 - 10)} = 10^{+6}$. The equilibrium favors products.

3.7 PROTON TRANSFER MECHANISMS

Mechanisms are our best guess of the steps occurring in a reaction. Mechanisms can never be proven but may be disproved by experimental data. Mechanisms are like a grammatically correct sentence; to make sense, we need to use known words properly. Our words are the electron flow paths and the grammar is the crosschecks for the path.

So far, we have one path (p.t.) and three crosschecks (media pH, most acidic H, and proton transfer K_{eq}). Before things get complicated with more paths and crosschecks, let's explore mechanisms.

3.7.1 Problem Space for Proton Transfer Mechanisms

The set of all possible steps that could occur in a problem is its problem space. It includes incorrect, partially correct, and correct routes. Our search for an answer occurs within this problem space. Often instructors present just the one correct route, leaving you to wonder whether your different route is OK or not. At this stage, we are concerned with a reasonable mechanism, not necessarily the absolute best. The problem space is often in the shape of a tree, with branches occurring at each decision point. We need an efficient way to search the problem space tree.

Consider the problem of driving from Los Angeles to San Francisco. The problem space is all the roads in the US. The first task in narrowing down the problem space is to decide in what direction the goal is. Since San Francisco is north of L.A. we can discard generally south routes, and the problem space reduces by half. The media pH crosscheck is a similar mechanistic check: is the reaction run in acid or base? We need to stay on the roads (our electron flow paths). Some routes branching north from L.A. may be interstate highways and others dirt roads. Other crosschecks would now come in to aid us in picking a good route. We would like a good route that does not require white-knuckle mountain roads and is reasonably direct (going via Chicago is out). Working backward from our goal can help a lot. We may find in mechanisms that a certain process, although reasonable, does not go on to product; it is a dead end. Backtrack to the last fork and try another route. Avoid the temptation to go off-road at the dead end, rather than going back to the main road.

3.7.2 Make Conscious Decisions

Being oblivious to forks in the road or deciding on impulse rather than using a map is an inefficient way to travel. Picking the first alternative that comes to mind is impulse, not thinking. Critical thinking is making a deliberate choice based on considering the worth of all the alternatives. Be willing to reconsider choices already made, and if you are really stuck, consider asking for directions. Show your instructor your scratch sheet for the problem, so you can see where you took a wrong turn. Learn from your mistakes.

3.7.3 What Happened?—Mapping Changes

A great help in decreasing the size of the problem space is to map what changes have occurred in the reaction. With mechanisms, and especially proton-transfer reactions, draw out all the hydrogen atoms near the areas that change (line structure can hide changes because the Hs are not drawn). It is much easier to crosscheck changes and your arrows if you always draw Lewis dot structures for any part of the molecule that comes near an arrow. You may have to number the carbon atoms in the reactant and in the product to identify what has changed. Ockham's razor (seek simplicity) guides us to make the fewest changes necessary when numbering our product.

3.7.4 Pruning the Tree—Crosschecks

Our crosschecks are tests of reasonableness. Our first check was north or south? Is the route going in the correct direction? Are we still on a good road? Are we climbing a ridiculously steep hill? For proton transfer mechanisms the equivalent questions are the following. Is the reaction in basic or acidic media, and is our mechanism consistent with that? Did we pick the correct position on the molecule to protonate or deprotonate? Did we get the p.t. path arrows correct? Is the K_{eq} of the process reasonable?

3.7.5 Example Proton Transfer Mechanism

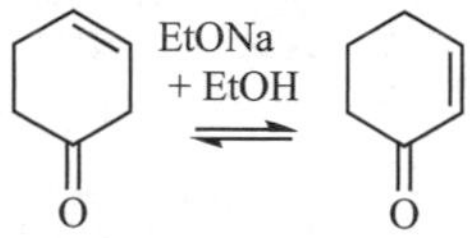
EtONa
+ EtOH
O
O

First, we need to figure out what has changed during the reaction. After drawing out the hydrogen atoms, we see that there are two ways we could number carbon atoms of the product, keeping the C=O as carbon number 1. The left one changes only carbons 2 and 4, whereas the right changes carbons 3, 4, 5, and 6. We choose the left, the simplest with the fewest changes. If we just picked the first numbering that popped into our head, there is a 50% chance of making the problem nearly impossible.

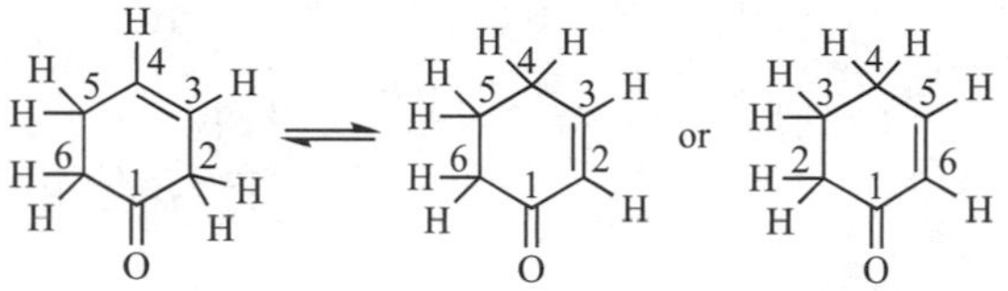
or

With the correct numbering we can see that carbon 2 lost an H and carbon 4 gained an H. On more complex problems, it is helpful to make a "to do" list of changes, but remember that list must be done in an order that makes chemical sense. The medium is basic, so acidic routes are out. Instead of protonating the molecule with an acid, we will deprotonate with a base. Both carbons 2 and 6 have acidic hydrogen atoms, since they are next to the polarized multiple bond, C=O. Although the H on carbon 2 is more acidic because it forms a more delocalized conjugate base, we should explore both possibilities. However good the deprotonation of carbon 6 may seem, it does not get us toward our goal. It is a very reasonable process that is a dead end. Since deprotonation of carbon 2 heads in the correct direction, we need to know if the K_{eq} is within the range of our base. The pK_a of our reactant is known to be 15.2, and the conjugate acid of our ethoxide base is pK_a 16, which gives a K_{eq} of $10^{(16-15.2)} = 10^{+0.8}$, passing the crosscheck, and actually downhill. The proton transfer electron flow path forms the Lewis structure of the product of the first step. The major resonance form is the middle one, but the right one gives us a hint as to why carbon 4 is basic. Therefore, if you are stuck, draw resonance forms.

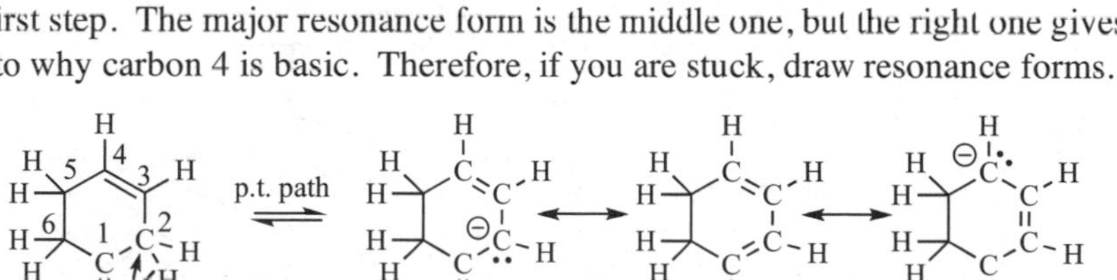
p.t. path
Reactant
:OEt
Extended enolate

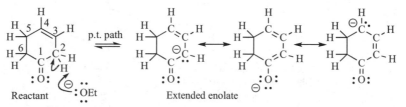

Our best acid in the solution is the solvent, ethanol. The proton transfer path again gives us the Lewis structure of the product. The pK_a of the product is known to be 17.6, and the pK_a of ethanol (our acid) is at 16, so that gives a K_{eq} of $10^{(17.6-16)} = 10^{+1.6}$, passing the crosscheck, and again downhill. Both proton transfer steps in this case were downhill because ethanol's pK_a was higher than that of the reactant and lower than that of the product. The product is downhill from the reactant because it is conjugated and the reactant is not. Both the reactant (pK_a of 15.2) and the product (pK_a of 17.6) deprotonate to give the same extended enolate anion, but since the product is more stable, it has more of a climb to get to the extended enolate, so its acidity is less, indicated by a higher pK_a.

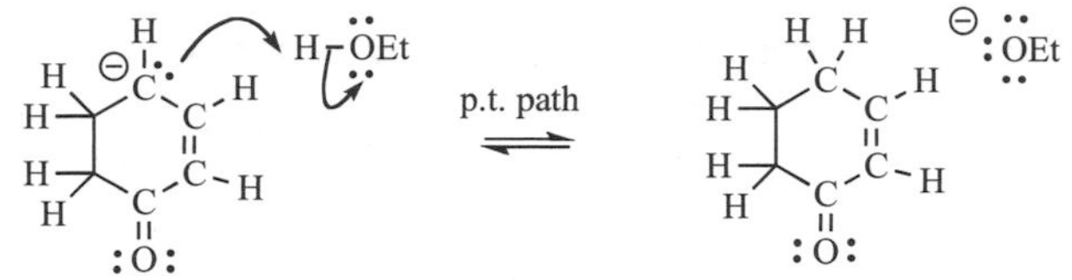
p.t. path
OEt

Now let's look at the problem space that we just navigated in Figure 3.6 to see all the branches and where else we could have gone under basic conditions. Two new enol structures from the protonation of the oxygen anion are very reasonable dead ends for this problem. A ΔH calculation will show that the enol is uphill from the reactant. The decisions and operations that we went through to navigate this example proton transfer mechanism problem space are flowcharted in Figure 3.7.

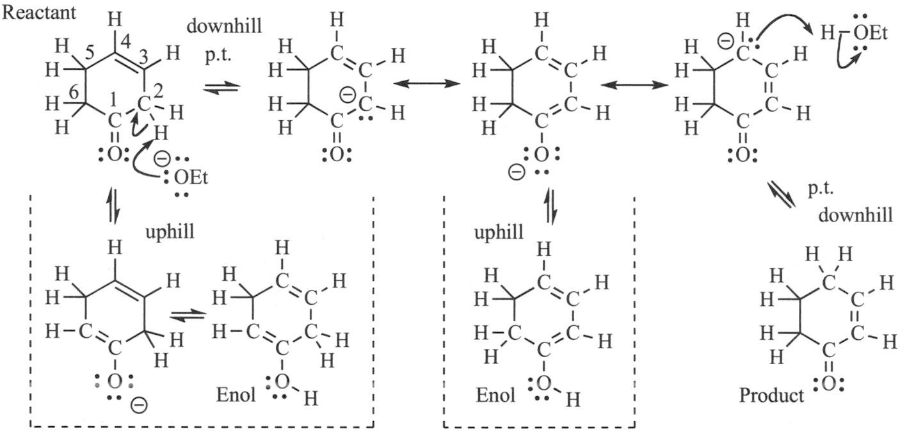
Reactant
downhill
p.t.
uphill
uphill
Enol
Enol
p.t.
downhill
Product

Figure 3.6 Reasonable basic problem space for the example proton transfer mechanism. Acid routes have been dropped. Only proton transfer routes are shown. Routes deprotonating carbons 3, 4, and 5 have been dropped because the proton transfer K_{eq} is less than 10^{-10}, and thus not useful.

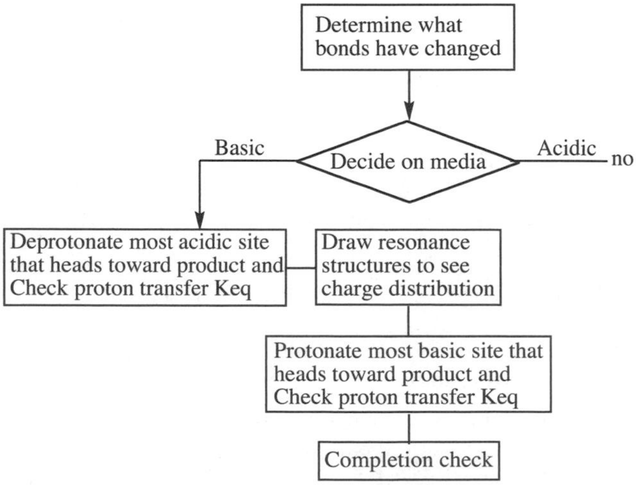
Determine what bonds have changed
Basic
Decide on media
Acidic
no
Deprotonate most acidic site that heads toward product and Check proton transfer Keq
Draw resonance structures to see charge distribution
Protonate most basic site that heads toward product and Check proton transfer Keq
Completion check

Figure 3.7 Flowchart for the 3.7.5 basic media example proton transfer mechanism.

3.8 COMMON ERRORS

Where could we have gone astray here? Most often, it is forgetting a crosscheck.

3.8.1 Failure to Stay on the Path

We could have done the previous mechanism in one step with four arrows, but that would assume a three-molecule collision, something that is very rare. The arrows would not involve the C=O, which is essential for the reaction to proceed. This reaction goes through a resonance-stabilized intermediate, the stability of which we can check. Made-up combinations of arrows are hard to check and may just shift the lines and dots of our Lewis structure without saying anything chemically understandable. For our mechanistic sentences to make sense we must use known words, the electron flow paths.

Biochemical Note: An enzyme's active site can have weak acids and bases close to our reactant. It is often possible to have proton transfer occurring in the same step as another electron flow path. These are discussed separately in Section 7.4.3.

3.8.2 Failure to Check K_{eq}

Often beginners will forget to check the K_{eq} for proton transfer and use a base to remove a proton that is not acidic. It should be assumed that the hydrogen is not acidic if you cannot find the appropriate or a related compound on the pK_a chart. As an illustration, you will not find diethyl ether, $CH_3CH_2OCH_2CH_3$, on the pK_a chart; none of its hydrogens is acidic. The carbonyl hydrogen of an aldehyde, RCHO, also is not acidic. If you cannot find the compound, or one with a closely related functional group, on the pK_a chart, then suspect that its hydrogens may not be acidic. Therefore, it is very important that you know how to use the pK_a chart.

3.8.3 Media pH Errors

In our mechanism example, we could have started off on an impulse to do an acidic mechanism in basic conditions. What would have caught us was our first K_{eq} crosscheck. Using ethanol as an acid (pK_a = 16), the most favorable K_{eq} we could have gotten would have been protonating the carbonyl lone pair at 10^{-23}, which is not usable. We would be saved by our crosscheck. Generally a good rule is: **In acid protonate the reactant, and in base deprotonate it.**

3.8.4 pK_a Span Errors

Beginners will sometimes try to do both acid-catalyzed and base-catalyzed processes at the same time, forgetting that a strong acid and base present in the same vessel would neutralize each other. **No medium can possibly be both strongly acidic and strongly basic.** The relative hydronium ion and hydroxide ion concentrations are defined by $K_w = [H^+][OH^-] = 10^{-14}$. Their p$K_a$ values span 17.4 pK_a units. In essence, when working in acidic media, avoid creating or using strong bases. In basic media, avoid creating or using strong acids.

3.8.5 Drop off H^+ Only Shorthand Notation

A shorthand notation you may see in some texts is to drop a proton off by itself or show a naked proton being picked up by a base. Under the conditions found in almost all reactions, the proton is transferred from acid to base and does not float off on its own. There is nothing wrong with using a shorthand notation, as long as you remember it is not reflecting reality.

3.8.6 Wrong pK_a Errors

Be careful when looking up a pK_a to determine a K_{eq}, because the wrong pK_a values will lead to an incorrect decision based on poor data. It is an unfortunate reality that many compounds can serve as either an acid or a base. A very common example is water. Water can act as a base to form hydronium or can act as an acid to form hydroxide. Since this book's pK_a chart lists both the acid and its conjugate base, water is on the pK_a chart twice, once as an acid (15.7) and once as a conjugate base for hydronium (−1.7).

3.8.7 Strained Intramolecular Proton Transfer Errors

Sometimes a proton can be transferred internally if the transition state is not too strained. Three- and four-membered transition states have bent bonds that make them energetically unfavorable. The optimum ring size for intramolecular proton transfer is a five- or six-membered transition state.

3.9 PROTON TRANSFER PRODUCT PREDICTIONS

The problem seems simple, "Draw a reasonable proton transfer reaction given the following conditions." However, several skills and correct decisions are required to get the correct answer. If any of these is done incorrectly, the final answer will be wrong, so let's pick this problem apart. A **proton transfer** is simply passing a proton from one site to another, from acid to base (Section 3.1). We need to decide whether the **conditions** are basic or acidic, since we tend to protonate substrates in acid and deprotonate substrates in base (Sections 3.2.1, 3.2.2). We need to identify the acid (Section 3.2.3) and decide where the most acidic hydrogen is located (Section 3.5). We also need to identify the base and its most basic site (Sections 3.2.4 and 3.4). For this we need to understand the acidity and basicity trends (Sections 3.3 and 3.4), and use a pK_a chart if one is available (Appendix). Once we have found the most acidic H and most basic site, then comes the test of thermodynamic **reasonableness** (Section 2.5): Is the K_{eq} greater than 10^{-10} (Section 3.6)? If that test is passed all we have to do is draw out the proton transfer correctly, using good Lewis structures (Section 1.3) and arrows (Section 1.4), and avoiding common mistakes such as charge balance errors. As you can see, this one "simple" problem makes use of much of what you've learned and shows the interrelatedness of organic chemistry. So let's try one!

Example problem

Draw a reasonable proton transfer reaction given the following conditions.

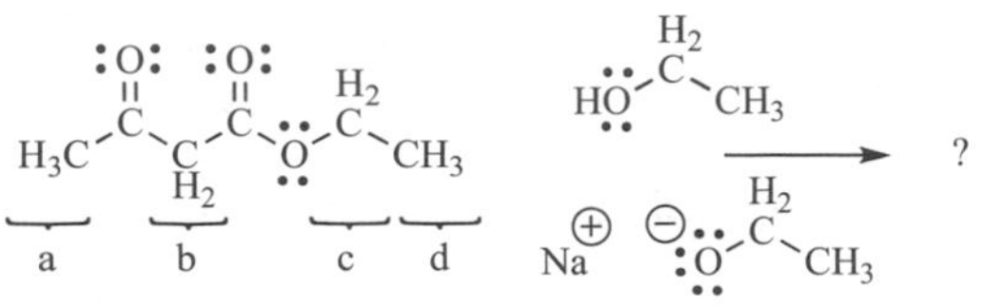
H$_3$C
H$_2$
a
b
c
d
HO
CH$_3$
Na
O
?

Answer: First decide whether the medium is acidic or basic. With both ethanol and ethoxide present, the media is highly basic (visualize Fig. 3.2 but with the pK_a of 16 of ethanol). Since ethoxide is much more basic than ethanol, ethoxide is our base. Now we must decide which is the most acidic H on the substrate: a, b, c, or d. Using acidity trends, we decide that the protons on carbon labeled "b" between the two electron-withdrawing groups will have the most delocalized conjugate base, and so are most acidic. We can check this with a pK_a chart: The H on "a" is adjacent to a ketone for a pK_a of about 19.2 ; the H on "b" is between a ketone and an ester for a pK_a of about 10.7; the Hs on "c" and "d" have no delocalization of the conjugate base and so are expected to be close to ethane at pK_a 50. Clearly the protons on "b" are most acidic because they have the lowest pK_a. The check for thermodynamic reasonableness is next: K_{eq} = $10^{(\mathrm{p}K_{\mathrm{abH}} - \mathrm{p}K_{\mathrm{aHA}})}$ or $10^{(16 - 10.7)} = 10^{+5.3}$ which is clearly a downhill process. After the check of K_{eq}, we can finally say that the following is a correctly drawn and **reasonable** proton transfer reaction. The arrows could be drawn to form any resonance form of the conjugate base. The conjugate base is a delocalized anion, and just the major resonance form is shown below.

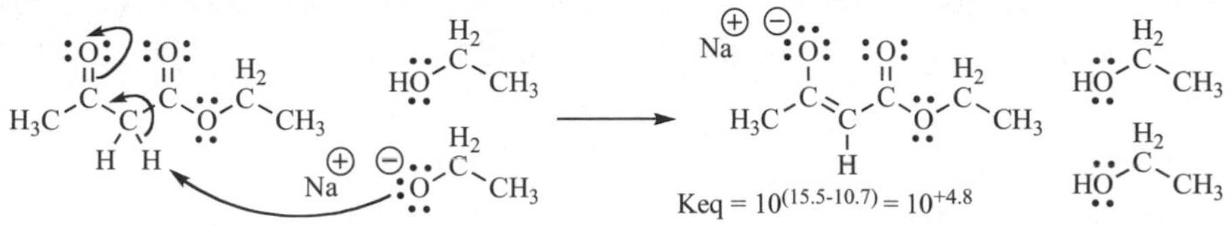
Keq = 10^(15.5-10.7) = 10^+4.8

The K_{eq} is greater than 1, indicating the proton transfer reaction favors products. Charge is balanced. Figure 3.8 shows each of the decisions in the problem space just navigated.

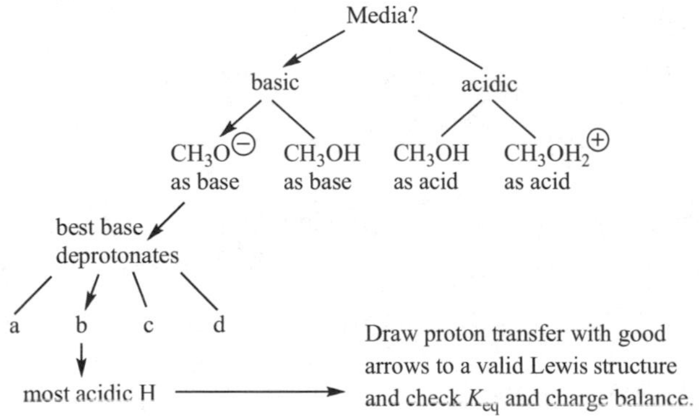
Media?
basic
acidic
CH3O⊖ as base
CH3OH as base
CH3OH as acid
CH3OH2⊕ as acid
best base deprotonates
a
b
c
d
most acidic H
Draw proton transfer with good arrows to a valid Lewis structure and check Keq and charge balance.

Figure 3.8 Problem space for predicting a reasonable proton transfer in basic media.

3.10 PROTON TRANSFER SUMMARY

The take-home message for this chapter is that structure determines reactivity. Figure 3.9 relates the major factors that influence the acidity of a compound.

- Strong acids have a low pK_a and have weak conjugate bases.
- Strong bases have a high pK_{abH} and have weak conjugate acids.
- Proton transfer $K_{eq} = 10^{(\mathrm{p}K_{\mathrm{abH}} - \mathrm{p}K_{\mathrm{aHA}})}$
- A useful lower limit for proton transfer is: K_{eq} equal to or greater than 10^{-10}.
- Always decide if the reaction is run in acid or base and stay within those bounds.
- Acidic Hs are found on heteroatoms and adjacent to polarized multiple bonds.
- In acid protonate the reactant, and in base deprotonate it.

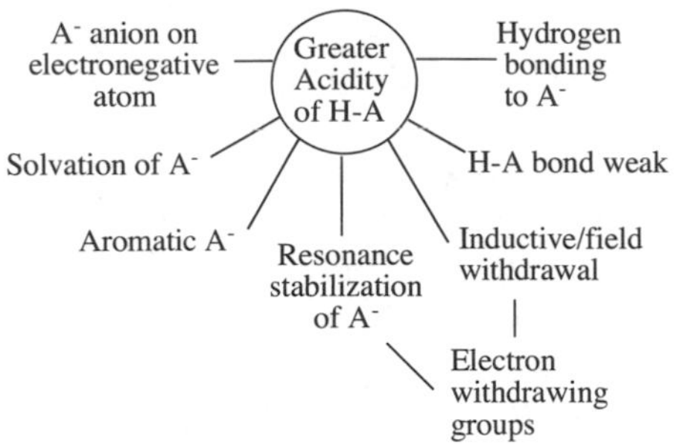
A⁻ anion on electronegative atom
Greater Acidity of H-A
Hydrogen bonding to A⁻
Solvation of A⁻
H-A bond weak
Aromatic A⁻
Resonance stabilization of A⁻
Inductive/field withdrawal
Electron withdrawing groups

Figure 3.9 Web of concepts for factors increasing acidity.

ADDITIONAL EXERCISES

3.1 Using the pK_a chart in the Appendix, calculate the numerical value of K_{eq} for these reactions. Predict whether it favors reactants or products and estimate $\Delta G°$ at 25°C.

(a) PhONa + PhSH $\rightleftharpoons$ PhSNa + PhOH

(b) 

(c) $(CH_3)_3COK$ + HC≡CH $\rightleftharpoons$ $(CH_3)_3COH$ + HC≡CK

3.2 Trends: With the help of the pK_a chart when needed, rank all species, beginning with the numeral 1 to designate:

(a) The most stable carbanion

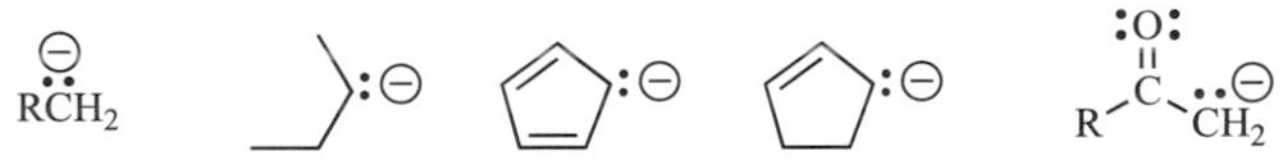

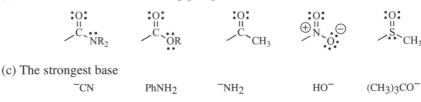

(b) The best electron-withdrawing group

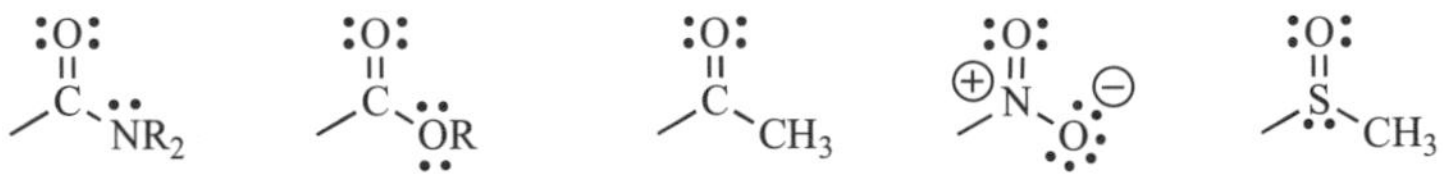

(c) The strongest base

^{-}CN $PhNH_2$ $^{-}NH_2$ HO^{-} $(CH_3)_3CO^{-}$

(d) The strongest acid

H_3O^+ HF CH_3OH HCN NH_3

3.3 With the help of the pK_a chart, circle the most acidic H in each of the following compounds.

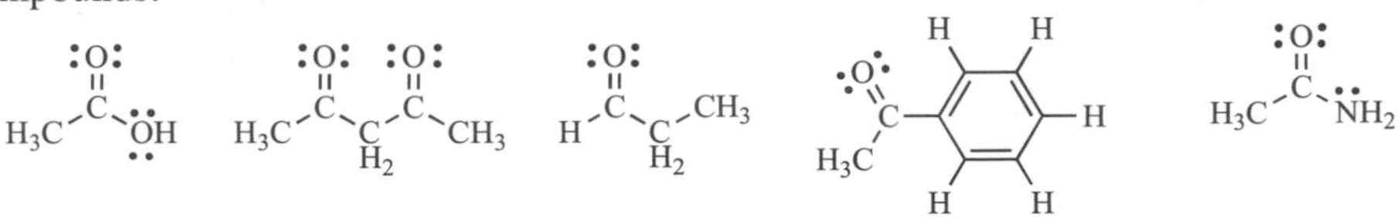

3.4 Use your knowledge of anion stability and resonance to decide which of the following two alternatives is the lower-energy process.

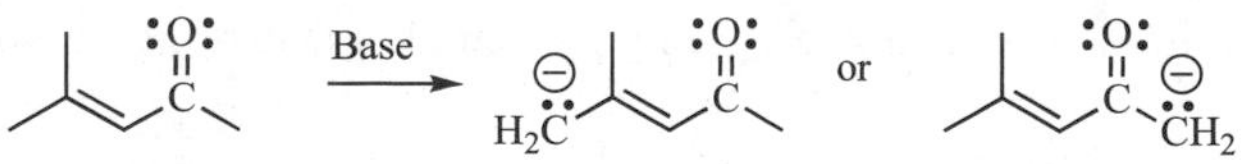
Base
or

3.5 Use your knowledge of proton transfer equilibria to decide which of the following two alternatives is the lower-energy process.

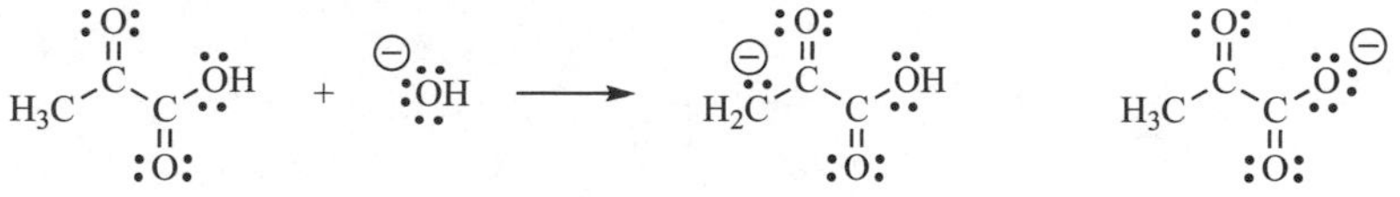

3.6 Ascorbic acid, vitamin C, has one very acidic H (pK_{a1} = 4.1), and all the others are not very acidic (pK_{a2} = 11.8). Use your knowledge of anion stability and resonance to find the acidic H and explain why it is so acidic. Draw all the resonance forms of the conjugate base.

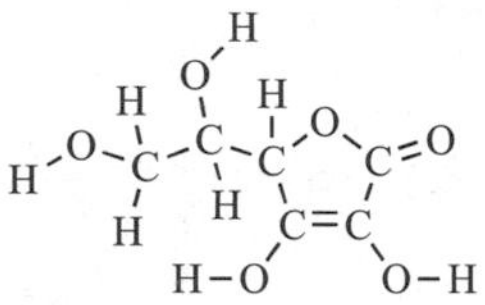

3.7 Using the pK_a chart, rank the following by acidity and explain the reason for the order you find.

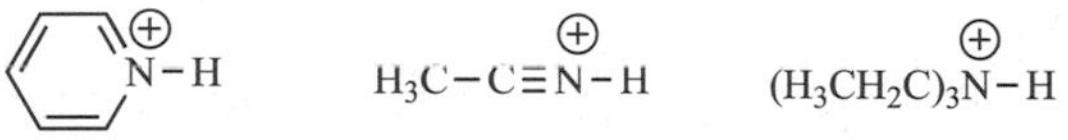

3.8 Succinic acid, $HOOCCH_2CH_2COOH$, has two pK_as, 4.2 and 5.6. Using crossing and break points, draw a species vs. pH graph for succinic acid, like Figures 3.2 and 3.3.

3.9 Draw a reasonable proton transfer mechanism for the following reaction catalyzed by sodium hydroxide in heavy water. D stands for deuterium. Treat D_2O just like H_2O.

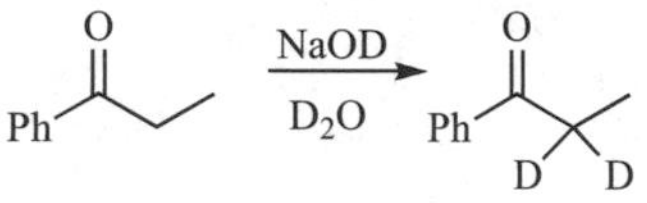
NaOD
D2O

3.10 Decide whether these conditions are acidic or basic and give an approximate pH.

(a) A water solution of 2 M NH_4Cl and 2 M NH_3

(b) A water solution of 1.0 M sodium acetate and 0.1 M acetic acid

(c) A water solution of 0.1 M PhONa and 1.0 M PhOH

(d) A water solution of 1 M KF and 1 M HF

3.11 Draw an energy diagram for the proton transfer reaction of $CH_3COOH + NH_3$. Calculate the proton transfer K_{eq} to get an approximate $\Delta G°$ for the reaction.

3.12 Provide an approximate pK_a for each set of hydrogens on the compound below.

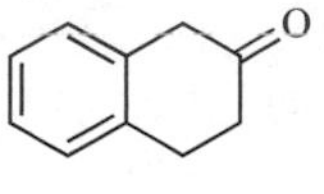

3.13 Fill in the blank for the missing partners in these proton transfer reactions.

(a) 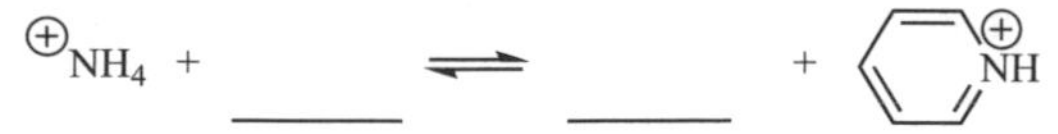

(b) 

(c) 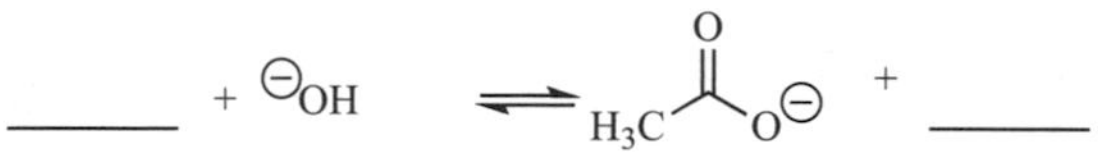

(d) 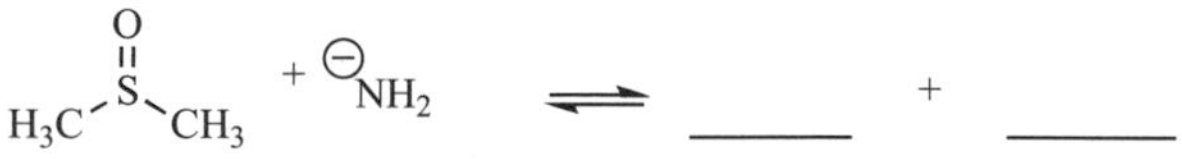

3.14 (a, b, c, d) Give K_{eq}, $\Delta G°$, and equilibrium direction of each problem 3.13a, b, c, d.

3.15 Predict the pK_as of all the Hs in the following compounds.

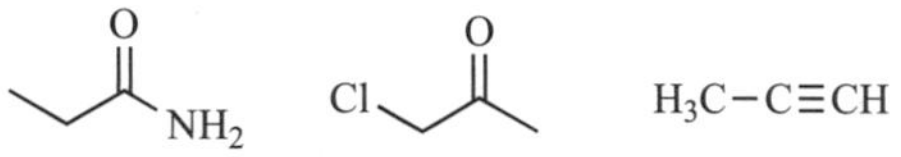

3.16 In each molecule below, circle the most acidic H.

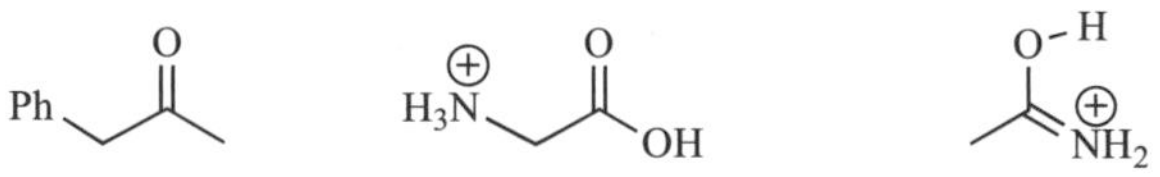

3.17 Trends: With the help of the pK_a chart when needed, rank all species, beginning with the numeral 1 to designate:

(a) The weakest base

^{-}CN $PhNH_2$ $^{-}NH_2$ HO^{-} $(CH_3)_3CO^{-}$

(b) The weakest acid

H_3O^+ HF MeOH HCN NH_3

(c) The least stable carbanion

$H_3C-\ddot{C}H_2^{\ominus}$ $NC-\ddot{C}H_2^{\ominus}$ $Ph-\ddot{C}H_2^{\ominus}$ $H_2C=\ddot{C}H^{\ominus}$ $O_2N-\ddot{C}H_2^{\ominus}$

3.18 Draw a reasonable proton transfer reaction given the following conditions.

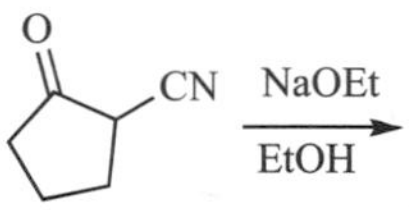

3.19 Draw a reasonable proton transfer reaction given the following conditions.

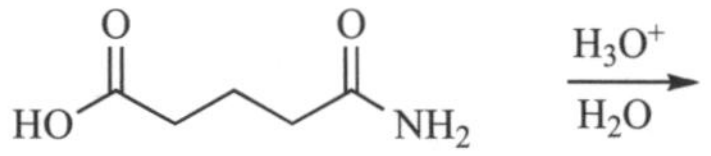

3.20 Provide a proton transfer mechanism for this isomerization of a prostaglandin. Watch out for a reasonable proton transfer that is a dead end!

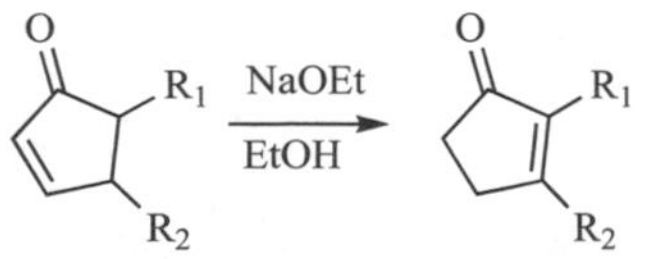

3.21 Draw the proton transfer products for the following reaction and using the pK_a chart in the Appendix, calculate the value of K_{eq} and estimate $\Delta G°$ at room temperature. Draw a qualitative energy diagram.

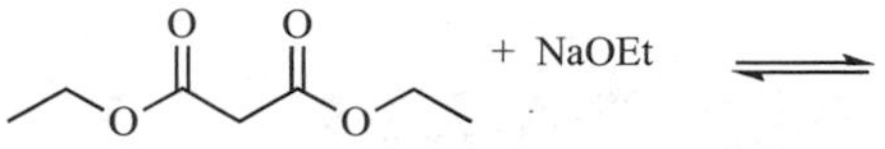
+ NaOEt

3.22 Carbonic acid, $HOCO_2H$, has two pK_as, 6.4 and 10.3. Using crossing and break points, draw a species vs. pH graph for carbonic acid, like Figures 3.2 and 3.3. Using this chart, decide whether the following conditions are acidic or basic and give an approximate pH: a water solution of 0.1 M Na_2CO_3 and 1.0M $NaHCO_3$.

3.23 Given that the K_{eq} for the reaction below converting the keto form into the enol is 10^{-8}, and that the pK_a for the keto form is 18.4, what is the pK_a of the enol? Both the enol and keto form have the same conjugate base, the enolate form.

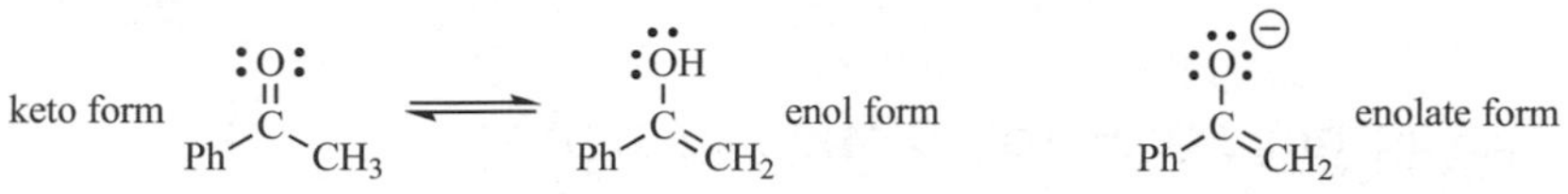
keto form
enol form
enolate form

3.24 For the species in problem 3.23, draw a qualitative energy diagram showing both curves for the enol and for keto forms reacting with ethoxide to produce the enolate form.

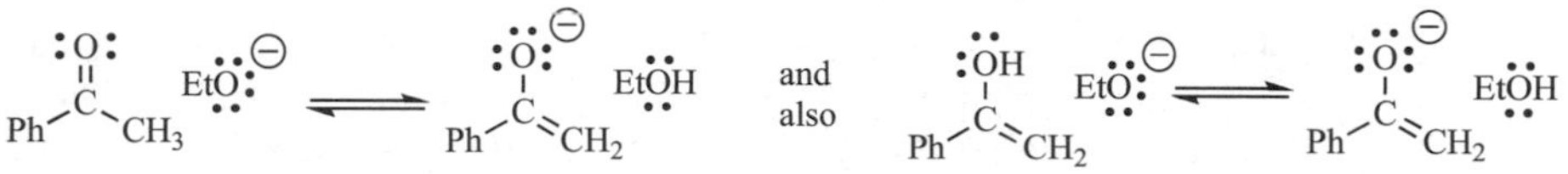
and also

3.25 Draw the proton transfer products and calculate the value of K_{eq} and $\Delta G°$ at room temperature. The cyanoester is not on the chart, but is easily estimated as halfway between the dicyano and the diester.

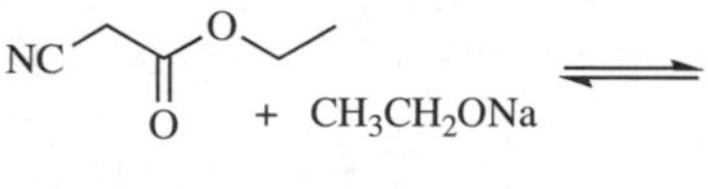
+ CH3CH2ONa

3.26 In cells, enolpyruvic acid is in equilibrium with pyruvic acid. In the lab in acidic media, the first step of this conversion is a protonation of enolpyruvic acid to give a resonance delocalized cation intermediate; the second step is a deprotonation to give pyruvic acid. Give a detailed mechanism for the conversion. Draw both resonance forms of the intermediate cation and decide which form is major. Use H_3O^+ to protonate and H_2O to deprotonate.

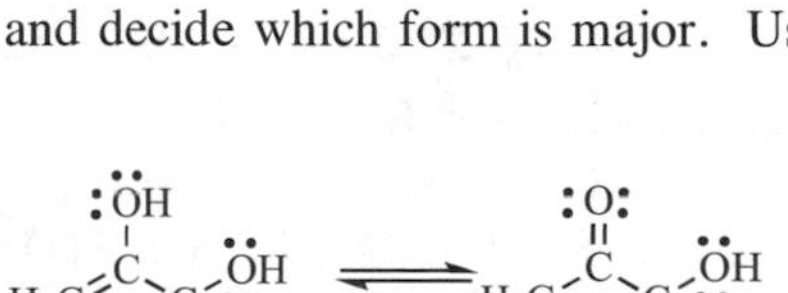

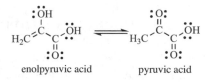

enolpyruvic acid   pyruvic acid

3.27 Provide a proton transfer mechanism for this isomerization in basic media. Which is favored, reactant or product?

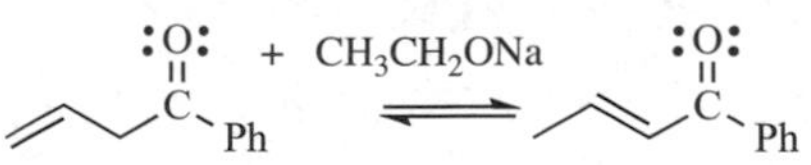
+ CH3CH2ONa

4

IMPORTANT REACTION ARCHETYPES

Electron Flow In Organic Chemistry: A Decision-Based Guide To Organic Mechanisms, Second Edition.
By Paul H. Scudder

4.1 INTRODUCTION TO REACTION ARCHETYPES

There are really only four organic reaction types (Fig. 4.1):

Substitution—trade a nucleophile for a leaving group

Elimination—lose a leaving group and usually a proton to form a pi bond

Addition—add a nucleophile or electrophile or both to a pi bond

Rearrangement—produce a constitutional isomer

The first three can occur intramolecularly (within the same molecule) or intermolecularly (via a bimolecular collision), whereas rearrangement is only an intramolecular process. If we understand idealized versions or archetypes of each of these four reaction types, we will have come a long way toward mastering the basics of most organic reactions. Figure 4.1 gives generic examples of each reaction type. Note that substitutions at a tetrahedral center are significantly different from substitutions at a trigonal planar center. Since trigonal planar center substitutions usually go via a two-step process of addition then elimination, it is easier to discuss them after those processes.

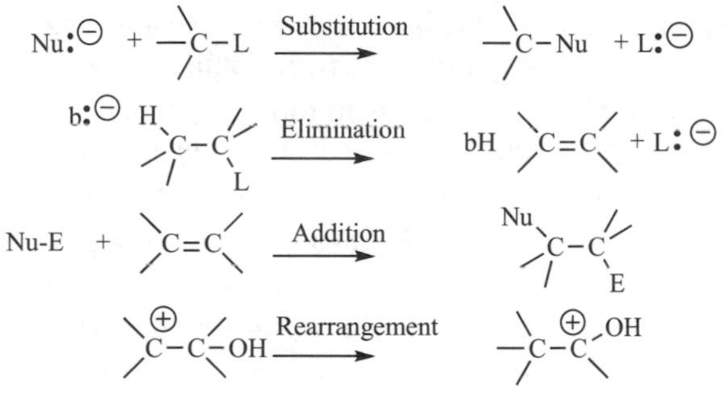
Substitution
Elimination
Addition
Rearrangement

Figure 4.1 Examples of the four different reaction archetypes

4.2 NUCLEOPHILIC SUBSTITUTION AT A TETRAHEDRAL CENTER

4.2.1 Substitution Electron Flow Paths

The energy surface in Figure 4.2 gives an overview of three possible routes to substitute a leaving group bound to a tetrahedral center. Our two new generic groups (a lone pair nucleophile and a leaving group on a tetrahedral carbon) are found as the reactants in the upper right corner. We will look at the reactivity of nucleophiles and leaving groups and what characterizes these two groups later in this section. The leaving group-carbon bond must break (the vertical axis) and the carbon-nucleophile bond must form (the horizontal axis) to reach the substitution products in the lower left. Three idealized routes can occur. The pentacovalent intermediate route is not available to second-row elements like carbon but is well established for third-row elements like phosphorus, capable of *d* orbital bonding.

There are two routes for carbon that differ only in the timing of when the nucleophile attacks and when the leaving group falls off: **S_N1** (substitution, nucleophilic, unimolecular): the leaving group departs first (rate-determining) by path D_N, producing a reasonably stable carbocation; the carbocation is then trapped by a nucleophile, path A_N. **S_N2** (substitution, nucleophilic, bimolecular): the simultaneous attack of nucleophile and loss of the leaving group, path S_N2. These three new electron flow paths are shown in

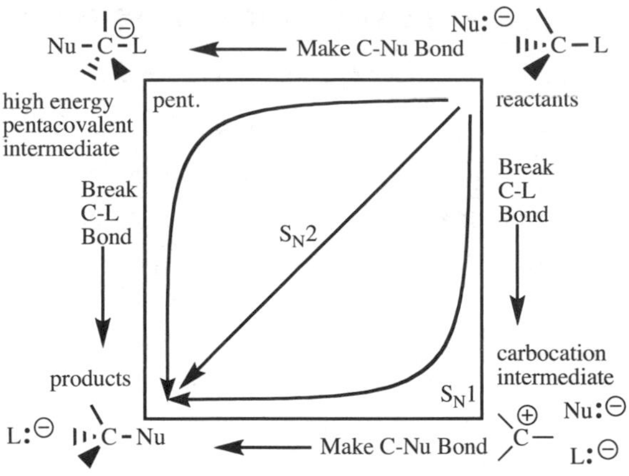

Nu - C - L
Make C-Nu Bond
Nu:
C - L
high energy pentacovalent intermediate
pent.
reactants
Break C-L Bond
Break C-L Bond
SN2
products
carbocation intermediate
SN1
Nu:
L:
C - Nu
Make C-Nu Bond
C
L:

Figure 4.2 Top view of a simplified energy surface for substitution at a tetrahedral center.

Figure 4.3. Path D_N (dissociation, nucleofugic) is the loss of a leaving group; the arrow starts from the C–L bond, breaking it, forming a carbocation and a new lone pair on the leaving group. Path A_N (association, nucleophilic) is the microscopic reverse, a lone pair nucleophile adds to the carbocation, to form a C–Nu bond. This two-step combination of paths D_N + A_N forms the S_N1 route. The two paths, combined into one step of A_ND_N, form the S_N2 route. But before we can discuss these routes, we must understand what makes a good nucleophile, a good leaving group, and a good carbocation.

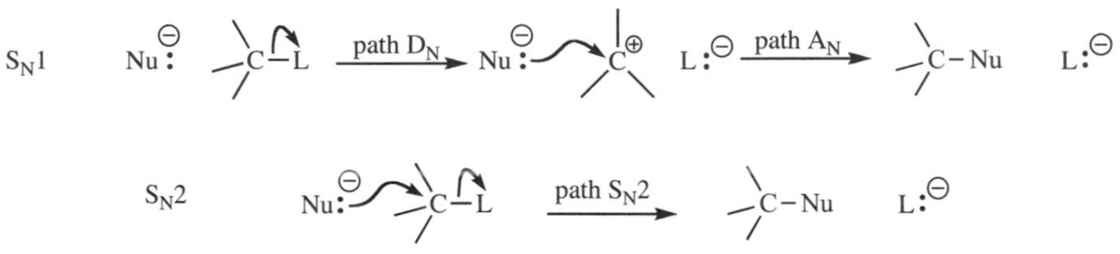

SN1
Nu:
C - L
path DN
Nu:
C
L:
path AN
C - Nu
L:
SN2
Nu:
C - L
path SN2
C - Nu
L:

Figure 4.3 The electron flow paths for substitution at a tetrahedral carbon.

4.2.2 Ranking of Nucleophiles

Nonbonding electron pairs are good electron sources, especially lone pairs of anions. The following list gives the relative nucleophilicity of many lone pair nucleophiles in an S_N2 reaction on methyl iodide. An anion is always a better nucleophile than its neutral counterpart; methoxide is almost 2 million times more nucleophilic than methanol.

We might suspect that the more basic species would also be more nucleophilic, but nucleophilicity does not parallel basicity exactly because softness must be considered. The harder ions are more tightly solvated in polar solvents and cannot serve as a nucleophile as well as a poorly solvated softer ion. To verify this, we need only to look at the following list and see that iodide at pK_{abH} −10 is over a thousand times more nucleophilic than the much more basic acetate, pK_{abH} 4.8. **For lone pairs of the *same element*, the more basic the lone pair is, the more nucleophilic it is.** To rank nucleophiles in this way, remember to look up the conjugate acid on the pK_a chart. The following is a ranked list of the pK_{abH} of only the oxygen nucleophiles from the previous list. The one exception to this trend is that nucleophiles with adjacent lone pairs like HOOH, NH_2NH_2, and NH_2OH are more nucleophilic than expected; this α-effect is most likely due to poorer solvent stabilization.

Relative nucleophilicity toward CH_3I (in CH_3OH as solvent)	
CH_3OH	1
F^-	500
CH_3COO^-	20,000
Cl^-	23,000
Et_2S	220,000
NH_3	320,000
PhO^-	560,000
Br^-	620,000
CH_3O^-	1,900,000
Et_3N	4,600,000
CN^-	5,000,000
I^-	26,000,000
Et_3P	520,000,000
PhS^-	8,300,000,000
$PhSe^-$	50,000,000,000

Steric hindrance decreases nucleophilicity. As the nucleophile becomes larger, its bulk tends to get in the way of its acting as a nucleophile, especially when the electrophile is also large. For this reason the larger *tert*-butoxide, $(CH_3)_3CO^-$, is a much poorer nucleophile than the smaller methoxide, CH_3O^-.

Most nucleophilic and most basic O	$\mathbf{pK_{abH}}$
CH_3O^-	15.5
PhO^-	10
CH_3COO^-	4.8
CH_3OH	−2.4
Least nucleophilic and least basic O	

Electron availability decreases as the orbital the electrons occupy is made more stable. As the hybridization of the nitrogen lone pair goes from sp^3 to sp^2 to sp, the lone pair gets less basic and less nucleophilic as indicated by the following trend.

Most nucleophilic and most basic N	
Et_3N:	pK_{abH} = 10.7; lone pair hybridization = sp^3
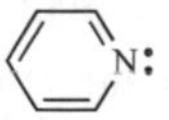	pK_{abH} = 5.2; lone pair hybridization = sp^2
$CH_3C{\equiv}N$:	pK_{abH} = -10; lone pair hybridization = sp
Least nucleophilic and least basic N	

An exception to this hybridization trend is that an ester's carbonyl oxygen sp^2 lone pairs are much more basic and more nucleophilic than its ether oxygen sp^3 lone pairs. When an ester's carbonyl lone pair is protonated, a delocalized cation is formed. When an ester's ether oxygen lone pair is protonated, the cation is less stable because it is not delocalized and the electron withdrawal by the adjacent carbonyl destabilizes it.

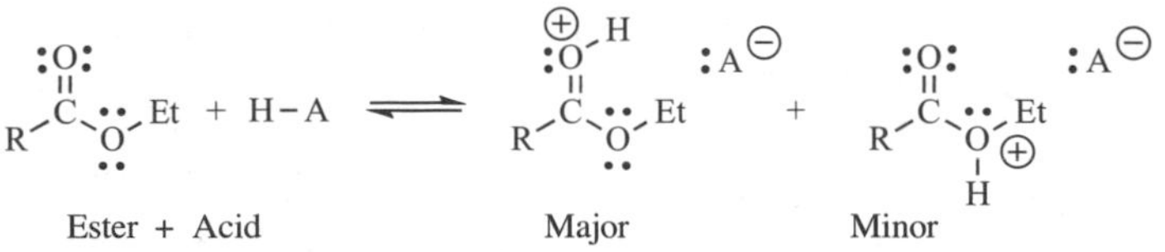
Ester + Acid
Major
Minor

Solvation is a hard–hard interaction, and therefore **solvent effects drastically alter nucleophilicity**. Small, hard, highly shielded ions must "break out" of the solvent cage to be available as nucleophiles, and thus nucleophilicity will vary greatly with solvation. Solvation, in fact, appears to be mostly responsible for the following halide nucleophilicity trend in methanol. The ion most strongly solvated is fluoride, the weakest iodide. In polar aprotic solvents like dimethyl sulfoxide, the halide anion is poorly solvated, and the nucleophilicity order reverses, being dominated by the strength of the bond formed (the C–F bond is strongest).

Best halide nucleophile in methanol $I^- > Br^- > Cl^- > F^-$ **Worst**

A p*K*a chart can be used as a reference for nucleophilicity only if the difference in softness is considered. A partially plus carbon atom is a much softer electrophile than a proton. Soft ions are more nucleophilic in protic solvents because tighter solvation greatly decreases the nucleophilicity of the hard ions.

We can restate the HSAB principle. Hard nucleophiles favor binding with hard electrophiles; soft nucleophiles favor binding with soft electrophiles. Most nucleophilicity charts show relative rates of nucleophilic attack with methyl iodide as the electrophile. A carbon–iodine bond is very soft because the electronegativity difference is nearly zero (Section 1.2). Therefore softer (less electronegative, more polarizable) atoms have lone pairs that are better electron sources toward soft electrophiles.

In summary, to rank the nucleophilicity of nonbonding electron pairs reacting in protic solvents with soft electrophiles such as R–X, rank first by softness, then by basicity (within the same attacking atom). However, for nonbonding electron pairs reacting with harder electrophiles such as a proton or a carbonyl, rank by basicity. Very reactive electrophiles like carbocations are not selective and react with the most abundant nucleophile (commonly the solvent).

Example problem

Which of the following, PhO^-, PhS^-, CH_3COO^-, is the best nucleophile to react with CH_3I in methanol? Which is the worst?

Answer: Since the electrophile is soft and the solvent is protic, the softness of the nucleophile must be considered. Because sulfur is softer than oxygen, the softest and most nucleophilic is PhS^-, thiophenoxide. The remaining oxygen anions are then ranked by basicity: Phenoxide (PhO^-), $pK_{abH} = 10$, is more basic than acetate (CH_3COO^-), at 4.8. Therefore the next best nucleophile is phenoxide, and the least is acetate.

4.2.3 Ranking of Leaving Groups

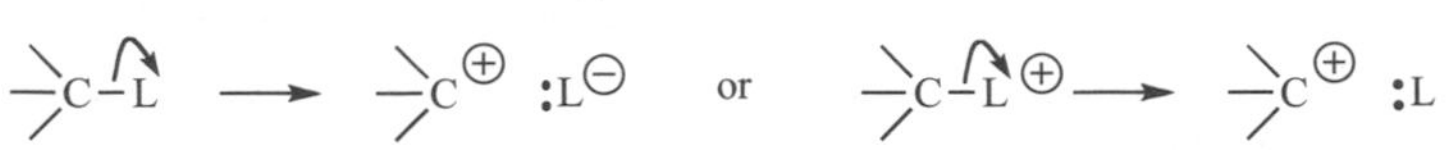
or

A group that separates from a compound and takes with it its bonding pair of electrons is called a *leaving group* (L) or nucleofuge, and can depart as an anion or as a neutral species. As the leaving group gets better, so does the substitution reaction to replace it. If a substrate has two different tetrahedrally bound leaving groups, then the better leaving group is usually the first to react. It is common to rank leaving groups by their conjugate acid pK_a, the pK_{aHL}. **A good leaving group, L, has a negative pK_{aHL}.** The pK_{aHL}s of poor leaving groups range from 13 to 30. The loss of a group like NH_2^- whose pK_{aHL} is greater than 30 is exceedingly rare.

Some groups do not fit this highly simplistic scheme for ranking leaving groups. If the L is highly electronegative, or if the carbon-leaving group bond is weaker or more polarizable, the bond is more easily broken. In ranking leaving groups by their pK_{aHL}, we are drawing a parallel between two reactions that are close but not identical:

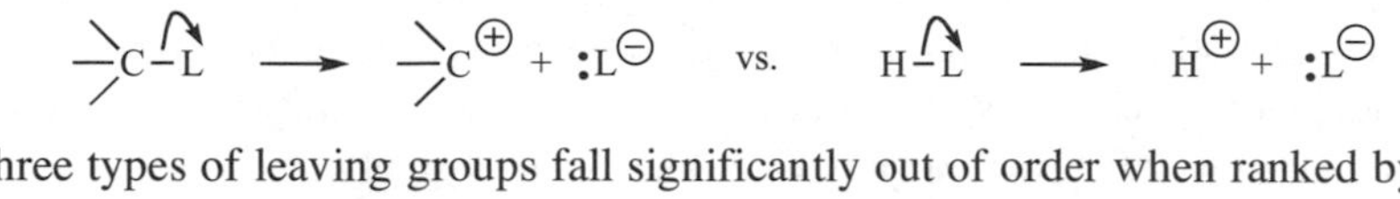
C–L
C⊕ + :L⊖
vs.
H–L
H⊕ + :L⊖

Only three types of leaving groups fall significantly out of order when ranked by pK_{aHL}. Fluoride and cyanide are much *poorer* leaving groups than predicted (due to a strong C–L bond). The charge delocalized alkyl or aryl sulfonates, RSO_3^- or $ArSO_3^-$, are much *better* leaving groups than predicted by their pK_{aHL}s (they are better than iodide); common representatives of these follow. Also, molecular nitrogen, N_2, is an excellent L.

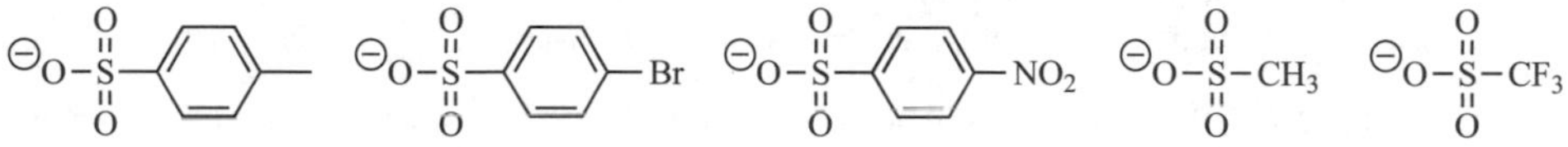
Br
NO2
CH3
CF3

Tosylate, OTs Brosylate, OBs Nosylate, ONos Mesylate, OMs Triflate, OTf

When using a pK_a chart to rank leaving groups, always look up the conjugate acid of the leaving group on the chart. If EtO^- departs, look up EtOH, pK_a +16; if EtOH departs, look up $EtOH_2^+$, pK_a −2.4. Protonation of a group before it falls off makes it cationic, more electronegative, and a much better leaving group. The following is a ranked list of common leaving groups and their conjugate acid pK_a. **Note that H^-, NH_2^-, and CH_3^-, each with a pK_{aHL} above 30 are not leaving groups.**

Excellent	**pK_{aHL}**
N_2	<−10
$CF_3SO_3^-$	−15
$ArSO_3^-$	−6.5
$CH_3SO_3^-$	−6

Good	**pK_{aHL}**
I^-	−10
Br^-	−9
Cl^-	−7
EtOH	−2.4
H_2O	−1.7
CF_3COO^-	+0.5

Fair	**pK_{aHL}**
$O_2NC_6H_4COO^-$	+3.4
$RCOO^-$	+4.8
$O_2NC_6H_4O^-$	+7.2
NH_3	+9.2
RS^-	+10.6
NR_3	+10.7

Poor	**pK_{aHL}**
HO^-	+15.7
EtO^-	+16
$RCOCH_2^-$	+19
$ROCOCH_2^-$	+26

Example problem

Which of these, R–Cl, R–Br, R–OH, has the best leaving group, and which has the worst?

Answer: The best leaving group is bromide, pK_{aHL} of –9. The next-best leaving group is chloride, pK_{aHL} of –7. The worst leaving group is hydroxide, pK_{aHL} = 15.7.

4.2.4 Carbocation Stability Trends

Figure 4.4 gives the geometries of the alkyl and vinyl carbocations. There are two common hybridizations for carbocations, sp^2 for alkyl carbocations and *sp* for vinyl carbocations. The carbon bearing the positive charge hybridizes to place the valence electrons in the most stable arrangement. Since hybrid orbitals with a higher percentage of *s* character (%*s*) are more stable, these are filled first, and an empty *p* orbital is left over. Alkyl carbocations can be relatively easy to form and are sp^2 and trigonal planar at the cationic center. Vinyl carbocations, however, are linear and more difficult to form because the cation is on a more electronegative *sp*-hybridized carbon.

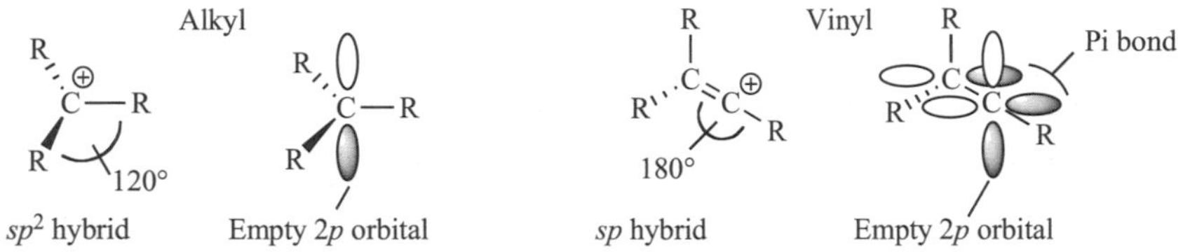
Alkyl
120°
sp^2 hybrid
Empty 2*p* orbital
Vinyl
180°
sp hybrid
Pi bond
Empty 2*p* orbital

Figure 4.4 The geometries of an alkyl carbocation (left) and a vinyl carbocation (right).

Carbocations are usually formed either by loss of a leaving group or by addition of an electrophile to a pi bond. Carbocation formation is very rare in nonpolar solvents that cannot stabilize the cation. Polar solvent molecules can cluster their lone pairs around the cationic center, forming an intermolecular donation of electron density that stabilizes the carbocation. Some carbocations are stable enough to be on the pK_a chart; the more stable ones are the weaker acids. Recall that when the pH equals the pK_a, the conjugate acid and base concentrations are equal. A protonated amide with a pK_a of –0.5 requires very acidic water before it is half-cation and half-neutral. A protonated ester with a pK_a of –6.5 is much less stable. However, since most carbocations are not on the pK_a chart, we need another method to rank carbocation stability.

A lone pair on an atom directly bonded to the cationic center serves as a pi-donor group and stabilizes the carbocation by donating some of the lone pair electron density. A pi bond is formed between the pi-donor lone pair and the empty 2*p* orbital of the carbocation (Fig. 4.5 top). The less electronegative the atom bearing the lone pair, the greater will be its capacity to donate. A lone pair in a nonbonding 2*p* orbital will be able to stabilize a carbocation better than a lone pair in a 3*p* or 4*p* orbital because of better overlap with the empty 2*p* carbon orbital. **The better the electron-releasing group, or the more of them attached to the cationic center, the more stable the carbocation is.**

Carbon–carbon double bonds can stabilize the carbocation by donating some electron density from the pi bond by resonance. The electrons in a pi bond are less available because they are usually in a lower-energy bonding molecular orbital compared to a nonbonding orbital of a lone pair. Resonance with a pi bond forms a conjugated system, distributing the charge over more atoms (Fig. 4.5 middle). Stabilization of a carbocation by one double bond gives the allylic cation.

Alkyl groups stabilize the carbocation to a much lesser extent by donating some electron density from the sigma framework via *hyperconjugation* (Fig. 4.5 bottom). The more substituted a carbocationic center is, the more stable it is (tertiary, R_3C^+ > secondary, R_2HC^+ > primary, RH_2C^+ > methyl, H_3C^+).

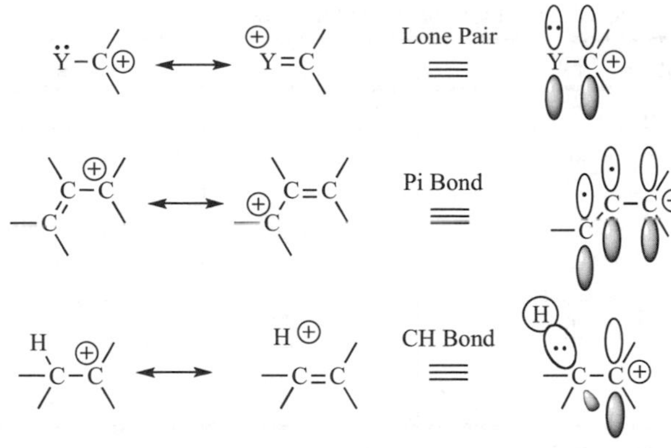
Lone Pair
Pi Bond
CH Bond

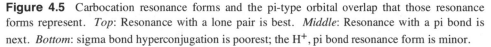

Figure 4.5 Carbocation resonance forms and the pi-type orbital overlap that those resonance forms represent. *Top*: Resonance with a lone pair is best. *Middle*: Resonance with a pi bond is next. *Bottom*: sigma bond hyperconjugation is poorest; the H^+, pi bond resonance form is minor.

Table 4.1 provides a ranking of carbocations by stability. We can state a general trend for carbocation stability: **Carbocations stabilized by resonance with a lone pair are more stable than those stabilized only by resonance with a carbon–carbon double bond, which in turn are more stable than those stabilized only by alkyl group substitution.** There are overlaps where three of a lesser type of stabilization are at least as good as one of the better type. For example, three R groups (tertiary alkyl) stabilize better than one pi bond (primary allylic). Vinyl carbocations are less stable than the corresponding alkyl carbocation of similar substitution (secondary vinyl cation is less stable than the secondary cation but more stable than the primary cation). Aromaticity greatly stabilizes carbocations; an example is tropylium cation, $C_7H_7^+$ (Section 1.9). Antiaromatic ions such as cyclopentadienyl cation (Section 1.9) are highly destabilized.

Example problem

Which of the following cations, $(CH_3)_2HC^+$, $Ph(CH_3)_2C^+$, $(CH_3)_3C^+$, is the most stable? The least stable?

Answer: The delocalized tertiary carbocation, $Ph(CH_3)_2C^+$, is the most stable. The *tert*-butyl carbocation, $(CH_3)_3C^+$, stabilized by three methyl groups, is of intermediate stability, whereas the isopropyl carbocation, $(CH_3)_2HC^+$, stabilized by only two methyl groups, is the least stable.

Carbocation Stabilization from a HOMO–LUMO Perspective
(A Supplementary, More Advanced Explanation)

Another way to look at the stabilization of a carbocation is to consider the interaction of the empty *p* orbital (LUMO) with the full donor orbital (HOMO). The best donors to an empty *p* orbital will be full orbitals close to the same energy (the stabilizing interaction is greater). For lone pair donors, as the atom bearing the lone pair becomes more electronegative, the pi-donor effect decreases because the energy of the HOMO drops. Stronger bonds have lower-energy HOMOs, farther away in energy from the empty carbon 2*p* orbital. A carbon–carbon pi bond will be a better electron-releasing group than a carbon–hydrogen sigma bond (Fig. 4.6).

Table 4.1 The Carbocation Stability Ranking

Carbocation Name	Structure	Comments on Stabilization
Stable carbocations		
Guanidinium cation	$(H_2\ddot{N})_3C^{\oplus}$	Three N lone pairs delocalize plus
Tropylium cation	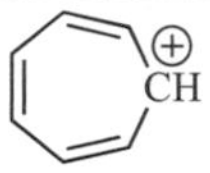	Stable by aromaticity
Protonated amide	$H_3C-C^{\oplus}(:\ddot{O}H)(:NH_2)$	Delocalized by N and O lone pairs
Moderately stable		
Protonated carboxylic acid	$H_3C-C^{\oplus}(:\ddot{O}H)(:\ddot{O}H)$	Delocalized by two O lone pairs
Triphenylmethyl cation	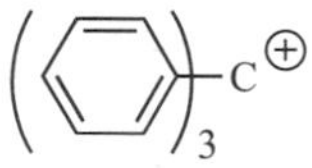	Delocalized by three phenyl groups
Protonated ketone	$H_3C-C^{\oplus}(:\ddot{O}H)(CH_3)$	Delocalized by one O lone pair
Diphenylmethyl cation	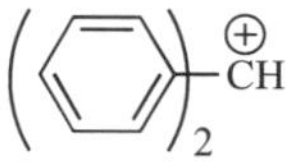	Delocalized by two phenyl groups
Average stability		
Tertiary alkyl cation (3°)	$(CH_3)_3C^{\oplus}$	Stabilized by three alkyl groups
Benzyl cation	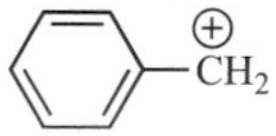	Delocalized by one phenyl group
Primary: allyl cation	$H_2C{=}HC-\overset{\oplus}{C}H_2$	Delocalized by one pi bond
Acylium cation	$H_3C-\overset{\oplus}{C}{=}\ddot{O}$	Lone-pair-delocalized vinyl cation
Moderately unstable		**No resonance delocalization**
Secondary: alkyl cation (2°)	$(CH_3)_2\overset{\oplus}{C}H$	Stabilized by two alkyl groups
Secondary: vinyl cation	$H_3C-\overset{\oplus}{C}{=}CH_2$	Plus on more electronegative *sp* C
Unstable		**(Rarely formed)**
Primary alkyl cation (1°)	$CH_3\overset{\oplus}{C}H_2$	Stabilized by one alkyl group
Primary vinyl cation	$H-\overset{\oplus}{C}{=}CH_2$	Plus on more electronegative *sp* C
Phenyl cation	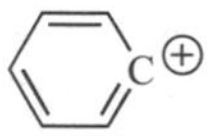	Bent vinyl cation & not delocalized
Methyl cation	$^{\oplus}CH_3$	No stabilization

The strained C–C sigma bonds of a three-membered ring have a HOMO much higher in energy than unstrained C–C sigma bonds and are the exception to sigma bonds being poor at stabilizing cations. A three-membered ring can stabilize an adjacent carbocation (a cyclopropylcarbinyl system, $C_3H_5–CH_2^+$) better than a phenyl can, $C_6H_5–CH_2^+$.

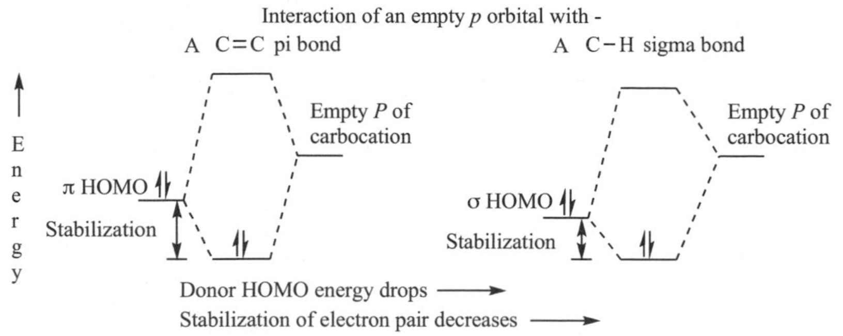

Interaction of an empty p orbital with -
A C=C pi bond
A C−H sigma bond
Energy
Empty P of carbocation
π HOMO
Stabilization
Empty P of carbocation
σ HOMO
Stabilization
Donor HOMO energy drops
Stabilization of electron pair decreases

Figure 4.6 Interaction diagrams of a carbocation with a C=C pi bond and a C–H sigma bond.

Nonclassical Carbocations

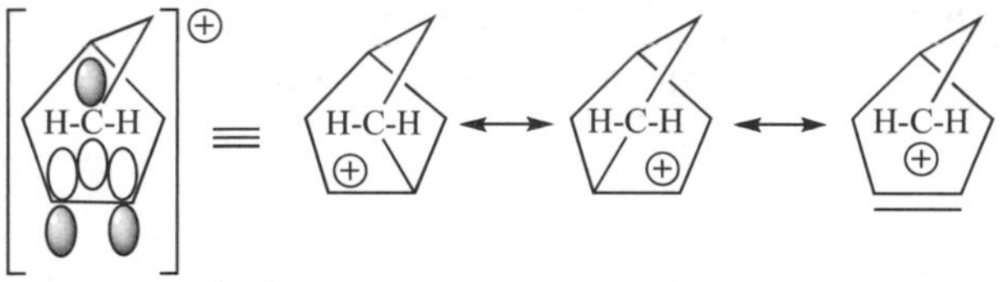

H-C-H
H-C-H
H-C-H
H-C-H

Figure 4.7 The 2-norbornyl nonclassical ion. The resonance forms on the right describe this unusual three-center, two-electron bonding molecular orbital.

Certain primary or secondary carbocations can enter into a type of bonding with a pi bond or strained sigma bond that involves three orbitals overlapping to give a bonding molecular orbital that is occupied by two electrons, similar to a pi-complex. These bridged species can be transition states, intermediates, or sometimes the lowest-energy structure of the carbocation. The 2-norbornyl cation (Fig. 4.7) is a nonclassical ion, but much experimentation was required to determine that the bridged ion was the lowest-energy species, not a transition state between two interconverting secondary carbocations.

To generalize, Figure 4.8 shows three ways that the **sigma overlap** of a lone pair, a pi bond, or a sigma bond can stabilize a carbocation. The most effective sigma-overlap stabilization comes from overlap of the empty orbital of the carbocation with a lone pair orbital forming an onium ion; for example, if X is Br, then it is called a bromonium ion. For comparison, Figure 4.5 gave the more common **pi overlap** stabilization of a carbocation by a lone pair, a pi bond, and a CH bond.

4.2.5 Ranking of Electron-Releasing Groups

The ranking of electron-releasing groups (erg) or electron donors is very important. The electron richness of electron sources will depend on this trend. The ranking is based on how available the electrons of the donor are; the more stabilized they are, the less available they are. Electrons in nonbonding orbitals are the most available. The best donors are anions that can donate an electron pair by resonance. The better donors all contain lone pairs that can form a strong pi bond to the empty *p* orbital of the cation, and are pi-donors for that reason. Table 4.2 is a ranked list of electron-releasing groups.

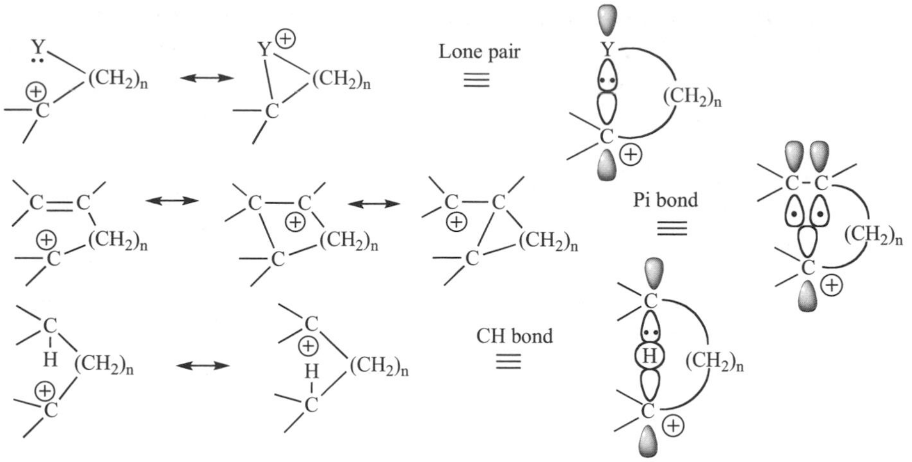
Lone pair
(CH2)n
Pi bond
CH bond

Figure 4.8 Summary of ways a filled orbital can stabilize a carbocation using sigma-type overlap. The arc symbolizes one or more CH_2 groups. Most common is the onium ion shown on the top.

Table 4.2 Ranking of Electron-Releasing Groups, ERG

Excellent donors	**Features of group**
$—\ddot{C}H_2^{\ominus}$	Anionic and least electronegative
$—\ddot{N}H^{\ominus}$ (two lone pairs)	Anionic and more electronegative than C
$—\ddot{O}\colon^{\ominus}$ (three lone pairs)	Anionic and more electronegative than N
Good donors	
$—\ddot{N}(CH_3)_2$	Neutral and less electronegative than O
$—\ddot{N}H_2$	Neutral and less electronegative than O
$—\ddot{O}H$ (two lone pairs)	Neutral and more electronegative than N
$—\ddot{O}CH_3$ (two lone pairs)	Neutral and more electronegative than N
$—\ddot{N}H—C(=O)CH_3$	Amide carbonyl decreases N lone pair availability
$—\ddot{S}CH_3$ (two lone pairs)	Sulfur 3*p* orbital has poor overlap with a 2*p* orbital
Poor donors	
—Ph	Delocalization through resonance with phenyl
—R	Hyperconjugation only, no lone pair or pi resonance
—H	No substituent at all
Very poor donors	
$—\ddot{O}—C(=O)CH_3$ (two lone pairs)	Electronegative, carbonyl decreases O lone pair availability
$—\ddot{C}l\colon$ (three lone pairs)	Electronegative and poor overlap with a 2*p* orbital

For first-row elements, as the atom containing the lone pair becomes more electronegative, it is less able to donate electron density. For this reason, carbon anions are better donors than nitrogen anions, which are in turn better donors than oxygen anions. Anions are the best donors because they donate both by resonance and by inductive/field effects. Neutral heteroatoms are more electronegative than carbon, so inductive electron withdrawal competes with resonance electron donation, making them poorer lone pair donors than anions.

An electron-withdrawing group attached to a lone pair donor decreases the electron availability of the lone pair. For example, the attachment of an electronegative carbonyl group to a nitrogen atom with a lone pair diminishes the electron donation ability of that lone pair; an amide nitrogen is a much poorer electron-releasing group than an amine nitrogen. Delocalization of the nitrogen lone pair into the carbonyl stabilizes the lone pair and makes the nitrogen partially positive and more electronegative.

The formation of a strong pi bond is responsible for the stabilization afforded by the pi-donor. However, pi bonds between orbitals from different shells necessarily have poor overlap and are weak. A sulfur atom is a poor pi-donor because there is poor overlap of the sulfur 3*p* orbital with the carbon 2*p* orbital. The two worst lone pair donors are poorer donors than alkyl groups because their high electronegativity is not compensated by their weak pi donation ability. For example, a chlorine atom is very electronegative and is a poor resonance pi-donor because of poor 2*p*-3*p* overlap.

Resonance effects will usually dominate over inductive/field effects. When the resonance effect is blocked, the only groups that will still be donors are anions and alkyl groups. The electronegativity of all the other groups will cause them to be inductive electron-withdrawing groups. Electron-withdrawing groups destabilize carbocations. Similarly, electron-releasing groups destabilize anions.

Example problem

Which of these, OCH_3, CH_3, NH_2, is the best electron-releasing group? The worst?

Answer: Both the ether and the amine are lone pair donors and so are better pi-donors than a methyl group, which is the weakest. The amine is the best electron-releasing group because the amine nitrogen is less electronegative than the ether oxygen.

4.2.6 Stereochemistry and Substitution Reactions

A very important implication of the three-dimensional nature of organic compounds is that they can be *chiral*, occurring in left- and right-handed forms. A compound is chiral if it does not superimpose on its mirror image (Fig. 4.9). A carbon atom with four different groups is called a chiral center.

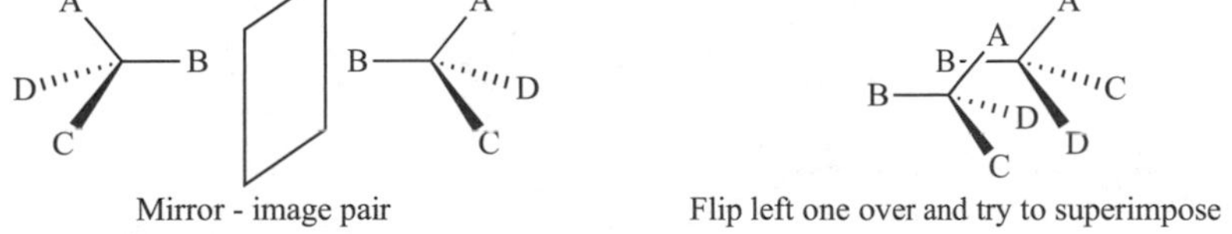
A
B
D
C
A
B
D
C
Mirror - image pair
A
A
B
B
C
D
D
C
Flip left one over and try to superimpose

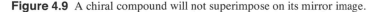

Figure 4.9 A chiral compound will not superimpose on its mirror image.

A molecule can be chiral without having a chiral center if it has a chiral shape such as a propeller or helix. A molecule is not chiral if it contains an internal mirror plane. A compound can contain chiral centers and not be chiral if there is an internal mirror plane. Whenever you are uncertain, build the model of the compound and try to superimpose it on a model of its mirror image. See the Appendix for how to name chiral compounds.

Enantiomers are mirror image pairs that are not superimposable (if the pair superimposes, then the two are the same). Enantiomers have identical physical and chemical properties; they can be distinguished only in a chiral environment, such as in the active site of an enzyme. A 1:1 mix of the two enantiomers is called a *racemic mixture* (symbolized by ±). Originally, most pharmaceuticals were synthesized as the racemic mixture because chiral synthesis was a real challenge. However, the enantiomers of many pharmaceuticals not only are often ineffective but also add to the toxicity, so great effort is now made to synthesize the effective enantiomer without its "evil twin." *Conformational isomers* just differ by single bond rotations, and therefore are interconvertable. *Stereoisomers* are isomers that have the same sequence of bonds and differ only in the arrangement of atoms in space. Stereoisomers that are not enantiomers are called *diastereomers*. For example, *cis* and *trans* double bond isomers are diastereomers. Diastereomers can often have different physical and chemical properties. Figure 4.10 shows the relationships between the different types of isomers.

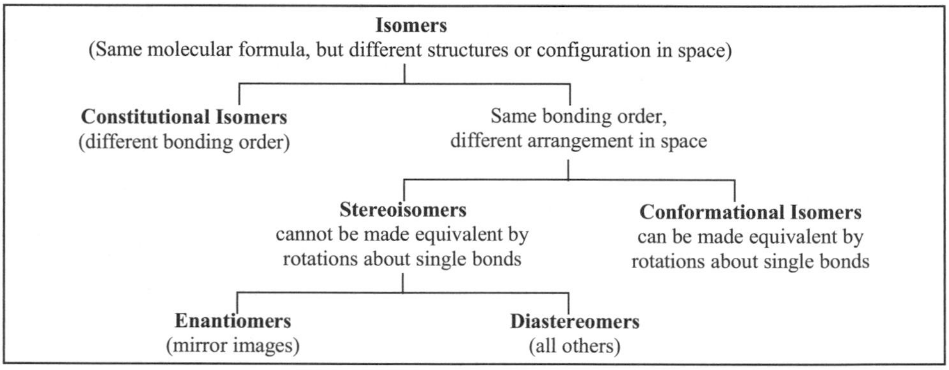
Isomers
(Same molecular formula, but different structures or configuration in space)
Constitutional Isomers
(different bonding order)
Same bonding order,
different arrangement in space
Stereoisomers
cannot be made equivalent by
rotations about single bonds
Conformational Isomers
can be made equivalent by
rotations about single bonds
Enantiomers
(mirror images)
Diastereomers
(all others)

Figure 4.10 The isomer family tree.

Figure 4.11 illustrates the three possible stereochemical outcomes of a substitution reaction. The S_N2 substitution gives inversion because the nucleophile must attack the carbon from the back side because the full nucleophile orbital must overlap well with the empty electrophile orbital (Section 4.2.7). The S_N1 substitution often gives racemization because the intermediate carbocation is trigonal planar, and the nucleophile can attack equally from top or bottom faces (Section 4.2.8).

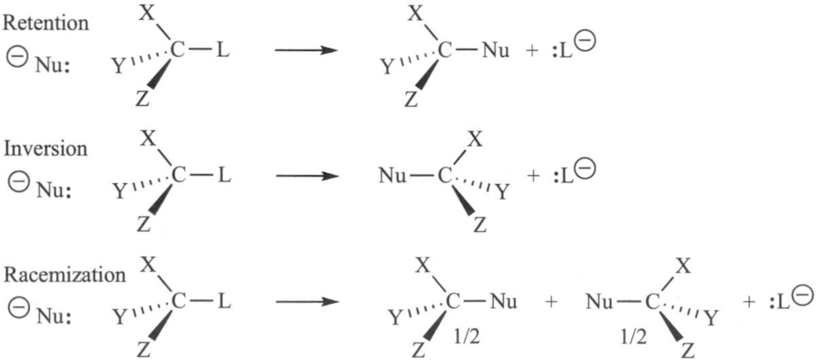
Retention
Inversion
Racemization
Nu:
1/2
1/2

Figure 4.11 The three possible fates of a chiral center undergoing a substitution reaction: retention, inversion, or racemization (producing a 1:1 mixture of enantiomers, usually shown by ±).

Example problem

What is the relationship between each of the following pairs of compounds?

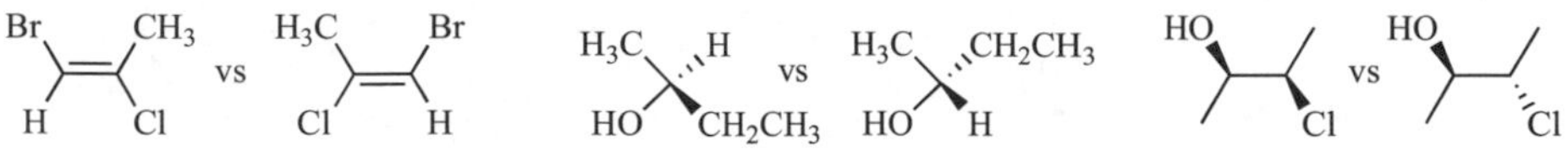
Br
CH3
H
Cl
vs
H3C
Br
Cl
H
H3C
H
HO
CH2CH3
vs
H3C
CH2CH3
HO
H
HO
Cl
vs
HO
Cl

A B C D E F

Answer: Compounds A and B are the same (flip B over onto A; they are both flat). Compounds C and D are enantiomers (mirror image pairs). Compounds E and F are diastereomers (they are not mirror images, but have different orientations in space).

4.2.7 S_N2 Electron Flow Pathway

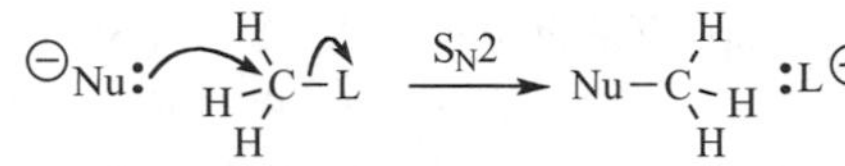
Nu:
H
H
C
L
H
SN2
H
Nu
C
H
H
:L

When the carbocation is unstable, the lowest energy path on Figure 4.2 is the S_N2 in which the nucleophile pushes out the leaving group. The first electron flow arrow starts from the nucleophile lone pair and ends at the carbon attacked. A second arrow breaks the carbon-leaving group bond and deposits the bonding pair of electrons as a lone pair on the leaving group.

S_N2 from a HOMO-LUMO Perspective

(A Supplementary, More Advanced Explanation)

When a nucleophile displaces a leaving group in an S_N2 substitution reaction, the nucleophile HOMO interacts with the LUMO of the leaving group bond as shown in Figure 4.12. Nucleophile attack produces the transition state wherein the nucleophile overlaps the largest lobe of the LUMO to get the greatest stabilization possible. The nucleophile displaces the leaving group from the back side, and therefore the configuration about the carbon attacked is inverted in this type of process. Front side displacement would involve much poorer interaction with the LUMO because of nearly equal positive and negative overlap.

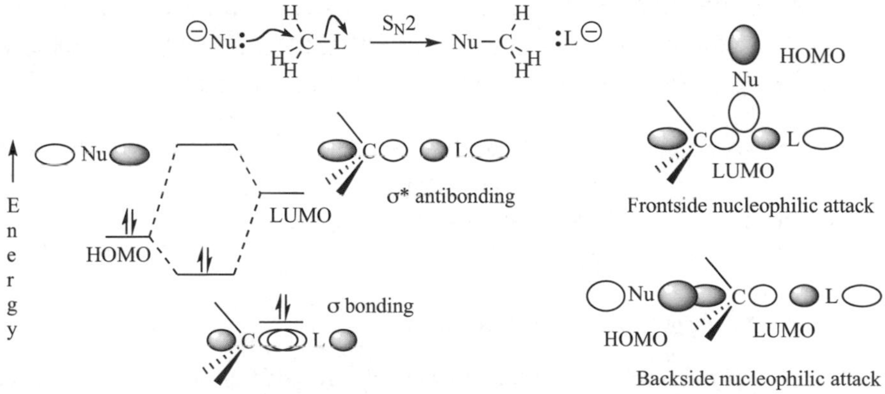
Nu:
C
L
SN2
Nu
C
:L
Energy
Nu
HOMO
LUMO
σ* antibonding
σ bonding
HOMO
Nu
LUMO
Frontside nucleophilic attack
HOMO
LUMO
Backside nucleophilic attack

Figure 4.12 The interaction of the HOMO of the nucleophile and the LUMO of the leaving group bond to carbon to produce a new sigma bond.

Stereochemistry of S_N2 reactions

Figure 4.11 illustrated the three possible stereochemical outcomes of a substitution reaction. The S_N2 reaction gives inversion because the nucleophile displaces the leaving group from the back side (pushing it out from the opposite side) as shown in Figure 4.12.

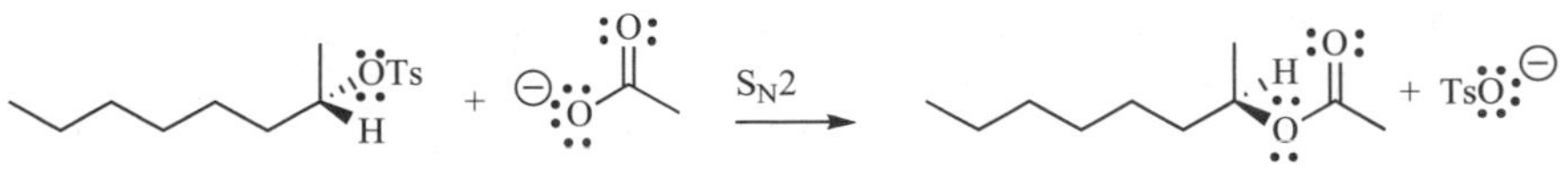

S_N2 Path Limitations

There are three requirements for the S_N2 reaction. In order to be pushed out, the leaving group must be at least fair, usually good. Section 4.2.3 gave a ranking of leaving groups; the pK_{aHL} cutoff for the S_N2 is usually no greater than 10. An exception to this is epoxides; strain relief of opening the three-membered ring makes up for the poorer alkoxide leaving group. The nucleophile also must be reactive enough to push out the leaving group. A list of nucleophile rankings for the S_N2 on methyl iodide in methanol solvent was given in Section 4.2.2. A poorer nucleophile can be compensated by a better leaving group and vice versa. Usually simple pi bond sources do not react by S_N2. Allylic sources like enolates, enols, and enamines do, since the attached electron-releasing group makes them more electron rich. Finally, the back side of the tetrahedral carbon attacked must be accessible to the nucleophile and not blocked by other groups.

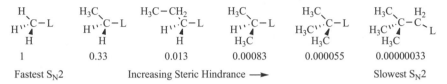

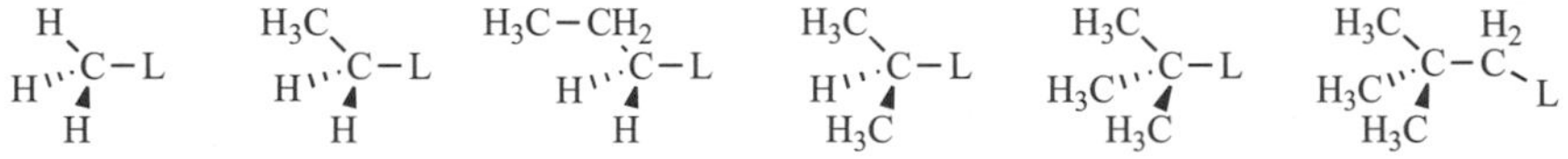
1
0.33
0.013
0.00083
0.000055
0.00000033
Fastest S_N2
Increasing Steric Hindrance
Slowest S_N2

Figure 4.13 Relative rates of S_N2 reactions as a function of steric hindrance of the site of attack.

Figure 4.13 gives the relative rates of S_N2 reactions as the groups around the carbon attacked get bigger. Methyl and primary are fastest; secondary is slow. Tertiary is so slow to S_N2 that other processes like S_N1 easily win out. Neopentyl is the slowest, as a methyl group of the *tert*-butyl group always blocks the back side nucleophilic attack.

Example problem

Which of the following, CH_3Cl, CH_3I, CH_3Br, will be the fastest S_N2 in methanol using hydroxide as a nucleophile? Which is the slowest?

Answer: The S_N2 on a better leaving group will be faster using the same nucleophile. The best leaving group is iodide, pK_{aHL} of −10, so is the fastest S_N2. The next-best leaving group is bromide, pK_{aHL} of −9. The worst leaving group and slowest S_N2 is chloride, $pK_{aHL} = -7$.

S_N2 Energetics and the ΔpK_a Rule

If the reactants and the products of a reaction are not charged, we can calculate the $\Delta H°$ (Section 2.5) and thereby know whether the reaction is uphill or downhill in energy. Unfortunately, this calculation is of limited use for mechanisms, since most reaction steps involve charged species as nucleophiles, electrophiles, or intermediates. We need a way to tell whether a step involving a charged species is energetically uphill or down.

For example, the S_N2 displacement of a leaving group by a nucleophile is microscopically reversible. Therefore a general rule is needed to determine whether the reaction will proceed to the left or to the right. We can use the relative basicity of the Nu and the L as a guide, because **usually the reaction will proceed to form the weaker**

base, as was seen in the proton transfer process. Sometimes a slightly stronger base can be formed if the bond broken is weaker than the one formed; the formation of the stronger bond would counterbalance the formation of a slightly stronger base. Even a 10-unit climb in pK_{abH}, corresponding to a 13.7 kcal/mol (57.3 kJ/mol) climb in energy, could occasionally be compensated for by bond formation. Similar to the proton transfer K_{eq} crosscheck in Section 3.6, we will state a general rule that will become one of our most important energetics crosschecks for polar reactions. **The ΔpK_a rule: Do not create a product or intermediate more than 10 pK_a units more basic than the incoming nucleophile.** The ΔpK_a rule can be used to predict feasibility and the reversibility of an S_N2 reaction. If the reverse reaction climbs more than 10 pK_a units, the reaction probably does not reverse. If the forward reaction climbs more than 10 pK_a units, the reaction probably does not go at all.

Example problem

Is the following S_N2 reasonable, and is it reversible?

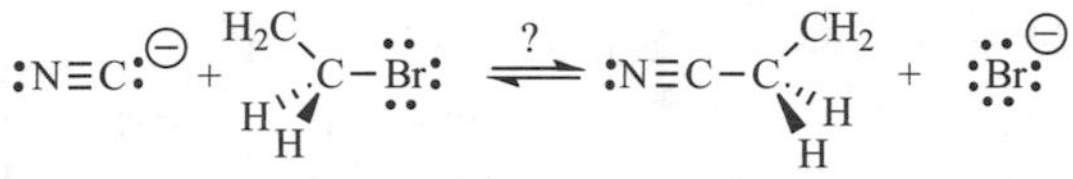

Answer: The cyanide nucleophile has a pK_{abH} = 9.2 and the bromide leaving group has a pK_{aHL} of −9. The forward direction drops 18.2 pK_a units, so is very favorable. The reverse reaction would have to climb the same 18.2 units, and so is ruled out by our ΔpK_a rule. The reaction goes forward, but is not reversible.

4.2.8 S_N1 Reaction: Electron Flow Pathways D_N + A_N

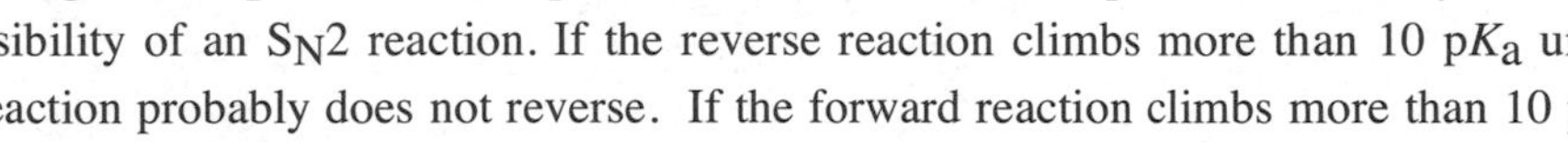

The S_N1 substitution reaction is a two-step process, a slow D_N step to break the carbon-leaving group bond forming a carbocation, followed by a fast A_N trapping of the carbocation to form the new bond. The D_N and A_N paths are just the reverse of each other. Carbocations have just three fates: They can be trapped by a nucleophile as discussed in this section; they can lose a proton to form the alkene (Section 4.3), or they can rearrange to another carbocation of equal or greater stability (Section 4.7).

Stereochemistry of S_N1 reactions

Figure 4.11 illustrated the three possible stereochemical outcomes of a substitution reaction. The S_N1 reaction often gives racemization because the trigonal planar carbocation can be trapped from either face equally. Sometimes S_N1 gives partial inversion (more inversion than retention) because the leaving group is ion-paired with the carbocation, and the back side is more open to nucleophilic attack because the leaving group is still blocking the front. Polar solvents aid the breaking up of ion pairs, so they are expected to favor complete racemization in the S_N1 reaction.

Example problem

For the compound below, draw an S_N1 substitution mechanism and show both enantiomers of the product that is formed.

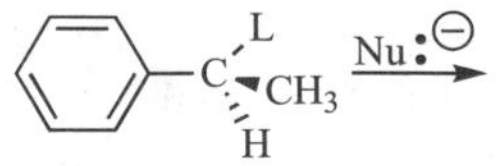
L
Nu:⊖
C
CH3
H

Answer: The S_N1 route first loses the leaving group through path D_N to form a trigonal planar carbocation that is trapped from either side by path A_N.

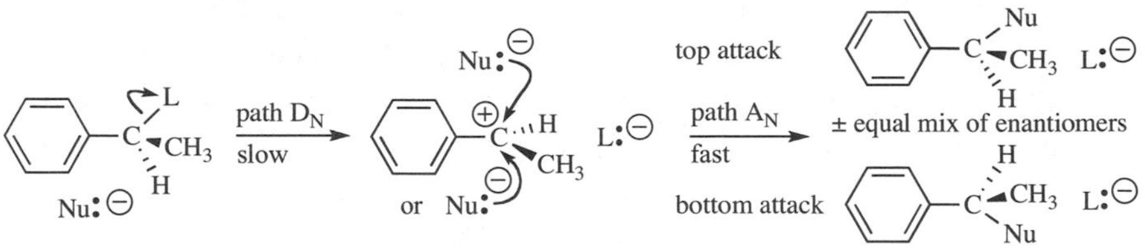
path DN
slow
Nu:⊖
or Nu:
L:⊖
top attack
path AN
fast
bottom attack
± equal mix of enantiomers
Nu
CH3
H

4.2.9 Ionization of Leaving Groups

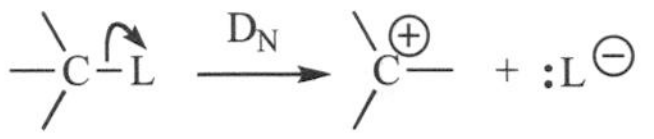
DN
+ :L⊖

The slow step in the S_N1 is the loss of the leaving group. The ionization of a leaving group depends primarily on three factors, the ability of the **solvent** to stabilize the charges formed, the stability of the **carbocation**, and the quality of the **leaving group**.

Table 4.3 gives conditions for ionization of leaving groups. **To summarize, definitely expect ionization when these three conditions are satisfied: a good leaving group ($pK_{aHL} < 0$), a polar solvent, and a carbocation of stability better than secondary.** Do not expect leaving group ionization to occur in a nonpolar solvent. If you do not have all three conditions satisfied, the S_N1 does not occur at a significant rate.

Table 4.3 The Ionization of Leaving Groups

Carbocation	pK_{aHL}	Solvent	Ionize?
1° Primary RCH_2^+	Any	Any	Almost never
2° Secondary R_2CH^+	Above 0	Any	Rarely
	Below 0	Polar	Slow
3° Tertiary R_3C^+	Any	Nonpolar	Rarely
	7 or higher	Polar	Rarely
	0 to 6	Polar	Medium
	Below 0	Polar	Fast

Example problem

Which of the following, $(CH_3)_3CBr$, Ph_2CHBr, $(CH_3)_2CHBr$, will be the fastest S_N1 in methanol? Which is the slowest?

Answer: The solvents and leaving groups are the same. The S_N1 reaction will be faster via the more stable carbocation. The most stable carbocation and fastest S_N1 reaction is via the diphenylmethyl carbocation; next is via the *tert*-butyl carbocation. The least stable isopropyl cation is the slowest, not expected to form at a significant rate.

4.2.10 S_N1/S_N2 Substitution Spectrum

It is best to consider a S_N1/S_N2 spectrum where there is competition between two rates: the rate of ionization of the leaving group and the rate of nucleophilic attack on the substrate or partially ionized substrate. Table 4.4 outlines the extremes of the spectrum.

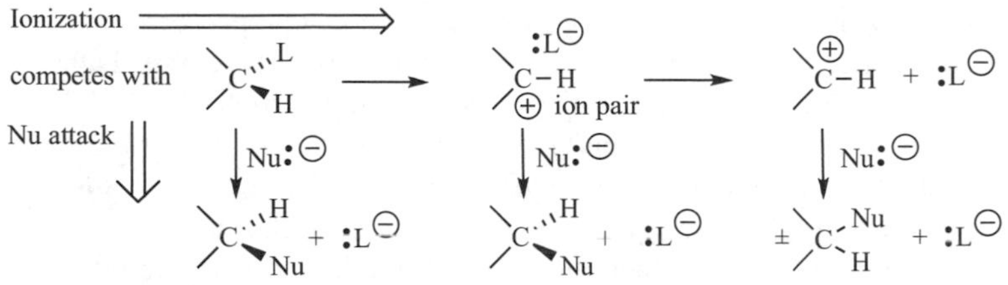

Ionization
competes with
Nu attack
ion pair
Nu:
:L

TABLE 4.4 The S_N1/S_N2 Spectrum Extremes

Variable	S_N1 (Path D_N then A_N)	S_N2 (Concerted A_ND_N)
Carbocation	Must be good (usually better than secondary) because it is an intermediate	Not formed, therefore carbocation stability is not a concern
Nucleophile	Weaker Nu (often neutral) tolerated since the sink is good	Good Nu (commonly anionic) required to push out the leaving group
Site of attack	Often hindered	Must be open
Stereochemistry	Racemization (or inversion on ion pair)	Inversion
Media	Often acidic	Often basic
Leaving group	Excellent	Average

Although the actual energy surface for the simplest substitution reaction is much more complex than the simplified surfaces drawn in the following figures, we can use these surfaces to understand how the changes in reaction conditions would affect the reaction path. The vertical axis of energy is added to Figure 4.2 to give Figure 4.14, where the front horizontal axis is still the degree of C–Nu bond making and the axis going back to front is the degree of C–L bond breaking. Reactants are in the back right corner at point R. The back left corner corresponds to only C–Nu bond making without C–L bond breaking, thus forming a high-energy pentacovalent intermediate since carbon

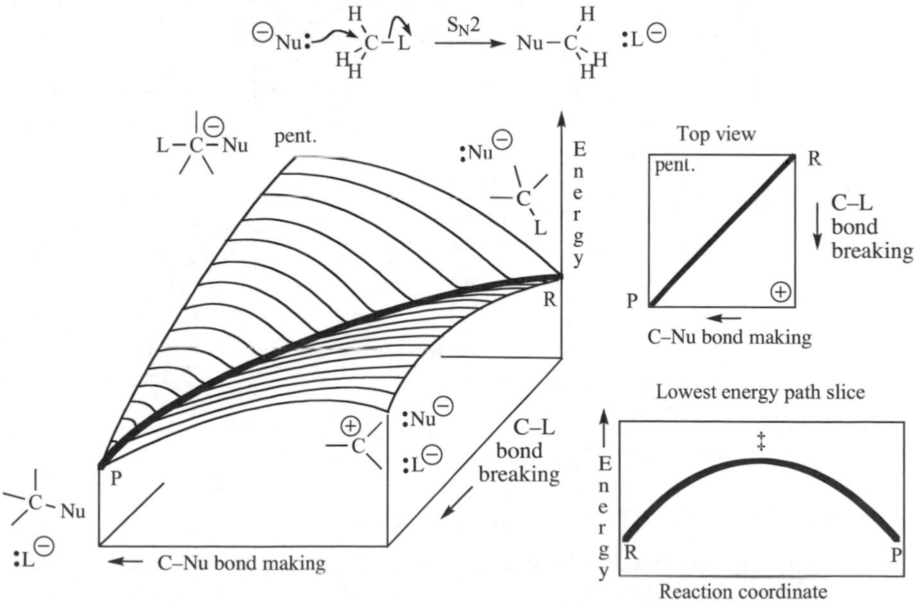

S_N2
pent.
Energy
R
P
C–L bond breaking
C–Nu bond making
Top view
Lowest energy path slice
Reaction coordinate

Figure 4.14 The S_N2 energy surface; C–L bond breaking and C–Nu bond making are concerted.

can't expand its valence shell. The front right corner corresponds to only C–L bond breaking and therefore is the carbocation; the front left corner is the substitution product P. The lowest-energy path on the surface is the dark bold line from R to P.

It is possible to understand the trends by watching how the substitution surface folds and/or tilts in response to a change. In Figure 4.14, the carbocation is poor; its corner is high in energy; since the pentacovalent corner is also high in energy, the surface folds down the middle. The reaction will go by the S_N2 process in which C–Nu bond making and C–L bond breaking occur to an equal extent, and the lowest-energy path will follow the *diagonal* through the "pass" between the two "mountains" created by the high-energy carbocation and high-energy pentacovalent intermediate.

A better nucleophile makes C–Nu bond making easier, and therefore lowers the entire left edge, since all points along that edge have the C–Nu bond made. Therefore, the effect of a better nucleophile is to lower the pentacovalent corner somewhat, favoring S_N2, and to lower the product, making the overall reaction more favorable.

On the other hand, if the carbocation is rather stable, the carbocation corner on the diagram will be lowered in energy, and now the lowest-energy path will go via the carbocation: the S_N1 substitution (Fig. 4.15). The highest barrier is the loss of the leaving group.

Steric hindrance raises the pentacovalent intermediate to an even higher energy, since it is the most crowded, with five groups around the central atom. The carbocation with only three groups around the central atom is least affected by steric hindrance. Therefore, increasing steric hindrance raises the pentacovalent intermediate greatly and the carbocation little, resulting in a tilting of the surface toward carbocation formation.

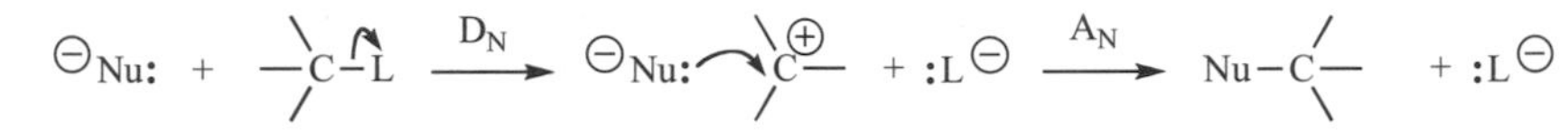
Nu:
L
D_N
Nu:
:L
A_N
Nu
:L

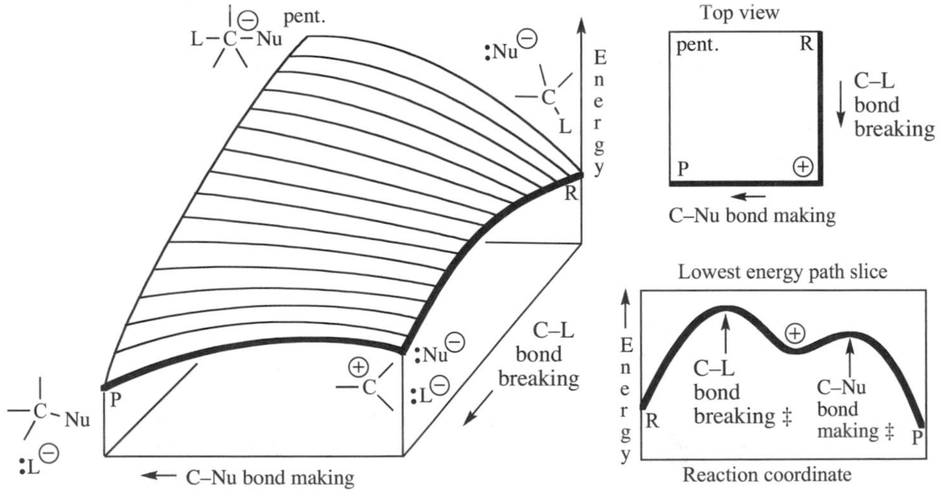
pent.
L–C–Nu
:Nu
L
Energy
R
P
Nu
:L
:Nu
:L
C–L
bond
breaking
C–Nu bond making
Top view
pent.
R
P
C–L
bond
breaking
C–Nu bond making
Lowest energy path slice
Energy
R
C–L
bond
breaking ‡
C–Nu
bond
making ‡
P
Reaction coordinate

Figure 4.15 The S_N1 energy surface. The C–L bond breaking occurs first to form a carbocation, and then the C–Nu bond is made.

If the leaving group is better, C–L bond breaking is easier, and therefore the entire front edge is lowered because all points along that edge have the C–L bond broken. The effect of a better leaving group lowers the carbocation corner, favoring S_N1, and lowers the product. This makes the overall reaction more favorable.

4.2.11 Substitution via a Pentacovalent Intermediate

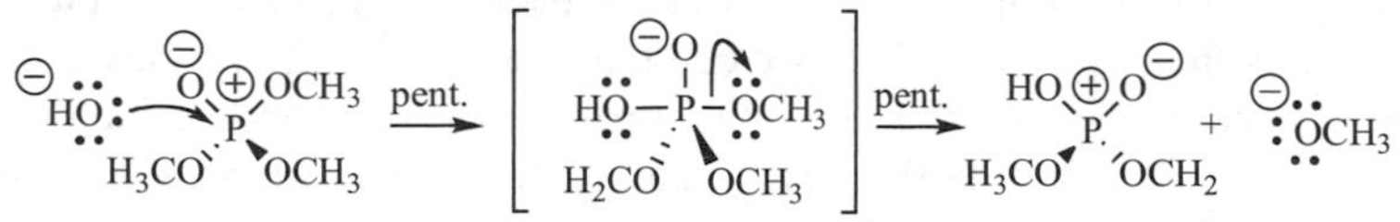
pent.
pent.

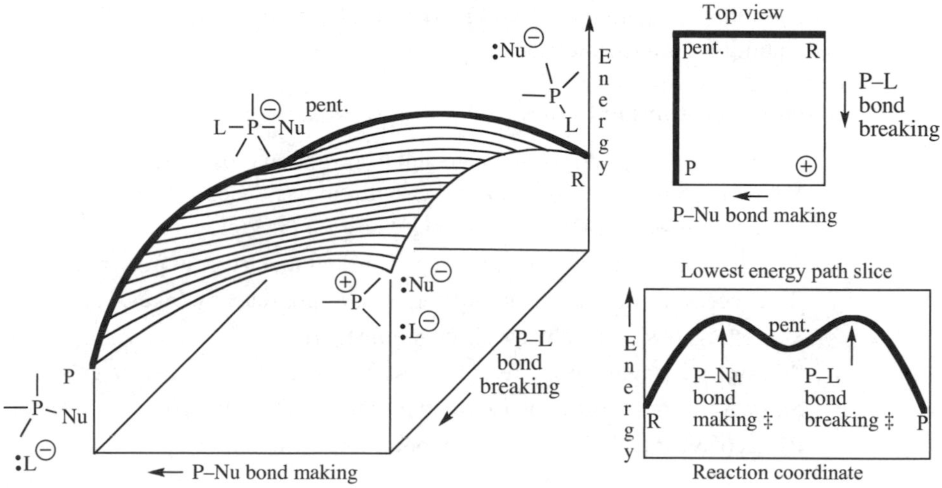
pent.
Energy
R
Top view
pent.
R
P
P–L bond breaking
P–Nu bond making
Lowest energy path slice
pent.
P–Nu bond making ‡
P–L bond breaking ‡
R
P
Energy
Reaction coordinate
P–L bond breaking
P
Nu
P–Nu bond making

Figure 4.16 The energy surface for substitution via a pentacovalent intermediate.

Third-row and higher elements can stabilize a pentacoordinate intermediate by *d*-orbital bonding. Their larger size also allows for a five-bonded intermediate to be less crowded. These factors lower the pentacovalent intermediate corner far enough so that the reaction can proceed through that intermediate (Fig. 4.16). Substitution at phosphorus commonly proceeds through a pentacovalent intermediate, via the pent. path, as shown in the above example. The pentacovalent intermediate shown in brackets is a trigonal bipyramid with the nucleophile in the apical position. By microscopic reversibility, the leaving group is also in the apical position. The better leaving groups lower the front edge of the diagram, and also lower the barrier for the second step, decreasing the lifetime of the intermediate. At the good leaving group extreme, the intermediate can last less than a vibration, and the route is difficult to distinguish from an S_N2 substitution.

4.2.12 Approaches to Substitution Mechanisms

With three substitution routes each with new crosschecks, our problem space has gotten more complex. The first task is always to map changes on a balanced reaction: What group was replaced, and what group replaced it? Is the medium acidic or basic? In acid, we can protonate leaving groups to improve them. In base we can often deprotonate the nucleophile to improve it. Remember, the medium can't be both acidic and basic. In essence, we check for possible proton transfer first. Next we decide the substitution route. Visualize that you are standing on the reactant "street corner" of the energy surface of Figure 4.2 and need to look at the "traffic lights" (energetics) to decide which way to cross. If the leaving group in the problem is on a carbon, then the pentacovalent route is out because it is too high in energy. Next check the three restrictions for the D_N step: a carbocation stability usually better than secondary, a polar solvent, a good leaving group with a $pK_{aHL} < 0$. If it fails on any of the three, the S_N1 route is also too high in energy. It is time to check S_N2 restrictions: good access, a decent nucleophile, and a decent leaving group. We are checking each branch of the problem space tree to see which is the best route before we commit to a direction to proceed. Once you decide the best route, make sure you draw the arrows correctly and keep track of charge balance. Use the known electron flow paths, and don't combine them to just save drawing an extra structure. Skipped or combined steps are difficult to cross-check. With practice, the consideration of alternatives and crosschecks become automatic. You will develop a good intuition for these reactions.

Common Substitution Errors

The most common errors are the result of forgotten crosschecks:
Mixed media errors (Don't do acidic routes in basic media or vice versa).
Proton transfer energetics errors (make sure K_{eq} is greater than 10^{-10}).
S_N2 does not occur on leaving groups attached to a trigonal planar center.
S_N2 access restrictions (a tertiary center does not react by S_N2)
S_N2 energetics error (always check the ΔpK_a rule)
S_N1 or S_N2 leaving group restriction (don't drop off poor leaving groups)
S_N1 carbocation restriction (don't make an unstable carbocation)
Pentacovalent path does not occur for second-row elements.

Example Substitution Mechanism

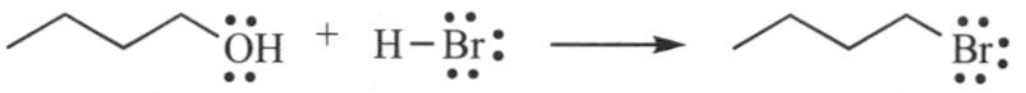

Often organic chemists don't bother to fully balance reactions, so that is your first task in order to be able to map the changes. With HBr present, the reaction is certainly in acidic medium. A substitution definitely has occurred; the OH has been replaced by Br. With a balanced reaction, we can see the leaving group was water.

We first check for favorable proton transfers. A proton transfer reaction not only improves the leaving group by protonation, but also produces a better nucleophile, the conjugate base of HBr, the bromide anion. The proton transfer passes a K_{eq} check: $\text{Log}K_{eq} = (pK_{abH} - pK_{aHA}) = (-2.4 - (-9)) = +6.6$, so $K_{eq} = 10^{+6.6}$ is very favorable.

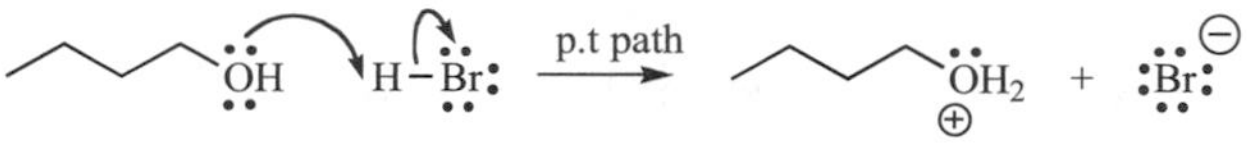
p.t path

It is time to decide on a substitution route. As the leaving group is on a second-row element, the pentacovalent path is out. A check of D_N restrictions shows the leaving group is good, and the solvent polar, but the carbocation would be primary, and therefore very unstable, so S_N1 is also out. A check of S_N2 access passes since the carbon bearing the leaving group is unhindered. The ΔpK_a rule has an acceptable climb of $pK_{aHL} - pK_{aHNu} = (-1.7 - (-9)) = +7.3$ pK_a units.

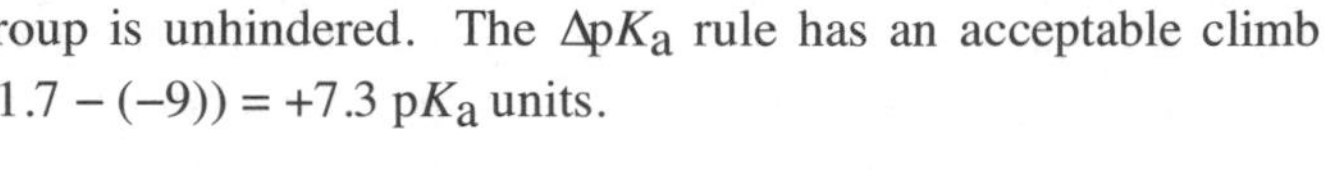
S_N2 path

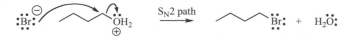

By using the crosschecks we have produced a reasonable mechanism and avoided the common errors that going with our first idea may have fallen prey to. Often it may seem that diving into a problem without considering all these restrictions is quicker, but much time is then lost exploring useless dead ends. A logical approach avoids those, and overall is faster and will get even faster with practice.

Nucleophilic Catalysis

Nucleophilic catalysis, often present in biochemical systems, is illustrated in the following example. The uncatalyzed reaction is slow because even though chloride is a decent leaving group, neutral water is a poor nucleophile. We expect an S_N2 reaction, not S_N1 because the carbocation would be primary, and the pentacovalent route is out because the leaving group is on a second-row element. This reaction is similar to the previous worked reaction mechanism, just with the conditions changed (excess water) to favor the reverse reaction.

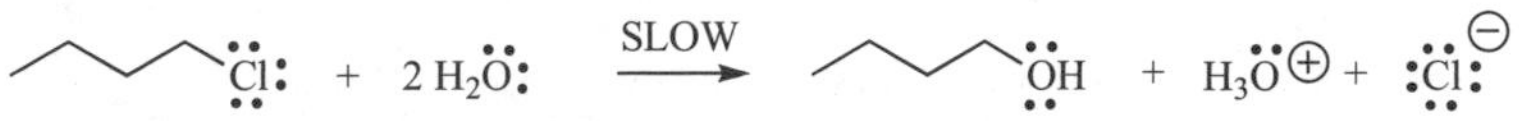
Cl
+ 2 H2O
SLOW
OH
+ H3O⊕ + Cl⊖

The reaction is impressively sped up by adding a small amount of KI. Potassium iodide is not an acid or a base, and KCl does not speed the reaction at all. What is happening here? It must be due to the iodide anion, acting as a nucleophilic catalyst.

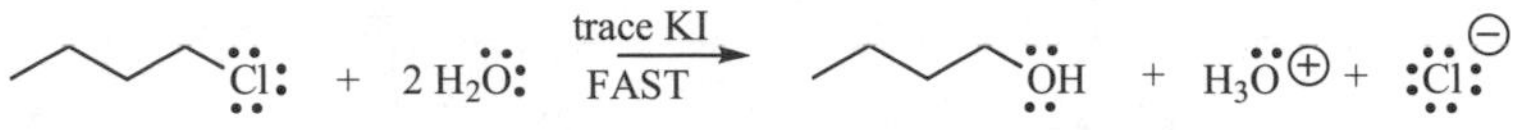
Cl
+ 2 H2O
trace KI
FAST
OH
+ H3O⊕ + Cl⊖

Iodide, being very soft and negatively charged, is a much better nucleophile than neutral water. Iodide can do an S_N2 substitution to replace chloride, giving the butyl iodide in a fast step. Since the C–I bond is much weaker than the C–Cl bond because the overlap is poorer, iodide is a much better leaving group than chloride. Water can now react with the butyl iodide in a fast S_N2 to give the butanol product, kicking out the iodide anion, which is now free to continue catalysis. Even though we have added an extra step, all the barriers are lower, and therefore the overall reaction is much faster.

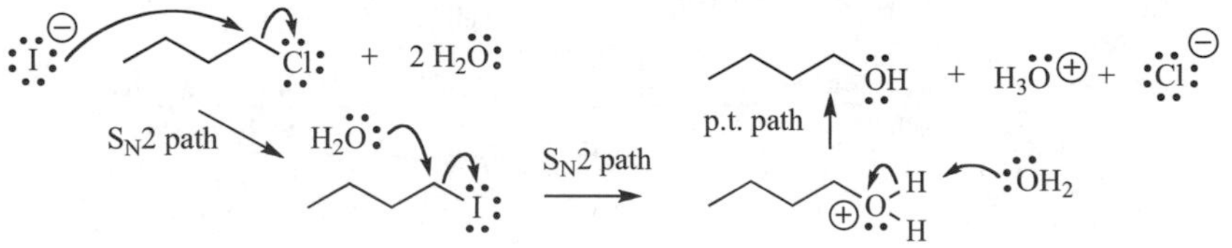
I⊖
Cl
+ 2 H2O
SN2 path
H2O
I
SN2 path
OH
+ H3O⊕ + Cl⊖
p.t. path
H
OH2
H

4.2.13 Substitution Summary

- For nucleophiles, rank first by softest as best then within the same atom, by most basic.
- For leaving groups, rank by the lowest pK_{aHL} with exceptions of sulfonates better, and fluoride and cyanide worse than expected.
- Carbocations are stabilized most by resonance with a lone pair, next by pi resonance, least by alkyl substitution. Three of the lesser is better than one of the better.
- Electron-releasing groups are best anionic, next the less electronegative pi-donors, alkyl substitution, and least, very electronegative pi-donors with poor $2p$-$3p$ overlap.
- Substitution version of the ΔpK_a rule: Never kick out a leaving group more than 10 pK_a units more basic than the incoming nucleophile.
- S_N2 does not occur on leaving groups attached to a trigonal planar center, or on a highly hindered site like a tertiary or neopentyl center.

• S_N2 must pass three tests: good access (primary or secondary), a leaving group with a pK_{aHL} of 10 or less, and a decent nucleophile.

• S_N1 must pass three tests: usually a carbocation that is usually more stable than simple secondary, a polar solvent, and a leaving group with a pK_{aHL} of zero or less.

• The pentacovalent path is for third-row and higher elements and must pass a ΔpK_a rule test.

• S_N2 goes via inversion at the site of nucleophilic attack, and S_N1 usually goes by racemization.

• Any route that goes via a carbocation like S_N1 may have rearrangement as a complication (Section 4.7)

The decision process flow chart for working a substitution problem is shown in Figure 4.17. More substitution examples can be found in Section 8.4.1.

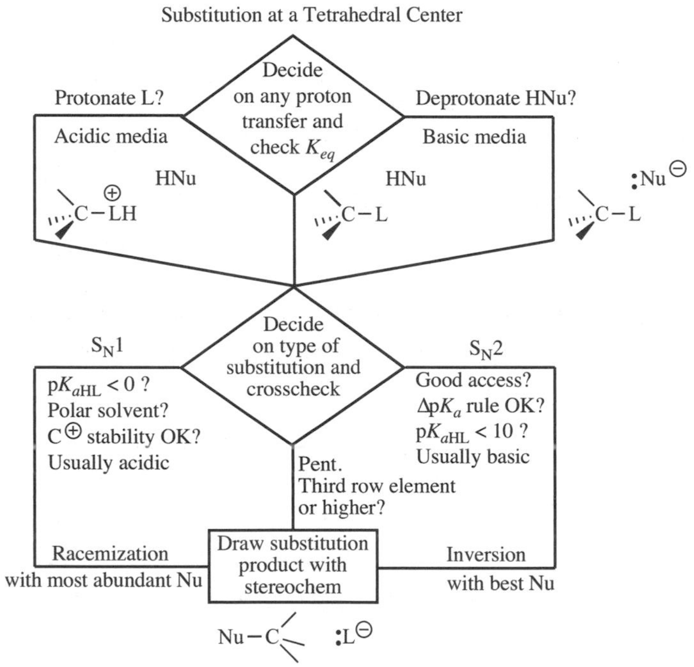
Substitution at a Tetrahedral Center
Decide on any proton transfer and check K_{eq}
Protonate L?
Acidic media
HNu
C–LH
Deprotonate HNu?
Basic media
HNu
C–L
:Nu⊖
C–L
Decide on type of substitution and crosscheck
S_N1
$pK_{aHL} < 0$?
Polar solvent?
C⊕ stability OK?
Usually acidic
S_N2
Good access?
ΔpK_a rule OK?
$pK_{aHL} < 10$?
Usually basic
Pent.
Third row element or higher?
Racemization
with most abundant Nu
Draw substitution product with stereochem
Inversion
with best Nu
Nu–C
:L⊖

Figure 4.17 The flowchart for substitution at a tetrahedral center.

4.3 ELIMINATION REACTIONS CREATE PI BONDS

4.3.1 Elimination Electron Flow Paths

Elimination reactions often compete with substitutions, but we will worry about that later. Substitution versus elimination is considered in depth in Section 9.5. There are no new generic classes of reactants to consider, but there are three new routes. The energy surface in Figure 4.18 gives an overview of three possible routes to eliminate a leaving group to form a pi bond. The reactants are found in the upper left corner. The leaving group–carbon bond must break (the vertical axis) and the carbon–hydrogen bond must break (the horizontal axis) to reach the elimination products in the lower right. Three idealized routes can occur.

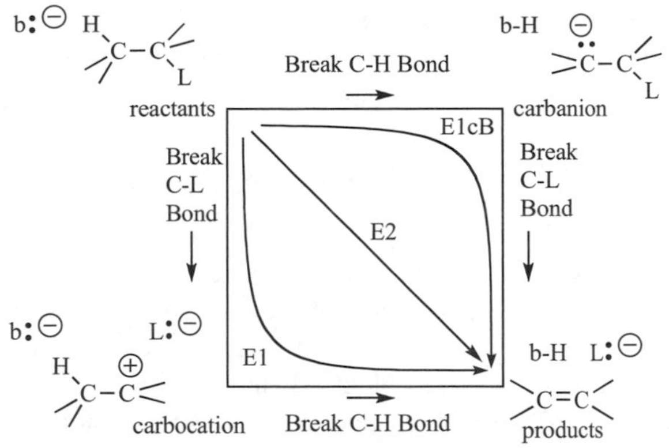
reactants
Break C-H Bond
carbanion
ElcB
Break C-L Bond
E2
Break C-L Bond
E1
carbocation
Break C-H Bond
products

Figure 4.18 Top view of the simplified energy surface for elimination.

The three pathways differ only in the timing of when the proton is removed and when the leaving group departs. **Generic E1** (elimination, unimolecular)—the leaving group departs first (rate-determining) by path D_N, producing a reasonably stable cation; a proton is then lost by path D_E, forming an alkene. **Generic E2** (elimination, bimolecular)—occurs by a concerted loss of a proton and the leaving group. **Generic ElcB** (elimination, unimolecular, conjugate base)—proton transfer forms an anion (the conjugate base) usually stabilized by an electronegative atom or group, and then the leaving group departs (rate-determining) by path E_β. Our two new electron flow paths are D_E (Dissociation, Electrofugic) and E_β (beta elimination from a lone pair). An *electrofuge* is a group that leaves **without** its bonding pair of electrons. The electron flow arrows for the three elimination routes are shown in Figure 4.19. Since we already know how to rank bases (Section 3.2), carbanions (Section 3.5.2), leaving groups (Section 4.2.3), and carbocations (Section 4.2.4), we can proceed directly to discuss each elimination route in detail.

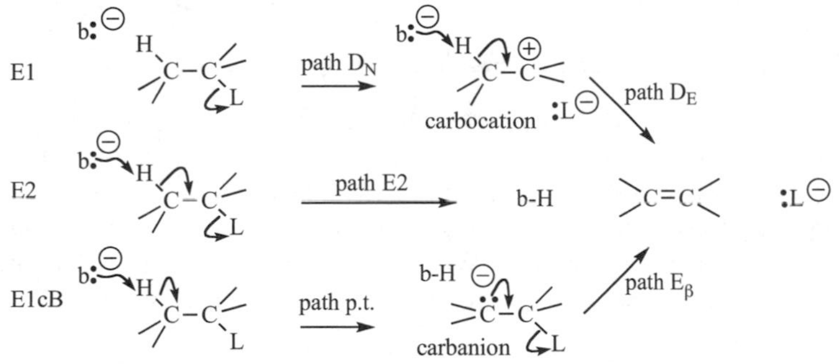
E1
path D$_N$
carbocation
path D$_E$
E2
path E2
b-H
ElcB
path p.t.
carbanion
path E$_\beta$

Figure 4.19 The electron flow paths for elimination.

4.3.2 The E1 Reaction: Electron Flow Pathways $D_N + D_E$

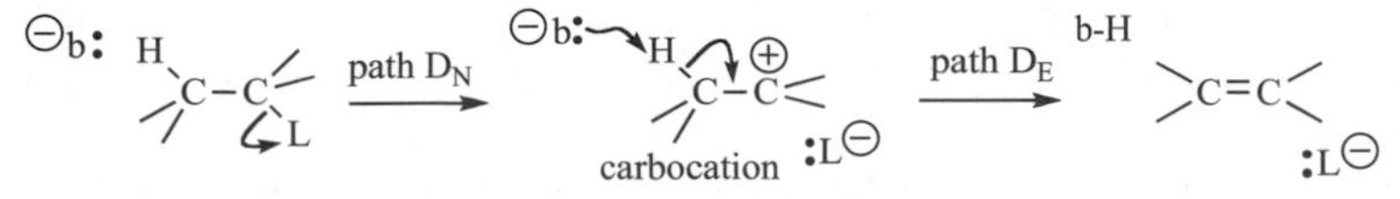
path D$_N$
carbocation
path D$_E$
b-H

The **E1 elimination** has the same rate determining first step, D_N, as the S_N1 substitution and is subject to the exact same **limitations: a carbocation stability usually**

better than secondary, a polar solvent, a good leaving group with a pK_{aHL} < 0. If it fails on any of the three, the E1 route is too high in energy. The second step, D_E, (Dissociation, Electrofugic) is a new electron flow path, and has the following form.

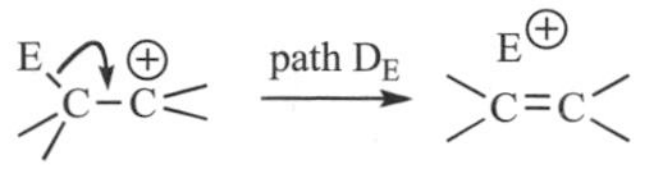
path D_E

The arrow starts from the C–E bond, breaking it and forming a new pi bond to the cationic center. The C–E bond and the carbocation empty orbital must be able to align in the same plane to form a pi bond, but because the C–C single bond between them can rotate, this is usually not a problem. If the electrofuge is not stabilized, then it will not fall off by itself, and must be removed by bonding to another species. This is the case in the E1 example shown previously, as a proton is never spat out into a vacuum, but is picked up by a basic species. This path is energetically downhill unless the electrofuge is less stable than the carbocation. We have now seen two of three paths available to carbocations: A_N trap a nucleophile, D_E lose an electrofuge, and yet to come, rearrange to another carbocation.

The Alkene Stability Trend

The following is a ranking of alkene stability. The most substituted alkene is the most stable. The *cis*-disubstituted double bond is more sterically hindered and less stable than the *trans*-disubstituted double bond. The *gem*-disubstituted double bond is about the same stability as the *cis*-disubstituted double bond. The monosubstituted 1-hexene is the least stable of the hexenes listed.

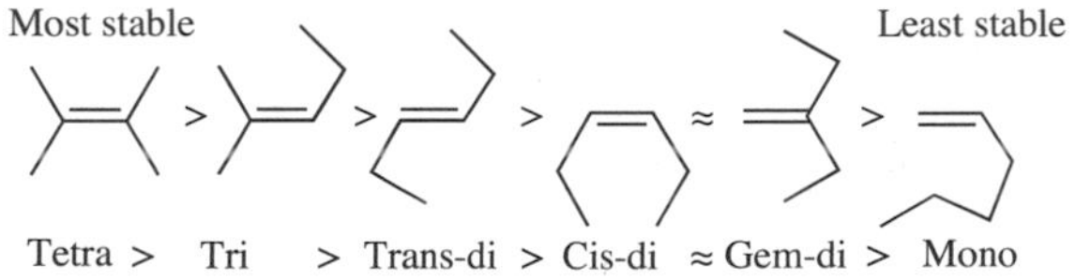
Most stable
Least stable
Tetra > Tri > Trans-di > Cis-di ≈ Gem-di > Mono

E1 Regiochemistry and Stereochemistry

Any H adjacent to the carbocation can be lost, often producing a different alkene. This loss is often reversible, so the E1 often produces an equilibrium mixture of alkenes, the most stable alkene predominating (Zaitsev's rule). When a reaction produces a mixture of different constitutional isomers, we use the term *regiochemistry* to refer to product ratios in that mixture. When the E1 produces a 1,2-disubstituted alkene, the trans stereoisomer predominates because it is less hindered, so more stable.

Example problem

For the compound below, draw an E1 elimination mechanism and show both major and minor products.

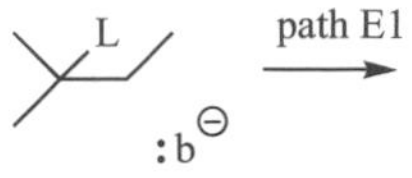
L
path E1
:b

Answer: The first step of the E1 is the loss of the leaving group by path D_N to give a carbocation, which is deprotonated by a base to form a pi bond by path D_E. Either the methyls or methylene adjacent to the carbocation can lose the H, but the more stable, more substituted product is major.

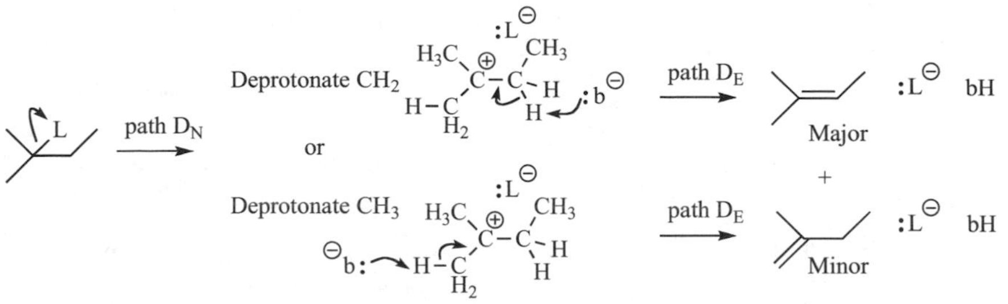
path D_N
Deprotonate CH_2
or
Deprotonate CH_3
path D_E
Major
Minor
bH

4.3.3 The E2 Electron Flow Pathway

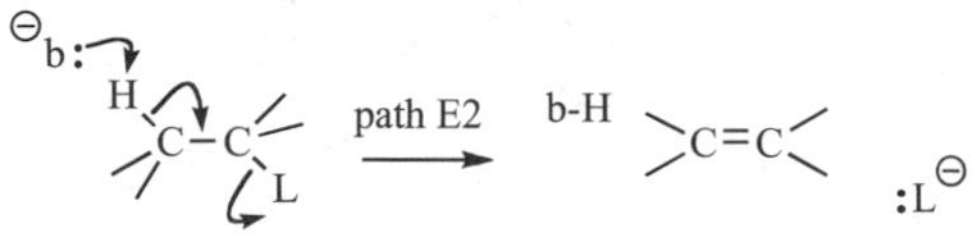
path E2
b-H

The E2 is a new electron flow path. The electron flow arrow starts from the base lone pair forming a bond to H. A second arrow breaks the C–H bond and forms a pi bond. The last arrow breaks the C–L bond and forms a lone pair on the leaving group. Since the *p* orbitals in the pi bond must be coplanar, and we are forming them in one step, then **the C–H and C–L bonds broken must be coplanar in the E2 transition state.** Those bonds can be in the same plane on the same side or on opposite sides. The staggered antiperiplanar arrangement shown, with the H and L coplanar but on opposite sides, is favored because the orbital overlap is better, and the base and leaving group have more space. Figure 4.20 illustrates several views of the E2 antiperiplanar transition state.

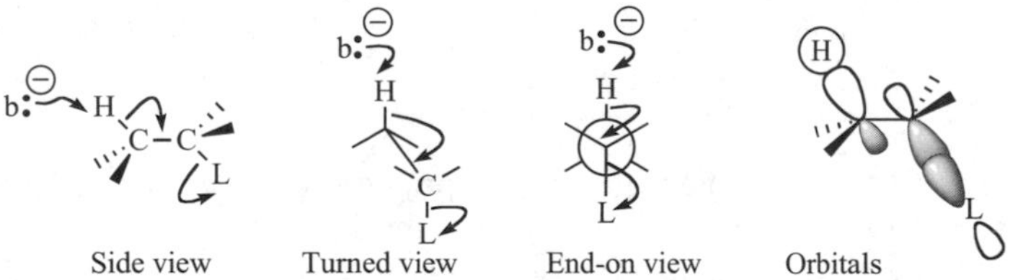
Side view
Turned view
End-on view
Orbitals

Figure 4.20 Views of the antiperiplanar transition state for the E2 elimination.

E2 Regiochemistry and Stereochemistry

Since the C–C single bond is free to rotate in open chain systems, the sterically less hindered E2 transition state will be the lowest energy. Because the E2 is rarely reversible, the lowest-energy transition state will determine the product. Both product stability and steric effects influence the energy of the transition state. In the E2 transition state the C–H bond and the C–L bond must lie in the same plane, preferably *anti* to each other for steric reasons. The protons on either carbon adjacent to the leaving group can be in proper alignment for reaction; thus different products are produced, as shown in Figure 4.21.

The following reactions show that the regiochemistry of the E2 elimination varies with the size of the base. The small base, methoxide, gives the *trans*-2-alkene as the major product because it is the most stable product with the least hindered transition state. If the base size is huge, like *tert*-butoxide, steric hindrance raises the energy of both the 2-alkene transition states, so the 1-alkene is preferred because the hydrogens on the end are more accessible to the large base (Hoffmann product). As the leaving group gets huge, sterically crowding the internal hydrogens, the 1-alkene is likewise favored.

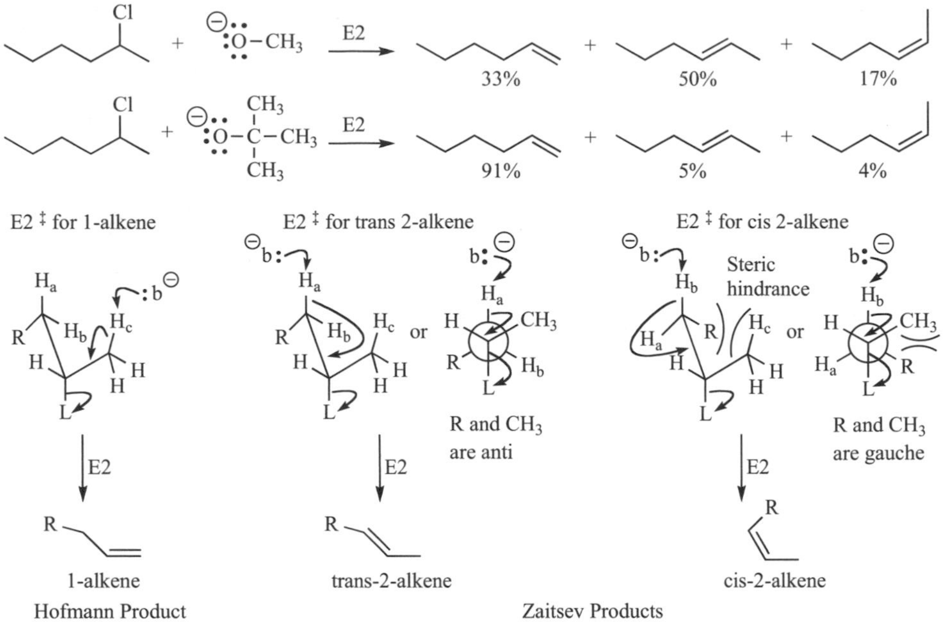
E2
33%
50%
17%
91%
5%
4%
E2 ‡ for 1-alkene
E2 ‡ for trans 2-alkene
E2 ‡ for cis 2-alkene
Steric hindrance
R and CH3 are anti
R and CH3 are gauche
1-alkene
Hofmann Product
trans-2-alkene
cis-2-alkene
Zaitsev Products

Figure 4.21 E2 elimination regioisomers and stereoisomers. The *cis*-2-alkene has the most crowded transition state, and the 1-alkene has the least crowded transition state.

4.3.4 The E1cB Reaction: Electron Flow Pathways p.t. + E$_\beta$

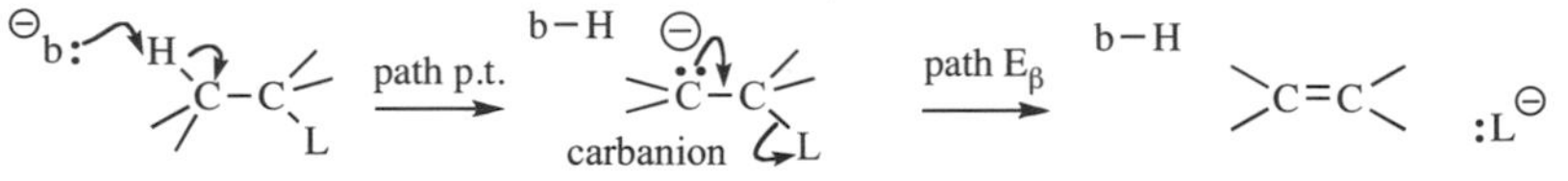
path p.t.
carbanion
path Eβ

The E1cB elimination starts with a proton transfer as its first step. It is the **conjugate base** (cB) of the reactant that kicks out the leaving group. The second step, E$_\beta$ (beta elimination from a lone pair), is a new electron flow path, and has the following general form. The arrow starts from the lone pair, forming a new pi bond. A second arrow starts from the C–L bond, breaking it and placing the electrons as a new lone pair on the leaving group. Whenever we make a new bond to an atom with a complete octet, we have to break one of its existing bonds, so as not to exceed its octet.

The K_{eq} of the proton transfer is the limiting factor for the E1cB. The E1cB is most often seen when an electron-withdrawing group increases the acidity of the C–H bond. **A useful K_{eq} of the proton transfer for E1cB is greater than 10^{-10}**.

E1cB Regiochemistry and Stereochemistry

The E1cB regiochemistry is solely determined by the position of the acidic hydrogen. Since the C–C single bond is free to rotate in open chain systems, the sterically less hindered product stereochemistry is usually formed. In the following example the carbonyl acts as an electron-withdrawing group to make the adjacent hydrogen acidic enough for the E1cB to proceed.

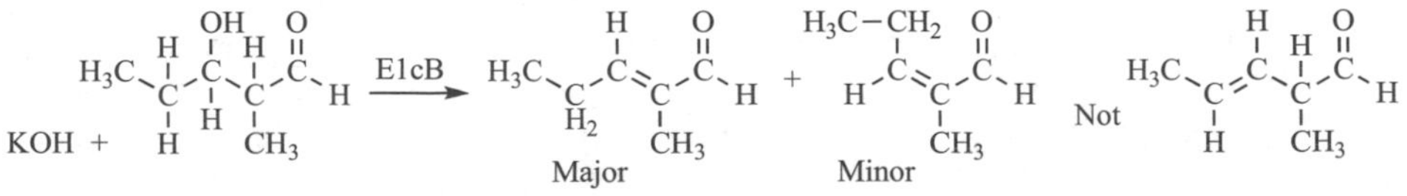
KOH +
E1cB
Major
Minor
Not

Example problem

For the reaction above run in basic water, draw an E1cB elimination mechanism and explain the regiochemistry and stereochemistry found.

Answer: The first step of the E1cB is the deprotonation of the acidic hydrogen next to the electron-withdrawing group by the p.t. path. This deprotonation determines the regiochemistry. The carbanion formed can then push out a poorer leaving group to form the alkene. The bond between the leaving group carbon and the carbanion can easily rotate. The major stereoisomer comes from the least crowded transition state (because the oxygen anion is surrounded by solvent) and is the more stable of the two possibilities.

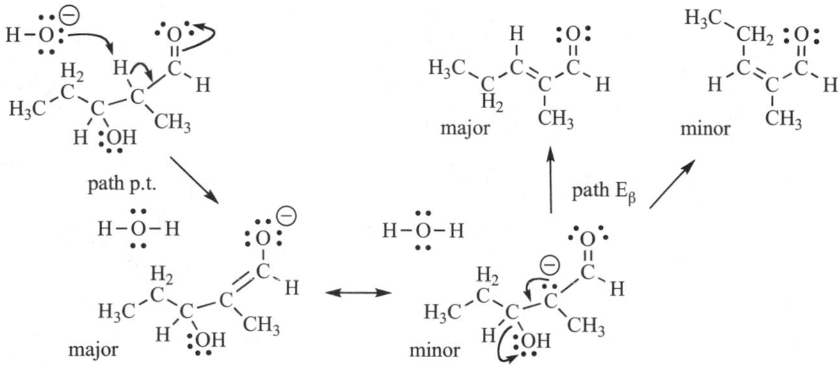
path p.t.
major
minor
path Eβ
major
minor

4.3.5 The E1/E2/E1cB Elimination Spectrum

The factors that decide which elimination path will occur are listed in Figure 4.22. For the E1 and the E1cB, solvation of the ionic intermediate is very important. The E1 and E1cB should be considered the two extremes of a spectrum of elimination mechanisms in which the E2 is in the center.

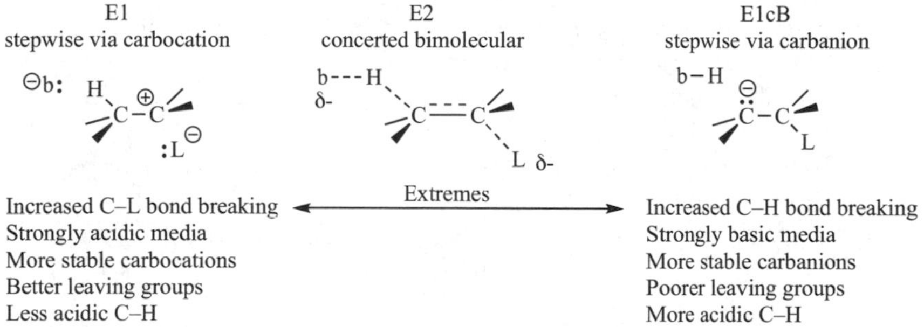
E1
stepwise via carbocation
E2
concerted bimolecular
E1cB
stepwise via carbanion
Extremes
Increased C–L bond breaking
Strongly acidic media
More stable carbocations
Better leaving groups
Less acidic C–H
Increased C–H bond breaking
Strongly basic media
More stable carbanions
Poorer leaving groups
More acidic C–H

Figure 4.22 The E1/E2/E1cB spectrum.

Mechanisms occur that do not cleanly fit the label of E1, E2, or E1cB. If the reaction is partway between E2 and E1, it is called E1-like, for the C–L bond is broken to a greater extent than the C–H bond, and a partial plus builds up on the leaving group carbon; this adjacent partial plus makes the C–H bond more acidic and therefore more easily broken. If the reaction is partway between E2 and E1cB, it is called E1cB-like, for the C–H bond is broken to a greater extent than the C–L bond, and a partial minus builds up on the carbon adjacent to the leaving group; this adjacent partial minus aids in pushing off the leaving group.

The elimination spectrum can be visualized as an energy surface (Fig. 4.23), where the front horizontal axis is the degree of C–H bond breaking, the axis going from back to front is the degree of C–L bond breaking, and the vertical axis is energy. Reactants are in the back left corner at point R; the back right corner corresponds to only C–H bond breaking and therefore is the carbanion; the front left corner corresponds to only C–L bond breaking and therefore is the carbocation; the front right corner, P, is the product alkene. The lowest-energy path is the dark line from R to P (shaded behind the surface).

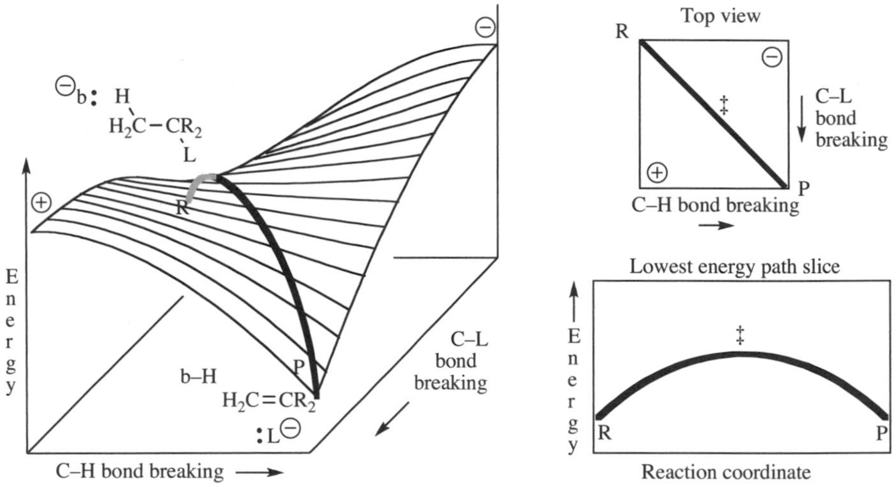
⊖b: H
H_2C-CR_2
L
⊕
R
⊖
Energy
b–H
P
$H_2C=CR_2$
:L⊖
C–H bond breaking
C–L bond breaking
Top view
R
⊖
‡
⊕
P
C–L bond breaking
C–H bond breaking
Lowest energy path slice
‡
Energy
R
P
Reaction coordinate

Figure 4.23 The E2 energy surface.

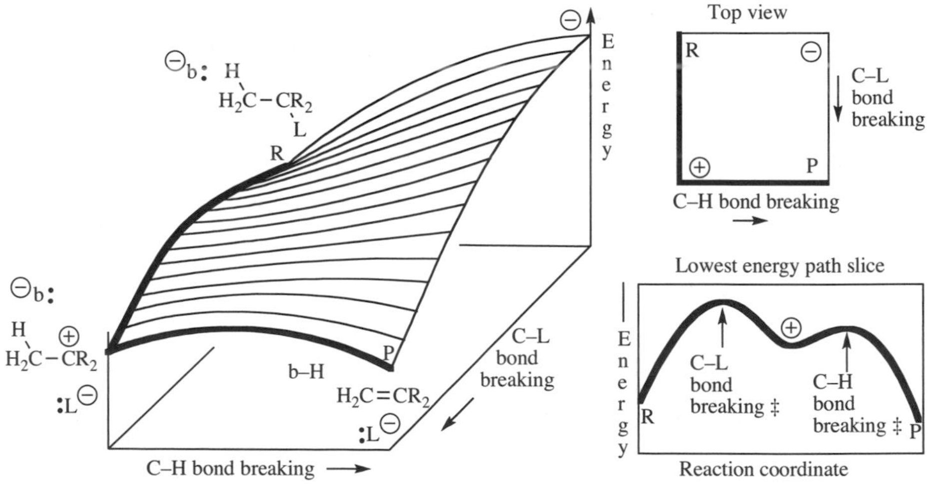
⊖
Energy
⊖b: H
H_2C-CR_2
L
R
⊖b:
H ⊕
H_2C-CR_2
:L⊖
b–H
P
$H_2C=CR_2$
:L⊖
C–H bond breaking
C–L bond breaking
Top view
R
⊖
⊕
P
C–L bond breaking
C–H bond breaking
Lowest energy path slice
Energy
⊕
C–L bond breaking ‡
C–H bond breaking ‡
R
P
Reaction coordinate

Figure 4.24 The E1 energy surface.

If both carbocation and carbanion are poor, the two corners corresponding to them will be high in energy; the surface folds down the middle. The reaction will go by the E2 process in which C–H bond breaking and C–L bond breaking occur to an equal extent, and the lowest-energy path will follow the diagonal through the saddle resulting from the high-energy carbocation and carbanion. If the carbocation is rather stable, the corner on the diagram corresponding to the carbocation will be lowered in energy, and now the lowest-energy path will go via the carbocation, the E1 elimination (Fig. 4.24).

If, on the other hand, the carbanion is reasonably stable and the carbocation is not, the corner on the diagram corresponding to the carbanion will be lowered in energy instead, and now the lowest-energy path will go via the carbanion, thus the E1cB elimination (Fig, 4.25).

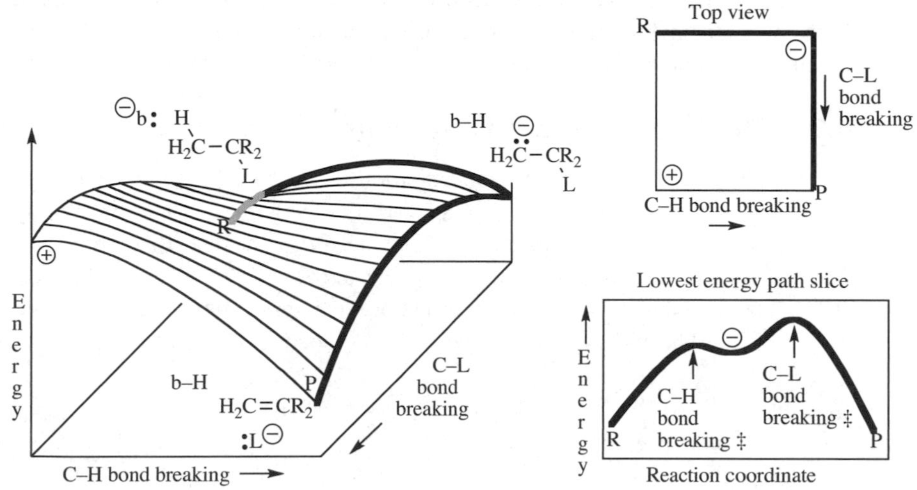

b: H
H2C−CR2
L
b–H
H2C−CR2
L
R
P
Energy
b–H
H2C=CR2
:L
C–L
bond
breaking
C–H bond breaking
Top view
R
C–L
bond
breaking
P
C–H bond breaking
Lowest energy path slice
Energy
C–H
bond
breaking ‡
C–L
bond
breaking ‡
R
P
Reaction coordinate

Figure 4.25 The E1cB energy surface.

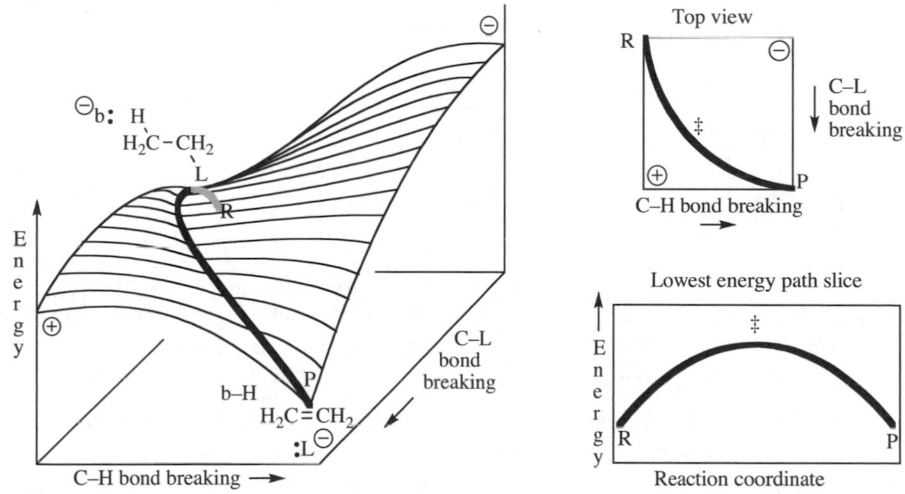

b: H
H2C−CH2
L
R
Energy
P
b–H
H2C=CH2
:L
C–L
bond
breaking
C–H bond breaking
Top view
R
‡
C–L
bond
breaking
P
C–H bond breaking
Lowest energy path slice
‡
Energy
R
P
Reaction coordinate

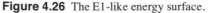

Figure 4.26 The E1-like energy surface.

Finally, if the carbocation is more stable than the carbanion but neither is very good, the two "mountains" on the diagram may be of different heights and will have a "pass" that is off to one side, not exactly on the diagonal (Fig. 4.26). In this case, the lowest-energy path to product takes a curve through that pass. More C–L bond breaking than C–H bond breaking occurs in going to the transition state "pass," a partial plus builds up on the leaving group carbon at the transition state, and the reaction is called E1-like. Had the carbocation been slightly more stable, the lowest-energy path would have fallen into the carbocation energy well, and the process would be E1, not E1-like.

In a similar manner, if the carbanion is more stable than the carbocation, the "pass" will be on the carbanion side and more C–H bond breaking than C–L bond breaking occurs in going to the transition state. A partial minus charge builds up on the carbon adjacent to the leaving group carbon at the transition state, helping to push out the leaving group, and the reaction is E1cB-like.

Also for the E1/E2/E1cB spectrum, similar to what was seen in the S_N1/S_N2 spectrum, a better leaving group will lower the front edge, making the overall process more favorable and tilting the surface toward the E1 path. If the H is more acidic or the base stronger (making deprotonation easier), the right-hand edge will be lowered, again making the elimination more favorable but this time tilting the surface toward the E1cB.

4.3.6 Eliminations in Cyclic Systems

Cyclohexane favors a chair conformation in which all the bonds are staggered. All the C–H bonds roughly in the plane of the ring are called equatorial; all those perpendicular are called axial. Cyclohexane can easily flip between two chair conformations, as shown in Figure 4.27; the barrier is only 10.8 kcal/mol (45 kJ/mol). All the axial positions become equatorial and vice versa upon ring flip. With substituted rings, the equilibrium favors the larger groups in the less crowded equatorial position.

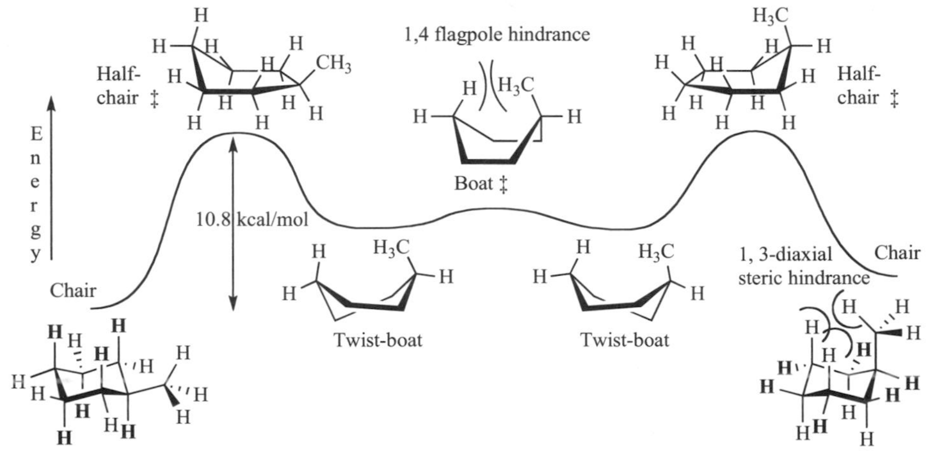
Half-chair ‡
1,4 flagpole hindrance
Boat ‡
Half-chair ‡
Energy
10.8 kcal/mol
Chair
Twist-boat
Twist-boat
1, 3-diaxial steric hindrance
Chair

Figure 4.27 The bold hydrogens on the left cyclohexane chair are axial and upon ring flip become equatorial. The adjacent axial hydrogens sterically hinder the axial methyl group in the chair on the right, which is 1.7 kcal/mol (7.1 kJ/mol) higher in energy than the chair on the left, with the equatorial methyl group. Most hydrogens in the twist-boat and boat are deleted for clarity.

In a freely rotating system it was relatively easy to achieve the orbital overlap requirements of the E2 elimination. Cyclic systems are rotationally constrained, so in the cyclohexane system, the coplanar overlap requirement is satisfied only if both C–H and C–L bonds are both axial. This *trans*-diaxial restriction affects the product distribution.

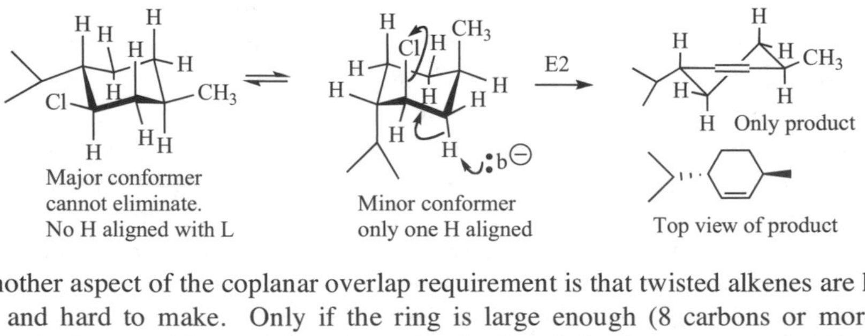
Major conformer
cannot eliminate.
No H aligned with L
Minor conformer
only one H aligned
E2
Only product
Top view of product

Another aspect of the coplanar overlap requirement is that twisted alkenes are high in energy and hard to make. Only if the ring is large enough (8 carbons or more) is a bridgehead alkene stable enough to be isolated. The following compound undergoes a *syn* coplanar E2 elimination and does not form the highly twisted bridgehead alkene.

H_a
L
H_b
H_a is coplanar, H_b is not
E2
Syn elimination
Only product
NOT
Highly twisted bridgehead
alkene not formed

4.3.7 Eliminations Yielding Carbonyls

The carbonyl group, C=O, is perhaps the most important functional group in organic chemistry. If we replace one of the carbons in our elimination surface with an oxygen atom, we get a related surface, shown in Figure 4.28.

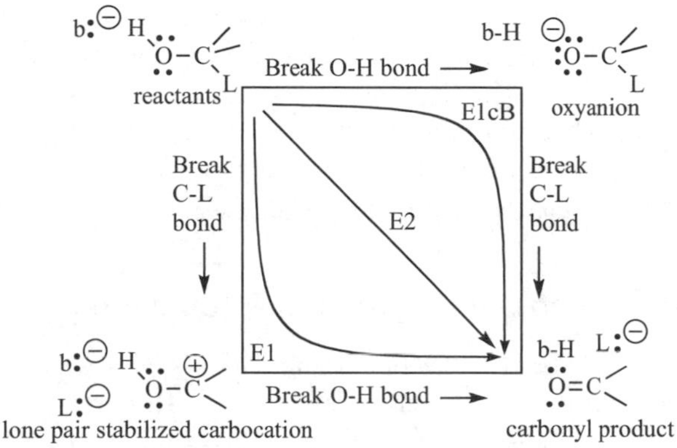

reactants
Break O-H bond
b-H
oxyanion
E1cB
Break
C-L
bond
E2
Break
C-L
bond
E1
b-H
lone pair stabilized carbocation
Break O-H bond
carbonyl product

Figure 4.28 Top view of the carbonyl-forming elimination surface.

All elimination routes are improved by this swap of oxygen for carbon (Fig. 4.29). The E1cB proton transfer step now creates a very stable oxyanion, an anion on a very electronegative oxygen atom. The slow step of the E1, loss of the leaving group, is also sped up because the oxygen lone pair can stabilize the carbocation formed. All routes have formed a very strong C=O bond. The primary determinant at this point, since all routes are reasonable, is the pH of the reaction medium. The E1cB is found in base because the oxyanion is basic (pK_{abH} is about 12 to 16). The E1 is found in acid because the lone pair stabilized carbocation is just a protonated carbonyl (pK_a is near −7). The E2 is found in more neutral media.

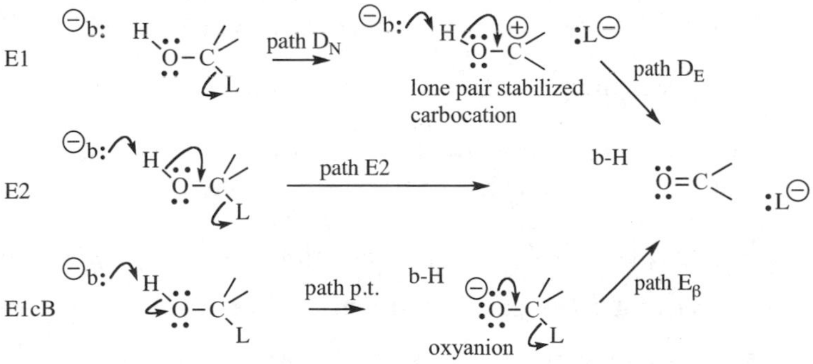

E1
path D_N
lone pair stabilized
carbocation
path D_E
E2
path E2
b-H
E1cB
path p.t.
b-H
oxyanion
path $E_β$

Figure 4.29 Three good carbonyl-forming elimination routes with the leaving group on carbon

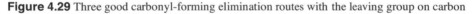

What would happen if we replaced the other carbon, the one bearing the leaving group, with oxygen? The three routes, shown in Figure 4.30, would have drastically different energetics. Loss of the leaving group in the first step of the E1 would produce a

highly unstable cation with an incomplete octet on a very electronegative atom. The E1 is not possible here. The deprotonation step in the E1cB creates an anion next to oxygen, which is destabilized by the adjacent lone pairs on oxygen. This leaves the E2 as the only reasonable route in this system. Placing a leaving group on oxygen and then proceeding with this E2 is the most common method to oxidize an alcohol to a ketone.

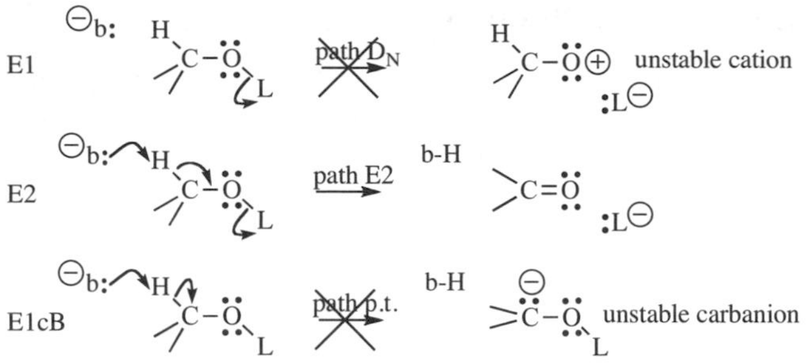

E1
path D_N
unstable cation
E2
path E2
b-H
E1cB
path p.t.
b-H
unstable carbanion

Figure 4.30 Only one carbonyl-forming elimination route with the leaving group on oxygen.

4.3.8 Approaches to Elimination Mechanisms

With the addition of the three elimination routes each with new crosschecks, our problem space has again gotten more complex. The first task is always to map changes on a balanced reaction. Is the medium acidic or basic? In acid, we can again protonate leaving groups to improve them. Like our approach to substitution mechanisms, we will first check for a favorable proton transfer before deciding which elimination route to take. Again visualize that you are standing on the reactant "street corner" of the elimination surface and need to look at the "traffic lights" (energetics) to decide which way to cross. Is there an appropriately positioned electron-withdrawing group making the C–H acidic? If not, our number of alternatives drops down to two, E1 or E2. Check the three restrictions for the D_N step: a carbocation stability usually better than secondary, a polar solvent, a good leaving group with a $pK_{aHL} < 0$. If it fails on any of the three, the E1 route is too high in energy. It is time to check E2 restrictions: a decent leaving group and the ability to get the C–H and C–L bonds in the same plane. Again, we check each branch of the problem space tree to find the best route before committing to a direction to proceed. Once you decide the best route, make sure you draw the arrows correctly, keep track of charge balance, and use the known electron flow paths.

Common Elimination Errors

The most common errors are the result of forgotten crosschecks. The following list shows the most common student errors on elimination reactions.

Mixed media errors (Don't do acidic routes in basic media or vice versa).

Proton transfer energetics errors (make sure K_{eq} is greater than 10^{-10}).

E2 energetics error (always check the ΔpK_a rule)

E1 carbocation restriction (don't make an unstable carbocation)

E1 leaving group restriction (the pK_{aHL} should be zero or below)

Charge Balance Errors Combined with Line Structure

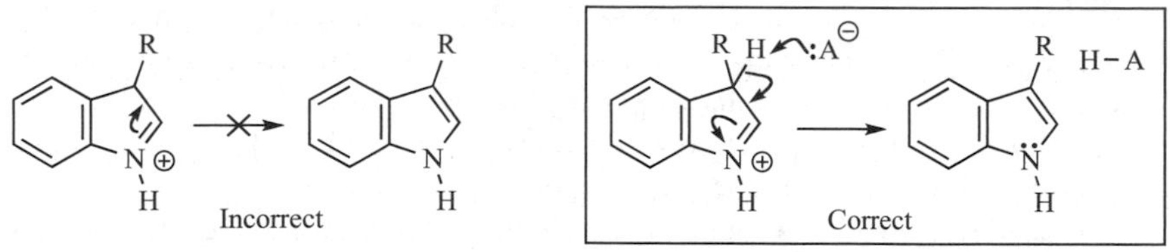
Incorrect
Correct

Line structure can be deceptive in eliminations because the important C–H bond is not explicitly drawn out, in this example, when there is a lone pair stabilized carbocation intermediate in the E1. The loss of a formal charge may be difficult to notice in reactions that drop off a proton because the proton is commonly not drawn as one of the products. Since the proton lost was not drawn in the line structure on the left, it was simply forgotten. The arrow made a double bond in the right place but flowed the electrons the wrong way: away from the positive center and not toward it. Besides the loss of the formal charge, the lone pair on nitrogen was lost; it was assumed in the product, but the incorrect arrow did not form it. The right side shows the correct electron flow path D_E, deprotonation of a carbocation to form a pi bond.

Internal Elimination Shortcut Errors

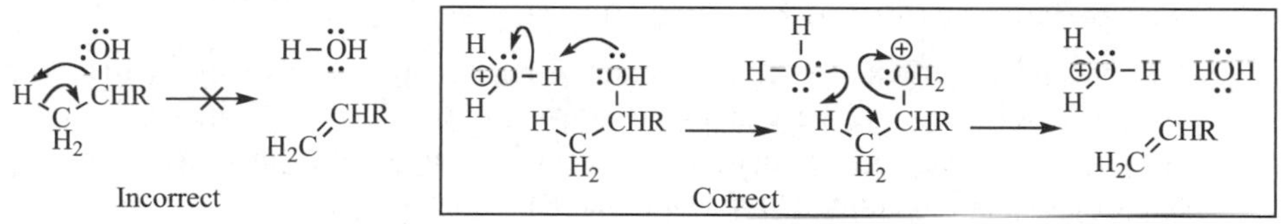
Incorrect
Correct

Alcohols do not do uncatalyzed eliminations of water at reasonable temperatures. The carbon–oxygen bond is not a base, and the carbon–hydrogen bond is not an acid. The process drawn on the left is a four-center, four-electron process, which with very few exceptions does not occur thermally. The most common route for water elimination is by acid catalysis, as shown on the right: path p.t., protonation of a lone pair, followed by the E2 elimination. If the carbocation is reasonably stable, the reaction may proceed via E1.

Orbital Alignment Errors

How much angular "slop" there is in a transition state is currently a debated topic. There is agreement that perpendicular orbitals do not interact. However, most calculations of the actual shape of orbitals show them to be much fatter than they tend to be drawn in beginning textbooks. It now appears that a 10° deviation from the optimum has a negligible effect.

The most important place to watch orbital overlap is in the formation of pi bonds. Pi bonds that have about a 30° twist can be made but are reactive. Since substitution competes with elimination, it does not take very much of a twist to tilt the balance away from elimination. There are many examples to show that changing the leaving group from approximately coplanar with the hydrogen to 60° out of alignment shuts off the E2 elimination process entirely. Be especially careful with eliminations in rigid systems.

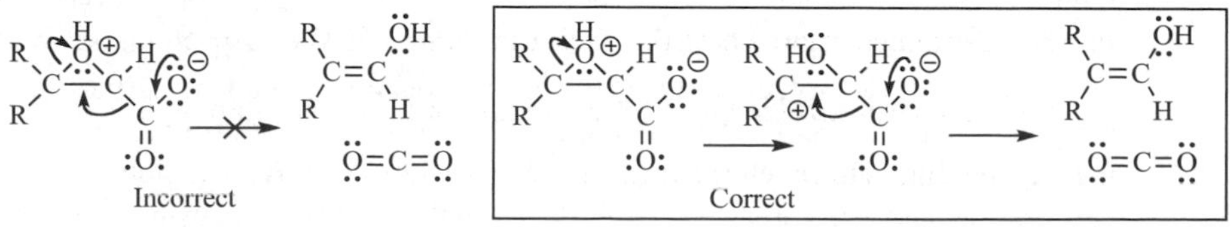
Incorrect
Correct

The loss of carbon dioxide (a decarboxylation reaction) on the left, written as a variant of the E2, contains a serious overlap error. The error is that the source and sink are about 60° out of being coplanar, clearly beyond the distortion limits of the E2 elimination (Fig. 4.31). We can propose a reasonable alternative elimination, the E1 process, that produces the same product. If the leaving group departs first, the carbocation formed can rotate about the single bond until the carbon–carbon bond of the source is coplanar with the empty *p* orbital. Then the loss of carbon dioxide can occur easily to form the pi bond of the enol product.

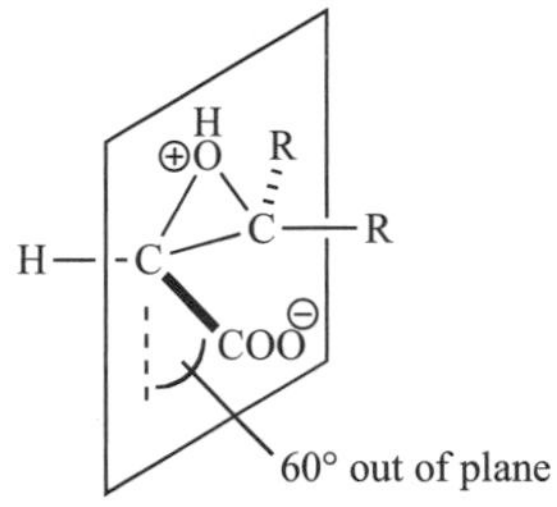
H
⊕O
R
C
R
H
C
COO⊖
60° out of plane

Figure 4.31 An illustration of improper orbital alignment. The C, C, and O of the three-membered ring define the plane. The carboxylate sticks out 60° from the plane of the ring.

Chemical structures drawn flat on a page may not reflect accurately the three-dimensional nature of the compound drawn. If you have any suspicion that a reaction step has poor orbital alignment, or is physically impossible (inaccessibility of a reactive site, for example), build the molecular model.

Example Elimination Mechanism—Dehydration of an Alcohol

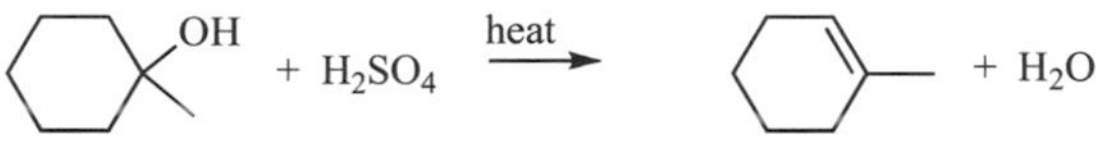
OH
+ H2SO4
heat
+ H2O

The first task in order to be able to map the changes is to check to see whether the reaction is balanced; it is, and we can see that the leaving group was water. With sulfuric acid present, the reaction is definitely in acidic medium. An elimination reaction has occurred since we have a new pi bond. We first check for favorable proton transfers. A proton transfer reaction will improve the leaving group significantly (hydroxide has a pK_{aHL} of 15.7, whereas water has a pK_{aHL} of −1.74). Although the listed pK_a of sulfuric acid is only approximate, the proton transfer easily passes a K_{eq} check: $\log K_{eq} = (pK_{abH} - pK_{aHA}) = (-2.4 - (-9)) = +6.6$, so $K_{eq} = 10^{+6.6}$, which is very favorable.

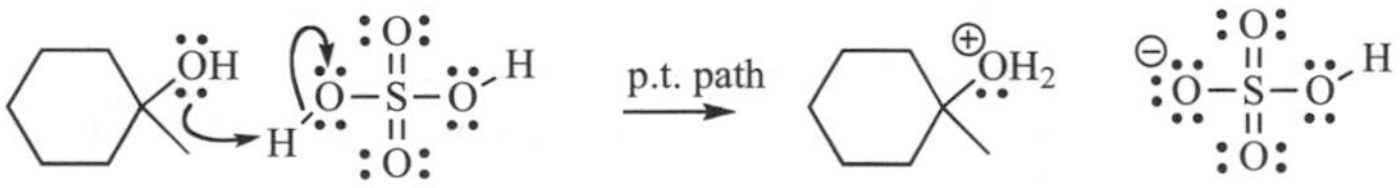
OH
O
O–S–O
H
H
O
p.t. path
⊕OH2
O
⊖O–S–O
H
O

It is time to decide on an elimination route. To aid in this decision, recall the elimination energy surface (Fig. 4.18). There is no electron-withdrawing group making the C–H adjacent to the leaving group carbon acidic, so the E1cB is out. The E1cB is rare in acidic media anyway. For the E1, a check of D_N restrictions shows the leaving group is good, the solvent polar, and the carbocation would be tertiary, and therefore sufficiently stable. Since the E1 passes all crosschecks, we do not have to worry about the E2 since the elimination surface is tilted toward E1. We consider E2 when both the carbocation and the carbanion corners are high in energy, causing the surface to fold down the middle, making the E2 our possible lowest-energy route.

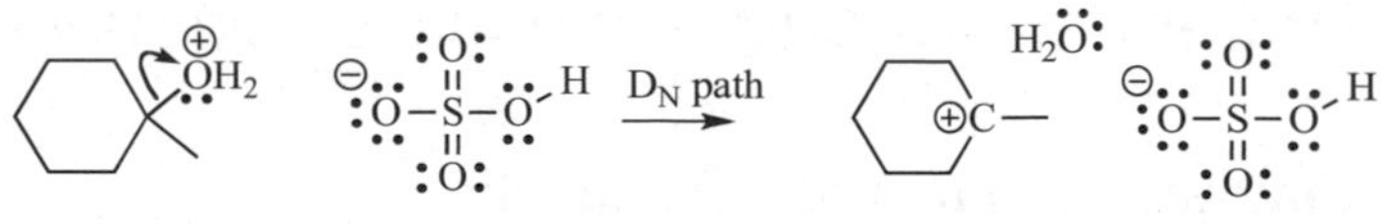
D_N path

Now we finish the E1 route by path D_E to produce the most stable trisubstituted alkene. The most basic group present is water, so it removes the proton from the carbocation. The protonated water would be expected to protonate the next molecule of alcohol, continuing the acid catalysis and freeing up water as a product.

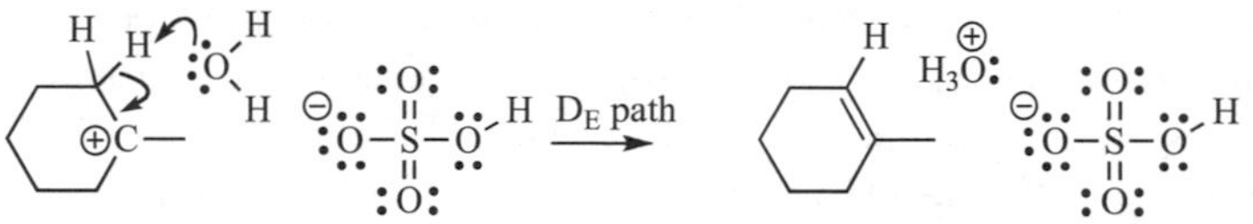
D_E path

4.3.9 Elimination Summary

• Elimination version of the ΔpK_a rule: Never kick out a leaving group more than 10 pK_a units more basic than the incoming base.

• E1 must pass three tests: a carbocation stability usually better than secondary, a polar solvent, and a leaving group with a pK_{aHL} of zero or less. Any route that goes via a carbocation like E1 may have rearrangement as a complication (Section 4.7).

• E2 occurs when the carbanion and carbocation are not very stable: The energy surface folds down the middle. The C–H and C–L eliminated must be close to coplanar.

• E1cB requires an acidic hydrogen and a good base. Check that the proton transfer K_{eq} is at least 10^{-10}. E1cB can tolerate poorer leaving groups.

More elimination examples can be found in Section 8.4.2 The decision process for working a elimination problem is shown in Figure 4.32.

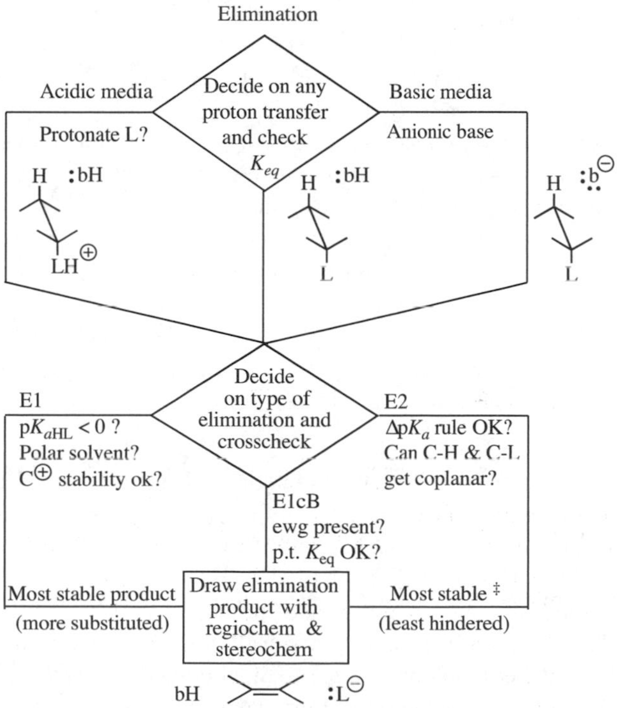
Elimination
Acidic media
Decide on any proton transfer and check K_{eq}
Basic media
Protonate L?
Anionic base
E1
$pK_{aHL} < 0$?
Polar solvent?
C⊕ stability ok?
Decide on type of elimination and crosscheck
E2
ΔpK_a rule OK?
Can C-H & C-L get coplanar?
E1cB
ewg present?
p.t. K_{eq} OK?
Most stable product
(more substituted)
Draw elimination product with regiochem & stereochem
Most stable ‡
(least hindered)

Figure 4.32 The flowchart for elimination.

4.4 ADDITION REACTIONS TO POLARIZED MULTIPLE BONDS

4.4.1 Introduction to Addition Reactions

An addition reaction is formally the reverse of an elimination reaction. We will see both electrophilic and nucleophilic additions in this section. There are few new generic classes of reactants to consider, and there are three new routes, but these are just the reverse of the elimination routes just covered. The new generic classes are discussed in much more detail in Chapter 5 and 6, but need to be introduced here. A carbon–carbon double bond can range from electrophilic to nucleophilic depending on what is attached to it (Fig. 4.33). Another way to make a double bond electrophilic is to replace one of its carbon atoms with an electronegative heteroatom like oxygen, C=O, a carbonyl.

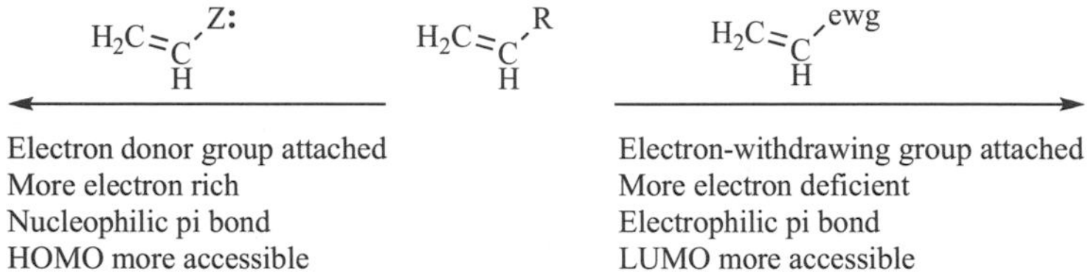
Electron donor group attached
More electron rich
Nucleophilic pi bond
HOMO more accessible
Electron-withdrawing group attached
More electron deficient
Electrophilic pi bond
LUMO more accessible

Figure 4.33 Electrophilic and nucleophilic pi bonds.

4.4.2 Addition Electron Flow Paths

The energy surface in Figure 4.34 gives an overview of three possible routes to add an electrophile and a nucleophile to a pi bond. The reactants are found in the upper left corner. In this example surface, the electrophile is a proton, delivered by an acid. The carbon–nucleophile bond must be made (the vertical axis) and the carbon–hydrogen bond must form (the horizontal axis) to reach the addition product in the lower right. Three idealized routes can occur.

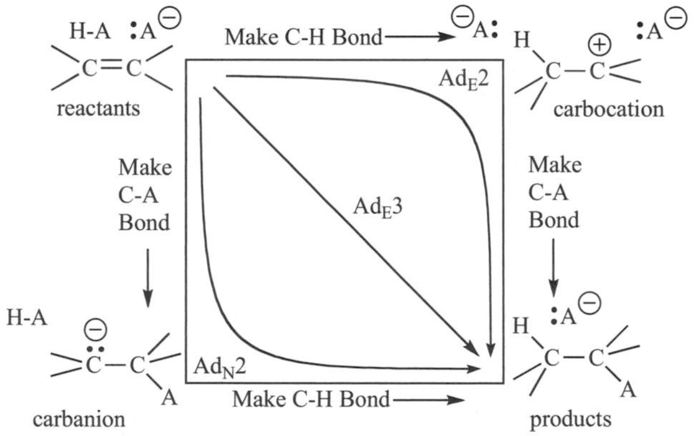
Make C-H Bond
reactants
carbocation
Make
C-A
Bond
Make
C-A
Bond
carbanion
Make C-H Bond
products

Figure 4.34 Top view of the simplified energy surface for addition to a carbon–carbon pi bond.

The three pathways differ only in the timing of when the electrophile is added and when the nucleophile is added. **Generic Ad$_E$2** (addition, electrophilic, bimolecular)—The electrophile adds first (rate-determining) by path A$_E$, producing a reasonably stable carbocation; a nucleophile traps the carbocation by path A$_N$, forming the addition product. **Generic Ad$_E$3** (addition, electrophilic, trimolecular)—The reaction occurs by a concerted addition of an electrophile and a nucleophile, path Ad$_E$3. **Generic Ad$_N$2** (addition, nucleophilic, bimolecular)—Nucleophilic addition (rate-determining) forms an anion usually stabilized by an electronegative atom or group, and then is protonated by path p.t. to give the addition product. Our three new electron flow paths are A_E (Association, Electrophilic), Ad_E3 (Addition, Electrophilic, Trimolecular) and Ad_N (Nucleophilic addition to a polarized multiple bond). The electron flow arrows for the three addition routes are shown in Figure 4.35. Since we already know how to rank all the intermediates, we can proceed directly to discuss each addition route in detail.

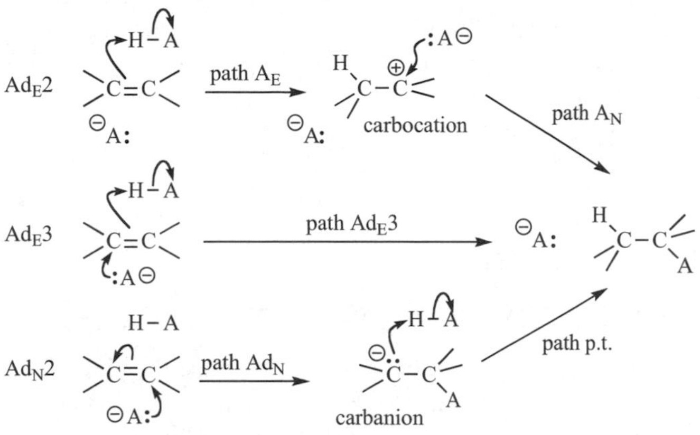

AdE2
path AE
carbocation
path AN
AdE3
path AdE3
AdN2
path AdN
carbanion
path p.t.

Figure 4.35 The electron flow paths for addition

4.4.3 The Ad_E2 Reaction: Electron Flow Pathways A_E + A_N

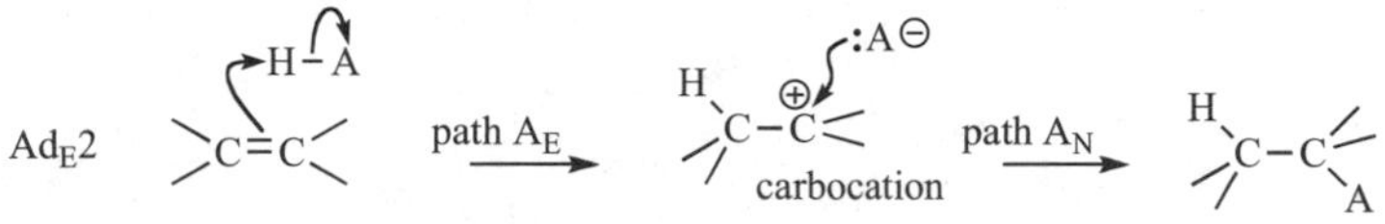

AdE2
path AE
carbocation
path AN

The Ad_E2 is the conceptual reverse of the E1 elimination. The first step, A_E, (Association, Electrophilic) is a new electron flow path and has the following general form. The arrow starts from the pi bond, breaking it and forming a new sigma bond to the electrophile, creating a carbocation next to the carbon–electrophile bond.

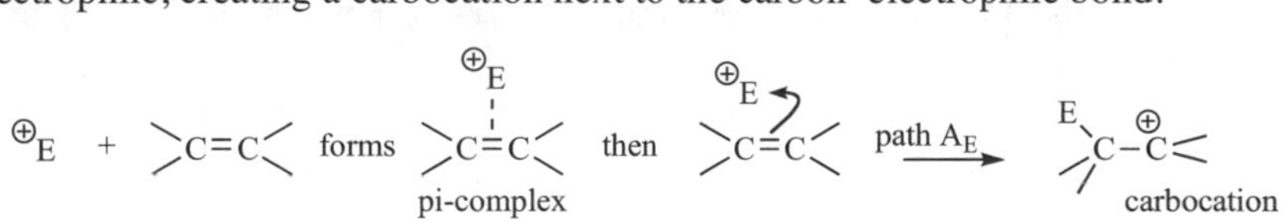

forms
pi-complex
then
path AE
carbocation

It is thought that the electrophile complexes to the electron cloud of the pi bond first, as in Figure 1.16, and then the sigma bond is formed. An energy diagram for a typical Ad_E2 is given in Figure 4.36. The stability of the carbocation (Table 4.1) dominates the energy diagram and is the major contributor to the reaction barrier.

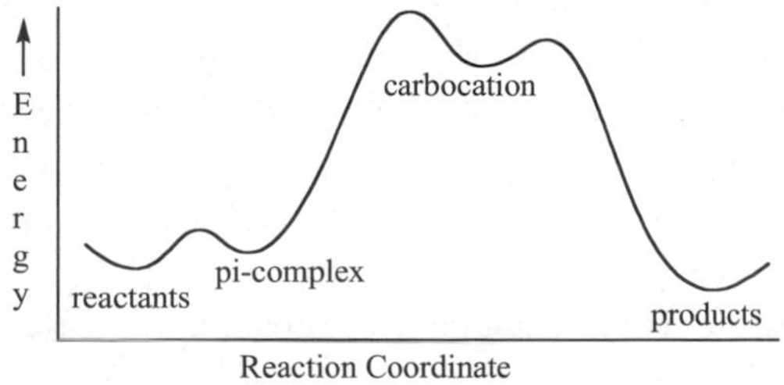

Energy
carbocation
reactants
pi-complex
products
Reaction Coordinate

Figure 4.36 The energy diagram for a typical Ad_E2 addition.

Additions to Alkenes from a HOMO-LUMO Perspective
(A Supplementary, More Advanced Explanation)

When an electrophile attacks an alkene (Fig. 4.37), it will approach the pi bond in a way to achieve the best overlap of its LUMO with the alkene HOMO. The attack is vertical, along the axis of the *p* orbitals. We have interacted an empty electrophile orbital with a full pi bonding orbital to achieve stabilization.

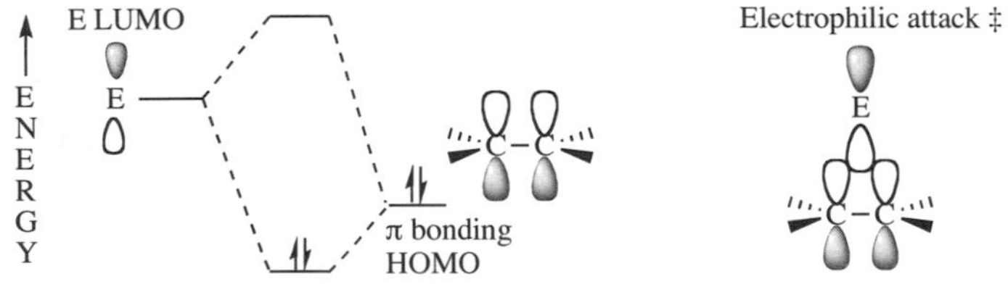
E LUMO
E
ENERGY
π bonding
HOMO
Electrophilic attack ‡
E

Figure 4.37 The interaction of the LUMO of the electrophile and the HOMO of the pi bond.

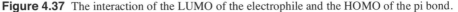

Ad$_E$2 Regiochemistry and Stereochemistry

There are two regiochemical possibilities since the electrophile could have added to either carbon of the pi bond. Since the addition reaction barrier is dominated by the stability of the carbocation formed, the reaction follows the lowest-energy route, namely, to create the more stable of the two possible carbocations. The modern statement of Markovnikov's rule is that the electrophile will add to the pi bond to form the more stable carbocation. The nucleophile adds equally to the top or bottom face of the carbocation.

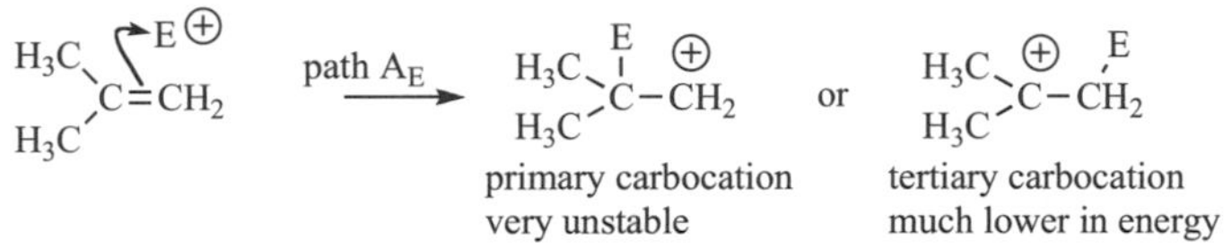
path A$_E$
or
primary carbocation
very unstable
tertiary carbocation
much lower in energy

In the following example, H_3O^+ is used to represent acidic water (which is almost all water, not H_3O^+). The pi bond is protonated to form a carbocation, resonance-stabilized by the phenyl. Water adds equally to either carbocation face to give a 1:1 mix of enantiomers of the protonated alcohol, which water then deprotonates to give the product.

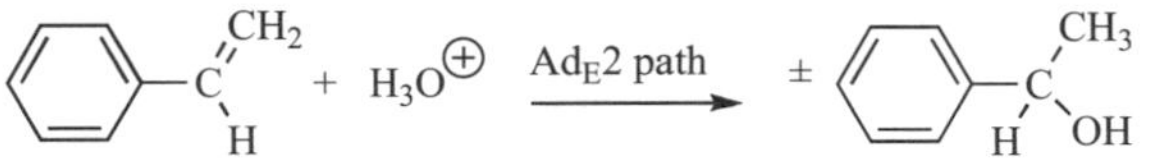
Ad$_E$2 path
±

Example problem

For the reaction above, using the explanation in the previous paragraph, draw an Ad$_E$2 addition mechanism and show both enantiomeric protonated alcohol products.

Answer: The first step of the Ad$_E$2 is the addition of the electrophile by path A$_E$ to give a carbocation, which is trapped by a nucleophile to form the product by path A$_N$. Deprotonation of the enantiomers shown below by water yields the neutral alcohol above.

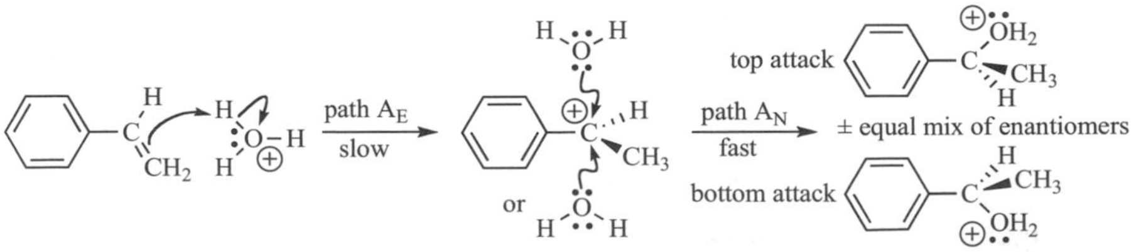
path A$_E$
slow
or
path A$_N$
fast
top attack
bottom attack
± equal mix of enantiomers

4.4.4 The Ad_E3 Electron Flow Pathway

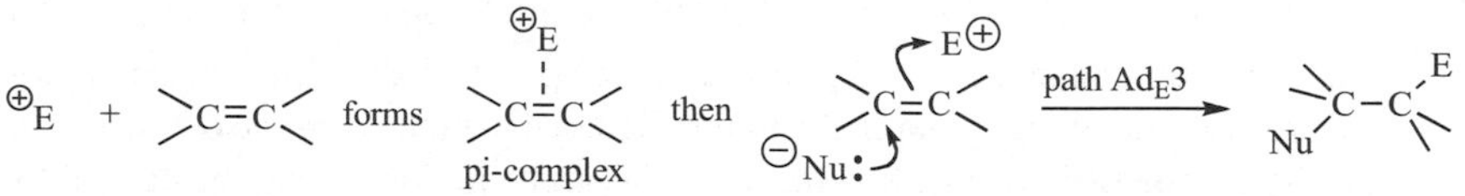

The Ad_E3 is conceptually the reverse of the E2 elimination, and a new electron flow path. Since three-molecule collisions are exceedingly rare, this reaction is thought to proceed by the nucleophile colliding with an electrophile/alkene pi-complex, making it just a two-body collision. The electrophile's pi-complexation polarizes the pi bond, making it susceptible to nucleophilic attack. The Ad_E3 avoids formation of an unstable carbocation. In the most general form, the electron flow arrow starts from the nucleophile lone pair and attacks the partial-positive end of the pi bond. The next arrow breaks the pi bond and forms a new sigma bond to the electrophile. In the case shown below, a third arrow is needed to break the H–A bond.

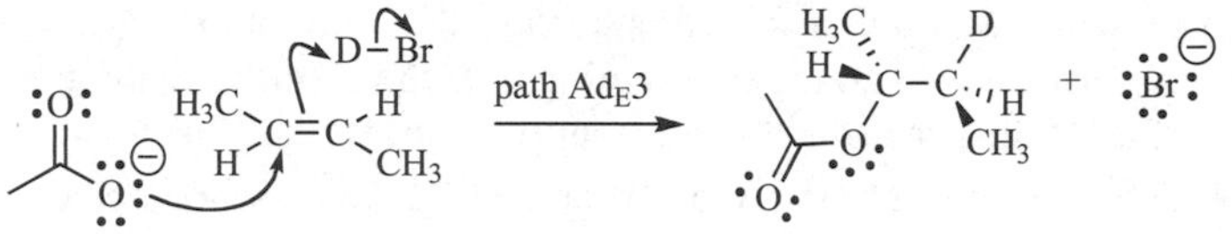

Ad_E3 Regiochemistry and Stereochemistry

We saw in the Ad_E2 that the electrophile first formed a pi-complex and then added to the pi bond to form the most stable carbocation. In the Ad_E3, the carbocation does not form, but the polarization of the pi bond by the pi-complex is similar. The nucleophile is attracted to the greatest partial plus of the complex, resulting in overall Markovnikov addition. Since the electrophile pi-complex blocks one face of the pi bond, the nucleophile attacks from the opposite face, giving overall *anti* addition. In the following example, the chloride nucleophile for the Ad_E3 comes from ionization of HCl.

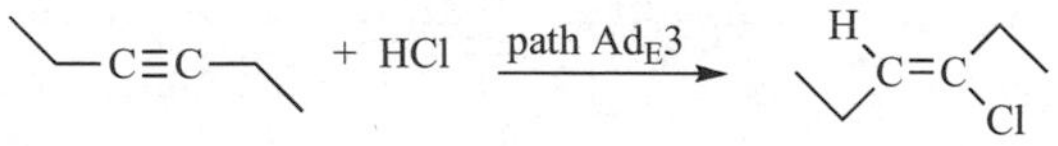

4.4.5 The Ad_N2 Reaction: Electron Flow Pathways Ad_N + p.t.

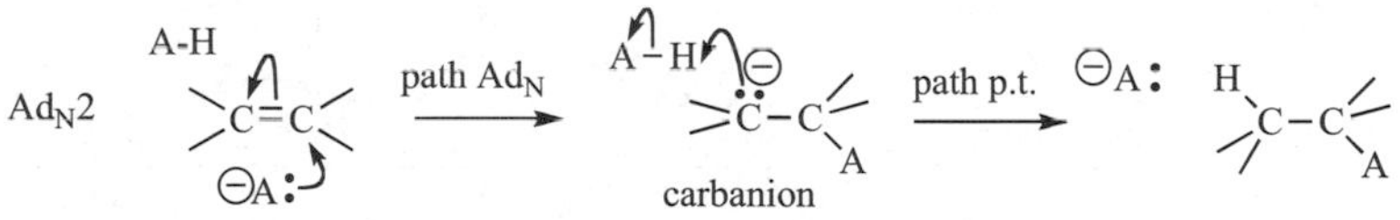

The Ad_N2 is the conceptual reverse of the E1cB elimination and has a new electron flow path, Ad_N, as its first step. The first arrow of the Ad_N starts from the nucleophile lone pair and makes a bond to the partial positive carbon of the pi bond. Next, an arrow starts from the pi bond, breaking it, and creates a carbanion next to the carbon–nucleophile bond. For the energetics of the Ad_N path to be reasonable, the carbanion needs to be stabilized by an adjacent electron-withdrawing group. **The best crosscheck for the Ad_N2 is the ΔpK_a rule (the carbanion formed should not be any more than 10 pK_a units more basic than the incoming nucleophile).**

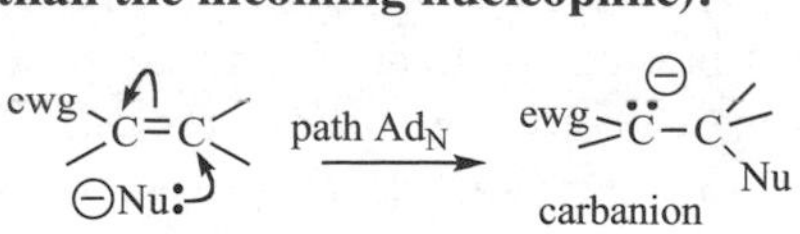

Ad_N2 Regiochemistry and Stereochemistry

The Ad_N2 regiochemistry is solely determined by the position of the electron-withdrawing group. The nucleophile attacks the double bond carbon not bearing the electron-withdrawing group. The stereochem is often a mixture because the carbanion can be protonated from either face. If the nucleophile is very basic and not tolerant of proton sources, then this protonation step occurs later in an acidic workup.

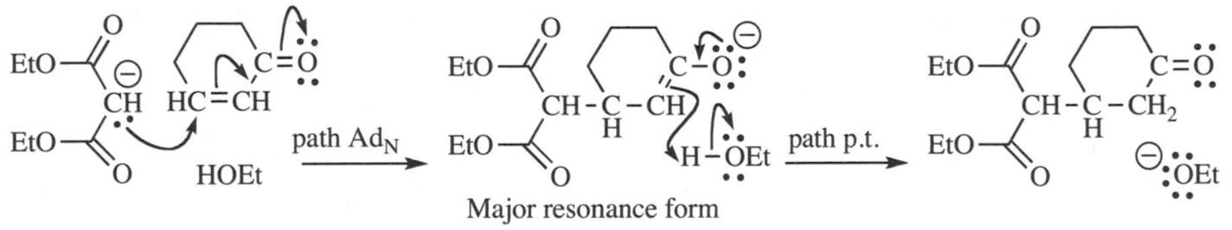

EtO
HOEt
path Ad_N
Major resonance form
H-ÖEt
path p.t.
:ÖEt

4.4.6 The Ad_E2/Ad_E3/Ad_N2 Addition Spectrum

The same factors that folded and tilted the elimination surface have a similar effect when the surface is tilted to favor addition. If the cation and anion are relatively unstable, the surface folds and the lowest-energy path is the Ad_E3 process. If an electron-withdrawing group stabilizes the carbanion, the Ad_N2 process is the lowest-energy route. If an electron-releasing group stabilizes the carbocation, the Ad_E2 is energetically best.

The Interrelationship of the Addition and Elimination Processes

Addition and elimination reactions, being complementary processes, occur on essentially the same energy surface. Conditions that lower the alkene corner favor elimination, whereas conditions that raise it or lower the addition product corner favor addition. A top view of the general addition/elimination energy surface is shown in Figure 4.38.

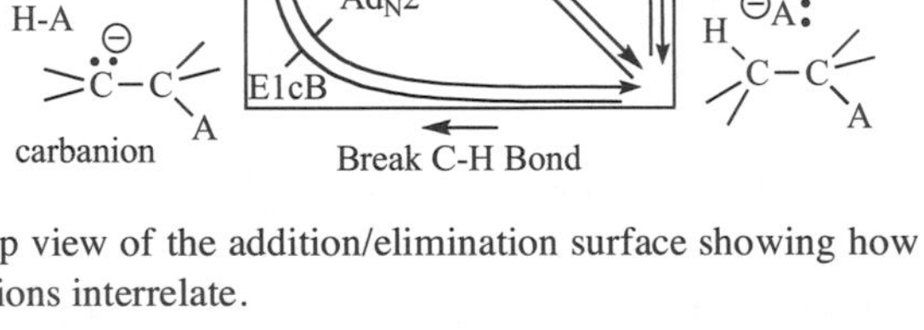

H-A
Make C-H Bond
carbocation
Ad_E2
E1
Make C-A Bond
Break C-A Bond
Ad_E3
E2
Ad_N2
E1cB
H-A
carbanion
Break C-H Bond

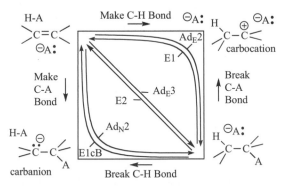

Figure 4.38 The top view of the addition/elimination surface showing how the different addition and elimination reactions interrelate.

4.4.7 Additions to Carbonyls

Additions to carbonyl compounds are perhaps the most important reactions in organic chemistry. If we replace one of the carbons in our addition surface with an oxygen atom, we get a related surface, shown in Figure 4.39.

The three routes in Figure 4.40 are similar to those for a carbon–carbon pi bond with the exception that the carbonyl lone pair is protonated instead of the carbonyl pi bond. This is due to the nonbonding electron pair on oxygen being energetically more accessible than a bonding pair of electrons. The major resonance form of the lone-pair stabilized carbocation is consistent with bond length data that the pi bond is still intact.

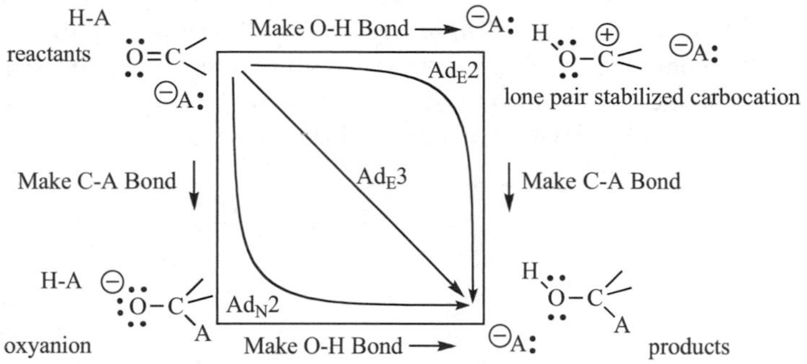
H-A
reactants
Make O-H Bond
lone pair stabilized carbocation
Make C-A Bond
Make C-A Bond
H-A
oxyanion
Make O-H Bond
products

Figure 4.39 Top view of the carbonyl addition surface.

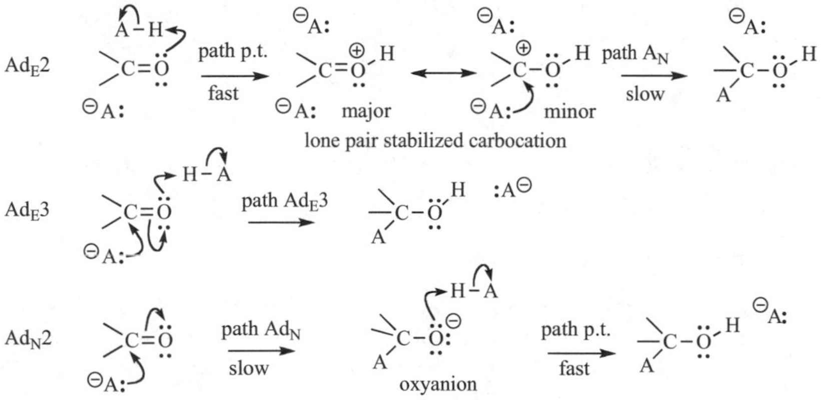
path p.t.
fast
major
minor
lone pair stabilized carbocation
slow
oxyanion
path p.t.
fast

Figure 4.40 Electron flow paths for addition to a carbonyl.

Additions to Carbonyls from a HOMO-LUMO Perspective

(A Supplementary, More Advanced Explanation)

When a nucleophile attacks a carbonyl group (Fig. 4.41), it will approach the carbonyl group so as to achieve the best overlap of its HOMO with the carbonyl's LUMO whose largest lobe is on carbon. The attack is not vertical, but close to the tetrahedral angle. We have interacted a full orbital with an empty one to achieve stabilization.

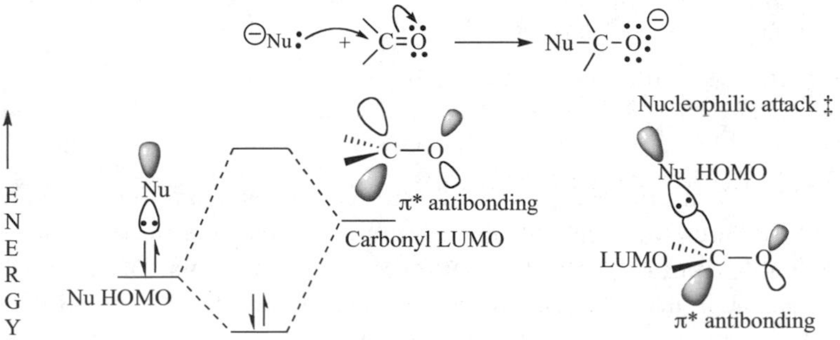
Nucleophilic attack ‡
ENERGY
Nu
Nu HOMO
π* antibonding
Carbonyl LUMO
Nu HOMO
LUMO
π* antibonding

Figure 4.41 The interaction of the HOMO of the nucleophile and the LUMO of the carbonyl to produce a sigma bond.

Likewise, when an electrophile attacks a carbonyl group, it will approach the carbonyl group so as to achieve the best overlap of its LUMO with the carbonyl's HOMO, the lone pair on oxygen (Fig. 4.42). The molecular orbitals are telling us what we already knew from the minor resonance form of the carbonyl group. Nucleophiles attack the δ+ end, and electrophiles attack the δ− end of the carbonyl group. The orientations of the orbitals tell us additional information on the position of attack.

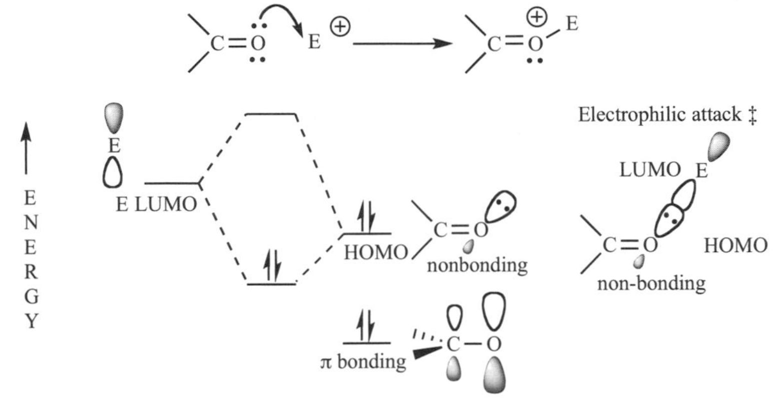
ENERGY
E
E LUMO
HOMO
nonbonding
π bonding
Electrophilic attack ‡
LUMO E
HOMO
non-bonding

Figure 4.42 The interaction of the LUMO of the electrophile and the HOMO of the carbonyl to produce a sigma bond. The carbonyl HOMO is the lone pair and not the pi bond.

Reversibility of Carbonyl Additions

The acidic media additions to carbonyls are often reversible. For example, addition of water to a carbonyl group usually favors the carbonyl. A $\Delta H°$ calculation gives the hydrate as slightly uphill of the carbonyl (two C–O bonds at 86 kcal/mol (172 kcal/mol) is weaker than one C=O at 177 kcal/mol).

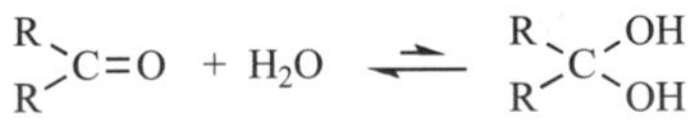

The Ad_N2 reaction on a carbonyl is often carried out in two separate and sequential reaction conditions if the nucleophile is a strong base (and would react with proton sources). The addition of the nucleophile occurs in basic aprotic media, followed by addition of a weak acid in the workup for the proton transfer step. The reversibility of the Ad_N2 can usually be predicted by the ΔpK_a rule, but in protic solvents if the nucleophilic attack forms a stronger base, a following irreversible proton transfer step may make the overall reaction favorable. This very favorable p.t. step is the driving force for the cyanohydrin formation reaction below.

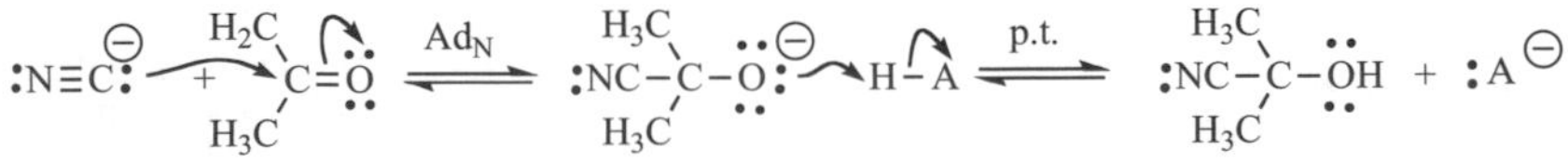
Ad$_N$
p.t.

A look at two energy diagrams for this reaction that differ only in the proton transfer step is instructive. The overall $\Delta G°$ from experiment is –2 kcal/mol (–8.4 kJ/mol). By estimating the pK_{abH} of the intermediate at about 13, we can get a value for the proton transfer K_{eq} that gives an approximate $\Delta G°$ value of –5 kcal/mol (–21 kJ/mol) for the following proton transfer. By subtraction, we get the $\Delta G°$ for the formation of the intermediate to be about +3 kcal/mol (+12.6 kJ/mol) and can construct an energy diagram, shown in Figure 4.43.

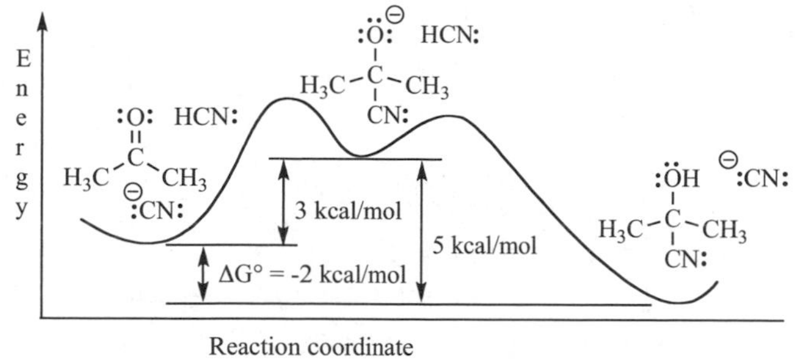
Energy
HCN:
H3C CH3
CN:
:O: HCN:
3 kcal/mol
5 kcal/mol
ΔG° = -2 kcal/mol
:ÖH :CN:
Reaction coordinate

Figure 4.43 The energy diagram for cyanohydrin formation.

If we change the acid from HCN to water, the first part of the diagram remains the same since the acid is just a spectator in the first step. With a weaker acid the proton transfer step is not favorable, and the equilibrium favors starting materials (Fig. 4.44).

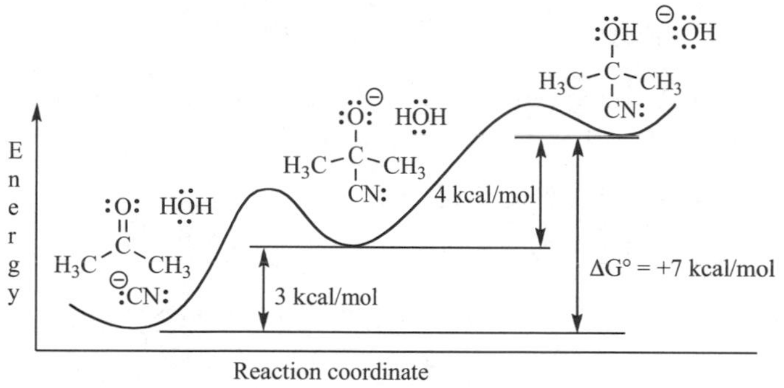
Energy
:ÖH :ÖH
HÖH
:O: HÖH
4 kcal/mol
3 kcal/mol
ΔG° = +7 kcal/mol
Reaction coordinate

Figure 4.44 The energy diagram for cyanohydrin cleavage.

4.4.8 Approaches to Addition Mechanisms

Although we have three new addition routes each with crosschecks, the similarity to the elimination routes should help. The first task is always to map changes on a balanced reaction. Is the medium acidic or basic? As always, we will first check for a favorable proton transfer before deciding which addition route to take. Again visualize the reactant "street corner" on the addition surface to decide which addition route to use. Is there an appropriately positioned electronegative atom or electron-withdrawing group making Ad_N2 viable? If not, our number of alternatives drops down to two, Ad_E2 or Ad_E3. For the Ad_E2, check the two possible carbocations for stability (better than secondary). If neither carbocation is reasonably stable, the Ad_E2 route is too high in energy, and the reaction will go Ad_E3. On either Ad_E2 or Ad_E3 remember to follow Markovnikov's rule when adding the electrophile. Check each branch of the problem space tree to find the best route before committing to a direction to proceed, then make sure you draw the arrows correctly, keep track of charge balance, and use the known electron flow paths.

Common Addition Errors

Besides the usual proton transfer errors, media pH errors, and media pH span errors, the most common addition error is forgetting to check the carbocation stabilities. A carbocation stability assessment is needed to decide between Ad_E2 and Ad_E3, and also to use Markovnikov's rule. For Ad_N2, a common error is to forget a ΔpK_a rule crosscheck.

Example Addition Mechanism

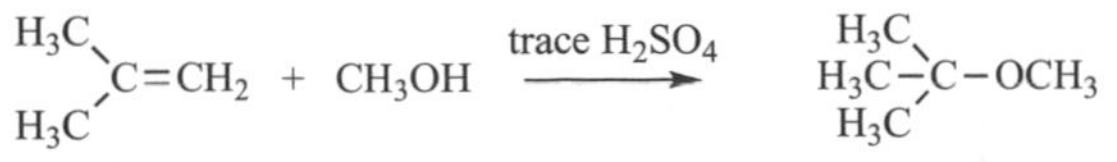

trace H_2SO_4

We are asked for the mechanism of formation for a controversial gasoline additive methyl *tert*-butyl ether or MTBE. With sulfuric acid present, the medium is certainly acidic. Mapping changes gives an H added to the pi bond methylene (CH_2) and the OCH_3 added to the pi bond quaternary carbon as expected from Markovnikov addition. The pi bond carries no electron-withdrawing group, so Ad_N2 is out. We need to decide between Ad_E2 and Ad_E3. Markovnikov addition of a proton forms a tertiary carbocation; the Ad_E2 route passes the carbocation stability check. We do not have to consider Ad_E3 if the surface is tilted toward Ad_E2. The sulfuric acid protonates the methanol solvent, so the acidic catalyst in this addition is protonated methanol.

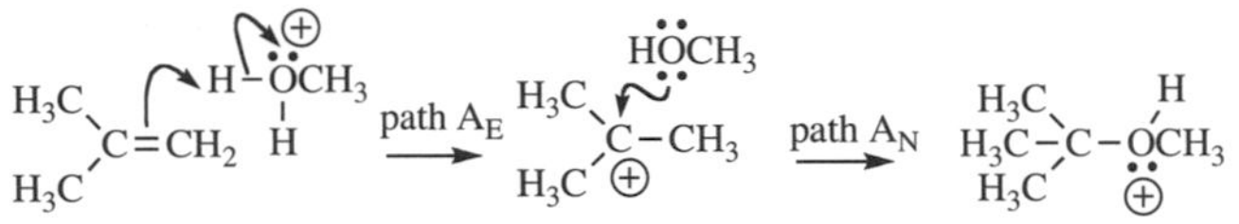

path A_E
path A_N

The Ad_E2 reaction gives the protonated product; deprotonation by methanol yields the neutral product and protonated methanol.

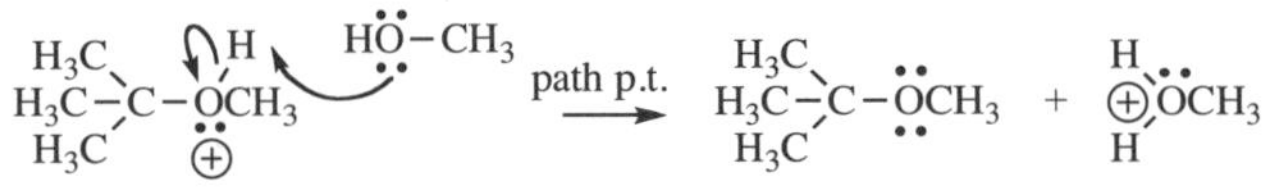

path p.t.

The decision process for working an addition problem is shown in Figure 4.45.

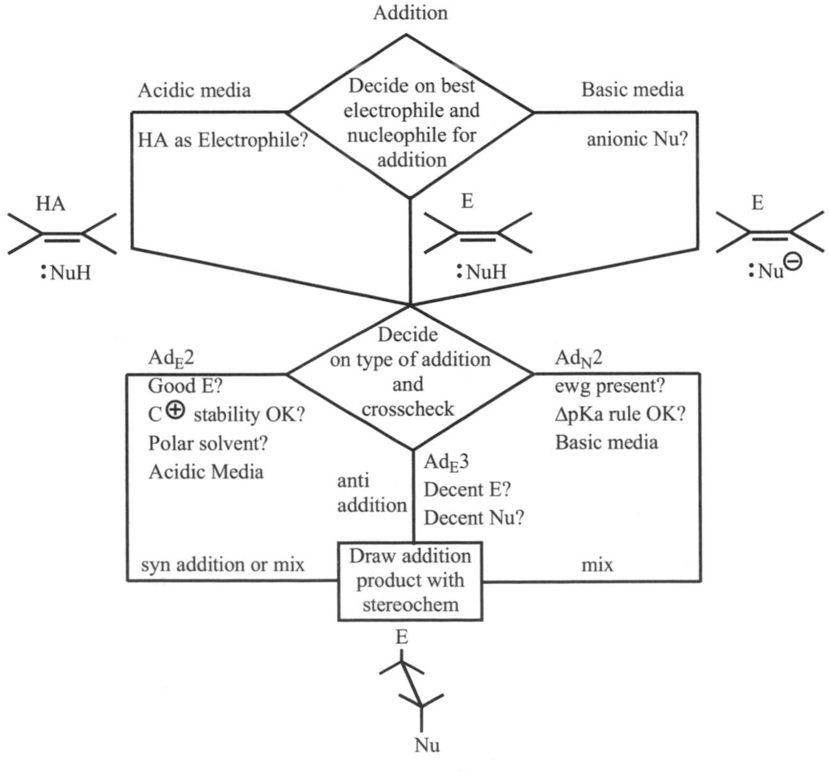

Addition
Acidic media
Decide on best electrophile and nucleophile for addition
Basic media
HA as Electrophile?
anionic Nu?
HA
:NuH
E
:NuH
E
:Nu⊖
Decide on type of addition and crosscheck
Ad_E2
Good E?
C⊕ stability OK?
Polar solvent?
Acidic Media
Ad_N2
ewg present?
ΔpKa rule OK?
Basic media
anti addition
Ad_E3
Decent E?
Decent Nu?
syn addition or mix
Draw addition product with stereochem
mix
E
Nu

Figure 4.45 The general flowchart for addition. A proton transfer step may follow.

4.4.9 Addition Summary

• Ad_E2 must form a carbocation of secondary or better, and tends to produce a mix of *syn* and *anti* addition. Markovnikov's rule is followed: The electrophile adds to form the most stable carbocation.

• Ad_E3 occurs when the carbanion and carbocation are not very stable: The energy surface folds down the middle. The electrophile and nucleophile end up on opposite faces of the pi bond (*anti* addition). Markovnikov's rule is followed.

• Ad_N2 follows the ΔpK_a rule: Never form an anion more than 10 pK_a units more basic than the incoming nucleophile. For this to happen, the anion formed must usually be on an electronegative atom or stabilized by an electron-withdrawing group.

• Species such as Y–L can serve first as the addition electrophile and leaving group can serve later as the nucleophile. More addition examples can be found in Chapter 8.

4.5 NUCLEOPHILIC SUBSTITUTION AT A TRIGONAL PLANAR CENTER

4.5.1 Addition–Elimination Electron Flow Paths

A leaving group on a trigonal planar center is almost exclusively replaced by an addition–elimination mechanism. The S_N2 has rarely been seen on a trigonal planar center, and the S_N1 is seen with this system in only very acidic media with great leaving groups. The group C=C–L is usually unreactive to nucleophiles, because the anion that results from nucleophilic addition is not stabilized. This ΔpK_a rule requirement for the nucleophilic attack means that addition–elimination reactions are commonly seen with ewg–C=C–L and Y=C–L, where Y is an electronegative heteroatom such as oxygen. The reactivity trends of carboxylic acid derivatives, O=C–L, are discussed in detail in Section 6.5. Carboxylic acid derivatives are useful and typical reactants for our next reaction archetype, addition–elimination.

We can modify our familiar addition and elimination surfaces to give us a combined simplified addition–elimination energy surface (Fig. 4.46). Although this system is further complicated by additional proton transfer reactions, we can get an overview of the problem space with this simplified surface as a map. The reactants are in the upper left corner.

In acidic media, the carbonyl can be protonated to give a more reactive electrophile, the lone-pair stabilized carbocation, shown on the top right of the figure. In acidic media the weaker nucleophile can add to the more reactive electrophile to give the neutral tetrahedral intermediate. Proton transfer interconverts the different charge types of the tetrahedral intermediate. The neutral tetrahedral intermediate undergoes elimination and loses the leaving group to head toward products (or the nucleophile to return to reactants). Common in acid is a protonation to improve the leaving group of the tetrahedral intermediate. Therefore, the most common acidic route is around the right side, the Ad_E2 followed by E1.

In base, there are no acids strong enough to protonate the carbonyl, but basic media usually contain much better nucleophiles, so protonation of the sink is not needed. In basic media, the nucleophile adds to the carbonyl to give the anionic tetrahedral intermediate, which kicks out the leaving group to form the product. In summary, the most common basic route is straight down the left side, paths Ad_N followed by E_β.

Reactions in neutral media or in enzymes can avoid the strongly acidic intermediates such as the protonated carbonyl by proceeding via the diagonals, Ad_E3 followed by E2.

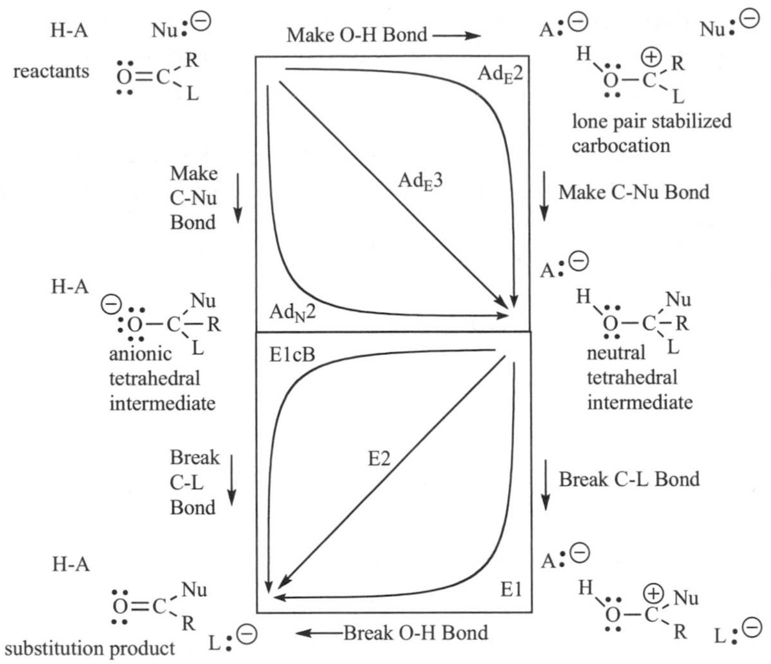
H-A
Nu:⊖
reactants
Make O-H Bond
AdE2
A:⊖
Nu:⊖
lone pair stabilized carbocation
Make C-Nu Bond
Make C-Nu Bond
AdE3
H-A
AdN2
A:⊖
anionic tetrahedral intermediate
E1cB
neutral tetrahedral intermediate
Break C-L Bond
E2
Break C-L Bond
H-A
E1
A:⊖
substitution product
L:⊖
Break O-H Bond
L:⊖

Figure 4.46 Top view of the simplified combined energy surface for addition–elimination.

The arrow pushing for the basic addition–elimination route for carboxyl derivatives follows. A very important crosscheck for this often-reversible route is the ΔpK_a rule. The reaction will tend to form the weaker base. If the leaving group is a lot less basic than the nucleophile, then the reaction may be irreversible (again the ΔpK_a rule).

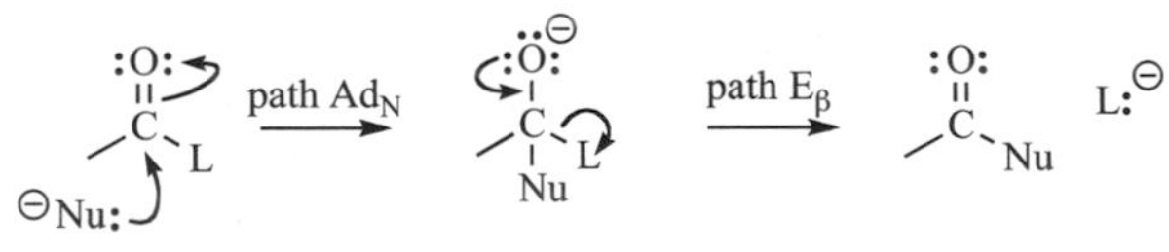
path AdN
path Eβ

The arrow pushing steps for the common acidic route (Ad_E2 then E1) is a bit longer. This route is usually reversible. In acidic media an additional proton transfer to improve the leaving group is common.

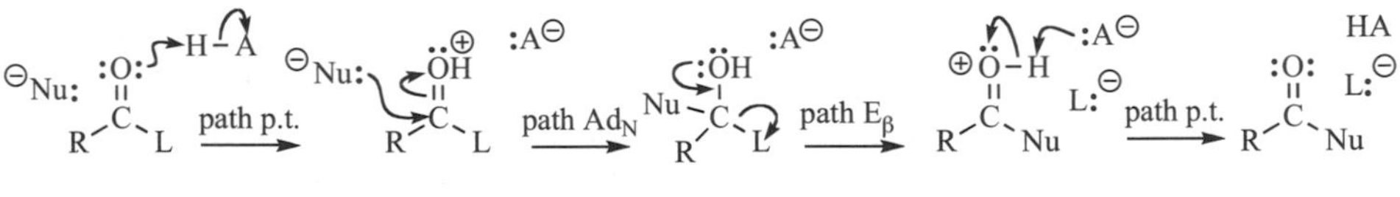
path p.t.
path AdN
path Eβ
path p.t.

4.5.2 Drawing Energy Diagrams and the ΔpK_a Rule

The ΔpK_a rule is exceedingly helpful in predicting the energetics of a reaction. We need some rule, no matter how approximate, to tell us whether a particular step or alternative is uphill or downhill in energy. The mechanism for basic hydrolysis of an ester can be used to illustrate the ΔpK_a rule and show how to draw an energy diagram.

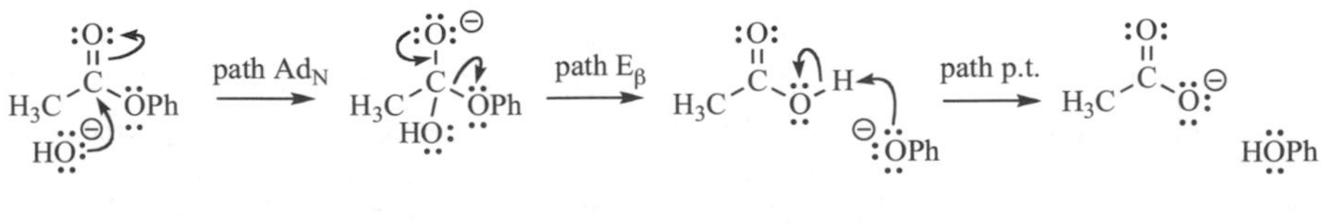
path AdN
path Eβ
path p.t.

The first step in drawing an energy diagram is to write out the mechanism, so you know how many intermediates there are. The reactants, each intermediate, and the products each go in an energy valley, or well. This energy diagram must have four wells. The first and third structures are both carboxylic acid derivatives, differing only in the nature of the anion, not in the number or type of bonds. The phenoxide (PhO^-) anion has a more distributed charge than hydroxide, and so is less basic than hydroxide. We can place the phenoxide and carboxylic acid well lower than the hydroxide and ester (reactants) well. The tetrahedral intermediate is placed at the highest energy because two C–O bonds are not as strong as one C=O bond, and we have lost entropy combining two molecules into one. The K_{eq} for proton transfer between phenoxide and acetic acid is calculated (Section 3.6) to be $10^{(10 - 4.8)} = 10^{+5.2}$ corresponding to a $\Delta G°$ at room temperature of approximately (5.2)(−1.36) = −7.1 kcal/mol (–29.6 kJ/mol), almost irreversibly downhill (Section 2.1). This last proton transfer step forming the resonance stabilized acetate ion is the driving force for the reaction, so we place the products' well lowest. Figure 4.47 shows the relative (and very qualitative) energy of the wells.

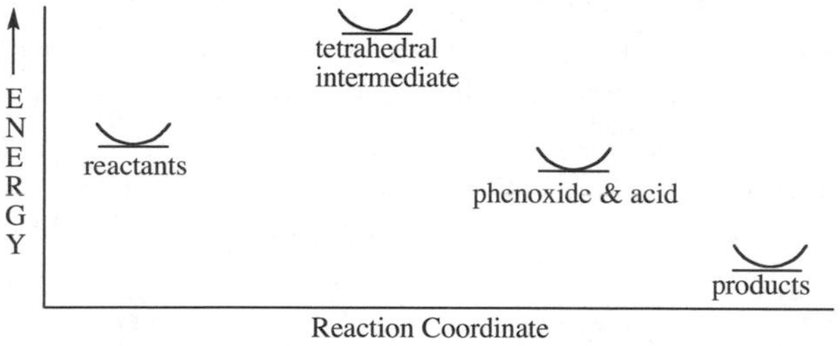
ENERGY
tetrahedral
intermediate
reactants
phenoxide & acid
products
Reaction Coordinate

Figure 4.47 Energy well positions for the basic hydrolysis of phenyl acetate.

The only thing left to do is decide the rate-determining step, which will usually be one side or the other of the highest well, the tetrahedral intermediate in this case. If we imagine sitting in the well of the tetrahedral intermediate, which way, forward or back, would be easiest? The three possible groups that could be lost from the tetrahedral intermediate are PhO^- (pK_{aHL} 10), HO^- (pK_{aHL} 15.7), or CH_3^- (pK_{aHL} 48), and clearly the first is the best. Methyl anion is not a leaving group at all. Since the best leaving group from the tetrahedral intermediate is PhO^-, the expulsion of the better leaving group should have a lower barrier (viewed from the well of the intermediate). The rate-determining step is the attack of hydroxide ion on the ester. Figure 4.48 gives a reasonable energy diagram for the reaction.

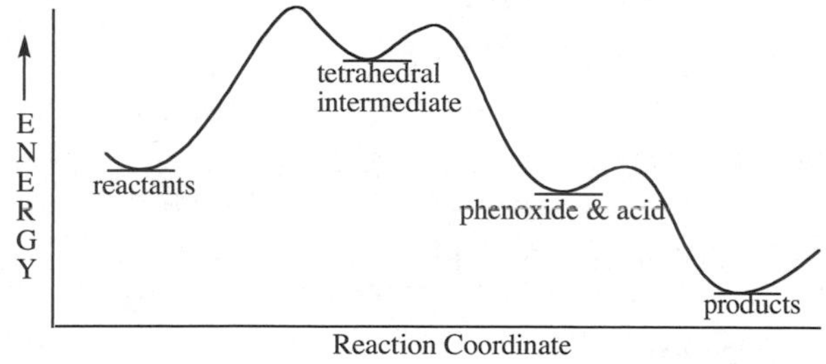
ENERGY
tetrahedral
intermediate
reactants
phenoxide & acid
products
Reaction Coordinate

Figure 4.48 Energy diagram for the basic hydrolysis of phenyl acetate.

The plot of pK_{abH}, the pK_a of conjugate acid of the most basic partner in each step versus the reaction coordinate, is shown in Figure 4.49 and has a profile similar to the energy diagram (Fig. 4.48). Several points can be made. Over the course of a reaction the pK_{abH} of the intermediates may increase slightly, but they will either hover about the initial pK_{abH} or very likely fall to yield a weaker base at the end of the reaction. The breaking of a weak bond and formation of a stronger bond can compensate for the formation of an intermediate with a higher pK_{abH}. **No reaction has huge jumps upward in the pK_{abH} of its intermediates.** There can be great downward jumps as strongly basic intermediates are neutralized.

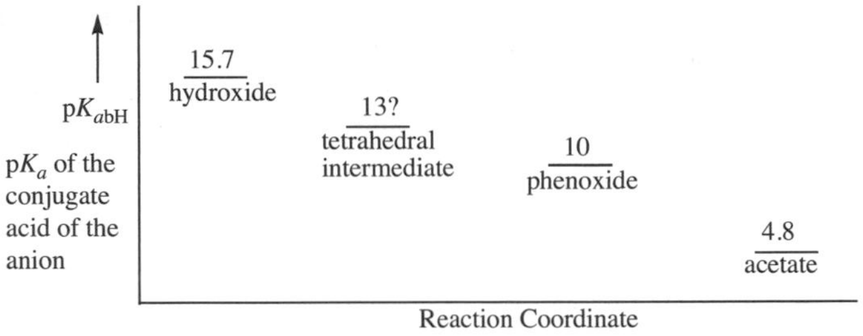
pK_abH
pK_a of the conjugate acid of the anion
15.7
hydroxide
13?
tetrahedral intermediate
10
phenoxide
4.8
acetate
Reaction Coordinate

Figure 4.49 The pK_{abH} of the most basic species in each step of the basic hydrolysis of phenyl acetate versus reaction coordinate. The pK_{abH} must be estimated for the tetrahedral intermediate.

Obviously, the match is not perfect. In the energy diagram (Fig. 4.48) the energy of all species in each step is taken into account, whereas in the pK_{abH} diagram only the most basic partner was considered. The loss of entropy when the two reactants join to form the tetrahedral intermediate causes a significant error. However, overall we can draw a rough association between a reaction proceeding energetically downhill and the fact that the pK_{abH} of the products is almost always less than that of the reactants. Consequently, the following is an extremely useful, although very crude, approximation. **The energy drops if the pK_{abH} drops significantly.** This gross generalization will allow us to choose which of two alternative reaction routes is probably of lower energy.

The Amazingly Useful pK_a Chart

The pK_a chart can be used for

- Ranking the strength of acids
- Ranking the strength of bases
- Calculating a proton transfer K_{eq}
- Ranking the stability of carbanions
- Approximate ranking of leaving groups
- Approximate ranking of electron-withdrawing groups
- Estimating the energetics of reaction.
- Checking the reversibility of a reaction (ΔpK_a rule on the reverse reaction)
- Helping to rank nucleophiles (in conjunction with softness) (Section 4.2.2)

Example problem

Does the following reaction favor products or reactants, and is it reversible?

Answer: Since the pK_{abH} of methoxide is 15.5 and that of chloride is −7, chloride is obviously the weaker base. The reaction proceeds to the right to form the weaker base and is not reversible because the reverse reaction would climb 22.5 pK_a units.

4.5.4 Approaches to Addition–Elimination Mechanisms

This time we are just applying known addition and elimination routes to a new reactant. The first task is always to map changes on a balanced reaction. Is the medium acidic or basic? The pH of the medium is very important in deciding which addition–elimination route to take, basic or acidic. Again visualize the reactant "street corner" on the addition surface to decide which way to go, then make sure you draw the arrows correctly, keep track of charge balance, and use the known electron flow paths.

Common Addition–Elimination Errors

Besides the usual proton transfer errors, the most common addition–elimination errors are switching media pH in midmechanism, media pH errors, and media pH span errors. A common error in basic media is to forget the important ΔpK_a rule crosscheck.

Errors from Shortcuts to Save Arrows or Compress Steps

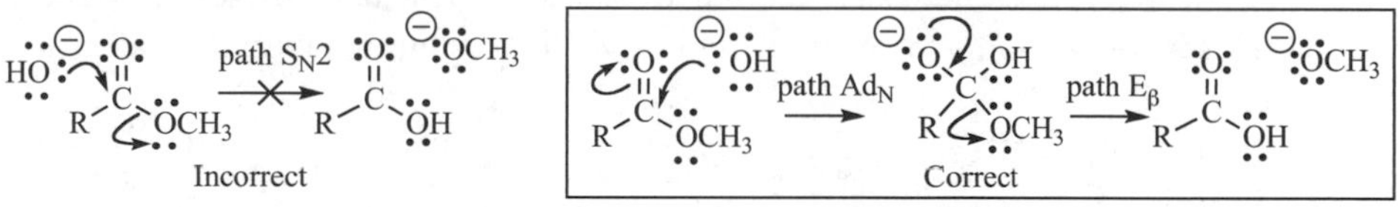
path S_N2
Incorrect
path Ad_N
path E_β
Correct

Although the S_N2 of leaving groups bound to a sp^2 center is a mechanism that has been searched for, with one or two rare exceptions there is little evidence in the literature that it has ever been found. The correct mechanism is path Ad_N, addition to a polarized multiple bond, followed by path E_β, beta elimination from an anion.

Mixed-Media Errors

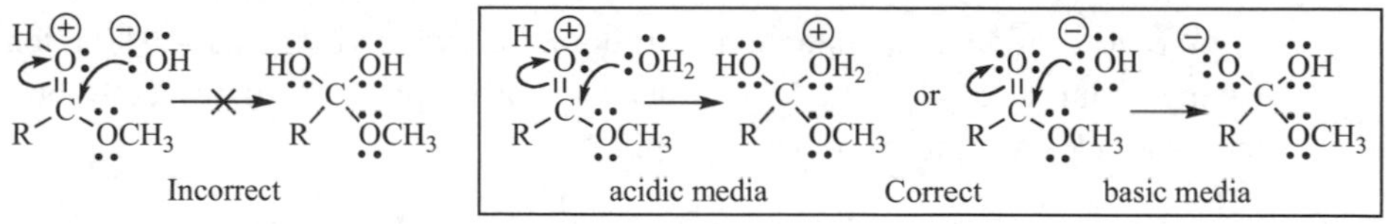
Incorrect
or
acidic media
Correct
basic media

Sometimes students will optimize the source and the sink and forget that those two species cannot exist in the same medium. Since the protonated ester pK_a is −6.5 and the pK_{abH} of hydroxide is 15.7, they span over 22 pK_a units. Therefore, any medium that would be acidic enough to have some concentration of the protonated ester would have essentially no concentration of hydroxide ion. The right examples show two correct alternatives: acidic medium with the protonated ester and neutral water as the nucleophile or basic medium with the neutral ester and hydroxide as the nucleophile.

Loss of a Poor Leaving Group Error

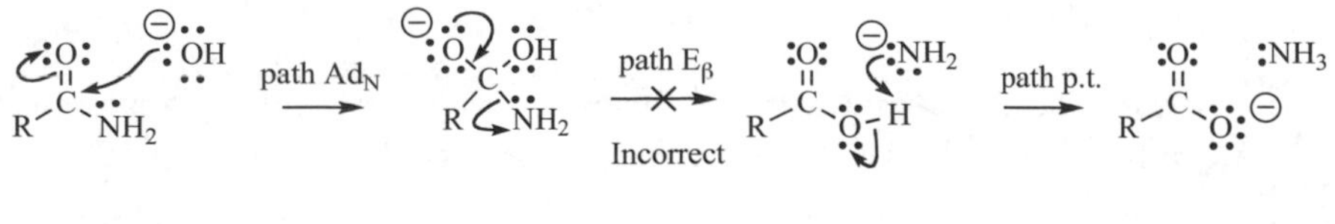
path Ad_N
path E_β
Incorrect
path p.t.

This is basically a ΔpK_a rule violation. Groups with a pK_{aHL} of over 30 are not leaving groups from tetrahedral intermediates, which don't get more basic than about a pK_{abH} of 16. Unfortunately these errors occur even in textbooks; for example, the basic amide hydrolysis mechanism shown above kicks out an NH_2^- anion of pK_{abH} 35 from a tetrahedral intermediate of about pK_{abH} 14, a climb of over 20 pK_a units. For a reasonable alternative, see the worked basic amide hydrolysis example in Chapter 10.

Example Addition–Elimination Mechanism

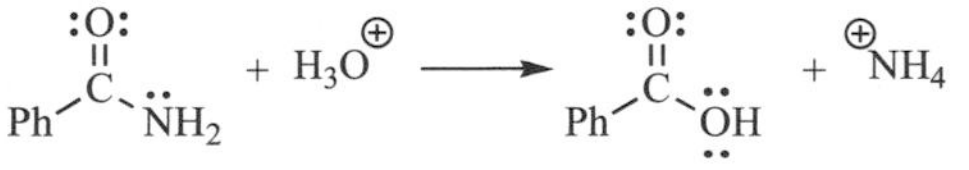

We are asked for the mechanism of acidic amide hydrolysis. The digestion of proteins is an amide hydrolysis. With acidic water present, we will choose the acid catalyzed addition–elimination route, Ad_E2 followed by elimination. We anticipate needing to add a proton transfer step to improve the leaving group, since it is currently very poor. The first step is to activate the carbonyl for attack by the mediocre nucleophile water. An alternative proton transfer to nitrogen is possible but unfavorable because this puts a plus charge next to the partial plus carbon of the carbonyl. Protonating the carbonyl instead makes a reactive resonance-stabilized electrophile. A crosscheck gives K_{eq} to be $10^{(-1.5-(-1.7))} = 10^{+0.2}$, slightly favoring products.

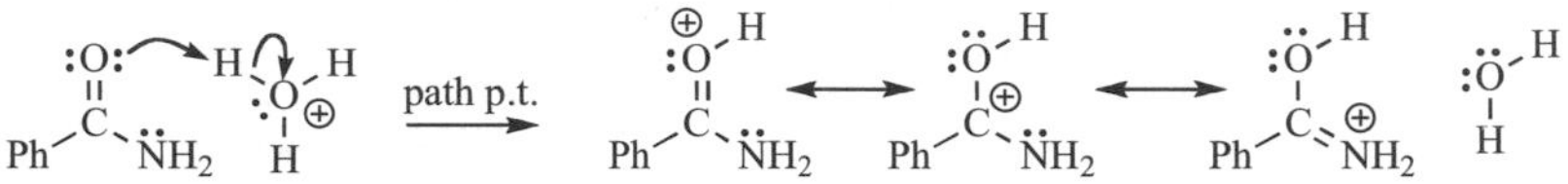
path p.t.

Water acts as a nucleophile on the protonated carbonyl and then loses a proton to give the neutral tetrahedral intermediate, completing the Ad_E2.

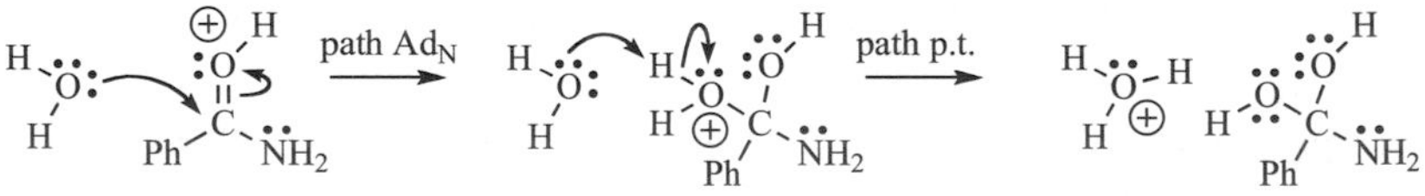
path Ad$_N$
path p.t.

We need to do an elimination to restore the carbonyl, but the NH_2 anion is not a leaving group at $pK_{aHL} = 35$. Protonation will improve it significantly, giving NH_3 neutral as a leaving group, at $pK_{aHL} = 9.2$.

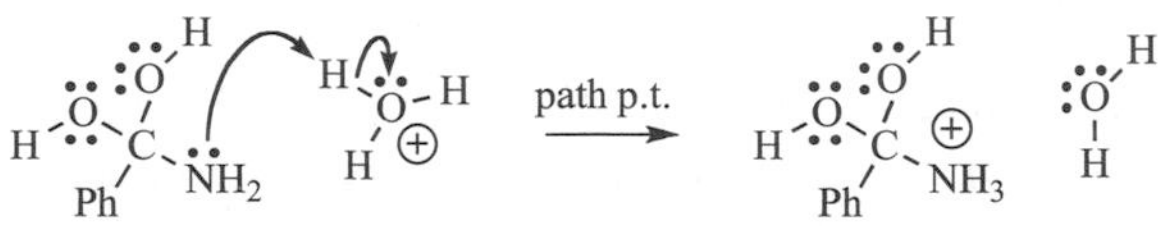
path p.t.

Now it is time to decide an elimination route. The E1 has a problem because although the carbocation would be stable and the water solvent is very polar, the leaving group is just not good enough. E1cB rarely occurs in acidic media because it needs a decent base for the initial deprotonation step. E2 remains a reasonable alternative.

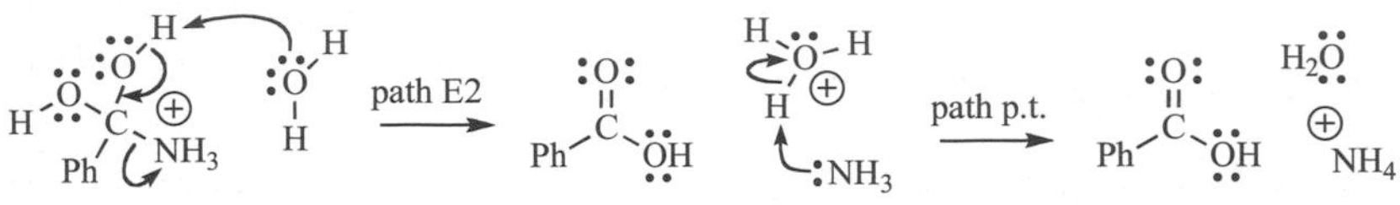
path E2
path p.t.

All we need to complete the mechanism is a very favorable proton transfer with a $K_{eq} = 10^{(9.2-(-1.7))} = 10^{+10.9}$, which makes the overall reaction irreversible.

4.5.5 Addition–Elimination Summary

The previously discussed limitations for both additions and eliminations apply. More addition–elimination examples can be found in Section 8.8. The decision process for working an addition problem is shown in Figure 4.50.

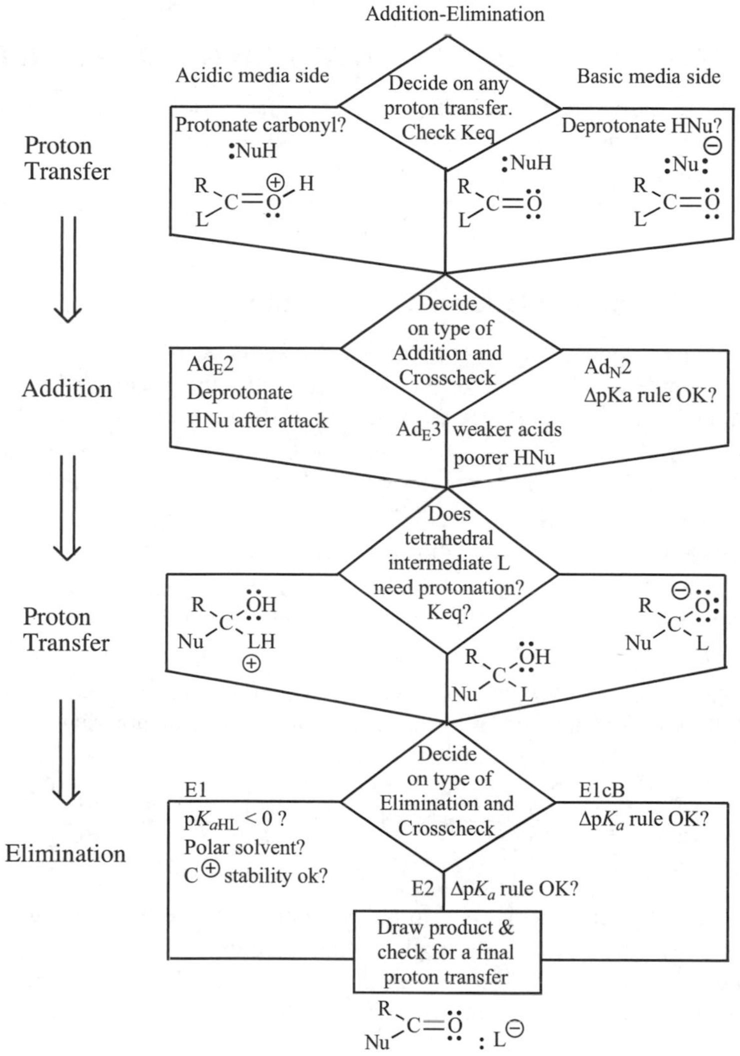
Addition-Elimination
Acidic media side
Decide on any proton transfer. Check Keq
Basic media side
Proton Transfer
Protonate carbonyl?
Deprotonate HNu?
Addition
Decide on type of Addition and Crosscheck
AdE2 Deprotonate HNu after attack
AdN2 ΔpKa rule OK?
AdE3 weaker acids poorer HNu
Proton Transfer
Does tetrahedral intermediate L need protonation? Keq?
Elimination
Decide on type of Elimination and Crosscheck
E1 pKaHL < 0 ? Polar solvent? C⊕ stability ok?
E1cB ΔpKa rule OK?
E2 ΔpKa rule OK?
Draw product & check for a final proton transfer

Figure 4.50 The flowchart for addition–elimination.

• Ad_E2 usually occurs in acid because it forms a very acidic protonated carbonyl.

• Ad_E3 occurs in buffered media, where the medium is not acidic enough to protonate the carbonyl and not basic enough to support strong nucleophiles.

• Ad_N2 occurs in base with good nucleophiles and follows the ΔpK_a rule: Never form an anion more than 10 pK_a units more basic than the incoming nucleophile.

• E1 tends to occur in acid because the leaving groups are often improved by being protonated and must pass three tests: a reasonably stable carbocation (usually better than secondary), a polar solvent, and a leaving group with a pK_{aHL} of zero or less.

• E2 often occurs in buffered media when the energy surface folds down the middle.

• E1cB usually requires basic media and an acidic hydrogen. Check that the proton transfer K_{eq} is more than 10^{-10}. Because of the anionic tetrahedral intermediate, E1cB can kick out poorer leaving groups. The E1cB follows the ΔpK_a rule: Never kick out a leaving group more than 10 pK_a units more basic than the tetrahedral intermediate.

4.6 ELECTROPHILIC SUBSTITUTION AT A TRIGONAL PLANAR CENTER

The most common example of electrophilic substitution at a trigonal planar center is electrophilic aromatic substitution, which will serve as our archetype. Aromaticity was covered in Section 1.9, and the reactivity trends of aromatics will be covered in detail in Chapter 5. Chapter 8 has additional reaction examples. Aromatic rings are usually poor nucleophiles; therefore excellent electrophiles are needed for the reaction to proceed.

4.6.1 Electrophilic Aromatic Substitution

There is only one route by which the aromatic rings react with electrophiles. The electrophile is believed to form a pi-complex with the pi cloud of the aromatic ring; then the electrophile attacks an electron-rich carbon of the ring by path A_E, creating a delocalized carbocation called the *sigma-complex*. The sigma-complex loses an electrophile, commonly a proton by path D_E to restore the aromaticity to the ring. This combination of the A_E then D_E paths has the mechanistic abbreviation S_E2Ar (substitution, electrophilic, bimolecular, aromatic). The arrow pushing for the electrophilic aromatic substitution is shown below. An arrow starts from a pi bond of the aromatic ring and creates the carbon electrophile bond in an A_E step. The first arrow of the D_E step starts from any slightly basic species (often the counterion of the electrophile) and *removes the proton at the site of electrophile attack.* A final arrow breaks the C–H bond and reforms the pi bond of the aromatic ring.

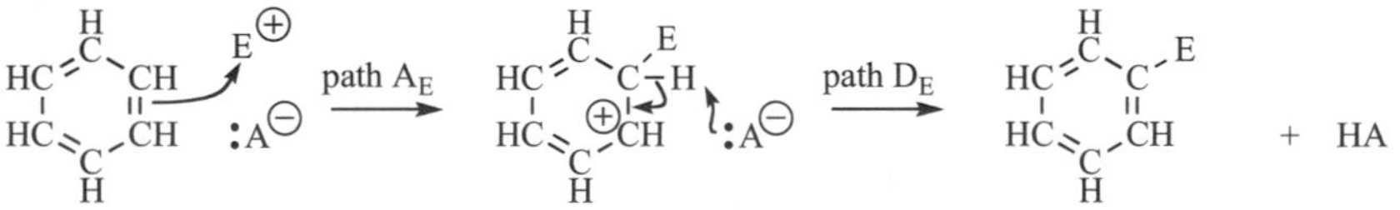
path A_E
path D_E
+ HA

Usually the attacking electrophile is more reactive than a proton, and therefore the rate-determining step is electrophile attack. Proton loss from the sigma-complex will be an easier process than loss of a reactive electrophile. Figure 4.51 shows a typical energy diagram for an electrophilic aromatic substitution.

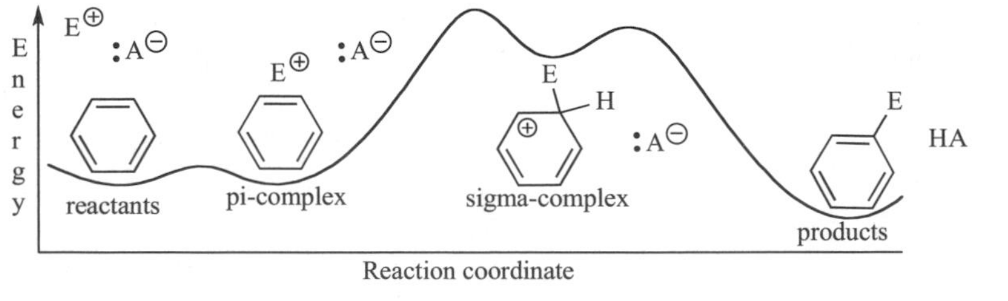
Energy
reactants
pi-complex
sigma-complex
products
HA
Reaction coordinate

Figure 4.51 A common energy diagram for electrophilic aromatic substitution.

Electrophilic Aromatic Substitution Regiochemistry

Groups present on the ring before the electrophile attacks dictate not only the reactivity of the ring but also the position of attack. Any group that has a resonance interaction with an aromatic ring has the greatest effect at the *ortho* and *para* positions. This is illustrated by the following resonance forms.

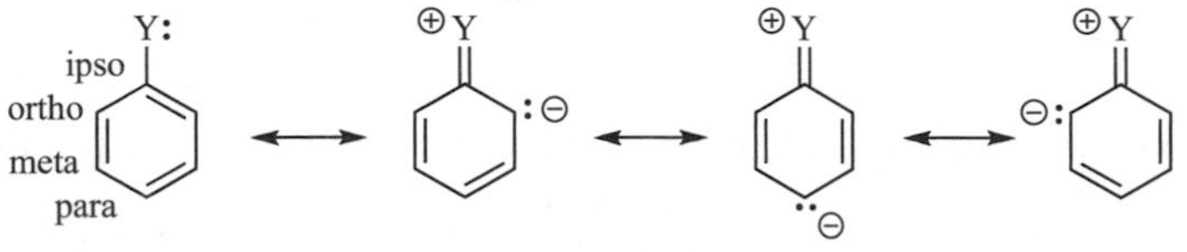
Y:
ipso
ortho
meta
para

Electron-releasing groups (erg) attached to the ring can stabilize the sigma-complex if the electrophile attacks either *ortho* or *para* to the group. In one resonance form of either sigma-complex, the positive charge is adjacent to the electron-releasing group. By stabilizing the sigma complex, the electron donors both increase the reactivity of the ring (by increasing the electron availability) and direct the incoming electrophile *ortho* or *para*. The *ortho* position is more hindered, so larger electrophiles usually go *para*.
Ortho attack:

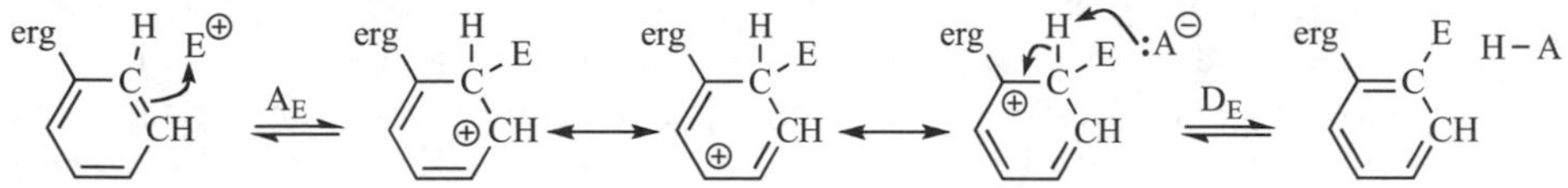

Para attack:

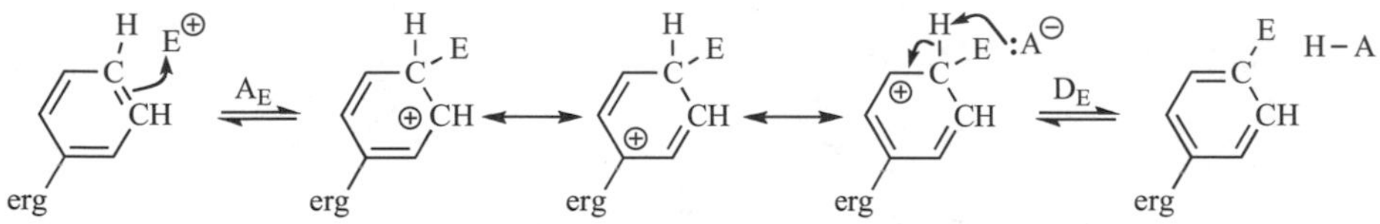

Electron-withdrawing groups (ewg) make the ring less electron rich, less reactive to electrophiles, and direct the incoming electrophile *meta*. The reactivity of the aromatic ring drops far enough so reactions with carbon electrophiles fail. The incoming electrophile is directed *meta*, going via the least destabilized sigma-complex, which has no resonance form with the positive charge adjacent to the electron-withdrawing group.
Meta attack:

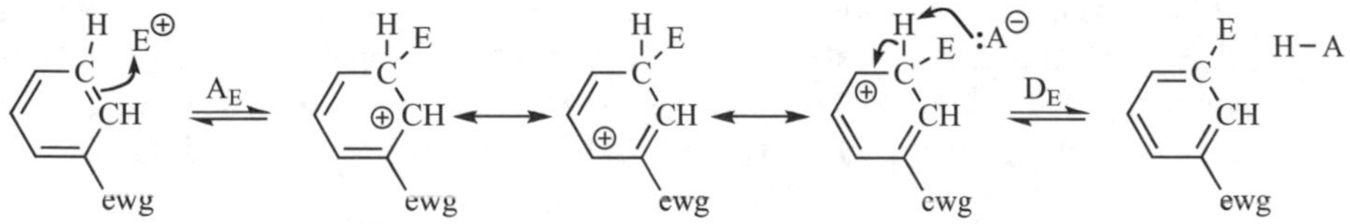

Occasionally the electrophile will attack the ring carbon that bears the group itself (*ipso* attack), but this site is definitely more hindered. If the group can form a cation equal to or more stable than the incoming electrophile, the group can depart from the sigma-complex as a cation (path D_E) and thus be replaced by the electrophile. Groups less electronegative than carbon, like silanes or metals, prefer the *ipso* substitution route.
Ipso attack:

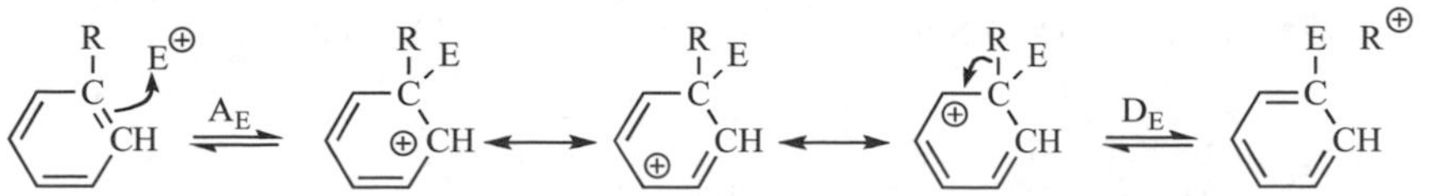

If more than one group is on the ring, the one that can stabilize the sigma-complex the most will be the one that directs electrophile attack. The reaction proceeds through the most stable of the possible sigma-complexes. For example, if two donors are on the ring, the electrophile attacks *ortho* or *para* to the best electron-releasing group.

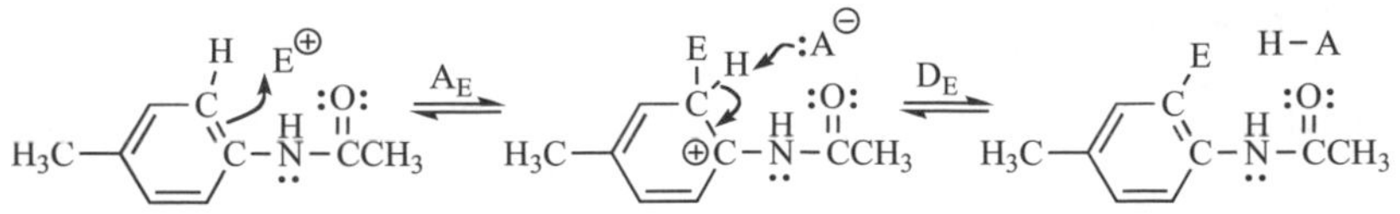

When two donors are *meta* to each other, attack at the position between them is extremely difficult because of steric hindrance.

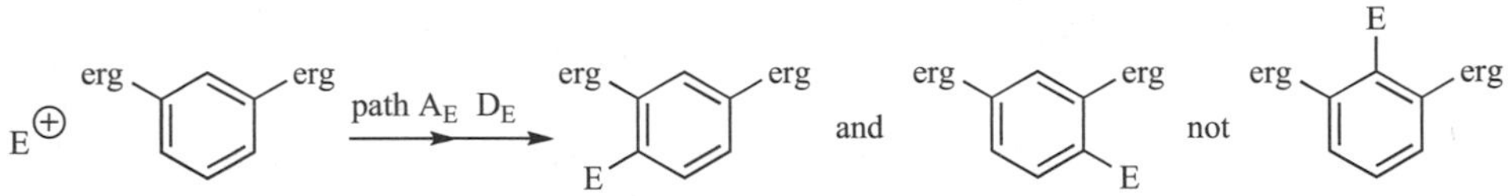

4.6.2 Approaches to Electrophilic Aromatic Substitution Mechanisms

Often the most difficult part of electrophilic aromatic substitution mechanisms is working out the generation of the reactive electrophile. The first task is always to map changes on a balanced reaction. The medium is almost always acidic because reactive electrophiles are present. The electrophilic addition to the aromatic ring is just a two-step process, A_E then D_E (usually a proton). Make sure you draw the arrows correctly, keep track of charge balance, and use the known electron flow paths.

Example Electrophilic Aromatic Substitution Mechanism

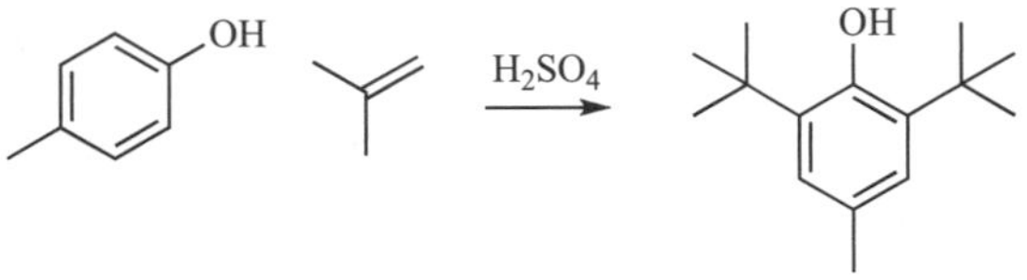

This is the synthesis of a somewhat controversial food preservative called BHT. It is an antioxidant that has been put in packaging to keep foods from becoming rancid from air oxidation. A balanced reaction would require two molecules of the alkene for each of the starting aromatic. The aromatic ring is relatively electron rich with two donors attached to it. With sulfuric acid present, the medium is definitely acidic. The first step is to generate the excellent electrophile needed for electrophilic aromatic substitution. For simplicity, let's symbolize sulfuric acid as H–A. The Markovnikov addition of a proton to isobutylene gives the *tert*-butyl carbocation, an excellent electrophile.

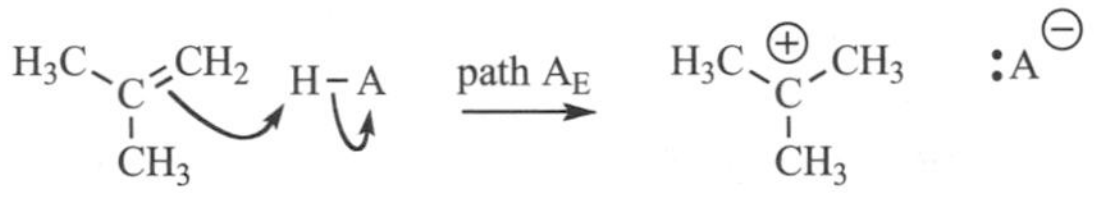

Now we have an excellent electrophile for the electrophilic aromatic substitution reaction. The hydroxyl lone pair is a much better electron-releasing group than the methyl, so the electrophile adds *ortho* to the best electron-releasing group, hydroxyl (the *para* position is already occupied by methyl). The D_E step removes the proton from the carbon that the electrophile attacked to rearomatize the ring.

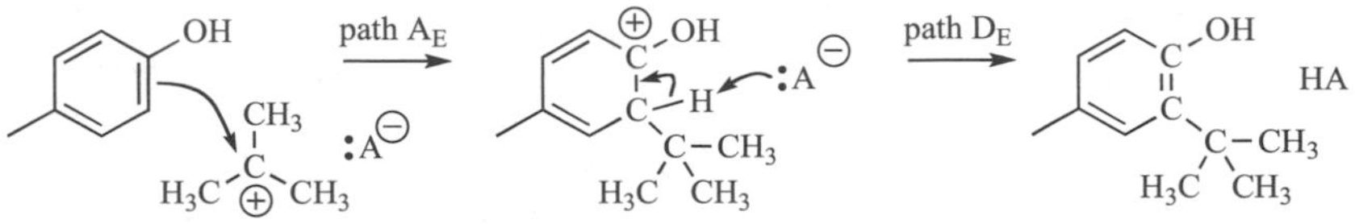

The electrophilic aromatic substitution reaction is repeated again on the other *ortho* position with the same electrophile to give BHT.

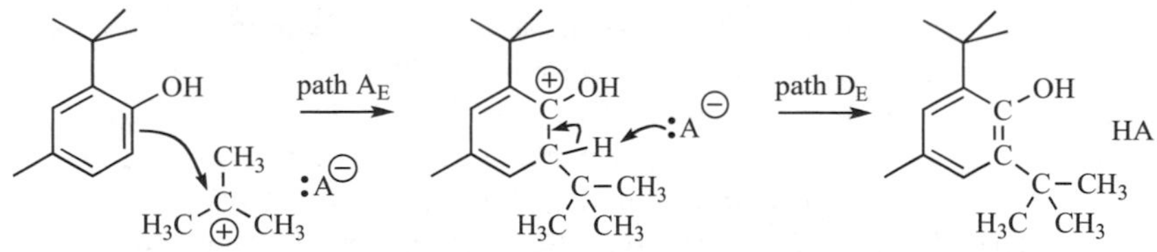
OH
path A_E
CH_3
H_3C CH_3
:A
path D_E
HA

Common Electrophilic Aromatic Substitution Errors

Some texts will draw a circle in a six-membered ring to symbolize an aromatic. A semicircle with a plus may be also used to symbolize a sigma-complex. Don't use the circle notation when doing mechanisms, because it is so easy to lose track of electrons.

Media pH errors and media pH span errors are common. **Since electrophilic aromatic substitution is almost exclusively an acidic media process, do not make any strong bases during the mechanism.** The proton on the aromatic ring becomes very acidic after the electrophile attaches and forms the carbocationic sigma-complex. However, before the electrophile attacks, the aromatic H is not acidic at all, pK_a = 43, so do not get your steps out of order and try to pull the H off first.

Line structure can set you up for a common beginner's error. You can forget about the H that needs to be pulled off in a D_E step because line structure does not show the hydrogens. **It is the proton at the site of electrophile attack that is lost, not any other.**

4.6.3 Electrophilic Aromatic Substitution Summary

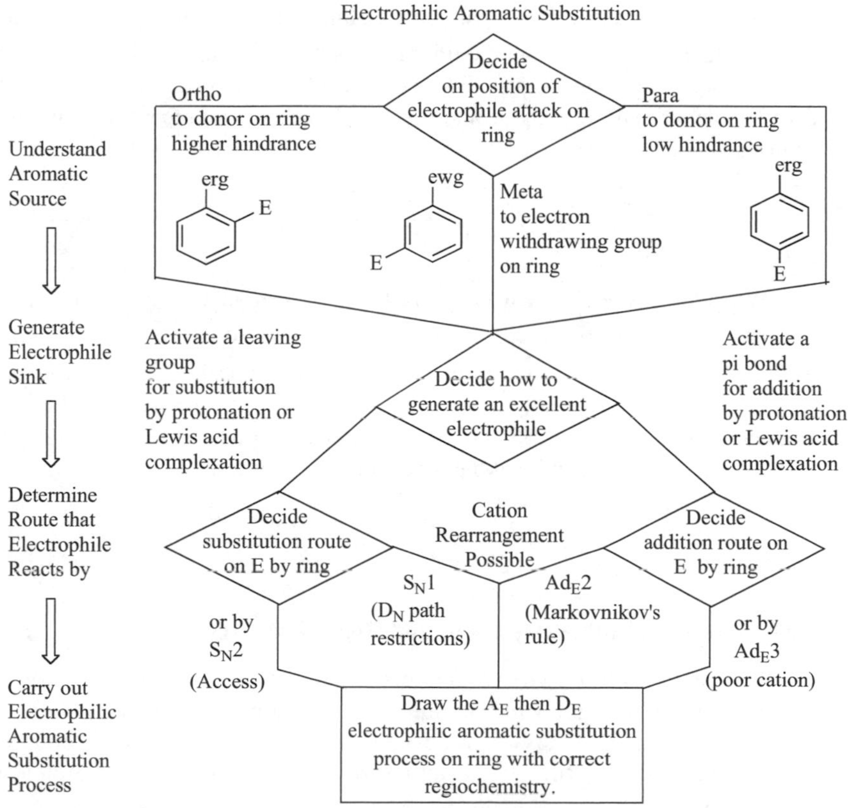
Electrophilic Aromatic Substitution
Understand Aromatic Source
Generate Electrophile Sink
Determine Route that Electrophile Reacts by
Carry out Electrophilic Aromatic Substitution Process
Decide on position of electrophile attack on ring
Ortho to donor on ring higher hindrance
Para to donor on ring low hindrance
Meta to electron withdrawing group on ring
erg
ewg
E
Activate a leaving group for substitution by protonation or Lewis acid complexation
Decide how to generate an excellent electrophile
Activate a pi bond for addition by protonation or Lewis acid complexation
Decide substitution route on E by ring
Cation Rearrangement Possible
Decide addition route on E by ring
S_N1 (D_N path restrictions)
Ad_E2 (Markovnikov's rule)
or by S_N2 (Access)
or by Ad_E3 (poor cation)
Draw the A_E then D_E electrophilic aromatic substitution process on ring with correct regiochemistry.

Figure 4.52 A general flowchart for electrophilic aromatic substitution.

The decision process for working an electrophilic aromatic substitution problem is shown in Figure 4.52. First we have to decide the best position on the ring to add the electrophile, and next we need to figure out how to generate the electrophile. Then we have to decide the process by which our electrophile reacts before adding it to the aromatic ring. Carbocation formation can have the possibility of rearrangement, our last archetype; Section 9.8 discusses the rearrangement decision. More examples of electrophilic aromatic substitution reactions can be found in Chapter 8.

4.7 REARRANGEMENTS TO AN ELECTROPHILIC CARBON

4.7.1 Introduction to Rearrangement Reactions

Rearrangements encompass so many different structural features and mechanisms that it is not possible to write a generic rearrangement example that would fit all. Instead, for our archetype, we will pick two competing rearrangements to an electrophilic carbon that complicate our other archetypes, primarily the S_N1, E1, and S_E2Ar. We have to be suspicious of the possibility of rearrangement occurring any time a compound is placed in strongly acidic medium.

4.7.2 Rearrangement Electron Flow Paths

The arrow pushing for the two rearrangement paths is shown below. When the carbocation is of sufficient stability, the leaving group will depart first (path D_N), and then rearrangement occurs in a path resembling the S_N1 process (1, 2 rearrangement, path 1,2R). The arrow starts from the migrating bond and points to the carbocation, indicating that the migrating group is just sliding over while maintaining orbital overlap between the carbon it leaves and the one it migrates to. This rearrangement usually creates an equal or more stable carbocation.

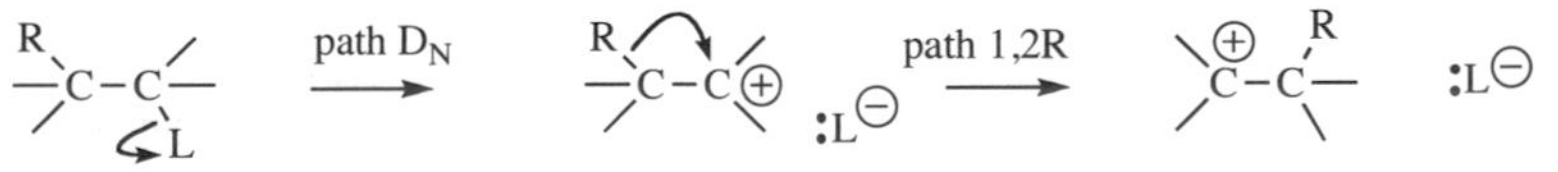

R
L
path D_N
R
:L⊖
path 1,2R
R
:L⊖

When the carbocation that would have formed in the D_N step is not very stable, rearrangement can still occur by a process similar to the S_N2. The migrating group displaces the leaving group from the back side. For this rearrangement with leaving group loss, the 1,2RL path, we start the first arrow from the bond of the migrating group and slide it over to the carbon bearing the leaving group. A second arrow starts from the C–L bond, breaking it and making a new lone pair on the leaving group.

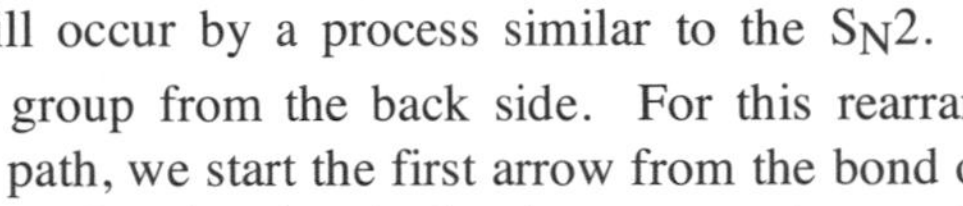

R
CH_2
L
path 1,2RL
R
H_2
:L⊖

Rearrangement Path Restrictions and Regiochemistry

The main path restriction on rearrangements is that the migrating group must maintain orbital overlap with both the atom it migrates *from* and the atom it migrates *to*. At the transition state, the migrating group is partially bonded to both atoms. If the leaving group departs first, the usual D_N restrictions apply (carbocation stability usually better than secondary, good leaving group, polar solvent). A specific example is useful. Isobutyl alcohol reacts with concentrated hydrobromic acid to give *tert*-Butyl bromide.

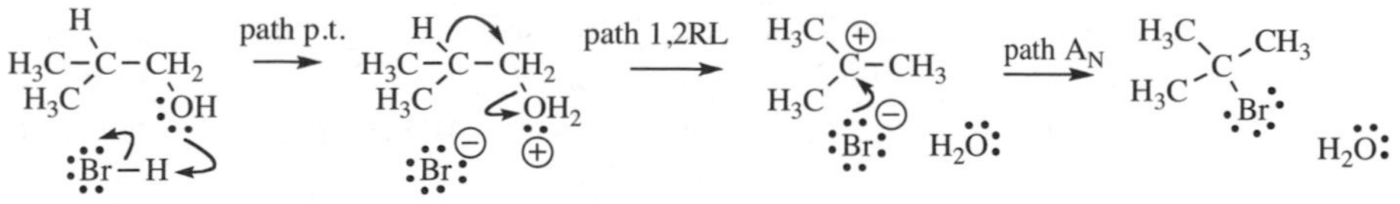

Our start is to check for any decent proton transfer reactions. Protonating the alcohol with HBr has a favorable $K_{eq} = 10^{(-2.4\ -\ (-9))} = 10^{+6.6}$, which greatly improves the leaving group. A check of the D_N restrictions finds a good leaving group, a polar solvent, but it would form an unstable primary alkyl carbocation, so we do not have all three criteria satisfied. To repeat, a highly unstable primary carbocation is not a reasonable alternative for a mechanism. The rearrangement can still go by the 1,2RL path. However, either a methyl or a hydride could migrate. If the methyl migrated, a secondary carbocation would be formed. The hydride migration produces a more stable tertiary carbocation. Since free rotation will line up either one, the lowest-energy route will form a more stable carbocation. The tertiary carbocation formed is trapped by path A_N, giving the product.

4.7.3 Approaches to Rearrangement Mechanisms

Predicting when rearrangement will occur is a bit of a challenge, covered more in Section 9.8; important factors are highly acidic media or strong Lewis acids, and a group that if it migrated would produce a compound of significantly greater stability. In mechanism problems, we have the distinct advantage of being given the product. If the product has a different bonding order from the reactant, then rearrangement has happened. We just have to find a reasonable route.

Common Rearrangement Errors

The most common error is to go with the first rearrangement route that you may think of and not consider important crosschecks like carbocation stability. Secondary alkyl carbocations are the tipping point. If your mechanism is going through a carbocation less stable than secondary alkyl, it is most likely incorrect. Orbital alignment needs to be watched in cyclic systems, where free rotation does not occur. Make sure the migrating group can maintain overlap in the migration; migrating groups usually move to the adjacent atom, and do not hop over anything.

Example Rearrangement Mechanism (Pinacol Rearrangement)

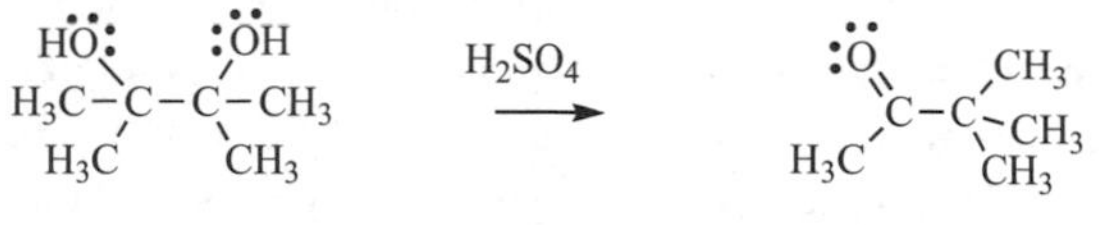

This is an example of a common type of acid-catalyzed rearrangement. A balanced reaction gives water as the other product. Initially, we explore proton transfer to improve the leaving group. Since the molecule is symmetrical, it matters not which hydroxyl is protonated. The proton transfer to the alcohol is downhill. If the conjugate acid is below (higher pK_a) the acid on the chart, then the proton transfer is favorable and there is no need to calculate the K_{eq}. Using H–A to represent sulfuric acid, we begin by protonating the hydroxyl group.

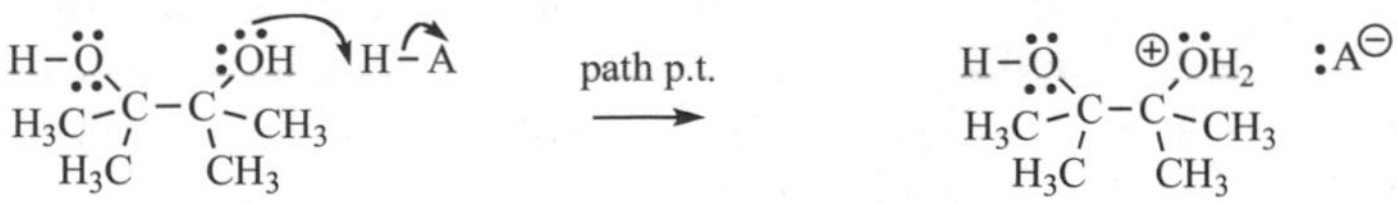

Now is the time to choose which of our two rearrangements to electrophilic carbon route to use. The D_N crosscheck passes, the pK_{aHL} of the leaving group is below zero, the carbocation is tertiary, and the solvent is polar. We can proceed with the D_N step.

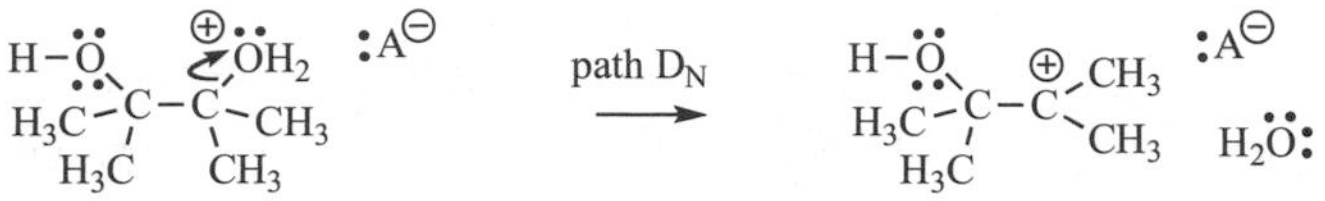

From the product, we see that a methyl has migrated; migrating the hydroxyl would just give the same connectivity as the reactant. The carbocation produced from the methyl migration is more stable than the one we started with because it is now resonance-stabilized by the lone pair on oxygen.

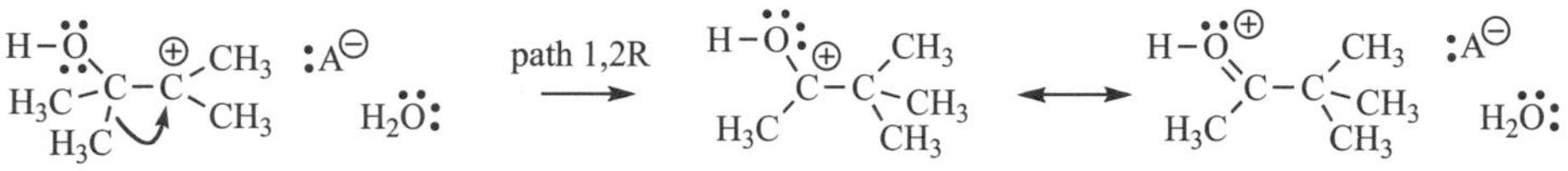

The rearrangement produces the protonated product, which is deprotonated by the strongest base in solution: neutral water.

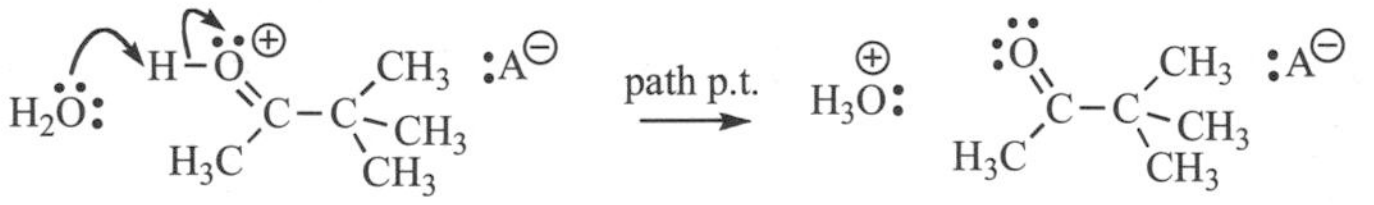

4.7.4 Rearrangement Summary

Rearrangement competes with substitutions, eliminations, and additions in acidic media and not in basic. If steric hindrance slows down the other routes, rearrangement can compete. Strongly acidic media or powerful Lewis acids encourage rearrangement. The following are some systems that commonly rearrange in acid. More rearrangement examples are in Section 8.11 and the decision whether to migrate or not is considered in Section 9.8.

4.8 REACTION ARCHETYPE SUMMARY

Proton transfer (Chapter 3) and the four reaction archetypes in this chapter, **substitution, elimination, addition,** and **rearrangement,** make up the core of an undergraduate organic course. Soon, we will be heading into more complex reactions that combine these archetypes or play one off against another. The archetypes and their mechanisms are so central to all reactions that rarely will we have to bring in a new electron flow path to explain a reaction. In the next two chapters, we discuss a way to categorize our reactants into generic electron sources and sinks so that the sheer mass of material does not become overwhelming. Then in Chapter 7, all of the electron flow pathways will be discussed in detail. We will be able to understand almost all organic reactions as consisting of selected sequences of the dozen most common electron flow pathways. Chapter 8 then maps the interaction of the electron sources with electron sinks, using these pathways to help you understand the vast majority of organic reactions. Chapter 9 deals with the major decisions like substitution versus elimination. Chapter 10 shows the analytical process for writing reasonable organic reaction mechanisms and for predicting the products of an organic reaction.

ADDITIONAL EXERCISES

4.1 Trends: rank all species, using the numeral 1 to designate:

(a) the most reactive for S_N1

(b) the most reactive for S_N2

(c) the most reactive nucleophile toward CH_3I

$PhS^{\ominus}$ $PhO^{\ominus}$ $CH_3O^{\ominus}$ $PhSe^{\ominus}$ CH_3OH

(d) the most stable carbocation

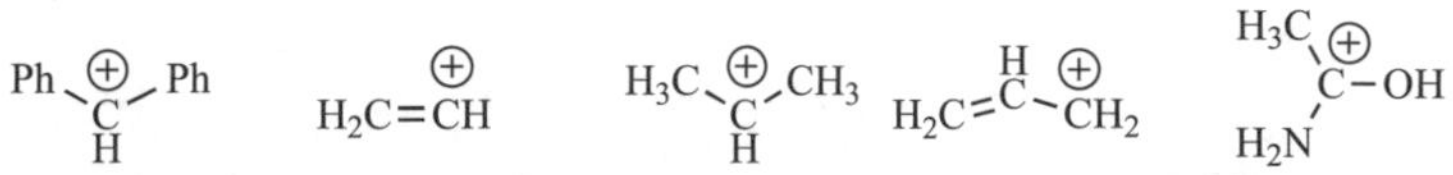

4.2 State what controls the regiochemistry of the E1, the E2, and the E1cB.

4.3 Using the energy surfaces in Section 4.2.10, predict how increasing the solvent polarity should affect the competition between S_N1 and S_N2.

4.4 Draw the substitution product and decide whether the reaction is S_N1 or S_N2.

(a) 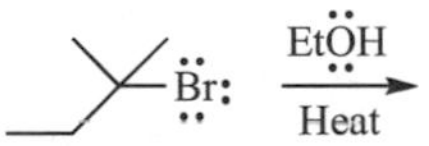

(b) 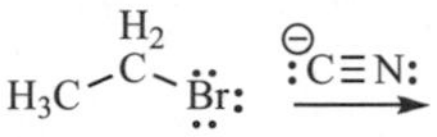

(c) $Ph_3P: + H_3C{-}\ddot{I}: \longrightarrow$

4.5 Draw the elimination product and decide whether the reaction is E1, E2, or E1cB.

(a) 

(b) 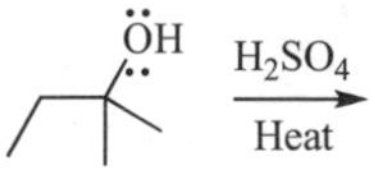

(c) 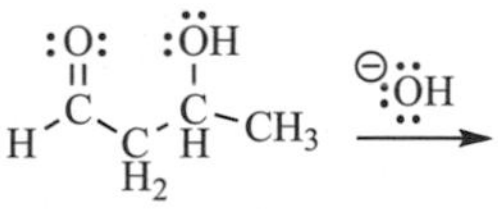

4.6 Draw the addition product. Decide whether the reaction is Ad_E2, Ad_E3, or Ad_N2.

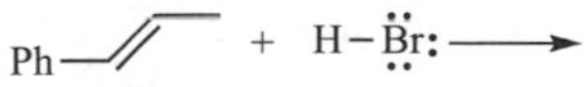

4.7 Draw the product addition–elimination. Decide whether the route is basic or acidic.

(a) 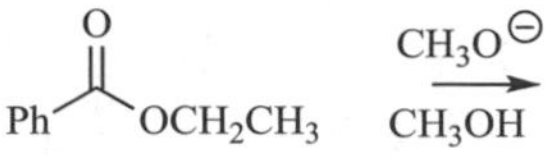

(b) 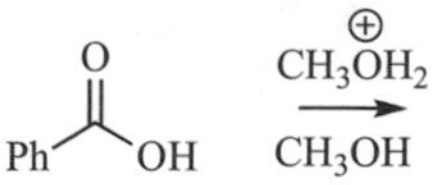

4.8 Draw the electrophilic aromatic substitution product below. The E adds just once.

(a) 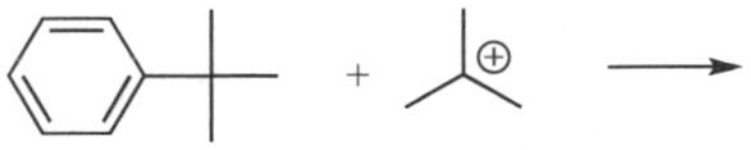

(b) 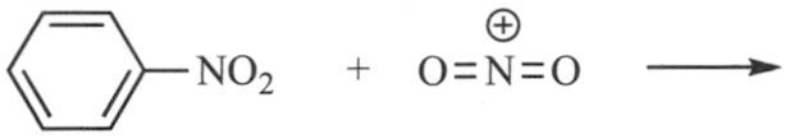

4.9 Use the appropriate electron flow pathway to rationalize the product of these reactions. First, decide which of the four reaction archetypes has occurred. These mechanisms have only one electron source and one sink present in the reaction mixture.

(a) 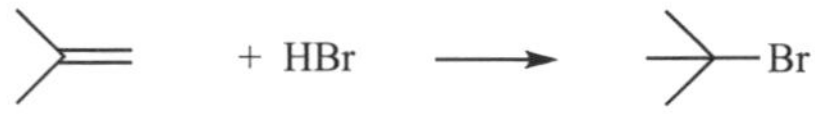

(b) 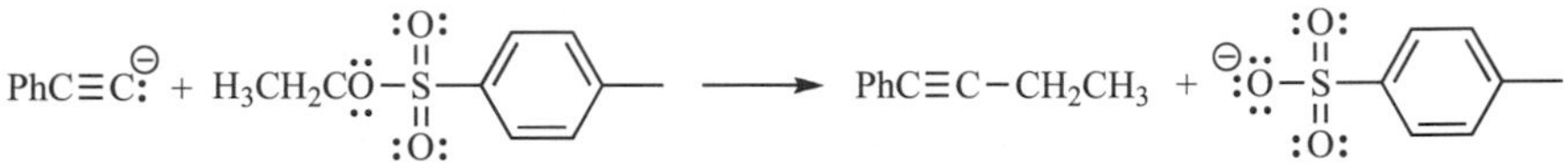

(c) 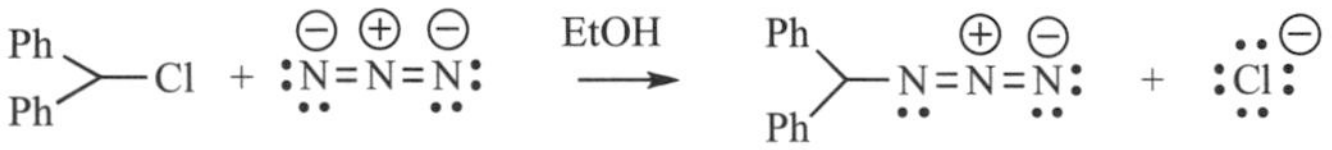

(d) 

(e) 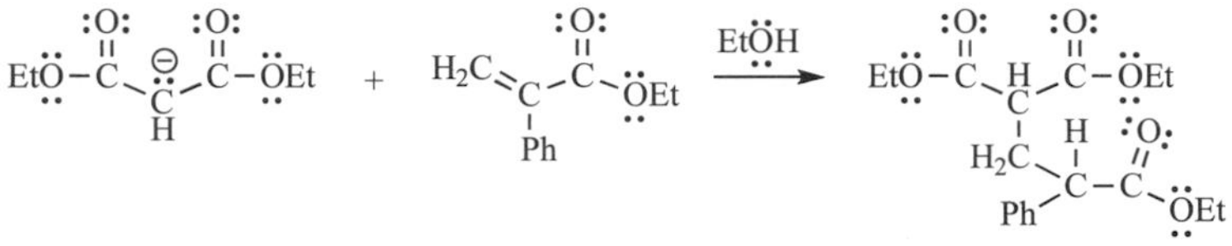

4.10 First decide which of the four reaction archetypes has occurred. Use the appropriate electron flow pathway to rationalize the product of these **two-step** reactions. The first step is a favorable proton transfer that generates the electron source for the reaction; there is only one electron sink.

(a) 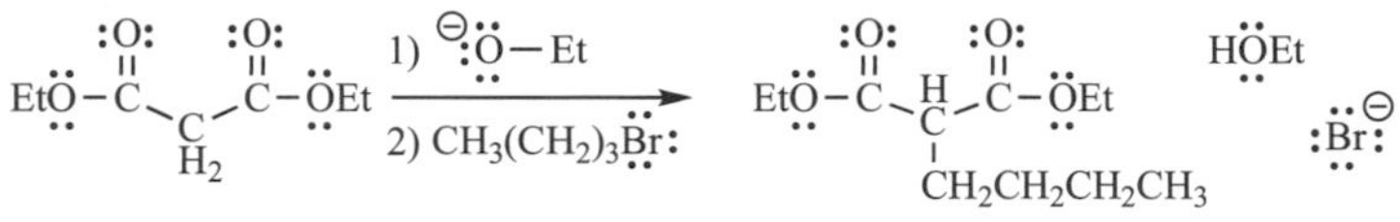

(b) 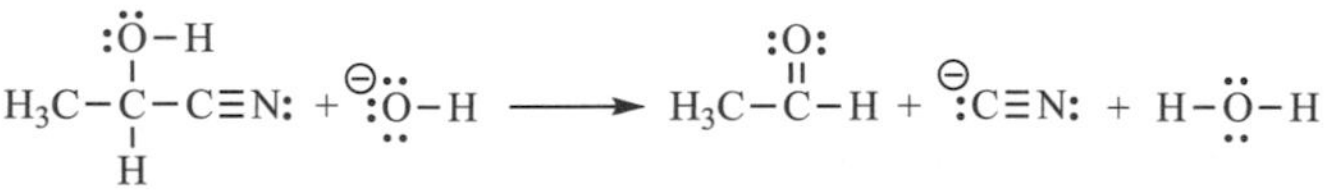

(c) $CH_3(CH_2)_{10}CH_2OH + HBr \rightleftharpoons CH_3(CH_2)_{10}CH_2Br + H_2O$

(d) 

4.11 Use the ΔpK_a rule to predict whether each of these reactions favors reactants or products.

(a) $H_3C{-}SH + I^{\ominus} \rightleftharpoons HS^{\ominus} + H_3C{-}I$

(b) 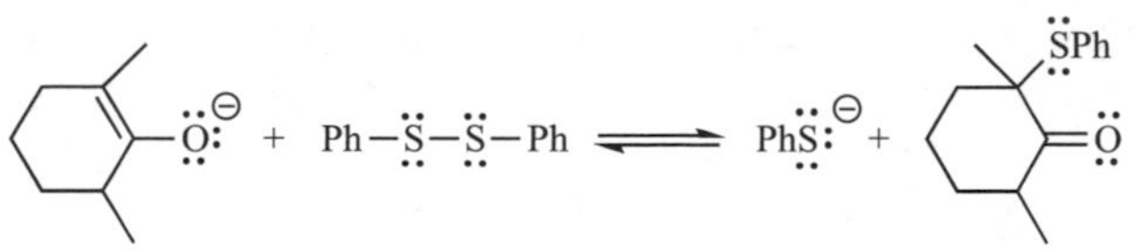

(c) 

(d) $PhH + LiOH \rightleftharpoons PhLi + H_2O$

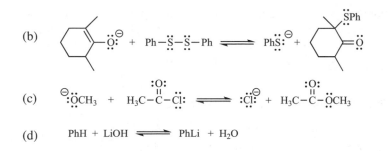

4.12 Use the ΔpK_a rule to determine which of the two following alternatives is the lower-energy process.

$CH_3COOEt + HO^-$ reacting to form $H_2O + {}^-CH_2COOEt$
or $CH_3COOEt + HO^-$ reacting to form $EtOH + CH_3COO^-$

4.13 Use the leaving group trend to decide which of the following two alternatives is the lower-energy process.

$ClCH_2CH_2CH_2Br + CN^-$ displacing bromide to form $ClCH_2CH_2CH_2CN$
or displacing chloride to form $BrCH_2CH_2CH_2CN$

4.14 Use the carbocation stability trend to decide which of the following two alternatives is the lower-energy process.

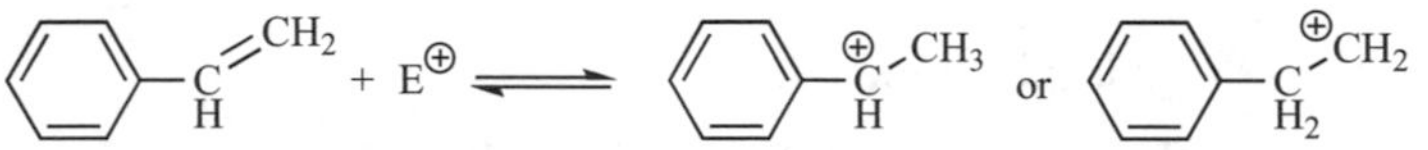

4.15 Decide which of the following two alternatives is the lower-energy process.

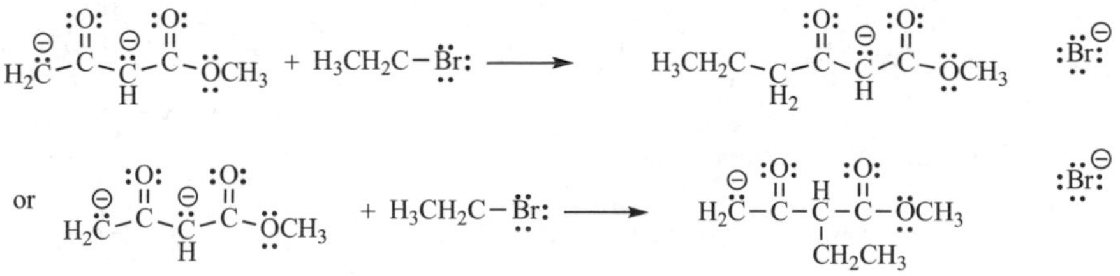

4.16 Decide which of the following two alternatives is the lower-energy process.

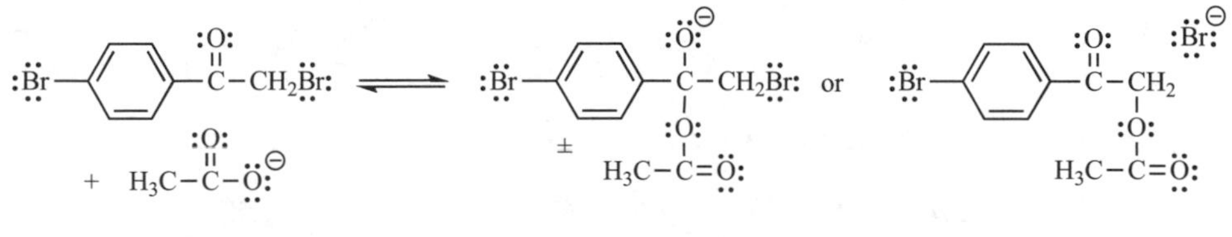

4.17 Account for the different rates in the ionization of the leaving group X. *Hint:* What factors influence the ease of ionization of a leaving group?

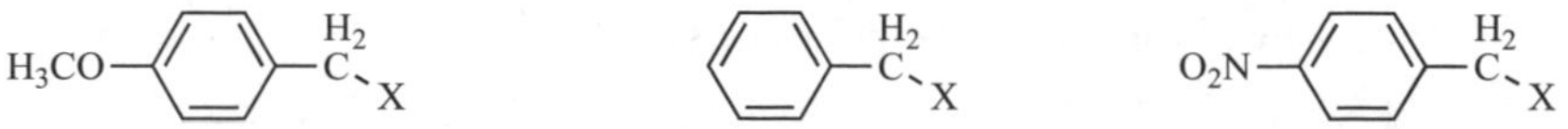

Fast Average Slow

4.18 Draw the problem space for the following reactant prediction problem. Select the best of two possible substitution alternatives. *Hint*: look at the proton transfer reaction.

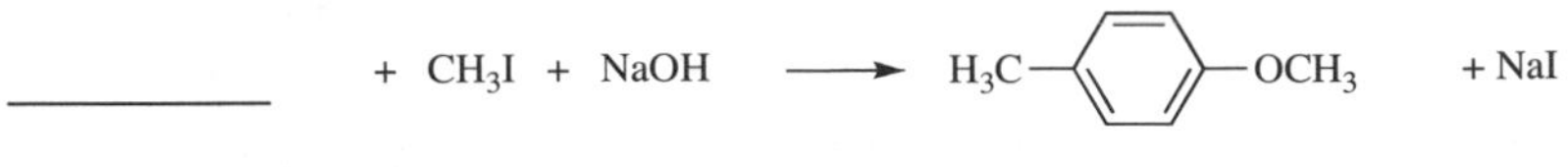

4.19 Which of the following structures are chiral?

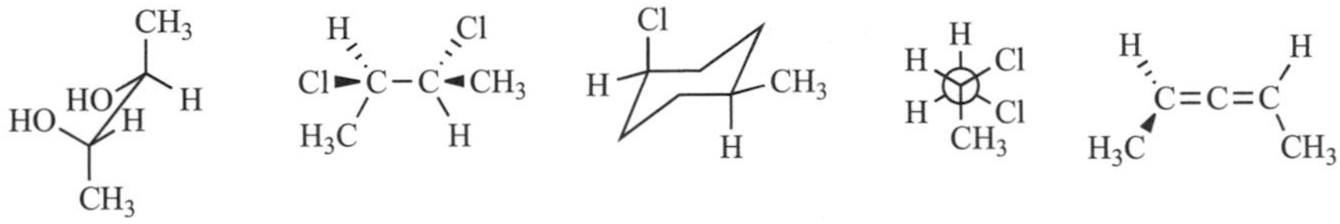

4.20 Explain why sodium sulfite below protonates on oxygen but alkylates with methyl iodide on sulfur. Draw Lewis structures for reactants and products.

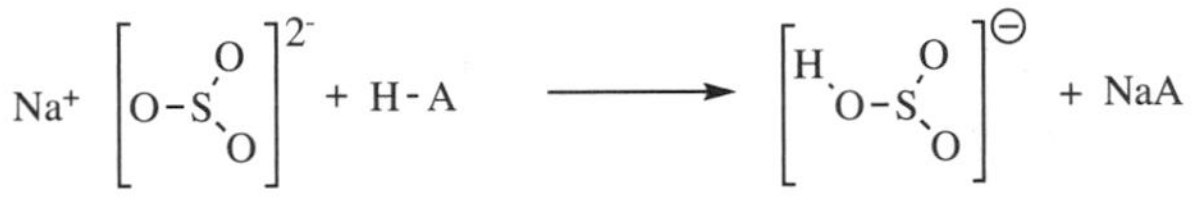

But

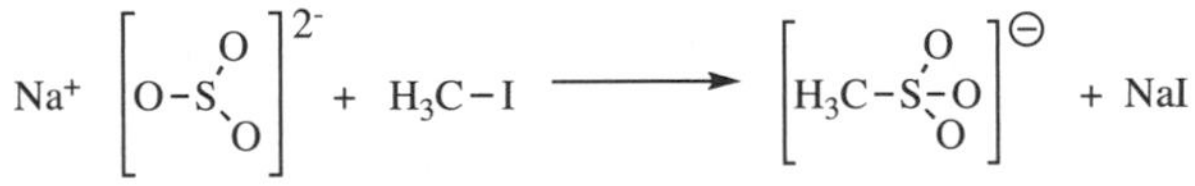

4.21 Trends: Rank all species, using the numeral 1 to designate:

(a) the best leaving group

—OCH_3 —Cl —OPh —OSO_2CH_3 —NH_3

(b) the most stable carbocation

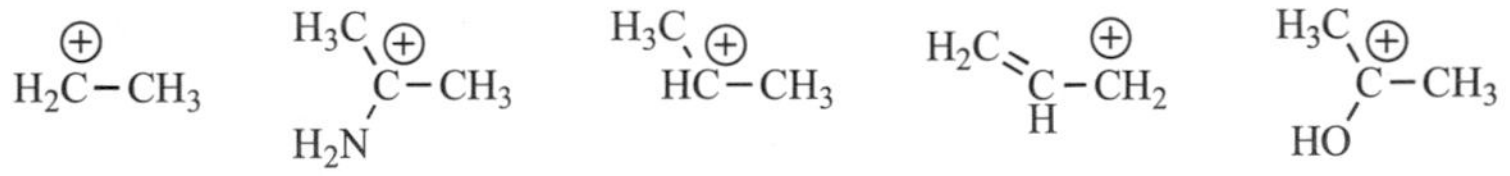

(c) the most stable carbanion

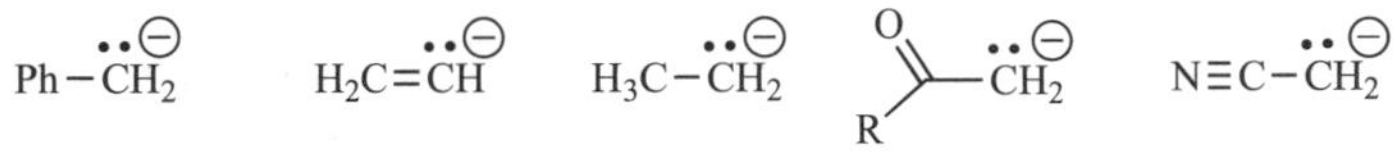

(d) the best electron-releasing group

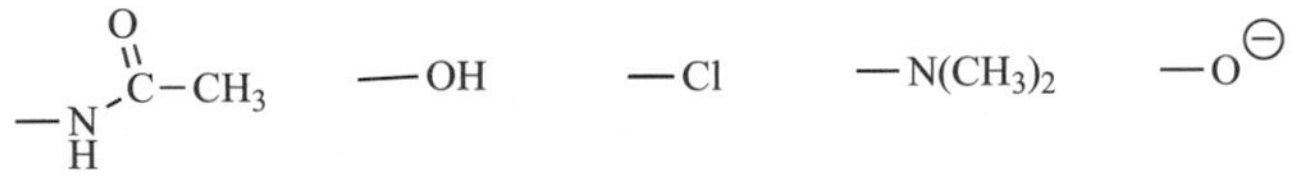

(e) the best electron-withdrawing group

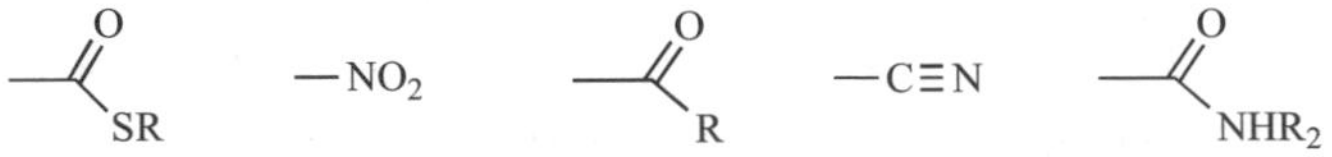

4.22 Explain the fact that thiocyanates react at sulfur whereas cyanates react at nitrogen. Draw Lewis structure for both products.

$$\text{R-Br} + \text{SCN}^{\ominus} \longrightarrow \text{RSCN} + \text{Br}^{\ominus}$$

$$\text{R-Br} + \text{OCN}^{\ominus} \longrightarrow \text{RNCO} + \text{Br}^{\ominus}$$

5

CLASSIFICATION OF ELECTRON SOURCES

5.1 GENERALIZED RANKING OF ELECTRON SOURCES

In this chapter we shall explore the classification and ranking of electron sources into the 12 common "generic" classes of electron sources. These generic classifications are based on chemical reactivity, and just like you do not expect a pig to fly, these classifications allow you to narrow down the chemical behavior of members of the class. A big problem in organic chemistry is the sheer magnitude of information, so efficient ways of organizing the material are very valuable. To repeat for emphasis, **there are only 12 generic classes from which a set of arrows, an electron flow, can start.** Some classes are quite rare, and others have several dozen examples. It is very important that you learn to classify according to generic class. **It is much easier to handle multiple examples of 12 familiar classes than hundreds of different special cases**. The first thing that you should do in learning a new reaction is to classify the electron source into one of these 12 classes. Three questions must be answered about each class: What characterizes the members of the class? What is the most electron-rich site? What determines reactivity within each class?

Before discussing the individual classes, we need to get an overview of all of them. Since most reactions have a hard–hard and a soft–soft component, it is important to notice both how much negative charge a source bears and how soft the source is. **Electrons that reside in more stable, lower-energy bonding orbitals are less available for use as electron sources in reactions.** We can divide good electron sources into three groups: nonbonding electrons, electron-rich sigma bonds, and electron-rich pi bonds.

The best electron sources are usually nonbonding electron pairs. They are electron rich, and no bonds need be broken to use them as electron sources. Other excellent electron sources are highly ionic sigma bonds and also pi bonds highly polarized by excellent electron-releasing groups.

The most stabilized electrons, such as those in strong sigma bonds or aromatic pi bonds, make poor electron sources. For example, carbon–carbon single bonds are generally poor sources. There are exceptions, of course; an aromatic bearing an excellent electron-releasing group is an improved but still mediocre source.

Although an electron flow path is shown for each source in this chapter, it is included only to point out the electron pair from which the electron flow must start. The detailed discussion of the electron flow paths is presented in Chapter 7.

5.2 NONBONDING ELECTRONS

5.2.1 Lone Pairs as Brønsted Bases

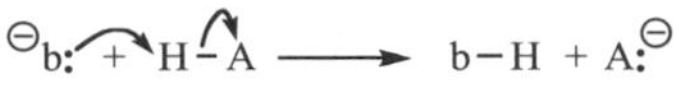

A nonbonding electron pair can serve as a Lewis base and attack an electron-deficient carbon or serve as a Brønsted base and pull off a proton. The decision of whether a lone pair serves as a base or a nucleophile will not concern us in this section, only how to recognize the generic class. As a Brønsted base, the electron flow starts from the lone pair of the base. Section 3.2 showed how to rank Brønsted bases. A synthetically useful proton transfer has a K_{eq} greater than 10^{-10}. See Section 3.6 for the calculation of the proton transfer K_{eq}. See Section 3.2.4 for common examples.

5.2.2 Heteroatom Lone Pairs as Nucleophiles

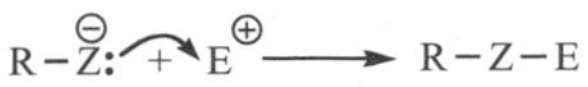

The electron flow starts from the lone pair. Nonbonding electron pairs are good electron sources, especially the lone pairs of anions. Nucleophilicity is a gauge of the rate of a Lewis base's attack on an electron-deficient carbon. A proton is much harder than an electron-deficient carbon, so the relative hardness of the nonbonding electron pair is important in any discussion of basicity versus nucleophilicity (see Section 9.5). To rank the nucleophilicity of a nonbonding electron pair, we must consider the major factors of charge, size, and solvation and the minor factors of electronegativity and strength of the bond that is formed. A detailed discussion of nucleophilicity is in Section 4.2.2, so just the major points are reviewed here.

- A charged anion is a better nucleophile than its neutral counterpart. Because softness is important, nucleophilicity does not parallel basicity exactly. For lone pairs of the *same element*, the more basic the lone pair is, the more nucleophilic it is.
- Steric hindrance decreases nucleophilicity. As the nucleophile becomes larger, its bulk tends to get in the way of its acting as a nucleophile, especially when the electrophile is also large.
- Electron availability decreases as the lone pair electrons become held more tightly; therefore anions on more electronegative atoms are less nucleophilic.
- Softness (less electronegative, more polarizable) is important because the more polarizable the attacking lone pair is, the more easily it can be distorted to form the new bond with the electrophile. Softness also affects how tightly solvated the nucleophile is, since solvation is a hard–hard interaction. Small, hard, highly shielded ions must "break out" of the solvent cage to be available as nucleophiles. Soft ions are more nucleophilic in protic solvents because tight solvation greatly decreases a hard ion's nucleophilicity.
- Nucleophilicity also depends on the electrophile. Hard nucleophiles favor binding with hard electrophiles; soft nucleophiles favor binding with soft electrophiles (HSAB principle, Section 2.4). Most nucleophilicity charts show relative rates of nucleophilic attack with methyl iodide, which is a very soft electrophile because the carbon–iodine electronegativity difference is nearly zero. A pK_a chart can be used as a reference for nucleophilicity only if the difference in softness is considered. A partially plus carbon atom is a much softer electrophile than a proton.

In summary, to rank the nucleophilicity of nonbonding electron pairs reacting in protic solvents with soft electrophiles such as R–L, first rank by softness, then by basicity (within the same attacking atom). However, for nonbonding electron pairs reacting with harder electrophiles such as a proton or a carbonyl, rank by basicity. Very reactive electrophiles like carbocations are not selective and react with the most abundant nucleophile (commonly the solvent).

5.2.3 Easily Oxidizable Metals

$$\text{M:} + \text{R}-\ddot{\underset{\cdot\cdot}{\text{Br}}}: \longrightarrow \text{M}^{2+} + \overset{\ominus}{\text{R:}} + :\ddot{\underset{\cdot\cdot}{\text{Br}}}:^{\ominus}$$

The valence electrons of a reactive metal such as lithium, sodium, magnesium, or zinc can be donated to another species to form an ionic bond. Reactions of oxidizable metals occur by *electron transfer* from the metal surface to the electron sink. This process oxidizes the metal, producing the metal ion, and reduces the electron sink. For example, sodium metal will lose one electron to achieve the noble gas configuration of neon in the sodium cation. In the case of magnesium and zinc that lose two electrons, the electron transfer very likely occurs one electron at a time (Chapter 11). In the *halogen metal exchange reaction* shown, the R–Br bond is broken in the reduction step, producing an ionic or partially ionic organometallic salt.

Metals differ greatly in how easily they are oxidized. The most reactive metals are those with a very negative standard electrode potential. Of the commonly used metals, the alkali metals lithium, sodium, and potassium are highly reactive; the next most reactive is magnesium, then zinc.

5.3 ELECTRON-RICH SIGMA BONDS

We can consider a partially ionic sigma bond as an already partially "broken" bond. An organometallic bond, for example, has electrons that are very available. Conversely, a carbon–carbon covalent bond is completely covalent; its electrons reside in a very stable sigma bonding orbital and are not available. The main exception to this is the C–C bonds of cyclopropane, which are destabilized by ring strain, and thus are more reactive.

5.3.1 Organometallics

The electron flow starts from the polarized sigma bond. Organometallics vary in their ability to act as electron sources. Very ionic organometallics are excellent electron sources, whereas the most covalent organometallics are poor sources. Therefore, checking the electronegativity difference between the metal and carbon will allow us to rank the reactivity of the organometallic when the R group is the same. **A good source will have a large electronegativity difference.** The following is a list of several organometallics ranked by their electronegativity differences (R is the same for all).

Electronegativity difference	**Most reactive**
1.62	R-Na
1.57	R-Li
1.24	R-MgX
0.90	R_2Zn
0.86	R_2Cd
0.65	$R_2Cu^-Li^+$
0.55	R_2Hg
0.22	R_4Pb
	Least reactive

To determine the reactivity of an organometallic on the basis of R when the metal is the same, compare the carbanion stability (Section 3.4.1). As the number of electron-donating alkyl groups on the carbanion increases, the carbanion gets less stable and more reactive. The alkyllithium general reactivity trend is tertiary > secondary > primary > methyl. The less stable the carbanion, the more reactive it is as an electron source. For carbanions that are resonance-stabilized by electron-withdrawing groups, the identity of the metal is of less importance than the stabilization of the carbanion.

Simple alkyl carbanions, like CH_3Li, are sp^3 hybridized at the carbanionic center and commonly exist as the tetramer, $(CH_3Li)_4$. This tetramer exists as a tetrahedron of lithium ions with a carbanion snuggled in between the three lithium atoms of each face.

Example problem

Which of the following, CH_3Li, CH_3MgBr, CH_3CdCl, is the most reactive organometallic? Which is the least?

Answer: The most ionic is the most reactive. The least electronegative metal in this group is lithium, so CH_3Li is the most reactive. The next most reactive is CH_3MgBr; the least reactive and the most covalent organometallic is CH_3CdCl.

5.3.2 Group 1A Metal Hydrides

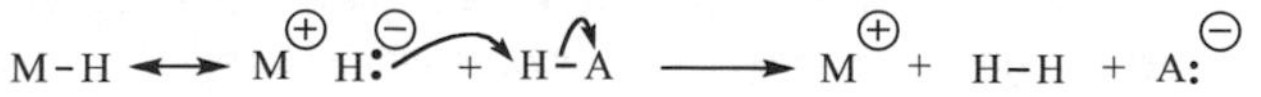

The electron flow starts from the polarized sigma bond. Metal hydrides can serve as good electron sources, but their behavior varies with the electronegativity of the metal. Alkali metal hydrides, NaH and KH, are ionic and function primarily as bases reacting with acidic protons to form hydrogen gas; little reduction occurs.

5.3.3 Complex Metal Hydrides

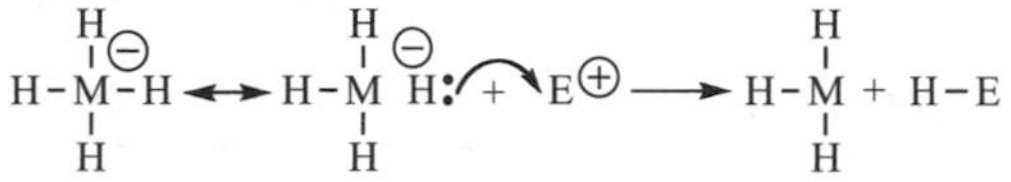

The softer complex metal hydrides, AlH_4^- and BH_4^-, ,donate hydride to reduce the substrate and are better nucleophiles than bases. To summarize, the harder alkali metal hydrides are good bases and poor nucleophiles, whereas the softer complex metal hydrides are good nucleophiles and weaker bases. As with the organometallics in the previous section, the electronegativity difference governs the reactivity of the metal hydrides. The metal hydride will become somewhat less reactive if the metal is made more electronegative by attaching an inductively withdrawing group. As the hydride source becomes less reactive, it also becomes more selective. The complex metal hydride reactivity trend is $\mathbf{AlH_4^- > HAl(OR)_3^- > BH_4^- > H_3BCN^-}$.

5.4 ELECTRON-RICH PI BONDS

5.4.1 Allylic Sources

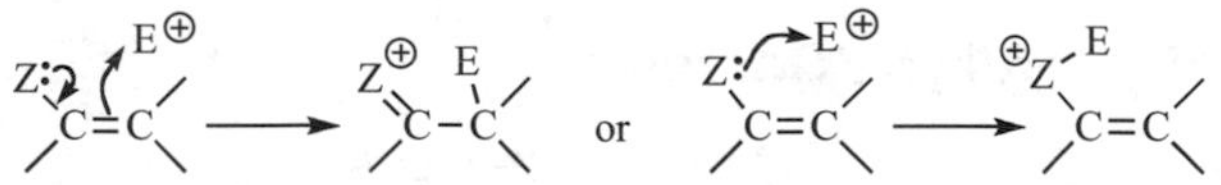

Pi bonds tend to be weaker than sigma bonds and are therefore usually easier to break. We can make a pi bond more reactive by attaching a good lone pair donor. The electron flow can come from either end of the allylic system. When pi bonds are used as electron sources, an attached lone pair electron-releasing group can greatly increase the availability of the pi bond's electrons and stabilize the cationic product usually formed upon an electrophilic attack on the double bond. We have in essence "extended" via delocalization with a double bond the properties of the lone pair donor group (vinylogy). These sources contain three adjacent *p* orbitals, and almost all are much more reactive than a simple alkene.

Resonance structures show us that either the double bond or the heteroatom can serve as an electron source. Because allylic sources can "bite" at two different sites they are called *ambident nucleophiles*. Usually the double bond is the attacking nucleophile; the decision about which end of the allylic source attacks the electrophile will be discussed in Section 9.4.

We will see later that some electron sinks are in equilibrium with a species that can serve as an allylic electron source, Z=C–C–H $\rightleftharpoons$ H–Z–C=C; this equilibrium is called *tautomerization* and is discussed in Chapter 7.3.8 under path combinations.

The rankings of allylic electron sources reflect the electron-releasing group trends already covered in Section 4.2.5 and thus do not present new trends for you to learn. **The better the donor is, the better the allylic electron source is.** The following are several allylic electron sources, ordered by their rank as sources:

Best 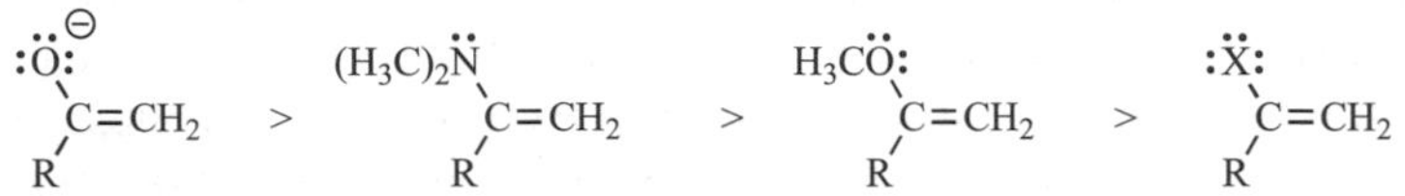 Worst

Enolate Enamine Enol ether Vinyl halide

The last one in the series, the vinyl halide, is actually a poorer electron source than a simple alkene because halogens are poorer donors than alkyl groups.

Example problem

Which of the following, CH_2=CH–O⁻, CH_2=CH–NR_2, CH_2=CH–Cl, is the most reactive allylic source? Which is the least?

Answer: The best donor is the oxygen anion, so the best allylic source of the three is the enolate, CH_2=CH–O⁻; the next best is the enamine, CH_2=CH–NR_2. The least is the vinyl chloride, CH_2=CH–Cl, because chlorine is a very poor donor.

Vinylogous allylic sources, :Z–C=C–C=C or :Z–C=C–C=O, commonly react at the center atom and therefore will be treated as a subset of this source. A few examples of these vinylogous allylic electron sources are, respectively, extended enolates, acetoacetates, and malonates.

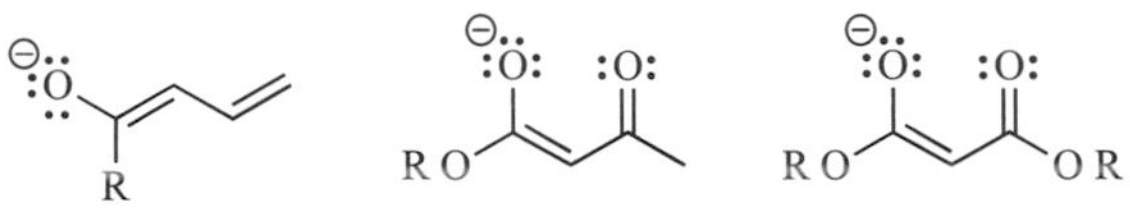

5.4.2 Allylic Alkyne Sources

$\ddot{Z}$–C≡C—

If we add a second pi bond perpendicular to an allylic system we create this rare electron source that should just be treated as a subset of the allylic sources.

5.5 SIMPLE PI BONDS

5.5.1 Alkene Sources

The pi bond is the electron source. For allylic sources, the electron pair donor stabilizes the resultant cation. The simple alkene reactivity trend reflects the stabilization of the resultant carbocation by alkyl substitution and delocalization with other double bonds. The ranking of simple double bonds as electron sources basically is: the more substituted, the more reactive.

Best 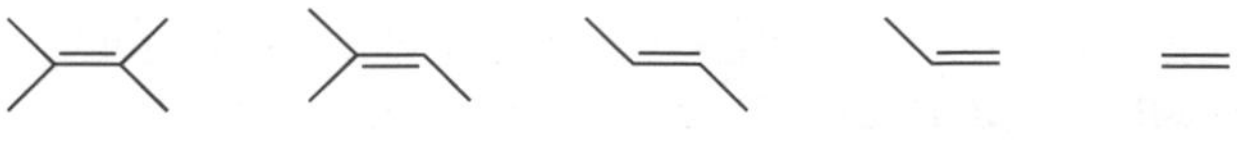 Worst

When a substituted double bond is attacked by an electrophile, the two possible carbocations formed can differ in stability. **The formation of the more stable carbocation will be the lower-energy process and will determine the site of electrophilic attack** (a modern version of Markovnikov's rule). Figure 5.1 shows that if the group on the double bond acts as an electron-releasing group, the carbocation adjacent to and stabilized by the electron-releasing group will be formed. Electron-releasing groups stabilize cations, and electron-withdrawing groups destabilize cations. Avoid forming cations adjacent to electron-withdrawing groups.

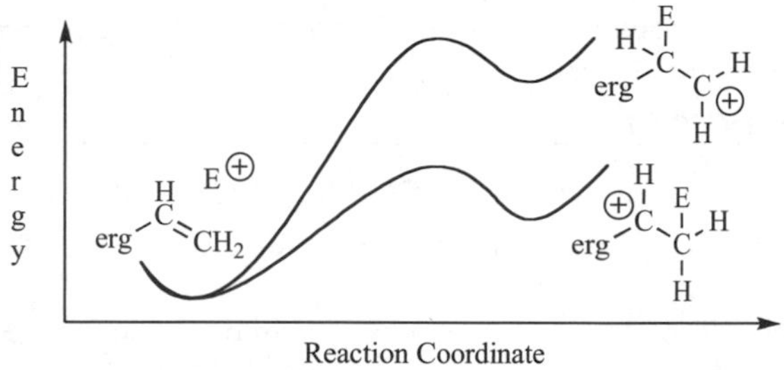
Energy
Reaction Coordinate

Figure 5.1 Energy diagrams for electrophilic addition to substituted alkenes. The cation adjacent to the electron-releasing group is stabilized.

5.5.2 Diene Sources

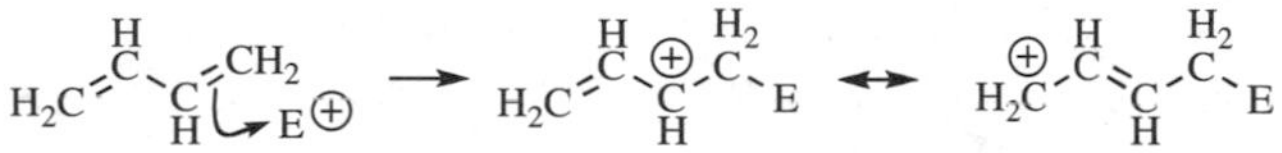

Dienes are conjugated systems of two pi bonds. Above, the simplest diene, 1,3-butadiene, adds the electrophile to the end to produce a resonance stabilized allylic carbocation. As with alkenes, the more substituted the diene is, the more reactive.

To make the choice on where to add an electrophile to a diene pi system, draw all possible carbocations and then pick the more stable one according to the trends you learned in Section 4.2.4. This is one of many times that we will use the trends to make a decision about the route a reaction will take.

Example problem

Predict where an electrophile would attack the following compound.

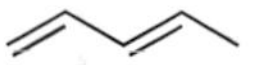

Answer: Draw out all four possible carbocations that would be formed upon electrophilic attack and rank their stability.

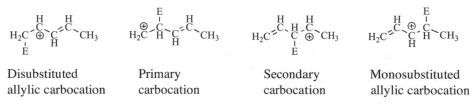

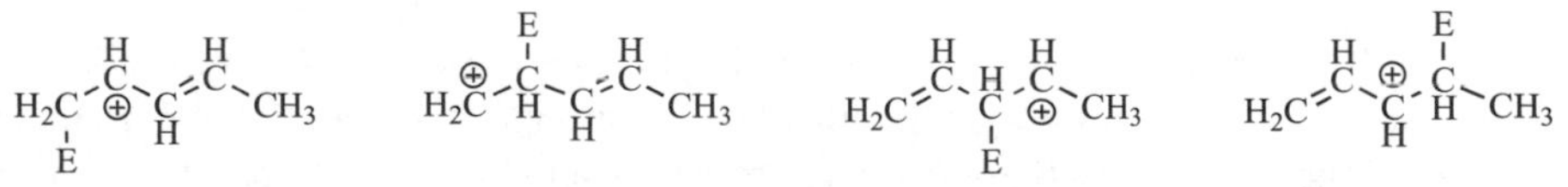

Disubstituted allylic carbocation	Primary carbocation	Secondary carbocation	Monosubstituted allylic carbocation

The ranking of these carbocations from most to least stable is disubstituted allylic > monosubstituted allylic > secondary > primary. Since the electrophile will add to form the most stable cation, the electrophile will add on the end to form the disubstituted allylic carbocation.

5.5.3 Alkyne Sources

The two perpendicular pi bonds of a triple bond can be considered separately, and thus this source becomes merely a subset of the double-bond sources. Alkynes are slightly poorer electron sources than alkenes because the vinyl cation formed upon electrophile attack is relatively unstable. Acetylide anions, $RC{\equiv}C^- M^+$, are treated as organometallic reagents (Section 5.3.1) and not as alkyne sources.

Example problem

Which of the following, $CH_3C{\equiv}CCH_3$, $RC{\equiv}C{-}NR_2$, $HC{\equiv}CH$, is the most reactive triple-bond electron source? Which is the worst?

Answer: The best electron-releasing group attached to the triple bond is the NR_2, so the best triple-bond electron source of the three is $RC{\equiv}C{-}NR_2$; the next best is $CH_3C{\equiv}CCH_3$ because methyl is a weak donor. The worst is $HC{\equiv}CH$, because the cation formed upon electrophilic attack lacks any form of stabilization.

5.5.4 Allene (Cumulene) Sources

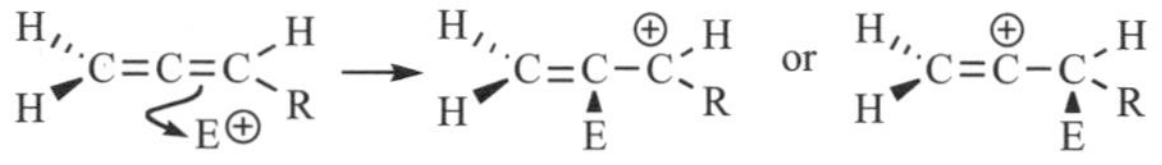

The least common of the simple pi bond sources are allenes, $R_2C{=}C{=}CR_2$, whose two perpendicular double bonds must again be considered separately. Allene reactions have a bit of a strange twist to them, literally. There are two possible sites for the electrophile to attack on the allene above. It would seem to be no contest between a resonance-stabilized allylic carbocation and the less stable vinyl carbocation. But a look at the orbital arrangement below for the reaction reveals that all is not right with the allylic system; in fact, it is twisted 90° out of alignment, so no resonance stabilization can occur at the transition state for addition.

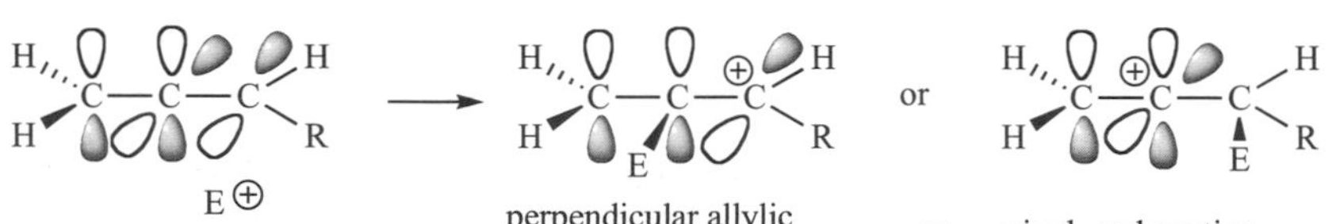

perpendicular allylic carbocation vs. vinyl carbocation

The resonance stabilization of an allylic carbocation is estimated to be 15 kcal/mol. The loss of this resonance stabilization means that the vinyl cation is of comparable stability. When the R group is hydrogen or alkyl, the electrophile adds to the end as shown below. If the R group on the allene is capable of additional resonance stabilization of the carbocation, then electrophile attack shifts to the middle carbon.

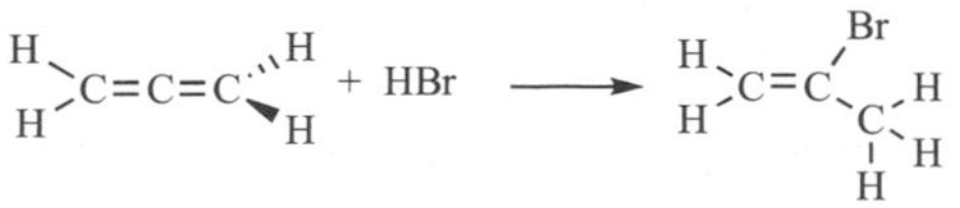

5.6 AROMATIC RINGS

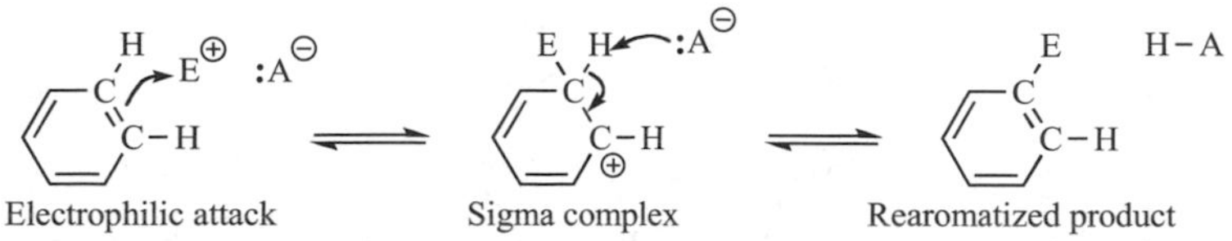
Electrophilic attack
Sigma complex
Rearomatized product

The pi cloud of the aromatic ring reacts with an electrophile to produce a cationic sigma complex that then loses an electrophile, commonly H^+, to restore the aromatic stabilization of the ring. Almost all aromatic rings are poor electron sources because the electrons in an aromatic ring are very stabilized (Section 1.9.3) and are therefore not very available. However, an aromatic ring can be made more reactive by the attachment of one or more electron-releasing groups.

To summarize Section 4.6, any group that has a resonance interaction with an aromatic ring has the greatest effect at the *ortho* and *para* positions. An electron-releasing group on the ring increases the electron availability at the *ortho* and *para* positions, increases reactivity, and stabilizes the carbocation formed upon attack at those positions. The less crowded *para* position is favored when the electron-releasing group is large or when the electrophile is large.

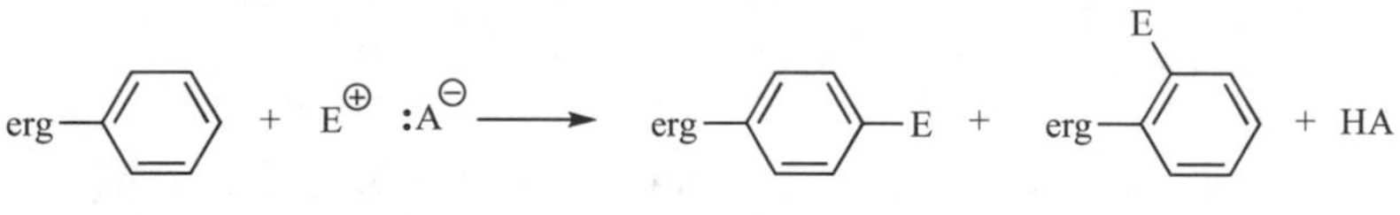

An electron-withdrawing group attached to the ring decreases the electron availability at the *ortho* and *para* positions (making the *meta* position the least electron poor), decreases reactivity, and destabilizes the carbocation formed upon attack at the *ortho* and *para* positions, so *meta* attack is preferred.

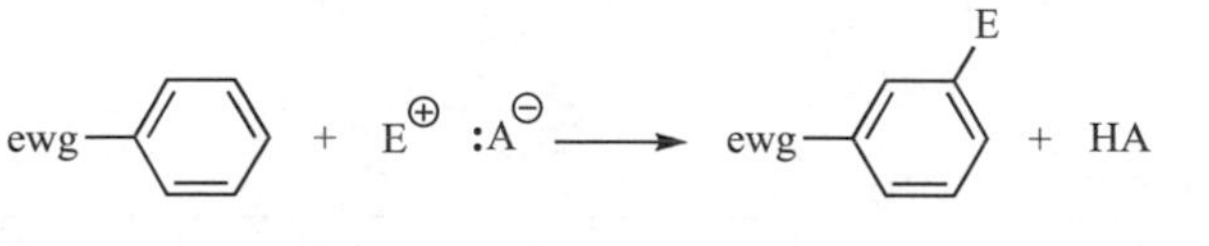

Therefore to rank the reactivity of a benzene ring to electrophilic attack, the better electron-releasing groups (Section 4.2.5) make the ring more reactive. The better electron-withdrawing groups (Section 3.5.1) make the ring less reactive. Combining these two trends back to back will give an overall reactivity trend for an aromatic ring substituted with a group, G, shown below.

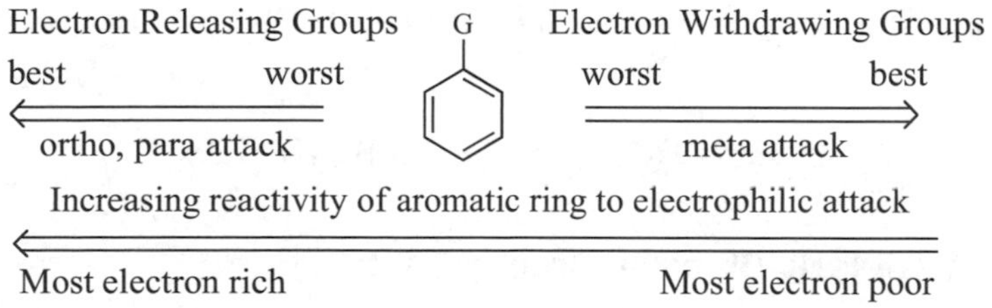
Electron Releasing Groups
G
Electron Withdrawing Groups
best
worst
worst
best
ortho, para attack
meta attack
Increasing reactivity of aromatic ring to electrophilic attack
Most electron rich
Most electron poor

Other Aromatics as Electron Sources

The stability of the sigma complex is the guiding principle when determining the site of electrophilic attack on any aromatic ring. The more approximately equal energy resonance structures the carbocation has, generally the more stable it is. Resonance forms that place the positive charge next to an ewg are poor and should not be counted.

Pyridine undergoes electrophilic attack *meta* to the nitrogen.

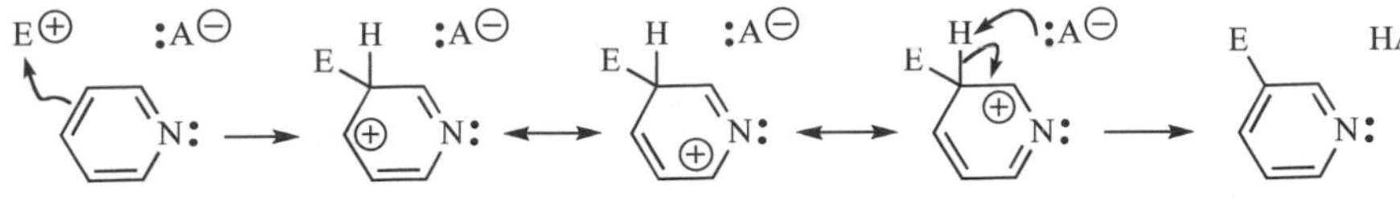

Sigma complex **resonance forms that have a heteroatom with an incomplete octet are especially bad**; no attack on the *ortho* or *para* positions of pyridine occurs.

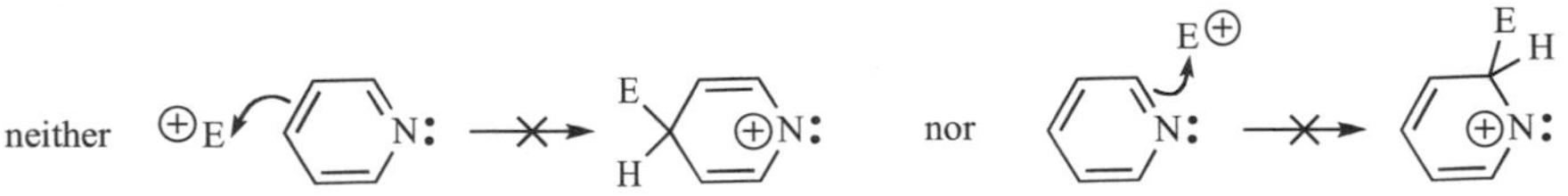

However, the attack of the electrophile on the nitrogen lone pair of pyridine is a competing reaction. Note that the nitrogen lone pair of pyridine is perpendicular to the pi system of the aromatic ring and therefore cannot be used to stabilize any sigma complex.

Five-membered ring heteroaromatics undergo electrophilic aromatic substitution at the position *ortho* to the heteroatom. The most stabilized sigma complex dominates. Furan example:

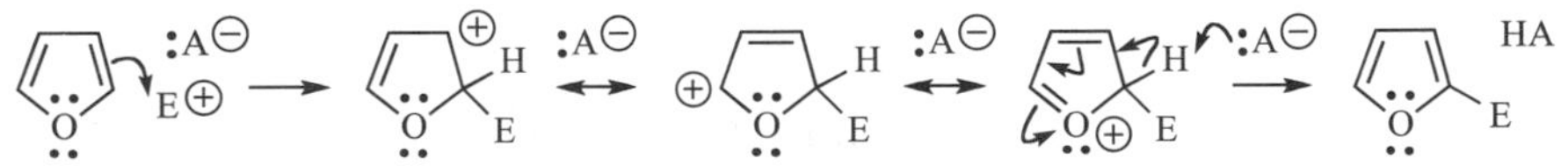

Naphthalene undergoes electrophilic aromatic substitution at the position next to the second ring. The sigma complex leading to the obtained product has more resonance structures, indicating a more delocalized charge than that of the alternative.

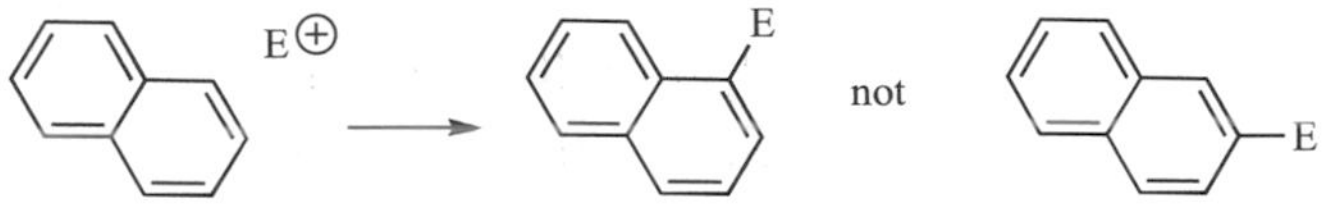

Example problem

Which of the following, $PhCH_3$, $PhNR_2$, PhCOR, is the most reactive aromatic ring toward electrophilic attack? Which is the least?

Answer: The best donor is the NR_2, so the most reactive ring of the three is $PhNR_2$; the next best donor is methyl, so the next most reactive ring is $PhCH_3$. The least reactive ring is PhCOR, because ketones are electron-withdrawing groups and deactivate the ring.

5.7 SUMMARY OF GENERIC ELECTRON SOURCES

5.7.1 Generic Electron Source Table

Table 5.1 lists the 12 generic electron sources from which an electron flow in a reaction may start. Conspicuous in their absence are covalent sigma bonds. **Regular sigma bonds are just too stabilized to act as electron sources** (only when destabilized in a highly strained three-membered ring may a carbon–carbon single bond behave as a source). Usually, the only single bonds that serve as sources are highly ionic ones, single bonds to metals.

Table 5.1 The Twelve Generic Electron Sources

Generic Class	Symbol	Examples
Nonbonding electrons (5.2)		
Lone pairs on heteroatoms (Nu / base dual behavior)	Z:	I^-, HO^-, H_2O, H_3N, *t*-BuO^- CH_3COO^-
Easily oxidizable metals	M:	Na, Li, Mg, Zn metals
Electron-rich sigma bonds (5.3)		
Organometallics	R–M	RMgBr, RLi, $R_2Cu^-Li^+$
Metal hydrides	MH_4^- (Nu) MH (Bases)	$LiAlH_4$, $NaBH_4$, NaH, KH
Electron-rich pi bonds (5.4)		
Allylic sources	C=C-Z:	Enols: C=C–OH Enolates: C=C–O^- Enamines: C=C–NR_2
Allylic alkyne sources	C≡C-Z:	Et–C≡C–NEt_2
Simple pi bonds (5.5)		
Alkenes	C=C	$H_2C=CH_2$
Dienes	C=C-C=C	$H_2C=CH-HC=CH_2$
Alkynes	C≡C	HC≡CH
Allenes	C=C=C	$H_2C=C=CH_2$
Aromatic rings (5.6)		
Aromatics	ArH	

5.7.2 Generic Electron Source Classification Flowchart

As noted in the beginning of this chapter, the reason for generic classification is so that you know how a compound is going to behave. Generic classification will allow you to predict what will happen when you encounter a compound you have not seen before. This powerful classification tool is useful only if it is actually employed to organize all new incoming organic reactions, and not relegated to just another thing to know.

The remaining task is to come up with a method for classification of a given structure into its most appropriate generic class. Often several functional groups are present in a reactant, and several reactants may be present in a reaction. One functional group may possibly fall into more than one class, so we need to pick the most applicable. We need to recognize the generic sources in a way such that we find the most reactive first. A flowchart of the classification process for electron sources is shown in Figure 5.2.

1. Oxidizable metals, organometallics, and metal hydrides are usually very reactive.

2. Next look for lone pair donors and the electron-rich pi bonds.

3. Then search for the simple pi bonds.

4. The least reactive, aromatics, are sorted out last.

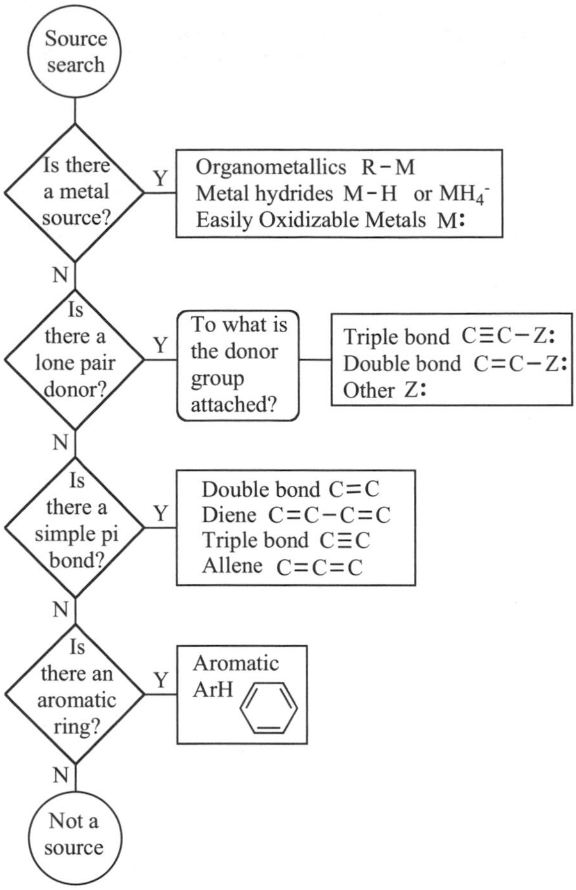
Source search
Is there a metal source?
Y
Organometallics R−M
Metal hydrides M−H or MH4-
Easily Oxidizable Metals M:
N
Is there a lone pair donor?
Y
To what is the donor group attached?
Triple bond C≡C−Z:
Double bond C=C−Z:
Other Z:
N
Is there a simple pi bond?
Y
Double bond C=C
Diene C=C−C=C
Triple bond C≡C
Allene C=C=C
N
Is there an aromatic ring?
Y
Aromatic
ArH
N
Not a source

Figure 5.2 Classification flowchart for electron sources

Example problem

To which generic classes do these belong: Ph–NH_2, CH_2=CH–Li, CH_2=CH–NH_2?

Answer: We notice that in Ph–NH_2 there is a nitrogen lone pair and an aromatic ring. Following our classification scheme, we drop it into the lone pair class, Z, rather than into the less reactive aromatic class. Likewise for CH_2=CH–Li , we classify it as an organometallic, R–M, rather than the much less reactive alkene. For CH_2=CH–NH_2, the nitrogen lone pair is conjugated with the double bond, giving an allylic source, C=C–Z. Since it is an ambident nucleophile, a decision is necessary as to which end of the allylic system to use as a source (Section 9.4). On each of these examples, the generic classification process gives us insight into how each compound will react. Understanding structure allows us to understand reactivity.

5.7.3 Common Sources Reactivity Summary

A much smaller set of sources is responsible for most reactions and is summarized in Table 5.2, which gathers examples, reactivity trends, common reactions, necessary decisions, and the related generic groups. Some have a decision discussed in Chapter 9.

Table 5.2 Summary Reactivity Table for Common Electron Sources

Symbol & Name	Common Examples in Order of Decreasing Reactivity	Reactivity Trend Principle	Common Reactions Decision?	Related Generic Groups
Z: Heteroatom lone pairs as nucleophiles	RS^- I^- CN^- RO^-	Softer anions are more nucleophilic. If same atom, more basic, more Nu	Substitutions & Additions. Nucleophile vs. base decision	See allylic sources
Z: Heteroatom lone pairs as Bases	R_2N^- *t*-BuO^- R_3N CH_3COO^-	The higher pK_{abH} is stronger	Proton transfer $K_{eq} > 10^{-10}$	NaH and KH as MH bases
R–M Organo-metallics	CH_3Li CH_3MgI $(CH_3)_2Cu^-\,Li^+$	The more ionic RM bond is more reactive	Substitutions & Additions. Deprotonates acidic H's	See enolates (allylic sources)
MH_4^- Complex metal hydrides	$LiAlH_4$ $LiAlH(OR)_3$ $NaBH_4$ $NaBH_3CN$	The more ionic MH bond is more reactive	Substitutions & Additions. Deprotonates acidic Hs on heteroatoms	NaH and KH as MH bases
C=C–Z: Allylic sources	Enolates $C{=}C{-}O^-$ Enamines $C{=}C{-}NR_2$ Enol ethers C=C–OR	The better donor on the pi bond is more reactive	Substitutions & Additions. Carbon vs. heteroatom decision	Extended enolates. Allylic alkyne sources
C=C Simple pi bonds	$R_2C{=}CR_2$ RHC=CHR $H_2C{=}CH_2$	The more stable the resultant carbocation, the more reactive the pi system.	Electrophilic Additions. Markovnikov's rule used for regiochemistry	Other pi systems: C=C–C=C C≡C C=C=C
ArH Aromatic rings	PhOR PhH PhCl PhCOR	A donor on the ring makes it more reactive; ewg makes ring less reactive.	Electrophilic aromatic substitution. Regiochemistry of addition	Hetero-aromatics, condensed aromatics

ADDITIONAL EXERCISES

5.1 Give the generic class of each of the following electron sources.

$LiAlH_4$	CH_3Li	NH_3	$CH_3OCH{=}CH_2$	Et_2NLi
Mg	CH_3COO^-	EtMgBr	$(CH_3)_2NCH{=}CH_2$	H_2O

5.2 Draw Lewis structures for the species in problem 5.1 and indicate on the structures which electron pair(s) would be the start of the electron flow in a reaction.

5.3 Give the generic class of each of the following electron sources.

NaH	$CH_3C{\equiv}CH$	$(CH_3)_2C{=}CH_2$	*t*-BuO$^-$	$PhCH_3$
$PhNH_2$	$^-C{\equiv}N$	Cl^-	$^-CH_2COOCH_3$	CH_3OH

5.4 Draw Lewis structures for the species in problem 5.3 and indicate on the structures which electron pair(s) would be the start of the electron flow in a reaction.

5.5 Give the generic class of each of the following electron sources.

$NaBH_4$	$^-O\text{-}CH{=}CH_2$	HO^-	Zn	NaOEt
$^-SC{\equiv}N$	$H_2C{=}C{=}CH_2$	PhSH	$H_2C{=}CHCH{=}CH_2$	F^-

5.6 Draw Lewis structures for the species in problem 5.5 and indicate on the structures which electron pair(s) would be the start of the electron flow in a reaction.

5.7 Circle the one atom on each molecule that is best for electrophilic attack.

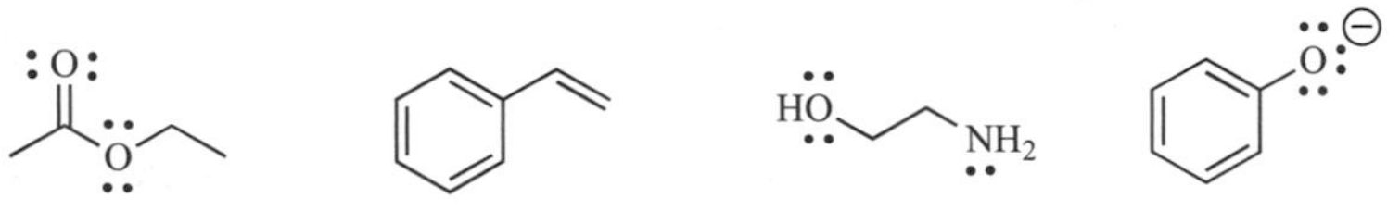

5.8 Rank all species, beginning with the numeral 1 to designate:

(a) The best nucleophile to react with CH_3I in CH_3OH

$PhSe^-$ PhO^- PhS^- CH_3COO^- CH_3OH

(b) The most reactive organometallic

$CH_3\text{-}K$ $CH_3\text{-}MgBr$ $(CH_3)_2Hg$ $(CH_3)_2Cd$ CH_3Li

(c) The best allylic electron source

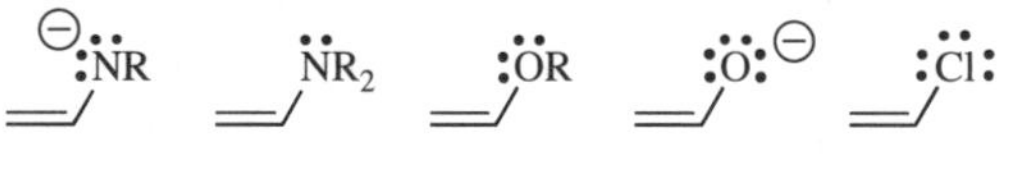

(d) The most reactive aromatic ring toward electrophilic attack

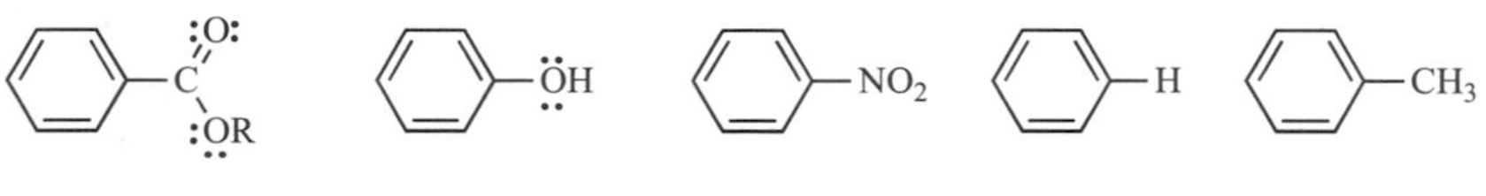

5.9 Rank which of the following pi bonds are most reactive to electrophilic attack.

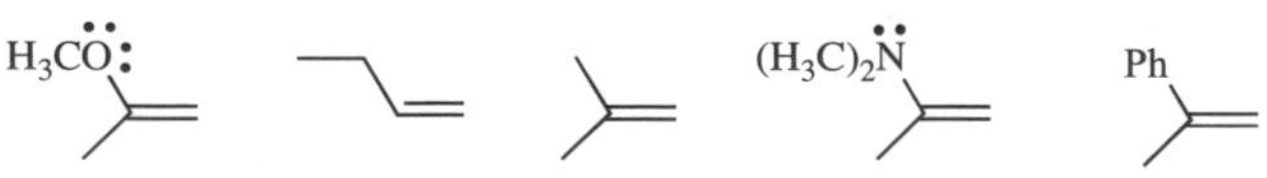

5.10 Use Markovnikov's rule to predict the site of electrophilic attack on each of the pi bonds in the compounds in problem 5.9.

5.11 At which position(s) on the aromatic ring would you expect electrophilic attack to occur for the following compounds?

$PhCF_3$ PhOH $PhNO_2$ PhCOOH PhCl

5.12 Use the nucleophilicity trend to decide which of the following two alternatives is the faster process in methanol solvent.

$CH_3CH_2CH_2Br + CH_3O^-$ displacing bromide to form $CH_3CH_2CH_2OCH_3$ or
$CH_3CH_2CH_2Br + CH_3S^-$ displacing bromide to form $CH_3CH_2CH_2SCH_3$

5.13 When an acyl chloride is added to aqueous ammonia two products are possible. Use the nucleophilicity trend to decide which of the following two alternatives is the faster process.

$CH_3COCl + H_2O$ displacing chloride to form CH_3COOH or
$CH_3COCl + NH_3$ displacing chloride to form CH_3CONH_2

5.14 Rank all species, beginning with the numeral 1 to designate:

(a) The strongest base

NH_2^- PhO^- CH_3COO^- CH_3O^-

(b) The most reactive complex metal hydride

$NaBH_3CN$ $LiAlH_4$ $NaBH_4$

5.15 Circle the one atom on each molecule that is best for electrophilic attack.

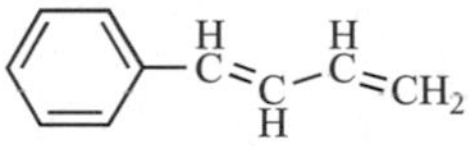
H
H
C
C
C
CH2
H
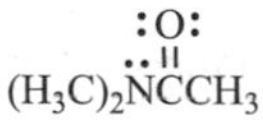
:O:
(H3C)2NCCH3
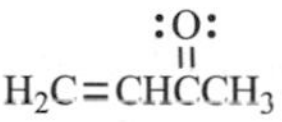
:O:
H2C=CHCCH3

$(H_3C)_2C{=}CH_2$

5.16 Decide whether the following aromatic rings would be attacked *ortho*/*para* or *meta*.

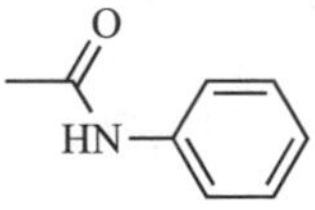
O
HN
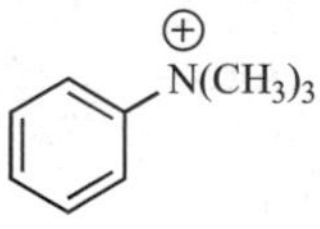
N(CH3)3
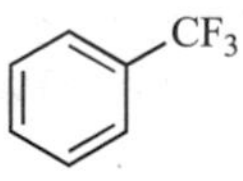
CF3
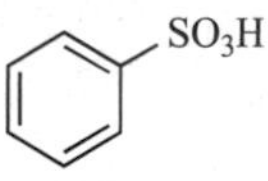
SO3H
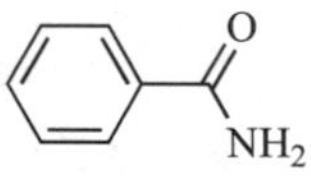
O
NH2

6

CLASSIFICATION OF ELECTRON SINKS

6.1 GENERALIZED RANKING OF ELECTRON SINKS

In this chapter we will study the 18 generic classes of electron sinks in which an electron flow ends. Because most reactions have a hard–hard and a soft–soft component, again it is important to notice both how much positive charge the sink bears and how soft it is. **The larger the partial plus is, the better the electron sink can attract a negatively charged nucleophile.**

The three general classes of electron sinks are an empty orbital, a weak single bond to a leaving group, and a polarized multiple bond. Excellent electron sinks are electron-deficient cationic species having an empty orbital, such as protons and carbocations, that can easily form a new bond with an electron source. Similarly, strong Lewis acids like $AlCl_3$ or BF_3 are also good electron sinks. Medium electron sinks are groups that accept

Electron Flow In Organic Chemistry: A Decision-Based Guide To Organic Mechanisms, Second Edition.
By Paul H. Scudder

the electron flow from the electron source by breaking a weak bond and forming a stable species. The bond that is broken is either a pi bond of a polarized multiple bond or a sigma bond to a leaving group. All of these electron sinks have a full, partial, or inducible plus charge on the atom that gets attacked by the electron source, the nucleophile. Here are a few examples listed with their bond polarization:

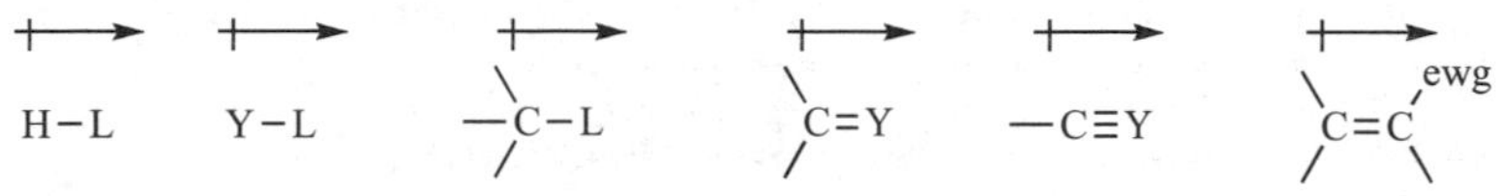

The poorer electron sinks would be any of the above species with poorer leaving groups, less electronegative atoms, or poorer electron-withdrawing groups.

Although an electron flow path is shown for each sink in this chapter, it is included only to point out the atom that would be attacked by the electron source. A detailed discussion of the electron flow paths is provided in Chapter 7.

6.2 ELECTRON-DEFICIENT SPECIES

6.2.1 Carbocations

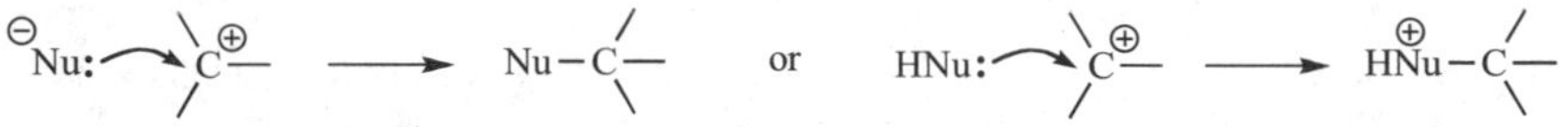

The best electron sinks, occurring almost exclusively in acidic media, are reactive cations such as carbocations (Section 4.2.4). Most are such good electron sinks that they can react with even very poor electron sources like aromatic rings. The most stable carbocations are the least reactive. Highly stabilized carbocations like $^+C(NH_2)_3$ are so stable that they can exist in basic media and make very poor electrophiles. The following are some of the more common reactive carbocations.

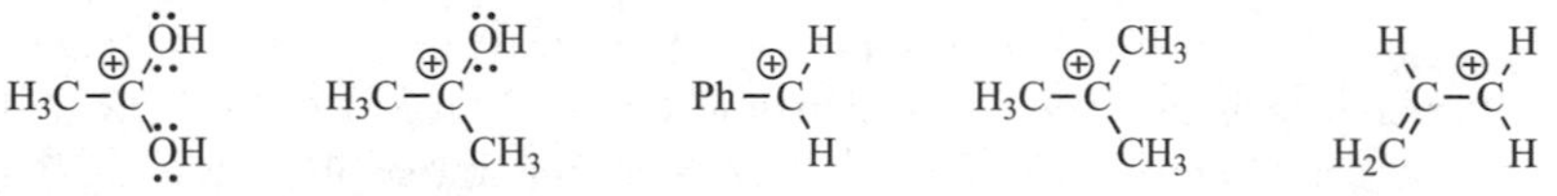

Protonated carboxyl — Protonated ketone — Benzyl cation — *t*-Butyl cation — Allylic cation

6.2.2 Inorganic Lewis Acids

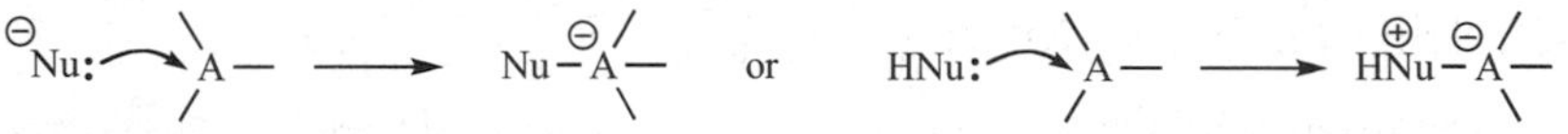

Electron-deficient Lewis acids like $AlCl_3$ or BF_3 behave much like carbocations. Like the $^+CR_3$ carbocation, BR_3 has an empty $2p$ orbital, exactly the same orbital configuration (*isoelectronic*), and will tend to react in a similar way. The strength of a Lewis acid varies with substituents; attached resonance donors reduce the Lewis acidity. A very crude ranking of Lewis acid strengths is $BX_3 > AlX_3 > FeX_3 > SbX_5 > SnX_4 > ZnX_2 > HgX_2$. Another strong Lewis acid is $TiCl_4$.

6.2.3 Metal Ions M^{2+}

Although not really electron deficient, metal ions may bear a high positive charge and can act as electrophilic catalysts because they can complex with many electron sinks and make them more positive, and thus more prone to nucleophilic attack. The extent of catalysis by the cationic counterion of an anionic nucleophile is sometimes underestimated. It is much more favorable to form an ion pair in most organic solvents than an isolated charge. Some metal ions that show electrophilic catalysis are Ag^+, Hg^{2+}, Li^+, and Mg^{2+} ions. Many biochemical reactions achieve catalysis by complexing a divalent metal ion such as Mg^{2+} or Zn^{2+} to activate the substrate.

6.2.4 Metal Ions as Oxidants Mn^{7+} and Cr^{6+}

Transition metal ions such as MnO_4^- and CrO_4^{2-} can serve as electron acceptors in oxidation reactions, for they can have many available oxidation states. Metal ions as electron sinks will be covered in more detail in Section 8.10

6.3 WEAK SINGLE BONDS

6.3.1 Acids

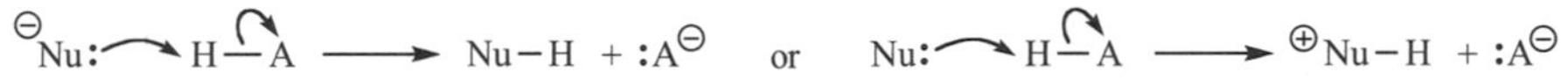

All Brønsted acids belong to this class. Two of four common reaction charge types are shown above (see Fig. 3.1). The more stable the conjugate base is, the stronger the acid is (see Section 3.2). Strong acids often will protonate the solvent; therefore the strongest acid that can exist in a solvent is the protonated solvent. See Section 3.6 for the calculation of the proton transfer K_{eq}, which should be greater than 10^{-10} to be useful.

6.3.2 Leaving Groups on Heteroatoms

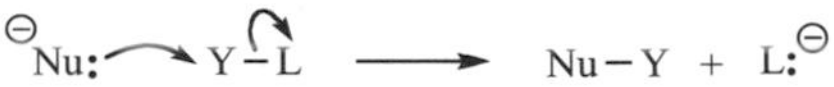

Bonds between heteroatoms and good leaving groups are generally weak. Lone pair–lone pair repulsion is responsible for weak bonds in RO–OR and RS–SR. These weak bonds are broken easily by an electron source that attacks Y and displaces the leaving group. The leaving group rarely departs first since an incomplete octet on heteroatom Y is very unstable. **Do not make any species that has an oxygen or nitrogen atom with an incomplete octet.** These reactions are often exothermic since the new bond formed is usually stronger than the weak Y–L bond broken. In unsymmetrical compounds like I–Cl, the δ+ end of the molecule gets attacked by the nucleophile (in this case iodine because Cl is more electronegative). A subset of the Y–L class that will be discussed in Sections 7.2.4 and 8.12 is the :Nu–L class in which Y bears a nucleophilic lone pair. Examples are halogens like Br_2 and Cl_2

6.3.3 Leaving Group Bound to an *sp*³ Hybridized Carbon

The sp^3-bound leaving groups can be directly displaced by the nucleophile as shown, or the leaving group can ionize off first to form a carbocation. As the leaving group gets better, the following equilibrium begins to produce more of the carbocation, an excellent electron sink. The weaker electron sources must wait until the carbocation is formed before the reaction can occur. As we saw in Chapter 4, the extent of this equilibrium depends heavily on the stability of the carbocation produced, the quality of the leaving group, in addition to solvent polarity. Remember that vinylic, C=C–L, sp^2-bound leaving groups behave much differently from sp^3-bound leaving groups.

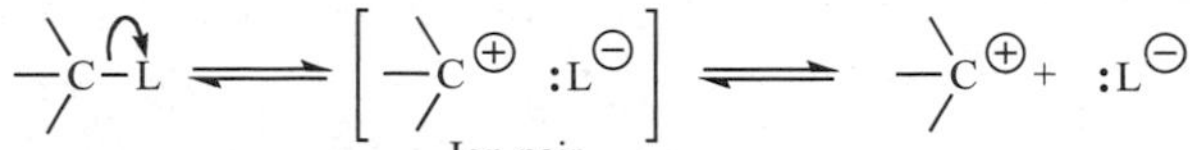
Ion pair

Any electron sink that contains a leaving group will become a much better sink as the leaving group improves. One common way to improve the quality of the leaving group (Section 4.2.3) is with acid catalysis. Acid catalysis is the complexation of a proton or strong Lewis acid (for example, BF_3 or $AlCl_3$) with an electron sink, which increases the electron sink's ability to accept electrons.

Poor leaving group
Good leaving group

In the previous acid catalysis example, the pK_{aHL} of the leaving group improves over 17 pK_a units (hydroxide loss at a pK_{aHL} of 15.7 compared to water loss at a pK_{aHL} of −1.7). In fact, no reaction at all would occur without catalysis because hydroxide is a poor leaving group. **With few exceptions, no reaction occurs when the electron sink bears a poor leaving group (pK_{aHL} greater than 12).** Because diethyl ether, $CH_3CH_2OCH_2CH_3$, is so unreactive, it is a common solvent for reactions involving very reactive nucleophiles. Strained ring ethers (three- and four-membered) are exceptions. Ring strain raises the energy of the reactant relative to a product in which the ring is broken, and the strain is relieved. The five-membered ring ether (tetrahydrofuran, THF) is not strained and is unreactive.

Reactive
Unreactive

The best sinks of the sp^3-bound leaving group class are allylic leaving groups, $CH_2=CH-CH_2-L$, and benzylic leaving groups, $Ph-CH_2-L$, because both are especially activated toward direct displacement (see Section 7.1.6) and ionization (see Section 7.1.2). Some additional examples of sp^3-bound leaving group sinks are:

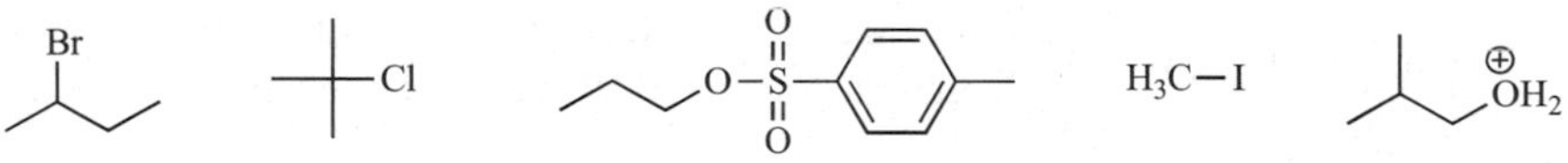

Compounds with more than one leaving group are just a subset of this class. Ketals, $R_2C(OR)_2$, and orthoesters, $RC(OR)_3$, are common examples.

Example problem

Which of the following, CH_3OCH_3, CH_3I, CH_3Cl, is the most reactive toward nucleophilic attack? Which is the least?

Answer: The most reactive bears the best leaving group. The most reactive is CH_3I, with a pK_{aHL} of −10. CH_3Cl is next, with a pK_{aHL} of −7. The least reactive is the ether, CH_3OCH_3, with a pK_{aHL} of 15.5.

6.4 POLARIZED MULTIPLE BONDS WITHOUT LEAVING GROUPS

6.4.1 Heteroatom–Carbon Double Bonds

The electronegativity of Y polarizes the double bond to put a partial plus charge on carbon, making it susceptible to nucleophilic attack. The electronegativity of Y stabilizes the anionic product of the nucleophilic attack. The atom Y is most commonly oxygen, occasionally nitrogen. Alkyl substitution decreases reactivity by weak electron donation into the carbonyl reducing the partial plus and by making the carbonyl less accessible; thus aldehydes are more reactive than ketones. Conjugation stabilizes the carbonyl and thus makes it less reactive; for example, $PhCOCH_3$ is less reactive than CH_3COCH_3.

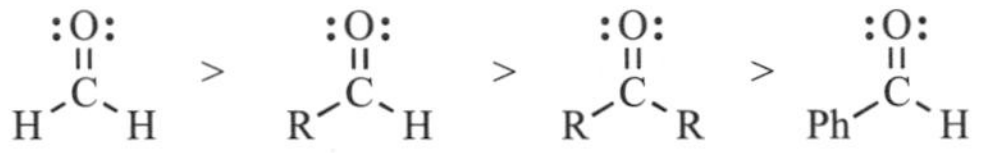

Acid catalysis, often by proton transfer or hydrogen bonding to Y, enhances the ability of this electron sink to accept electrons. Since the lone pair on an sp^2 nitrogen is easier to protonate than a lone pair on an sp^2 oxygen (see pK_a chart, Appendix), acidic catalysis occurs much more readily with nitrogen. The carbocation thus formed is a much better electron sink. Additional examples of this sink are:

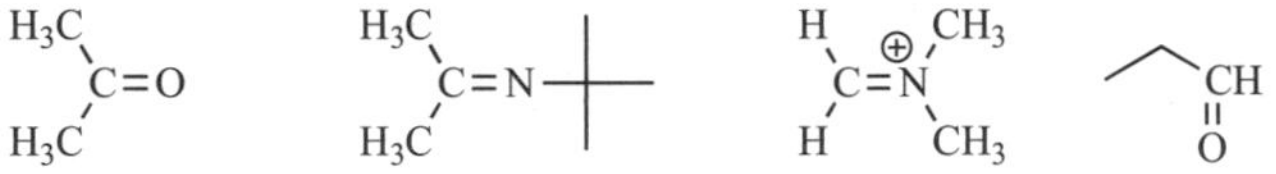

Example problem

Which of the following is the most reactive toward nucleophilic attack? Which is the next most reactive? Which is the least reactive?

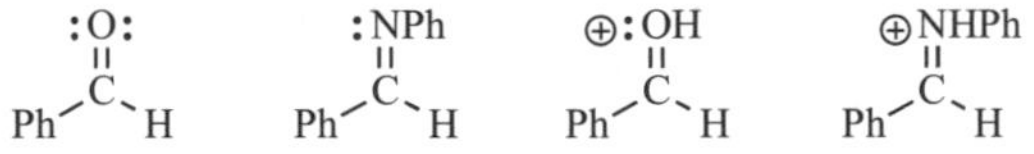

Answer: The most reactive is the protonated aldehyde. The protonated imine is next, for it is slightly more stable. The least reactive is the neutral imine because the anion formed upon nucleophilic attack is more basic than the anion formed upon attack of the neutral aldehyde.

6.4.2 Heteroatom–Carbon Triple Bonds

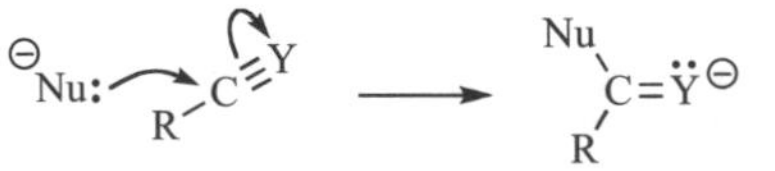

The most common representatives of this electron sink are the nitriles with Y as N. Nitriles are much less reactive than ketones. The electronegativity of Y polarizes the triple bond so that carbon bears a partial plus charge. The two perpendicular double bonds can be treated separately; thus attack by a nucleophile is identical to the doubly bonded electron sink, C=Y. Some examples are Ph–C≡N, and acylium, $CH_3–C≡O^+$.

6.4.3 Heterocumulenes

The heterocumulene derivatives, C=C=Y or Z=C=Y, are a minor variation of the polarized multiple-bond sink and undergo an identical nucleophilic attack at the C=Y bond. The only difference is that the intermediate is a resonance hybrid that protonates to form the most stable product. The product with the stronger bonds predominates.

p.t.

Some examples of heterocumulene sinks are:

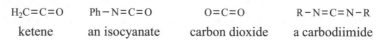

$H_2C{=}C{=}O$	$Ph{-}N{=}C{=}O$	$O{=}C{=}O$	$R{-}N{=}C{=}N{-}R$
ketene	an isocyanate	carbon dioxide	a carbodiimide

6.4.4 Conjugate Acceptors

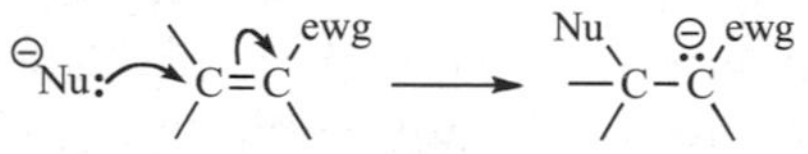

The electron-withdrawing group polarizes the carbon–carbon double bond in a way similar to the C=Y electron sink and makes this sink a subset of that group (with Y equaling C–ewg). The electron-withdrawing group must be present for the double bond to accept electron density from a nucleophile. The attached electron-withdrawing group makes the C–ewg carbon more electronegative. **Trading C–ewg for Y is simply exchanging an electronegative carbon atom for an electronegative heteroatom. The better the electron-withdrawing group is, the better the electron sink will be.** More than one good electron-withdrawing group on a double bond, $C{=}C(ewg)_2$, gives a very reactive electron sink because the resulting anion is stabilized to a much greater extent.

A complication arises in that the nucleophile now has a choice; it can add in a conjugate fashion, or it can add to the electron-withdrawing group directly. Section 9.6, Ambident Electrophiles, discusses this problem in depth. Examples of these sinks are:

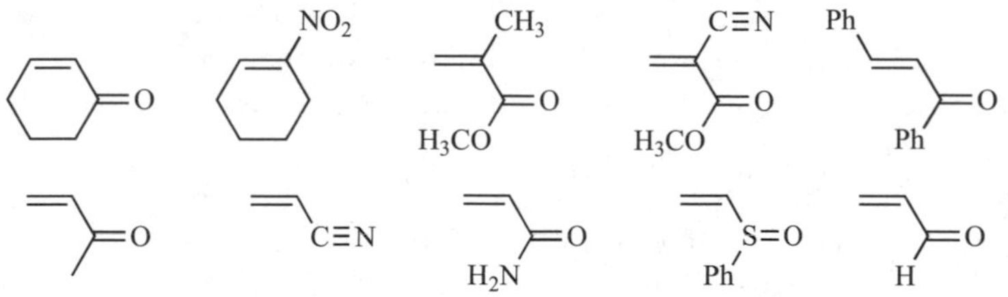

6.4.5 Triply Bonded and Allenic Conjugate Acceptors

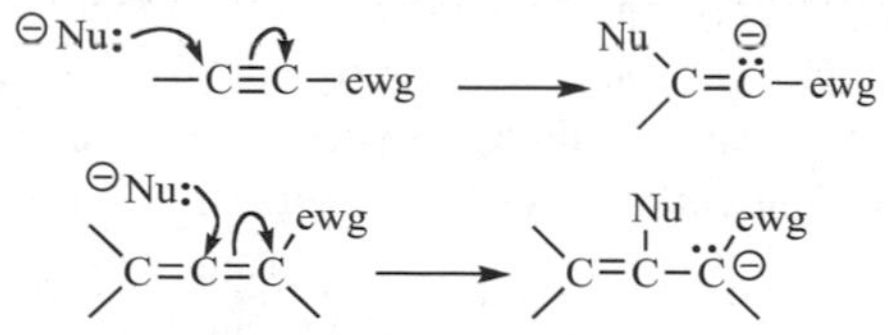

These two electron sinks are the least common of the six that make up the set of multiple-bonded electron sinks with no attached leaving group. They behave very similarly to the conjugate acceptors previously discussed. The electronegative C–ewg group replaces the electronegative Y in accepting the electron flow from the nucleophile.

6.5 POLARIZED MULTIPLE BONDS WITH LEAVING GROUPS

6.5.1 Carboxyl Derivatives

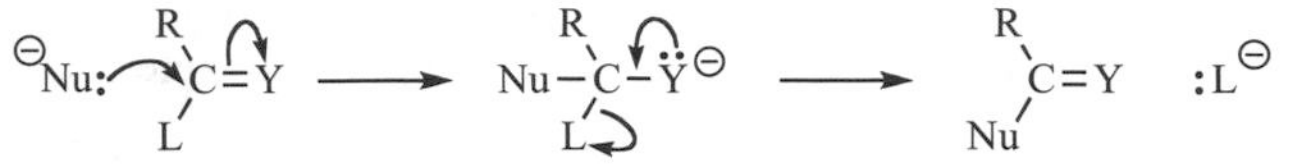

The most common representatives of this group are the carboxylic acid derivatives with Y as oxygen. As we saw in Section 4.5, replacement of the leaving group is a two-step process: First nucleophile addition occurs, then the leaving group is lost. The rate-determining step is almost always the nucleophilic attack; thus the rate depends on the ability of this electron sink to attract and accept electron density from a nucleophile.

If the leaving group is a good electron donor it will delocalize and diminish the partial plus on the carbon of the carbonyl, decreasing its hardness and therefore decreasing its ability to attract a negatively charged nucleophile. Conjugation of a leaving group lone pair tends to make these systems more stable and less reactive (Fig. 6.1). The barrier to reaction increases since this delocalization of the reactant is not possible in the tetrahedral intermediate. In an acyl chloride, the most reactive carboxyl derivative, this delocalization of the leaving group lone pair into the carbonyl is very minor because of poor $3p$–$2p$ overlap between the chlorine and the carbonyl carbon. **As L becomes a better donor, carboxyl derivatives become less reactive.** The donor ranking was covered in Section 4.2.5.

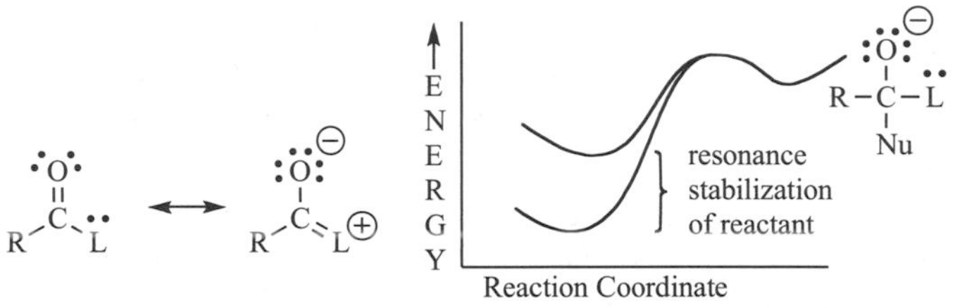
ENERGY
resonance stabilization of reactant
Reaction Coordinate

Figure 6.1 Reactant stabilization relative to the intermediate increases the reaction barrier.

Carboxylic acids are doubly poor sinks, for they can protonate the nucleophile (thus making it useless) and form the very unreactive carboxylate anion. Anions rarely attack anions because of the repulsion of like charges. **Only R–Li and $LiAlH_4$ are reactive enough to attack the carboxylate anion** (which is ion paired with a metal cation to cancel the negative charge repulsion of the Nu). The less reactive nucleophiles will react only with the most reactive carboxyl derivatives. Using the donor trend, we can place ketones as more reactive than esters. Acyl halides, with the halide being both a poor donor and a good leaving group, are the most reactive carboxyl derivatives. A carboxyl derivative with a good leaving group is very reactive because the loss of the leaving group from the intermediate is often the driving force for the overall process. Acyl chlorides are more reactive than anhydrides for this reason. In general, the better the leaving group on the carboxyl derivative, the more reactive it will be. The better donor group deactivation and the better leaving group activation of carboxyl derivatives do not often conflict, so either trend can be used to judge the reactivity of this electron sink.

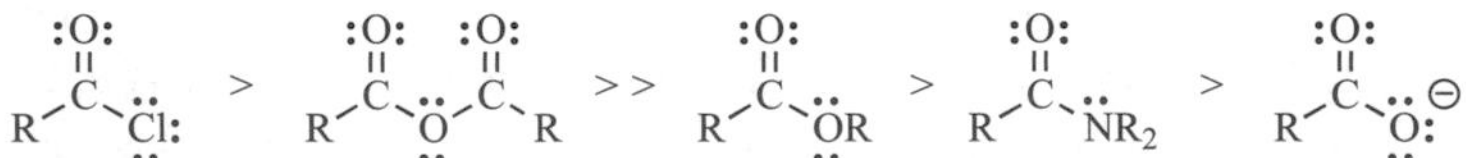
Best electron sink
Worst electron sink

One minor variation of this group is the carbonate derivatives, which have two leaving groups attached to the carbonyl instead of just one. These systems can be quite unreactive since the partial plus on the carbonyl is often diminished by two donor groups. Both leaving groups can be replaced by a nucleophile. If the leaving groups are different enough (for example, Cl and OEt), the monosubstitution product can be isolated.

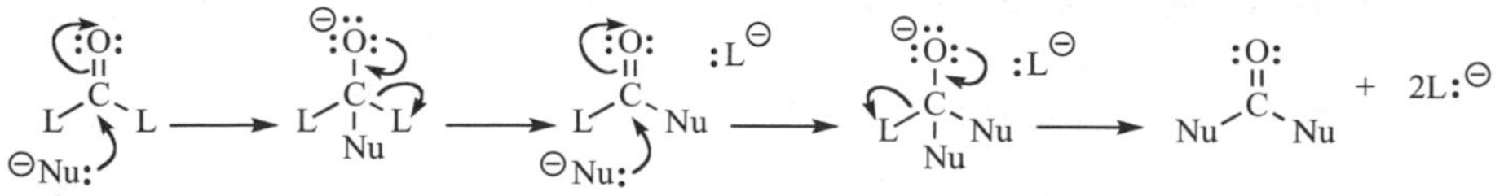

Example problem

Which of the following is the most reactive toward nucleophilic attack? Which is the least reactive?

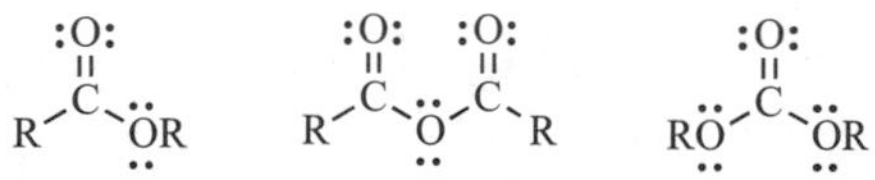

Answer: The most reactive bears the weakest donor. The anhydride is the most reactive. The ester is the next most reactive; least reactive is the carbonate.

6.5.2 Vinyl Leaving Groups

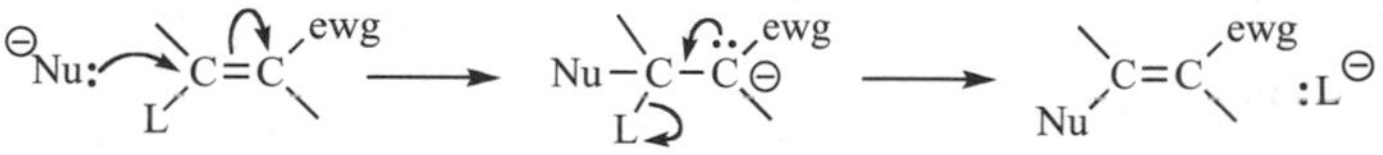

This less common group L–C=C–ewg bears the same relationship to the carboxyl derivatives L–C=Y discussed in Section 6.5.1 as the conjugate acceptors C=C–ewg bear to the C=Y series. Again we have merely substituted C–ewg for the electronegative Y atom in the group. To illustrate how similar the reactivities are, if the ewg is C=Y, then the L–C=C–C=Y group is merely a vinylogous L–C=Y group. Aromatic rings that bear electron-withdrawing groups and a leaving group can be included in this classification. Again, more than one leaving group can be attached, thereby increasing the reactivity.

The electron-withdrawing group is absolutely necessary for the acceptance of electron density from a nucleophile. Simple olefinic halides, RCH=CH–X, are very unreactive because there is no good place to put the electrons from the nucleophile.

6.5.3 Leaving Groups on Polarized Triple Bonds

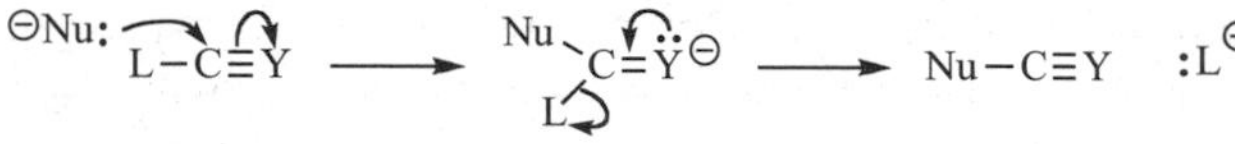

Two very rare electron sinks, **L–C≡Y** and **L–C≡C–ewg**, the first represented by the acid bromide of HOCN, Br–C≡N, react as expected for a multiply bonded electron sink with attached leaving group. The C–ewg again replaces Y in Br–C≡C–COOEt.

6.6 SUMMARY OF GENERIC ELECTRON SINKS

6.6.1 Reactivity Comparison Between Various Classes

All the electron sinks are summarized in Table 6.1; however, the general reactivity trends are not as clear-cut as they were for electron sources. Generally, look first for electron-deficient species, then for the others. The reactivity trend within each class is more important than the class to which a species belongs. For example, **acyl chlorides,** the top end of the reactivity spectrum for polarized multiple bonds with leaving groups, **are more reactive than most alkyl halides**; however, **esters are less reactive than most alkyl halides. Acyl chlorides are more reactive than aldehydes. Nitriles tend to be about as reactive as amides.**

6.6.2 Generic Electron Sink Table

Table 6.1 The 18 Generic Electron Sinks

Generic class	Symbol	Examples
Electron-deficient species (6.2)		
Carbocations	$-\overset{\oplus}{C}<$	$^{+}C(CH_3)_3$
Lewis acids	$-A<$	BF_3, BH_3, $AlCl_3$
Metal ions	M^{2+}	$Hg(OAc)_2$
Weak single bonds (6.3)		
Acids	H–L (or H–A)	HCl, $ArSO_3H$, CH_3COOH
Leaving groups on a heteroatoms	Y–L	RS–SR, HO–OH, Br–Br, HO–Cl
Leaving groups bound to sp^3 carbon	$-\!\!\!>$C–L	CH_3CH_2–Br, $(CH_3)_3C$–Br, $(CH_3)_2CH$–Cl, CH_3–I Ketals $R_2C(OCH_3)_2$ (2 poor Ls)
Polarized multiple bonds without leaving groups (6.4)		
Heteroatom-carbon multiple bonds	>C=Y	Aldehydes: RHC=O Ketones: R_2C=O Imines: R_2C=NR
	–C≡Y	Nitriles: RC≡N
Conjugate acceptors	C=C–ewg	Enones: C=C–C=O, Conjugated esters: C=C–COOR
	C≡C–ewg	RC≡C–COOR
Heterocumulenes	C=C=Y	Ketenes: R_2C=C=O
	Z=C=Y	Ph-N=C=O, RN=C=NR, O=C=O
	C=C=C–ewg	$(CH_3)_2C$=C=$CHCOOCH_3$
	ewg–C=C=C–ewg	PhCOCH=C=CHCOPh
Polarized multiple bonds with leaving groups (6.5)		
Carboxyl derivatives	>C=Y (with L)	Acyl chlorides: RCOCl Anhydrides: RCOOCOR Esters: RCOOEt Carbonate derivatives: ClCOCl
Vinyl leaving groups	L–C=C–ewg	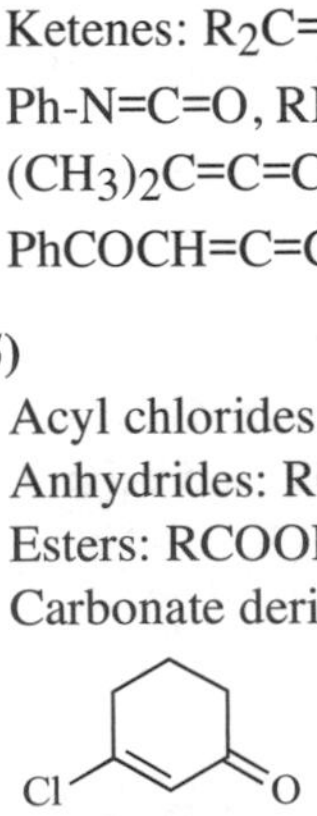
Leaving groups on polarized triple bonds	L–C≡Y L–C≡C–ewg	Br–C≡N Br–C≡C–COOR

Cl
O

6.6.3 Generic Electron Sink Classification Flowchart

Classification of a given structure into its most appropriate generic class relies on your ability to recognize electron-deficient species, leaving groups, and polarized multiple bonds. Figure 6.2 shows a flowchart for classification of electron sinks.

1. Carbocations and Lewis acids are usually highly reactive, and thus are found first.
2. Next look for any leaving groups. Is the leaving group bound to carbon and, if so, what is the hybridization of that carbon? Then sort into the two general classes: weak single bonds and polarized multiple bonds with leaving groups.
3. Search for any polarized multiple bonds without leaving groups.
4. Finally, metal ions as sinks are sorted last since they tend to be the least reactive.

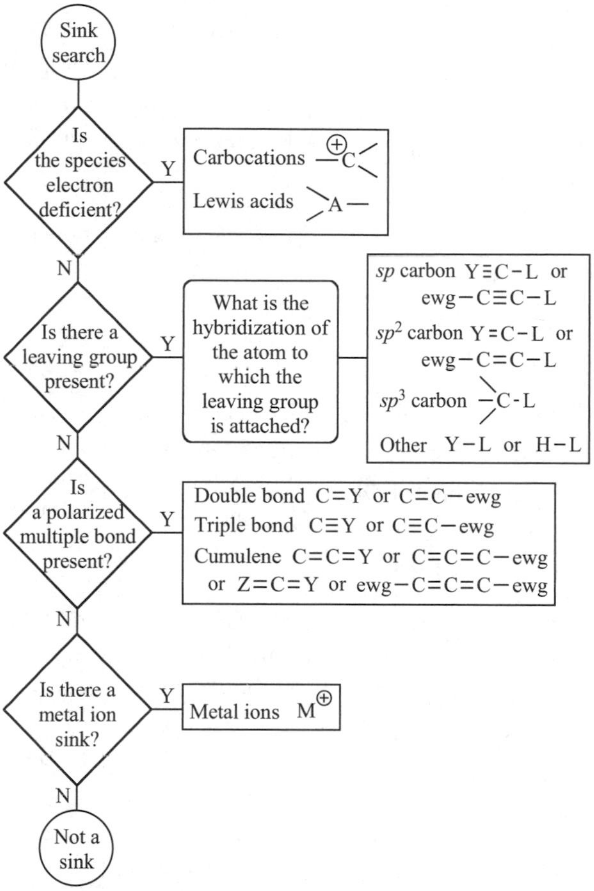

Sink search
Is the species electron deficient?
Y
Carbocations
Lewis acids
N
Is there a leaving group present?
Y
What is the hybridization of the atom to which the leaving group is attached?
sp carbon Y≡C−L or ewg−C≡C−L
sp² carbon Y=C−L or ewg−C=C−L
sp³ carbon C-L
Other Y−L or H−L
N
Is a polarized multiple bond present?
Y
Double bond C=Y or C=C−ewg
Triple bond C≡Y or C≡C−ewg
Cumulene C=C=Y or C=C=C−ewg or Z=C=Y or ewg−C=C=C−ewg
N
Is there a metal ion sink?
Y
Metal ions M⊕
N
Not a sink

Figure 6.2 Classification flowchart for electron sinks

Example problem

To which generic classes do these belong, PhCOOH, CH_2CHCHO?

Answer: We have two problems: translating the typed formula into a recognizable structure and doing a correct classification. The first compound is a carboxylic acid, which should tip us off to put it in the acids, the H–L or H–A class. The next compound is an aldehyde conjugated to a pi bond. Since an aldehyde is a polarized multiple bond and also an ewg, we would put this compound in the conjugate acceptor class, C=C–ewg.

6.6.4 Common Sinks Reactivity Summary

A much smaller set of sinks is responsible for most reactions and is summarized in Table 6.2, which gathers reactivity trends, common reactions, and related generic groups.

Table 6.2 Summary Reactivity Table for Common Electron Sinks

Symbol & Name	Common Examples in Order of Decreasing Reactivity	Reactivity Trend Principle	Common Reactions. Decision?	Related Generic Groups
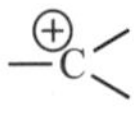 Carbocation	$(CH_3)_2CH^+$ $(CH_3)_3C^+$ Ph_3C^+ $(H_2N)_3C^+$	More stabilized, less reactive. Donors stabilize Lone pairs > pi bonds > R > H	C^+ trap Nu. Rearrange to equal or more stable C^+? Drop off H^+?	Inorganic Lewis Acids
H-L or H-A Acids	H_2SO_4 HBr HCl	The lower pK_a is stronger	Proton transfer $K_{eq} \geq 10^{-10}$	Carbon acids H_3C-ewg
Y-L Leaving group on a heteroatom	Iodine I_2 Bromine Br_2 Chlorine Cl_2 Peroxides ROOR	A weaker bond is more reactive	Often Substitution Nu attacks δ+ end of Y-L if different	Leaving groups on phosphorus or silicon
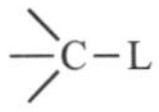 Leaving group on sp^3 carbon	Sulfonates $R\text{-}OSO_2R$ Alkyl iodides R-I Protonated alcohols ROH_2^+ Alkyl bromides R-Br Alkyl chlorides R-Cl	The better L is more reactive	Substitution versus Elimination Decision Section 9.5	2 Ls on C R_2CL_2 3 Ls on C RCL_3
 Polarized multiple bond without L	Protonated carbonyls $R_2C{=}OH^+$ Iminium ions $R_2C{=}NR_2^+$ Aldehydes RHC=O Ketones $R_2C{=}O$ Imines $R_2C{=}NR$	The larger partial plus carbon is more reactive	Addition. Following Substitution or Elimination? (8.5.1)	Polarized triple bonds R-C≡Y without leaving groups
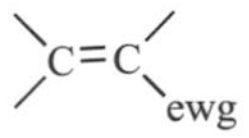 Conjugate acceptor	Conj. nitros $C{=}C\text{-}NO_2$ Enones C=C-COR Conj. esters C=C-COOR Conj. nitriles C=C-CN	The better the ewg, the more reactive	Addition 1,2 vs. 1,4 ? Conjugate addition or direct? (9.6)	Triply bonded conjugate acceptors C≡C-ewg
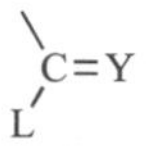 Polarized multiple bond with L	Acyl halides RCOCl Anhydrides RCOOCOR Thioesters RCOSR Esters RCOOR Amides $RCONR_2$	The poorer donor L is, the more reactive	Addition-Elimination Add a second Nu after add-elim? (9.2)	Carbonate derivatives $L_2C{=}Y$ L on triple bonds L-C≡Y

ADDITIONAL EXERCISES

6.1 Give the generic class of each of the following sinks.

PhNCO   BF₃   HOBr   (CH₃)₂CHBr   (CH₃)₂CO

(CH₃CO)₂O   CH₃Br   AlCl₃   CH₂=CHCOCH₃   CO₂

6.2 Draw Lewis structures for all the species in problem 6.1 and designate which atom(s) on each would be attacked by the electron source in a reaction.

6.3 Give the generic class of each of the following sinks.

CH₃CN   (CH₃)₂NCH₂⁺   CH₃CH₂I   CH₃COCl   (CH₃)₃C⁺

H₂SO₄   PhSO₃CH₃   CH₃SO₂Cl   SOCl₂   CH₂=CHCN

6.4 Draw Lewis structures for all the species in problem 6.3 and designate which atom(s) on each would be attacked by the electron source in a reaction.

6.5 Give the generic class of each of the following sinks.

COCl₂   CH₃COOH   H₂O₂   CH₃COOEt   CH₃OSO₃CH₃

PhCHO   EtOCOCl   Cl₂   Cl−C≡N   H₂C=C=O

6.6 Draw Lewis structures for all the species in problem 6.5 and designate which atom(s) on each would be attacked by the electron source in a reaction.

6.7 Circle the best atom on each molecule for nucleophilic attack.

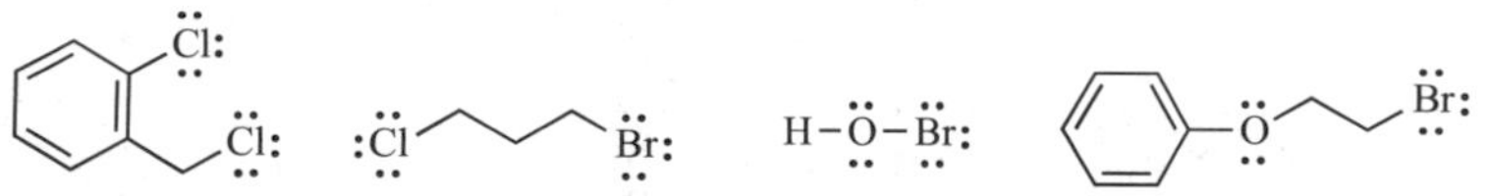

6.8 Trends: rank all species; use the numeral 1 to designate:

(a) The most reactive toward nucleophilic attack

CH₃−OCH₃   CH₃−Br   CH₃−N(CH₃)₂   CH₃−I   CH₃−Cl

(b) The most reactive conjugate acceptor

NO₂   COR   CO₂R   CONR₂   SO₂R

(c) The most reactive toward nucleophilic attack

$$\underset{R}{\overset{\overset{\displaystyle :O:}{\|}}{C}}\!-\!\ddot{O}\!:^{\ominus} \quad \underset{R}{\overset{\overset{\displaystyle :O:}{\|}}{C}}\!-\!\ddot{O}R \quad \underset{R}{\overset{\overset{\displaystyle :O:}{\|}}{C}}\!-\!\ddot{O}\!-\!\underset{R}{\overset{\overset{\displaystyle :O:}{\|}}{C}} \quad \underset{R}{\overset{\overset{\displaystyle :O:}{\|}}{C}}\!-\!\ddot{C}l\!: \quad \underset{R}{\overset{\overset{\displaystyle :O:}{\|}}{C}}\!-\!\ddot{N}R_2$$

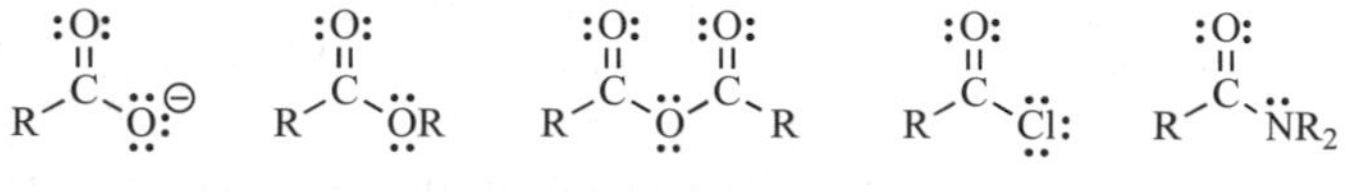

6.9 Explain in general terms why ketals, R₂C(OR)₂, fail to react (act as electron sinks) in basic media but react rapidly in acidic.

6.10 Circle the best atom on each molecule for nucleophilic attack. The comparison is between different classes of sinks.

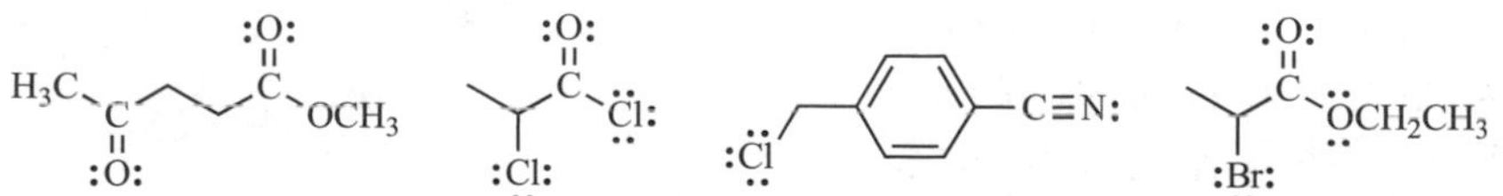

6.11 Sodium borohydride reduces ketones quickly but reacts very slowly to reduce esters. Comparatively, how quickly would you expect an aldehyde to react?

6.12 Thioesters, common in living systems, are more reactive than esters. Explain.

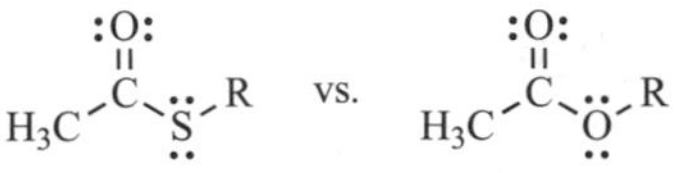

6.13 Circle the best atom on each molecule for nucleophilic attack.

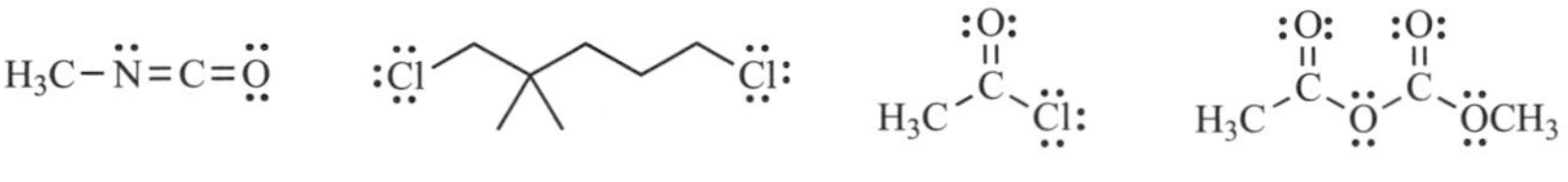

6.14 Trends: rank all species; use the numeral 1 to designate:

(a) The most reactive toward nucleophilic attack

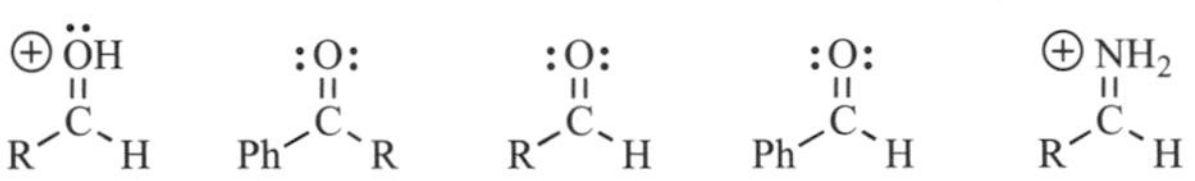

(b) The most reactive carbocation

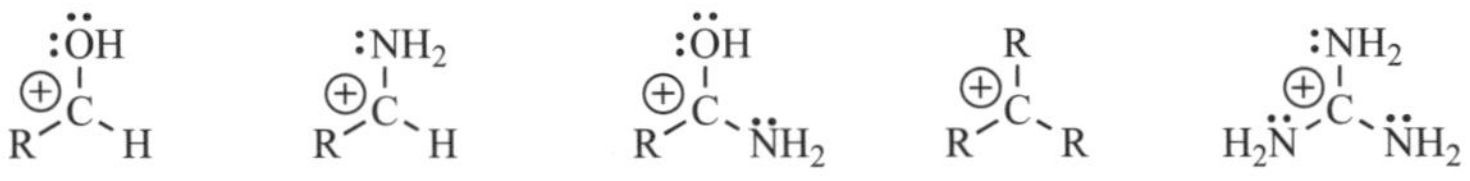

(c) The most reactive toward nucleophilic attack

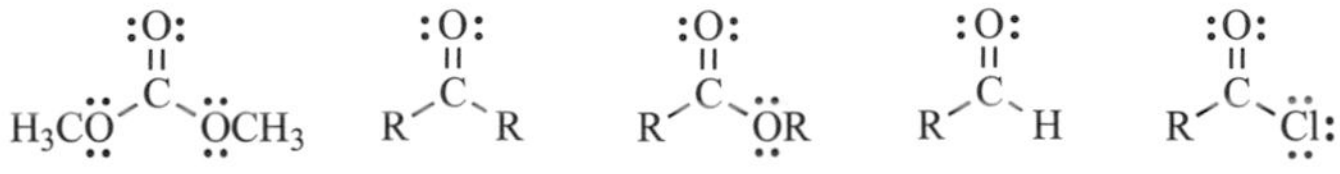

(d) The strongest acid

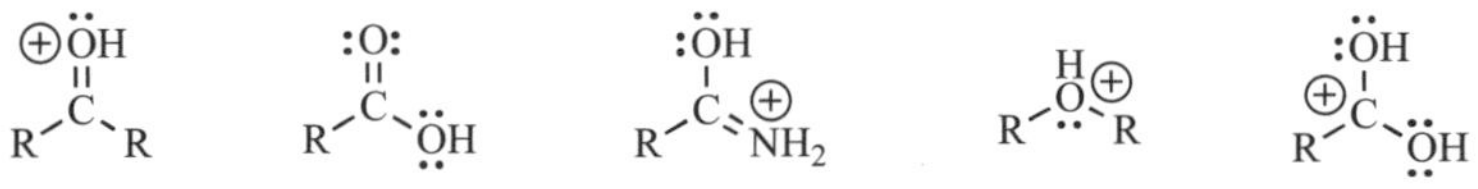

6.15 Carbamic acids are unstable and fall apart easily. The salt is stable. Explain.

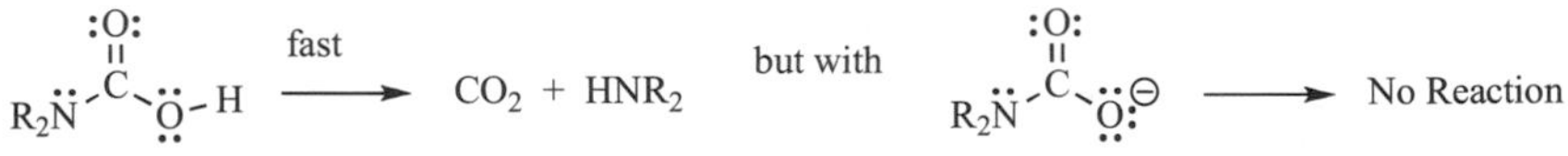

6.16 Some carbocations are stable enough to appear on the pK_a chart. How does their reactivity correspond to their acidity? Rank the following, use numeral 1 to designate the most reactive, and explain the trend. The pK_a is listed below the structures.

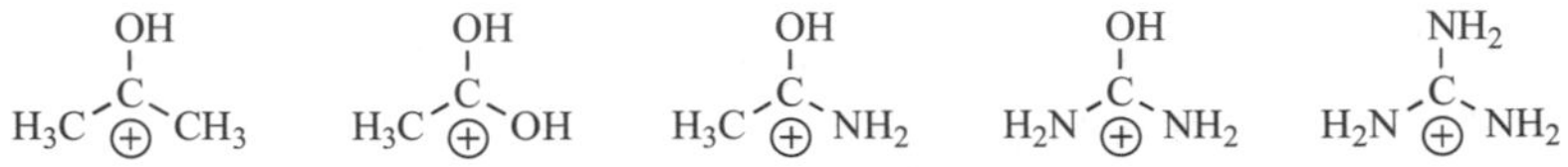

-7 -6 -0.5 0.18 13.6

7

THE ELECTRON FLOW PATHWAYS

7.1 THE DOZEN MOST COMMON PATHWAYS

Your Mechanistic Vocabulary:

p.t.	Proton Transfer to a Lone Pair
$\mathbf{D_N}$	Ionization of a Leaving Group
$\mathbf{A_N}$	Trapping of an Electron-Deficient Species
$\mathbf{A_E}$	Electrophile Addition to a Multiple Bond
$\mathbf{D_E}$	Electrofuge Loss from a Cation to Form a Pi Bond
$\mathbf{S_N2}$	The S_N2 Substitution
E2	The E2 Elimination
$\mathbf{Ad_E3}$	The Ad_E3 Addition
$\mathbf{Ad_N}$	Nucleophilic Addition to a Polarized Multiple Bond
$\mathbf{E_\beta}$	Beta Elimination from an Anion or Lone Pair
1,2R	Rearrangement of a Carbocation
1,2RL	Rearrangement with Loss of Leaving Group

7.2 SIX MINOR PATHWAYS

pent.	Substitution via a Pentacovalent Intermediate
6e	Concerted Six-Electron Pericyclic Reactions
Ei	Thermal Internal *Syn* Elimination
NuL	Nu–L Additions
4e	Four-Center, Four-Electron Reactions
$\mathbf{H^- t.}$	Hydride Transfer to a Cationic Center

Crosschecks for Suspected Additional Minor Paths

7.3 COMMON PATH COMBINATIONS

S_N1 (Substitution, Nucleophilic, Unimolecular), $D_N + A_N$

Ad_E2 (Addition, Electrophilic, Bimolecular), $A_E + A_N$ and Hetero Ad_E2, p.t. + Ad_N

E1 (Elimination, Unimolecular), $D_N + D_E$ and Lone-Pair-Assisted E1, E_β + p.t.

S_E2Ar Electrophilic Aromatic Substitution, $A_E + D_E$

E1cB (Elimination, Unimolecular, Conjugate Base), p.t. + E_β

Ad_N2 (Addition, Nucleophilic, Bimolecular), Ad_N + p.t.

Addition–Elimination, $Ad_N + E_\beta$

Tautomerization, H–C–C=Y $\rightleftharpoons$ C=C–Y–H

Chapter 3 covered the proton transfer electron flow path and reviewed the factors that contribute to acidity. Chapter 4 introduced all the rest of the major electron flow paths along with the four reaction archetypes, substitution, elimination, addition, and rearrangement. This chapter gathers together all the major electron flow paths, introduces a few minor paths, and reviews common path combinations. Section 7.4, Variations on a Theme, shows how the 18 electron flow paths might be reasonably extended and modified.

7.1 THE DOZEN MOST COMMON PATHWAYS

Although there are thousands of different organic reactions, they can be explained by mechanisms that are a combination of relatively few pathways for electron flow. The following are twelve of the most common generic electron flow pathways. These pathways should become a very important part of your mechanistic "vocabulary." You will need to have an excellent command of these dozen pathways to be able to combine them to postulate a reasonable mechanism for almost any reaction.

These electron flow pathways have been shown to be reasonable by many detailed mechanistic studies. Therefore if we confine our use of arrows to stepwise combinations of these pathways, taking into account each pathway's restrictions, we are pretty much assured that whatever we write will also be judged reasonable if not correct.

Beware of shortcuts that combine steps to save redrawing a structure. Any sequence of arrows on one structure implies that that particular electron flow occurs in one step, which in turn demands that all the interacting orbitals be properly lined up. Any sequence of more than three arrows may have major $\Delta S^{\ddagger}$ problems. There are examples in published work of absolute nonsense written with arrows, so always be suspicious of any set of arrows that does not fit a known pathway.

All of the electron sources and sinks previously discussed can react by at least one of these pathways or a simple sequence of them. We must learn what sort of functionality is required for each pathway and also notice the limitations on each pathway. Eventually we will be trying to decide among alternate routes, and the limitations are important so that we may narrow down the possibilities. Each of these pathways has four charge types since the source and/or sink can be charged or neutral: Nu with E, Nu^- with E, Nu with E^+, and Nu^- with E^+. Although any combination of E with Nu can be viewed as either an electrophilic attack or a nucleophilic attack, for consistency we will call it nucleophilic attack if the organic reactant is the sink, and electrophilic attack if an inorganic reactant is the sink.

Because of microscopic reversibility, many of these pathways are merely the reverse of others. The transition states for each related pair of paths are similar (they need not be identical, for the reaction conditions are often slightly different). A figure that illustrates the **approximate** orbital alignment and the transformation of the orbitals of the reactants into those of the product is given for each pair of paths.

7.1.1 Path p.t., Proton Transfer to a Lone Pair

Proton transfer can occur from any acidic to any basic groups or to and from the solvent. A common shortcut in writing mechanisms is to draw just the proton rather than the acid or the protonated solvent, but you should remember that a naked proton will never be floating free in solution. Always check the proton transfer K_{eq} (Section 3.3).

Proton transfer is either a deprotonation of the reactant by a base (symbolized by either b^- or A^-),

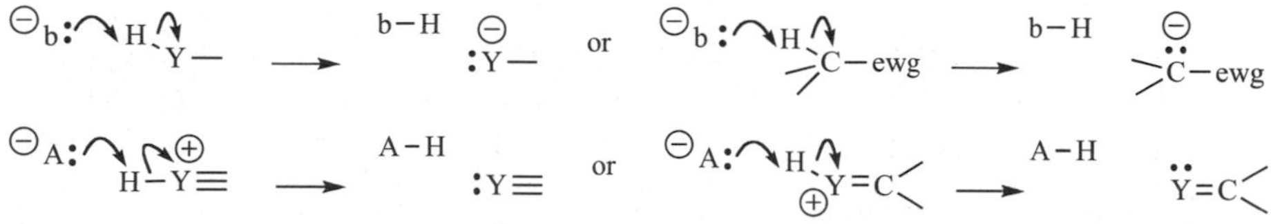

or the microscopic reverse, protonation of the reactant (four charge types of each exist).

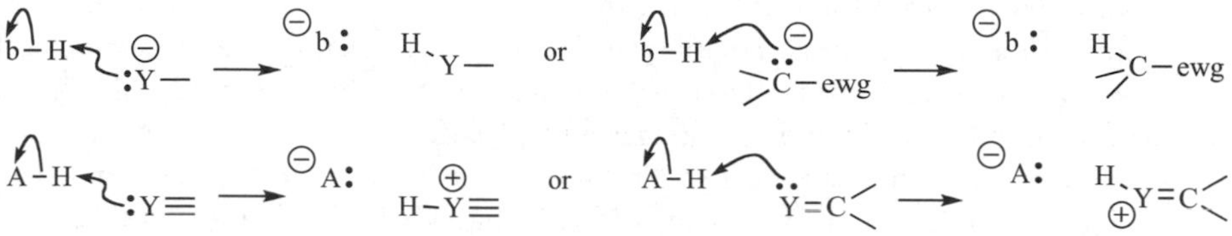

The nonbonding lone pair electrons of the carbonyl group, C=O, are more available than the electrons of the pi bond; proton transfer occurs to and from the lone pairs rather than to and from the carbonyl pi bond.

Proton transfer can occur in any medium, depending on the charge type. **A solution that contains equal concentrations of a base and its conjugate acid has a pH that is equal to the pK_{abH}** (Section 3.2.1). The pH of the medium will be close to the pK_{abH} of the anion needed for reaction, because the anion must have a high enough concentration to react at a reasonable rate. As an illustration, if a ketone enolate, pK_{abH} = 19.2, is needed for reaction, the medium will definitely be very basic.

Protons on electronegative heteroatoms (H–Y) are usually rather acidic because the electronegativity of the heteroatom can stabilize the conjugate base. An electronegative carbon atom, C–ewg, can replace an electronegative heteroatom, Y.

Path Crosschecks (See Chapter 3 for more discussion)

Energetics: The proton transfer K_{eq} should be greater than about 10^{-10}; although more uphill proton transfers occur, there are usually more probable competing processes.

Path Details

Overlap: (viewed as a reactant deprotonation) With H–C–ewg, the C–H bond and the adjacent *p* orbital of the pi system of the ewg must be roughly coplanar (Fig. 7.1). For H–Y, little angular dependence is expected since the base is overlapping with a spherical hydrogen 1*s* orbital and forms a lone pair on the heteroatom.

Media: acidic, basic, or neutral. The pH is often near the pK_{abH} of the conjugate base formed.

Solvent: often protic or polar to stabilize anion formed. The strongest acid possible in a solvent is the protonated solvent. The strongest base is the deprotonated solvent.

Steric: proton must be accessible to base.

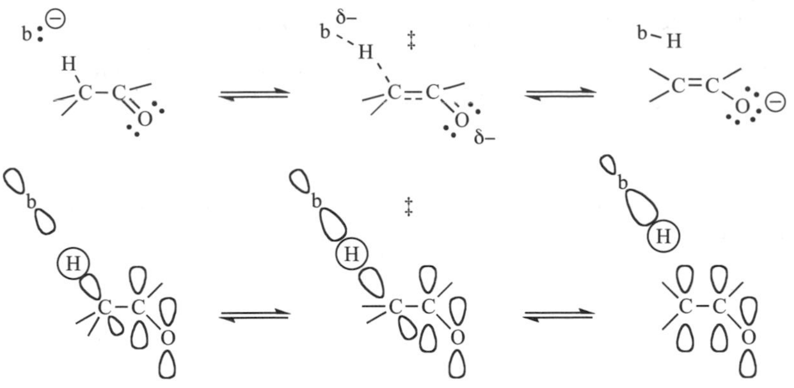

Figure 7.1 Loss of a proton adjacent to a carbonyl with the transition state shown in the center.

Kinetics: Proton transfer catalyzes many reactions. Proton transfer between heteroatom lone pairs is very fast, often at the diffusion-controlled limit. Under reversible (equilibrium) conditions, the most acidic proton is removed preferentially. However, if the deprotonation is done under irreversible conditions, the proton removed is determined by kinetics, not thermodynamics (Section 9.3). Anion basicity always competes with nucleophilicity. Proton transfer is slow enough between organometallics and protons adjacent to carbonyls (carbon bases with carbon acids) that addition of the organometallic to the carbonyl is the dominant process, path Ad_N.

7.1.2 Path D_N, Ionization of a Leaving Group

(Dissociation, Nucleofugic)

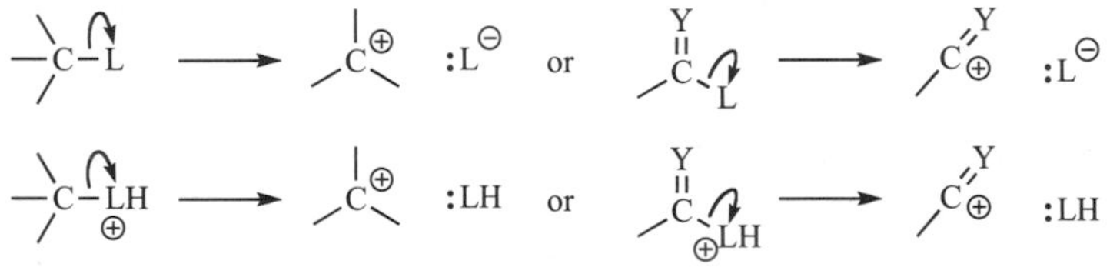

Path Crosschecks

Energetics: often uphill since a bond is broken and usually charged pieces form; to be favorable, both pieces should be reasonably stable ($pK_{aHL} < 0$ and a carbocation better than secondary) and the solvent should be polar to stabilize charges (Section 4.2.9).

Path Details

Overlap: Leaving group departs along the axis of the bond (Fig. 7.2). (Cleavage is a simple extension of a bond stretching vibration.)

Cation: Usually better than 2° in stability; 1° carbocations, RCH_2^+, are very rare.

Leaving group: For ⋟C–L, the L must be good, usually with a $pK_{aHL} < 0$. However, for Y=C–L or even more so for C=C–L, the L must be excellent (improved by Lewis acid complexation or protonation) to compensate for forming a less stable carbocation.

Media: commonly acidic, occasionally neutral, rarely basic. The formation of unstable cations like the vinyl cation occurs only in strong acid.

Solvent: polar to stabilize cation.

Steric: Sterically crowded leaving groups tend to leave a little faster because of the strain released in going to the planar carbocation.

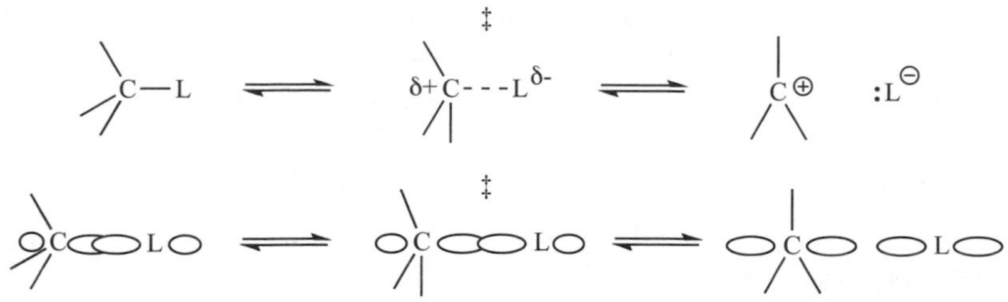

Figure 7.2 Ionization of a leaving group with the transition state in the center.

7.1.3 Path A_N, Trapping of an Electron-Deficient Species
(Association, Nucleophilic, the Reverse of the Previous Reaction)

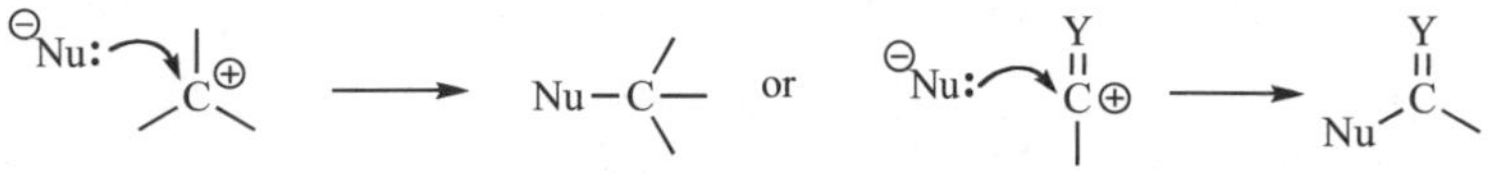

Included in this group are any of the isoelectronic Lewis salt-forming reactions.

Path Crosschecks

Energetics: usually downhill since charge is neutralized, but may be reversible by path D_N (Section 7.1.2) if D_N path crosschecks are met.

Path Details

Overlap: same as path D_N; Nu must approach the empty *p* orbital along its axis so that good overlap occurs (Fig. 7.2, but view the figure from right to left).

Media: commonly acidic, occasionally neutral, rarely basic.

Solvent: polar to stabilize cation.

Steric: relatively little problem because access to the cationic center is usually very good because the cation is flat.

7.1.4 Path A_E, Electrophile Addition to a Multiple Bond (Association, Electrophilic)

An alkene is an average electron source, and an aromatic compound is usually worse; therefore to get electrophilic addition to alkenes and aromatic compounds to occur one needs a good electron sink. Often a loose association of an electrophile with thc pi electron cloud (called a pi-complex) occurs before the actual sigma bond formation step. The best electrophiles, carbocations, add easily. For an overview of electrophilic additions to alkenes, see Section 4.4.

Path Crosschecks

Energetics: usually uphill and dominated by the stability of the carbocation. **The electrophile adds to give the most stable of the possible carbocation intermediates** (Markovnikov's rule). See Table 4.1 for the ranking of carbocation stability. If the electrophile is the most stable of the possibilities, one expects the reverse reaction to be rapid and the equilibrium to favor reactants. Usually the carbocation formed should be more stable than secondary, or the Ad_E3 addition path will be favored.

Path Details

Overlap: Electrophile must approach along the axis of the *p* orbital attacked, not from the side (Fig. 7.3).

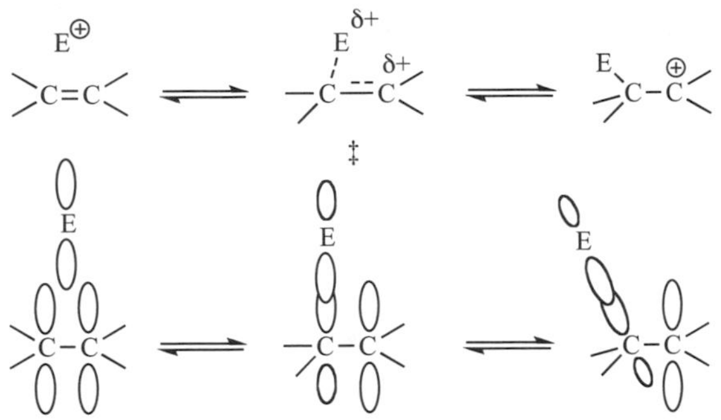

Figure 7.3 Electrophile addition to a double bond with the transition state shown in the center.

Media: most commonly acidic since good electrophiles are necessary for reaction. Occasionally neutral, rarely basic.

Solvent: often polar to stabilize carbocation.

Steric: will direct electrophile attack only if there is no difference in the two possible carbocations formed.

Variations: The electrophile is commonly a proton. The reaction is usually uphill in energy unless the acid is very strong or the cation is very stable, and can easily reverse since most carbocations are very strong acids. As the carbocation becomes less stable, the equilibrium shifts toward reactants. This equilibrium can be shifted to form more of the carbocation by making the acid stronger. For protonation of a pi bond see Figure 7.4 (but view the figure from right to left).

7.1.5 Path D_E, Electrofuge Loss from a Cation To Form a Pi Bond

(Dissociation, Electrofugic, the Reverse of the Previous Reaction)

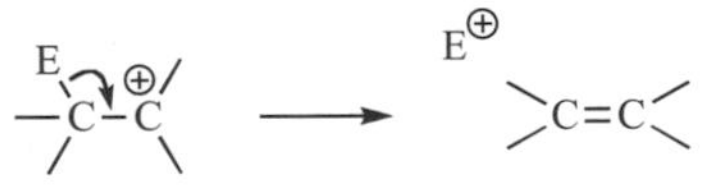

Path Crosschecks

Energetics: downhill unless the electrophile is less stable than the carbocation. *Overlap*: Since a pi bond is formed, the C–E bond broken must be roughly coplanar with the carbocation empty *p* orbital so a good pi bond is formed (Fig. 7.3, right to left).

Path Details

Media: acidic or neutral.

Electrophile lost (electrofuge): must be reasonably stable compared to carbocation.

Solvent: polar to stabilize cation.

Steric: little problem.

Variations: The electrophile lost is commonly a proton. This reaction is downhill in energy unless the A–H is a very strong acid. Figure 7.4 shows the orbital overlap.

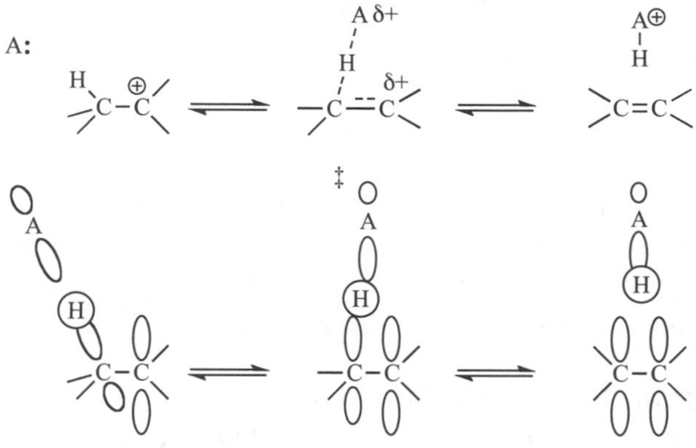

Figure 7.4 Loss of a proton from a carbocation with the transition state shown in the center.

7.1.6 Path S_N2, The S_N2 Substitution

(Substitution, Nucleophilic, Bimolecular)

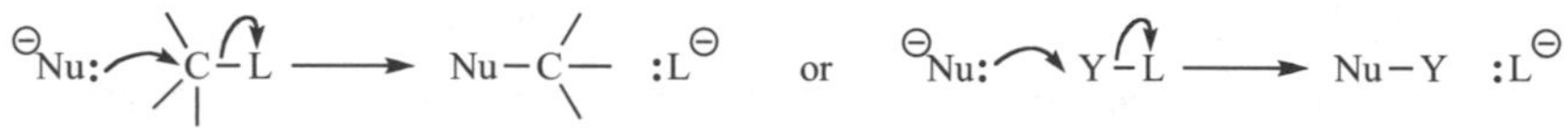

The leaving group is forced out by the nucleophile, as introduced in Section 4.2.7. The transition state is a five-coordinate carbon. **The tetrahedral configuration at the carbon atom is inverted**. The transition state has a carbon *p* orbital partially bonded to the nucleophile on one side and the leaving group on the other (back side attack, see Fig. 7.5). The "2" in the name S_N2 indicates that there are two reactant molecules involved in the rate-determining step (the nucleophile and the molecule attacked).

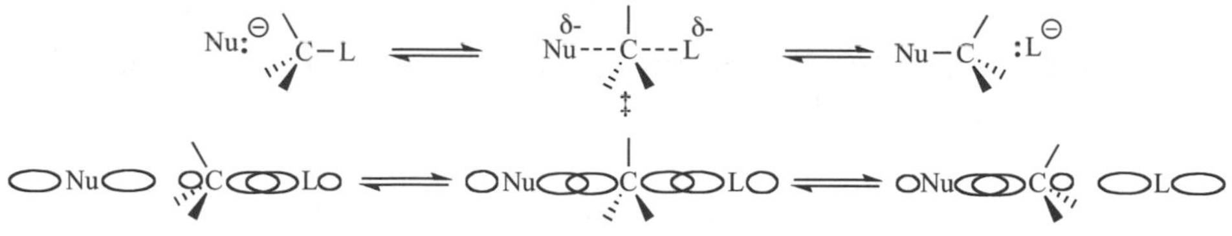

Figure 7.5 The S_N2 substitution with the trigonal bipyramidal transition state shown in the center.

Path Crosschecks

Energetics: do not kick out a leaving group more than 10 pK_a units more basic than the incoming nucleophile. This substitution version of the ΔpK_a rule can also be used to decide whether the forward or reverse reaction is favored, since the reaction tends to form the weaker base.

Leaving group: good or better, occasionally fair (pK_{aHL} ≤ 10). An exception is epoxides, where the strained three-membered ring is destabilized relative to ring opening.

Steric: good access to carbon attacked is critical. Methyl, CH_3–L, and primary, CH_3CH_2–L, are best; secondary, $(CH_3)_2CH$–L, is possible if unhindered. **Tertiary**, $(CH_3)_3C$–L, **does not react by this pathway**. Neopentyl, $(CH_3)_3C$–CH_2–L, also fails to react by this path because a methyl group always blocks the nucleophile's line of approach. **Leaving groups on a trigonal planar center like aromatic, Ar–L, and vinylic, C=C–L, leaving groups do not react via the S_N2 process.**

Path Details (See Section 4.2.7 for more discussion)

Overlap: Nu, carbon attacked, and L are usually close to collinear; distortions are tolerated, for three-membered rings form easily.

Media: basic or neutral, occasionally acidic.

Solvent: polar or medium polarity.

Special substrates: Allylic, H_2C=CH–CH_2–L, benzylic, $PhCH_2$–L, and :Z–CH_2–L react even more quickly than methyl. The overlap of the *p* orbital (of the double bond, aromatic, or lone pair) with the adjacent *p* orbital of the S_N2 transition state forms a pi-type bond that lowers the energy of the transition state, speeding up the reaction.

7.1.7 Path E2, The E2 Elimination

(Elimination, Bimolecular, Concerted Loss of a Proton and Leaving Group)

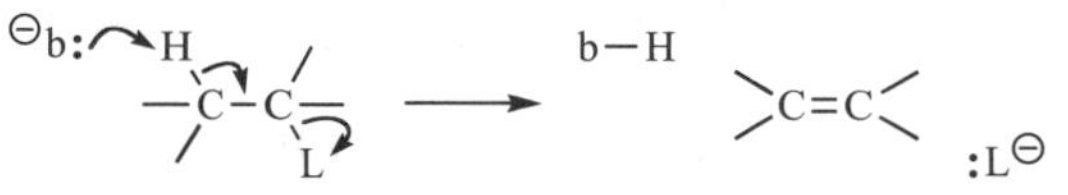

⊖b:
H
C
C
L
b–H
C=C
:L⊖

Path Crosschecks (See Section 4.3 for an overview of elimination)

Energetics: use elimination version of the ΔpK_a rule that the leaving group should not be more than 10 pK_a units more basic than the incoming base.

Overlap: Both the C–H bond and the C–L bond must be roughly coplanar (usually achieved by free rotation), shown in Figure 7.6, to form an untwisted pi bond. Forming an unstable, twisted pi bond takes more energy, so other processes such as substitution can successfully compete.

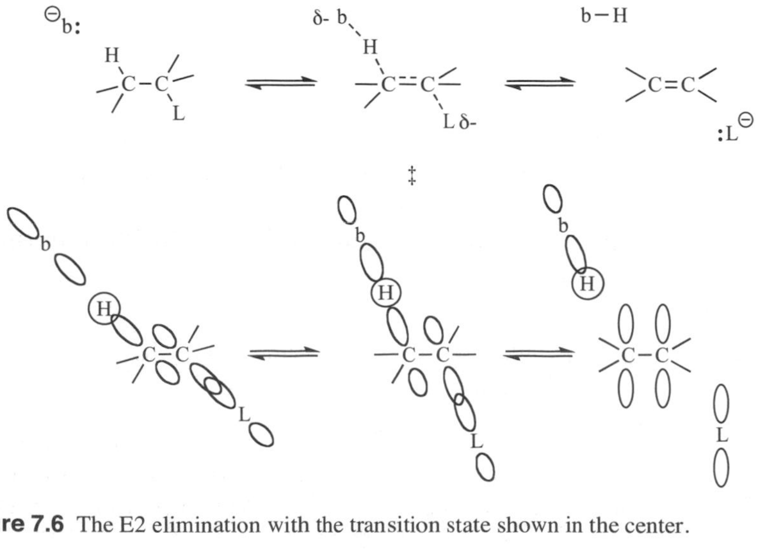

⊖b:
δ- b
b–H
H
C-C
L
C=C
L δ-
C=C
:L⊖
‡
b
H
L

Figure 7.6 The E2 elimination with the transition state shown in the center.

Path Details

Leaving group: fair or better.
Media: normally basic, occasionally neutral, rarely acidic.
Solvent: medium polarity.

Steric: Base must be able to approach the C–H bond easily. For this reason the *anti* elimination, H and L on opposite sides, is preferred over the *syn*, H and L on the same side. The leaving group crowds the incoming base in the *syn*.

Variations: A heteroatom can replace either carbon so that the elimination forms a carbon–heteroatom pi bond (Section 4.3.7). For other variations, see Section 7.4.1.

7.1.8 Path Ad_E3, The Ad_E3 Addition

(Addition, Electrophilic, Trimolecular, the Reverse of the Previous Reaction)

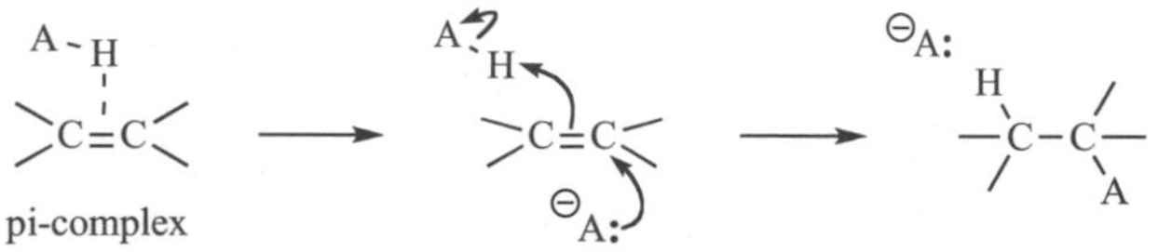
pi-complex

Pi complexation or hydrogen bonding serves further to polarize and activate the double bond toward an otherwise slow nucleophilic attack. (A dotted line indicates complexation only, does not change the electron count, and should not be confused with a solid line describing a bonding pair of electrons.)

Path Crosschecks

Energetics: Favorable in acidic media because the ΔH is usually exothermic. The H–A is usually pi-complexed or hydrogen bonded to the pi bond to activate it for nucleophilic attack. A three-molecule collision of HA, pi bond, and Nu would be rare. Expect Ad_E3 when other addition routes like the Ad_E2 are disfavored. (Section 4.4.2)

Path Details

Overlap: The H attacked, the pi bond, and the A^- must be roughly coplanar. Figure 7.6, viewed from right to left, with the substitution of A for both b and L, approximates the overlap requirements. The A^- attacks the largest partial plus of the pi-complex. Markovnikov's rule is followed even though there is no full positive charge formed.
Media: normally acidic, occasionally neutral.

Steric: The nucleophile is often the conjugate base of the acid, A^-, and must be able to approach the pi bond easily. For this reason *anti* addition is preferred over the *syn*. The H–A complexation crowds the incoming nucleophile in the latter.

7.1.9 Path Ad_N, Nucleophilic Addition to a Polarized Multiple Bond

(Addition, Nucleophilic)

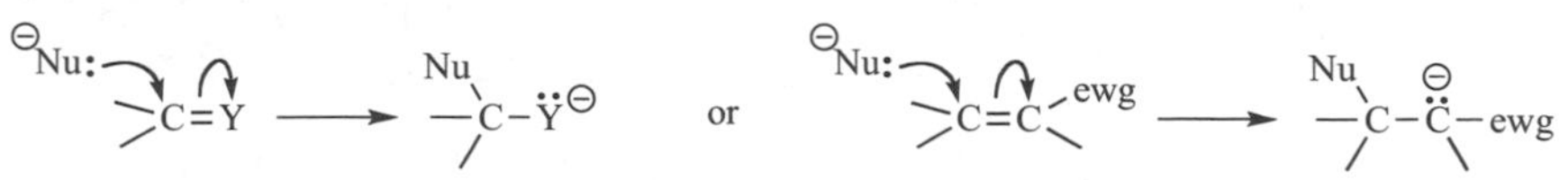

In this pathway electrons flow from the source to the multiply bonded carbon, break the pi bond, and produce a stable anion. An electronegative carbon atom, C–ewg, can replace an electronegative heteroatom, Y, and the electron flow does not change. When the electron sink is a carbon–carbon multiple bond with an electron-withdrawing group, the reaction is called *conjugate addition*. The electron-withdrawing group is essential, for without a means to stabilize the carbanion formed, electron flow cannot occur.

Path Crosschecks (See Section 4.4.5 for more discussion)

Energetics: utilize the ΔpK_a rule. The anion formed must usually be more stable than the Nu or the reaction will reverse by path E_β, the microscopic reverse of this path. However, in a protic solvent the initially formed anion may be rapidly and irreversibly protonated before the reverse reaction can occur. In acidic media, protonation of Y can also occur prior to or simultaneous with nucleophilic attack (see Section 7.4.3, Extent of Proton Transfer Variations).

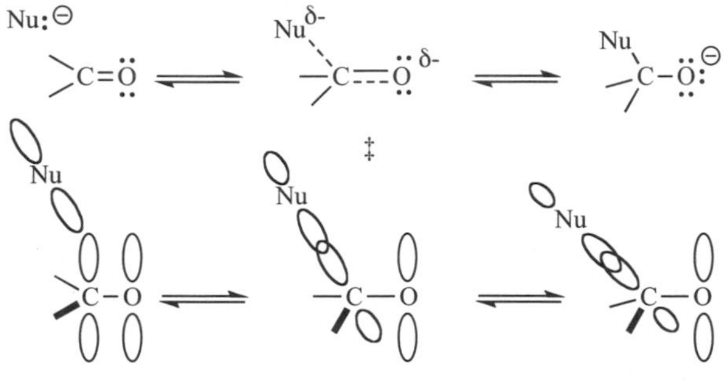

Figure 7.7 The addition to a polarized multiple bond with the transition state shown in the center.

Path Details

Overlap: Nu must approach the δ+ carbon attacked from behind and in the plane of the *p* orbitals of the pi bond (Fig. 7.7).

Media: no limitation; however, if the Nu gets protonated it's no longer a good Nu.

Solvent: often polar to stabilize the anion formed.

Steric: The approach of the Nu to the sink is not usually blocked; the Nu attacks the planar carbonyl from top or bottom. However, rate differences are common. Example: Aldehydes are much more reactive than ketones toward Nu attack.

Variations: Because for R–C≡Y, the two pi systems of the triple bond are perpendicular to each other, they behave independently. The pathway is the same as for double bonds with few path differences:

Overlap: not a problem because of the cylindrical shape of the triple bond.

Resultant anion: slightly more stable being sp^2 hybridized (for the same Y).

Steric: usually not a problem.

7.1.10 Path E_β, Beta Elimination from an Anion or Lone Pair

(The Reverse of the Previous Reaction)

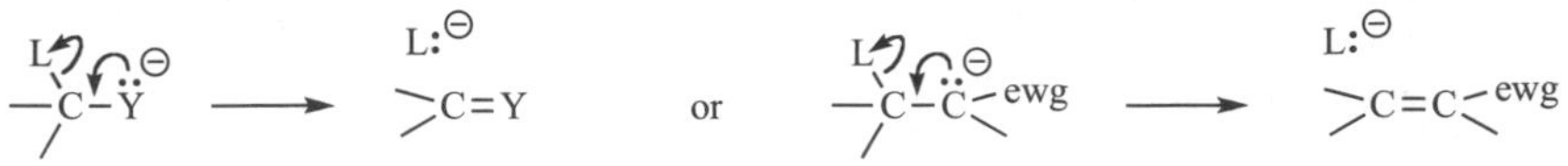

Path Crosschecks

Energetics: use ΔpK_a rule (Sections 3.6 and 4.5.2). The leaving group should not be more than 10 pK_a units more basic than the anion pushing it out.

Overlap: Anion *p* orbital and C–L bond must be roughly coplanar to form an untwisted pi bond (usually not a problem if free rotation is possible). View Figure 7.7 from right to left.

Path Details

Leaving group: can be fair to poor if the pushing anion is less stable. This is a case of a powerful source being able to force out a poorer leaving group. Use ΔpK_a rule here.

Media: no limitation. However, if the anion or lone pair becomes protonated, the electron pair needed to push out the leaving group is lost.

Solvent: polar to stabilize anion.

7.1.11 Path 1,2R, 1,2 Rearrangement of a Carbocation

Rearrangement will usually occur whenever a full positive charge is formed on a carbon that has an adjacent group capable of shifting over to it to form a more stable cation. If the carbon bears only a partial plus, the tendency to rearrange is less. Rearrangement of alkyl groups to an anionic or radical center does not occur.

Path Crosschecks

Energetics: An energetically favorable rearrangement produces a more stable cation (further discussion in Section 4.7). Carbocations of similar stability can also be formed by migration, but rarely is a significantly less stable carbocation formed this way.

Overlap: The most important path restriction to rearrangements is that **the migrating group must maintain orbital overlap with both the atom it migrates from and the atom it migrates to**. At the transition state the migrating group orbital will overlap **both** atoms as shown in Figure 7.8. Usually a group migrates to the neighboring atom from the one it starts out on—a 1,2 shift.

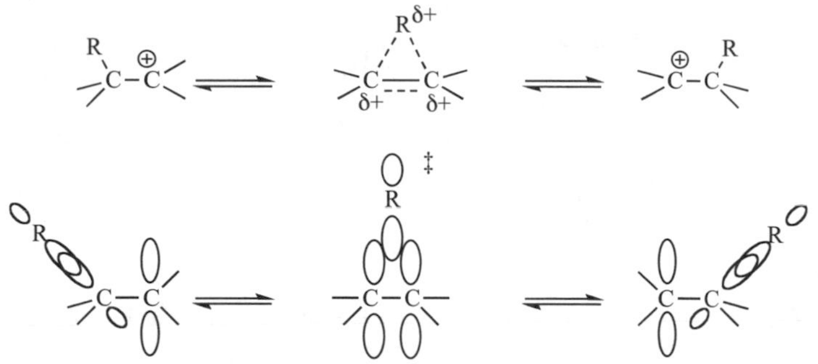

Figure 7.8 The 1,2 rearrangement of a carbocation. The transition state is in the center.

Path Details

Media: commonly acidic, occasionally neutral, rarely basic.

Solvent: polar to stabilize cation.

Migratory aptitudes: At the transition state the migrating group bears a partial positive charge. In freely rotating systems, the groups that can best tolerate this partial plus migrate the best; the order is $-H > -Ph > -CR_3 > -CHR_2 > -CH_2R > -CH_3$.

However, in rigid or conformationally restricted systems, the migrating group is the one that is best able to achieve proper alignment with the p orbital of the carbocation, regardless of the previous trend.

7.1.12 Path 1,2RL, 1,2 Rearrangement with Loss of Leaving Group

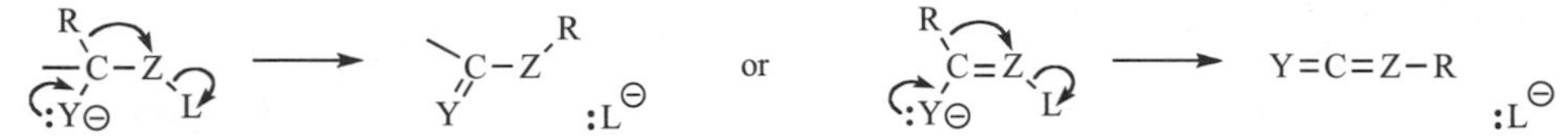
R
C–Z
:Y⊖
L
R
C–Z
Y
:L⊖
or
R
C=Z
:Y⊖
L
Y=C=Z–R
:L⊖

Often rearrangement occurs as shown with a path reminiscent of the generic S_N2, a back side displacement of the leaving group by the migrating group (further discussion in Section 4.7). There are some rearrangements in which it appears that the leaving group departs first, as in the S_N1 rather than the S_N2, so that the electron sink the group migrates to is the carbocation (we would then call it path D_N followed by 1,2R).

Path Crosschecks

Energetics: This path avoids forming an unstable cation and therefore is a useful alternative to path D_N followed by path 1,2R. Expect to use this path if a D_N path fails its crosschecks because of cation instability, for instance, if it would have formed a primary carbocation, or formed an electronegative heteroatom with an incomplete octet.

Overlap: as shown by Figure 7.9, the migrating bond and the leaving group bond must lie in the same plane for good overlap. Again, the migrating group slides from the starting orbital to the neighboring orbital, maintaining overlap as it goes.

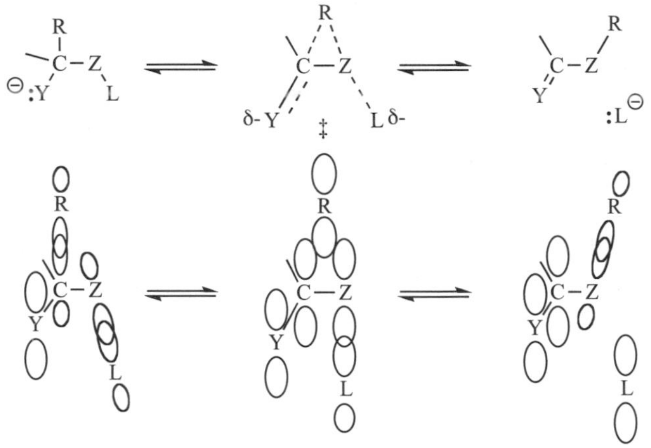
R
C–Z
⊖:Y
L
R
C—Z
δ-Y
‡
L δ-
R
C–Z
Y
:L⊖
R
C–Z
Y
L
R
C—Z
Y
L
R
C–Z
Y
L

Figure 7.9 Rearrangement with loss of a leaving group showing the transition state in the center.

Path Details

Leaving group: excellent to fair. With excellent leaving groups, the pushing Y^- group may not be needed and can be replaced by R. Heteroatom Z can be replaced by a carbon. See Section 4.7.2.

Solvent: usually polar.

Variations: Sometimes the leaving group can be replaced by a polarized multiple bond as the electron sink, shown in the following reaction. In essence, rearrangement can proceed with any of our three general classes of electron sinks: an empty p orbital, a weak single bond to a leaving group, or a polarized multiple bond.

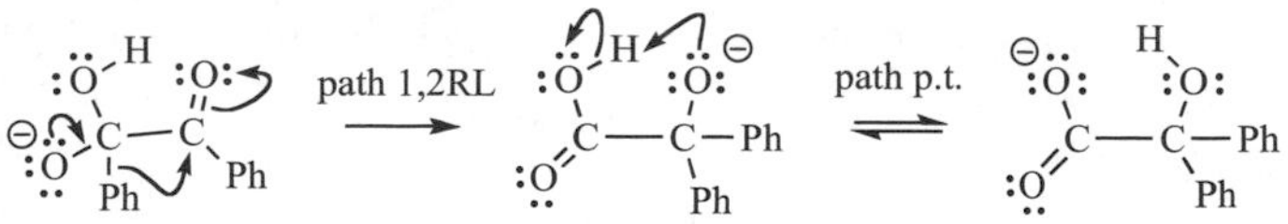

7.2 SIX MINOR PATHWAYS

7.2.1 Path pent., Substitution via a Pentacovalent Intermediate

(For Atoms Capable of Pentacoordinate Bonding Only, Not for Carbon)

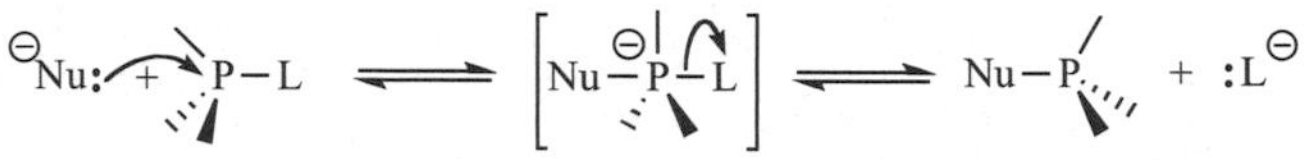

A nucleophile adds, forming a trigonal bipyramidal pentacovalent intermediate that then ejects the leaving group in the microscopic reverse of the nucleophilic addition reaction. The pentacovalent intermediate is often short-lived, and this is shown by enclosing it in brackets. The pentacovalent intermediate bonds that are colinear are called axial (or apical) bonds, whereas the three bonds that lie in the plane perpendicular are called equatorial. Three common elements that can react by this pathway are silicon, phosphorus, and sulfur. Hydrolysis of phosphate esters like RNA occurs by this path.

Path Crosschecks

Energetics: Pentacoordinate intermediates are possible for third-row or higher elements, like Si, P, and S, that are larger and have available *d* orbitals. Do not use this path for second-row elements like carbon.

Path Details

Overlap: Attacking Nu forms an axial bond; therefore, by microscopic reversibility, the departing L must leave from an axial position. The orbital overlap is similar to that shown for the S_N2 (Fig. 7.5), but the structure in the center of the figure is now an intermediate rather than a transition state.

Leaving group: good to fair. If the leaving group is very good, the intermediate would be expected to have a short lifetime and may become just a transition state (S_N2).

Media: acidic, basic, or neutral.

Solvent: commonly polar or medium polarity.

Steric: access usually not a problem since bond lengths are longer.

7.2.2 Path 6e, Concerted Six-Electron Pericyclic Reactions

The largest class of thermal concerted reactions involves six-electron cyclic transition states. Reactions include rearrangements and cycloadditions (two pieces forming a ring). Although the following reactions may all look different, they all involve six electrons (three arrows) going around in a circle (pericyclic). These pericyclic reactions are discussed in more detail with their crosschecks in Chapter 12.

Thermal rearrangements:

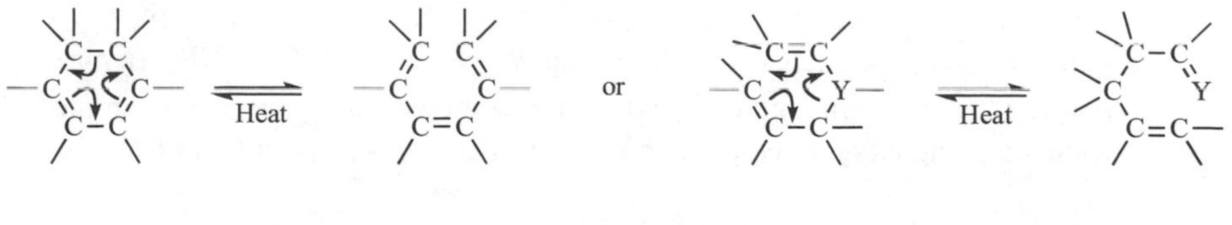
Heat
or
Heat

Thermal cycloadditions or cycloreversions (the reverse reaction):

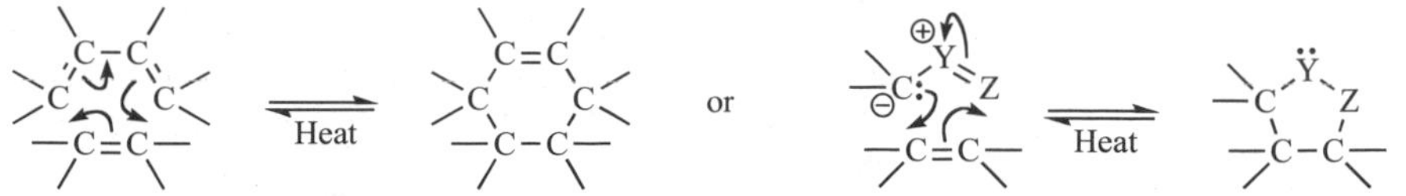

Metal-chelate-catalyzed additions:

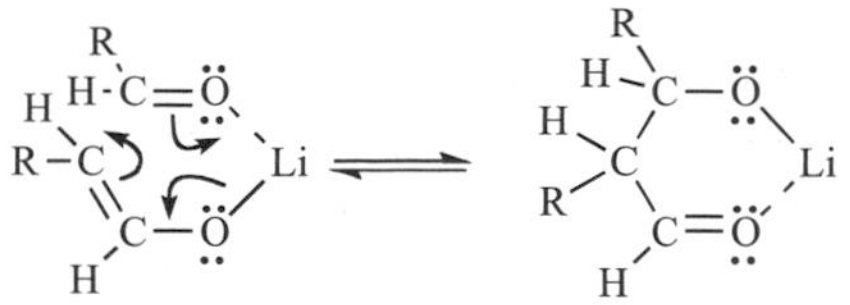

The transition state for these reactions has been described as aromatic since it has six electrons in orbitals that form a loop. In the neutral cases the direction that the arrows are drawn does not matter. The reverse reactions of all of these examples also occur. The major path limitation is to achieve the proper alignment for the cyclic transition state. There is a large diversity of compounds that react by this pathway; in many of the reactions, there are examples of reactions where a carbon atom has been replaced by a heteroatom or vice versa.

There are many reactions in which a metal ion is present and serves to hold the reacting partners together by complexation. Some of the metals for which this process is common are aluminum, magnesium, and lithium. In the following hydride transfer example, in addition to serving as a counterion for the $R_2HC{-}O^-$, the aluminum acts as a Lewis acid catalyst complexing the carbonyl and making it a better electron sink.

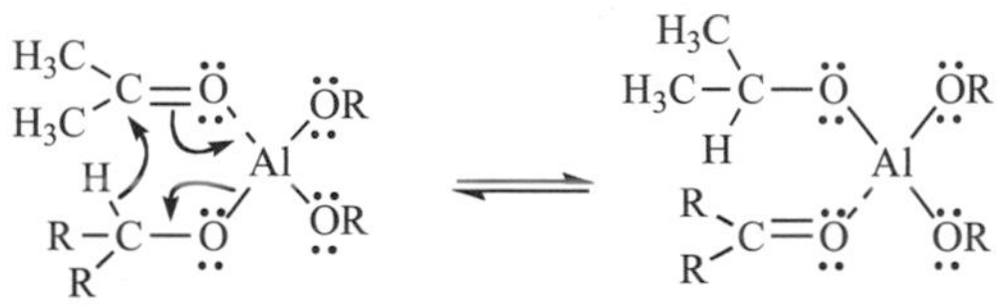

Path Crosschecks and Details

Overlap: These six-electron pericyclic reactions have very ordered transition states, so achieving that arrangement is necessary for the success of the reaction.

Energetics: These reactions often have a highly favorable ΔH of reaction but a negative ΔS of reaction because the product has fewer degrees of freedom than the reactants. For this reason the ΔH term can be overwhelmed by the $-T\Delta S$, so the reaction ($\Delta G = \Delta H - T\Delta S$) can reverse at higher temperatures.

7.2.3 Path Ei, Thermal Internal *Syn* Elimination

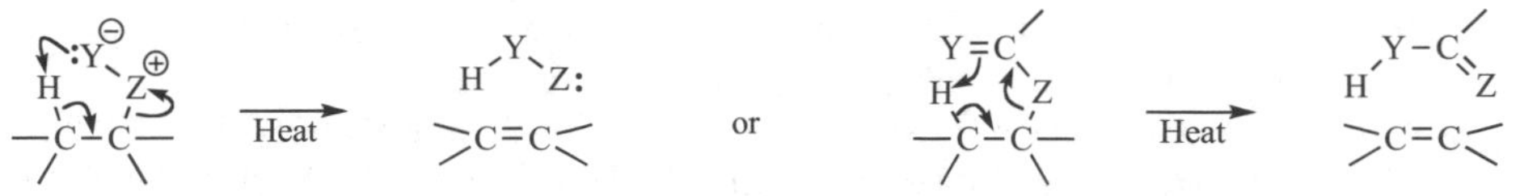

This reaction is just a thermal internal *syn* E2 elimination, and a relative of path 6e. There are many possible reactants for this internal *syn* elimination path. Five-membered transition state examples include Y equals oxygen and Z is $^+NR_2$ (amine oxides), SPh (sulfoxides), or SePh (selenoxides). Six-membered transition state examples include both Y and Z being oxygen (esters), or Y is sulfur and Z is oxygen (xanthates).

Path Crosschecks

Overlap: like the E2, the C–H bond and the C–Z bond involved in the elimination must be roughly coplanar for strong, untwisted pi bond formation (Fig. 7.10).

Energetics: Poorer leaving groups require a higher temperature. Selenoxides eliminate easiest, sulfoxides next, and amine oxides require the highest temperatures.

Path Details

Anion: average.

Leaving group: the weaker the C–Z bond strength, the better the leaving group.

Media and solvent: no limitation; can occur in the gas phase.

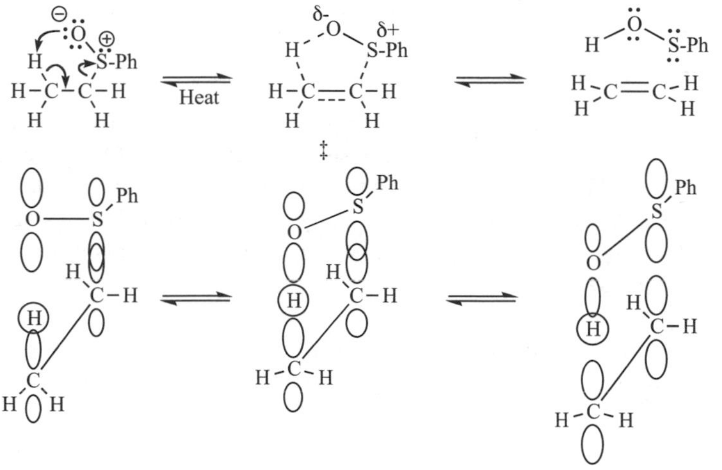
Heat
δ-
δ+
S-Ph
Ph
‡

Figure 7.10 Thermal internal *syn* elimination with the transition state shown in the center.

Variations: An additional six-membered transition state example is the cyclic decarboxylation. The overlap limitation for the new carbon–carbon pi bond requires the breaking C–C bond to align coplanar with the carbon *p* orbital of the carbonyl.

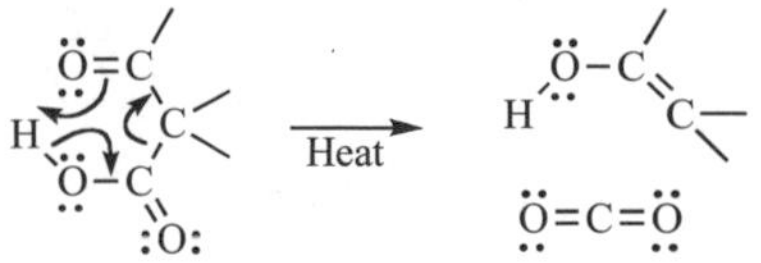
Heat

7.2.4 Path NuL, Nu–L Additions (Three-membered ring formation)

This pathway shows some variation on the timing of when the leaving group falls off. Occasionally the addition may be concerted as shown (with bromine, for example), but more often the leaving group loss and the nucleophilic attack on the pi bond are separate steps. The leaving group may fall off before attack (a carbene if Nu is carbon).

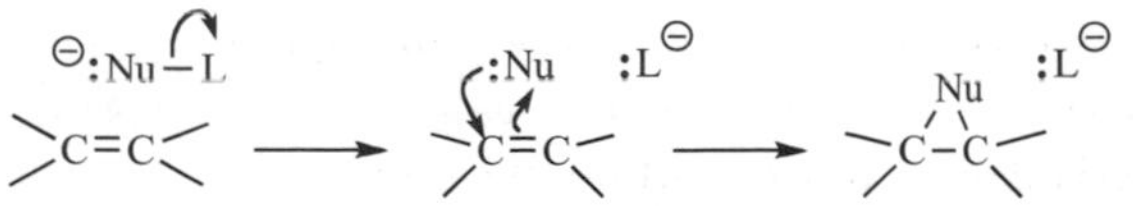

If the multiple bond is capable of stabilizing a carbocation, as erg–C=C is, then the leaving group may depart with nucleophilic attack.

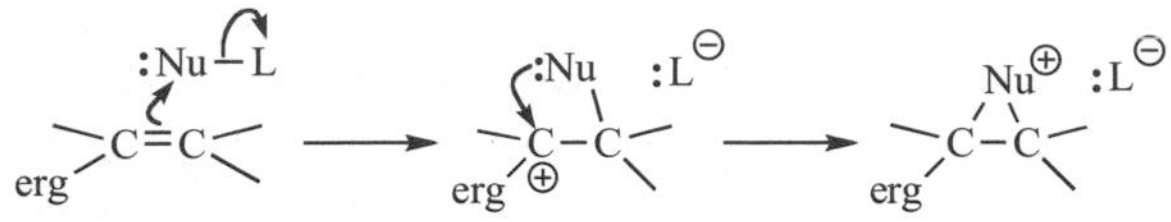

If the multiple bond is capable of stabilizing an anion, as C=Y is, then the leaving group may fall off after nucleophilic attack (for a thorough discussion, see Section 8.12).

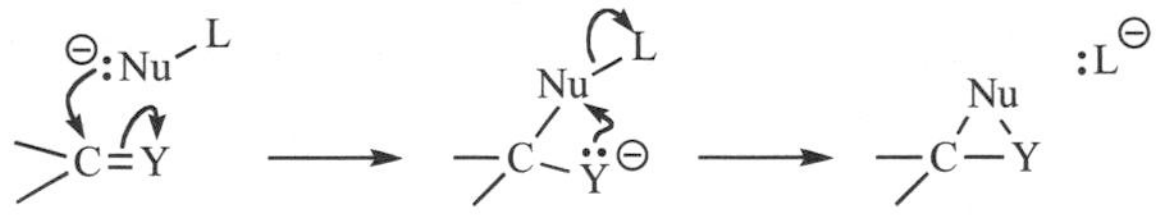

Path Details (Concerted)

Overlap: Nu–L must approach the pi bond from above or below and not from the side. The Nu, the L, and the double-bond carbons should be close to coplanar, as shown in Figure 7.11. The nucleophile and leaving group orbitals are close to perpendicular, so they react independently. A concerted example is shown, but most reactions are stepwise.

Leaving group: good to fair, may leave before addition occurs.

Steric: usually no problem.

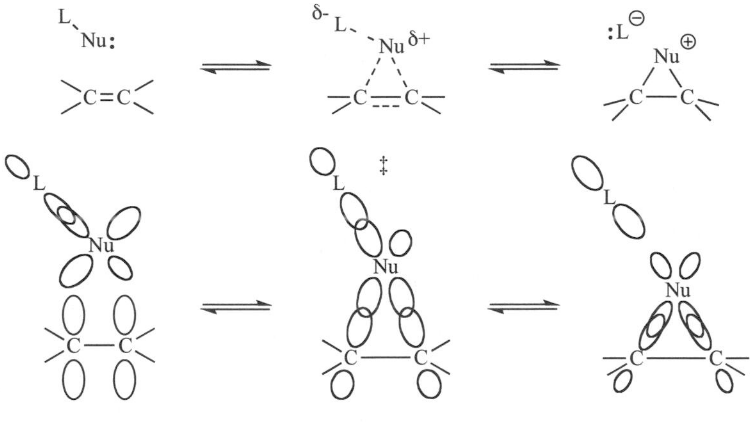

Figure 7.11 The addition of a Nu–L species to form a three-membered ring with the transition state shown in the center.

7.2.5 Path 4e, Four-Center, Four-Electron

Four-center, four-electron processes do not occur thermally. The transition states for these unfavorable reactions have been described as having antiaromatic destabilization because they have four electrons in a normal closed loop. There are **three exceptions** that go by this path; all have some unusual orbital arrangement to allow them to bypass the problem of antiaromatic destabilization of their transition states.

The addition of boranes to pi bonds occurs by this pathway (also the less common but similar trivalent aluminum hydrides, such as diisobutylaluminum hydride, $[(CH_3)_2CHCH_2]_2AlH$, add by this pathway). The overlap path limitation is that all four atoms must be roughly coplanar (Fig. 7.12). A pi-complex of the empty *p* orbital on boron with the double bond forms first, then collapses by this four-electron process to the product. Boron is the electrophile, and hydrogen is the nucleophile (boron is less electronegative than hydrogen). Because the nucleophile orbital and electrophile orbital are perpendicular (and therefore do not interact), the loop is not closed. The direction of electrophilic addition is determined by the formation of the most stable partial plus.

Alternatively, the two bond formation steps may not be synchronous but occur so close together in time that the intermediate has an insignificant lifetime. Electron density from the double bond flows into the empty orbital on boron, creating an electron-deficient center on a carbon to which the now electron-rich boron can donate a hydride. The addition is *syn*; both boron and hydrogen add to the same face of the double bond.

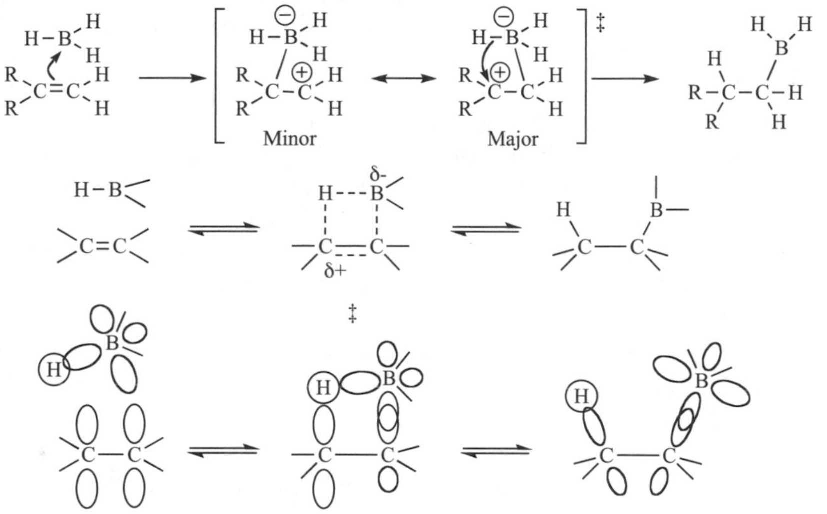
Minor
Major

Figure 7.12 Addition of a borane to a double bond with the transition state shown in the center.

A second exception occurs with the elimination of phosphine oxides from oxaphosphetanes. The pentacoordinate phosphorus serves as the electron source in an internal elimination of the oxygen bound to phosphorus.

The last exception has an unusual twist, literally. The transition state is a strange loop with a half-twist, a Möbius loop. In contrast to a normal loop, Möbius loops are predicted to be stable with 4*n* electrons in them. Given enough heat, the cyclobutene sigma bond twists open to form a diene. This and other pericyclic reactions are discussed in more detail in Chapter 12.

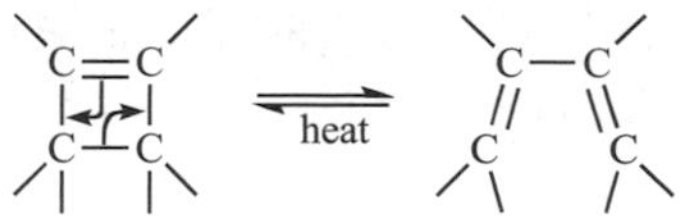
heat

7.2.6 Path H^- t., Hydride Transfer to a Cationic Center

Hydride has the electronic configuration of helium, but still is relatively reactive. Salts like sodium hydride are very basic because hydrogen, the conjugate acid, has a pK_a of 36. With that high a pK_{abH}, hydride makes a very poor leaving group. However, a carbocation can abstract a hydride to form an equal or more stable carbocation.

Path Crosschecks

Overlap: Only that the carbocation must approach within bonding distance to the hydride transferred. This pathway is related to path 1,2R, rearrangement of a carbocation, in which a group migrates with its pair of bonding electrons to an adjacent carbocation. In this minor path, a hydrogen with its bonding electron pair is passed to a nonadjacent carbocation; the empty *p* orbital of the carbocation must get close enough to allow the hydride to be partially bonded to both atoms at the transition state, as in Figure 7.13.

Path Details

Carbocation formed: of equal or greater stability than starting carbocation.

Media: acidic, occasionally neutral.

Solvent: commonly polar or medium polarity.

Steric: not a problem because the carbocation is flat.

Energetics: An energetically favorable transfer forms a more stable carbocation. Similar-stability carbocation formation is also facile, but not formation of a less stable carbocation.

Related path: If the carbocation is replaced by a partial plus center, then a pushing Y^- is required on the hydride donor, and the reaction tends to go by path 6e (last example in Section 7.2.2).

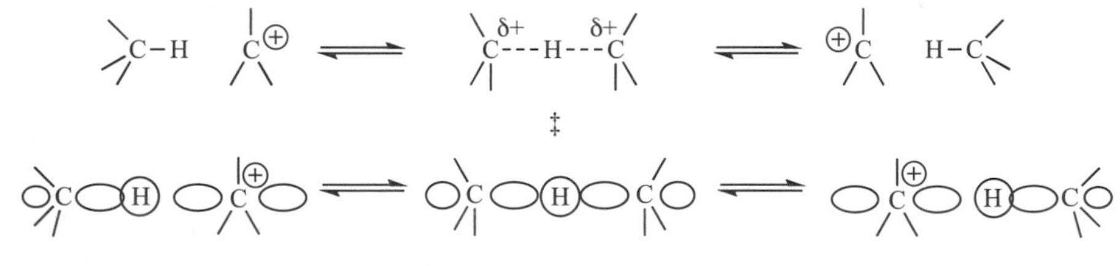
C−H
C⊕
δ+
δ+
C---H---C
‡
⊕C
H−C
C
H
C
H

Figure 7.13 Hydride abstraction by a cation with the transition state in the center.

Biochemical Example: NADH Reduction of Acetaldehyde

NADH (nicotinamide adenine dinucleotide) is a biochemical source of hydride. In the following example NADH reduces acetaldehyde to ethanol via minor pathway H^- t., hydride transfer to a cationic center. A Zn^{2+} ion acts as a Lewis acid to polarize the acetaldehyde carbonyl (similar to protonating the carbonyl). The Lewis acid makes the carbonyl a better electron sink by increasing the partial positive charge on carbon. In fact, the electrophilic catalysis by 2+ and 3+ metal ions can accelerate additions to carbonyls by over a million times. The formation of the aromatic pyridinium ring in the NAD^+ product helps balance the energetics of this easily reversible reaction.

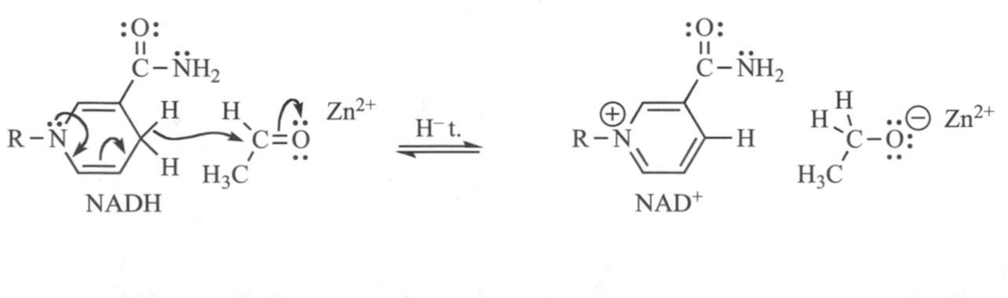
:O:
C−NH2
R−N
H
H
Zn2+
C=O
H3C
NADH
H−t.
R−N
H
NAD+
H
C−O:
Zn2+

Crosschecks for Suspected Additional Minor Paths

Any suspected new electron flow path should be well tested before it is accepted as a new pathway. It may be just a combination of (Section 7.3) or a variation on (Section 7.4) already known paths. Watch for ΔS problems of too many things happening at once. Look for steric and strain problems and check the orbital alignment with molecular models—orbitals that will become double bonds in the product must be able to get close to coplanar in the starting material. Check any intermediates for stability, especially if charged. Check that the electronics fit HOMO–LUMO and HSAB theory. Check the energetics with a ΔH calculation or the ΔpK_a rule, and be skeptical anyway, it's healthy.

7.3 COMMON PATH COMBINATIONS

Several pathway combinations occur so frequently that they have been given a name. Most were introduced and discussed in Chapter 4 and are grouped here as a summary. **These "phrases" in your mechanistic vocabulary are the next step toward your eventual construction of a grammatically correct mechanistic "sentence."**

7.3.1 S_N1 (Substitution, Nucleophilic, Unimolecular), $D_N + A_N$

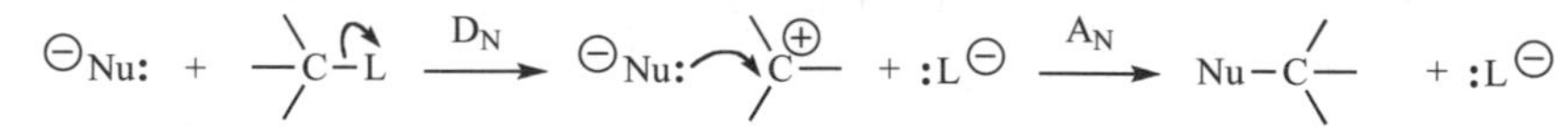

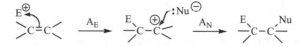

The rate-determining ionization of a good leaving group, path D_N, produces a carbocation-leaving group ion pair that may or may not dissociate before the carbocation is trapped by a nucleophile, path A_N. Substrates that would make a poor cation, CH_3–L, and Ar–L, do not react via the S_N1 process. The nucleophile attacks the carbocation or ion pair and gives the product in a fast second step. If the leaving group were on a chiral center, a racemic mixture can result because the nucleophile can attack the free carbocation from either the top or bottom face equally. Often, however, nucleophilic attack on the ion pair results in inversion, since the leaving group partially blocks one face. If the nucleophile is the solvent, this process is called solvolysis. (See Section 4.2.1 for more discussion of the various substitution types.)

7.3.2 Ad_E2 (Addition, Electrophilic, Bimolecular), $A_E + A_N$

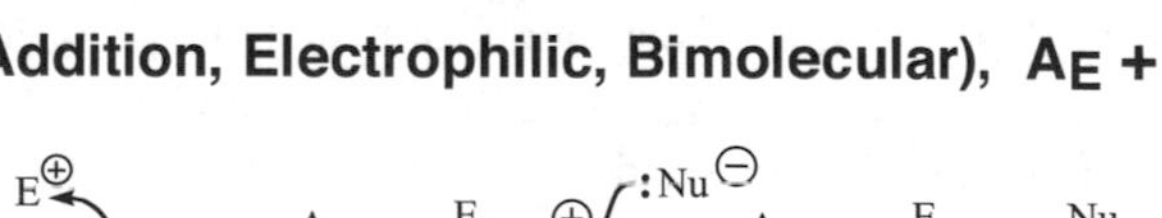

The Ad_E2 is the major addition pathway of alkenes and dienes. Electrophile addition to a pi bond or protonation of a pi bond, path A_E, produces the most stable cation, which is then trapped by a nucleophile, path A_N. (Section 4.4.2 discusses the various addition types.) If the electrophile is a proton, this reaction is the reverse of the E1 reaction. The reaction commonly produces a mixture of *syn* and *anti* addition.

Hetero Ad_E2, p.t. + Ad_N

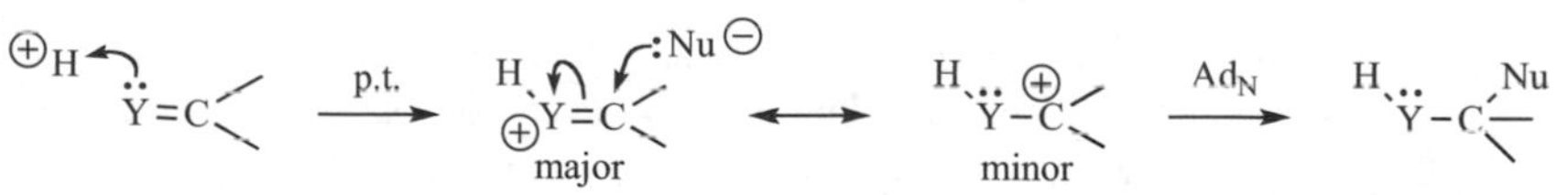
major
minor

This reaction is a minor variant of the Ad_E2 in which the electrophile, usually a proton, attacks the lone pair rather than the less available heteroatom–carbon pi bond. The lone pair is protonated, path p.t., to give a highly polarized multiple bond that can be viewed as a stabilized carbocation. The nucleophilic attack on this carbocation could also be viewed as path A_N, trapping of a cation by a nucleophile, instead of path Ad_N.

7.3.3 E1 (Elimination, Unimolecular), $D_N + D_E$

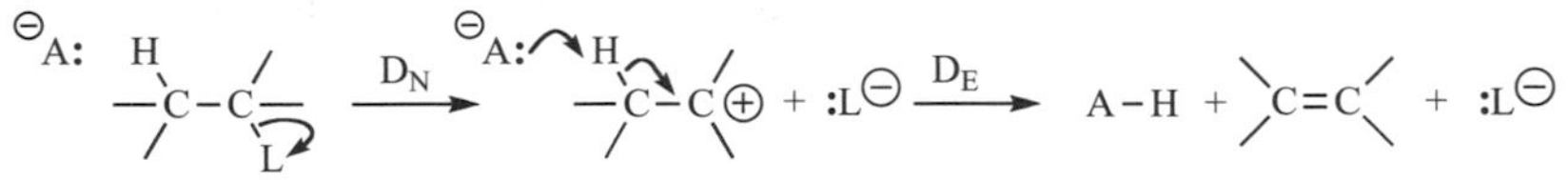

After rate-determining ionization of a good leaving group, path D_N, the cation is deprotonated by path D_E to produce a pi bond. This elimination is the reverse of the Ad_E2 addition. As might be expected, the E1 process competes with the S_N1 process. **If A^- acts as a base, E1 occurs; if it acts as a nucleophile, S_N1 occurs.** Section 9.5 will discuss such decisions in detail.

Lone-Pair-Assisted E1, E_β + p.t.

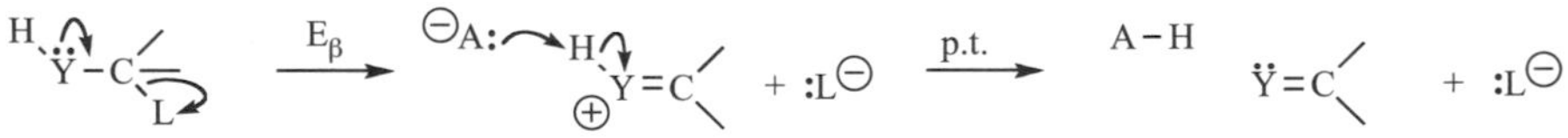

This reaction is the reverse of the hetero Ad_E2 reaction. The lone-pair-assisted E1 uses a properly aligned lone pair to expel the leaving group, path E_β. The resultant cation is then deprotonated, path p.t.

7.3.4 S_E2Ar Electrophilic Aromatic Substitution, $A_E + D_E$

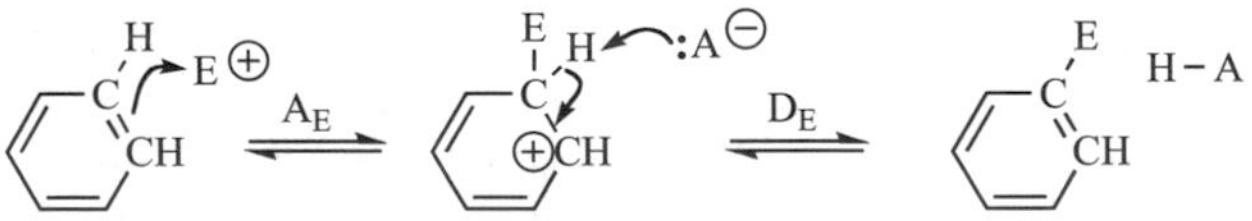

This reaction was introduced in Section 4.6, Electrophilic Substitution at a Trigonal Planar Center. The electrophile adds to the pi bond of the aromatic ring, path A_E, followed by deprotonation of the cation formed, path D_E, restoring aromatic stabilization. See Section 5.6 Aromatic Rings, for a discussion of electrophilic aromatic substitution on heteroaromatics like pyridine and on condensed aromatics like naphthalene.

7.3.5 E1cB (Elimination, Unimolecular, Conjugate Base), p.t. + E_β

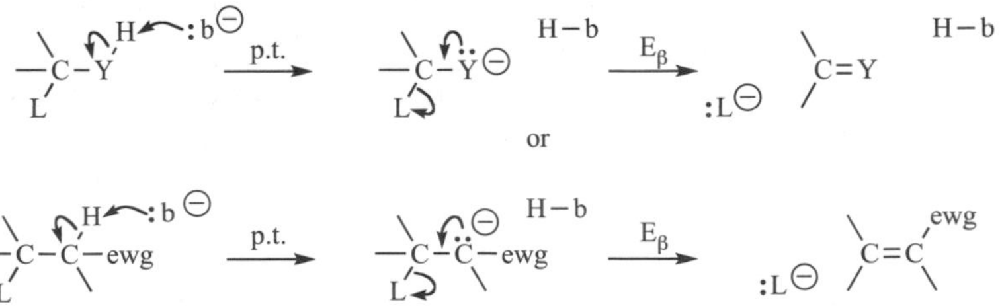

A stabilized anion forms first via path p.t., and then a fair leaving group departs in the slow step via the beta elimination from an anion path E_β. (See Section 4.3 for more discussion of the various elimination types.)

7.3.6 Ad_N2 (Addition, Nucleophilic, Bimolecular), Ad_N + p.t.

This reaction is the reverse of the E1cB elimination above. A rate-determining addition of a nucleophile to the polarized multiple bond occurs by path Ad_N, producing a stabilized anion that is then protonated by path p.t. (See Section 4.4 for more discussion of the various addition types.)

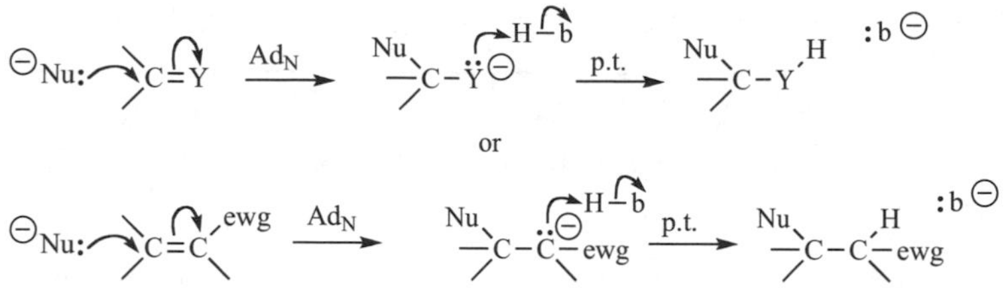

7.3.7 Addition–Elimination, $Ad_N + E_\beta$

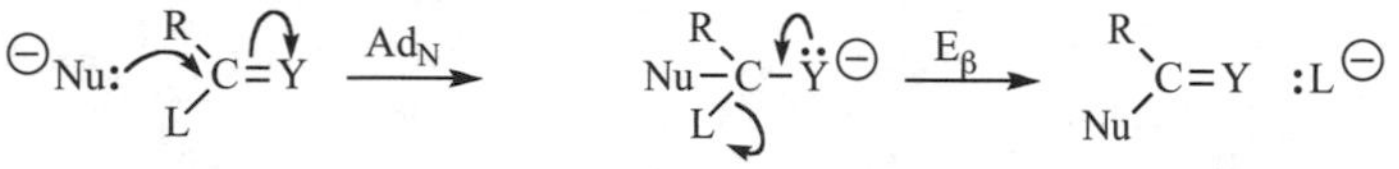

This is the almost exclusive route in basic media for replacement of a leaving group bound to a double bond. This reaction was introduced in Section 4.5, Nucleophilic Substitution at a Trigonal Planar Center. The most common substrates for this reaction are the carboxyl derivatives. A nucleophile adds first via path Ad_N, and then the leaving group departs via the beta elimination from an anion path E_β.

7.3.8 Tautomerization, taut.

$$H-C-C=Z \rightleftharpoons C=C-Z-H$$

Tautomerization is the shift of an H from a carbon adjacent to a carbon–heteroatom double bond to the heteroatom itself (and the reverse). It is an acid- or base-catalyzed equilibrium. Two examples are the keto/enol pair (Z = oxygen) and the imine/enamine pair (Z = nitrogen). Base catalysis goes via the enolate anion.

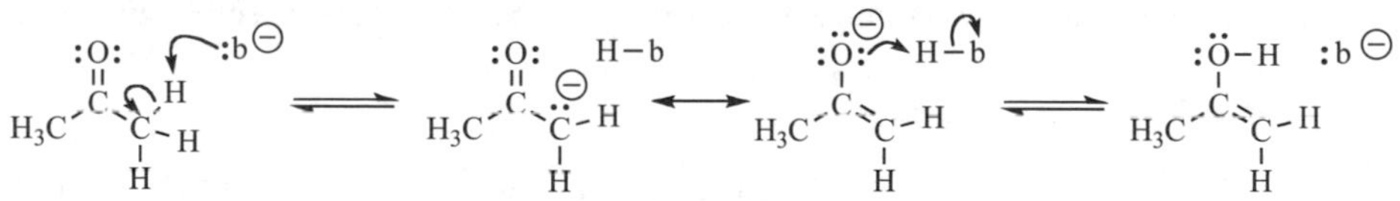

Keto Enolate Enol

Acid catalysis goes via the lone-pair-stabilized carbocation.

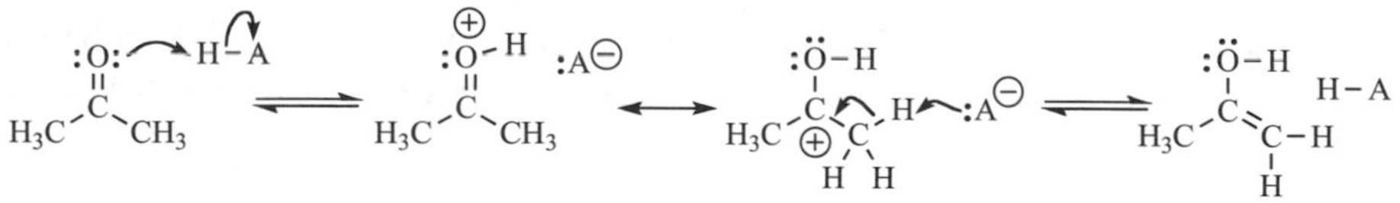

Keto Carbocation Enol

Common Error

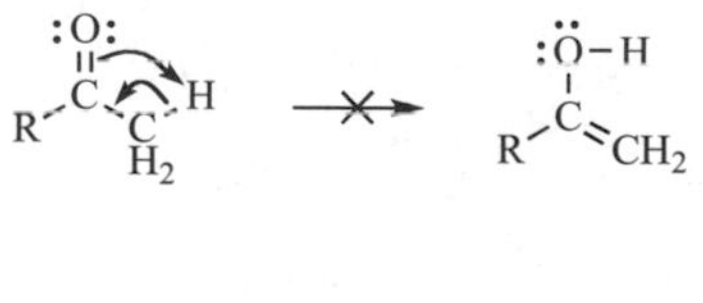

This shortcut certainly saves writing many structures and arrows, but there is good experimental evidence that this tautomerization definitely needs to be catalyzed by acid or base and will not proceed on its own. A four-center, four-electron process with very few exceptions (minor path 4e) does not occur thermally (see Section 7.2.5). The correct process is either the base- or acid-catalyzed process shown above, depending on pH. A proton-transfer variation (Section 7.4.3) is the less common push-pull route (Fig. 7.18).

Exercise

Using the previous structures as a guide, draw the arrows for the tautomerization of the enol form back to the keto form for both acidic and basic media. What pathways did you use?

Answer: The acidic route from the enol to the keto form is path A_E then p.t. Note that this backward route uses the reverse of each path.

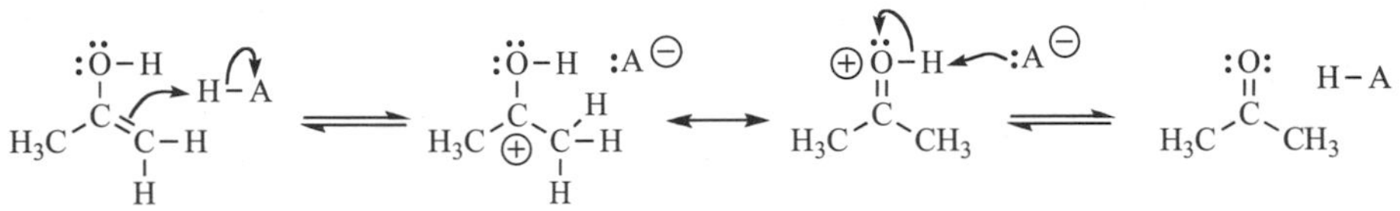

Enol Carbocation Keto

The basic route from the enol to the keto form is via proton transfer, path p.t. twice.

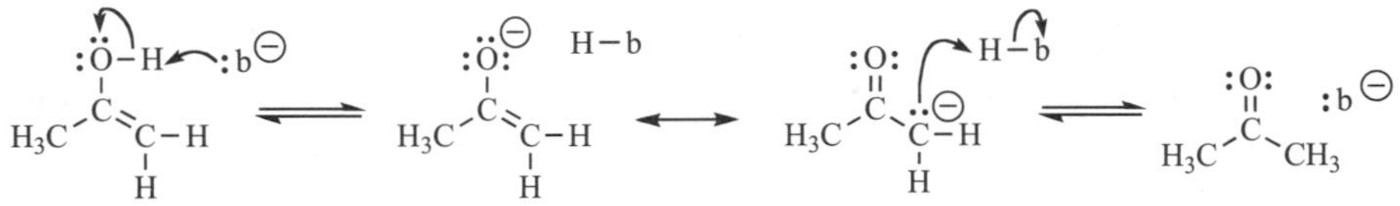

Enol Enolate Keto

7.4 VARIATIONS ON A THEME

The variety of the acids, bases, electrophiles, nucleophiles, and leaving groups may sometimes make an otherwise familiar reaction look unrecognizable. It is very important that you be able to recognize and classify groups into their respective generic classes.

7.4.1 Atom Variations

The pathways just discussed should be considered as generic; the route for electron flow is important, but the identity of the atoms in each path can easily change. For example, all of the electron flow paths were written generally, with Nu standing for any nucleophile, b for any base, L for any leaving group, and Y for any electronegative atom like oxygen, nitrogen, sulfur, or even C–ewg, an electronegative carbon atom. This section gives some examples of how the atoms can vary while the process remains essentially the same.

E2 Variants

Besides having several charge types, the E2 pathway has many variants, a few of which will be discussed in this section as illustrations of how the identity of the atoms can change while the electron flow stays the same. As was shown in Section 4.3.7, a heteroatom can be replaced for carbon and not affect the process at all:

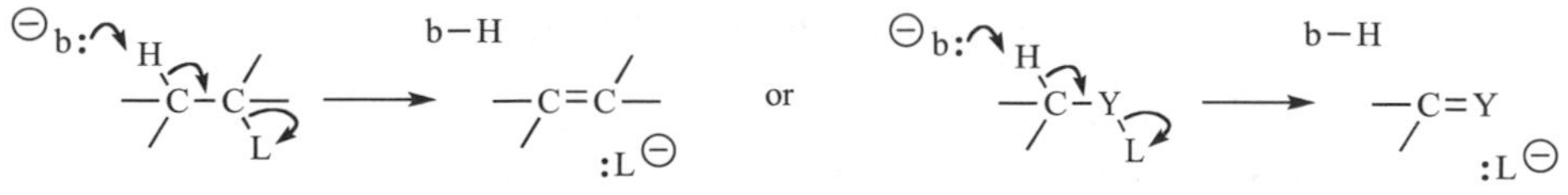

The dehalogenation reaction is a variant of the E2 where the H and L are now halogens, commonly bromine, and the base has been replaced by iodide. The orbital alignment restrictions are the same as for the E2 reaction:

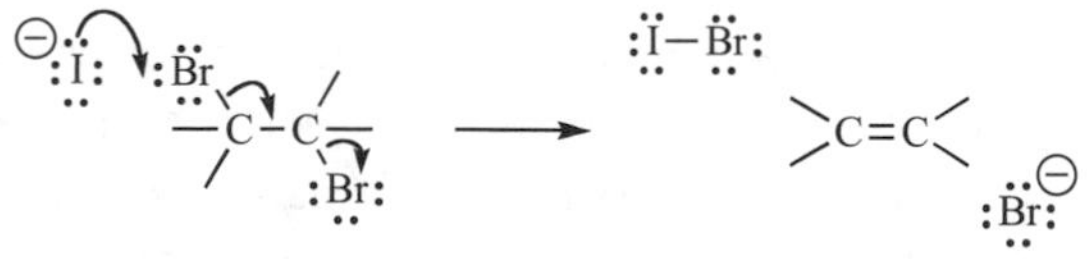

A fragmentation reaction is a reaction that breaks a molecule apart, usually into at least two pieces, by breaking a carbon–carbon bond. A fragmentation can be considered to be a type of elimination, similar to the E1cB or the E2, in which a weakened carbon–carbon single bond is broken. A fragmentation has the same path restrictions as the elimination it resembles. The fragmentation reaction below is an E2 elimination with another electron source replacing the base acting on a C–H bond. All orbitals that form double bonds must be coplanar and usually must be arranged as shown. The source can also be carboxlyate, in which Y is oxygen, and then the fragmentation would be called a decarboxylation and produce CO_2.

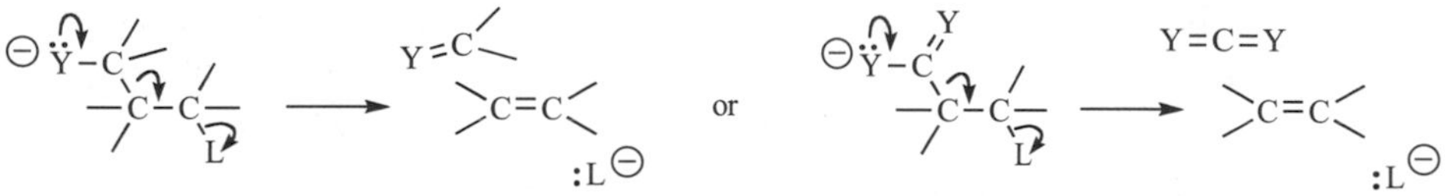

Enolates as Leaving Groups

The reaction shown on the right in the following structures (if Y is oxygen) also produces CO_2. It can be considered a variant of deprotonation adjacent to a polarized multiple bond. Again, we have changed the identity of the atoms but the electron flow remains the same. The overlap limitation for deprotonation, shown previously in Figure 7.1, similarly requires that the breaking C–C bond must align with the *p* orbital of the C=Z pi bond. Occasionally an electron flow can be viewed in more than one way; this decarboxylation reaction could also be considered an elimination in which the leaving group is an enolate.

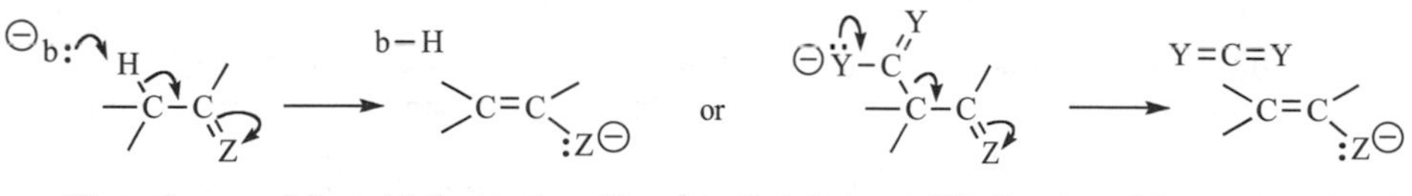

The reverse of the aldol reaction (Section 8.5.5) is an E1cB type of fragmentation in which the leaving group is an enolate:

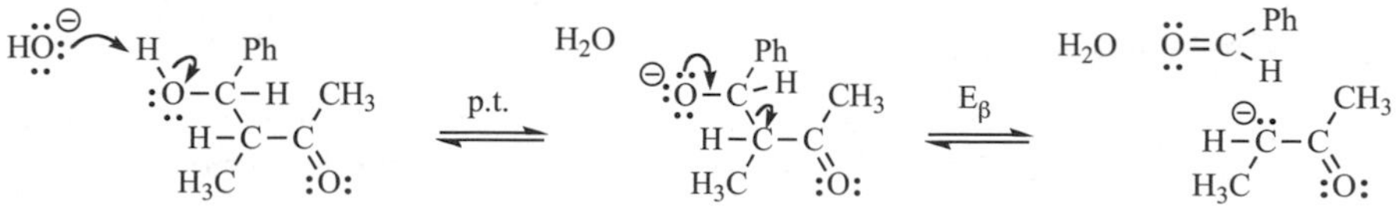

7.4.2 Vinylogous Variations

Additional conjugation may not change the system significantly (the principle of vinylogy, Section 1.9.2). Vinylogous paths are often denoted with a "prime" mark.

Vinylogous Substitutions at a Tetrahedral Center

The S_N2' reaction is just a vinylogous S_N2 reaction, where the nucleophile attacks a pi bond adjacent to the leaving group:

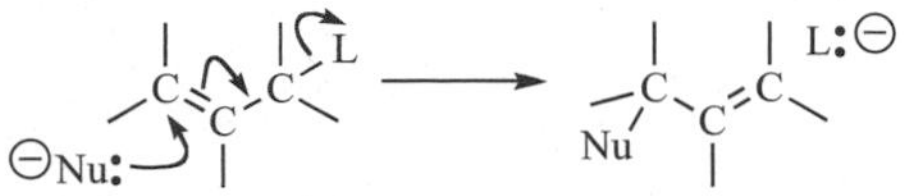

However, the overlap restrictions now include two more atoms and therefore may be more difficult to achieve. For the S_N2' the Nu, *p* orbital of carbon attacked, and C–L bond should be coplanar (often *syn*). This variant is found mostly in rigid systems where the C=C–C–L is locked into the proper alignment. Freely rotating systems go by the normal S_N2 reaction unless the S_N2 site is sterically blocked, and the S_N2' site is open. Triply bonded systems C≡C–C–L (propargyl leaving groups) do not have as critical an alignment problem; one of the two pi bonds will be close to the correct position.

The S_N1 also has a vinylogous version. In the following reaction, loss of the leaving group produces a resonance-delocalized allylic carbocation. Trapping of the carbocation on the other end of the allylic system gives an S_N1' reaction, a vinylogous S_N1. Trapping of the carbocation on the carbon that held the leaving group competes (the normal S_N1).

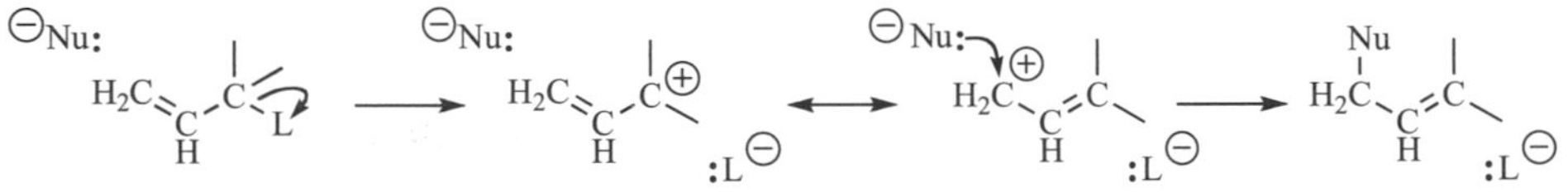

Vinylogous Substitutions at a Trigonal Planar Center

These electron sinks are much less common. If the ewg is a carbonyl and the leaving group is chloride, this electron sink is a vinylogous acid chloride (Cl–C=C–C=O). Just like all L–C=Y these groups substitute by addition-elimination.

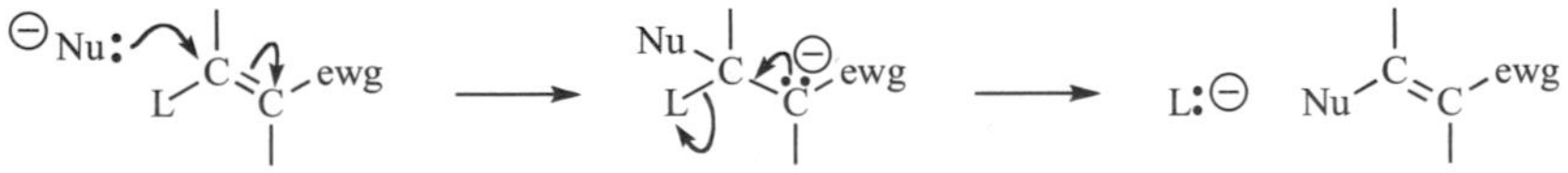

Vinylogous Eliminations

These vinylogous eliminations with one double bond between the C–H and C–L reactions are sometimes called 1,4-eliminations; a vinylogous E2 example follows. As in the E2, orbital alignment is important: both C–H and C–L must be coplanar with their adjacent *p* orbital on the pi bond. Freely rotating systems have little problem with this.

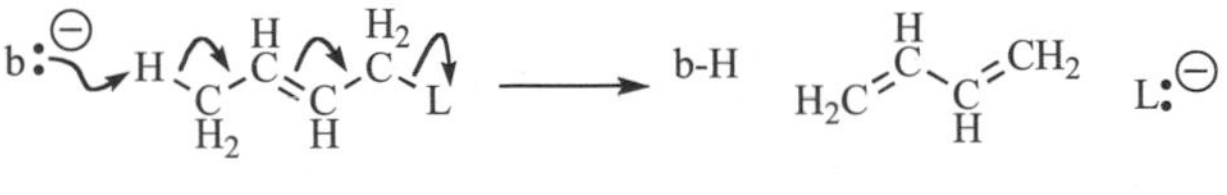

Vinylogous Additions

These vinylogous additions are the reverse of the vinylogous eliminations and are sometimes called 1,4-additions. The electrophile adds to form the allylic carbocation that can add the nucleophile at either end. The overall process, a vinylogous Ad_E2, often produces a mixture of 1,2 and 1,4 addition products (see Section 8.3.6 for details).

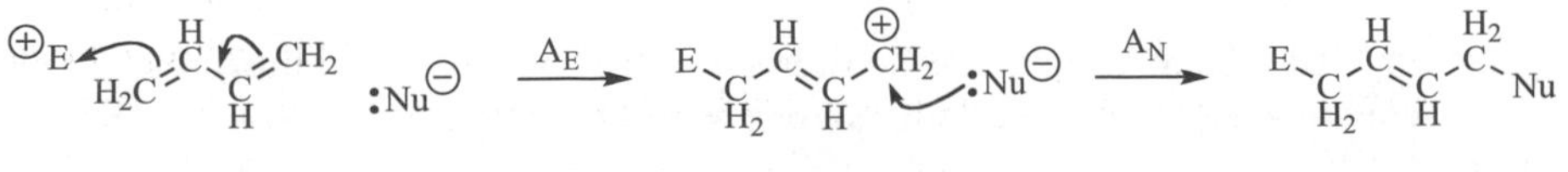
AE
AN

7.4.3 Extent of Proton Transfer Variations

Another topic that is appropriate under the variations-on-a-theme rubric is the extent of proton transfer between groups in protic solvents. If the rate of reaction catalyzed by an acid or a base depends on the pH of the medium and not the identity of the acid or base, then the reaction is called *specific acid* or *specific base catalyzed*. In the case of water, the specific acid is the protonated solvent, hydronium, and the specific base is the deprotonated solvent, hydroxide. Specific-acid-catalyzed reactions do not show a rate change if the identity or concentration of the acid is changed as long as the pH remains the same. *Specific-acid catalysis* indicates that the proton-transfer equilibria are fast and not involved in the rate-determining step. For similar base-catalyzed reactions, the term *specific-base catalysis* is used. However if the proton transfer is part of the rate-controlling step, then the concentration and pK_a of the acid matter, and the reaction is called *general acid* or *general base catalyzed*. If the presence of weaker proton donors, the [HA], contributes to the rate, the reaction is *general acid catalyzed*. *General-acid catalysis* indicates a reaction mechanism that has proton transfer occurring in the rate-determining step (for example, attack on a species hydrogen bonded to HA). For similar base-catalyzed reactions, the term *general-base catalysis* is used.

We have already seen one general acid-catalyzed electron flow path, the hetero Ad_E3, in Chapter 4. Here it is in more detail along with other general acid- and general base-catalyzed processes.

General Acid-Catalyzed Addition to a Polarized Multiple Bond

The hetero Ad_E3 (Fig. 7.14) involves nucleophilic attack on a hydrogen-bonded complex between the heteroatom lone pair and a weak acid and is a general acid-catalyzed addition. Note that the stereoelectronic requirements are different since the hydrogen-bonded lone pair lies in the pi nodal plane, and the nucleophilic attack occurs perpendicular to this plane. The reverse of this reaction is similar to the E2.

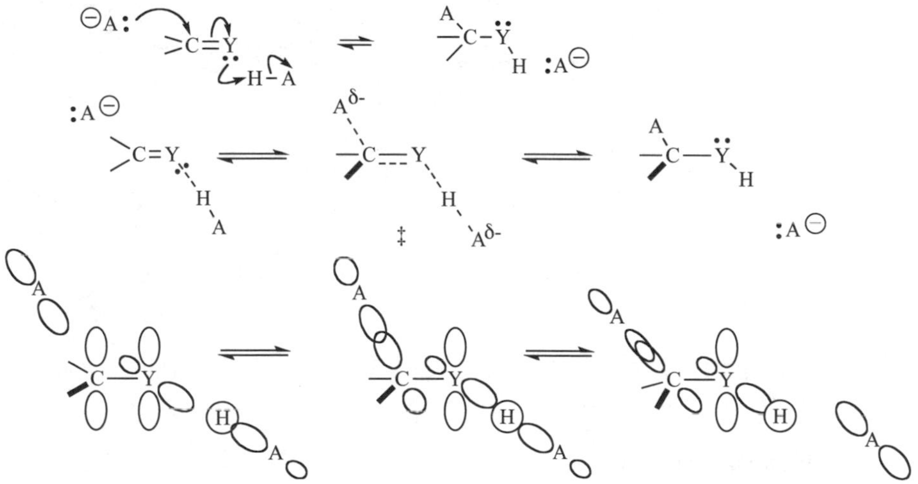

Figure 7.14 The hetero Ad_E3 addition with the transition state in the center.

Common Errors

The following set of arrows is a commonly drawn shortcut, and although it gets all the lines and dots in the right places, it ignores the fact that the hydrogen bond is usually to the lone pair and not to the less available pi bond.

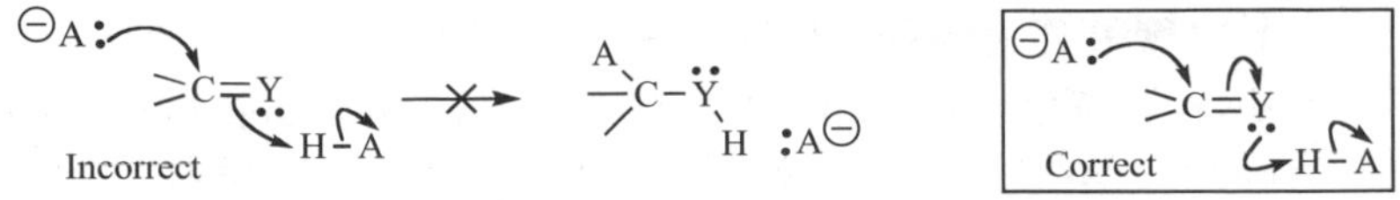
Incorrect
Correct

General Base-Catalyzed Addition to a Polarized Multiple Bond

This pathway (Fig. 7.15) is very similar to path Ad_N, with the difference that the nucleophile is poorer and is hydrogen bonded to a base when this pair collides with the polarized multiple bond. In this pathway the electron flow comes from the base to break the Nu–H bond, which in turn enhances the nucleophilicity of the nucleophile's lone pair. This lone pair attacks the multiply bonded carbon, breaks the pi bond, and produces a stable anion similar to path Ad_N.

Two processes are occurring to a varying extent: The hydrogen is being pulled off by the base, and the nucleophile lone pair is attacking the polarized multiple bond. If the nucleophile is especially poor, proton removal would need to be complete before a good enough nucleophile would be generated for the attack to proceed; this would then be path p.t. followed by path Ad_N. If the nucleophile is very good, proton removal would be unnecessary, and the reaction would go by path Ad_N. This path is used when there is a weak base present like acetate, $pK_{abH} = 4.8$.

An electronegative heteroatom, Y, can be replaced by an electronegative carbon atom, C–ewg. The electron-withdrawing group is extremely necessary, for without a means to stabilize the carbanion formed, the electron flow cannot occur. The enhancement of nucleophilicity by hydrogen bonding may occur in a protic solvent whenever the nucleophile bears a reasonably acidic hydrogen.

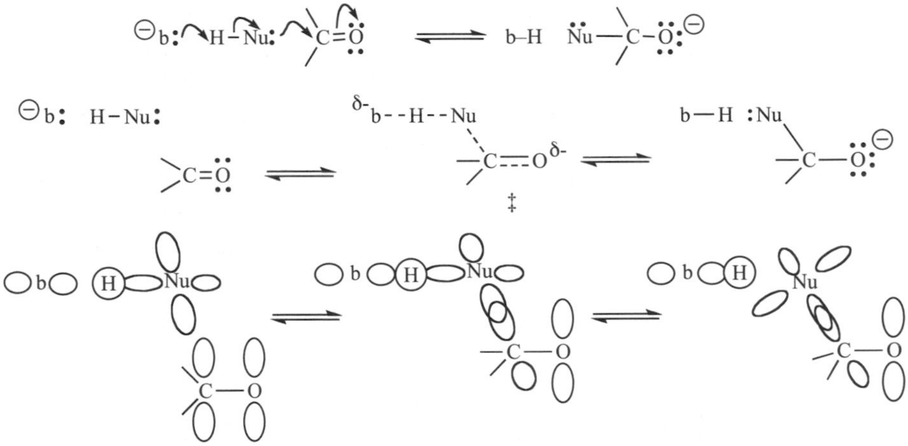

Figure 7.15 The general base-catalyzed addition to a polarized multiple bond with the transition state shown in the center.

Path Details

Overlap: Nu must approach the δ+ carbon from behind and close to the pi plane.
Media: A weak base is needed; the medium is usually weakly basic to neutral.
Solvent: Protic and polar to stabilize the anion formed.
Steric: The approach of the Nu to the sink is usually not a problem.
Energetics: Anion formed should be more stable than base or the reaction reverses.

Common Errors

The following set of arrows is the frequently drawn shortcut; it ignores the role of the lone pair and incorrectly implies that the sigma bond is acting as a nucleophile. The four-arrow description is more rigorous and agrees better with known enzymatic catalysis.

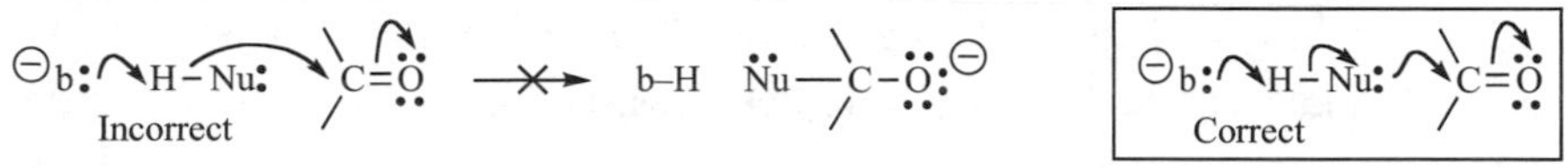

Occasionally an arrow in the middle of the flow is forgotten, quite often the breaking of a C–H or O–H bond. In the left example, the origin of the middle arrow is unclear; it appears to come from a lone pair position, but the lone pairs have been omitted. The correct path is the general base-catalyzed addition to a polarized multiple bond.

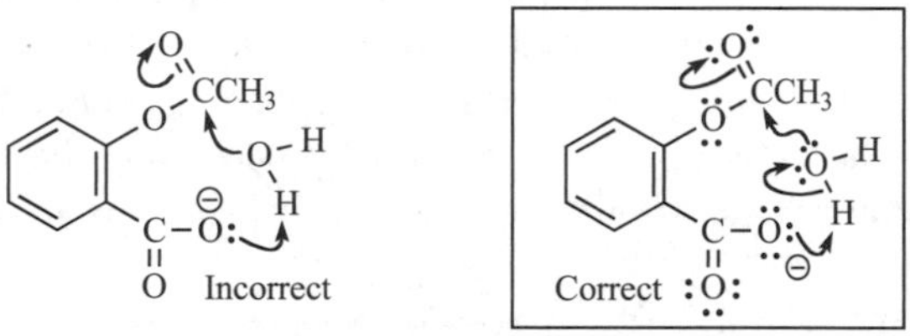

General Acid-Catalyzed Beta Elimination

(The Reverse of the Previous Reaction)

The difference between this path and path E_β is that the leaving group is poorer and must hydrogen bond to a weak acid (like RCOOH) to be good enough to depart. The hydrogen bond enhances the quality of the leaving group. This is the middle of three paths that involve different degrees of proton transfer. Catalysis by hydrogen-bonded species may occur any time a reasonably basic leaving group is lost in a protic solvent. If the leaving group were very poor, it would need to be fully protonated before it would be good enough to depart; this would be path p.t. followed by path E_β. If the leaving group is good, then proton transfer is unnecessary and the reaction would proceed by path E_β.

Path Details

Overlap: Anion *p* orbital and C–L bond close to coplanar (Fig. 7.15, right to left).
Leaving group: Average, but needs a lone pair to hydrogen bond to the weak acid.
Media: A weak acid is needed; the medium is usually weakly acidic to neutral.
Solvent: Protic and polar to stabilize anion.
Energetics: Use the ΔpK_a rule.

Common Errors

To save drawing an extra arrow, chemists may draw the following set of arrows, a shortcut that does get all the lines and dots in the right places but has some problems that are more obvious in this direction than they were in the microscopic reverse.

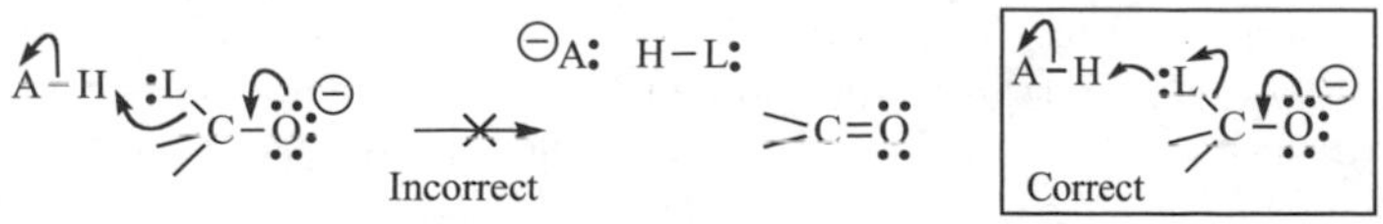

The problems with the three-arrow shortcut are that it ignores the necessary role of hydrogen bonding to the lone pair of the leaving group, that it protonates the unavailable sigma bond electron pair, and that it does not quite agree with intramolecular and enzymatic studies. This process is more rigorously drawn with four arrows to emphasize the role of the lone pair.

Summary of Polarized Multiple Bond Reactions

Figure 7.16 shows the top view of an energy surface that relates all three **basic media** mechanisms for addition of a nucleophile to a polarized multiple bond. The reactants are in the upper left corner, and the products are in the lower right corner. If the base is weak and the nucleophile is good, the nucleophile adds (path Ad_N) and later gets deprotonated (path p.t.). If the nucleophile is poor or the base very strong, deprotonation occurs first (path p.t.), followed by addition to the polarized multiple bond (path Ad_N), and this is called specific base catalysis. If the nucleophile is mediocre and a weak base is present, then a general base-catalyzed process (the diagonal route) occurs in which partial proton removal from the nucleophile enhances its ability to attack the C=Y.

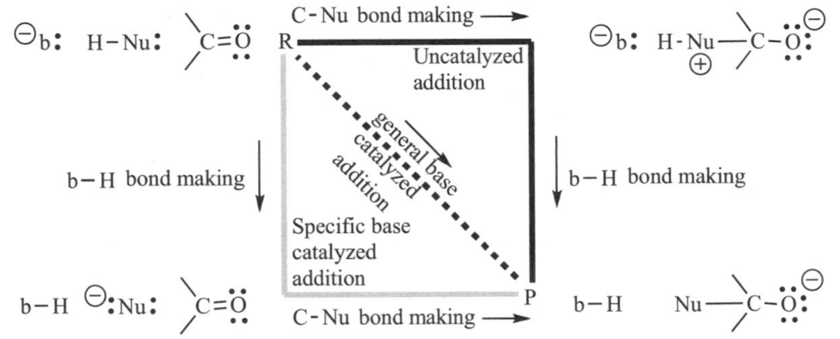
C-Nu bond making
R
Uncatalyzed addition
general base catalyzed addition
b-H bond making
b-H bond making
Specific base catalyzed addition
P
C-Nu bond making

Figure 7.16 Polarized multiple bond addition/elimination mechanisms in basic media. Reactants are in the upper left. The black route is an uncatalyzed addition. The gray route is the specific base catalyzed, and the dashed diagonal is the general base-catalyzed process.

Figure 7.17 is the top view of an energy surface of three **acidic media** mechanisms that can occur depending on the acidity of the media. Specific acid catalysis, protonation followed by nucleophilic attack (path p.t. then Ad_N) can occur in strong acid with weak nucleophiles. General acid catalysis, (diagonal) protonation along with nucleophilic attack (path Ad_E3) occurs in weak acids. Nucleophilic attack followed by protonation (path Ad_N then p.t.) would occur with good nucleophiles in weakly acidic media.

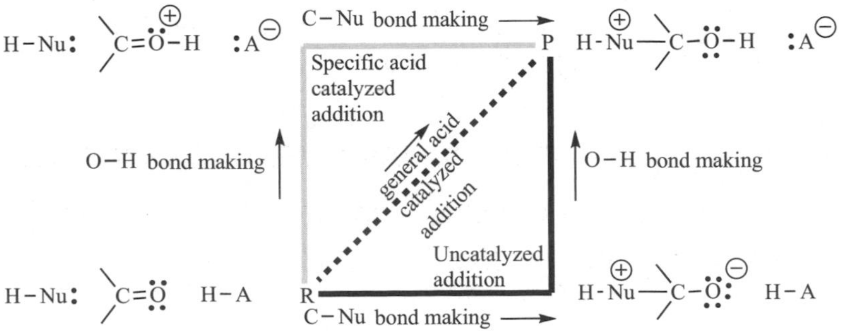
C-Nu bond making
P
Specific acid catalyzed addition
O-H bond making
general acid catalyzed addition
O-H bond making
Uncatalyzed addition
R
C-Nu bond making

Figure 7.17 Polarized multiple bond addition/elimination mechanisms in acidic media. Reactants are in the lower left. The black route is an uncatalyzed addition. The gray route is the specific acid catalyzed, and the dashed diagonal is the general acid-catalyzed process.

Push–Pull Catalysis of Enolization

As mentioned in Section 2.4, enzymes can have several catalytic groups in the cavity of their active site. When the reactant fits into that active site, these catalytic groups may function together in a push–pull catalysis. For example, the top view of the energy surface for enolization is shown in Figure 7.18. The reactants are in the lower left corner, and the products are in the upper right. The specific acid-catalyzed process forms the O–H bond before the C–H bond breaks (up, then right), whereas the specific base-catalyzed process breaks the C–H bond before the O–H bond forms (right, then up). The push–pull catalysis is the diagonal route in which the C–H bond breaks as the O–H bond forms.

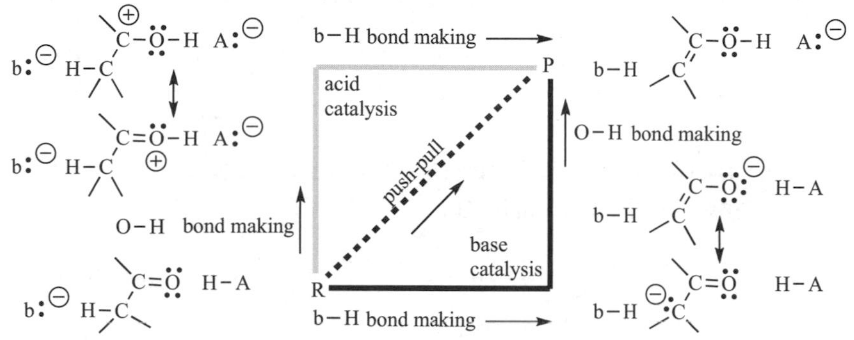
b–H bond making
P
acid
catalysis
O–H bond making
push-pull
O–H bond making
base
catalysis
R
b–H bond making

Figure 7.18 The energy surface for enolization. Reactants are in the lower left. The gray path is acid catalyzed, the black path is base catalyzed, and the dashed diagonal path is the push–pull path.

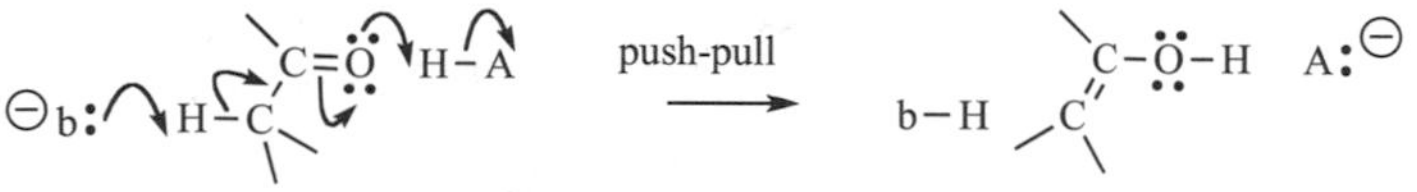
push-pull

We have now seen general acid- and general base-catalyzed processes for polarized multiple bond additions, eliminations, and enolizations. **There is good reason to expect that all reaction types, additions, eliminations, substitutions, and rearrangements have general acid- or general base-catalyzed routes under the correct conditions.** We can also expect that enzyme active sites with appropriately positioned acidic and basic groups can easily do general acid, general base, and push–pull catalysis.

Charge Types

This section should close with a reminder about charge types. Almost all of the pathways have several charge types. Nucleophiles can be anionic or neutral; electrophiles can be cationic or neutral; and leaving groups can depart as anions or neutrals. Reactions catalyzed by a Brønsted acid can often be similarly catalyzed by a Lewis acid or by a positively charged metal ion.

7.5 TWELVE MAJOR PATHS SUMMARY AND CROSSCHECKS

p.t., Proton Transfer to and from an Anion or Lone Pair

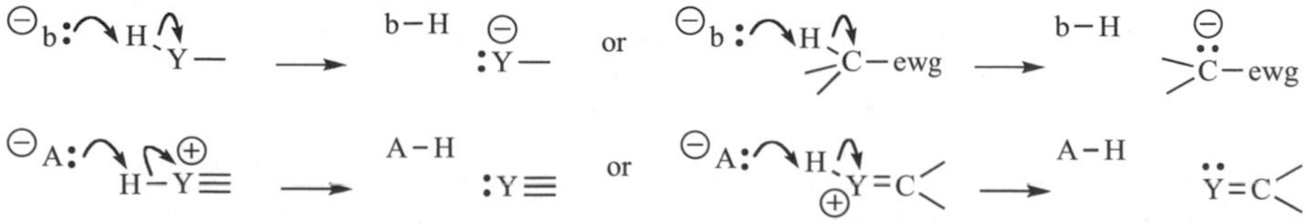

Both deprotonations (above) and protonations (below) Crosscheck: K_{eq} or ΔpK_a rule

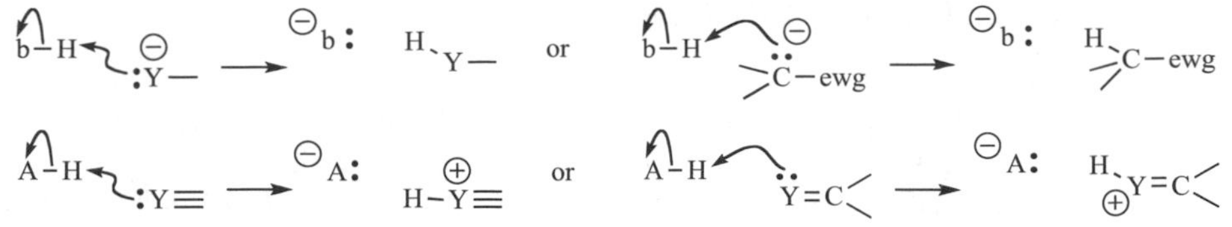

D_N, Ionization of a Leaving Group

Crosscheck: ΔpK_{aHL} less than zero, usually more stable than 2° cation, polar solvent

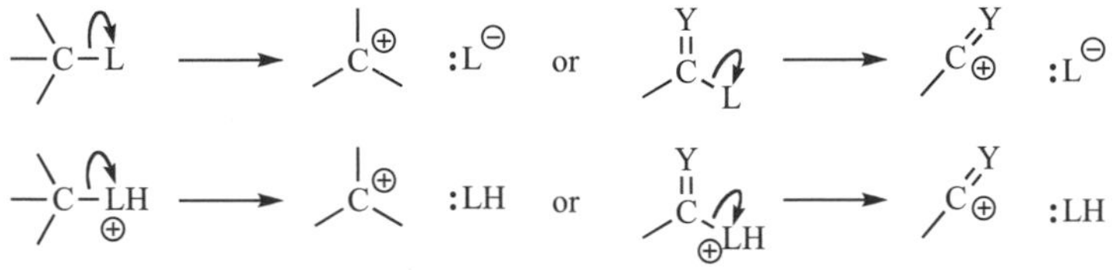

A_N, Trapping of an Electron-Deficient Species

Crosscheck: none, usually OK

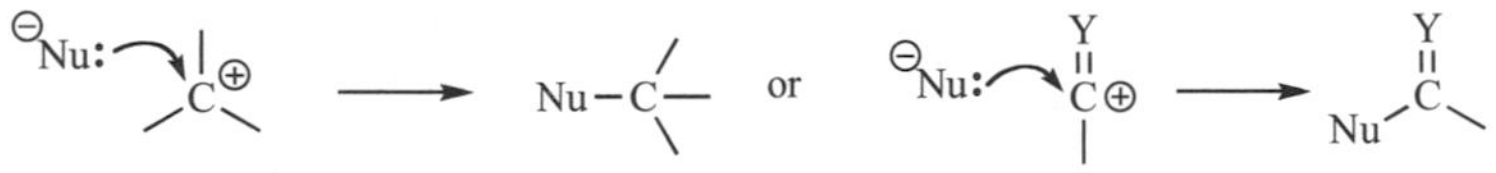

A_E, Electrophile Addition to a Multiple Bond

Crosscheck: Markovnikov's rule—make the most stable carbocation

D_E, Electrofuge Loss from a Cation To Form a Pi Bond

Crosscheck: Electrofuge lost should be reasonably stable

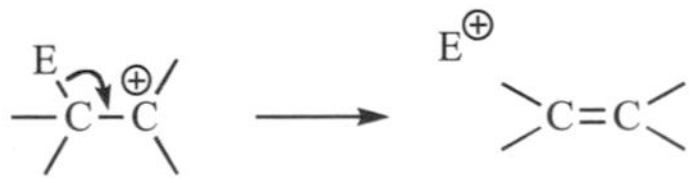

S_N2, The S_N2 Substitution

Crosscheck: ΔpK_a rule on Nu and L, $\Delta pK_{aHL} \leq 10$, good access to C attacked

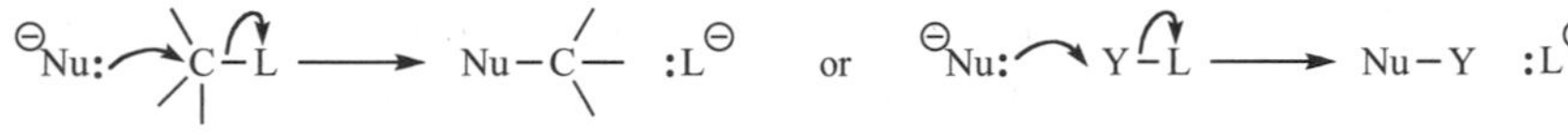

E2, The E2 Elimination

Crosscheck: ΔpK_a rule on base and L, C–H bond and C–L nearly coplanar

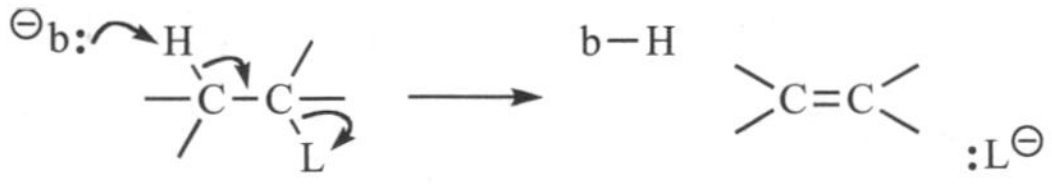

Ad_E3, The Ad_E3 Addition

Crosscheck: Markovnikov's rule followed because of polarization of pi-complex

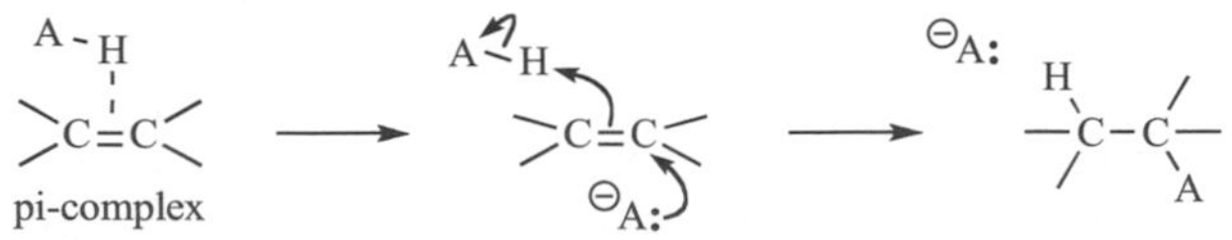
pi-complex

AdN, Nucleophilic Addition to a Polarized Multiple Bond

Crosscheck: ΔpK_a rule on Nu and anion formed

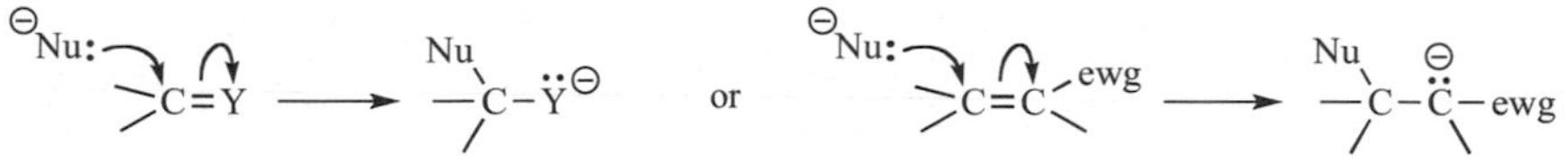

Eβ, Beta Elimination from an Anion or Lone Pair

Crosscheck: ΔpK_a rule on L and anion reactant

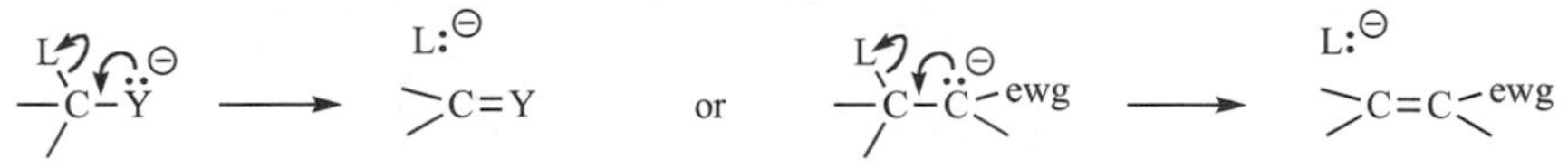

1,2R, Rearrangement of a Carbocation

Crosscheck: usually forms carbocation of equal or greater stability

Path 1,2RL, 1,2 Rearrangement with Loss of Leaving Group

Crosscheck (if no pushing Y): good L, usually better than 2° cation, polar solvent

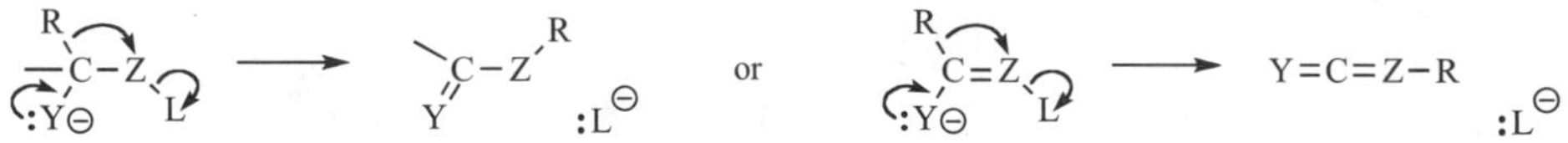

7.6 SIX MINOR PATHS SUMMARY

Path pent., Substitution via a Pentacovalent Intermediate

(Not for carbon, only for third-row and higher elements)

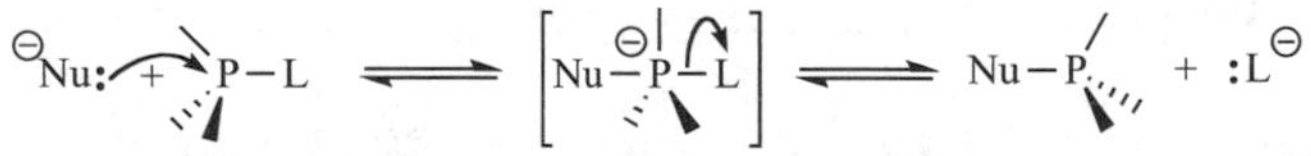

Path 6e, Concerted Six-Electron Pericyclic Reactions

Thermal rearrangements:

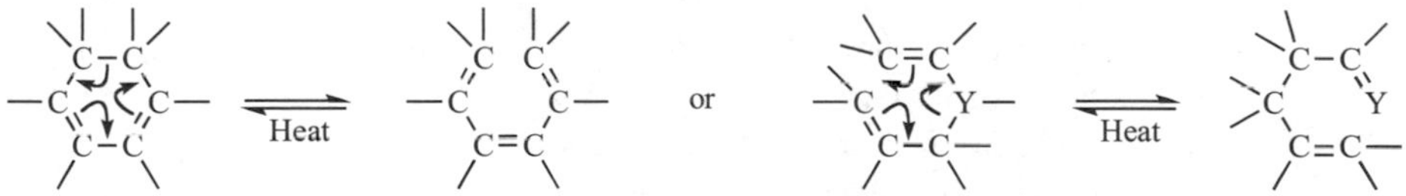

Thermal cycloadditions or cycloreversions (the reverse reaction):

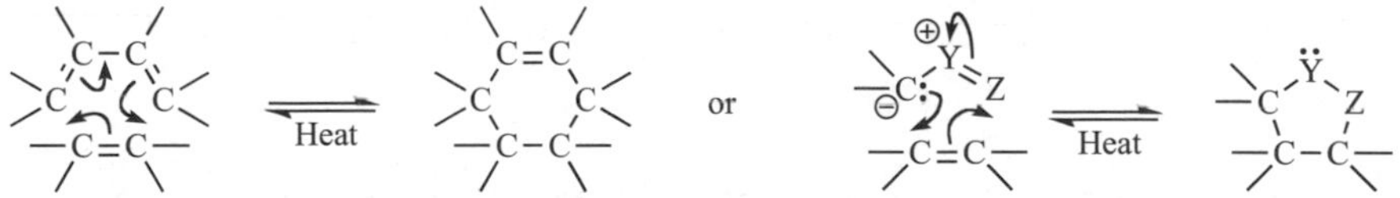

Metal-chelate-catalyzed additions:

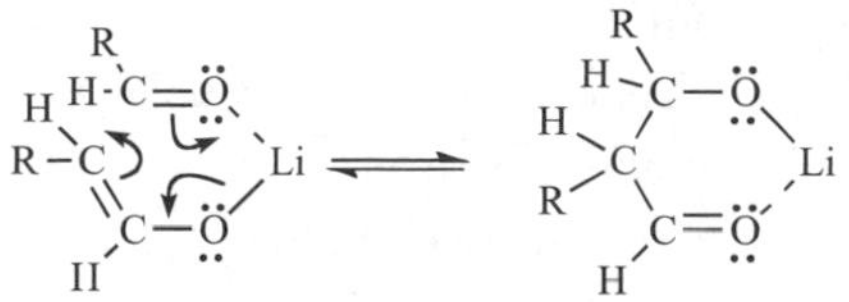

Path Ei, Thermal Internal *Syn* Elimination

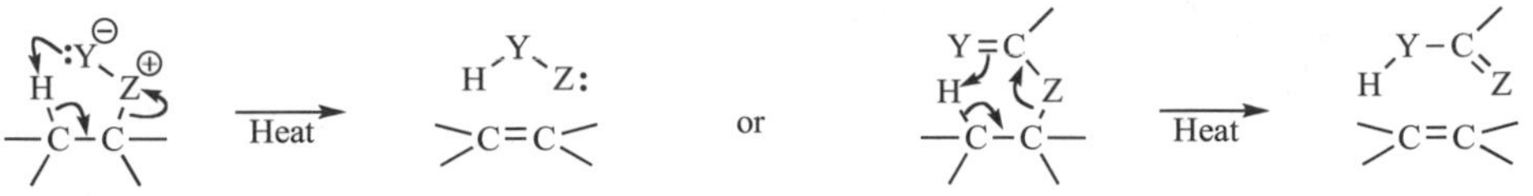

Path NuL, Nu–L Additions (three-membered ring formation)

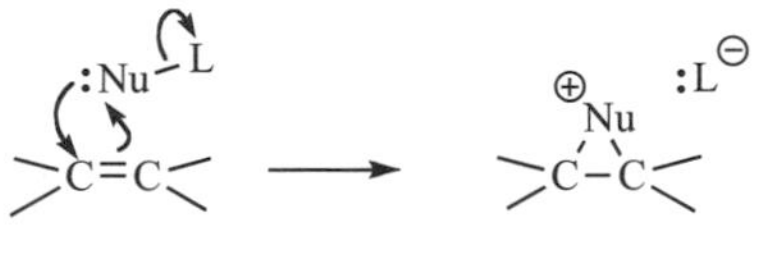

Path 4e, Four-Center, Four-Electron (three cases only)

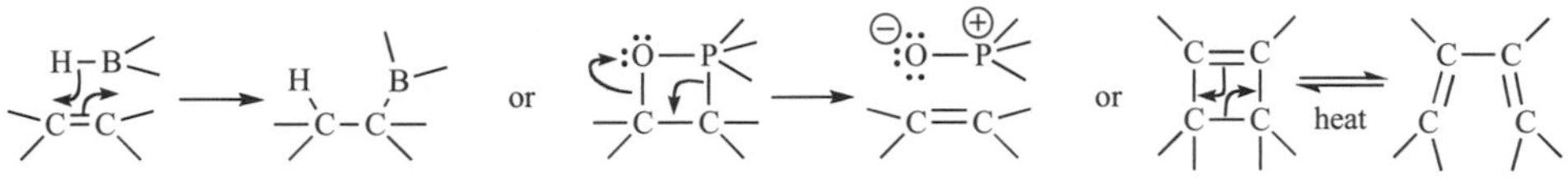

Path H⁻ t., Hydride Transfer to a Cationic Center

ADDITIONAL EXERCISES

7.1 Look at problem 1.5, and place the name of the path over each transformation arrow.

7.2 Look at problem 1.6, and place the name of the path over each transformation arrow.

7.3 Look at problem 4.9, and determine which path(s) were used in each mechanism.

7.4 Look at problem 4.10, and determine which path(s) were used in each mechanism.

7.5 Draw a general example of each of these most common pathways.
(a) path p.t., Proton transfer to a lone pair
(b) path D_N, Ionization of a leaving group
(c) path A_N, Trapping of a cation by a nucleophile
(d) path A_E, Electrophile addition to a multiple bond
(e) path D_E, Electrofuge loss from a cation to form a pi bond
(f) path S_N2 substitution

7.6 Draw a general example of each of these most common pathways.
(a) path E2 elimination
(b) path Ad_E3 addition
(c) path Ad_N, Nucleophilic addition to a polarized multiple bond
(d) path E_β, Beta elimination from an anion
(e) path 1,2R, Rearrangement of a carbocation
(f) path 1,2RL, Rearrangement with loss of a leaving group

7.7 Draw a general example of each of these common path combinations.
(a) S_N1 substitution
(b) Ad_E2 addition
(c) E1 elimination
(d) Electrophilic aromatic substitution

7.8 Draw a general example of each of these common path combinations.
(a) E1cB elimination
(b) Ad_N2 addition
(c) Addition–Elimination route of carboxyl derivatives
(d) Tautomerization

7.9 Sort the path combinations in problems 7.7 and 7.8 into those that prefer predominantly acidic media, basic media, or both.

7.10 A few major electron flow pathways have critical requirements. What are the requirements for the following paths that, if not met, the path will probably not happen?
(a) p.t., Proton transfer to a lone pair
(b) D_N, Ionization of a leaving group
(c) S_N2 substitution
(d) E2 elimination
(e) 1,2R, Rearrangement of a carbocation

7.11 Many reactions give product mixtures. Give the product of an S_N2 and also the product of an S_N2' on the following compound. Use Nu to symbolize the nucleophile.

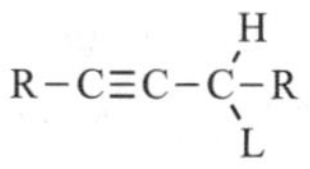
H
R−C≡C−C−R
L

7.12 From biochemical studies, a general acid catalyst is most effective if its pK_a is close to the pH of the medium. Explain this by examining the extremes. What would happen if the acid catalyst were very acidic, a much lower pK_a; then consider if the acid catalyst were not very acidic, a much higher pK_a?

7.13 Classify the reaction below into one of the four archetypes, and list the possible paths. Using the path restrictions, pick the only path or common path combination that fits for this reaction. Notice how this analysis process turns what appears to be an open-ended problem into a multiple-choice problem. Write out all the steps of the mechanism.

(CH3(CH2)3)3P:
+
I
(CH3(CH2)3)3P
I

7.14 Classify the reaction below into one of the four archetypes, and list the possible paths. Using the path restrictions, pick the only path or common path combination that fits for this reaction. Notice how this analysis process turns what appears to be an open-ended problem into a multiple-choice problem. Write out all the steps of the mechanism.

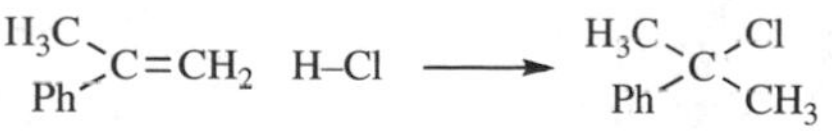
H3C
C=CH2
Ph
H–Cl
H3C
Cl
C
Ph
CH3

7.15 Classify the **base-catalyzed** (sodium hydroxide) reaction below into one of the four archetypes, and list the possible paths. Using the path restrictions, pick the only path or common path combination that fits for this reaction. Notice how this analysis process turns what appears to be an open-ended problem into a multiple-choice problem. Write out all the steps of the mechanism.

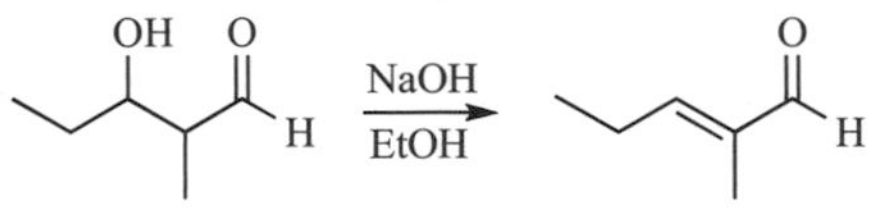
OH
O
H
NaOH
EtOH
O
H

7.16 The mechanism for this reaction under **acidic** conditions is: p.t., substitution; give the product and the mechanism with all its steps. Remember to choose between the two types of substitutions for carbon, S_N1 or S_N2.

$$(CH_3)_3COH + HCl \longrightarrow$$

7.17 The mechanism for the **acid-catalyzed** (sulfuric acid) Fischer esterification reaction below is: addition, p.t., p.t., elimination; write out the mechanism with all its steps. Remember to choose between the three types of additions, Ad_E2, Ad_N2, or Ad_E3, then choose between the three types of eliminations, E1, E2, or E1cB.

O
Ph
O
H
+ ROH
H—A
O
Ph
O
R
+ H2O

7.18 The mechanism for the following reaction in methanol is: p.t., addition, then elimination. A **base-catalyzed** addition (sodium hydroxide) is followed by an **acid-catalyzed** elimination (hydrochloric acid). Write out the product and the mechanism with all its steps. Note the aldehyde has no acidic hydrogens to remove. Remember to choose between the three types of additions, Ad_E2, Ad_N2, or Ad_E3, then choose between the three types of eliminations, E1, E2, or E1cB.

$$PhCHO + CH_3NO_2 \xrightarrow[\text{2) HCl}]{\text{1) NaOH}}$$

7.19 The mechanism for the **base-catalyzed** (sodiun ethoxide) Claisen condensation reaction below is: p.t., addition, elimination, p.t.; write out the mechanism with all its steps. Remember to choose between the three types of additions, Ad_E2, Ad_N2, or Ad_E3, then choose between the three types of eliminations, E1, E2, or E1cB.

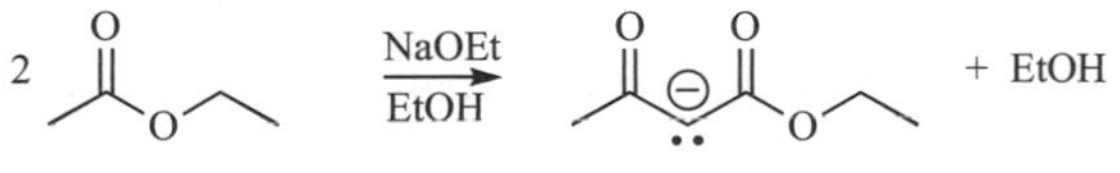
2
O
O
NaOEt
EtOH
O
O
O
+ EtOH

7.20 The mechanism for this **base-catalyzed** (sodiun ethoxide) reaction in ethanol is: p.t., substitution, p.t., substitution; write out the mechanism with all its steps. Remember to choose between the two types of substitutions for carbon, S_N1 or S_N2.

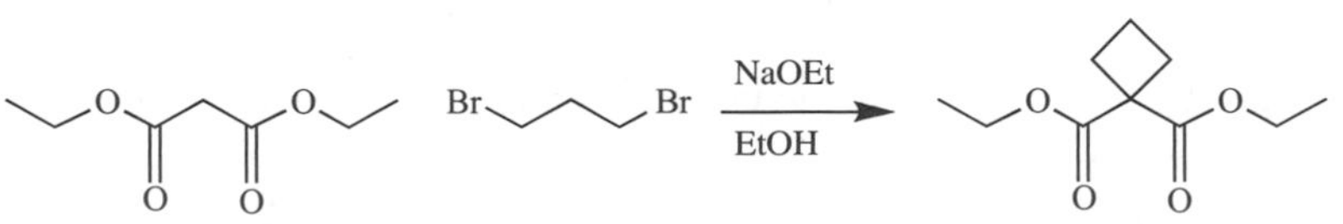
O
O
O
O
Br
Br
NaOEt
EtOH
O
O
O
O

8

INTERACTION OF ELECTRON SOURCES AND SINKS

Electron Flow In Organic Chemistry: A Decision-Based Guide To Organic Mechanisms, Second Edition.
By Paul H. Scudder

The flow of electron density, symbolized by a set of arrows, comes from the generic electron sources discussed in Chapter 5, via the pathways of Chapter 7, and ends in one of the generic electron sinks covered in Chapter 6. This chapter links common sources and sinks with appropriate pathways and discusses specific examples of each process.

8.1 SOURCE AND SINK CORRELATION MATRIX

A correlation matrix is the best way to display all the simple interactions between two groups so that no interactions are accidentally left out. In Chapters 5 and 6, we identified the electron sources and sinks that will be reaction partners. Those interactions of the generic sources with sinks are displayed with a matrix, each cell of the matrix corresponding to a specific combination of source and sink. Almost all of the reactions in organic chemistry can be described as simple bimolecular collisions between one of the 12 generic electron sources (Section 5.7) and one of the 18 generic electron sinks (Section 6.6). Many reaction mechanisms are multistep processes in which the product of the initial collision of source and sink then collides with another reactant and continues on. A giant matrix containing all the electron sources and sinks can be drawn, but it would be 12 × 18 and thereby contain 216 individual cells. This large matrix contains many cells in which no reaction occurs and many cells that have been little researched.

For simplification, a much smaller 7 × 7 matrix (Table 8.1) will be used that contains representatives of the more common sources and sinks. Within each cell of the matrix is the subsection in Sections 8.2 to 8.8 that covers the appropriate pathways for electron flow for that particular combination of source and sink. Any important combinations not included in the simplified matrix are covered in Sections 8.9 to 8.12.

Table 8.1 Correlation Matrix for the More Common Electron Sources and Sinks: An Index to Sections 8.2 through 8.8

	Common Electron Sources						
Electron Sinks	Lone pair Z:	Base and Adjacent CH	Complex Metal Hydride MH_4^-	Organo-metallic R–M	Allylic :Z–C=C	Simple Pi Bonds C=C C≡C	Aromatics 
H–A	**8.2.1**	**8.2.2**	**8.2.3**	**8.2.4**	**8.2.5**	**8.2.6**	**8.2.7**
Y–L	**8.3.1**	**8.3.2**	**8.3.3**	**8.3.4**	**8.3.5**	**8.3.6**	**8.3.7**
sp^3 C–L	**8.4.1**	**8.4.2**	**8.4.3**	**8.4.4**	**8.4.5**	**8.4.6**	**8.4.7**
C=Y	**8.5.1**	**8.5.2**	**8.5.3**	**8.5.4**	**8.5.5**	**8.5.6**	**8.5.7**
C≡Y	**8.6.1**	**8.6.2**	**8.6.3**	**8.6.4**	**8.6.5**	**8.6.6**	**8.6.7**
C=C–ewg	**8.7.1**	**8.7.2**	**8.7.3**	**8.7.4**	**8.7.5**	**8.7.6**	**8.7.7**
L–C=Y	**8.8.1**	**8.8.2**	**8.8.3**	**8.8.4**	**8.8.5**	**8.8.6**	**8.8.7**

8.2 H–A SINKS REACTING WITH COMMON SOURCES

8.2.1 Lone Pair Sources Reacting with Acids

Acids react by simple protonation of the electron source via proton transfer (see Section 3.6 for the calculation of the K_{eq}).

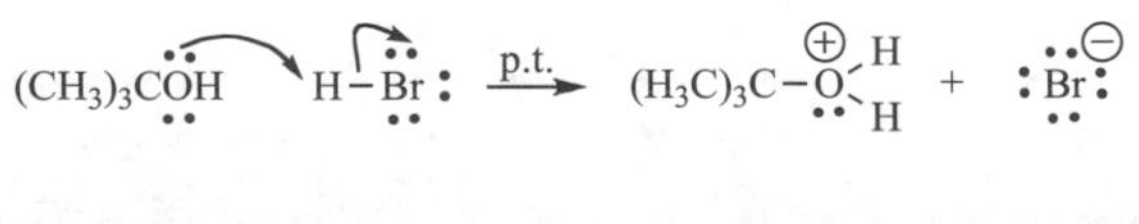
$(CH_3)_3COH$
H–Br:
p.t.
$(H_3C)_3C-O$
H
H
+
:Br:

8.2.2 Bases Reacting with Acids Having an Adjacent CH

This simple neutralization reaction always favors the weaker base. Thus the most acidic H is deprotonated, even if the acid has an adjacent CH. Because of the higher electronegativity of heteroatoms, protons on heteroatoms are usually lost before protons on carbon. The base *n*-butyllithium deprotonates the nitrogen of diisopropyl amine. The adjacent CH is about 14 orders of magnitude less acidic. Always check the K_{eq} to be sure your proton transfer is energetically reasonable. The K_{eq} for the reaction below is highly favorable, $10^{(50-36)} = 10^{+14}$.

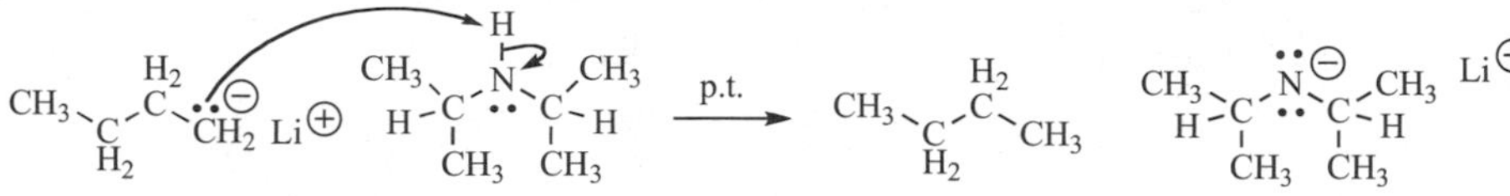

8.2.3 Complex Metal Hydrides Reacting with Acids

Aluminum hydrides react violently with protic solvents and acids (path p.t.) to produce hydrogen gas and can be used only in *aprotic* media, like diethyl ether. The borohydrides, BH_4^- are weaker sources and react much more slowly with protic solvents than do the aluminum hydrides. After the first step, the trivalent aluminum or boron species is now electron deficient and can react with a lone pair donor to form a tetravalent Lewis acid–base salt (path A_N) that can donate hydride again. This repeats until all hydrogen–metal bonds have been used.

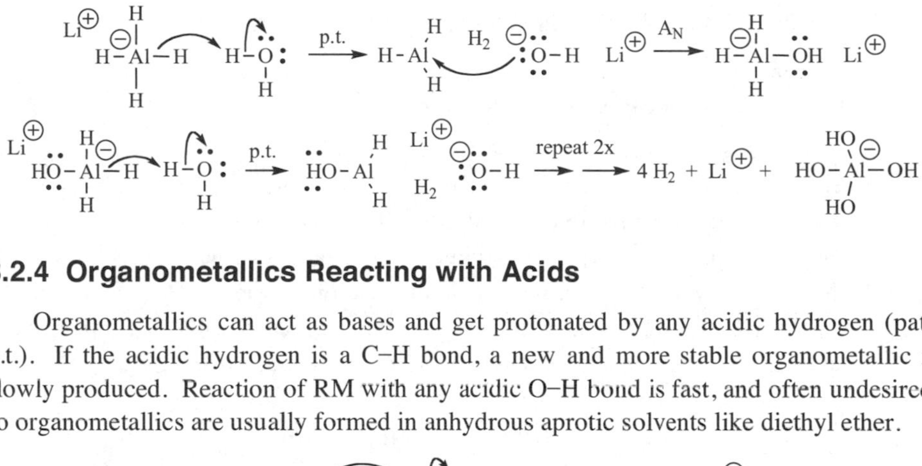

8.2.4 Organometallics Reacting with Acids

Organometallics can act as bases and get protonated by any acidic hydrogen (path p.t.). If the acidic hydrogen is a C–H bond, a new and more stable organometallic is slowly produced. Reaction of RM with any acidic O–H bond is fast, and often undesired, so organometallics are usually formed in anhydrous aprotic solvents like diethyl ether.

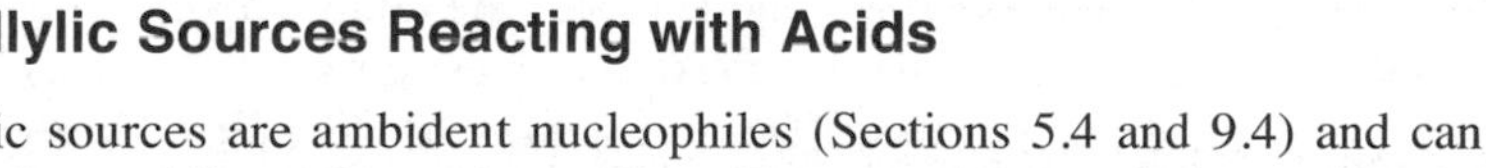

8.2.5 Allylic Sources Reacting with Acids

Allylic sources are ambident nucleophiles (Sections 5.4 and 9.4) and can therefore attack an electrophile at either of two sites. Since proton transfer is commonly reversible and rapid, an equilibrium mixture is quickly achieved. When the source is anionic, the protonation occurs most rapidly on the heteroatom (hard–hard, path p.t.) but can also occur on carbon. Equilibration to the more stable product occurs by proton transfer (Section 7.3.8, tautomerization). A ΔH calculation (Section 2.5) will verify that the carbon-protonated species is the more stable product.

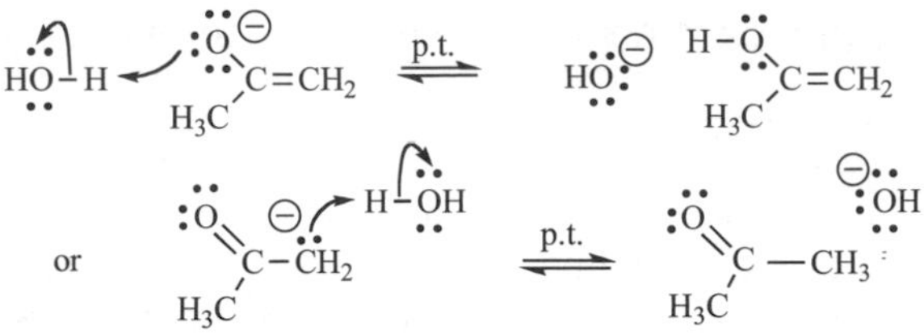

When the source is neutral, protonation can occur to produce the heteroatom-protonated species or the delocalized heteroatom-substituted carbocation. Protonation on the more basic heteroatom lone pair is the lower-energy process, but the heteroatom-protonated species usually just returns to reactants.

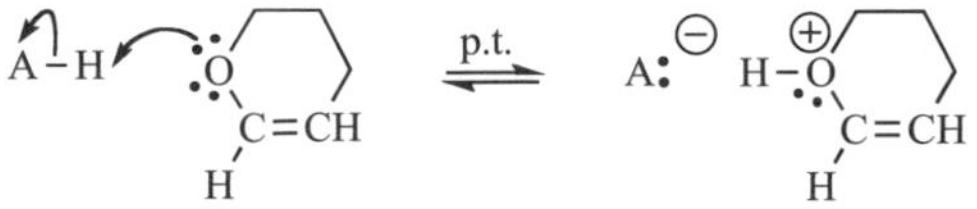

However, protonation on carbon creates a delocalized heteroatom-substituted carbocation that can easily be captured (path A_N) by solvent or another nucleophile.

Ad_E2 example (with follow-up proton transfer to solvent):

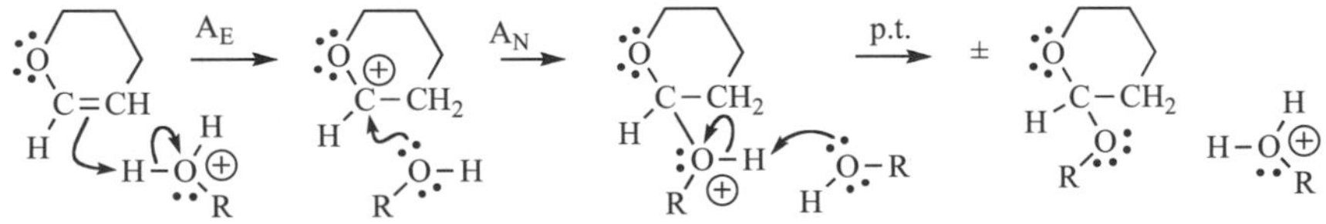

8.2.6 Simple Pi Bonds Reacting with Acids

Electrophilic Addition of H–A to Alkenes

As was discussed in detail in Sections 4.4.1 through 4.4.4, here are two mechanisms by which acids add to alkenes, the Ad_E2 and Ad_E3 processes. The reaction is carried out in the dark to minimize free radical side reactions (Chapter 11). In the Ad_E2, the proton adds by pathway A_E to produce the more stable carbocation intermediate (Markovnikov's rule); in a second step the cation is trapped by a nucleophile, path A_N. The Ad_E2 often gives a mixture of *syn* and *anti* addition since the nucleophile can approach the carbocation from either top or bottom face. Ad_E2 example:

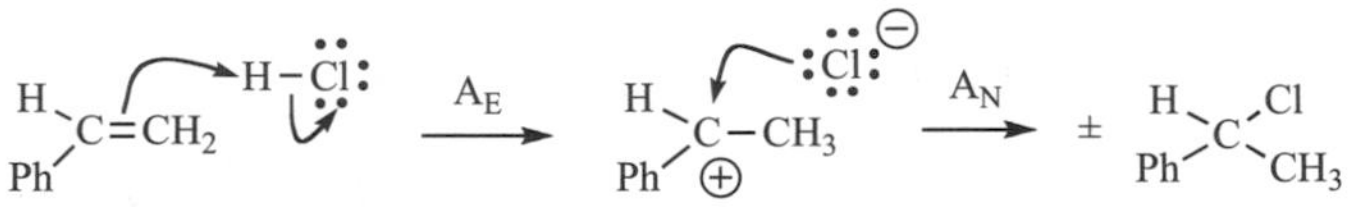

The Ad_E3 occurs when the cation is not as stable. Instead of proton transfer, a pi-complex forms, which polarizes and activates the double bond. The nucleophile attacks the largest partial plus of the complex and gives the same Markovnikov orientation of product as the Ad_E2 process. The pi-complex usually blocks one face of the double bond, therefore the nucleophile adds to the opposite face, giving overall *anti* addition as the predominant process. Ad_E3 Example:

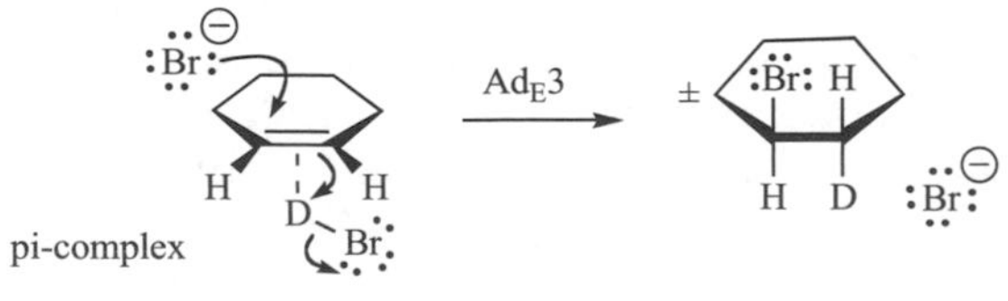

Electrophilic Addition of H–A to Alkynes

The two perpendicular pi bonds of an alkyne react independently of one another. Alkynes commonly undergo Markovnikov *anti* addition of acids by the Ad_E3 process (Section 7.1.8) because the Ad_E2 would go via a less stable vinyl cation intermediate. Slightly more vigorous conditions will favor the double-addition product.

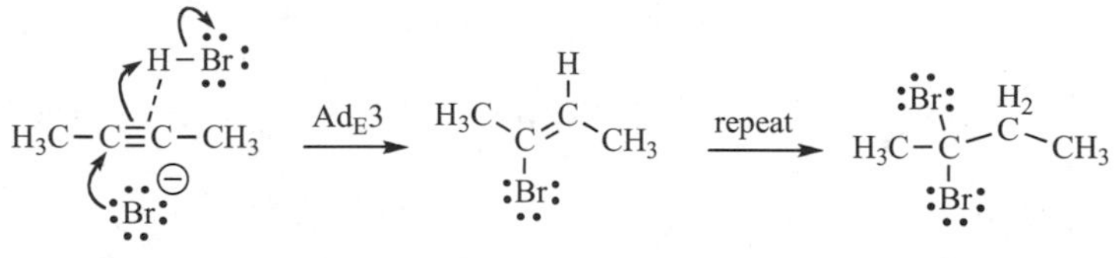

Ad_E3
repeat

Electrophilic addition can be followed by tautomerization (Section 7.3.8).

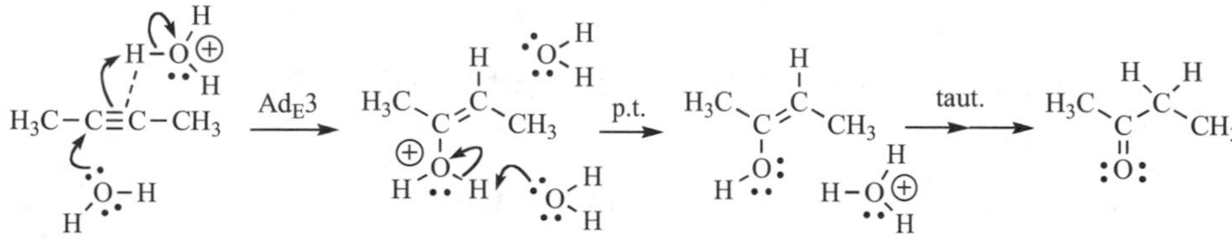

Ad_E3
p.t.
taut.

Electrophilic Addition of H–A to Dienes C=C–C=C

An electrophile will attack a diene to produce the delocalized allylic cation. The partial charge on each end of the allylic system will be different if the substitution at each end of the allylic unit is different. Since the attraction of unlike charges contributes greatly to the bringing of nucleophile and electrophile together (hard–hard), **the larger the partial plus is on an atom, the greater a negatively charged nucleophile is attracted to it.**

For the simplest diene, 1,3-butadiene, electrophile attack creates an allylic cation where the greatest partial plus is on the end bearing the methyl donor group.

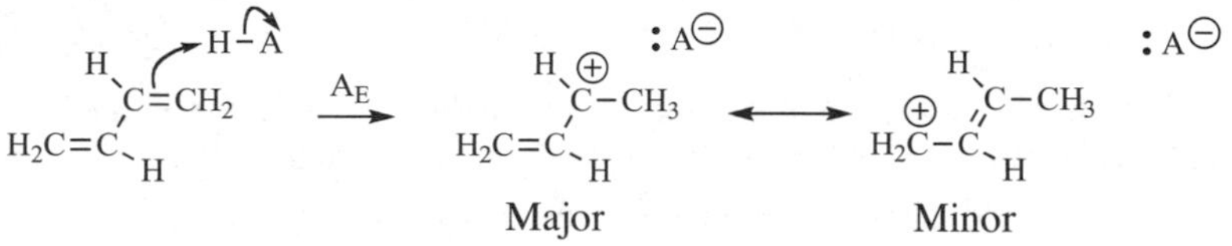

A_E
Major
Minor

If the rate of nucleophile attack governs the product distribution (the nucleophilic attack is not reversible at the given temperature), the major product will be the 1,2-addition product that results from Nu attack at the greatest partial plus (kinetic control, Section 2.6). However, if the addition of the nucleophile is reversible (higher temperature or the Nu is a good L), then the 1,4-addition product will be formed because the more substituted double bond is the more stable product (thermodynamic control, Section 2.6). Suspect kinetic control if the reaction temperature is significantly below 0°C. Diene example (Ad_E2):

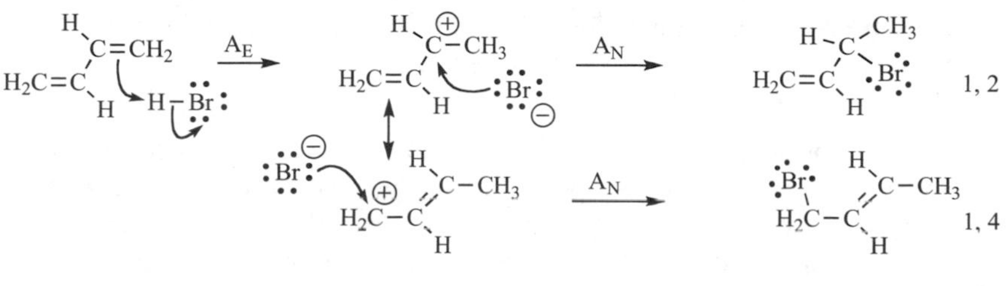

A_E
A_N
A_N
1, 2
1, 4

8.2.7 Aromatics Reacting with Acids

Aromatics Reacting with H–A

This reaction is noticeable only if *ipso* attack triggers rearrangement, or loss of a group on the ring, or if a deuterated acid is used. One possible mechanism for desulfonation of an aromatic ring is shown below.

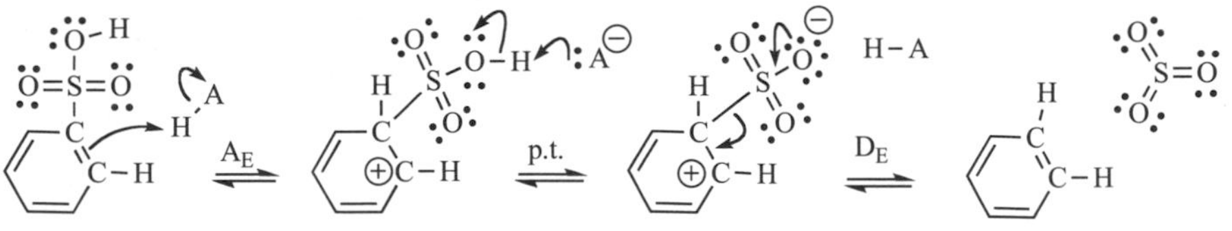

8.3 Y–L SINKS REACTING WITH COMMON SOURCES

8.3.1 Lone Pair Sources Reacting with Y–L

Since Y often represents an electronegative atom, Y^+ is usually a poor cation, so S_N1 is ruled out, commonly leaving S_N2. The ΔpK_a rule is helpful in determining the position of equilibrium in the displacement of a leaving group by a nucleophile.
Example (with follow-up proton transfer):

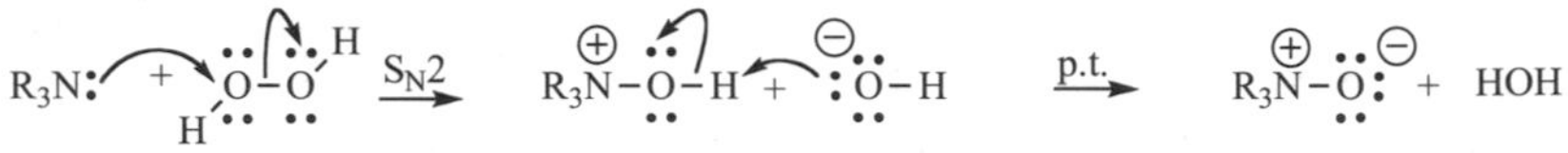

Sulfur example—formation of a tosylate from toluenesulfonyl chloride and an alcohol:

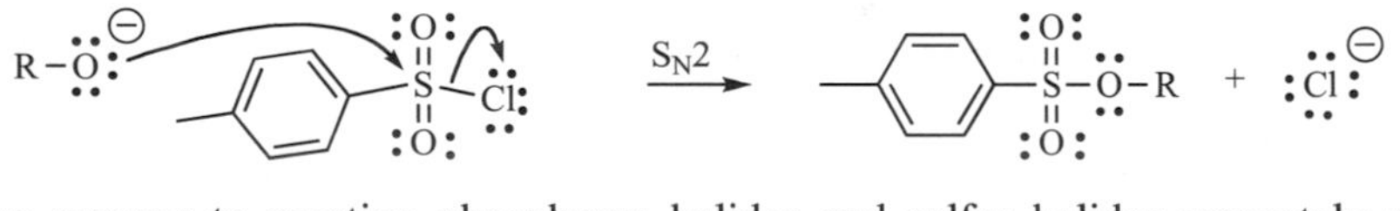

Two reasons to mention phosphorus halides and sulfur halides separately are that although sulfur probably goes by the S_N2, most likely the substitution at phosphorus goes via the pentacovalent intermediate (path pent.). Second, the phosphorus-oxygen bond is so strong that follow-up reaction often occurs in which O–P serves as a leaving group. Overall, an OH group is converted to a better leaving group, then is displaced by a nucleophile, commonly the leaving group on the Y–L. Phosphorus example—with follow-up S_N2 substitution, covered in Section 8.4.1:

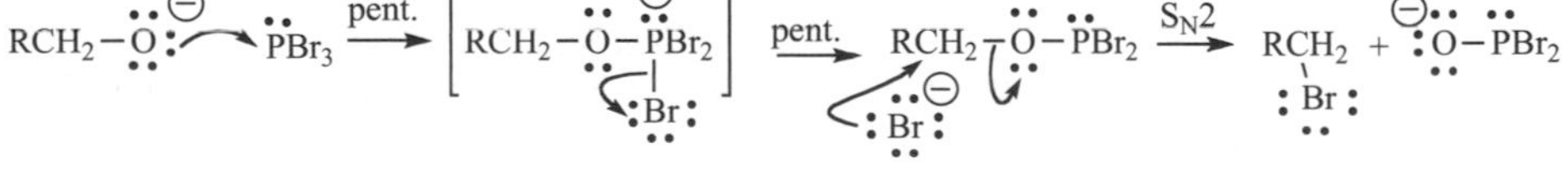

A follow-up elimination to the alkene sometimes occurs (see next subsection).

Biochemical Example

Adenosine triphosphate, ATP^{4-}, is the body's energy currency and a Y–L sink. It activates compounds by phosphorylation. The mechanism of the enzyme glycerol kinase (Fig. 8.1) is postulated to go by a pentacovalent phosphorus intermediate. A base in the active site makes the oxygen lone pair more nucleophilic by deprotonating it as it attacks (general base catalysis, Section 7.4.3). The magnesium ion acts as an electrophilic catalyst (Section 6.2.3), improving the leaving group, ADP^{3-}.

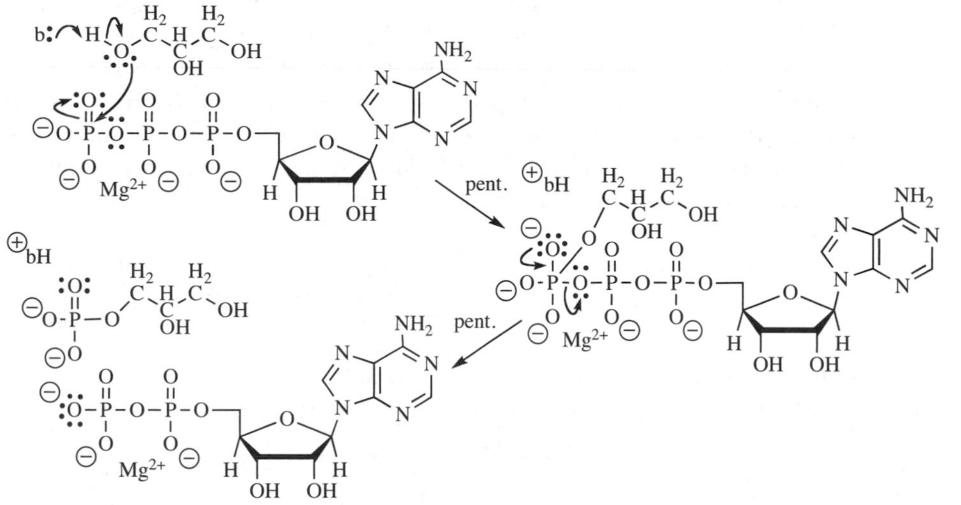

Figure 8.1 Phosphorylation of glycerol by ATP^{4-}.

8.3.2 Bases Reacting with Y–L Having an Adjacent CH, Oxidations

Since Y often represents an electronegative atom, Y^+ is usually a poor cation. Therefore, the E1 path is out (Section 4.3.7). The bases used in this reaction are usually weak, leaving the E2 or the Ei as the common route. For examples of this reaction see Section 8.12.3, Nu–L Reacting with sp^3 C–L, and Section 8.10.2, Metal Ions as Oxidants.

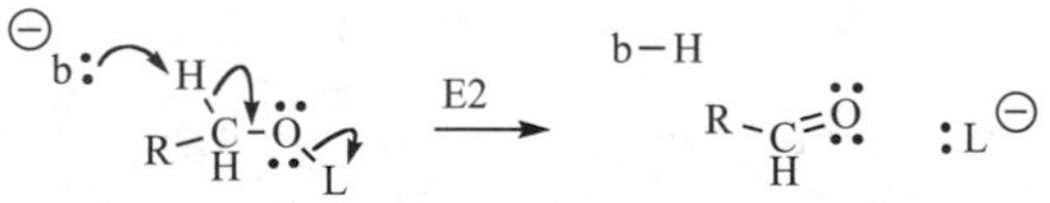

8.3.3 Complex Metal Hydrides Reacting with Y–L

The substitution reaction occurs in neutral or basic media via the S_N2 pathway to replace the leaving group with hydrogen. Often the metal cation can form a complex with the leaving group to aid in its leaving.

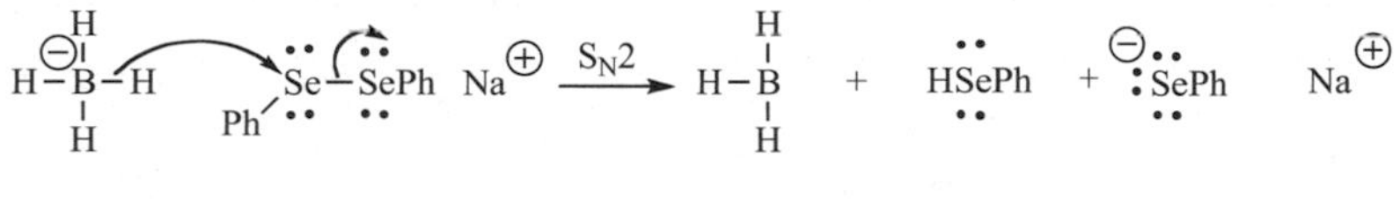

8.3.4 Organometallics Reacting with Y–L

Since organometallics react with acids (Section 8.2.4), they can be used only in basic or neutral media and in aprotic solvents. The most common reaction is S_N2 substitution.

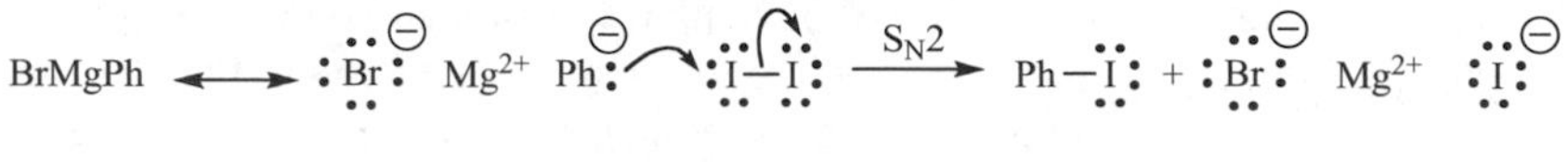

8.3.5 Allylic Sources Reacting with Y–L

An allylic electron source can be considered to be a vinylogous lone pair. Sometimes the lone pair itself is the nucleophile and can undergo reactions similar to the lone pair sources in the first column of the matrix. However, most often the nucleophile is the end carbon atom of the pi bond. The heteroatom lone pair is much harder than the pi bond, and the decision of which to use as the nucleophile is based on the hardness of the electrophile (HSAB principle) and will be discussed at greater length in Section 9.4. The Y–L reaction is almost exclusively by the S_N2 pathway because Y^+ is usually a poor cation. However, the substitution at phosphorus and silicon usually goes via the pentacovalent intermediate (path pent.). The following examples show the dual reactivity of the same allylic source, the enolate ion. Z as Nu (hard E, Si–Cl bond polarized because of a large electronegativity difference):

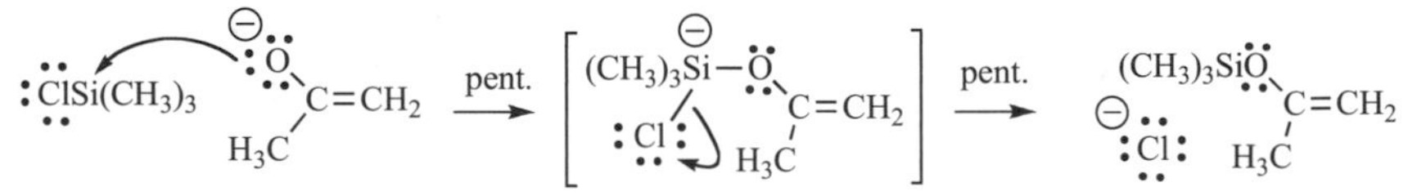

C as Nu (soft E, Br–Br bond not polarized at all):

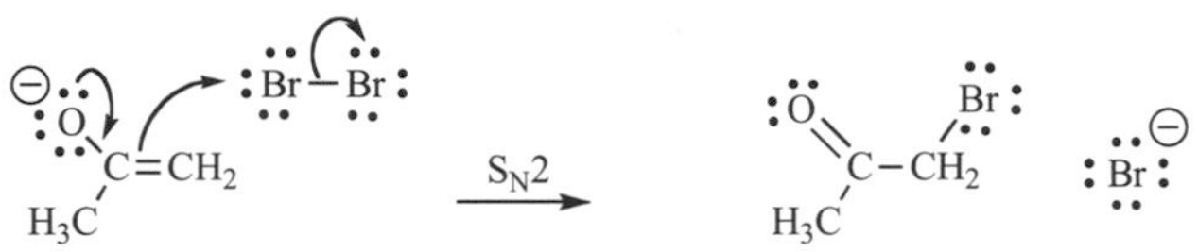

8.3.6 Simple Pi Sources Reacting with Y–L

Electrophilic Addition of Y–L to Alkenes

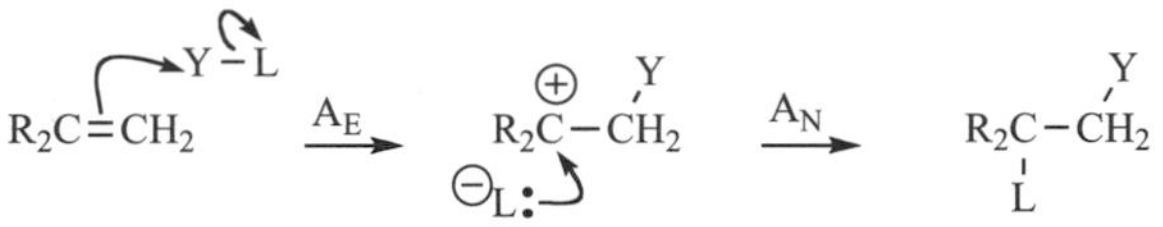

This reaction and the electrophilic addition of acids compose almost all of the ionic reactions of simple alkenes as electron sources. When Y bears a lone pair, the intermediate carbocation can be further stabilized by sharing the lone pair electrons with the cationic center in a bridge to form a three-membered ring, resonance form B in the following example. Although the Ad_E2 process (A_E, followed by A_N) was drawn above, it would often be proper to use the Nu–L path (Section 7.2.4) to get to the bridged onium ion directly, followed by path S_N2 to open the bridge. An example is bromination, in which the bromide leaving group traps the cation when run in nonnucleophilic solvent.

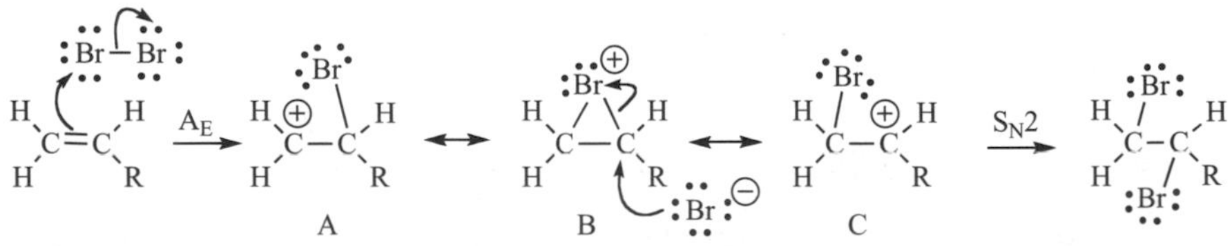

Bridging gives all atoms an octet, and some energy is released in forming a new bond, but this is at the expense of creating ring strain and putting a partial positive charge on the heteroatom. Highly electronegative elements like fluorine and oxygen are relatively poor at bridging. Less electronegative elements have a greater tendency to share their lone pairs and bridge. Stabilization of the carbocation by electron-donor groups favors the open form. Bridging is favored if the cation is poorly stabilized.

The second step in the electrophilic addition of Y–L to alkenes is the attack of a nucleophile on the cationic intermediate by the original leaving group or the solvent. The bridging group is displaced by an S_N2 path, so the nucleophile comes in from the opposite face, giving overall *anti* addition. The site of attack will depend on which resonance form, A or C, contributes more to the overall hybrid. The nucleophile will usually attack the carbon with the greater partial plus. Since the bromine could have attacked the pi bond from either the top or bottom face, the product is racemic. Solvent trap example (same reaction run in water solvent):

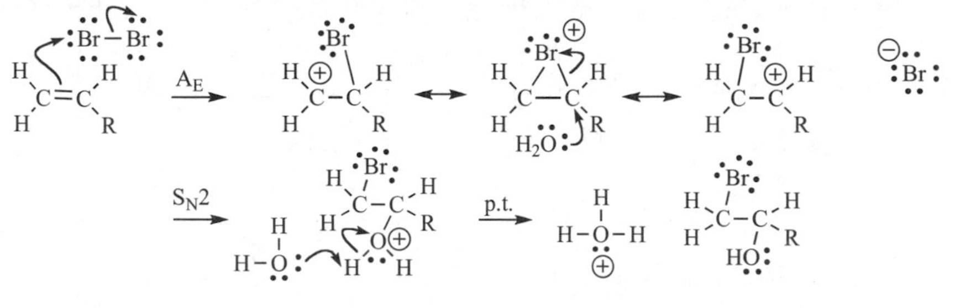

Electrophilic Addition of Y−L to Alkynes

Alkynes undergo the same set of reactions as alkenes but are slightly less reactive because the intermediates involved are less stable. For the Ad_E2 process, the vinyl cation intermediate formed is less stable than the alkyl carbocations formed when electrophiles attack alkenes. Addition to the vinyl cation produces a mixture of *syn* and *anti* addition. Stabilizing the vinyl cation by bridging is less favorable since the bridged ion is more strained and may have some antiaromatic character.

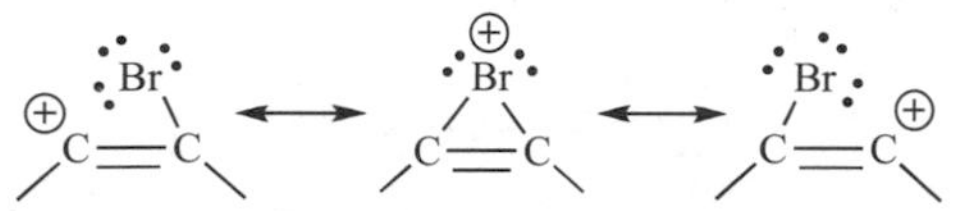

The Ad_E3 process competes with the Ad_E2 and can make the overall addition *anti*. The Ad_E2 process is seen when there is some additional stabilization of the vinyl cation. Double additions are frequent. Often the product of an electrophilic addition is less reactive than the original alkyne, and thus the reaction can be stopped before a second addition occurs (if run with an equimolar amount of electrophile and alkyne). Ad_E3 example (note the *anti* addition):

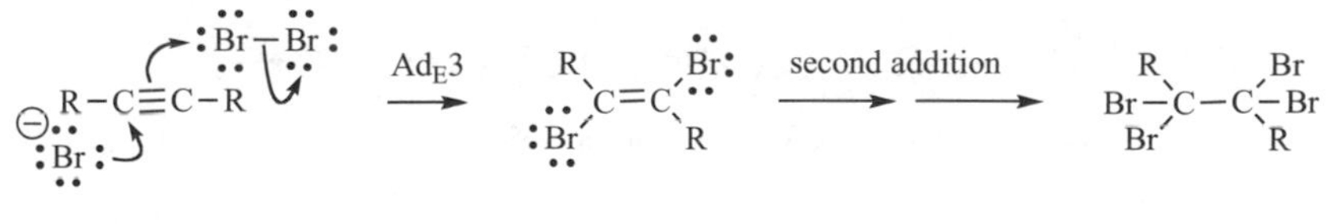

Electrophilic Addition of Y−L to Dienes

Since the allylic cation formed upon addition of an electrophile to an alkene is relatively stable, dienes add Y−L by the Ad_E2 process. Both 1,2 and 1,4 addition can occur. At higher temperatures, the more stable 1,4 product predominates.

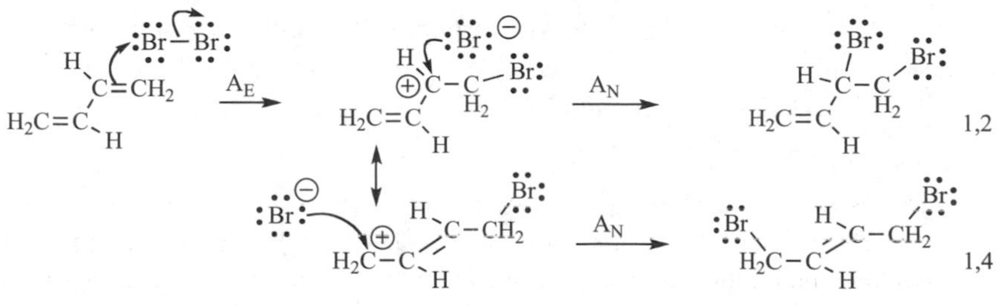

8.3.7 Aromatic Sources Reacting with Y–L

Because of the lower reactivity of aromatic sources, many of the electrophiles are cationic, and their generation and use are similar to an S_N1 process, as in the following nitration reaction. Protonation on nitric acid's more basic oxygen atoms is reversible and does not produce the electrophile. Only when nitric acid's OH is protonated can water be lost to give the resonance-delocalized NO_2^+ electrophile needed for the reaction.

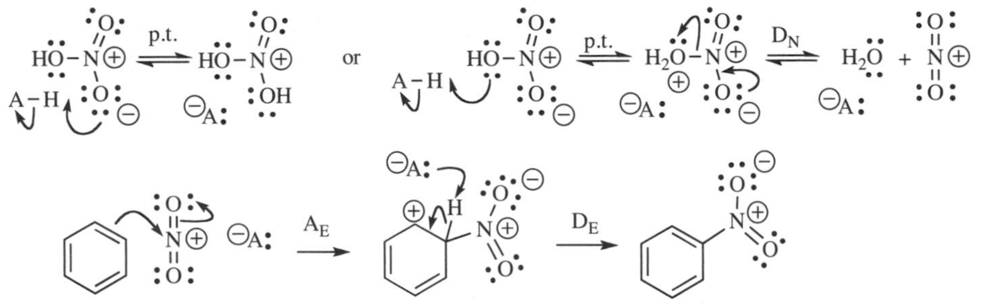
p.t.
or
p.t.
D_N
A_E
D_E

Other reactions merely require that a Lewis acid catalyze the departure of the leaving group similar to an S_N2 process on the electrophile, as in the following chlorination reaction, catalyzed by aluminum chloride.

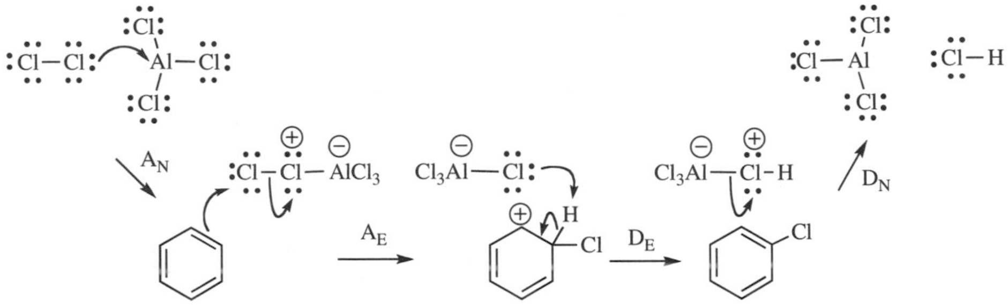
A_N
A_E
D_E
D_N

8.4 *sp*3 C–L SINKS REACTING WITH COMMON SOURCES

8.4.1 Lone Pair Sources Reacting with *sp*3 C–L

The substitution versus elimination decision is considered in depth in Section 9.5.

Alkylation

As was covered in Section 4.2, there are two routes that differ only in the timing of when the nucleophile attacks and when the leaving group falls off: **Generic S_N1**—Leaving group departs first by path D_N, producing a reasonably stable cation; the cation is then trapped by the nucleophile, path A_N. **Generic S_N2**—Simultaneous attack of nucleophile and loss of the leaving group. It is best to consider a S_N1/S_N2 spectrum where there is competition between the rate of ionization of the leaving group and the rate of nucleophilic attack on the substrate or partially ionized substrate.

S_N2 example (poor cation):

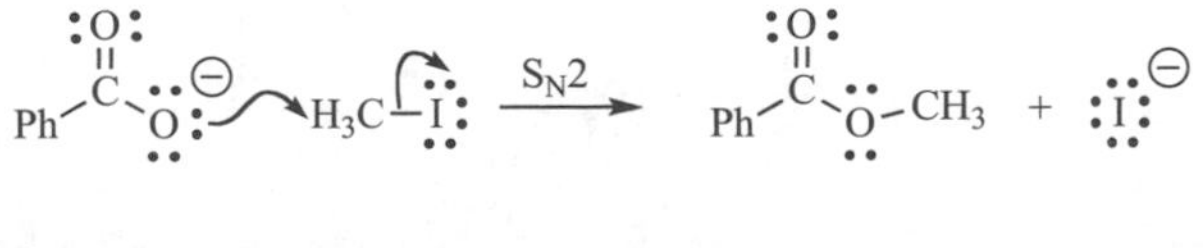
S_N2

S_N1 example (hindered site) with follow-up proton transfer:

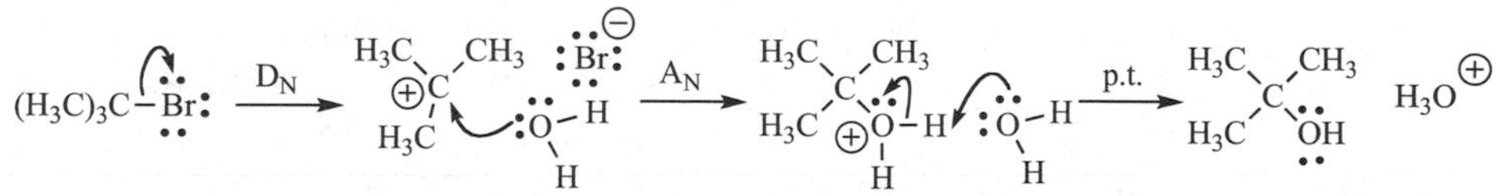
$(H_3C)_3C-Br$
D_N
A_N
p.t.
H_3O

S_N2 Biochemical Example

S-adenosylmethionine is the primary biochemical methyl group donor and provides an excellent example of a biological S_N2 reaction (Fig. 8.2).

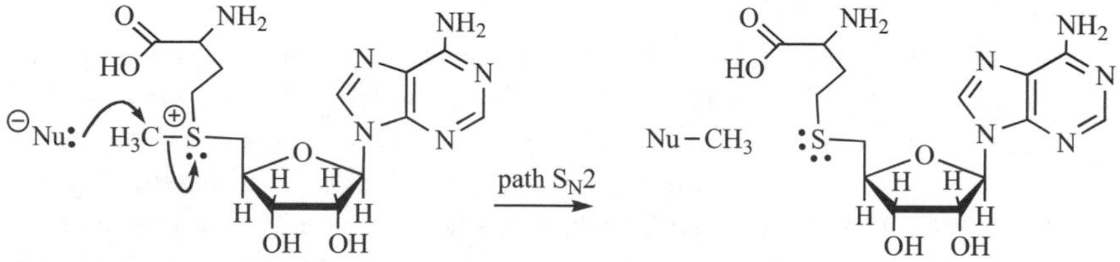
path S_N2
Nu$-CH_3$

Figure 8.2 Methylation of a nucleophile by *S*-adenosylmethionine.

S_N1 Biochemical Example

Glycogen phosphorylase mobilizes the body's stored glycogen to glucose-1-phosphate for use in glycolysis, the metabolic sequence in producing energy from glucose. The first step is a proton transfer to improve the leaving group, another sugar. A D_N step starts the S_N1 to give an oxygen lone pair stabilized carbocation that is trapped by phosphate in an A_N step to give glucose-1-phosphate (Fig. 8.3).

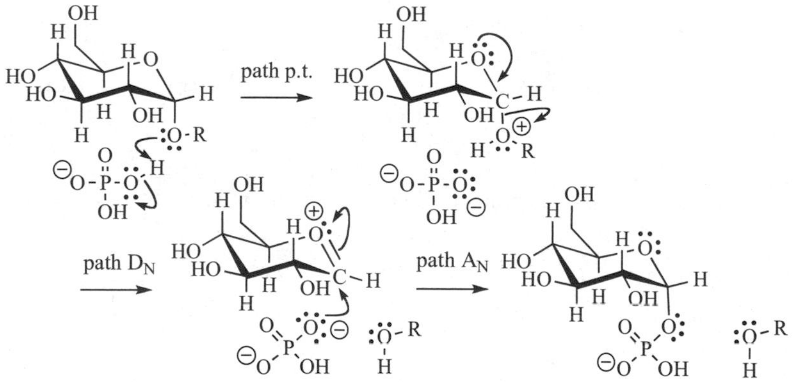
path p.t.
path D_N
path A_N

Figure 8.3 Phosphorylation of glycogen.

8.4.2 Bases Reacting with sp^3 C–L Having an Adjacent CH, Eliminations

As covered in Section 4.3, there are three pathways that differ only in the timing of when the proton is pulled off and when the leaving group falls off: **Generic E1**—The leaving group departs first by path D_N, producing a reasonably stable cation; the proton is then lost by path D_E, forming the alkene. **Generic E2**—Simultaneous loss of proton and leaving group. **Generic E1cB**—Proton transfer forms an anion stabilized by an electronegative atom, and then anion pushes out the leaving group by path E_β. Path selection between substitution and elimination is considered in depth in Section 9.5.

Alkene Formation via the E1/E2/E1cB Spectrum

The different elimination paths often produce different alkene constitutional isomers as products (*regiochemistry*). For the E2 reaction the C–H bond and the C–L bond must lie in the same plane, preferably *anti* to each other for steric reasons. The protons, shown in bold, on either carbon adjacent to the leaving group can be in proper alignment for reaction; thus different products are produced (the *cis* product is also formed).

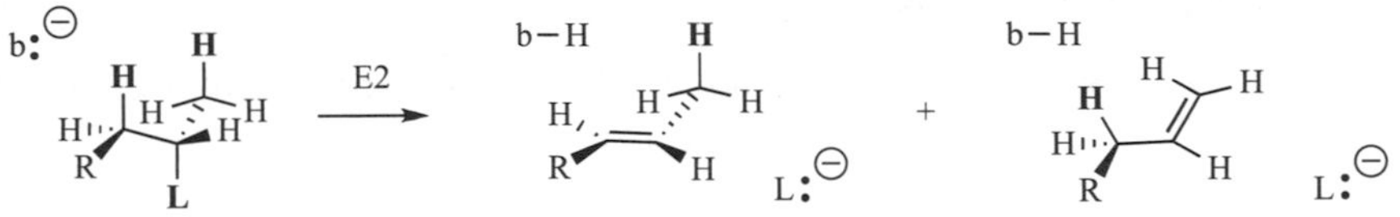

The regiochemistry of the elimination depends on the type of elimination process that occurs. **The E1 process favors the formation of the more substituted alkene** because reversible protonation of the double bond occurs and creates an equilibrium mixture that favors the more stable product. **The E2 regiochemistry is controlled by the need to minimize steric interactions in the transition state**; the size of the base is important because one proton may be more accessible than another, as in Figure 4.21. **The E1cB regiochemistry is determined by the loss of the most acidic proton.** Elimination reactions can produce different stereoisomers, for example, *cis* and *trans* alkenes. Since the *trans* isomer is usually of lower energy because of steric reasons, it usually predominates over the *cis* isomer in the product mixture.

E1 example—good L and cation, forms most substituted alkene (Zaitsev's rule):

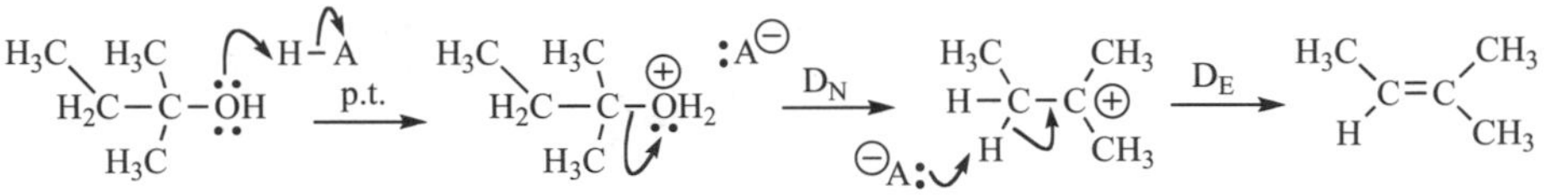

E2 example—concerted:

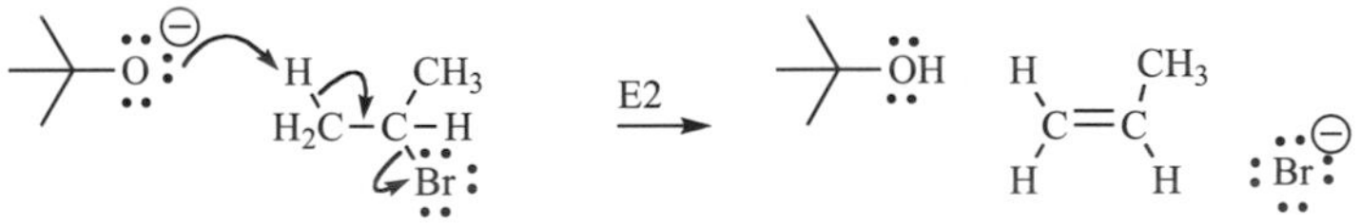

E1cB example—ewg makes C–H acidic; anion needed to kick out poor L:

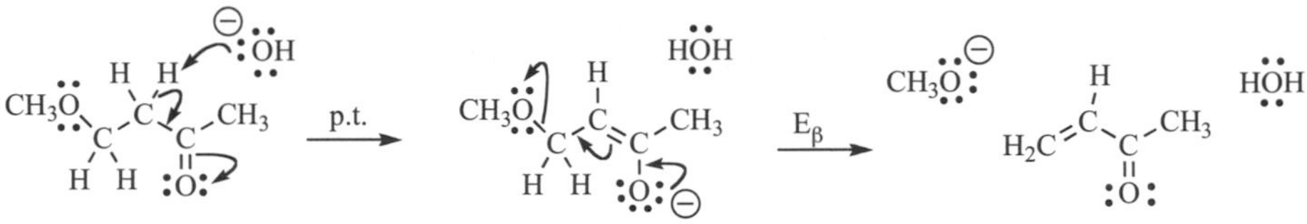

Heteroatom Variations (sp^3 C–L Having an Adjacent YH)

Anionic eliminations of the following type will also be considered as belonging to the E1cB class of eliminations. Since protons on heteroatoms are usually rather acidic because of the electronegativity of the heteroatom, the first step, deprotonation, can occur with a rather weak base. The second step is beta elimination from an anion, E_β, the reverse of nucleophile addition to a polarized multiple bond.

Hetero E1cB example (see Figs. 4.43 and 4.44 for an energy diagram):

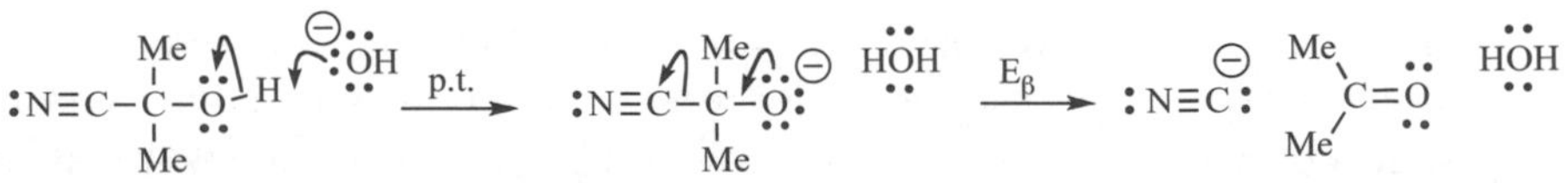

Likewise, a lone pair on a heteroatom can aid the loss of the leaving group in the E1 process. The following example shows first the protonation of the leaving group followed by the lone-pair-assisted E1, path E_β followed by proton transfer.

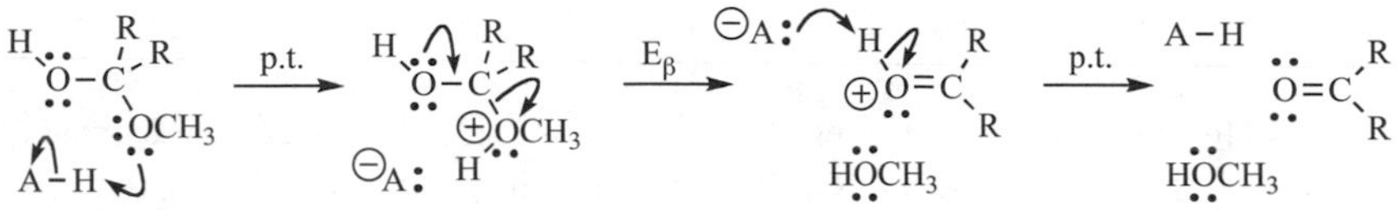

A hetero E2 example was discussed in Section 7.4, Variations on a Theme.

E1cB Biochemical Example

The synthesis of fats is an important way an organism stores the energy from foods. The dehydration step in fatty acid biosynthesis is an example of an E1cB sequence in which the conjugate acid of the initial base serves to protonate the leaving group on departure. This second step is a general acid-catalyzed beta elimination (Section 7.4.3).

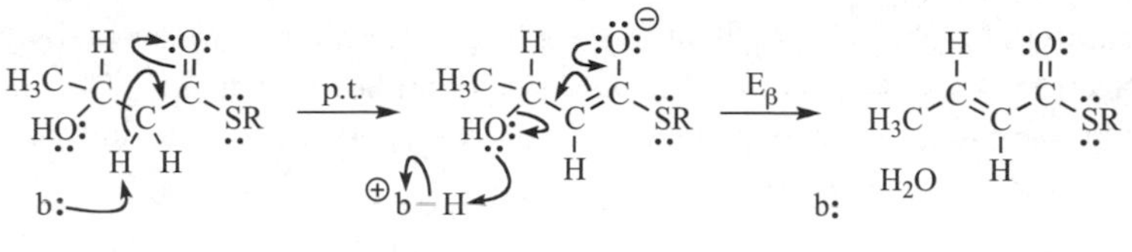

8.4.3 Complex Metal Hydrides Reacting with sp^3 C–L

The substitution reaction occurs in neutral or basic media and in aprotic solvents via the S_N2 pathway to replace the leaving group with hydrogen. Often the metal cation can form a complex with the leaving group to aid in its leaving.

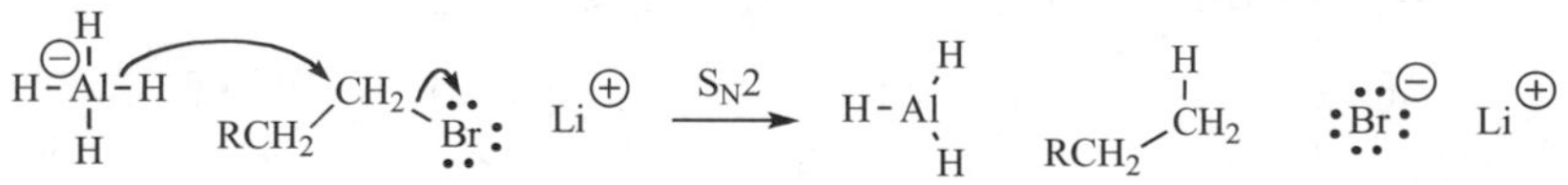

8.4.4 Organometallics Reacting with sp^3 C–L

Since organometallics react with acids (Section 8.2.4), they can be used only in basic or neutral media and in aprotic solvents. The most common reaction is substitution via S_N2. Substitution is favored by sulfonate leaving groups.

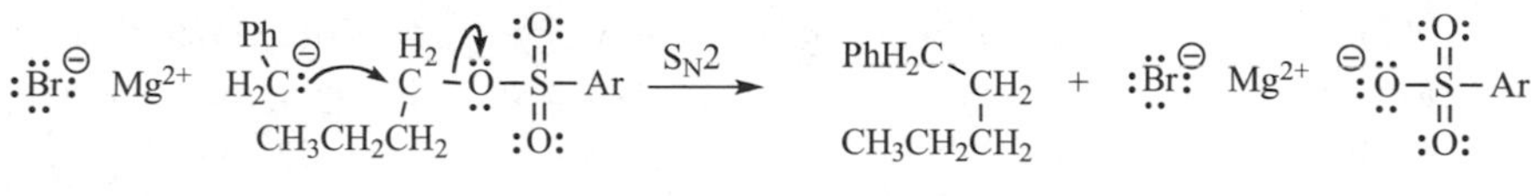

8.4.5 Allylic Sources Reacting with sp^3 C–L

An allylic electron source can be considered to be a vinylogous lone pair. Sometimes the lone pair itself is the nucleophile and can undergo reactions similar to the lone pair sources in the first column of the matrix. However, most often the nucleophile is the end carbon atom of the pi bond. The heteroatom lone pair is much harder than the pi bond, and the decision of which to use as the nucleophile is based on the hardness of the electrophile (HSAB principle) and will be discussed at greater length in Section 9.4.

Enamines, N–C=C, usually alkylate on carbon with this sink, and produce an iminium ion, +N=C–C–R, which is normally hydrolyzed to the ketone in an acidic water workup (the reverse of imine formation, Section 8.5.1). Enamine alkylation example:

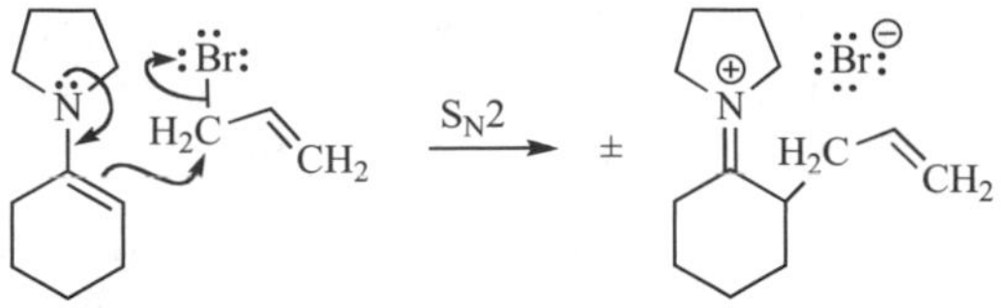
:Br:
N
H2C
CH2
SN2
±
:Br:
N
H2C
CH2

However, with simple enolates, $^{-}$O–C=C, alkylation can be a messy reaction. The alkylation product and the enolate can usually proton transfer, producing new electron sources and leading to undesired products. The best way to run this reaction is to add the enolate source into an excess of sink and hope that the alkylation proceeds faster than proton transfer.

Allylic Electron Sources with Additional Delocalization

One can speed up alkylation relative to elimination by stabilizing an enolate anion with an additional ewg, making the anion softer and less basic. For example, the acidic CH_2 of malonates, $CH_2(COOEt)_2$, or acetoacetates, CH_3COCH_2COOEt, deprotonates easily and makes an excellent nucleophile. The ester can later be hydrolyzed to the acid (Section 8.8.1) and removed by decarboxylation via path Ei. This ester is a detachable ewg that makes the enolate less basic and softer, favoring substitution over elimination.

Acetoacetate example:

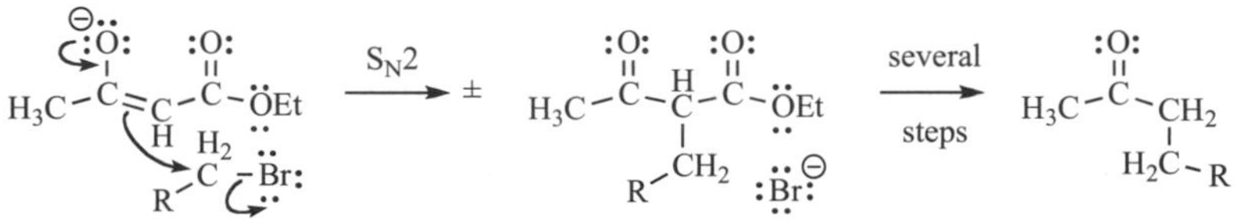
:O:
:O:
H3C
C
C
C
OEt
H
H2
C
Br:
R
SN2
±
:O:
:O:
H3C
C
C
C
OEt
H
CH2
R
:Br:
several
steps
:O:
H3C
C
CH2
H2C
R

8.4.6 Simple Pi Sources Reacting with *sp*3 C–L

Simple pi bonds are usually not good enough nucleophiles to react with alkyl halides. If a Brønsted acid or a Lewis acid is added to improve the electron sink, then addition can happen, but it is often complicated with polymerization.

Biochemical Example

The synthesis of terpene natural products begins with the following coupling reaction (Fig. 8.4). The magnesium ion acts as a Lewis acid and complexes to the pyrophosphate (diphosphate) leaving group (OPP) to activate its departure. The delocalized allylic carbocation adds to the pi bond of isopentenyl pyrophosphate, which then loses a proton to form the coupling product geranyl pyrophosphate. The coupling can be repeated with another isopentenyl pyrophosphate, adding another five-carbon piece.

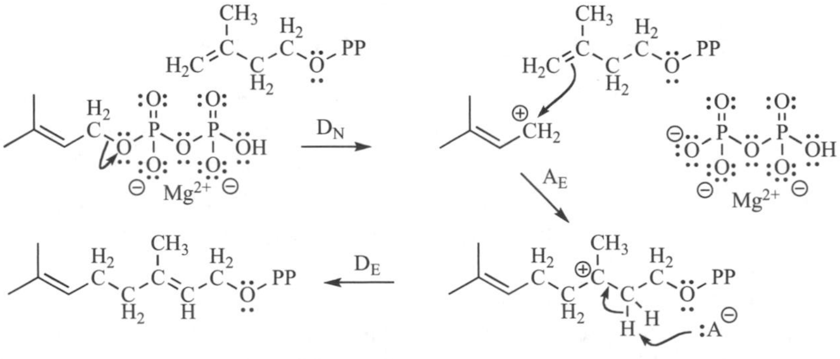
CH3
H2
H2C
C
C
C
O
PP
H2
:O:
:O:
P
P
O
OH
Mg2+
DN
CH2
AE
DE
H
:A

Figure 8.4 Biosynthesis of geranyl pyrophosphate from dimethylallyl pyrophosphate.

8.4.7 Aromatic Sources Reacting with sp^3 C–L

Friedel–Crafts Alkylation

Aromatic rings are usually not good enough nucleophiles to react with alkyl halides. If Lewis acids are added to improve the electron sink, then electrophilic aromatic substitution occurs via the carbocation if it is more stable than secondary (an S_N1 viewed from the electrophile). With Lewis acids and methyl, primary, or secondary alkyl halides, electrophilic aromatic substitution usually occurs via direct displacement of the leaving group (an S_N2 viewed from the electrophile).

This reaction has several limitations and problems. First, the reaction fails for deactivated rings. Second, if it can, the carbocation almost always rearranges to a more stable carbocation. Third, once the alkyl group is attached, it activates the ring for further attack so sequential electrophile additions to give multiple substituted aromatics are possible. The electrophiles are generated in a variety of ways:

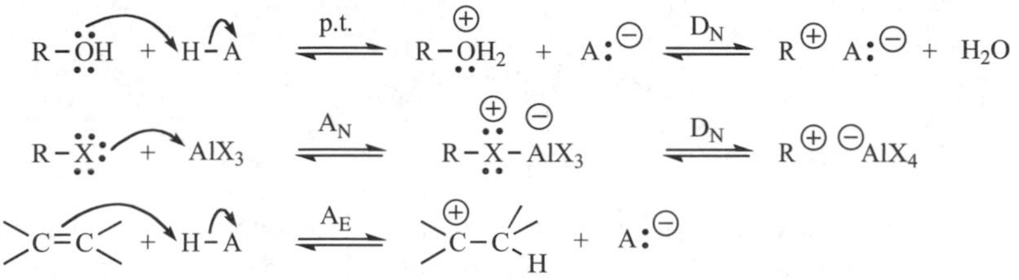

Alkylation example (an S_N2 viewed from the electrophile):

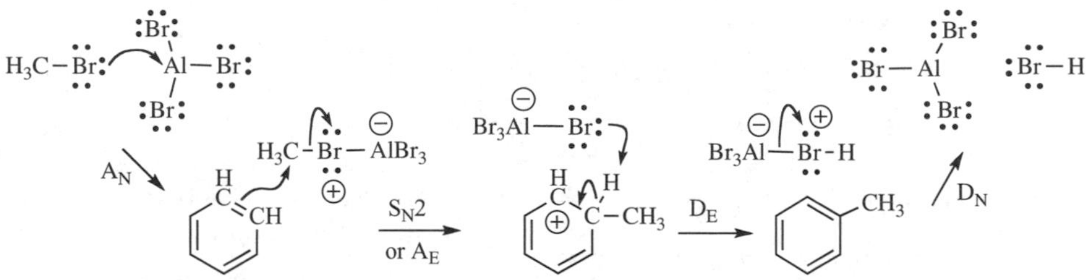

Alkylation example (an S_N1 viewed from the electrophile):

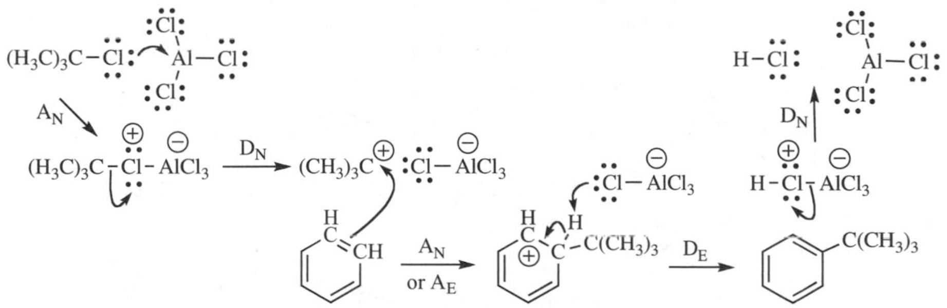

8.5 C=Y SINKS REACTING WITH COMMON SOURCES

Section 4.47 introduced the theory of additions to carbonyls, C=O, the most common example of C=Y sinks. All three routes, Ad_N2, Ad_E2, and Ad_E3 were found to occur.

8.5.1 Lone Pair Sources Reacting with C=Y

These reactions are very common (see Figs. 7.16 and 7.17 for the interrelationship of the available paths in acidic and in basic media). **In basic media the attack of the nucleophile on a polarized multiple bond forms an anion that in the workup of the reaction is usually protonated (path combination Ad_N2).** In protic solvents if the nucleophilic attack forms a stronger base, a following irreversible proton transfer step may make the overall reaction favorable (see Figs. 4.43 and 4.44).

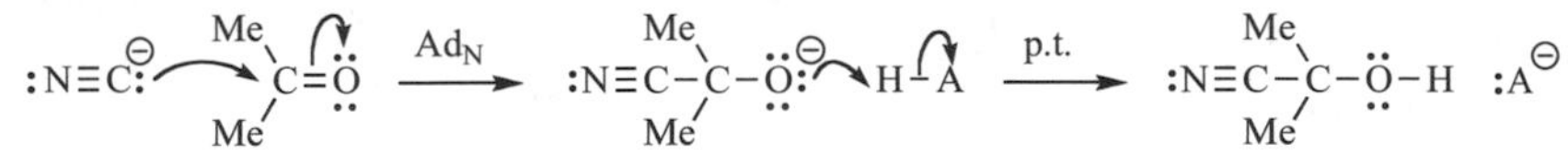

In acidic media, polarized multiple bonds often undergo acid catalyzed addition, and **a common mode of addition is the Ad_E2.** Deprotonation of the nucleophile by solvent gives the neutral compound. Common examples of this easily reversible Ad_E2 reaction are the formation of *hydrates* (NuH is H_2O) and, if NuH is ROH, *hemiacetals* (from aldehydes) and *hemiketals* (from ketones). Usually this reaction favors reactants.

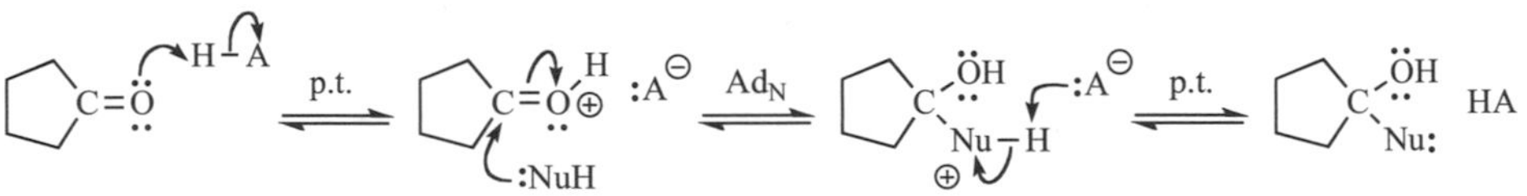

A follow-up reaction often occurs depending on the nucleophile and conditions. Six new functional groups can be easily made from the aldehyde or ketone (Table 8.2).

Table 8.2 New functional groups formed from aldehydes and ketones reacting with water, alcohols, and amines

Nucleophile	Mechanistic Process	Product Name	Structure
H_2O	Ad_E2	Hydrate	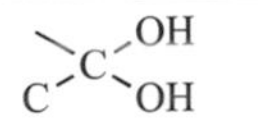
ROH	Ad_E2	Hemiacetal from aldehyde Hemiketal from ketone	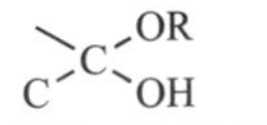
ROH	Ad_E2 then S_N1	Acetal from aldehyde Ketal from ketone	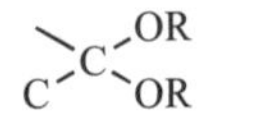
ROH	Ad_E2 then E1	Enol ether	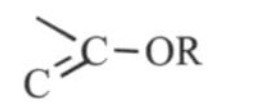
RNH_2	Ad_E2 then E1	Imine	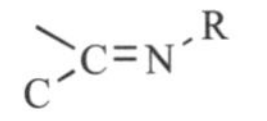
R_2NH	Ad_E2 then E1	Enamine	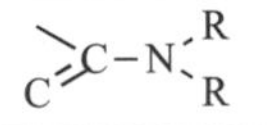

Biochemical Example

Five- and six-membered ring hemiacetals and hemiketals are common configurations of many sugars. Ribose, which forms a part of the structure of RNA, closes to form a five-membered ring preferentially.

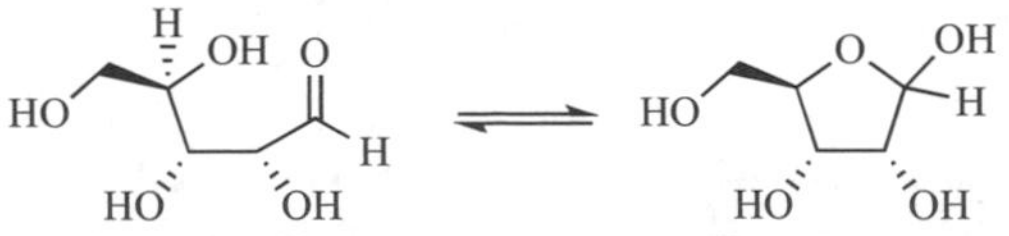

ribose open form ribose hemiacetal

If the following equilibria are **driven by the removal of water**, a follow-up reaction can occur in acid in which the OH group gets protonated (path p.t.), then is lost as water (path D_N), and the carbocation formed either is trapped (S_N1) or loses a proton (E1). The identity of the nucleophile is used to determine the route that the reaction will take.

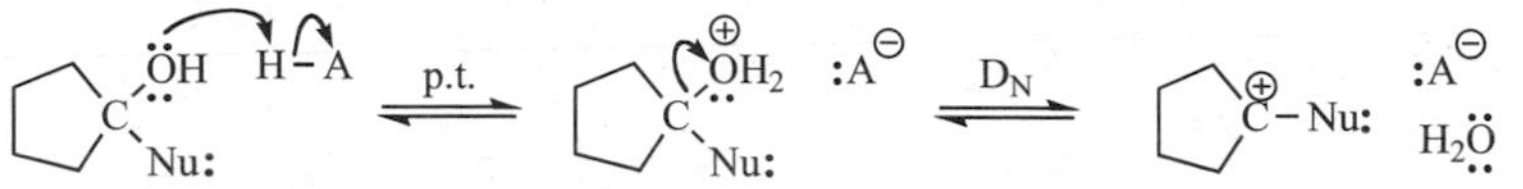

The carbocation can add a second nucleophile (path A_N followed by path p.t.).
This is common for alcohol sources (NuH is ROH), in acidic media, with removal of water to form *ketals* and *acetals*.

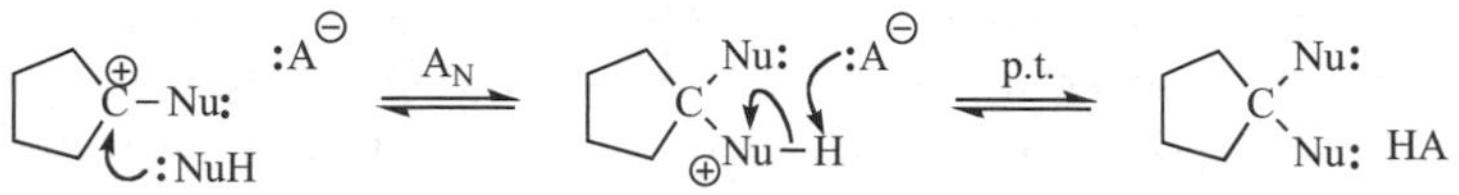

In compounds with multiple functional groups, ketals can be used to protect a ketone from basic nucleophiles and can later be removed by acidic water.
Cyclic ketal example:

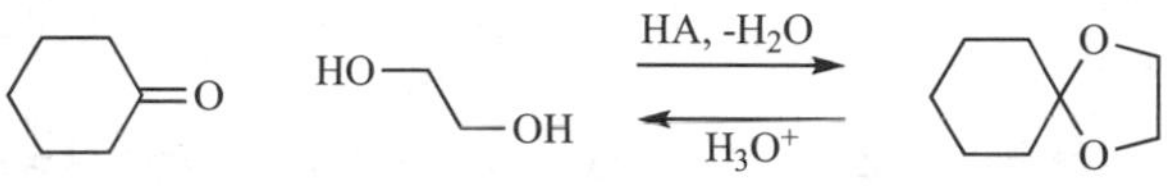

Alternatively, the C–H bond next to the carbocation can be deprotonated (path D_E). This is common for secondary amine sources (NuH is R_2NH), in acidic media, with removal of water to form *enamines*.

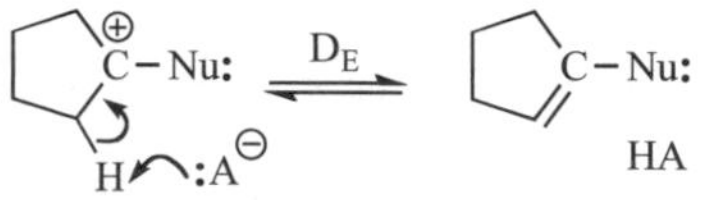

Common example (cyclic secondary amine):

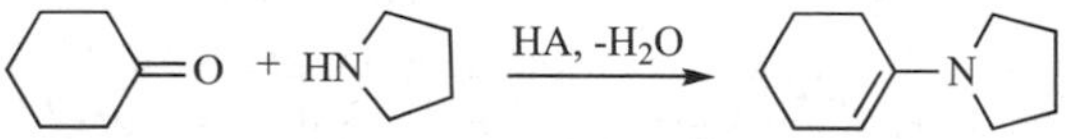

However, if there were two Hs on the original nucleophile (for example RNH_2), we could now remove that second H on the nucleophile to yield a stronger heteroatom–carbon double bond (path D_E). This is common for primary amine sources (NuH_2 is RNH_2) in acidic media, with removal of water to form *imines* (Section 10.5.2).

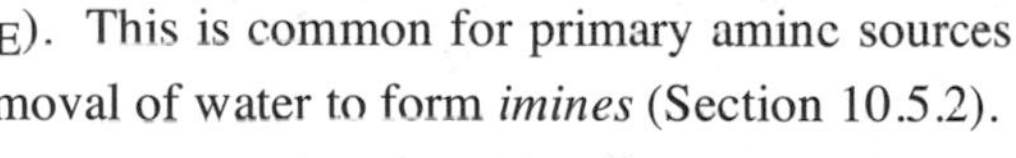

Finally, if the NuH_2 is $HONH_2$ the reaction forms an oxime (N is a better Nu than O).

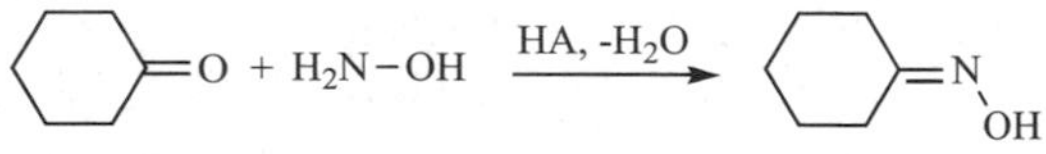

Biochemical Example

Pyridoxal (vitamin B_6) forms imines (Schiff bases) with amino acids as the first step in catalyzing their transformation (for one example, see Section 2.2). After pyridoxal has performed its particular catalysis on the attached amino acid, the imine can readily hydrolyze by the reverse of the formation reaction (shown in the worked mechanism example 10.5.2) to release the altered amino acid.

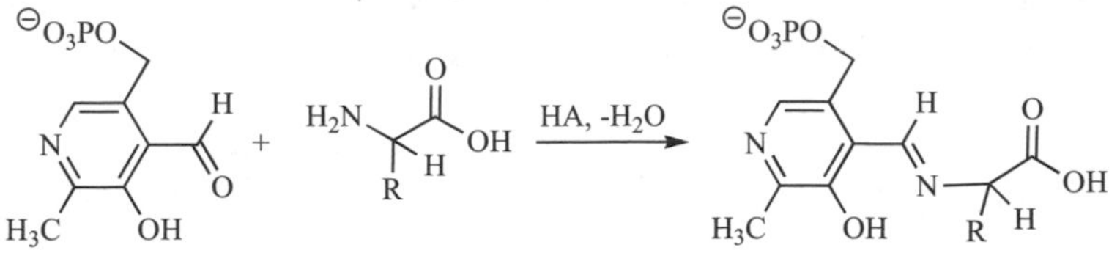

8.5.2 Bases Reacting with C=Y Having an Adjacent CH, Enolates

The polarized multiple bond makes adjacent C–H bonds acidic; deprotonation generates the enolate, an allylic electron source. Often there is more than one possible regiochem or stereochem of the enolate, which can be selected by reaction conditions (see Section 9.3).

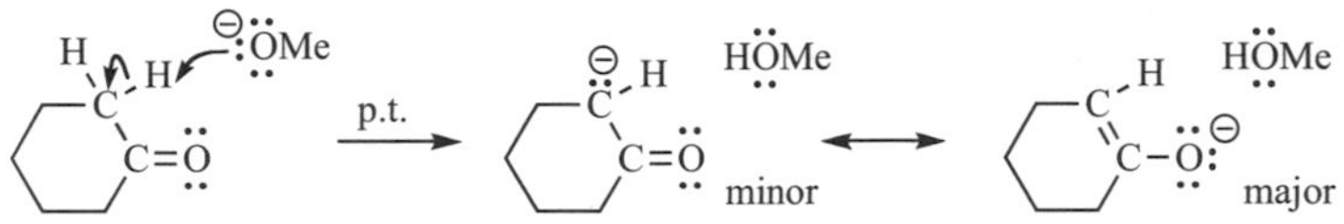

8.5.3 Complex Metal Hydrides Reacting with C=Y

Hydride is added to the multiple bond (path Ad_N) as expected. Usually the metal cation complexes with the Y heteroatom; nucleophilic attack on the polarized multiple bond will then yield a product that is stabilized by ion pairing.

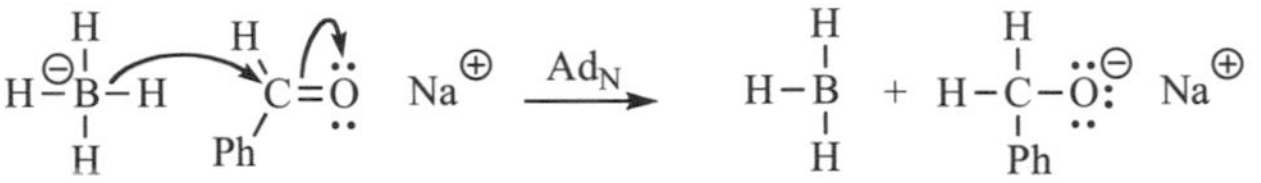

Chiral Lewis acids combined with hydride sources like borohydride can reduce carbonyls to give one enantiomer selectively. The chiral Lewis acid complex with the carbonyl makes one face of the carbonyl more accessible to the hydride nucleophile. The chirality of the Lewis acid determines the chirality of the product.

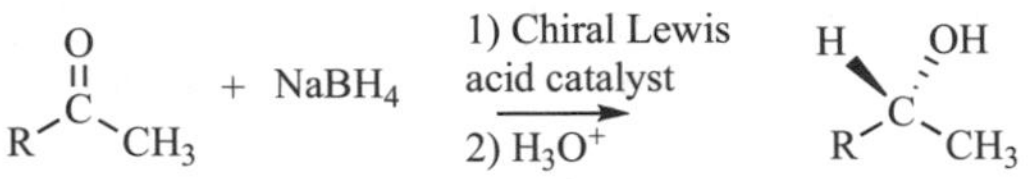

8.5.4 Organometallics Reacting with C=Y

Organometallics react with this sink by addition to the multiple bond (path Ad_N). The more covalent, less reactive organometallics, like R_2Cd, react very slowly with almost all of these sinks, whereas organomagnesiums, RMgX, and organolithiums react quickly. Complexation of the metal ion to the Y heteroatom catalyzes this reaction. Organometallics react much faster as nucleophiles with polarized multiple bonds than as bases with the adjacent C–H bonds. (carbon–acid, carbon–base proton transfer is slow).

C=Y example:

Chiral Lewis acids combined with less reactive organometallics like organotins (organostannanes) can add to carbonyls to give one enantiomer selectively. The chiral Lewis acid complex with the carbonyl makes one face of the carbonyl more accessible to the organometallic nucleophile. The product chirality is set by the chiral catalyst.

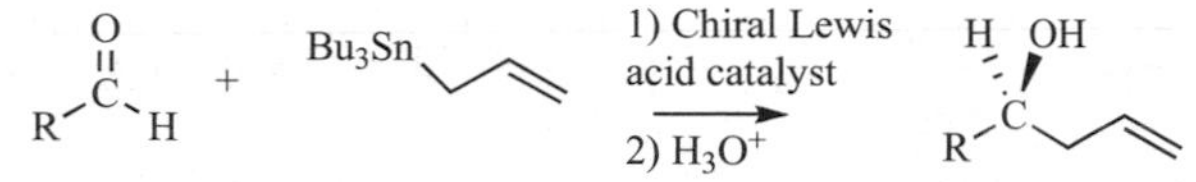
Bu3Sn
1) Chiral Lewis acid catalyst
2) H3O+

8.5.5 Allylic Sources Reacting with C=Y

C=Y as Sink, Aldol Reaction

Under equilibrium conditions (thermodynamic control), the allylic source adds to the polarized multiple bond (path Ad_N). However, the allylic source can also serve as a base and may deprotonate the sink, creating a mixture of sources and sinks and thus a messy statistical mixture of products. Clean products result if the source is just the deprotonated sink or if the sink has no acidic protons. With ketones, the equilibrium of the attack step favors the starting materials, and therefore the reaction goes to completion only if driven by a following elimination. In the next Ad_N2 example, the source is the deprotonated sink. The product is an aldehyde-alcohol, or aldol, a name now used for the general process of an enol (acidic media) or enolate (basic) reacting with an aldehyde or ketone.

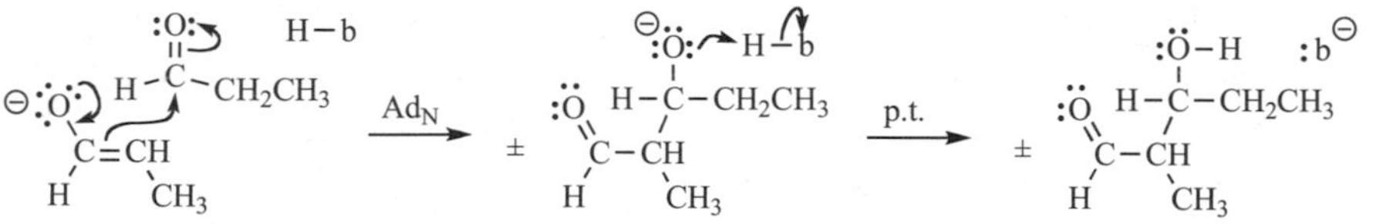
AdN
p.t.

If the reaction is heated, elimination usually follows via the E1cB path combination, because the carbonyl is an electron-withdrawing group.

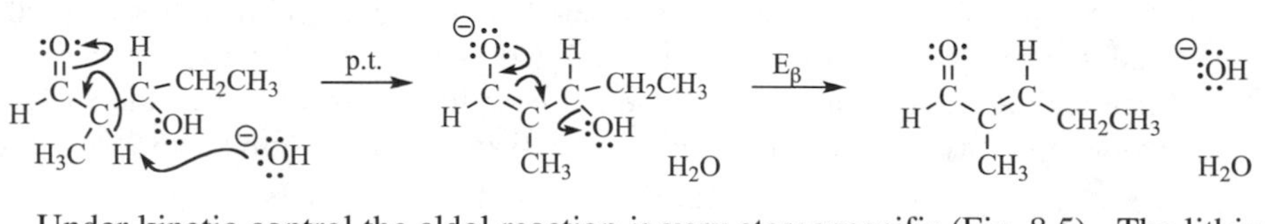
p.t.
Eβ

Under kinetic control the aldol reaction is very stereospecific (Fig. 8.5). The lithium enolate is generated in an aprotic solvent, and then the carbonyl compound is added. The reaction proceeds via the metal-chelated minor path 6e. The minimization of steric effects in the chair transition state and the stereochemistry of the enolate (Section 9.3) determine the stereochemistry of the product.

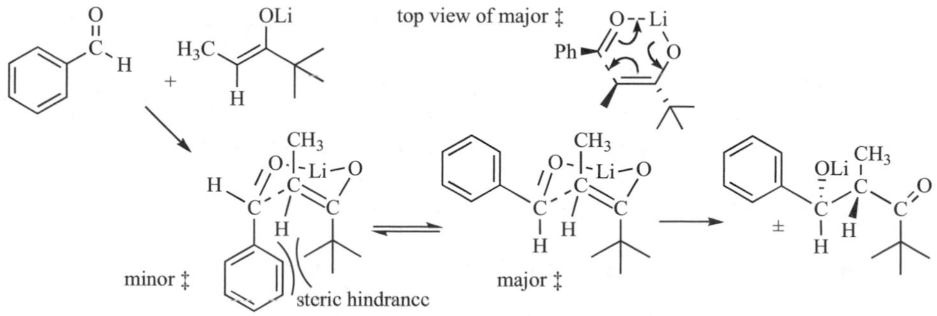
top view of major ‡
minor ‡
steric hindrance
major ‡

Figure 8.5 The stereospecific aldol reaction goes via the least hindered chair transition state.

The following is an example of an aldol reaction with a nonenolizable sink. The additional conjugation with the aromatic ring makes the follow-up elimination favorable.

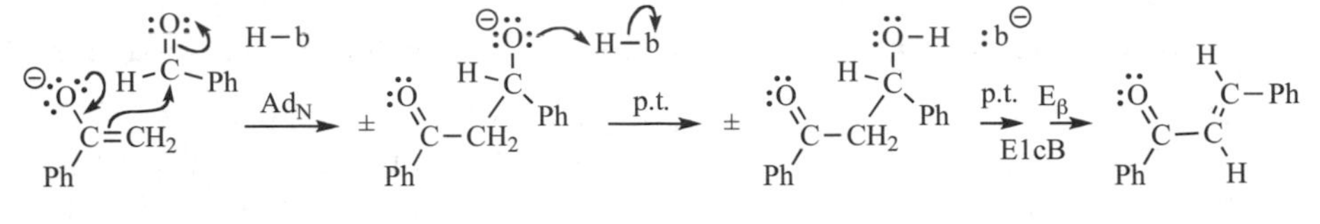
AdN
p.t.
p.t. Eβ
E1cB
Ph

Biochemical Example

Fats and carbohydrates are metabolized down to carbon dioxide via an acetyl unit, $CH_3C{=}O$, which is attached to a coenzyme, HSCoA, as a thioester called acetyl CoA. Acetyl CoA enters the citric acid cycle and eventually is converted to two molecules of carbon dioxide. The first step in the citric acid cycle is the aldol of acetyl CoA with oxaloacetate (Fig. 8.6). What is so elegant about this aldol is that the acidic and basic groups within the enzyme's active site provide a route that avoids any strongly acidic or basic intermediates. The enzyme accomplishes an aldol reaction at neutral pH, without an acidic protonated carbonyl or basic enolate intermediate via push–pull catalysis (Section 7.4.3).

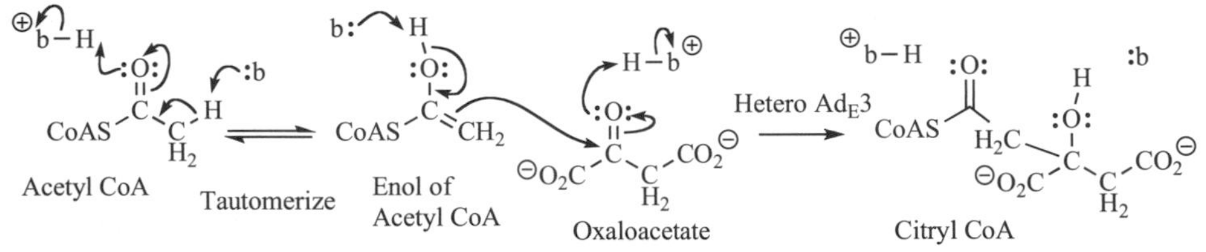
Acetyl CoA
Tautomerize
Enol of
Acetyl CoA
Oxaloacetate
Hetero AdE3
Citryl CoA

Figure 8.6 The biosynthesis of citryl CoA.

8.5.6 Simple Pi Bond Sources Reacting with C=Y

Simple pi bonds are usually not good enough nucleophiles to react with polarized multiple bonds. If a Brønsted acid or a Lewis acid is added to improve the electron sink, then addition can occur via the lone-pair-stabilized carbocation as the sink. Figure 8.7 shows a mechanistic example from a short synthesis of the human hormone estrone.

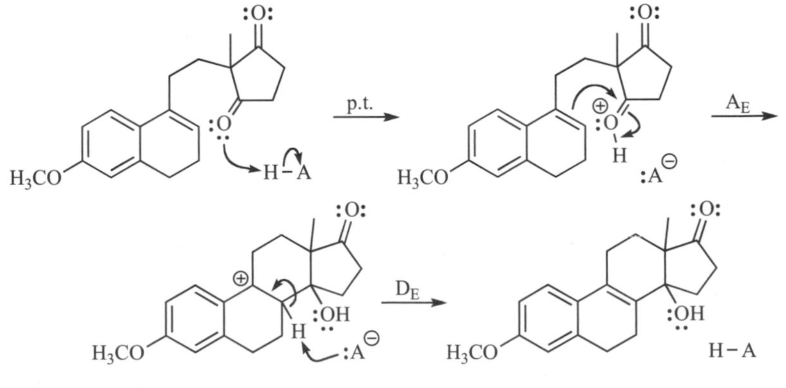
p.t.
AE
DE

Figure 8.7 The electrophilic ring closure step in a synthesis toward estrone.

8.5.7 Aromatic Sources Reacting with C=Y

Aromatic rings are usually not good enough nucleophiles to react with polarized multiple bonds. If a Brønsted acid or a Lewis acid is added to improve the electron sink, then electrophilic aromatic substitution occurs, usually via the lone-pair-stabilized carbocation as an electrophile.

Friedel–Crafts—Aromatic Rings with Lone-Pair-Stabilized Carbocations as Sinks

The electrophile is often a carbonyl that is protonated or complexed to a Lewis acid. The product can enter into a second Friedel–Crafts alkylation so that often the product is a result of two attacks. The synthesis of the insecticide DDT (Fig. 8.8) is an example.

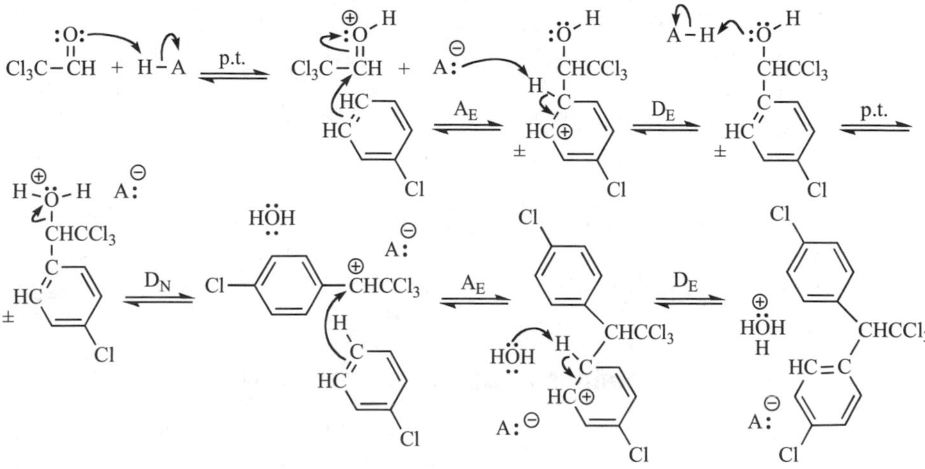

Figure 8.8 The synthesis of DDT catalyzed by sulfuric acid.

8.6 R–C≡Y SINKS REACTING WITH COMMON SOURCES

8.6.1 Lone Pair Sources Reacting with R–C≡N

The two pi bonds in a triple bond are perpendicular to each other, so they act independently. In the next example, the attack of the Nu is aided by protonation from the protic solvent, the hetero Ad_E3 path. Tautomerization then gives the more stable amide.

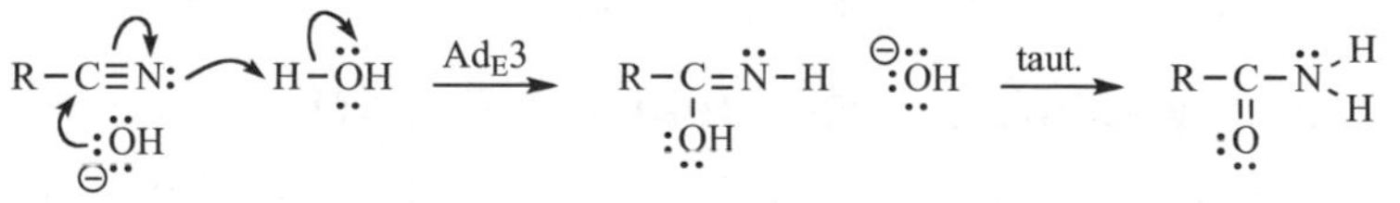

8.6.2 Bases Reacting with R–C≡N Having an Adjacent CH, Nitrile Enolates

The polarized multiple bond of the nitrile makes the adjacent CH bonds acidic (pK_a = 25); deprotonation by a strong base gives the nitrile enolate, an allylic electron source.

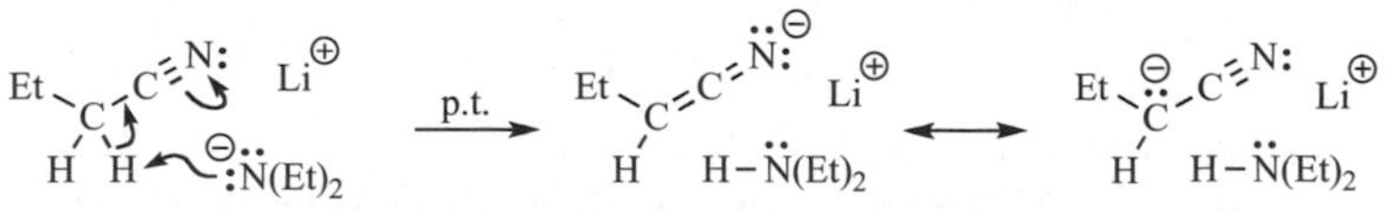

8.6.3 Complex Metal Hydrides Reacting with R–C≡N

With $LiAlH_4$ a second attack on the nitrile occurs. The product of the first addition can form a Lewis salt, which (most likely after acting as a hydride source) catalyzes the second addition. The product is the amine, after the acidic water in the workup protonates the nitrogen and dissociates it from the metal. This double addition can be prevented by using one equivalent of a less reactive aluminum hydride like $LiHAl(OEt)_3$ or by the use of one equivalent of $[(CH_3)_2CHCH_2]_2AlH$, diisobutyl aluminum hydride (see Section 9.2 for the multiple addition decision of nucleophiles to nitriles).

Double addition example (path Ad_N, then A_N, followed by the Lewis salt serving as a hydride source, then Ad_N again):

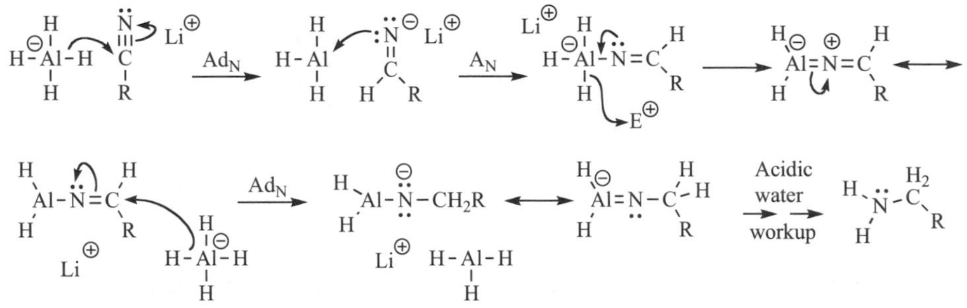
Ad

8.6.4 Organometallics Reacting with R−C≡N

Organometallics react with this sink by addition to the multiple bond (path Ad_N). The more covalent, less reactive organometallics, like R_2Cd, react very slowly with almost all of these sinks, whereas organomagnesiums, RMgX, and organolithiums react quickly. Complexation of the metal ion to the Y heteroatom catalyzes this reaction. Organometallics react much faster as nucleophiles than bases (carbon–acid, carbon–base proton transfer is slow). The resulting nitrogen anion is protonated in the acidic workup to give the imine, which then is hydrolyzed to the ketone. The mechanism of this acidic water workup is the reverse of imine formation, discussed in Section 10.5.2.

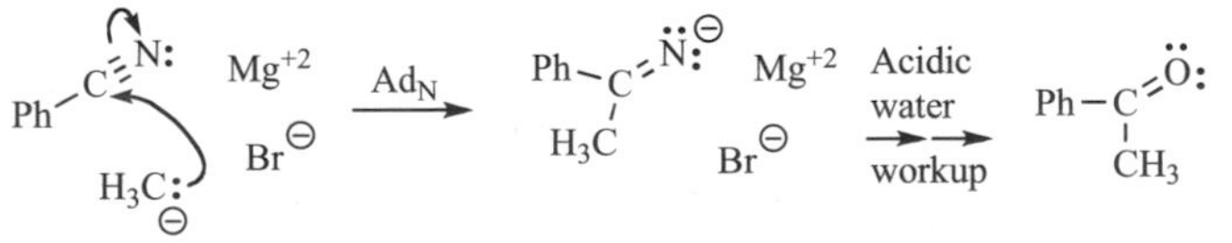
Acidic
water
workup

8.6.5 Allylic Sources Reacting with R−C≡N

The following example is an intramolecular reaction similar to the aldol reaction. A deprotonated nitrile (nitrile enolate) acts as a nucleophile and adds via an Ad_N path to another nitrile. The acidic water workup is the reverse of imine formation, Section 10.5.2.

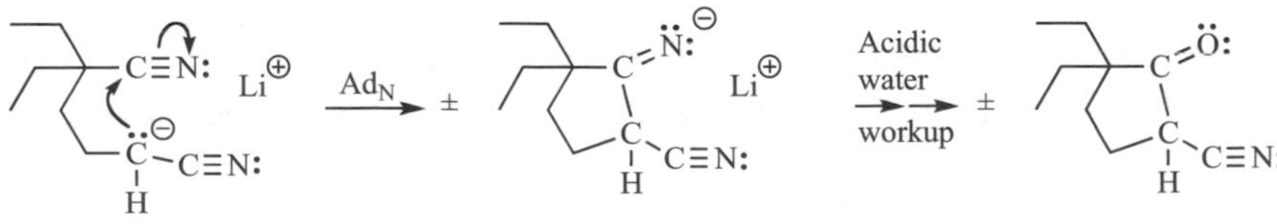
Acidic
water
workup

8.6.6 Simple Pi Bond Sources Reacting with R−C≡N

Simple pi bonds are not good enough nucleophiles to react with nitriles. It is basically a case of little push and little pull, so no electron flow occurs. If the pi bond is made more nucleophilic by addition of a pi donor (making it an allylic source) then the reaction is similar to the one above. Alternatively, a Brønsted acid or a Lewis acid could be added to improve the electron sink, but few examples of this reaction are known.

8.6.7 Aromatic Sources Reacting with R–C≡N

Aromatic rings are usually not good enough nucleophiles to react with nitriles. If a Brønsted acid or a Lewis acid is added to improve the electron sink, then electrophilic aromatic substitution can occur. The acidic water workup first protonates the aluminum salt of the addition product to form the imine, which is then hydrolyzed to the ketone by the reverse of imine formation discussed in Sections 8.5.1 and 10.5.2.

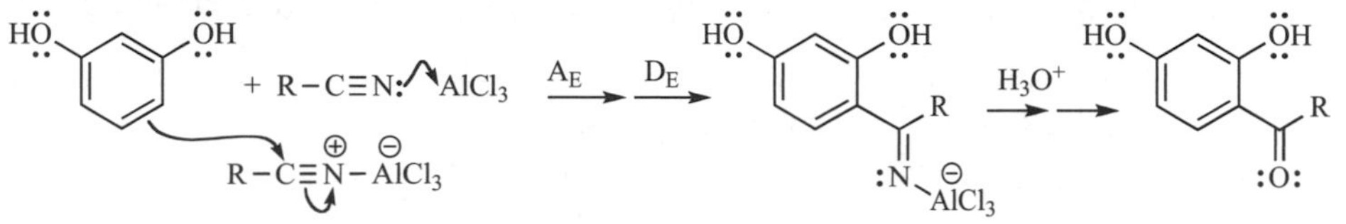

HO
OH
+ R–C≡N:
AlCl3
R–C≡N–AlCl3
AE
DE
R
:N
AlCl3
H3O+
:O:

8.7 C=C–EWG SINKS REACTING WITH COMMON SOURCES

8.7.1 Conjugate Addition by Lone Pair Sources to C=C–ewg

When an electronegative carbon atom replaces the electronegative Y atom, the Ad_N process is called *conjugate addition*. Competition between direct addition to the ewg and conjugate addition is discussed in Section 9.6 (the conjugate addition site is softer). This reaction goes by the Ad_N2. A rapid, irreversible proton transfer from the protic solvent is the driving force for this example.

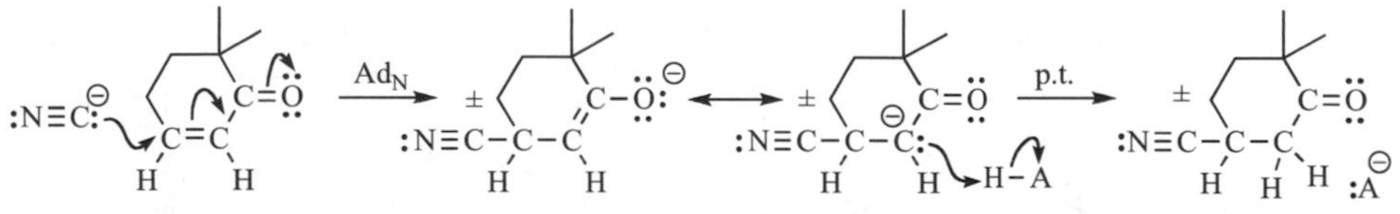

:N≡C:
AdN
p.t.
:N≡C–C–C
H–A
:A

8.7.2 Bases Reacting with C=C–ewg Having an Adjacent CH, Extended Enolates

These deprotonations generate extended enolates, which are vinylogous allylic sources. Deprotonation can also occur on the methyl next to the carbonyl in the following example. However, the extended enolate is more delocalized. Section 9.3 covers the regiochemistry and stereochemistry of enolate formation in more detail.

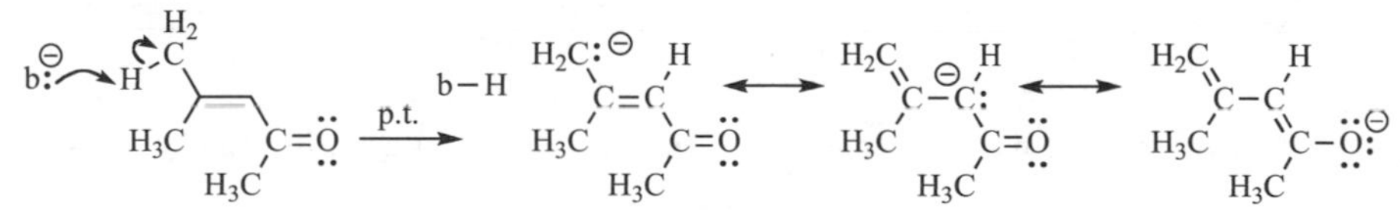

b:
H2
H3C
p.t.
b–H
H2C:

8.7.3 Conjugate Addition by Complex Metal Hydrides to C=C–ewg

Aluminum hydride, AlH_4^-, is relatively hard; therefore conjugate addition is usually only a minor side reaction, as most of the hydride addition occurs to the harder electron-withdrawing group and not the double bond (direct addition, Section 9.6). Borohydride, BH_4^-, since it is softer, has a much greater tendency toward conjugate addition but is often difficult to predict (see Section 9.6). Reagents and conditions have been found to give the conjugate addition.

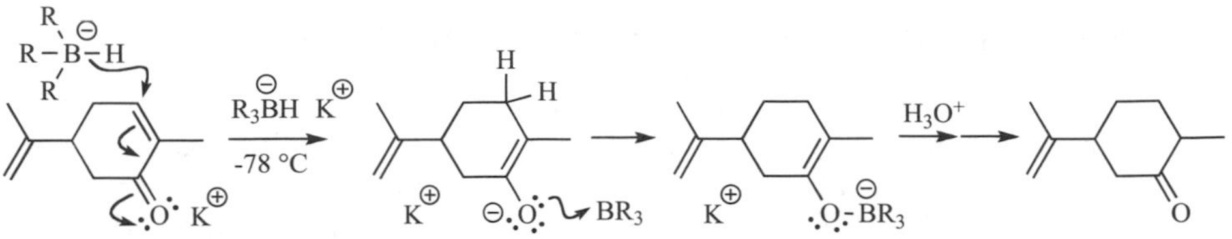

8.7.4 Conjugate Addition by Organometallics to C=C–ewg

Since the δ+ at the conjugate position is less, the carbon atom is softer, and softer organometallics tend to favor conjugate attack (see Section 9.6).

Example (the cuprate is a soft organometallic):

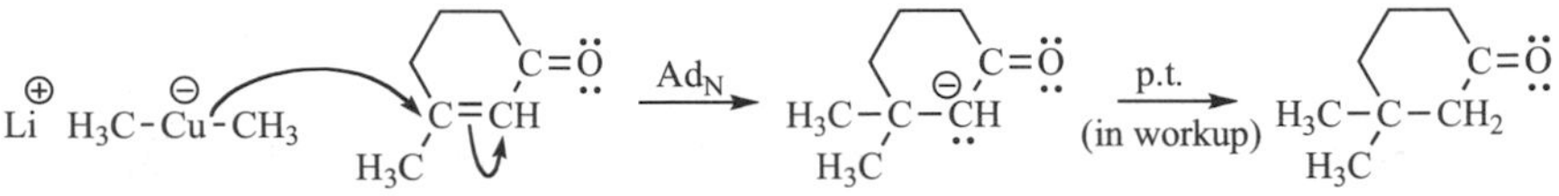

8.7.5 Conjugate Addition by Allylic Sources to C=C–ewg

The reaction is Ad_N2 addition to the polarized multiple bond (Ad_N, then p.t.). This reaction is easy to predict because the most stable product is formed under these equilibrium conditions. HSAB theory predicts that the combination of the two softest sites is favored. This conjugate addition is often called a Michael addition and forms the first step in the Robinson annulation diagramed in Figure 8.9. The later steps are an aldol addition (Ad_N2) to form the new ring followed by elimination by E1cB to give the bicyclic product.

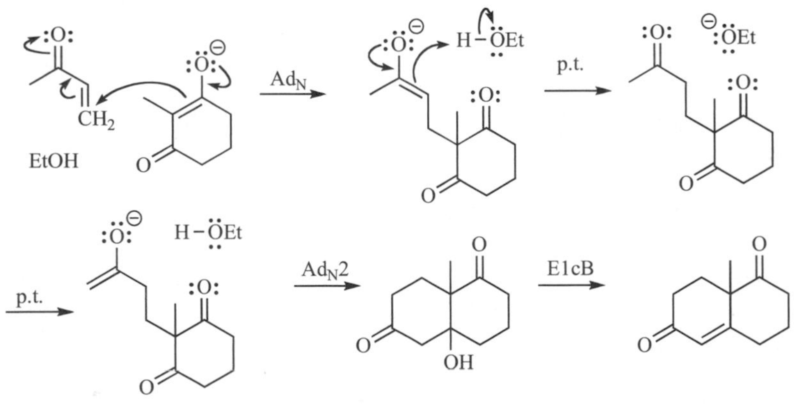

Figure 8.9 The Robinson annulation, conjugate addition followed by an aldol and an elimination.

8.7.6 Conjugate Addition by Simple Pi Sources to C=C–ewg

Simple pi bonds are not good enough nucleophiles to react with conjugate acceptors. It is basically a case of little push and little pull, so no electron flow occurs. If the pi bond is made more nucleophilic by adding a pi donor (making it an allylic source), then the reaction is similar to the one above. Alternatively, a Brønsted acid or a Lewis acid could be added to improve the electron sink; few examples of this reaction are known.

8.7.7 Conjugate Addition by Aromatic Sources to C=C−ewg

Aromatic rings are usually not good enough nucleophiles to react with conjugate acceptors. If a Brønsted acid or a Lewis acid is added to improve the electron sink, then electrophilic aromatic substitution can occur.

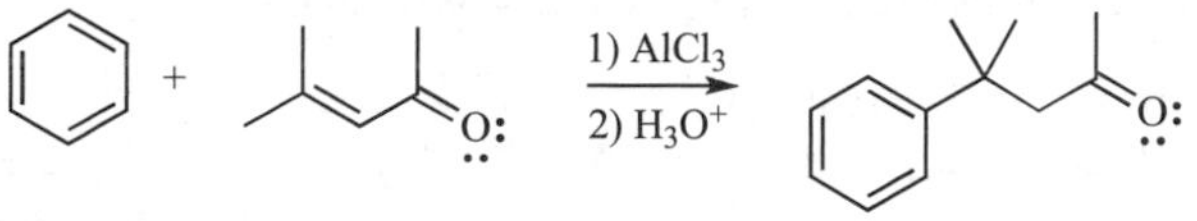

1) $AlCl_3$
2) H_3O^+

8.8 L−C=Y SINKS REACTING WITH COMMON SOURCES

8.8.1 Lone Pair Sources Reacting with L−C=Y

The most common representatives of the L−C=Y class of electron sinks are the carboxyl derivatives with Y equal to oxygen. **In basic media there is only one pathway: the addition–elimination path, path Ad_N + E_β** (see Section 4.5.1). The leaving group should be a more stable anion than the nucleophile, or the reaction will reverse at the tetrahedral intermediate. A follow-up reaction of a second addition to the polarized multiple bond occasionally occurs. With lone pair sources a second addition is rare because the nucleophile is usually a relatively stable species; the second tetrahedral intermediate tends to kick it back out (see Section 9.2).
Basic media addition–elimination example (path Ad_N + E_β):

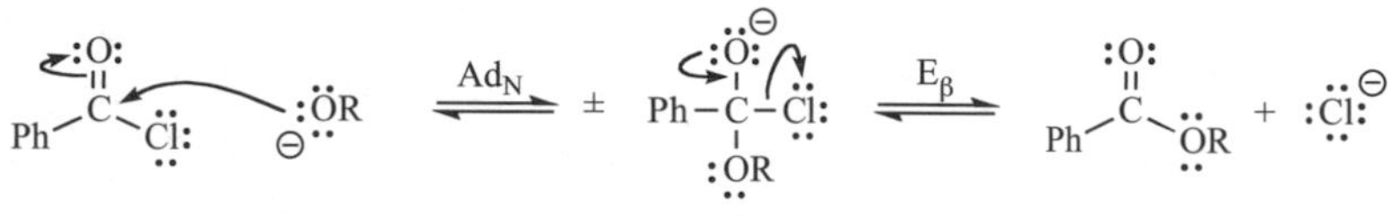

Ad_N
E_β

In acidic media, the carbonyl lone pair can get protonated to produce a much better electron sink. This better electron sink can be attacked by weaker nucleophiles. Then proton transfer to the leaving group makes it a better leaving group. The reaction is reversible; the position of equilibrium is determined by mass balance.
Acidic media example (Ad_E2, p.t. from Nu, p.t. to L, E1):

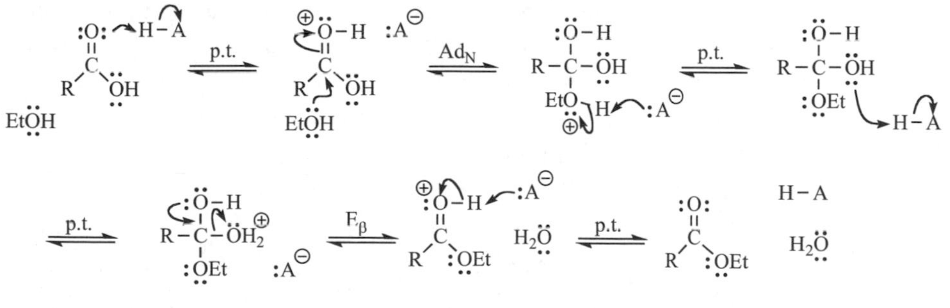

p.t.
Ad_N
p.t.
p.t.
E_β
p.t.

8.8.2 Bases Reacting with L−C=Y Sinks Having an Adjacent CH, Enolization

These enolates are usually generated by deprotonation below 0°C. At higher temperatures the enolate ejects the leaving group via path E_β to form a reactive heterocumulene (usually an undesirable side reaction). Enolization example:

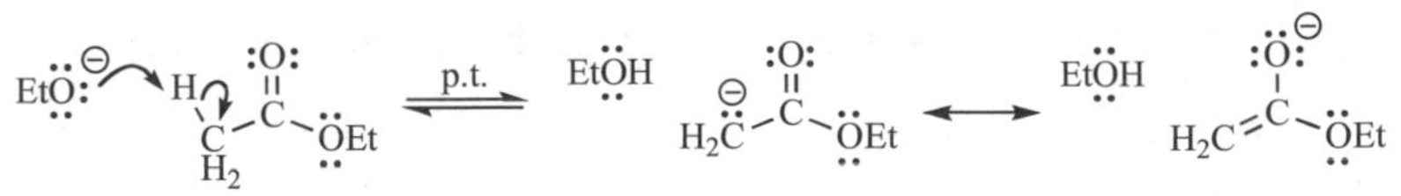

8.8.3 Metal Hydrides Reacting with L–C=Y

The product of the initial reduction is most often more reactive than the starting material; therefore a second addition is very common (Ad_N, then E_β, then Ad_N, covered in Section 9.2). An Ad_N then E_β product can be obtained with acyl halides and one equivalent of a less reactive metal hydride at low temperature. Borohydrides selectively react with aldehydes and ketones in the presence of less reactive esters and amides.

Second-addition example (path Ad_N then E_β, Lewis salt formation with the aluminum species to create the nucleophilic aluminate, followed by a second Ad_N):

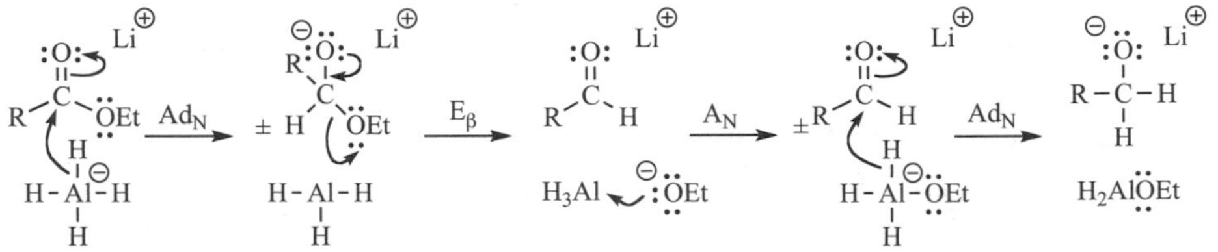

If the L is poor as in amides, Lewis salt formation allows an aluminum oxide to be kicked out instead of the amine (path Ad_N, then A_N, followed by the Lewis salt serving as a hydride source, then E_β, and then Ad_N).

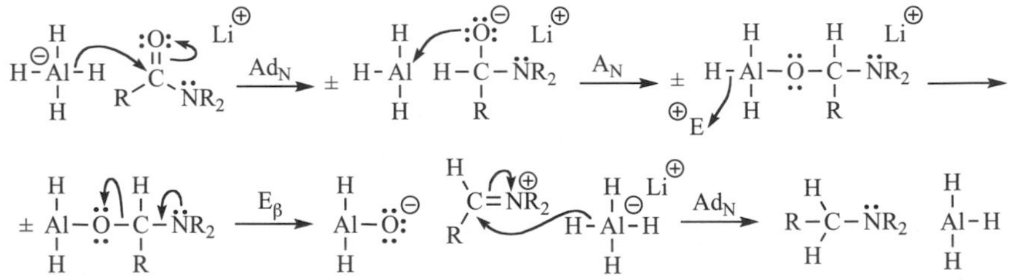

8.8.4 Organometallic Sources Reacting with L–C=Y Sinks

Organometallics follow the addition–elimination pathway ($Ad_N + E_\beta$) characteristic of this sink. However, the products of the addition–elimination path often undergo a second attack by the organometallic because the $Ad_N + E_\beta$ product is often more reactive than the original substrate (see Section 9.2). The reactivity ranking of the electron sinks and sources is important in this decision.

Second-addition example (path Ad_N, then E_β, then Ad_N again):

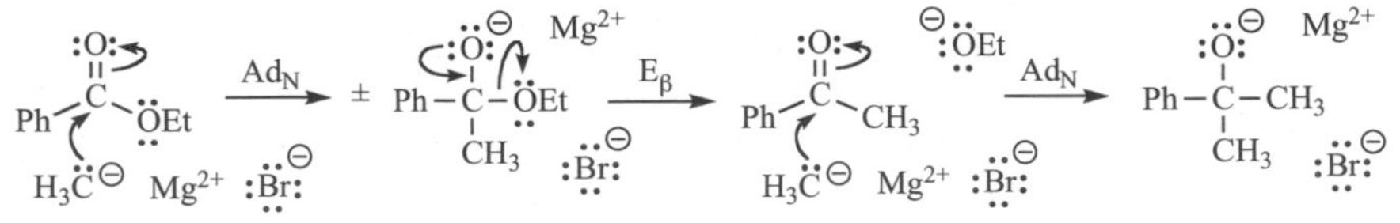

Single-addition example (path Ad_N, then E_β):

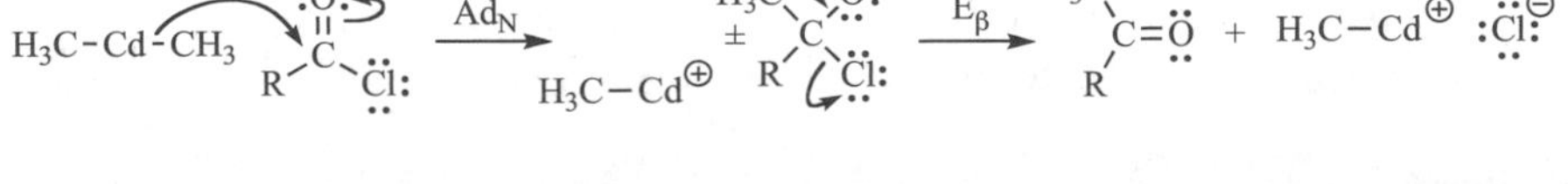

8.8.5 Allylic Sources Reacting with L–C=Y Sinks, Acylation

Allylic sources attack carboxyl derivatives such as esters almost exclusively on carbon by the path Ad_N, then E_β. As in Section 8.5.5, the enolate source should be the deprotonated sink or the sink should not be enolizable. A final proton transfer to give a resonance delocalized stable anion is frequently the driving force for the reaction.
Claisen condensation example (source is deprotonated sink), details in Section 10.7.2:

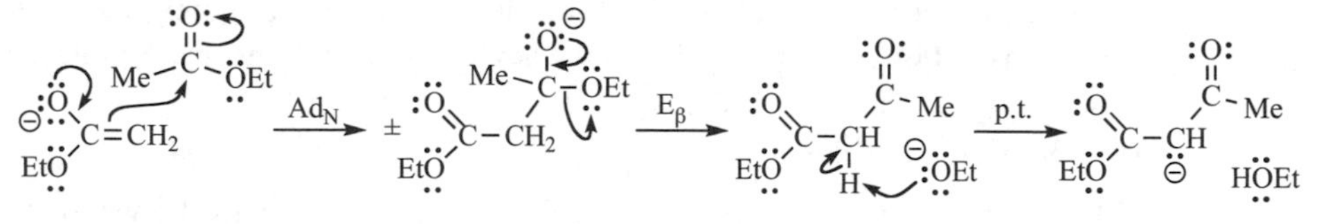

Biochemical Examples

Fats are metabolized down to a thioester called acetyl CoA. Acetyl CoA enters the citric acid cycle and eventually is converted to two molecules of carbon dioxide. The cleavage step in metabolizing long-chain fatty acids is the reverse of the Claisen reaction of the previous section. A thiol group attached to the enzyme is the nucleophile for general base-catalyzed addition, starting the reverse Claisen.

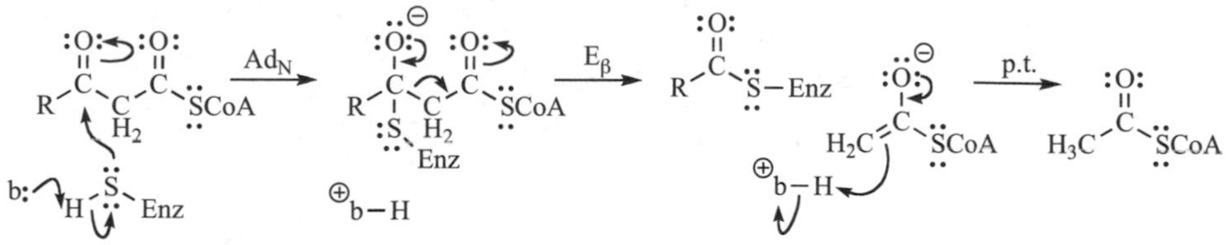

The release of the shortened fatty acid from the enzyme is an example of a lone pair nucleophile attacking a L–C=O sink as in Section 8.8.1. Again, a general base catalyzes the addition to the thioester.

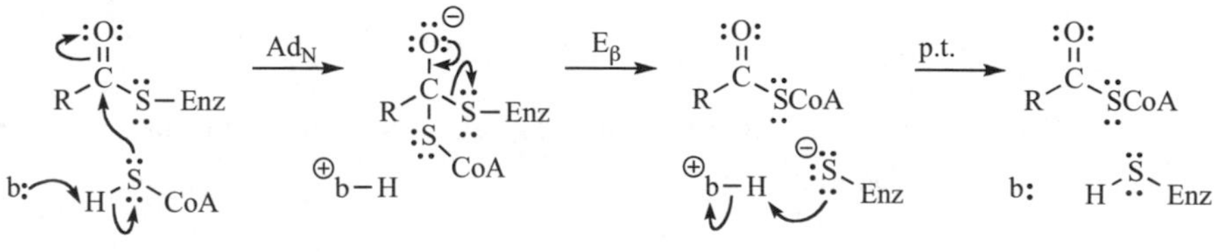

8.8.6 Simple Pi Bond Sources Reacting with L–C=Y Sinks

Simple pi bonds are usually not good enough nucleophiles to react with this sink. If a Brønsted acid or a Lewis acid is added to improve the electron sink, then addition can happen but it is often complicated with other reactions. We can improve the electrophilicity of an acyl chloride by complexing it with the Lewis acid aluminium chloride. The $AlCl_3$ can complex with either oxygen or chlorine lone pairs to form a reactive complex. Also, the leaving group can depart to give the reactive acylium ion.

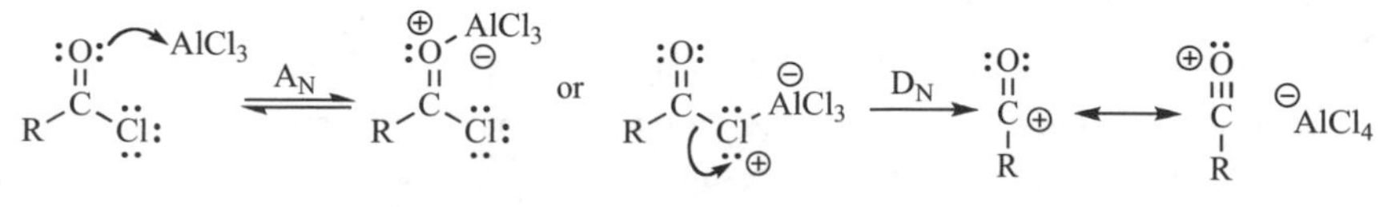

Lewis acid complexes Acylium ion

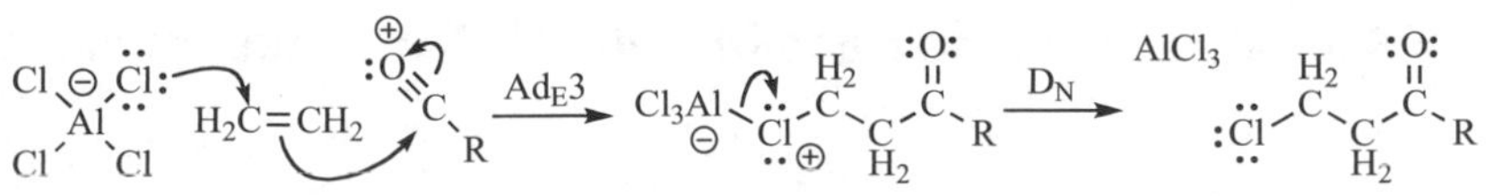

8.8.7 Aromatic Sources Reacting with L–C=Y Sinks

Aromatic rings are usually not good enough nucleophiles to react with this sink. If a Brønsted acid or a Lewis acid is added to improve the electron sink, then electrophilic aromatic substitution occurs, usually via the acylium cation as an electrophile.

Friedel–Crafts Acylation

The electrophile is the acylium ion, $R–C≡O^+$, generated by Lewis acid-catalyzed ionization of a leaving group (path D_N) from acyl halides or acid anhydrides (shown in the previous section). The proton that is lost comes from the same carbon that the electrophile attacked. The reaction fails for deactivated rings (Ar–ewg, *meta* directors). After the electrophile adds it deactivates the ring toward further attack. No rearrangement of the electrophile occurs.

Acylation example:

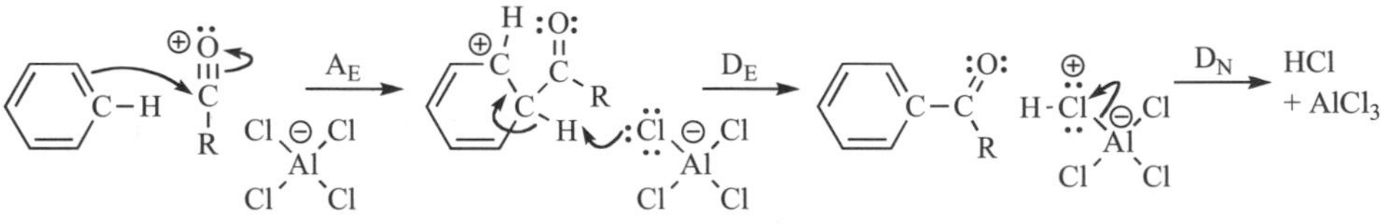

8.9 MISCELLANEOUS REACTIONS

8.9.1 Electron-Deficient Species as Electron Sinks

Lone Pairs as Sources

The trapping of a cation by a nucleophile, A_N, can be considered the electrophilic addition of a cation to a lone pair.

$$Me_3C^{\oplus} \quad :\overset{\ominus}{\underset{..}{Br}}: \quad \xrightarrow{A_N} \quad Me_3C–\overset{..}{\underset{..}{Br}}:$$

The electrophilic addition of a Lewis acid to a lone pair to form a Lewis salt is an isoelectronic reaction. An example is the first step of borate ester hydrolysis:

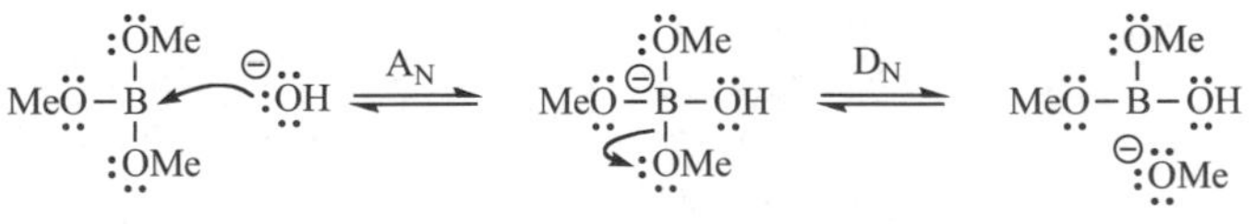

Pi Bonds as Sources

Since the product of the carbocation addition to an alkene via path A_E is also a carbocation that can rearrange and/or attack another alkene molecule (polymerization), unwanted product mixtures can result.

$$(CH_3)_3C^{\oplus} \quad C=C(CH_3)_2 \quad \xrightarrow{A_E} \quad \overset{(H_3C)_3C}{\underset{H_2C–\overset{\oplus}{C}(CH_3)_2}{}}$$

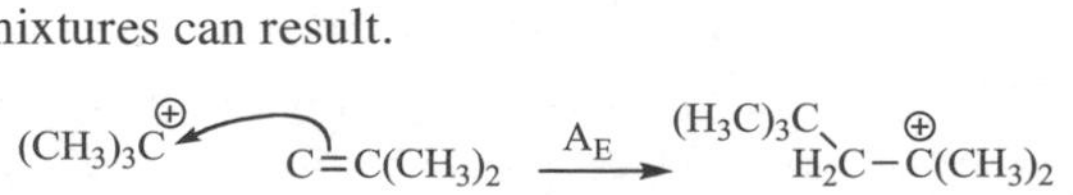

Borane, $2BH_3 \rightleftharpoons B_2H_6$, isoelectronic with carbocations, adds exclusively *syn* to multiple bonds. Initially a pi-complex is believed to form, followed by rapid hydride donation to the largest ∂^+ atom (4e path, Fig. 7.12). The reaction repeats to produce dialkyl and occasionally trialkylboranes, R_3B.

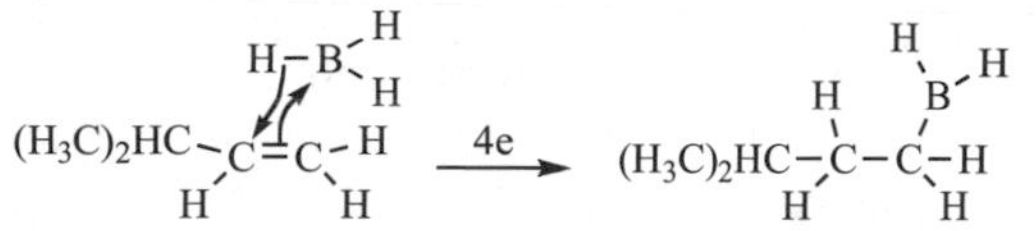

One equivalent of R_2AlH at low temperature will add once to nitriles or many carboxylic acid derivatives to give aldehydes after an acidic aqueous workup (see Section 9.2.2).

8.9.2 Carbonate Derivatives as Electron Sinks

These are represented by the following structures and can undergo (depending on the quality of the leaving group) single, double, and triple additions by nucleophiles via the addition–elimination path. The tendency toward a single addition–elimination is highest with haloformates (center of the following structures) because one leaving group is much better than the other. This is multiple addition decision is discussed in Section 9.2.

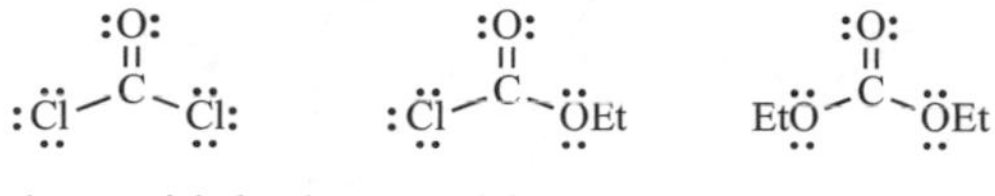

Carbonate derivatives with both a good leaving group and an OH are not stable with respect to an E1cB or E2 elimination reaction to produce carbon dioxide.

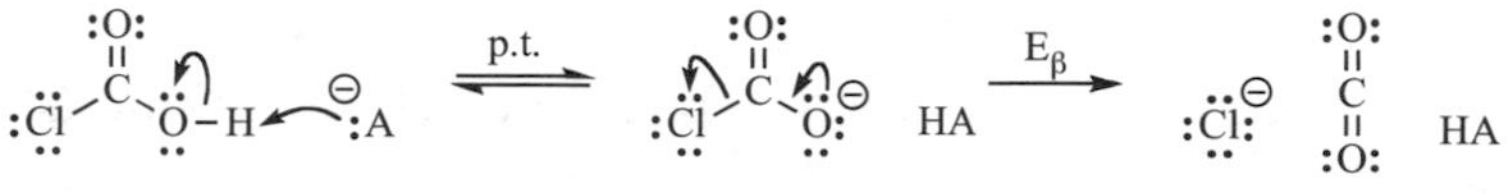

8.9.3 Heterocumulenes as Electron Sinks

Heterocumulenes are attacked at the central carbon. In acidic media, proton transfer may occur first. This is a common way to activate an amino acid COOH for laboratory peptide synthesis (R = cyclohexyl). Lone pair source Ad_E2 example:

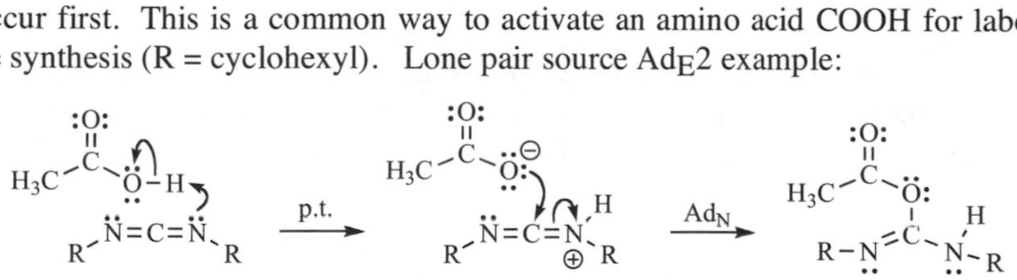

In basic media, the resultant anion from the Nu attack is protonated in the workup. Organometallic Ad_N2 example (acidic water workup protonates the carboxylate anion):

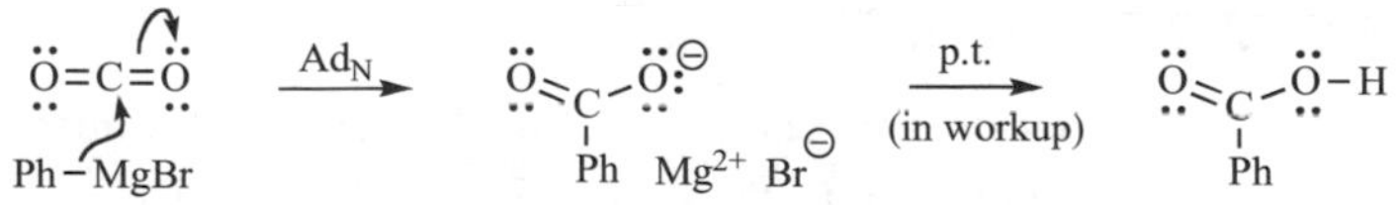

8.9.4 Nucleophilic Aromatic Substitution

There are three routes available for this very poor electron sink, Ar–L . If the leaving group is good—for example, chloride—but there are no electron-withdrawing groups on the ring, temperatures over 300°C will be required to get reaction. Nitro groups and other good electron-withdrawing groups on the ring lower the temperature for reaction by stabilizing the intermediate carbanion.

$Ad_N + E_\beta$ example:

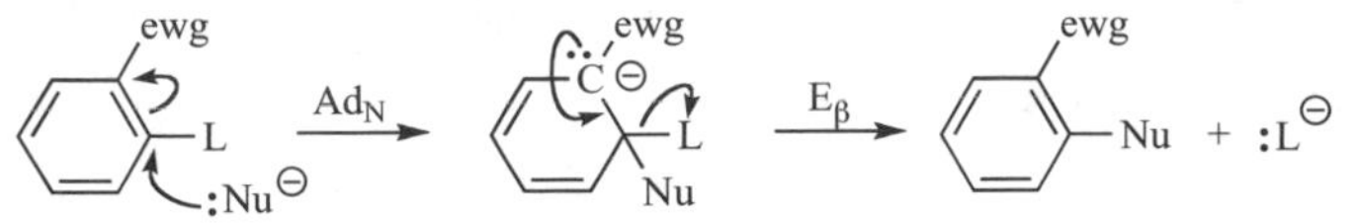

An unusual reaction occurs with aromatic halides and very strong bases; the substitution reaction can go via elimination to the highly reactive benzyne followed by addition, path E2 then Ad_N2.

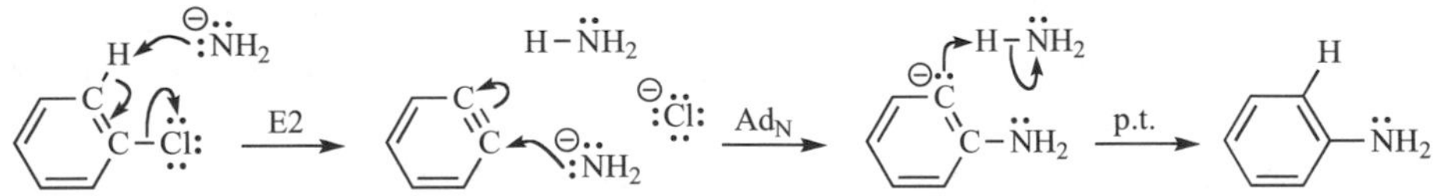

If the leaving group is excellent, namely N≡N in diazonium ions, an S_N1-like path may occur. The many mechanisms for substitution on these diazonium ions are complex; some involve electron transfer, radical intermediates, or copper catalysis (Chapter 11).

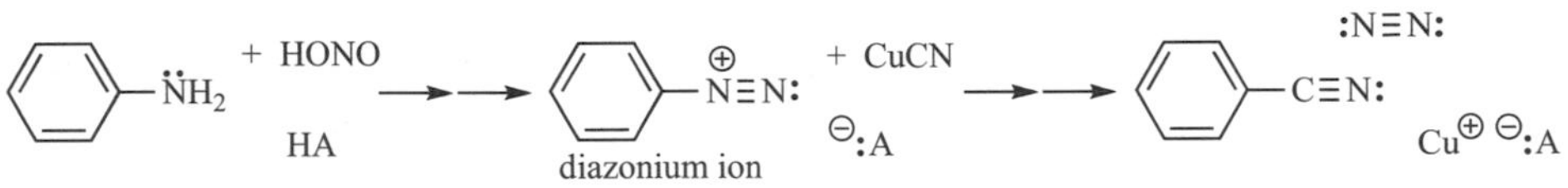

8.10 METAL IONS AS ELECTRON SINKS

8.10.1 Metal Ions as Electrophiles and Electrophilic Catalysts

The role of metal ions as electrophilic catalysts is often underestimated. The formation of good ion-pair stabilization in the transition state in an anionic nucleophilic attack lowers the energy of the transition state. Even a powerful hydride nucleophile such as $LiAlH_4$ will not add to a carbonyl if the lithium ion is kept from complexing with the carbonyl.

A good example of a metal ion as an electrophile is the oxymercuration reaction. The nucleophile is often solvent, usually water or an alcohol. The charge is distributed, so little rearrangement occurs. The mercury is removed in a subsequent reduction step.

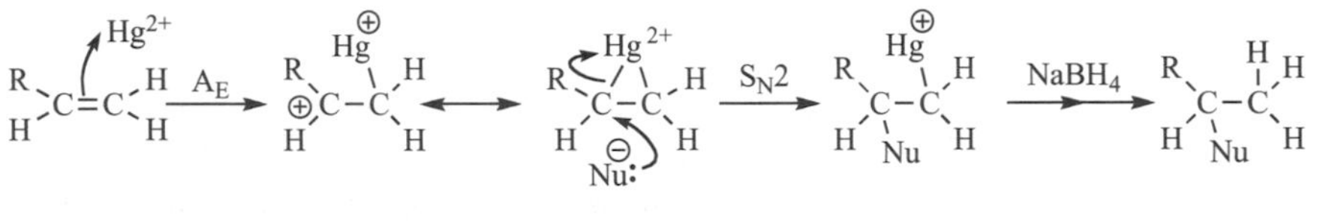

8.10.2 Metal Ions as Oxidants

Chromium trioxide and chromic acid, H_2CrO_4, are commonly used to oxidize alcohols. The alcohol adds to the chromium species to give a chromate ester, which undergoes elimination by an E2 process.

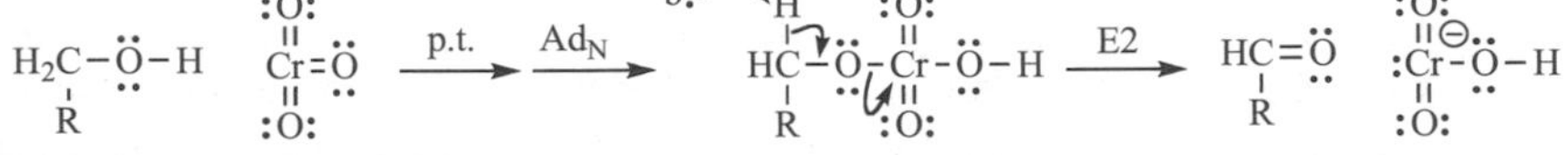

If water and a little acid are present, the initially formed aldehyde adds water to form a hydrate, which is further oxidized to the carboxylic acid.

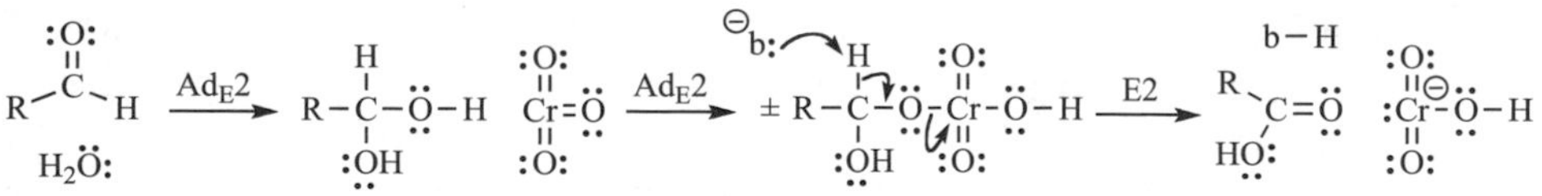

8.10.3 Hydrogenation via Transition Metal Catalysts

The exploration of transition metals as redox catalysts is a large and very active area of research. Transition metal mechanisms are quite different from those discussed so far. Catalytic hydrogenation of pi bonds is an important reaction for organic synthesis, and will serve as a good example. The mechanism proposed for catalytic hydrogenation by platinum, palladium, nickel, rhodium, iridium, and ruthenium is shown in Figure 8.10.

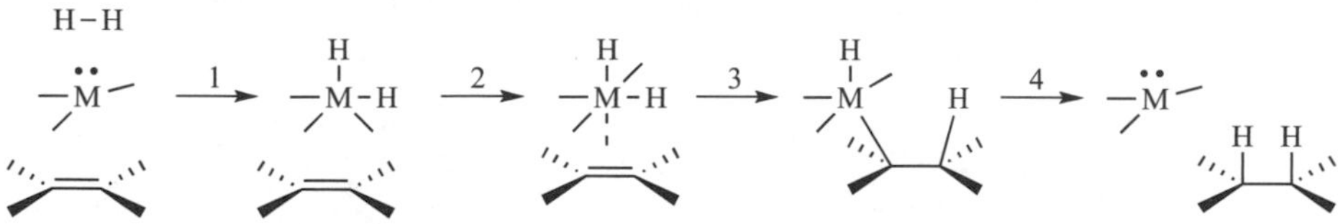

Figure 8.10 The proposed mechanism for catalytic hydrogenation.

Step 1 is the *oxidative addition* of hydrogen; the metal loses two electrons to form a bond to each of the two hydrogen atoms. Step 2 is the *complexation* of the alkene to the metal, activating the pi bond. Complexation usually occurs from the least hindered face of the pi bond. Step 3 is the *insertion* of the metal hydride bond into the pi bond, forming a metal-carbon bond. Step 4 is the *reductive elimination* (the reverse of oxidative addition), which returns the metal to its original oxidation state and releases the alkane. The reductive elimination is usually very fast. Both hydrogens are added to the same face of the pi bond, a *syn* addition.

8.11 REARRANGEMENTS TO AN ELECTROPHILIC CENTER

Rearrangements to an electrophilic carbon were briefly introduced in Section 4.7, and the electron flow paths were covered in Sections 7.1.11 and 7.1.12. To recap, the most important path restriction on rearrangements is that orbital overlap must be maintained by the migrating group with both the atom it migrates from and the atom it migrates to. In the transition state, the migrating orbital overlaps the orbitals of both. Rearrangement can proceed with any of our three general classes of electron sinks: an empty *p* orbital, a weak single bond to a leaving group, or a polarized multiple bond. The migration to an empty *p* orbital on carbon was the worked mechanism example in Section 4.7.3. Migration to polarized multiple bond (benzil to benzilic acid rearrangement):

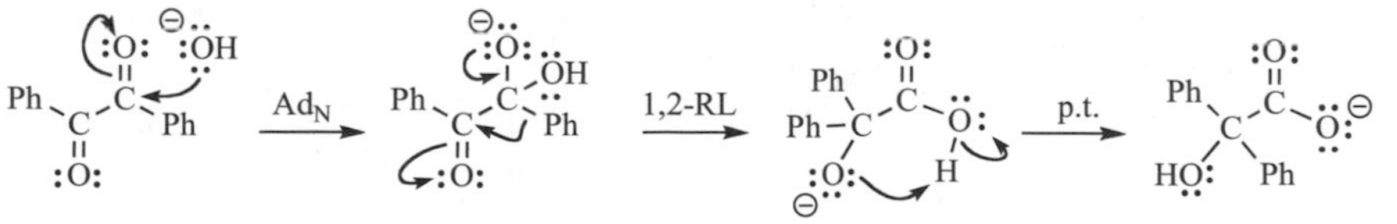

With a weak single bond to a leaving group, rearrangement can occur by a path (1,2RL) reminiscent of the generic S_N2. Good orbital alignment is crucial; the group that is best lined up to displace the leaving group from the back side is the one that preferentially migrates. Occasionally the leaving group will depart first (path D_N), and then rearrangement occurs (path 1,2R) in a path resembling the S_N1 process. Examples of migration to oxygen and migration from boron will be given in Section 8.12, Nu–L Pathways. The migration can occur to a vinylic center as the following examples show.

Migration to nitrogen, basic media example (in this example, the migration is followed by Nu attack on the heterocumulene, Section 8.9.3, and then decarboxylation, Section 8.9.2):

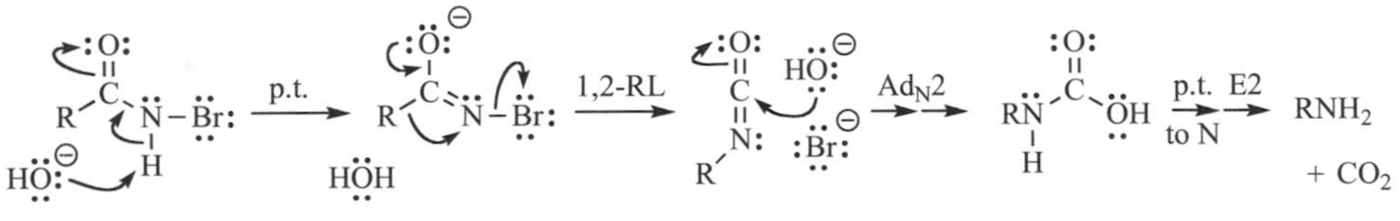

Migration to nitrogen, acidic media example (in this example, the leaving group returns as a nucleophile to trap the heterosubstituted vinyl cation):

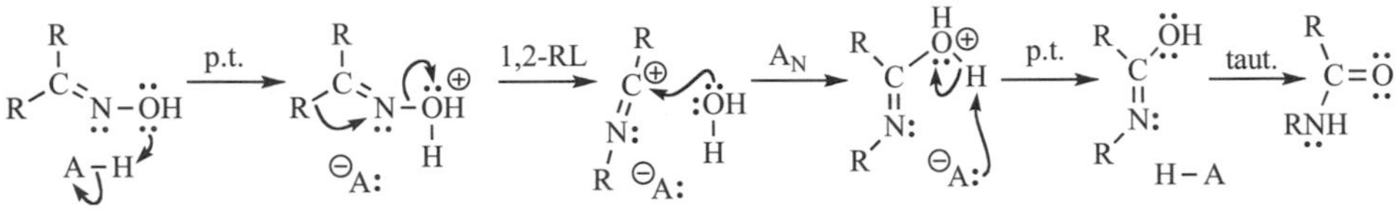

More examples of rearrangements to an electrophilic center are shown in the next section.

8.12 Nu–L REACTIONS

The Nu–L class of reagents can be recognized by a nucleophilic center to which a leaving group is directly bonded. **This unusual arrangement in which the same atom can serve both as a sink and a source is characteristic of the Nu–L class**, and thus their reactions are reasonably similar, most often three-membered ring formation or addition followed by rearrangement. Below are several examples of the Nu-L class.

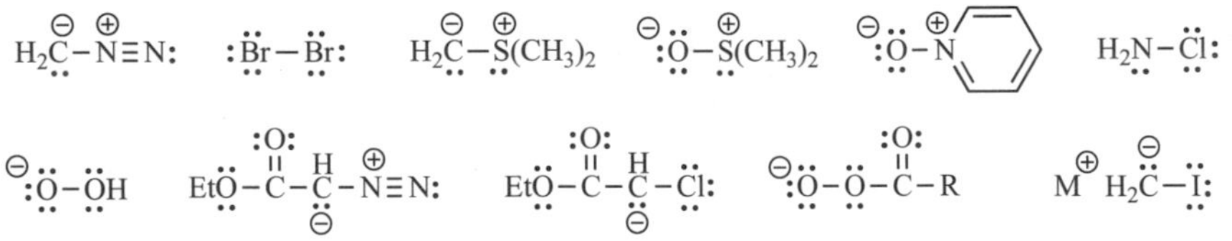

8.12.1 Nu–L Reacting with Trialkylboranes

Trialkylboranes, R_3B, react with Nu–L reagents to form Lewis acid–base salts that rearrange by migration of the alkyl group to the Nu with loss of the leaving group (path 1,2RL). The process usually repeats until all the alkyl groups have reacted. The borate that is formed can be hydrolyzed to give the free RNuH (see Section 8.9.1). This is a very general and useful set of reactions.

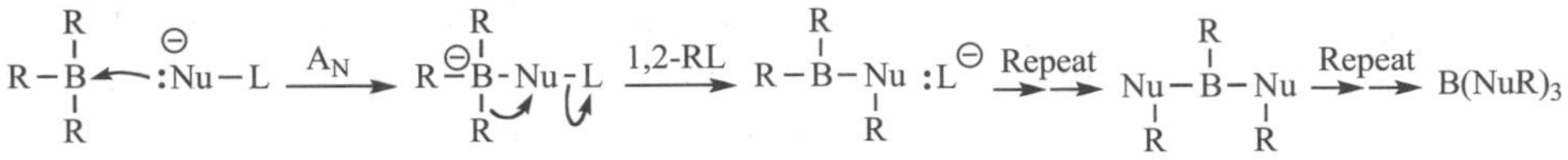

8.12.2 Nu–L Reacting with Acids

The Nu portion of the Nu–L can be protonated with acids, producing a reactive species with which the conjugate base of the acid may react.

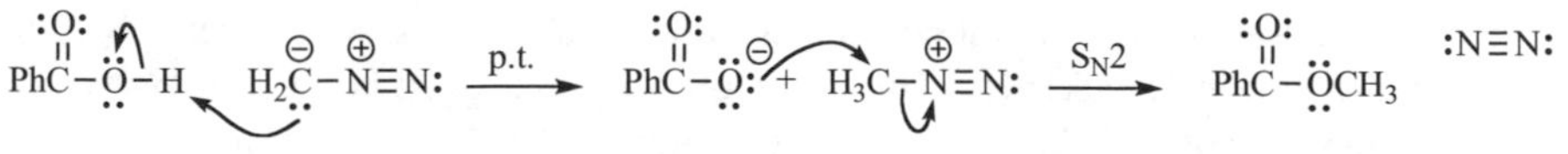

8.12.3 Nu–L Reacting with sp^3 C–L

The Nu–L can substitute the leaving group on carbon and can be followed by an elimination reaction.

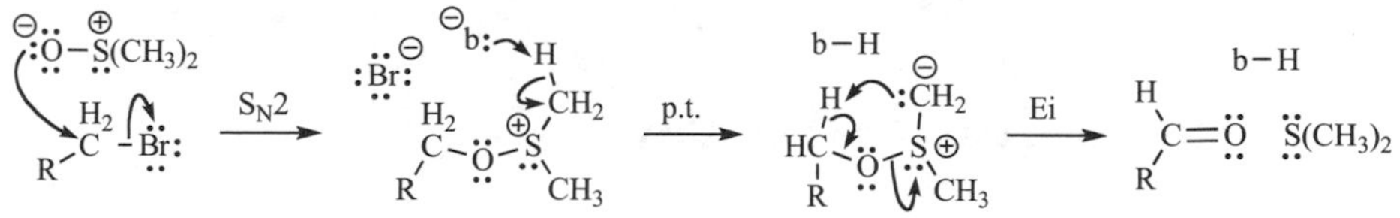

8.12.4 Nu–L Reacting with C=C

Leaving Group Drops Off First: Carbenes

Carbenes bear a resemblance to carbocations in that there is an empty *p* orbital that can behave as an electron sink. However, a full orbital that can serve as an electron source is on the same atom. Trichloromethyl anion loses chloride, forming the reactive dichlorocarbene, a neutral, electron-deficient, electrophilic intermediate. Stabilization in dichlorocarbene results from the interaction of the full lone pair orbitals of chlorine with the empty *p* orbital of the carbene (Fig. 8.11). If the donors on the carbene are good enough, the carbene becomes nucleophilic. With few exceptions, carbenes react stereospecifically with double bonds to produce three-membered rings.

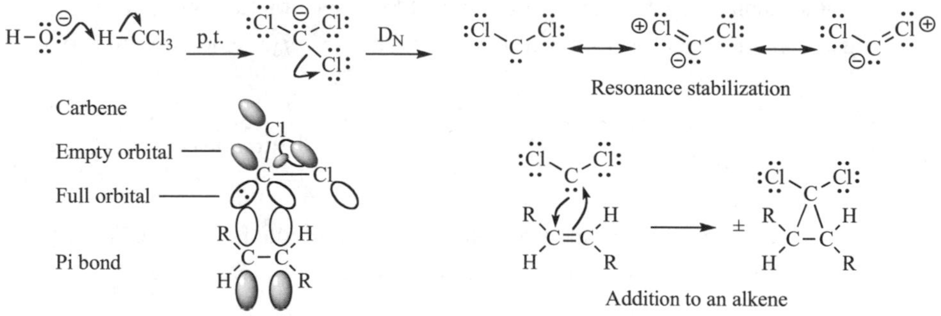
Resonance stabilization
Carbene
Empty orbital
Full orbital
Pi bond
Addition to an alkene

Figure 8.11 Dichlorocarbene addition.

Leaving Group Loss Concerted with Attack (path NuL)

These reactions are characteristically stereospecific. The formation of a bromonium ion from attack of bromine on a pi bond was the first example of this path (Section 8.3.6).

The organometallics of this subset are called carbenoids because of their similar reactivity to carbenes. Carbenoids are generated by the reaction of a halide with a metal by halogen–metal exchange. The presumed sequence is shown in the following example. The structure of the organometallic is currently under debate.

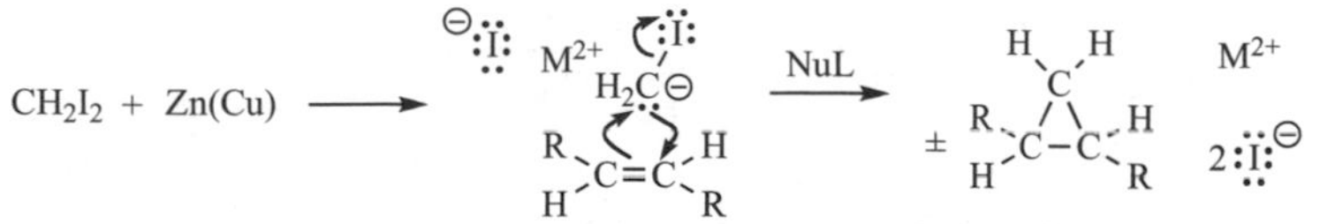

Although they are not considered carbenoids, peracids are mechanistically similar in their additions to alkenes. The intramolecular hydrogen bond (shown by two resonance forms) partially breaks the O–H bond, making it a better nucleophile.

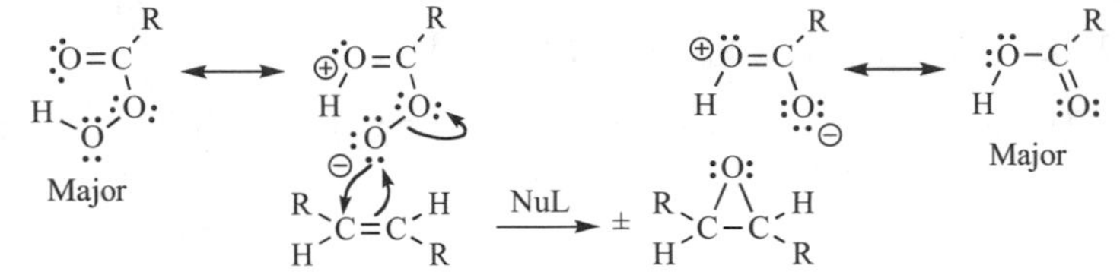

8.12.5 Nu–L Reacting with C=Y

Leaving Group Drops Off After Attack

This mechanism implies a new intermediate, one that stabilizes the anion formed upon nucleophilic attack. The Nu–L species acts first as an electron source, then in a second step functions as an electron sink. Common members of this subset, the ylides, have adjacent plus and minus charges, $^{-}$Nu–L^{+}. This mechanism is a combination of the addition to a polarized multiple bond (path Ad_N) followed by an internal S_N2.

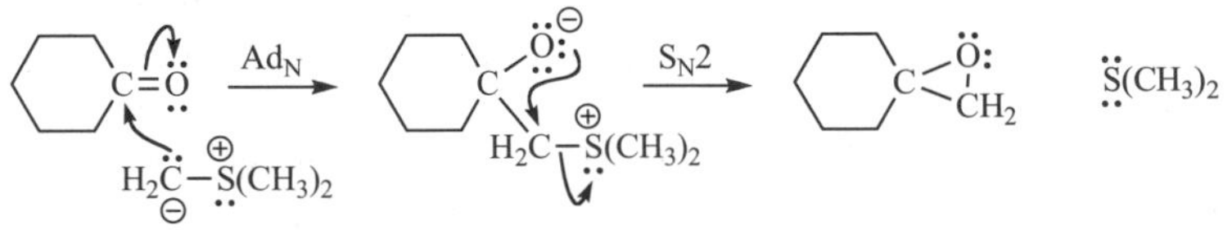

Phosphorous ylides are almost the only representative of a mechanistic variant. Rather than being displaced in an internal S_N2 reaction, the phosphorus bonds to oxygen and then undergoes an elimination (path 4e) as the very stable phosphine oxide.

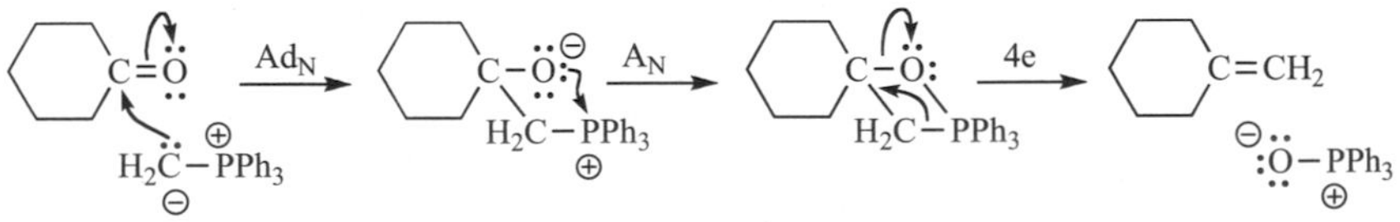

The Nu–L species can attack the polarized multiple bond (path Ad_N), then rearrange (path 1,2RL) with loss of the leaving group, discussed in Section 8.11.

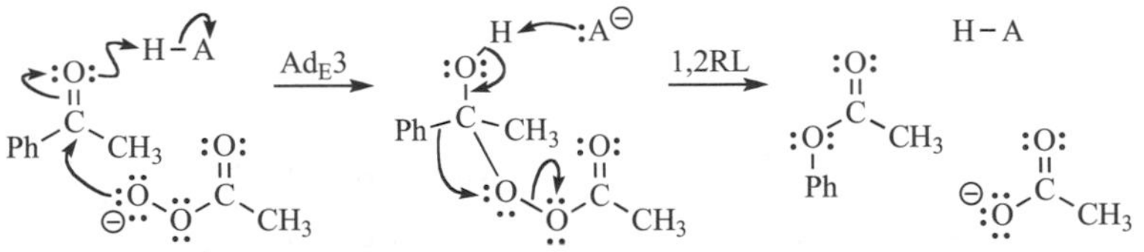

8.12.6 Nu–L Reacting with C=C–ewg

The allylic source, formed by the conjugate addition of peroxide ion to the polarized pi bond, acts as a nucleophile in an internal S_N2.

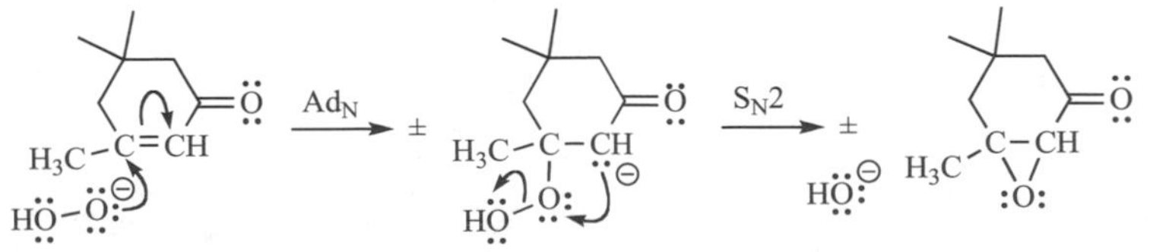

8.12.7 Nu–L Reacting with L–C=O

The expected addition–elimination path $Ad_N + E_\beta$ occurs.

Example (which gives another Nu–L species after proton transfer):

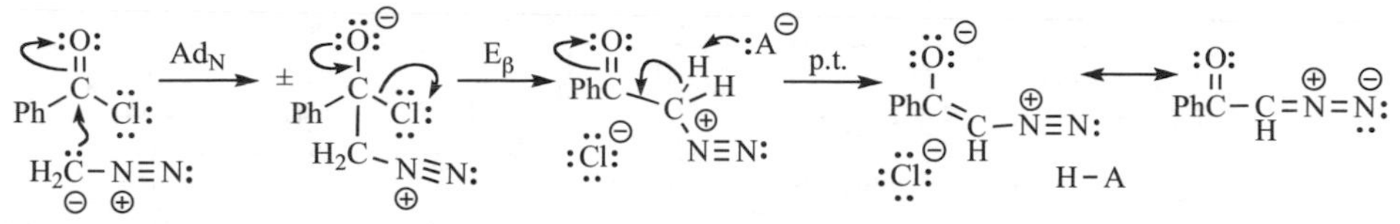

Azide ion addition–elimination is followed by a rearrangement with loss of nitrogen (path 1,2RL) in the following example.

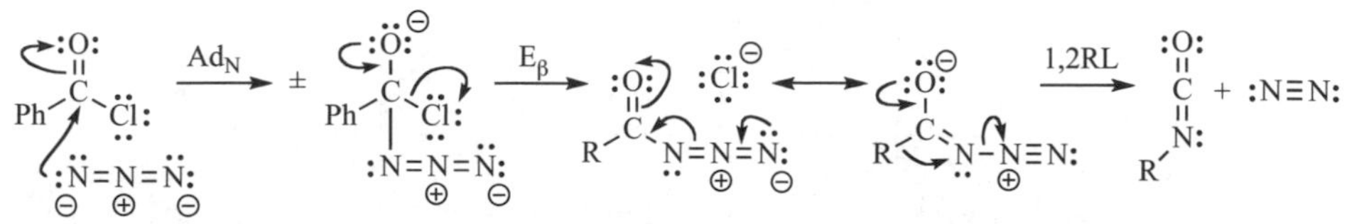

8.12.8 Thiamine-Catalyzed Decarboxylation of Pyruvate

Biochemical Example

Thiamine is an amazing coenzyme unique in biochemistry. Thiamine is an elegantly optimized catalyst that serves as an attachable/detachable electron-withdrawing group to allow electron flow when it would otherwise be impossible. One reaction thiamine catalyzes is the decarboxylation of pyruvate, CH_3COCOO^-, to acetaldehyde, CH_3CHO (Fig. 8.12). Initially, a deprotonation forms the nucleophilic thiamine carbene. This thiamine carbene adds to the more electrophilic ketone carbonyl of pyruvate. A fragmentation step via an E_β path (in which the leaving group is an enamine) can now occur because the thiamine accepts the electron flow from the carboxylate anion (discussed in Section 7.4, Variations on a Theme). The electrons flow back out of the thiamine to pick up a proton (or other electrophile). Then thiamine detaches as a leaving group by E2 to start the process all over again. Without the thiamine coenzyme, the decarboxylation of pyruvate does not occur at all, for the reaction would have to produce the highly unstable acyl anion, $[CH_3–C{=}O]^-$.

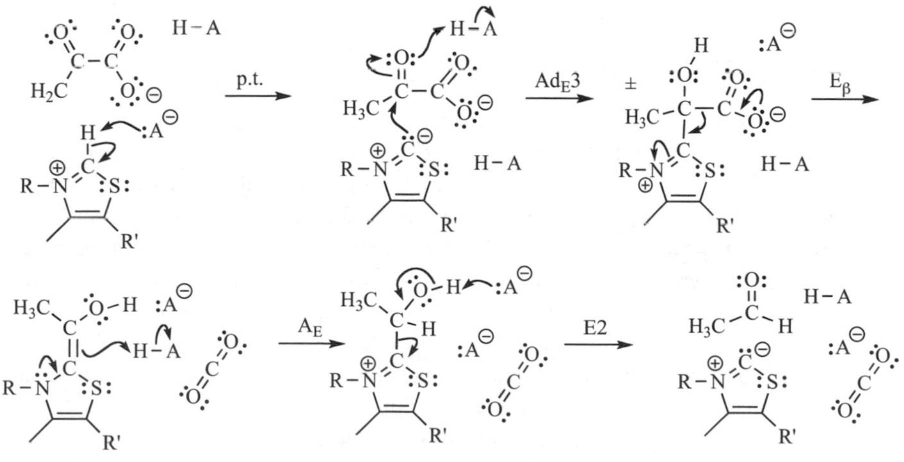

Figure 8.12 The thiamine (vitamin B1)-catalyzed decarboxylation of pyruvate.

8.13 PRODUCT MATRIX SUMMARY

Table 8.3 Product matrix

Source⇒ Sink⇓	Lone pair Nu Z:⊖	Base b:⊖ and adjacent CH	Metal hydride $MH_4^{\ominus}$	Organometal. R–M	Allylic[a] C=C–Z:⊖	Simple pi bonds C=C	Aromatic ArH
H–A	p.t. Z–H	p.t. b–H	p.t. H–H	p.t. R–H	p.t.[a] HC–C=Z	H add. A C–C	H exchange Ar–H
Y–L	subst. Z–Y	elim. C=Y	subst. H–Y	subst. R–Y	subst.[a] Y–C–C=Z	Y add. L C–C	Ar subst.[c] Ar–Y
C–L	subst. C–Z	elim. C=C	subst. C–H	subst. C–R	subst.[a] C–C–C=Z	usually no reaction[d]	Ar subst.[c] C–Ar
C=Y	add. Z–C–Y⊖	p.t. ⊖C–C=Y	add. H–C–Y⊖	add. R–C–Y⊖	add.[a] ⊖Y–C–C–C=Z	usually no reaction[d]	Ar subst.[c] Ar–C–YH
—C≡Y	add. Z–C=Y⊖	p.t. ⊖C–C≡Y	add. H–C=Y⊖	add. R–C=Y⊖	add.[a] C=Y⊖ C–C=Z	usually no reaction[d]	Ar subst.[c] Ar–C=YH
C=C–ewg[e]	1,4 add.[e] Z–C–C⊖–ewg	p.t. C=C–ewg C⊖	1,4 add.[e] H–C–C⊖–ewg	1,4 add.[e] R–C–C⊖–ewg	1,4 add.[a,e] C–C⊖–C–ewg C–C=Z	usually no reaction[d]	Ar subst.[c,e] Ar–C–C–ewg
L–C=Y	add.-elim. Z–C=Y	p.t. L–C=Y ⊖C	add.-elim.[b] H–C=Y	add.-elim.[b] R–C=Y	add.-elim.[a] C=Y C–C=Z	usually no reaction[d]	Ar subst.[b, c] Ar–C=Y

[a] Ambident source, Section 9.4. [b] Multiple addition possible, Section 9.2. [c] With Lewis or Bronsted acid catalyst. [d] Reacts with strong acid catalyst but may give polymers or mixtures. [e]Ambident sink, Section 9.6.

The product matrix (Table 8.3) serves as a summary of most of Chapter 8. It is the source and sink matrix that acted as an index to Sections 8.2 through 8.8 filled in with the most common reaction product in generic form. It cannot be stressed too much that **everything depends on your ability to recognize the generic class of the sources and sinks** that are present in the reaction mixture. However, sometimes a species has a dual reactivity and therefore may fit into more than one generic class. A common example of this is that most anions can behave as a nucleophile or as a base. Chapter 9 discusses the common major decisions.

ADDITIONAL EXERCISES

Ever notice how the book problems within a chapter appear easier than the ones at the end of the chapter? And perhaps the ones on an exam may appear harder still? It might just be a matter of "mental priming" of your analysis process. Within the chapter, you know that the problem has something to do with what you just read; you do not have to search for the principles to apply to solve the problem. At the chapter's end, you have a broader area to search, but usually the problems relate somehow to the chapter just covered. On an exam that covers several chapters, you may have difficulty trying to figure out which part of what you have learned applies to the problem. Therefore a crucial part of problem analysis is determining what information is relevant. We have learned how to classify sources and sinks and to use that to predict reactivity. The real reward of classifying sources and sinks is in priming your analysis of a problem you have never seen before. As was done in this chapter, focus on the sink first, then find a source. (See: Hints to Selected Problems from Chapters 8, 9, and 10, Appendix, if you get stuck.)

8.1 In the following problem, identify the electron sink and look at the product to identify the electron source and predict the pathway(s) to the product.

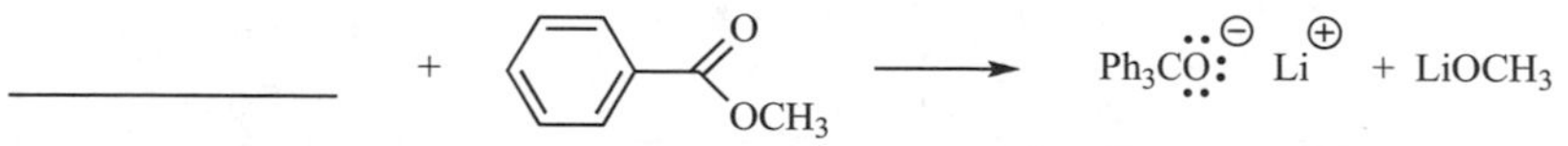
+
O
OCH3
Ph3CO: Li
+ LiOCH3

8.2 In the following problem, identify the electron sink and electron source and predict the pathway(s) to the product.

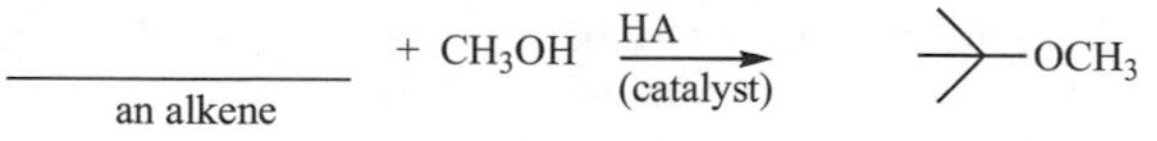
an alkene
+ CH3OH
HA
(catalyst)
OCH3

8.3 In the following problem, identify the electron source and look at the product to identify the electron sink and predict the pathway(s) to the product.

(CH3)CuLi
+
O

8.4 In the following problem, identify the electron sink and electron source and predict the elimination product.

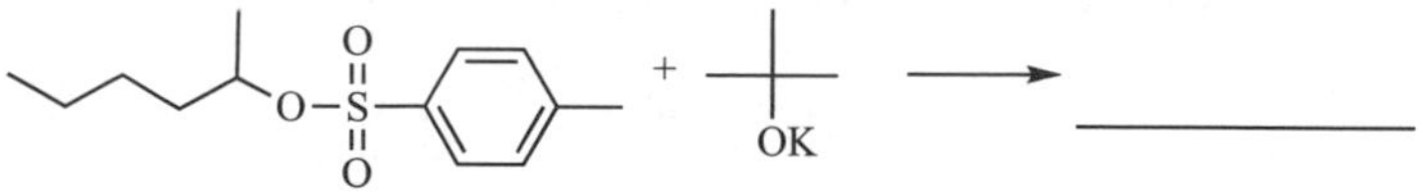
O
O–S
O
+
OK

8.5 In the following problem, identify the electron sink and electron source and predict the product and the pathway(s) to the product. The solvent is ethanol.

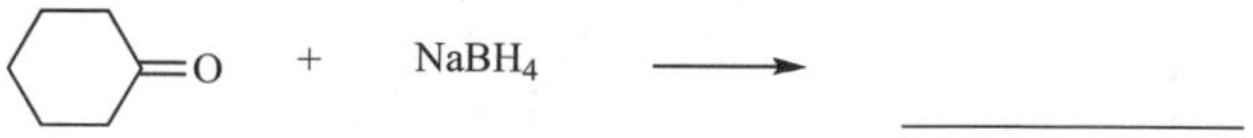
O
+
NaBH4

8.6 In the following problem, identify the electron sink and look at the product to identify the electron source and predict the pathway(s) to the product.

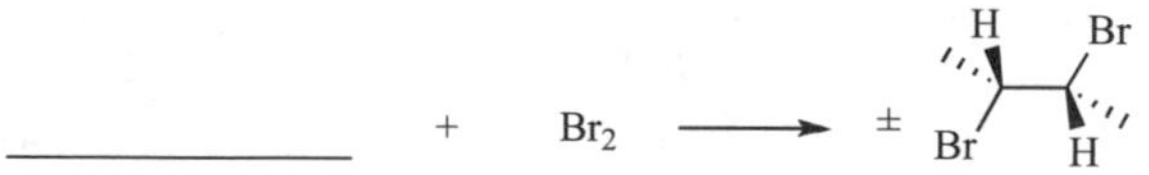
+
Br2
±
H
Br
Br
H

8.7 In the following problem, identify the electron sink and look at the product to identify the electron source and predict the pathway(s) to the product.

Br
+

8.8 In the following problem, identify the electron sink and electron source and predict the pathway(s) to the amide product. The solvent is water.

Ph–C≡N + KOH

8.9 In the following problem, identify the generic electron sink and electron source and predict the products.

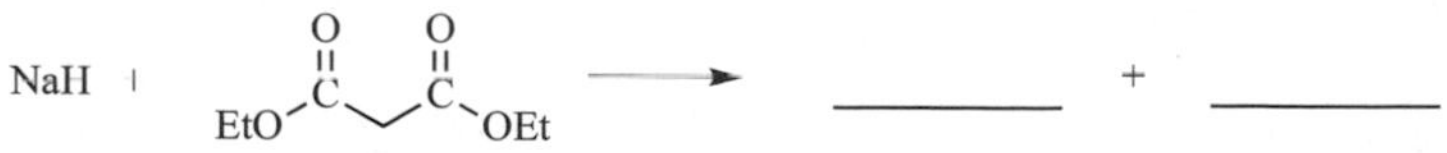
NaH +
O
O
C
C
EtO
OEt
+

8.10 In the following problem, identify the electron source and look at the product to identify the electron sink and predict the pathway(s) to the product.

N
+
N
O
Cl

8.11 In the following problem, identify the electron sink and electron source and predict the product and the pathway(s) to the product.

+ HBr

8.12 In the following problem, identify the electron sink and electron source and predict the product and the pathway(s) to the product.

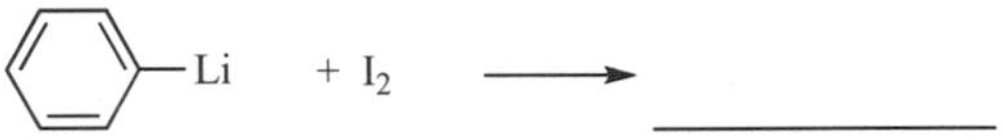
Li
+ I2

8.13 In the following problem, identify the best electron sink and electron source and predict the product and the pathway(s) to the product.

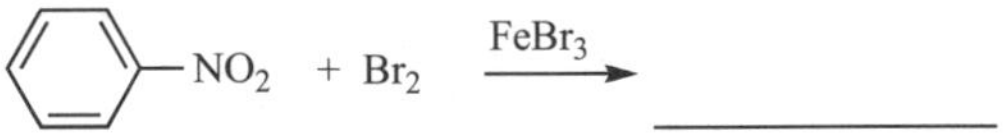
NO2
+ Br2
FeBr3

8.14 In the following problem, identify the electron sink and electron source and predict two possible products. A Chapter 9 decision is required to decide which one is favored.

Br
+ CH3ONa
or

8.15 The carbon-carbon bond-forming step of the acid-catalyzed aldol reaction has an enol (allylic source) attacking a protonated carbonyl (which is just a lone-pair-stabilized carbocation). With those hints, give a mechanism for the acid-catalyzed aldol reaction.

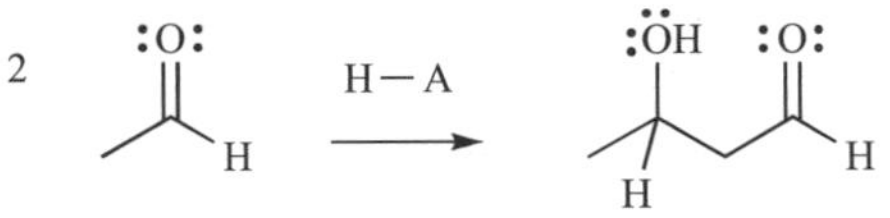
2
:O:
H
H−A
:ÖH
:O:
H
H

8.16 In the following problem, identify the best electron sink and electron source and predict the product and the pathway(s) to the product.

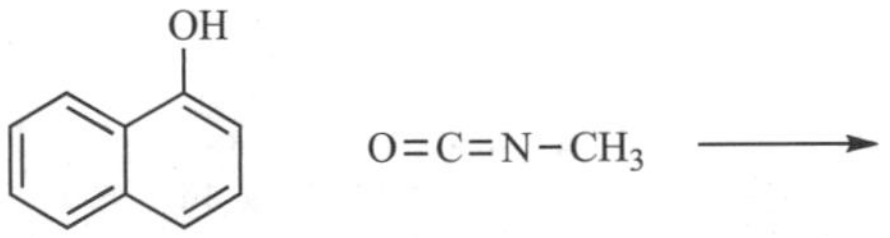
OH
O=C=N−CH3

8.17 In the following Lewis acid-catalyzed ($AlCl_3$) problem, identify the electron sink and electron source and predict the product and the pathway(s) to the product.

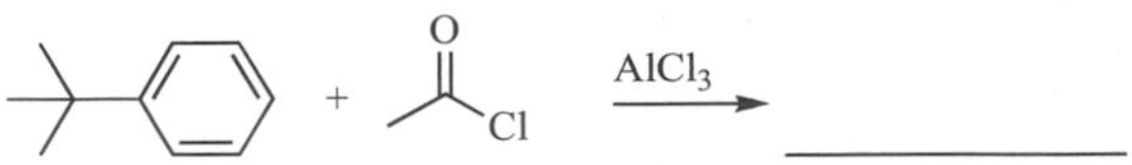
+
O
Cl
AlCl3

9

DECISIONS, DECISIONS

Electron Flow In Organic Chemistry: A Decision-Based Guide To Organic Mechanisms, Second Edition.
By Paul H. Scudder

9.1 DECISION POINT RECOGNITION

In Chapter 4 the decision to ionize or not to ionize the leaving group was part of your ability to predict whether the first step in an E1 or S_N1 would proceed. Several other similar decisions are collected in this chapter. The most difficult choice is between a species serving as a nucleophile or as a base (Section 9.5). Other decisions are basically a choice of regiochemistry (Sections 9.3, 9.4, 9.6). One (Section 9.2) is a question of extent of reaction: stop or keep going. Another (Section 9.7) is the competition between internal reactions and external ones. The last (Section 9.7) is the competition between nucleophilic trapping and cation rearrangement.

The most important question that you must continually ask yourself when examining any step in a process that has known and unknown branch points is, What else can happen? This may seem to add complexity, but it is the essence of critical thinking. Obvious routes often do not work and therefore lead to frustration and the question, "Where did I go wrong?" The answer is usually, "You forgot to watch for alternative routes and proceeded with the first idea that occurred to you."

Draw out every step of the process and look for alternative routes from each step; try to find at least one alternative for each step. For hints check resonance forms, proton transfer, alternative sources and sinks, less likely combinations, and higher-energy species. Specifically look for the more obvious decision points covered in this chapter.

If you find that you habitually miss a decision point or two, then a good idea would be to make up a checklist of reminders until new habits form. Refer to this checklist whenever you are working a problem. Continue to personalize this checklist by adding reminders to watch for common errors that you find yourself repeating.

9.2 MULTIPLE ADDITIONS

9.2.1 Nucleophilic Addition to L–C=O

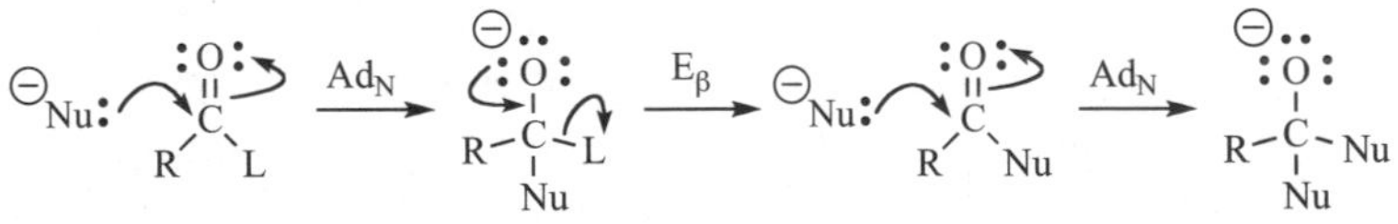
Nu
R
L
Ad_N
Nu
E_β
Nu
R
Nu
Ad_N
R
Nu
Nu

The products of an addition–elimination path very often undergo a second attack by the nucleophile (Ad_N + E_β then a second Ad_N) because the product of the first addition–elimination is also a reactive electron sink. If the addition–elimination product is more reactive than the original starting material, a second addition almost certainly will occur. A ranking of the reactivity of electron sinks and sources is very important for path selection. For each reaction, the useful limit must be found for each reactivity spectrum. How far down the reactivity trend for the electron source can we go and still have the reaction with a particular electron sink occur at a useful rate? There are four ways that the reaction may stop after the initial Ad_N + E_β:

1. **Unfavorable Equilibrium:** The Nu is such a weak base that the second addition easily reverses and the equilibrium favors the more stable carbonyl. Heteroatom lone pair sources almost never add twice.

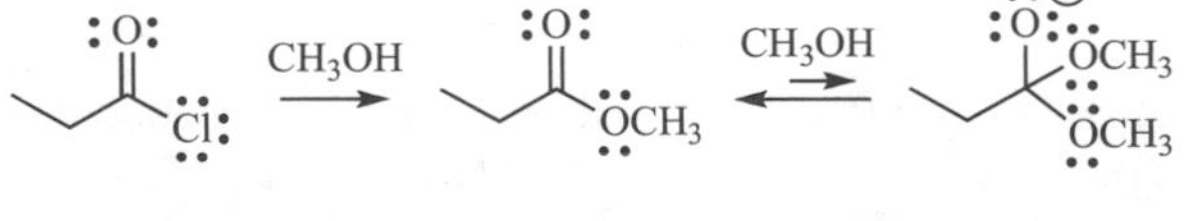
CH_3OH
OCH_3
CH_3OH
OCH_3
OCH_3

2. **Selective Nucleophile:** The initial electron sink must be more reactive than the product of the addition, and the electron source, although reactive enough to add once, is not reactive enough to make a second addition to the less reactive compound. Organometallics in the reactivity range of organocadmiums, organozincs, and organocoppers (metal electronegativities are 1.69, 1.65, 1.90, respectively) add quickly to the more reactive acyl chlorides but slowly to the ketone products.

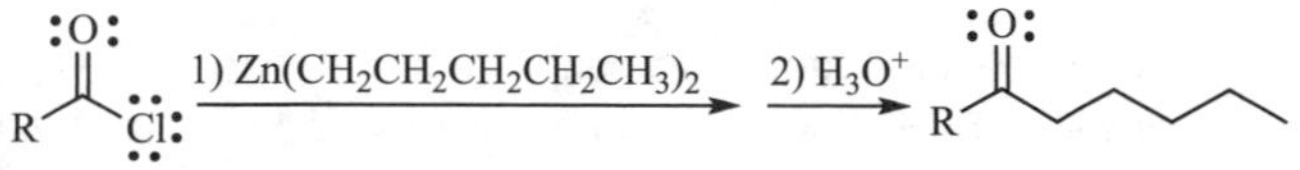
1) Zn(CH2CH2CH2CH2CH3)2
2) H3O+
R
Cl
R

3. **Poor Leaving Group:** The leaving group is so poor (in amides, $pK_{aHL} = 36$, for example) that it cannot be ejected easily from the tetrahedral intermediate. Since the carbonyl does not reform, there is no site for nucleophilic attack. The carbonyl forms in the acidic workup (no Nu left then) by protonating the nitrogen, making it into a better L.

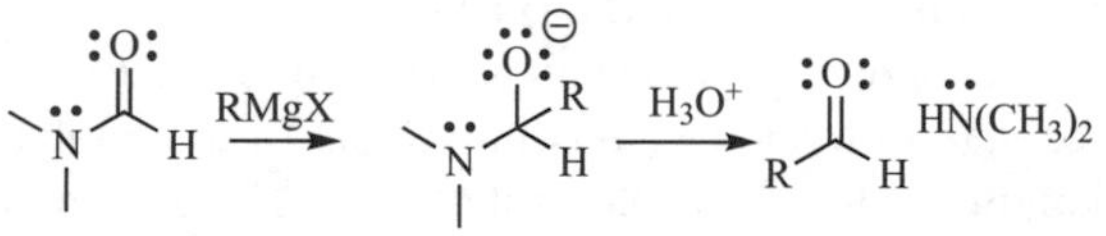
RMgX
H3O+
HN(CH3)2

4. **Low Temperature:** One equivalent or less of the electron source was added to the sink at a low enough temperature (usually –78°C). Since the addition step usually has a lower barrier than the elimination step, the loss of the leaving group from the tetrahedral intermediate by elimination path E_β does not occur to a significant extent before the electron source is used up. Tetrahydrofuran is the preferred solvent.

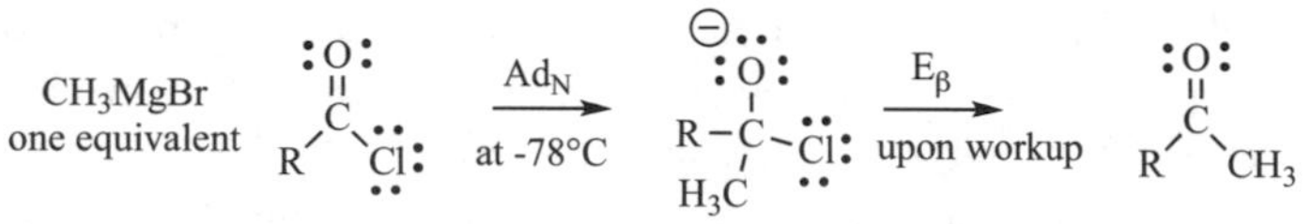
CH3MgBr
one equivalent
AdN
at -78°C
Eβ
upon workup
H3C
CH3

9.2.2 Additions of Nucleophiles to Nitriles

With few exceptions, organometallics add only once to nitriles (Section 8.6.4). Only if there is a very strong Lewis acid present to activate the anionic intermediate of the first nucleophilic attack will a second addition take place. An example of this second addition is the reduction of nitriles with $LiAlH_4$, which proceeds all the way to the amine (Section 8.6.3). The AlH_3 formed from the initial reduction acts as a Lewis acid to catalyze the second addition. This second addition can be prevented by using one equivalent of a less reactive aluminum hydride like $LiAlH(OEt)_3$ or diisobutylaluminum hydride, $[(CH_3)_2CHCH_2]_2AlH$. The reactions in the acidic water workup are the reverse of imine formation (Section 10.5.2).

One equivalent of $LiAlH(OEt)_3$ example:

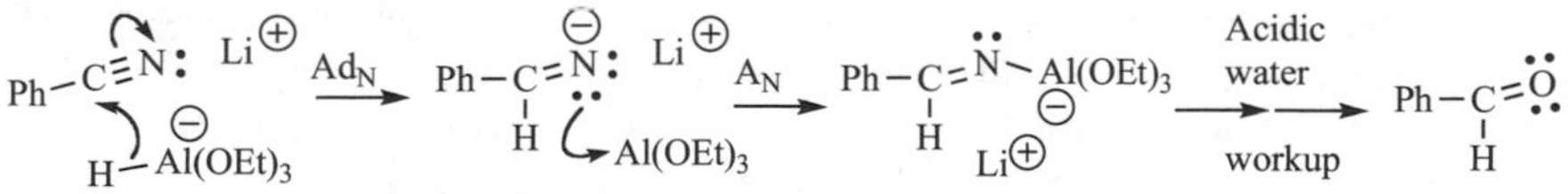
AdN
AN
Al(OEt)3
Acidic
water
workup

One equivalent of $[(CH_3)_2CHCH_2]_2AlH$ example:

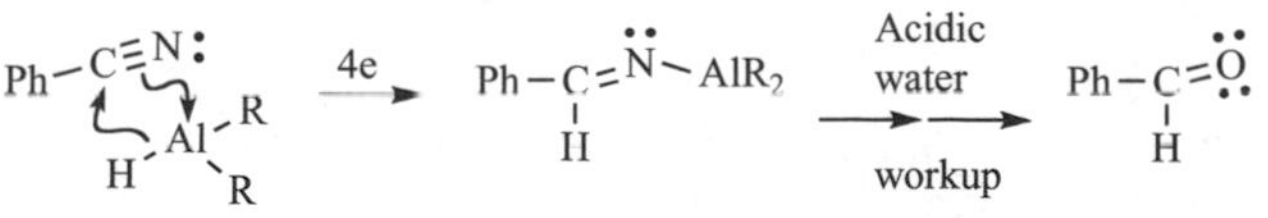
4e
AlR2
Acidic
water
workup

9.3 REGIOCHEMISTRY AND STEREOCHEMISTRY OF ENOLATE FORMATION

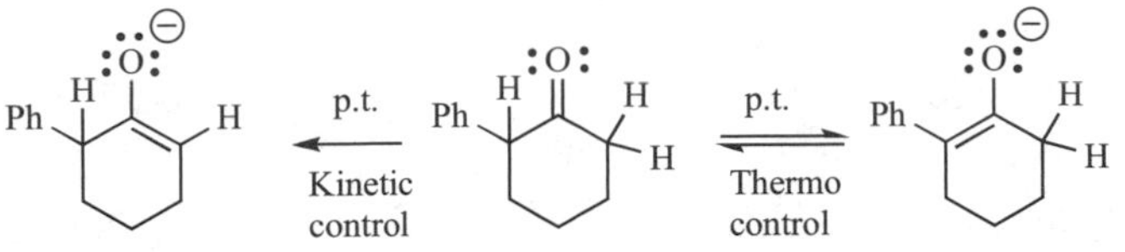
Ph
H
H
p.t.
Kinetic
control
Ph
H
H
H
p.t.
Thermo
control
Ph
H
H

An unsymmetrical ketone can deprotonate to form two possible regioisomeric enolates, and the choice of which enolate forms depends on whether the deprotonation was carried out under kinetic or thermodynamic control. The thermodynamic enolate is the most stable of the possible enolates. In general, the more substituted enolate is the more stable (reflecting the double-bond stabilities); however, exceptions are common. The kinetic enolate is usually the result of deprotonating the most accessible acidic hydrogen. The three hydrogens on a methyl outnumber and are more accessible than a single hydrogen on a more substituted center.

Deprotonation of the ketone must be fast, complete, and irreversible for kinetic control of enolate formation. No equilibration of the enolates can be allowed to occur. Optimum conditions for kinetic control of deprotonation are: Add the ketone slowly to an excess of very strong base (usually $i\text{-Pr}_2\text{NLi}$, the anion of diisopropyl amine, $\text{p}K_{\text{abH}} = 36$) in an aprotic solvent (such as dry tetrahydrofuran or dimethoxyethane). Since the K_{eq} for deprotonation of a ketone with this base is $10^{(36\,-\,19.2)} = 10^{+16.8}$, the reaction is complete and irreversible.

Any equilibrium will produce the thermodynamically most stable enolate. The most stable enolate will have the greatest charge delocalization. In the above example, the thermodynamically favored enolate is conjugated; the kinetically favored enolate is not. Common conditions for thermodynamic control are to use average bases (like sodium ethoxide or potassium *tert*-butoxide, $\text{p}K_{\text{abH}}$s 16 to 19) in alcohol solvents. Proton transfer equilibria rapidly occur among base, solvent, ketone, and enolate. Sodium hydride or potassium hydride in an ether solvent are also thermodynamic reaction conditions that allow equilibration between the ketone and the enolate. Enones have two possible enolates; weaker bases give the thermodynamically more stable extended enolate, whereas kinetic conditions produce the cross-conjugated enolate.

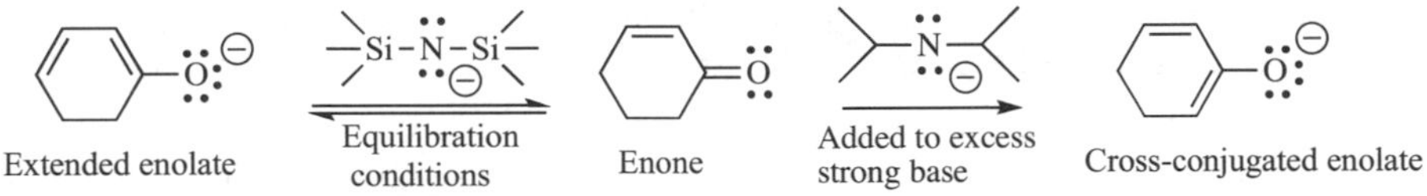
Si
N
Si
Extended enolate
Equilibration
conditions
Enone
N
Added to excess
strong base
Cross-conjugated enolate

In addition to regiochemistry, acyclic carbonyl compounds can produce two possible stereoisomeric enolates, E or Z, as shown above. Steric interactions determine the favored enolate stereochemistry. Under reversible conditions, Z enolates are more stable than E as they minimize steric interactions, especially if R is large. Z enolates are also usually favored under irreversible conditions in polar aprotic solvents like HMPA that complex cations well and break up ion pairing, effectively reducing the bulk around the oxygen anion. Under irreversible conditions in ether solvents, the E enolate is often favored because the steric size of the base/cation aggregate around the oxygen dominates, especially if R is smaller, as with esters.

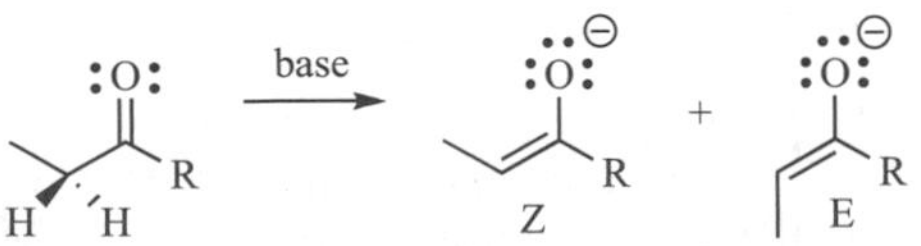
base
R
H
H
R
Z
+
R
E

9.4 AMBIDENT NUCLEOPHILES C=C–Z:

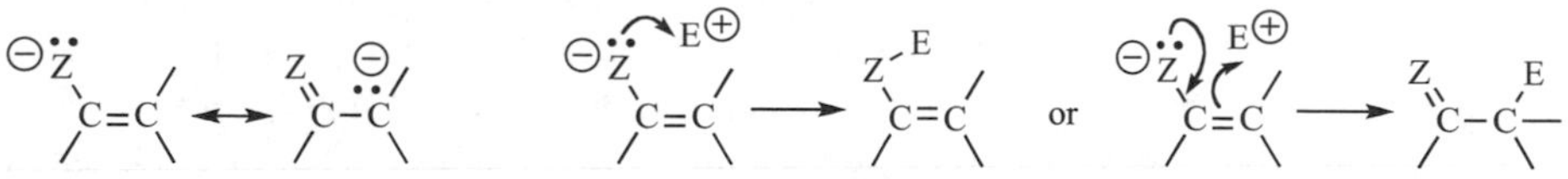

All allylic sources have the capability to act as a nucleophile on either δ– end of the source. Since the two ends of the allylic source commonly differ in electronegativity, the charge on each end differs. The two ends commonly differ in polarizability also. Therefore the Z end is usually much harder than the softer C end. For soft electrophiles, the soft–soft component is most important; therefore the atom with the greatest polarizability will be the best nucleophile. For hard electrophiles, the hard–hard component is most important, therefore the atom with the largest partial minus will be the best nucleophile (charge control). Solvation is also a hard–hard interaction, and the tighter the solvation around the Z end, the more hindered and poorer the Z nucleophile is.

However, a ΔH calculation usually predicts the C-reacted compound to be thermodynamically more stable than the Z-reacted compound (mainly because of the greater C=Z bond strength in the C-reacted product compared to the C=C in the Z-reacted). However, this does depend on the relative C–E vs. O–E bond strength. It is important to determine which is the dominant effect, product formation based upon product thermodynamic stability or upon kinetic direction from HSAB theory. To do this we need to determine whether the reaction is under kinetic or thermodynamic control. Figure 9.1 gives a flowchart for the decision for a common ambident nucleophile, an enolate anion (Z equals oxygen).

Use the ΔpK_a rule on the reverse reaction to determine if the overall reaction is reversible.

If irreversible, then hard with hard, soft with soft, since the E stays where it first attaches.

Polar aprotic solvents leave the oxygen anion poorly solvated and free to serve as a nucleophile

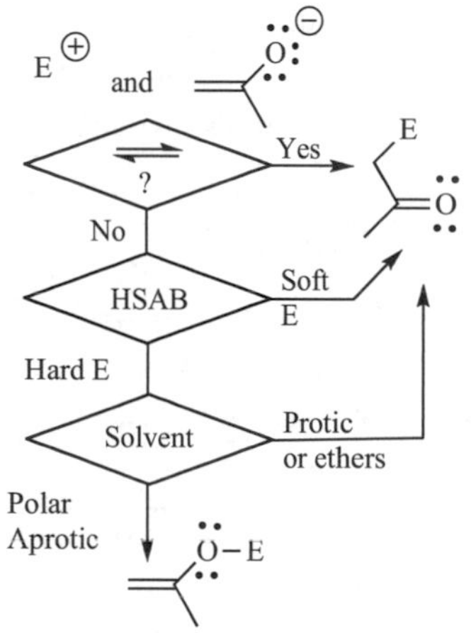

E and
?
Yes
No
HSAB
Soft E
Hard E
Solvent
Protic or ethers
Polar Aprotic
O–E

Figure 9.1 Flowchart for the ambident nucleophile decision of an enolate.

9.4.1 Alkylation of Enols, Enolates, and Enamines with sp^3 C–L

Enolate general example:

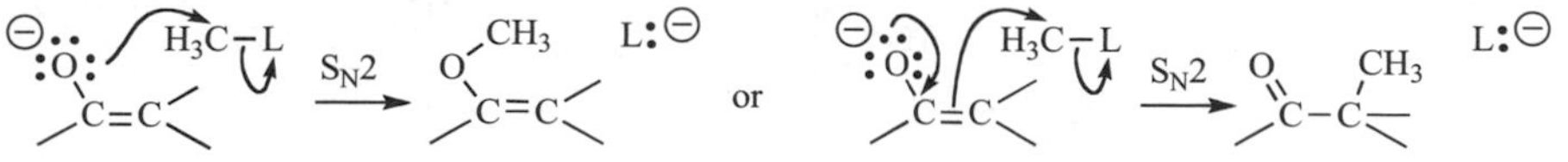

Alkylation on oxygen　　　　Alkylation on carbon

Most of these alkylations are not reversible, so whatever process is faster dominates (kinetic control). The oxygen atom is much harder than the carbon atom of the enolate and bears more of the charge because of its greater electronegativity. The decision whether to use the heteroatom or the carbon atom as the nucleophile is based on the hardness of the electrophile and the solvent. Soft electrophiles like R–Cl, R–Br, and R–I will alkylate on the softer carbon of the enolate. Hard electrophiles will attack the heteroatom (hard with hard) in polar aprotic solvents that poorly solvate the heteroatom. Since carbocations are harder electrophiles (greater charge) than sp^3 C–L, the tendency to alkylate on the heteroatom will increase as the reaction mechanism shifts from S_N2 toward S_N1. With sp^3 C–L, as the leaving group gets more electronegative, the partial plus on carbon increases, and the electrophile gets harder; therefore the amount of alkylation on the heteroatom lone pair increases—for example, oxonium ions, $(CH_3)_3O^+$, sulfate esters, $CH_3OSO_2OCH_3$, and sulfonate esters, $ArSO_2OCH_3$, tend to alkylate on the heteroatom in DMF, DMSO, and HMPA. A special case and not really considered an enolate, phenoxide usually alkylates on oxygen because alkylation on carbon would interrupt the aromaticity of the ring:

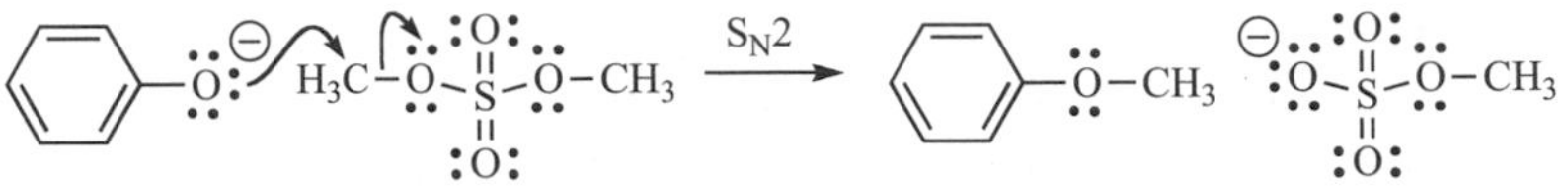
H3C
O
S
CH3
SN2

With anionic allylic sources, highly polar aprotic solvents increase the amount of lone pair alkylation because poor solvent stabilization of the anion leaves the heteroatom end less hindered by solvent. Conversely, groups bound to the heteroatom increase the steric hindrance about it, and therefore decrease the tendency toward heteroatom alkylation. For example, enamines (R_2N–C=C) tend to alkylate on carbon rather than nitrogen (as shown in Section 8.4.5).

Since alkylation is usually not reversible, the products are the result of kinetic control. However, iminium ions, $CH_2{=}N(CH_3)_2{}^+$, are stabilized electrophiles and reversibly add to enols to give the thermodynamically more stable C-alkylated product.

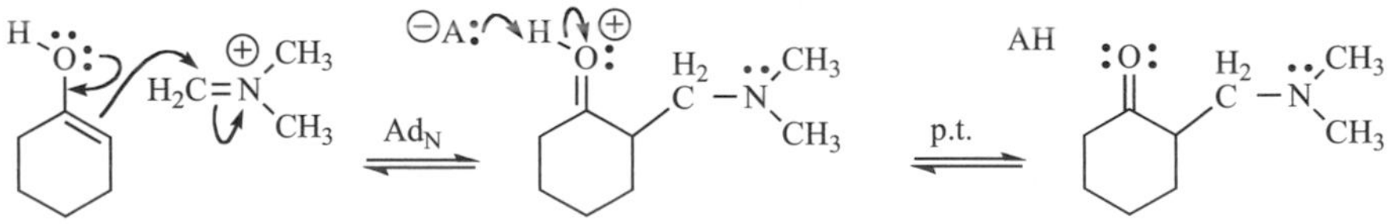
H2C=N
CH3
AdN
AH
p.t.

In the following list are the extremes for the alkylation reaction of allylic sources; however, product mixtures occur rather often.

Alkylation on Z (less common)	Alkylation on C (much more common)
Hard electrophiles	Soft electrophiles
Highly polar aprotic solvent	Ether or alcohol solvent
Heteroatom accessible	Carbon atom accessible
Irreversible reactions	Reversible reactions

9.4.2 Reaction of Enols and Enolates with Y–L

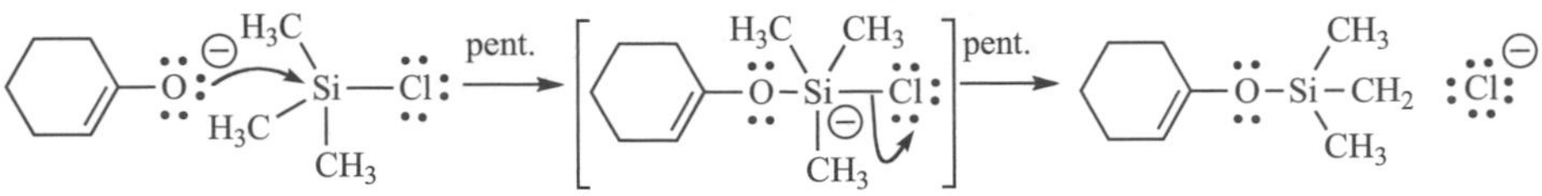
H3C
Si
Cl
CH3
pent.
pent.
CH2

In the above section the attacking electrophile was a carbon atom of varying hardness, and since the C–C and C–O bonds were about the same strength, the C=C vs. C=O bond strength dominated the thermodynamic choice. With the Y–L sink, the difference in the Y–C vs. Y–O bond strength can be controlling. With enolates for example, the Si–O bond is 36 kcal/mol stronger than the Si–C bond, so the O-silated product is both thermodynamic and kinetic favored. As a result trimethylsilyl chloride, $(CH_3)_3Si–Cl$, having a highly polarized silicon–chlorine sigma bond, alkylates exclusively on the oxygen of enolates. Likewise the P–O bond is over 70 kcal/mol stronger than the P–C bond, so all P–L species go on the oxygen of enolates also. On the other hand, the Br–C bond is 20 kcal/mol stronger than the Br–O bond, making the C-brominated product the thermodynamic one. The soft Br_2 brominates enolates on carbon.

9.4.3 Acylation of Enols, Enolates, and Enamines with O=C–L

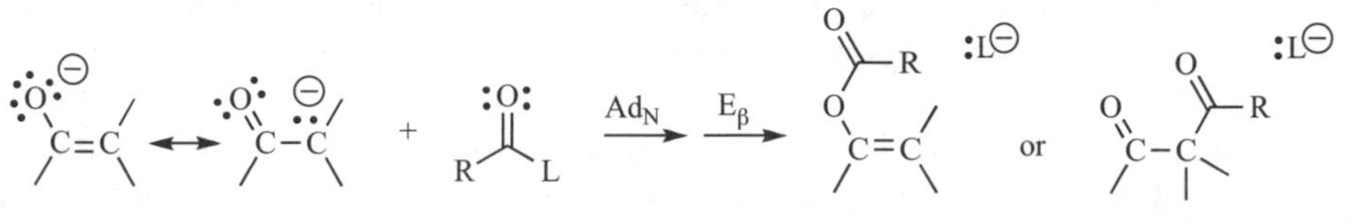

Acylation on oxygen or Acylation on carbon

For the most common conditions, **basic media**, acylation almost always goes on carbon. Acylation on the heteroatom produces a product that can be attacked by another molecule of the allylic source to produce the C-acylated compound. The product of any equilibrium is the thermodynamic C-acylated product.

The basic media exceptions can be easily understood if we invoke HSAB theory and realize that the kinetic and thermodynamic products are different. As L becomes a poorer donor, the partial plus on the acyl carbon increases, making it harder. Acylation on the heteroatom of the allylic source is fast for acyl halides and anhydrides where the acyl carbon is harder (greater partial plus) than the acyl carbon of esters. If the reaction is under kinetic control (allylic source added to an excess of acyl halide or anhydride), the Z-acylated product is formed; however, if equilibration occurs (excess of allylic source), the product will be the C-acylated, thermodynamic product.

In **acidic media**, the electron sink is most often the carbocation produced from protonating the acylating agent, and therefore the sink is very hard. Attack by the Z end (harder end) of the allylic source is very fast. For enols, the Z-acylated kinetic product can be isolated. Since the Z-acylated enol is itself an allylic source (but weaker), it can be forced by more vigorous conditions to equilibrate to the more stable C-acylated product. For enamines, the Z-acylated enamine is a good acylating agent; any excess of enamine will attack it and equilibrate it to the more stable C-acylated product.

In summary, acylation on Z is achieved with hard electrophiles (acyl halides) under kinetic control; acylation on C is achieved with any good O=C–L electrophile under thermodynamic control. As with alkylation, C acylation is more common.

9.4.4 Amides and Amidates

In amides the oxygen is the better nucleophile because it is partially minus and the nitrogen is partially plus. The partial plus on the nitrogen repels any like charged electrophile. However, in the amidate anion both nitrogen and oxygen share the negative charge, and therefore the softer nitrogen is commonly the better nucleophile.

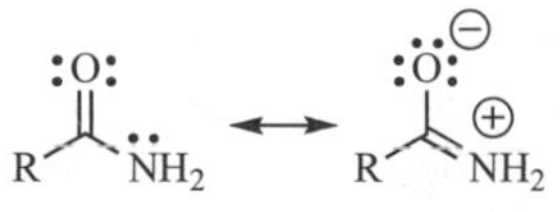

Amide resonance forms

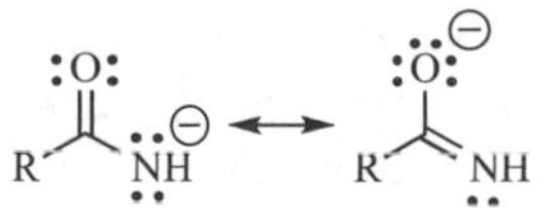

Amidate resonance forms

9.5 SUBSTITUTION VS. ELIMINATION

Nucleophilicity vs. basicity is perhaps the most difficult decision to make because almost any anion can serve as either a base (elimination) or a nucleophile (substitution):

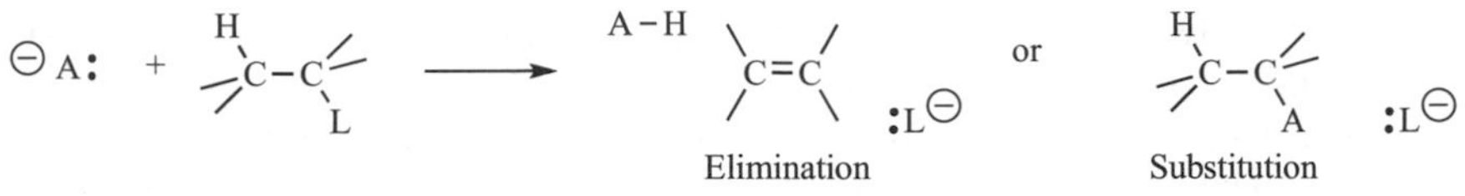
Elimination
or
Substitution

Again HSAB theory is useful. In general, the C–H bond is significantly harder than the C–L bond. Therefore the softer anions will tend toward substitution, and the harder anions will tend toward elimination. However, the situation is not that simple, for the accessibility of the substitution site and the size of the anion play a major role also. If the anion is hindered, it is a poor nucleophile. If the site is so hindered that the nucleophile cannot attack it, the balance tilts toward elimination. **The substitution/elimination decision becomes a function of three major variables: nucleophilicity, basicity, and steric hindrance.**

The substitution energy surface can be placed adjacent to the elimination surface since they share the C–L bond-breaking axis (Fig. 9.2). Now the factors that tilted each of the surfaces can be used to understand the substitution vs. elimination competition.

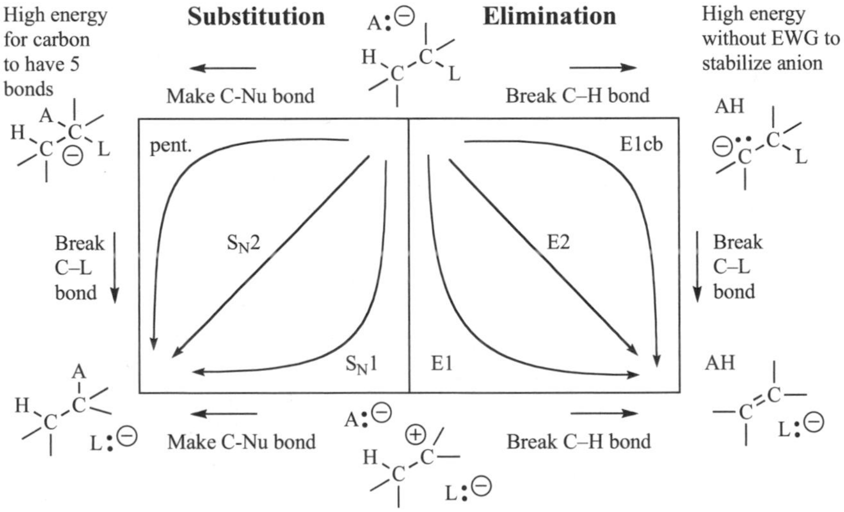
High energy for carbon to have 5 bonds
Substitution
Elimination
High energy without EWG to stabilize anion
Make C-Nu bond
Break C–H bond
pent.
E1cb
Break C–L bond
S_N2
E2
Break C–L bond
S_N1
E1
Make C-Nu bond
Break C–H bond

Figure 9.2 Top view of a simplified substitution/elimination surface. Reactants are in top center.

S_N1 vs. E1 Competition

With a reasonably stable carbocation and energy as a vertical axis, we get Figure 9.3, the S_N1 vs. E1 surface. In acidic media the S_N1 competes with the E1 process; the first step of both processes is to lose the leaving group to form the carbocation. Carbocations are excellent electron sinks and tend to react quickly with low selectivity. Equilibrium thermodynamics is therefore the best way to bias this surface. The C–Nu corner is raised if the Nu is poor and also is a good leaving group so that it falls off again, returning to the carbocation. The alkene corner is now the lowest point on the surface and is favored by any equilibrium. An example is elimination of alcohols by concentrated phosphoric acid.

Attempts to bias this surface toward substitution present a slight problem. The C–Nu corner cannot be lowered very much by making the nucleophile better because the acidic media that the ionization reaction prefers will protonate any good nucleophiles. The best way to increase substitution is to use a low-basicity nucleophile and to shift the equilibrium by mass balance: Make the nucleophile the solvent if possible (solvolysis).

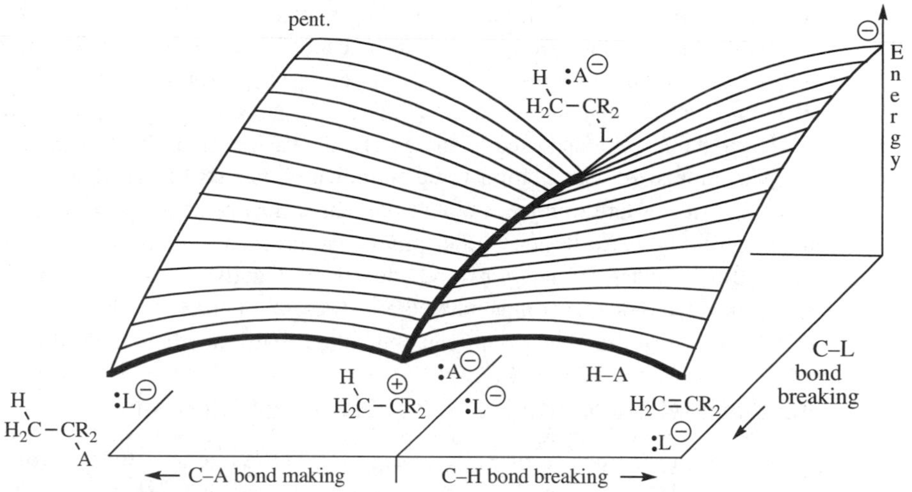
pent.
H :A⊖
H₂C−CR₂
L
⊖
Energy
H
H₂C−CR₂
A
:L⊖
H ⊕
H₂C−CR₂
:A⊖
:L⊖
H–A
H₂C=CR₂
:L⊖
C–L
bond
breaking
← C–A bond making
C–H bond breaking →

Figure 9.3 The simplified $S_N1/E1$ energy surface. Reactants are in top center.

S_N2 vs. E2 Competition

There normally is competition in basic media between the S_N2 and E2. In contrast to the previous surface, Figure 9.4 shows that the split in paths occurs much earlier because the carbocation is much higher in energy and not an intermediate. The choice of reaction path depends heavily on kinetics because the system is not usually reversible.

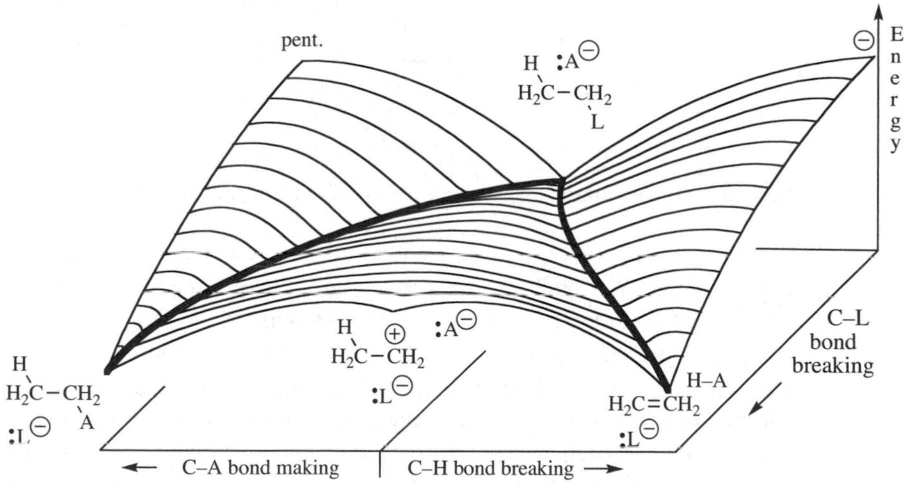
pent.
H :A⊖
H₂C−CH₂
L
⊖
Energy
H
H₂C−CH₂
:L⊖ A
H ⊕ :A⊖
H₂C−CH₂
:L⊖
H–A
H₂C=CH₂
:L⊖
C–L
bond
breaking
← C–A bond making
C–H bond breaking →

Figure 9.4 The simplified $S_N2/E2$ energy surface. Reactants are in top center.

To bias this surface toward elimination, one makes the S_N2 a higher-energy route by using hard, highly hindered strong bases, which are poor nucleophiles. Strong bases make the deprotonation energetically more favorable and thus lower the energy of the right edge. Poor nucleophiles make the substitution reaction more difficult and thus raise the energy of the left edge of the diagram. The entire surface then tilts toward elimination. Highly hindered strong bases with low nucleophilicities, $(CH_3)_3CO^-K^+$ or $(i\text{-Pr})_2N^-Li^+$, are used to optimize elimination.

Conversely, good, unhindered, soft nucleophiles of low basicity bias the surface toward substitution by tilting it exactly the opposite way. Poor bases raise the energy of the right edge; good nucleophiles lower the energy of the left edge; the entire surface tilts toward substitution. As mentioned above, the groups around the C–L site make a large difference in whether substitution or elimination occurs. As the C–L site gets more hindered, the left side is raised in energy and the surface then tilts toward elimination.

To summarize, substitution versus elimination is a multivariable decision that breaks down into major and minor variables. The three major variables are C–L site hindrance, nucleophilicity, and basicity; the minor variables are reaction temperature (higher temperature favors elimination) and the electronegativity of the leaving group (more electronegative leaving groups make the C–L carbon harder). A less common variable is that electron-withdrawing groups can make the C–H so acidic that elimination dominates.

3D Correlation Matrix for Substitution vs. Elimination

We need to consider each different type of C–L site as a function of the other two major variables, nucleophilicity and basicity. A three-dimensional correlation matrix is the best way to do this (Fig. 9.5).

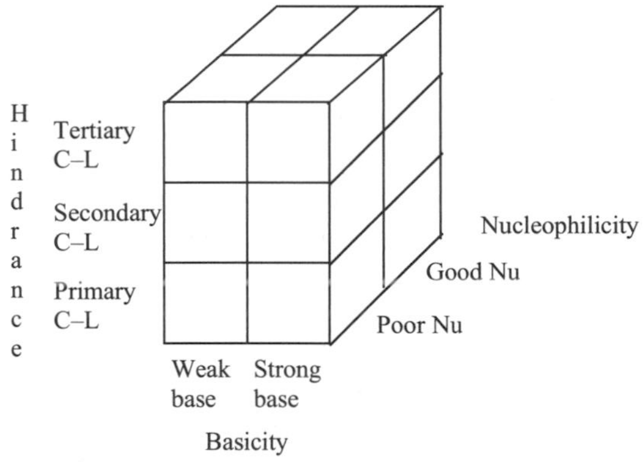
Hindrance
Tertiary C–L
Secondary C–L
Primary C–L
Nucleophilicity
Good Nu
Poor Nu
Weak base
Strong base
Basicity

Figure 9.5 A three-dimensional correlation matrix for substitution versus elimination.

We divide each axis into convenient portions corresponding to the most common cases (although we actually have a continuum in three dimensions). Hindrance can vary with adjacent substitution as noted below. An approximate tipping point for basicity is roughly $pK_{abH} = 10$, but this varies with the system. For nucleophilicity, the poor nucleophiles are often highly hindered anions or are neutral solvents with poor polarizability (Section 4.2.2). To make the individual cells in this matrix easier to view, we will consider each vertical layer separately, and show them as individual two-dimensional matrices in Figure 9.6.

The unhindered primary layer has the strong base, poor nucleophile box as its mixture quadrant with all others as substitution. Notice how the mixture quadrant in the primary matrix moves around to the strong base, good nucleophile quadrant in the secondary matrix, then again to the weak base, good nucleophile quadrant in the tertiary matrix, each time leaving behind an elimination quadrant.

For each mixture quadrant more information is required to be able to make a decision on substitution vs. elimination. The major variables have balanced out, and therefore the minor variables can tilt the balance. In the primary mixture quadrant, if the leaving group is a halide, elimination is the major product. If, however, the leaving group is a sulfonate, its electronegativity increases the partial plus on carbon (making it harder); substitution is then the major product. In the secondary mixture quadrant, elimination is common for any reaction that is heated. Elimination has a slightly higher energy barrier than substitution and is therefore favored by higher temperatures. In the tertiary mixture quadrant, a fine-tuning of basicity is important, for even mild bases like acetate cause elimination to occur.

Substitution vs. Elimination Decision

Example Reagents

Weak base	Strong base	
$I^{\ominus}$, $RS^{\ominus}$	$HO^{\ominus}$, $EtO^{\ominus}$	Good nucleophile
H_2O, EtOH	$(CH_3)_3CO^{\ominus}$, $R_2N^{\ominus}$	Poor nucleophile

Unhindered Primary

Weak base	Strong base	
Subst. by S_N2	Subst. by fast S_N2	Good nucleophile
No rxn. or very slow S_N2	Mixture (depends on L)	Poor nucleophile

Secondary

Weak base	Strong base	
Subst. by S_N2	Mixture (depends on temp.)	Good nucleophile
No rxn. or slow S_N2 or S_N1	Elimin. by fast E2	Poor nucleophile

Tertiary

Weak base	Strong base	
Mixture (depends on basic.)	Elimin. by fast E2 or E1	Good nucleophile
Subst. by S_N1 (polar solvent)	Elimin. by fast E2 or E1	Poor nucleophile

Figure 9.6 Substitution/elimination correlation matrix layers from Figure 9.5 sorted by hindrance. The matrix in the upper left gives some common examples of each type of reagents.

For the weak base, poor nucleophile quadrant, substitution occurs by the S_N2 path for primary and secondary substrates, but because the electron source is poor by definition, the reaction can be very slow if the leaving group is not very good. Tertiary substrates in this quadrant substitute by the S_N1 mechanism.

Even if a group is somewhat removed from the site, it can still hinder the approach of a nucleophile. Although $R{-}CH_2{-}L$ is a general description of a primary site, the size of R is very important because it can change which hindrance matrix best applies. The matrix for hindered primary leaving groups, like isobutyl, $(CH_3)_2CH{-}CH_2{-}L$, is similar to the matrix for secondary leaving groups. The matrix for hindered secondary leaving groups is similar to that of tertiary leaving groups. Vinyl leaving groups, $H_2C{=}CH{-}L$, do not substitute under normal conditions and with a strong base eliminate to the alkyne. For $CH_3{-}L$ systems, substitution occurs in all quadrants because elimination is not a structural possibility.

We treat as an exception our less common major variable, the presence of an electron-withdrawing group. Leaving groups beta to an electron-withdrawing group, ewg$-CH_2CH_2-$L, commonly eliminate because the increased acidity of the H makes the C$-$H easier to break, thereby favoring elimination by E1cB.

All multivariable decisions are messy, and if this presentation appears too clean to hold universally, you are correct in your suspicion. We are predicting the predominant product in a product mixture; rarely is the predicted product the only product. There are examples where a major variable joined with the other minor variables (and an experimental technique or two) can outweigh the other two major variables. Alcohol dehydration is such a case. Alcohols are heated in concentrated sulfuric or phosphoric acid, and the alkene is distilled out of the reaction mixture into another flask. The elimination is forced: The medium is highly acidic and dehydrating; sulfate and phosphate are weak bases, poor nucleophiles, and good leaving groups; heating not only favors elimination but also helps remove the product, thus displacing any equilibrium.

9.6 AMBIDENT ELECTROPHILES C=C–EWG

9.6.1 Conjugated Ketone Systems—Enones

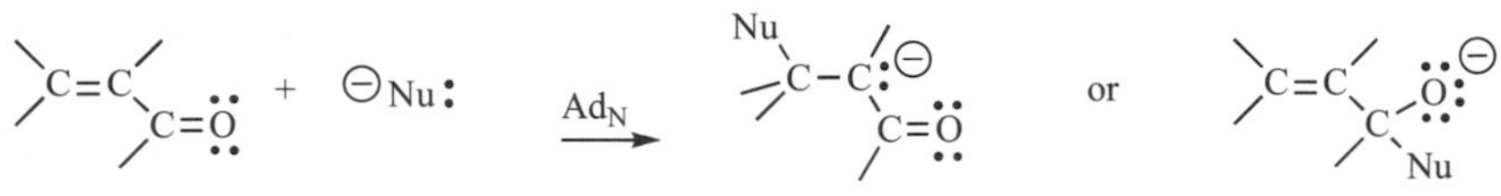

Nu:
Ad_N
or

Enone 1,4 Attack or 1,2 Attack

A conjugate acceptor system with two δ+ sites can be attacked at either site. The thermodynamic product will be the most stable of the possible products. A ΔH calculation shows that the thermodynamic product is the **1,4 or conjugate addition product**, primarily because of the greater bond strength of the C=O bond than the C=C bond. Any equilibrium would produce the conjugate addition product.

Because the effect of an electronegative group diminishes with distance, the carbonyl carbon in this system will have the greatest partial plus and therefore will be harder and attract a hard, negatively charged nucleophile best; the **1,2 product or normal addition product** will be the kinetic product for hard nucleophiles. Soft nucleophiles therefore prefer conjugate attack. Finally, if one site is very sterically hindered, attack at the more open site will dominate.

In summary, highly reactive, harder organometallics add irreversibly to an enone and produce the 1,2 product. The softer, more stable delocalized anions add reversibly and therefore produce the 1,4 product. Organomagnesium reagents give 1,2 additions except when the carbonyl of the enone is hindered (by a phenyl or larger group). Several nucleophiles and their predominant site of attack on an enone are presented in Table 9.2 and diagramed in Figure 9.7.

Table 9.2 The Mode of Attack of Various Nucleophiles on Enones

Nucleophile	Favored Attack	Rationale
R-Li	1,2	Very hard organometallic
R-Mg-X	1,2 (usually)	Hard organometallic
R_2CuLi	1,4	Soft organometallic
R-Mg-X/CuX	1,4	Makes soft organometallic by transmetallation
Enolates	1,4	Soft Nu, reversible addition
Lone pair Nu	1,4	Stable Nu, reversible addition
$LiAlH_4$	1,2 (mainly)	Hard hydride source
$NaBH_4$ or $KBHR_3$	1,4 (mainly)	Softer hydride source

9.6.2 Miscellaneous Ambident Electrophiles

As the carbonyl becomes more reactive, 1,2 attack will become more predominant; in contrast to what was found for enones, borohydride attacks 1,2 on conjugated aldehydes. All organomagnesium reagents add 1,2 on conjugated aldehydes. With lone pair nucleophiles (often very hard), conjugated acyl halides undergo 1,2 attack (hard carbonyl, often irreversible 1,2 addition), whereas conjugated esters undergo 1,4 attack (softer carbonyl, reversible 1,2 addition). Only 1,2 attack occurs on phenyl-substituted carbonyls like $PhCOCH_3$, because conjugate attack would disrupt the aromaticity of the aromatic ring.

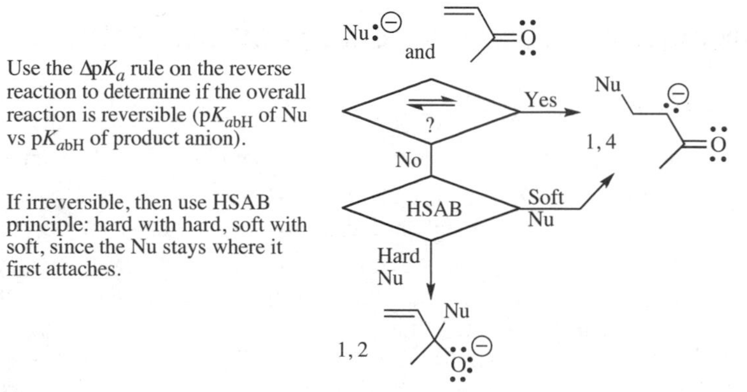
Use the ΔpKa rule on the reverse reaction to determine if the overall reaction is reversible (pKabH of Nu vs pKabH of product anion).
If irreversible, then use HSAB principle: hard with hard, soft with soft, since the Nu stays where it first attaches.
Nu:
and
?
Yes
No
HSAB
Soft Nu
Hard Nu
1,4
1,2
Nu

Figure 9.7 Flowchart for the ambident electrophile decision of an enone.

9.7 INTERMOLECULAR VS. INTRAMOLECULAR

One of the determining factors in reaction rates is how frequently the reaction partners collide. If collisions are more frequent, the reaction rate will be faster. Increasing the concentration of a reactant invariably increases the collision frequency and therefore the reaction rate. If the nucleophile and electrophile are part of the same molecule, they may collide much more often than is possible in even the most concentrated solutions. Therefore intramolecular reaction rates can easily exceed intermolecular rates. The energy of the cyclic transition state for an intramolecular process depends on the rigidity and size of the loop of atoms in the cyclic transition state and on the orbital alignment restrictions of the process involved.

For the S_N2 reaction shown in the following example, the determining factor for ring closure in a freely rotating chain is its size. The formation of very small rings requires the bending of sigma bonds, creating significant ring strain. In the formation of large rings (rings greater than eight atoms), the ends that must be brought together to form the ring may collide so infrequently that bimolecular reactions can easily compete. The optimum ring size for closure is a five-membered ring; six is not as good; seven starts to get too large. A three-membered ring forms easily, for although there is much ring strain, the number of effective collisions is great since the ends are so close. Four-membered ring formation is poor because the ring strain has not dropped off appreciably, but the number of effective collisions has.

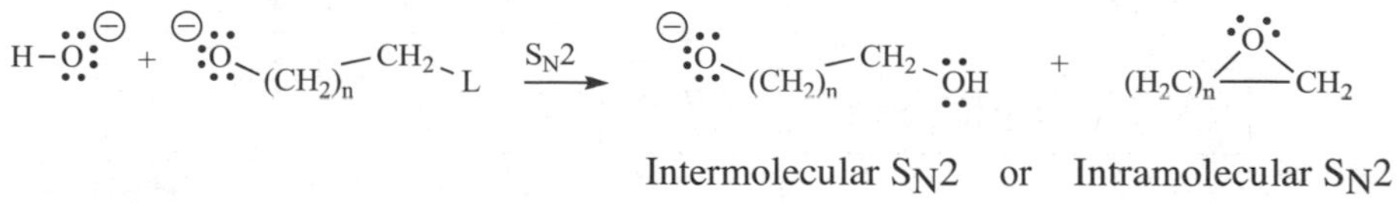
H–O:
(CH2)n
CH2
L
SN2
OH
(H2C)n
CH2
Intermolecular SN2 or Intramolecular SN2

Although there is some variation among systems, the general order is 5 > 6 > 3 > 7 > 4 > 8 to 10-membered transition states. In more concentrated solutions intermolecular reactions may be competitive with four-membered ring formation. In summary, **any intramolecular process with a 5-, 6-, 3-, or 7-membered transition state is almost always faster than the corresponding intermolecular process.**

In forming a ring on a substrate that has some conformational rigidity, you must be sure that the ends can reach each other. In fact, any intramolecular process should be checked with molecular models to see whether the reacting partners can interact as needed for the reaction.

9.8 TO MIGRATE OR NOT TO AN ELECTROPHILIC CENTER

Predicting when migration to an electrophilic center will occur is not an easy task. The main three reactions of carbocations are to trap a nucleophile (path A_N), to lose an electrofuge like a proton and form a pi bond (path D_E), or to rearrange to a carbocation of equal or greater stability (path 1,2R) (Section 4.7). Often it is a matter of kinetics. Can the carbocation survive long enough to rearrange, or does A_N or D_E happen first? Carbocation rearrangements are favored by strongly acidic media, precisely because those conditions have very poor nucleophiles and weak bases, so that carbocation has time to rearrange (a good base or nucleophile would get protonated in strong acid). Less stable carbocations are rarely intermediates. With few exceptions, tertiary carbocations will not rearrange to secondary; secondary carbocations will not rearrange to primary.

In order to rearrange, the migrating group orbital must align with the empty *p* orbital of the carbocation. Another way to look at it is that the migrating group is participating in hyperconjugation with the carbocation before migration. As Figure 9.8 shows, the migrating group maintains bridging-type bonding through the transition state between the atom migrated from and the atom migrated to (see the 1,2R path, Section 7.1.11 and the 1,2RL path, Section 7.1.12).

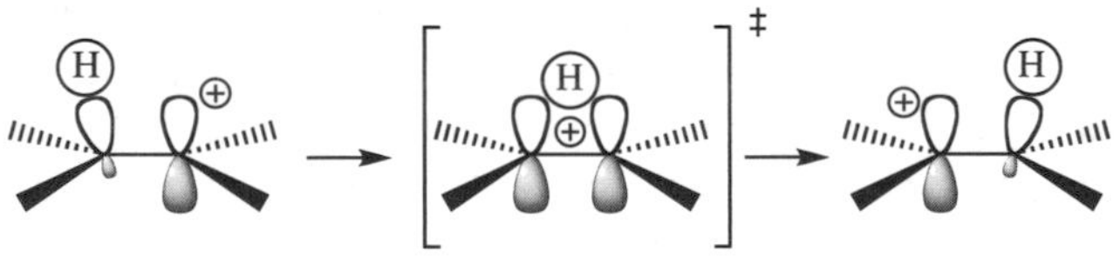
H
H
H
‡

Figure 9.8 The migration of a hydride to a cationic center maintains overlap throughout the move.

Rearrangement commonly is found with paths D_N and A_E, which make up path combinations, S_N1, E1, Ad_E2. Figure 9.9 shows an energy surface map of the spectrum between the diagonal one-step and two-step rearrangement processes.

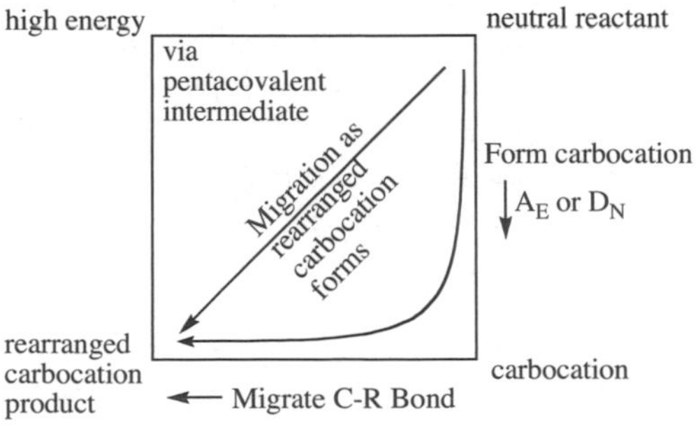
high energy
neutral reactant
via
pentacovalent
intermediate
Migration as
rearranged
carbocation
forms
Form carbocation
A_E or D_N
rearranged
carbocation
product
Migrate C-R Bond
carbocation

Figure 9.9 Top view of a simplified energy surface for carbocation rearrangements.

With tertiary or more stable carbocations, we would have one extreme, with the carbocation corner low enough in energy to be the two-step intermediate. Secondary carbocations seem to be the tipping point. The following arrows are for the corresponding two-step processes via an intermediate carbocation.

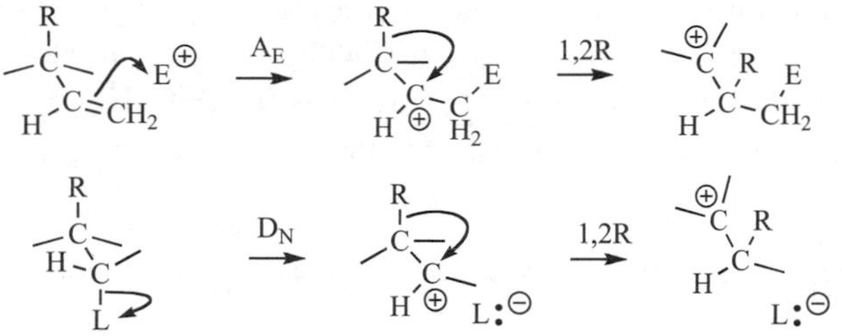

At the other extreme, the one-step rearrangement (the diagonal on Fig. 9.9 that bypasses the initial carbocation) would predominate when that carbocation would be less stable. As in the spectrum for S_N1 and S_N2, as the carbocation that would be formed by the D_N or A_E becomes less stable, the corresponding carbocation corner of the rearrangement surface begins to rise in energy. If the carbocation is very poor, like primary, the surface folds down the middle and the lowest-energy route is the diagonal. The following arrows are for the corresponding one-step diagonal processes.

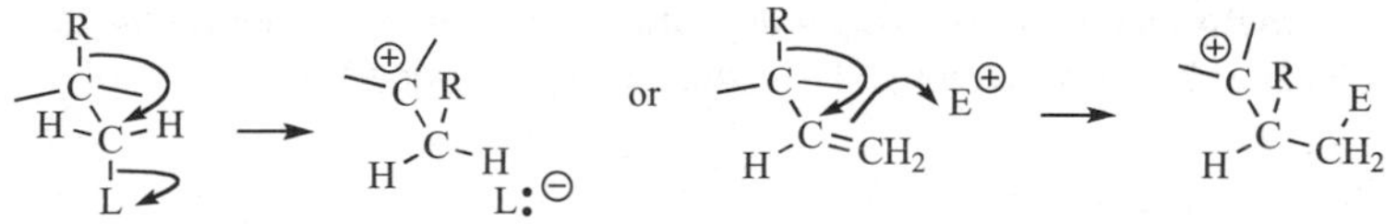

In between, with a secondary carbocation we would expect that the lowest-energy path would begin to break the C–L bond, creating a partial plus, but before the C–L bond is fully broken, the R would migrate. The lowest-energy route would curve on the surface between the two extremes.

The faster carbocation rearrangements tend to make a more stable carbocation. Certain systems that are more prone to fast rearrangement are shown below.

neopentyl isobutyl strained ring pinacol

Example problem

To which carbocations would the above compounds rearrange when the leaving group is lost?

Answer: The neopentyl system would migrate a methyl to give a tertiary carbocation. Isobutyl systems migrate an H to give the *tert*-butyl carbocation. The strained ring compound produces the 1-methylcyclopentyl carbocation. (Relief of severe

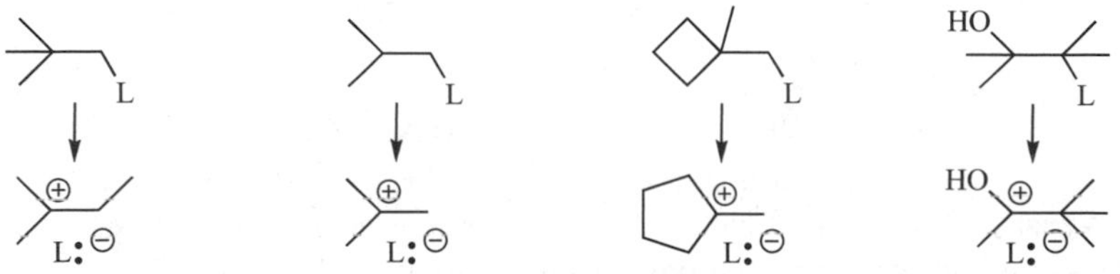

ring strain can often be enough to create a less stable carbocation, like secondary from tertiary.) Migration of a methyl in the pinacol system creates a very stable carbocation whose resonance form is a protonated carbonyl.

One last common system that rearranges by a different mechanism is shown below. Unlike the previous rearrangements that go through a planar carbocation that could be trapped on either side, this process replaces the leaving group with retention because it is two sequential S_N2 reactions. This double inversion process gives overall retention. However, the phenonium ion does have two different CH_2 carbons that could be attacked by the nucleophile to open the strained three-membered ring.

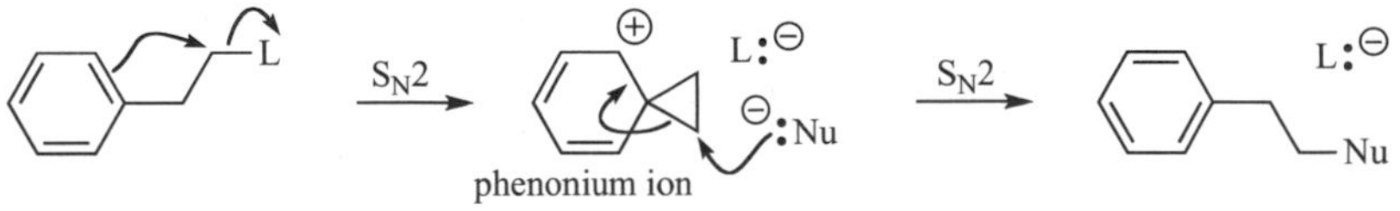
L
SN2
L:⊖
⊖:Nu
phenonium ion
SN2
L:⊖
Nu

9.9 SUMMARY

In summary, a caution: In complex multivariable decisions like those covered in this chapter the best we can hope for is to be able to predict correctly in straightforward cases. Many of these reactions produce mixtures of products, and we are often just trying to predict the major product.

Finally, a common error is building a rationale on kinetic control when the system is under thermodynamic control (or vice versa). Check for reversibility with the ΔpK_a rule.

ADDITIONAL EXERCISES

(See the Appendix: Hints to Selected Problems from Chapters 8, 9, and 10 if you need assistance.)

9.1 For the following reactions decide whether substitution or elimination dominates.

	Substrate	Reagent	Temperature
(a)	$CH_3CH_2CH_2CH_2OH$	HBr	(assume 25°C if not given)
(b)	$CH_3CH_2CH_2CH_2OTs$	t-BuO$^-$	
(c)	$CH_3CH_2CH_2CH_2Br$	t-BuO$^-$	
(d)	$(CH_3)_2CHCH_2Br$	I^-	
(e)	$(CH_3)_2CHBr$	EtOH	
(f)	$(CH_3)_2CHBr$	EtO$^-$	55°C
(g)	$(CH_3)_2CHCH_2CH_2Cl$	$[(CH_3)_2CH]_2N^-$	
(h)	$(CH_3)_2CHBr$	CH_3COO^-	
(i)	$(CH_3)_2CHBr$	EtS$^-$	
(j)	$(CH_3)_2CHCHBrCH_3$	CH_3COO^-	55°C
(k)	$CH_3CH_2CBr(CH_3)_2$	CH_3OH	
(l)	$(CH_3CH_2)_3CCl$	CH_3O^-	
(m)	$(CH_3)_3CBr$	CH_3COO^-	

9.2 a-g For each of the following reactions a decision discussed in this chapter is required. Choose which product would be preferred.

(a) 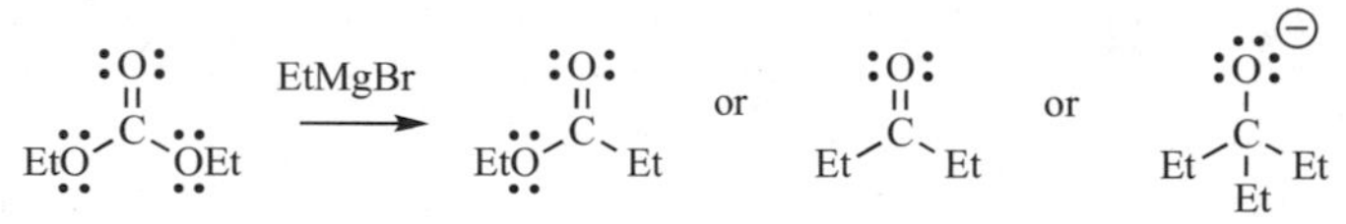

(b) 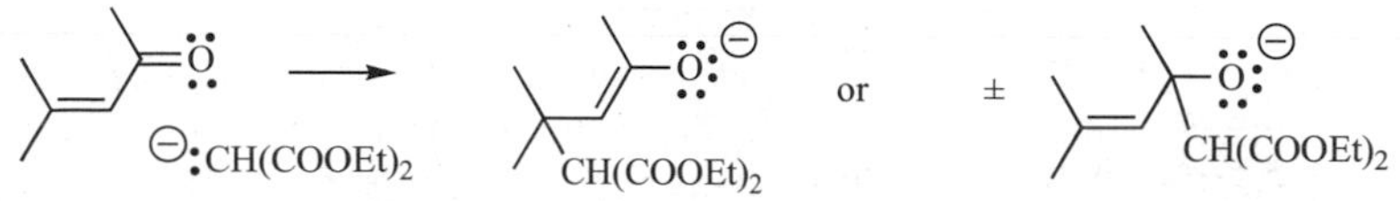

(c) 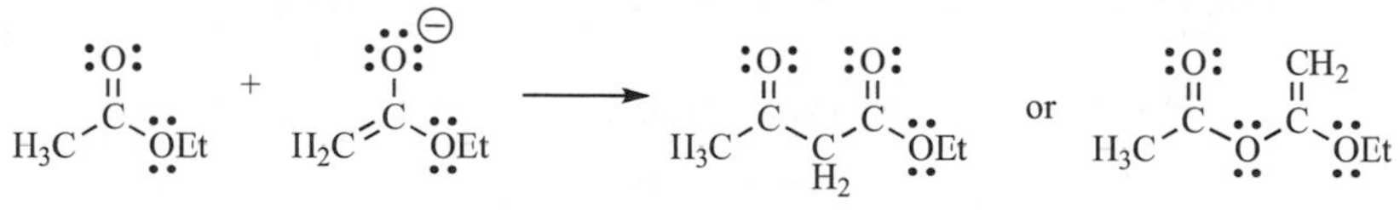

(d) 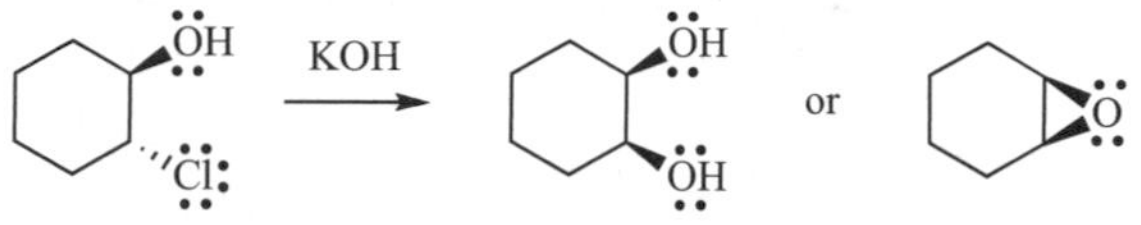

(e) 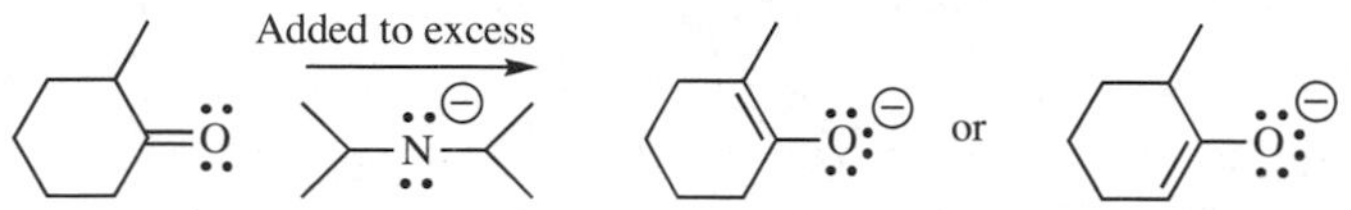

(f) 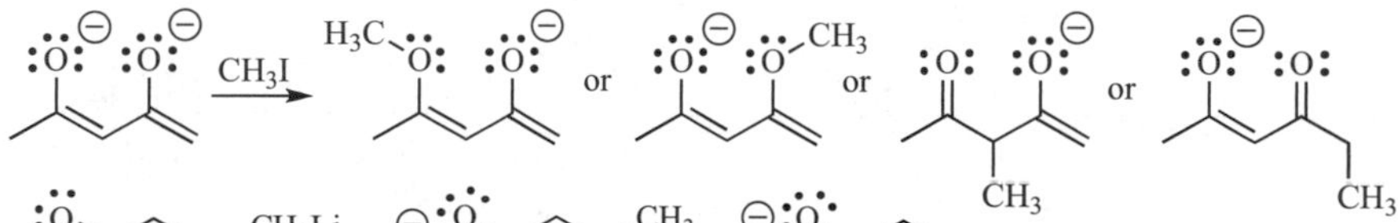

(g) 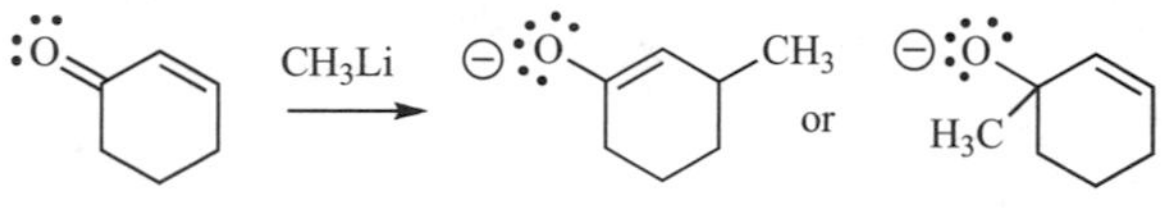

(h) 

9.3 a-g Provide a mechanism for the formation of the preferred products in problem 9.2. Combine the source and sink with the appropriate pathway(s).

9.4 a-g Each of the following reactions has only one electron source and sink. Give the product of the following reactions; a decision discussed in this chapter is required. Assume a mildly acidic workup for the organometallic reactions that just protonates any anionic species but goes no further.

(a) 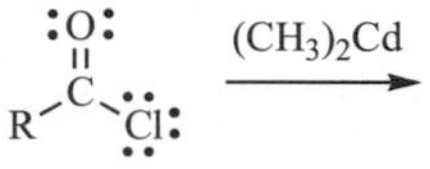

(b) 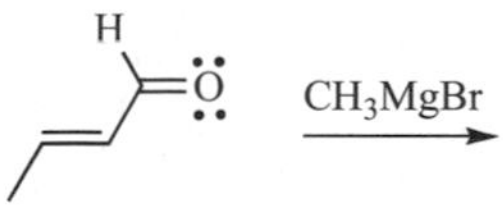

(c) 

(d) 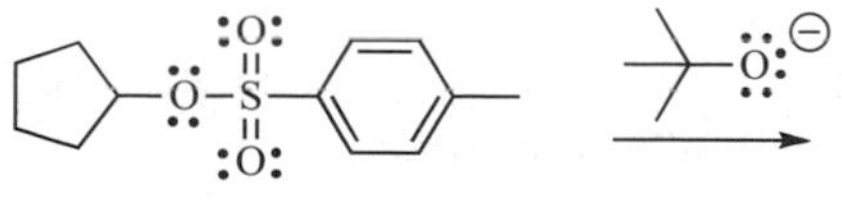

(e) 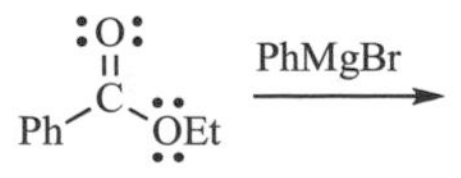

(f) 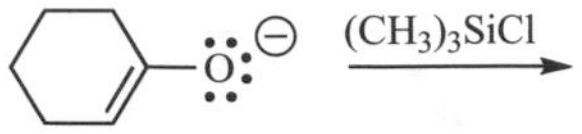

9.5 a-g Provide a mechanism for the formation of the preferred products in problem 9.4. Combine the source and sink with the appropriate pathway(s).

9.6 The following compounds rearrange under the reaction conditions; give the product.

(a) 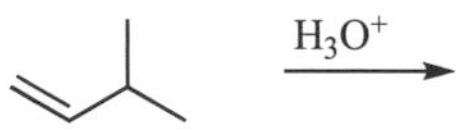

(b) 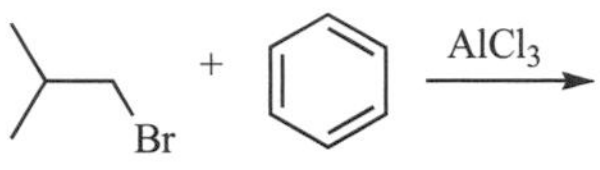

(c) 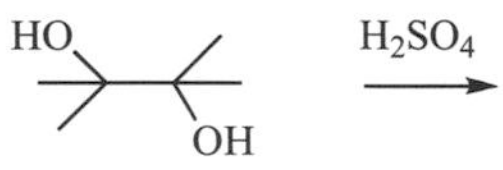

9.7 In the following problem, identify the electron sink and electron source and predict two possible products. When one equivalent of the hydride source is added at low temperature, which product would be preferred?

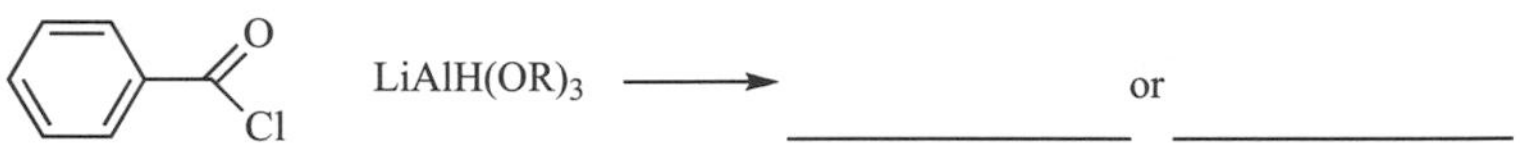

9.8 In the following reaction, run under strong acid catalysis, two enol acetates could form. Predict which would be the major product.

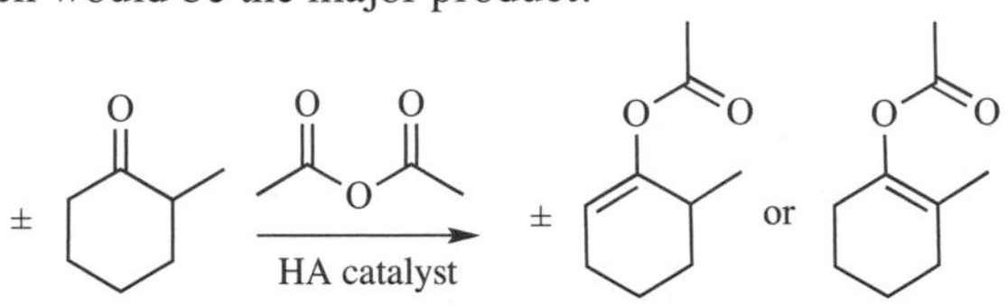

9.9 For each of the following reactions a decision discussed in this chapter is required. Choose which product would be preferred.

(a) 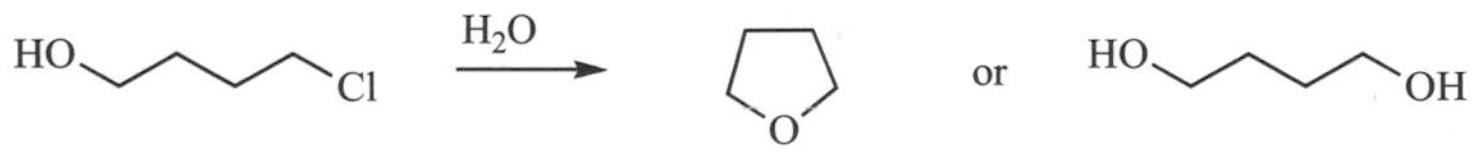

(b) 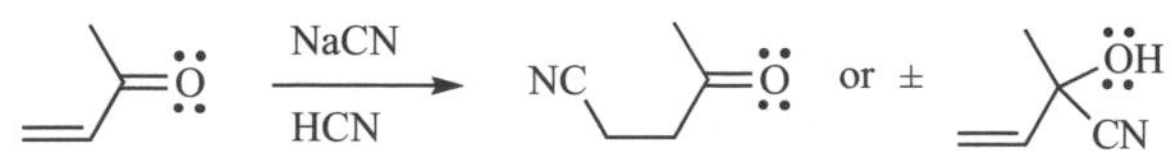

(c) 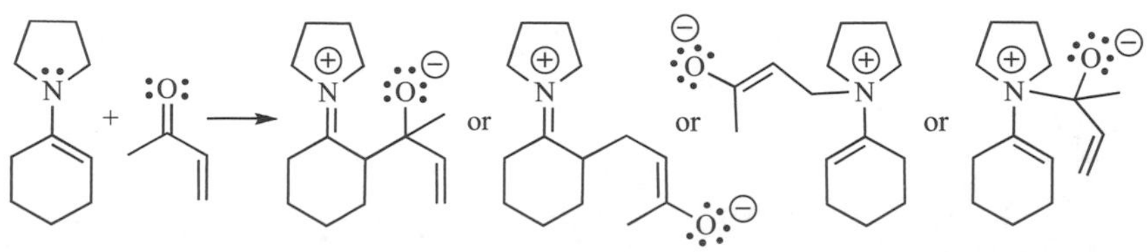

10

CHOOSING THE MOST PROBABLE PATH

Electron Flow In Organic Chemistry: A Decision-Based Guide To Organic Mechanisms, Second Edition.
By Paul H. Scudder

10.1 PROBLEM SOLVING IN GENERAL

The following hints are intended to help you avoid the most common "traps" or conceptual blocks encountered in solving a problem in organic chemistry or actually in any field. Specific decisions were covered in Chapter 9, and procedures for various problem types are covered later in this chapter. All of these procedures are extra work when they are compared to "just having the answer pop into your head," but they make the problem-solving process more systematic and more reliable; they make it easier to retrace your thinking to check the answer and simpler to "debug" when things go wrong.

The more complex a problem is, the less likely you will be able to see the answer from the start. Therefore you must begin work on the problem without any clear idea of how to work it. If so far you have always seen the answer from the start, working on a complex problem is, in a way, like flying on instruments through a cloud bank when you are used to visual navigation. Often students will just stare at the problem and wait for the answer to hit them. With difficult problems this can be a very long wait. These tips allow you to pursue the answer actively rather than waiting for it to come to you.

Study the Material Before Attempting the Problems

Study with a pencil to underline important ideas; make margin notes about things that do not make sense; write down questions to ask the instructor; make a written summary of what you have read. Understand the trends, vocabulary, and principles that are needed to solve the problem. Target your study to concentrate on weak areas as indicated by exams and homework. Patch your conceptual holes! Because organic chemistry is cumulative, like a foreign language, any important concepts you miss in the beginning will come back to bite you in the end.

Establish an Informational Hierarchy While You Study

You tend to remember the things that you use the most and those that are most recent. The only way you can raise the mental awareness of the important concepts over the more recent insignificant fact is to **review the most important material more often**. As any course of study progresses, the incoming facts and important concepts must be put in their place in the hierarchy. You manage an information overload by allowing the little fact to be forgotten after it has served its purpose of being an example for a more important principle. The older, important concepts must be continually raised above the newer facts by additional review (even though you feel you know the concepts very well; you are just reinforcing their importance by keeping them active). Continuously review the classification of the sources and sinks and the electron flow pathways.

Collect a "Toolbox" of Commonly Needed Problem Solving Items

As a start toward this, the Appendix gathers together important tools: a pK_a chart, a bond-strength table, a list of sources and sinks, the electron flow pathways, trends, general rules, and other useful items. Continue to customize your toolbox with any valuable tools that you need.

Read the Problem Carefully—Understand the Problem

Be sure to know what is asked. Can you rephrase the question in your own words? Make a model or draw a figure. Can you break the problem into separate, smaller, more easily solved units? Can you give a rough estimate of the overall process?

Map Changes—Understand Bonds Made and Broken

For molecules drawn with skeletal structures, draw out all the atoms near any changes in the structure. It is easy to forget about hydrogen atoms that are not drawn. Number the atoms of the skeleton if it is not obvious. Pick the simplest numbering scheme, the one with the fewest changes. What bonds were broken? What bonds have to be made? Is the system set up at the beginning to do any of these on your list? Would a proton transfer get you started? Keep an open mind over which process occurs first.

Do Not Ignore What You Do Not Understand

If you do not know what a reagent in a reaction is doing, don't ignore it. It may be critical for the solution of the problem. Solvents can often be ignored, so know the common abbreviations for them.

Gather All Applicable Information About the Problem First

Look to see that there is enough information to solve the problem. Search for relationships between the data and what is wanted. Write down any chemical equations that relate to the problem. What are the assumptions and limitations that come with the problem? What principle(s) was the problem designed to illustrate?

Classify Into Generic Groupings

Back off from the problem and get an overview of it. Have you seen a problem of the same generic class before? Often changes in the hydrocarbon "grease" around a reactive site will disguise the similarities to reactions that you've seen before: Don't slip on the grease! What are the *real* similarities and differences? Decide what parts of that related problem are useful. Have you seen a problem with a similar start or finish point?

Recognize Possible Intermediate Goals

For a mechanism question, number the atoms in the starting material and try to find those atoms in the product. Find out what bonds need to be broken or made. On what atoms do you need to generate nucleophiles or electrophiles? What are all the possible starting points? Are there any possible last steps? Middle steps?

Always Write Down Any Possibility That You Consider

Work on paper, not "in your head." The biggest problem-solving trap is to shoot down a possibility "in the air" without writing it down first. **Writing down something that you are considering forces you to take more time to think about it** and produces a written record that it has been checked out.

Have a Systematic Method to Your Answer Search

Random trial and error does not work on complex problems. A systematic search for the answer may seem like a long and tedious process, but it is much quicker and more reliable than random shots in the dark.

Look For Alternatives—Generate All Paths, Then Select the Best

Pick the best path and check it out, but always keep in mind the other alternatives in case the first one does not work out. In writing down the alternatives, you may get other ideas. Don't waste a lot of time at an apparent dead end before going back to check out other possibilities.

Recognize the Generic Form of Each Step

Classify each step or group of steps as a proton transfer, substitution, elimination, addition, or rearrangement. Then check for other alternative forms of the same process that may fit the reaction conditions better. For example, your first impulse in a particular problem might be to write some sort of elimination because it seems to get you closer to the answer. If you recognize that what you need is a type of elimination, you would know that eliminations could go by E1, E2, E1cB, or Ei mechanisms. You can then pick the most appropriate one, rather than remaining with the first one that occurred to you.

Don't Skip Steps—Look for Any Hidden Decision Points

One common trap is to skip a step rather than take the time to write out the intermediate. If there is another route from the unwritten intermediate, the decision point will be missed and difficult to locate on a crosscheck. Be continually on the lookout for the major decisions discussed in Chapter 9. The most common and also the most commonly missed decision is the nucleophile versus base decision.

Work Carefully and Cross-Check the Work as You Go

Simple errors can lead to some rather absurd predictions. Arrow pushing is a form of electron bookkeeping; therefore the arrows and Lewis structures must correlate and be accurately drawn for the results to make sense. The total charge on both sides of the transformation arrow should be the same. Pay attention to detail.

Beware of Memorization

Memorization is never a satisfactory or reliable substitute for understanding what you are doing. Beware of mindlessly "turning the crank" on a problem; it might not be connected to anything at all.

Make a Scratch Sheet Into an Idea Map

Keep your scratch sheets neat. Be organized on the scratch paper so you can trace your thoughts and be able to go back to other ideas and check your thinking **without getting lost**. Remember, you are exploring unknown territory; draw yourself a map of the problem space as you go.

Watch for Bad Habits

Do not stop at the first answer you find but continue to look for better solutions. Make up a list of your own problem-solving bad habits and refer to it when you work a problem. For example, if you usually forget to check for nucleophilicity vs. basicity, write yourself a reminder. Know the common "potholes" on the route to your answer.

Stay on the Pathways

Although it is tempting to get the problem over with by using a burst of arrows and as few steps as possible, our objective is to write a **reasonable** hypothesis, not necessarily the shortest one. **Construct your mechanistic "sentences" with known, tested "words," the electron flow pathways.** Mechanistic steps that you invent as a beginner may not be reasonable. Selecting from known pathways turns an apparent open-ended question into a simpler multiple-choice question.

Don't Force the Answer!

Be patient. The correct answer will seem to have a natural flow to it; nothing will have to be forced. Some dead ends can seem frustratingly close to the final answer, and there is a tendency, born out of this frustration, to "hammer home" the "last step" even though you know it is not right. Put the hammer away when working a jigsaw puzzle.

If Stuck, Don't Just Stare at the Page, Draw Something on It!

Reread the problem, then draw resonance forms, three-dimensional diagrams, other equilibria, less probable possible paths, anything to search actively for that crucial hint that is needed. **Blank paper gives no hints.**

Recognize the Limits Placed by the Reaction Conditions

Is the reaction in acidic or basic media, in protic or aprotic solvents? Classify the electron sink, and then sort out those pathways that do not use that sink to help you reduce the number of paths you need consider. Is the reaction site accessible, or do large groups block it? Are there constraints on orbital alignment that need to be met?

Beware of Limits That You Place on the Problem

Sometimes we have a tendency to make an initial guess as to the region in which the solution will be found, and later when the solution is not found within those limits to forget that those limits were not imposed by the problem. Without being aware of it, we can artificially confine our search to an area that does not contain the answer.

If Stuck, Examine the Other Possibilities at Each Decision Point

Force yourself to consider other alternatives. One tends to be reluctant to cross-check a hard decision, especially if it was a hassle to make. When cross-checking decisions, make sure that you **go back all the way to the first decision** at the start. Check all decisions, especially those that seem obvious.

When Done, Always Go Back and Check the Answer

Can you see the answer now at a glance? Is there a second way to get the same result? Does it make sense? Use the law of microscopic reversibility to check alternatives on reversible steps. Is the assumed causality correct?

Practice, Practice, Practice

Only by working many problems and many different types of problems can you get good at problem solving. Just reading the material is not sufficient. Recognizing the correct answer is usually easy; you must be able to **use** what you've learned. You need to work actively with the material to build a network of interrelated concepts. Working problems allows you to find your weak areas before an exam reveals them.

What Can Be Learned from the Problem and the Methods You Used to Solve It That Would Be Applicable to Other Problems?

Learn not only from your mistakes but also from your successes!

10.2 GENERAL MECHANISTIC CROSSCHECKS

For a mechanism to be considered reasonable it must fit into the body of knowledge that has accumulated over the years. The following are some points that need to get checked against any mechanistic proposal in order to judge whether or not the mechanism is reasonable.

Electron Flow Pathway Check

The most important check is to make sure that each mechanistic step is a valid electron flow path. The mechanistic "sentence" you write must use known "words" to be understandable. There is a temptation, arising out of frustration, to rearrange the lines and dots of reactant into the lines and dots of product with a mechanistically meaningless barrage of arrows just to get the problem over with. **It is important that you assemble your mechanistic hypothesis from known steps, the 18 electron flow pathways.** Check that the electron flow starts at a lone pair or bond that is a good electron source. Make sure that the flow continues without interruption and ends at a good electron sink.

Did you consider alternate pathways that achieve the same overall transformation (the energy surface street corner analogy) and then pick the most appropriate?

Completeness Check

When you generate the possible paths, make sure that you have not left out any important alternatives, then pick the most probable and check that decision. A common problem-solving error is to race off with the first possibility that looks reasonable, failing to check all the possibilities. **This completeness check is especially important for the first step, where the initial direction that you set out to explore is determined.** If your first step is in error, you may be reluctant, several steps later, to go back to the beginning and reconsider. Rather than retreat to the start, the typical student is more likely, as frustration builds, to force an incorrect answer just to be done with the problem.

Media Check

An important crosscheck is the media restriction. Paths that form reactive cations almost exclusively occur in acidic media. Likewise, paths that form reactive anions are the domain of basic media. No medium can be both a strong acid and a strong base; it would neutralize itself. **The reactive species in equilibrium-controlled reactions have a limited range of acidities.** For example, in neutral water the hydronium ion concentration and the hydroxide ion concentration are both 10^{-7} mol/L. Their relative concentrations are defined by $K_w = [H^+][OH^-] = 10^{-14}$. Their p$K_a$ values span 17.4 pK_a units. Would a reaction mechanism proposal be reasonable if it required both hydronium and hydroxide ions? At what point does the span of pK_a values become unreasonable because the needed species have too low a concentration to react at a usable rate?

To attempt to answer these questions, let's return to the example given in Section 3.3, the deprotonation of an ester (pK_a 25.6) by ethoxide ion (pK_{abH} 16); the pK_a span is 9.6. This small concentration of the ester enolate can then enter into a reaction with the original ester (see the product prediction in Section 10.6.2 for a discussion of the entire surface). If we try to run this reaction catalyzed by phenoxide anion (pK_{abH} 10) instead of ethoxide, the reaction fails; no product is isolated, only unreacted starting materials. There are two ways that we could interpret this result: first that the base was too weak, and no appreciable concentration of ester enolate was formed. A second interpretation is that there no longer was sufficient driving force for the reaction since the anion of product is now a stronger base (only by 0.7 pK_a units) than phenoxide; the reaction is no longer forming a weaker base. The pK_a span with phenoxide anion was 15.6 pK_a units.

Therefore as a first approximation, **a pKa span of 10 units is acceptable, but a span of 15 pKa units is probably not.** In your mechanisms, make sure that the span of pK_as of any species that are in equilibrium do not stray too far from the acceptable range. If several steps occur in the same medium, check the pK_a span of all species within those steps. In a way, we are just restating the ΔpK_a rule (which set a limit of 10 for a single step) and applying it to all species in equilibrium throughout the reaction. Certainly the pK_a span at which the reaction rate becomes too slow is open to debate.

Exceptions are expected to arise when the driving force for a reaction is an irreversible proton transfer at the diffusion-controlled limit in water or a similar protic solvent. With a maximum bimolecular rate constant of 10^{10} L/mol-s and a solvent concentration of 55.5 mol/L (pure water), a reactive intermediate could conceivably have a concentration as low as 10^{-16} mol/L and the reaction still proceed at a reasonable rate.

Energetics and Stability Check

We have three ways to check the energetics of a process. For proton transfer reactions we can calculate a K_{eq}. A K_{eq} greater than 1 indicates that products are favored and that the $\Delta G°$ is negative. If neutral compounds are transformed into neutrals, we can calculate the $\Delta H°$ for the reaction from the bond strengths. If the reactants or products are ionic or neutral, we can use the ΔpK_a rule. The most reliable of these three is the K_{eq}, next probably the $\Delta H°$, and least the ΔpK_a rule. A common error is to forget to check the proton transfer K_{eq}. Use the trends to gauge the stability of any intermediates formed in the reaction. Common errors include failure to judge the stability of carbocations, carbanions, and leaving groups.

Charge and Typo Check

Make sure that the Lewis structures and their formal charges are correct. An error on a formal charge can be especially dangerous since the attraction of opposite charges is used to predict reactive sites. The charge must balance on both sides of the reaction arrow. Don't accumulate charges on your intermediates as the reaction progresses. Avoid generating any species with adjacent like charges. Look for line-structure errors, particularly for the "vaporization" of an H that was not drawn in the line structure.

10.3 THE PATH-SELECTION PROCESS

One can view the reaction process as a wandering over an energy surface via the lowest-energy path toward the lowest accessible point on the surface. The reacting partners have a limited amount of energy; only certain transition states and intermediates can be achieved, and the reaction will end up at the **lowest accessible** point on the surface, which may not necessarily be the lowest point on the surface.

The most useful mental models approximate some physical reality. Since we will be exploring an energy surface with peaks, passes, and valleys unknown to us, the best mental model would be a mountainous wilderness, but one luckily crossed with trails. Imagine being dropped off in the middle of this wilderness at a trail junction, assured that one of the many trails leads down and out. We know that the desired route will be the lowest-energy route. We certainly do not want to panic and get off the trails into real unknown territory or to guess randomly at a direction and wander off. We must carefully and logically explore the nearby wilderness, drawing a map as we go.

Armed with the tools that we have learned in the previous chapters, we can predict where the lowest-energy path may go. The **electron flow pathways are the trails to guide us** in this wilderness of an unfamiliar energy surface. Some trails may dead-end in high valleys that have no reasonable exit other than the one we came in. Others may lead us to a low point (side product) on the energy surface but not the lowest accessible point (major product). The reacting partners are also exploring these dead ends, and very few reactions produce only one product.

What do we need in order to succeed? We need the ability to draw a good map (if you don't want to repeat some paths or wander around in circles), the ability to recognize whether we are going uphill or down (energy awareness), the ability to recognize all the paths and when they branch (you can't stay on the trails if you can't recognize them), and finally the ability to cross-check what you have done (was the hike to a certain point a silly thing to do or is it on the right track?).

The exploration process is really quite simple. The hard part is to stick with it when things are not going well and you are feeling lost; it's easy to get discouraged and do something rash just to be done with the struggle. The other time that it is hard to stick with it is when things are going so well that the end is in sight, and it feels that it might be fine to abandon the trails and just make a straight run for the end. Both arise out of the natural tendency to rely on insight because the alternative, rigorous logical analysis, is a lot of hard work.

So what is the search process? First understand where you are. Observe your surroundings; take inventory; classify what you have to work with. Second, start drawing your map by marking the start point down and note all the paths that lead from it. Find a downhill path or at best the least uphill path. Follow that path to the next point and mark that on your map. Again observe your surroundings. Now cross-check that going to this new point is indeed a reasonable thing to have done (here is where you make use of all those trends). Cross-check that the energy of the route is reasonable (with the ΔpK_a rule or K_{eq} calculation or ΔH calculation—whichever is appropriate). Now look for all the new paths leading off this new point, and start the process again. Always keep in mind your starting point and do not climb too high before returning to a lower point to explore other routes.

How do you go about learning this method of exploration? First learn the needed skills (Chapters 1–9), and then try them out on easy problems with the help of an experienced guide. On these guided tours we will explore the entire surface, including all its dead ends, so that you feel you know the territory. We will gradually explore more difficult surfaces until you feel confident enough to venture out on your own. Let's look at each step of the exploration process (Fig. 10.1) in more detail.

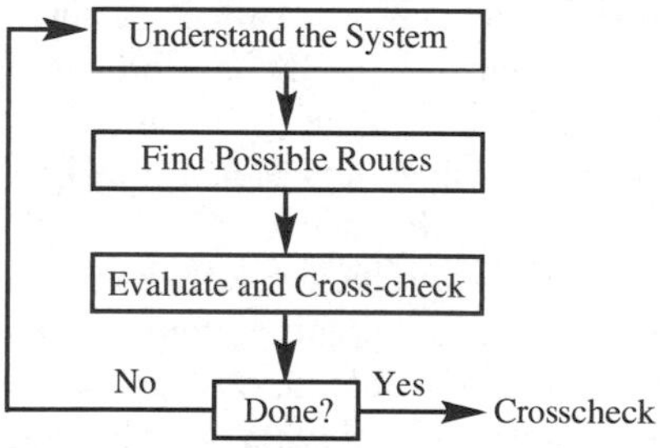
Understand the System
Find Possible Routes
Evaluate and Cross-check
No
Done?
Yes
Crosscheck

Figure 10.1 The overall generate and select process used throughout this text.

Understand the System

Write down the Lewis structure of the reactants, complete with formal charges, and draw any major resonance forms. Look for leaving groups, polarized single and multiple bonds, acids and bases. Classify into generic sources and sinks and then rank them. **The reaction usually occurs between the best source and sink.** Above all, note if the medium is acidic or basic. In basic media, find the best base, and then locate any acidic hydrogen within range (not more than 10 pK_a units above the pK_{abH} of the base). In acidic media, identify the best sites for protonation. Likewise, do not create a species that is more than 10 pK_a units more acidic than your acid. Understand what bonds have been made or broken, but do not lock into an arbitrary order as to which occurred first.

Find Possible Routes

Generate all the possible paths. The trails are few, since only those pathways that have the same sources and sinks as the reacting partners need be considered. The media will also restrict paths. Do not use acid media paths when the medium is definitely basic and vice versa. Draw the Lewis structure of the product. Never discard a route without first taking a small amount of time to write down the result (this will force you to consider it more carefully). Again, use scratch paper to draw a map of this new problem space that you are exploring so that you can keep track of where you have been. Note branches in the path that you have passed but not explored (and may want to come back to), and write down all the possibilities branching from the spot that you are currently considering. Remember the "street corner" analogy that we used in Chapter 4 when exploring the common reaction archetypes. Drawing such a map may seem like it would slow down the process of reaching the goal, but actually it speeds it up because much less time is spent lost, wandering aimlessly, taking random shots at the answer.

Evaluate and Cross-Check

So that we do not have to explore every possible path, but only the most probable ones, the trends allow us to gauge the relative height of the passes and high valleys that we have to choose among. Especially cross-check the media restriction and the stability of intermediates. We can often use either the ΔpK_a rule on our alternatives or the calculation of the ΔH or the K_{eq} to make sure the energetics of the process are reasonable. **Generate and select** the best path. As with any problem for which you cannot see the route to the answer at the start, you must be very careful at the beginning to select the best route, not just the first one you see. Double-check any major decision.

Repeat the Process

With mechanism problems, the final product is given, so the end point is unambiguous. The problem-solving thought process resembles a deliberate search from the starting point, which eventually triggers an insight coming from the goal. A novice can learn the methods of deliberate searching; insight comes by being familiar with the territory and will get better with practice.

With product-prediction problems, there will always be some doubt as to whether you are at the lowest accessible point on the surface or are in the energy well of a side product. Hopefully, the end should be obvious because you will definitely be downhill from your starting point, and it will seem that all alternatives for several steps around are definitely uphill. Again, knowing when you are done also improves with practice.

10.4 REACTION MECHANISM STRATEGIES

Mechanism problems are easier than product-prediction problems because helpful information can be derived from the product. Given the starting and ending points you need only supply the lowest-energy path between them. You may not be able to see the answer to a long mechanism problem at the start; just let a logical analysis carry you to the answer. Don't panic and try to force a quick (and incorrect) answer.

Commonly, organic equations are not balanced, so make sure the chemical equation balances. This may give you a hint in that it may identify another product; for example, water may have been lost. If there is a small piece lost in the reaction, try to find out where it came from. Next identify the carbons that belong to the starting materials within the product (number the carbons if you have trouble) so that you can tell what bonds have been made or broken. If more than one numbering scheme is possible, go with the scheme that requires the fewest changes. If the starting material structure just will not fit into the product structure (taking into consideration any small pieces lost), rearrangement may have occurred.

Look at each new bond and then at the polarization of the corresponding sites in the starting materials so that you can decide what the original nucleophile–electrophile pairs might have been for those bonds. If more than one new bond has been made, consider all possible sequences of bond formation, not just the first one that occurs to you. Look at adjacent functionality to get a hint as to whether it would stabilize an electrophile's or nucleophile's partial or full charge. Ask how proton transfer might have generated the needed nucleophiles and electrophiles. Consider any reasonable proton transfer and always check the K_{eq} value (at best $\geq 10^{-10}$).

Use the general trends to guide the decision process. At every decision point, **generate** all the possible paths, and then **select** the best one on the basis of trends. For example, if there is a choice of making two possible cations, rank their stability and pick the route that goes through the most stable cation. If nothing looks reasonable, return to any decision point where the trends indicated that an alternative was also feasible. Use "hindsight" (knowing the final product) as a guide through any difficult decisions.

Finally, do not combine steps to save drawing out the structure again. Keep to the pathways to avoid the trap of merely rearranging the lines and dots of the structures on the page. Be as rigorous in your own practice as you would be on an exam; you are trying to develop good intellectual habits. The prime directive is this:

If it does not make sense, it's wrong.

10.5 WORKED MECHANISM EXAMPLES

The problems at the end of Chapters 8 and 9 were short and simple mechanism- and product-prediction problems. Chapter 8 problems had only one source and sink and no major decisions to be made, and those in Chapter 9 were the same except that one major decision was required. We are now going to cover problems of gradually increasing complexity: more than one source and sink, longer problems, more decisions, and more alternate routes. For each example, cover each step and try to provide the expected information; compare yours with what is given, then move on to the next portion.

10.5.1 Alkyne Hydrobromination

Give a reasonable mechanism for the following reaction.

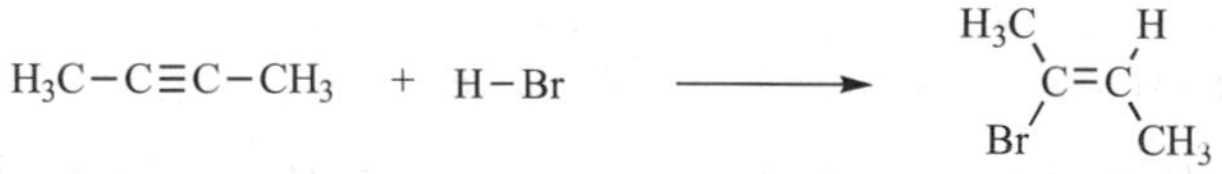
H3C−C≡C−CH3
+ H−Br
H3C
H
C=C
Br
CH3

Understand the System

Balanced? Yes. Generic process? An addition has occurred. Medium? Strongly acidic, HBr pK_a is −9. Sources? Triple bond and the bromide lone pair. Sinks? The H–Br bond is easily broken because hydrogen bromide is a strong acid. Bromide anion is also expected to be present as a conjugate base from any proton transfers. Stereochemistry of addition is *anti* since the H and the Br end up on opposite sides of the double bond.

Find Possible Routes

We are at a well-known trail junction, the starting point of the addition surface from Section 4.4.2. Figure 10.2 is a simplified addition surface serving as a problem space map to guide our decisions and to remind us of alternative possibilities, because the first answer we think of may not be the best.

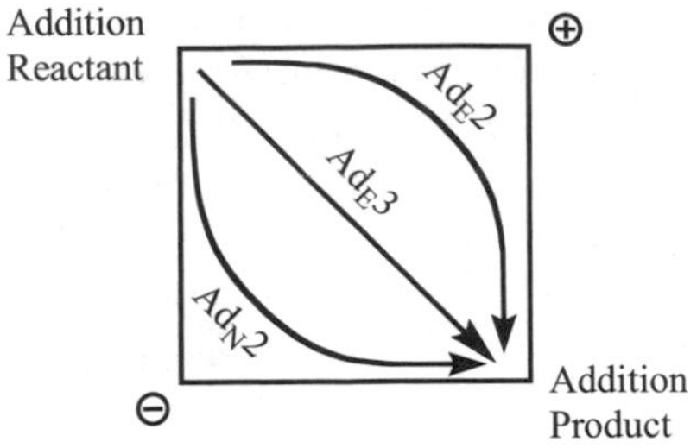
Addition Reactant
⊕
AdE2
AdE3
AdN2
⊖
Addition Product

Figure 10.2 Simplified addition surface serving as a problem space map for example 10.5.1.

From the surface, there are three possible routes that we must decide between: the Ad_N2 addition (path Ad_N followed by p.t.),

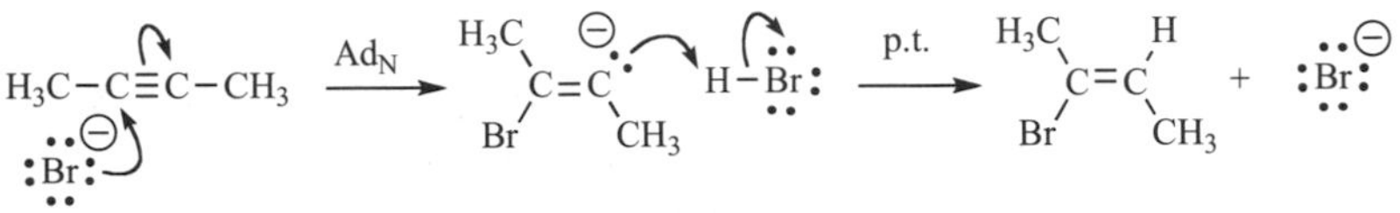

the Ad_E2 addition (path A_E followed by A_N),

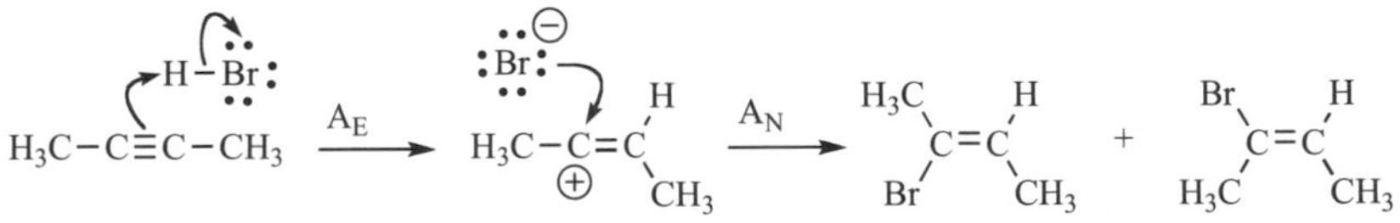

and the Ad_E3 addition.

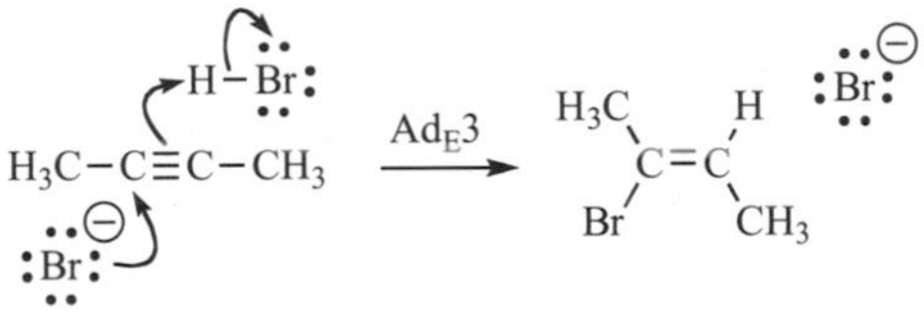

Evaluate and Cross-Check

We must judge the possible routes as to how reasonable they are to the sources, sinks, and the media. The Ad_N2 is not appropriate for acidic media because it produces a highly basic anion, pK_{abH} 44. The ΔpK_a rule would likewise throw out the Ad_N2 since the incoming bromide nucleophile at pK_{abH} −9 would never be expected to form a product anion of pK_{abH} 44. This route is so uphill that it would never occur. In this example to illustrate the sorting process, we showed all the possibilities, but in later mechanism examples we will consider only those routes that are contenders for the lowest-energy route.

The Ad_E2 and Ad_E3 processes are appropriate for the acidic medium. The Ad_E2 goes via an unstable secondary vinyl carbocation. A second problem is that that carbocation is expected to be linear and could therefore trap bromide from either the top or the bottom face to produce a mixture of stereoisomers.

The Ad_E3 avoids the unstable carbocation and rationalizes the stereochemistry of the product. The attack is *anti*; the bromide attacks the bottom face since the pi complex of the alkyne with HBr blocks the top attack. In this case the addition surface has folded so that the Ad_E3 is the lowest-energy route proceeding up through a pass between the two high corners of an unstable anion and an unstable cation. For an example of this surface, see the E2 surface (the reverse reaction) (Fig. 4.23). The Ad_E3 route explains the stereochemistry and is reasonable for the alkyne substrate and strongly acidic media.

10.5.2 Imine Formation

Give a reasonable mechanism for the following biochemically important reaction.

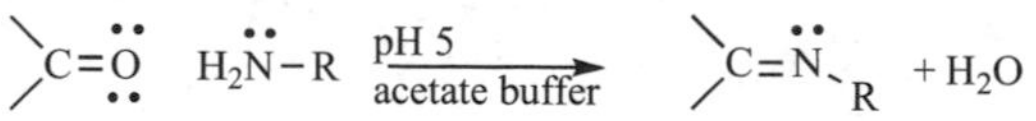

pH 5
acetate buffer

Understand the System

Balanced? Yes. Medium? Mildly acidic. A common acetate buffer is created from almost equal concentrations of acetic acid and sodium acetate. Sources? The amine lone pair and the carbonyl lone pair. Sinks? The ketone is a C=Y. Acidic Hs? None within range of pH 5. Basic sites? An oxygen and/or nitrogen lone pair could become protonated. Leaving groups? None. Resonance forms? The carbonyl charge separated form, $^{+}C{-}O^{-}$, shows the carbonyl carbon to be a good sink.

Bond changes? The amine nitrogen has to bond to the carbonyl carbon first, and then we have to create a new pi bond to nitrogen by removing the oxygen as water. Some proton shuffling has occurred because nitrogen has lost two Hs and oxygen has gained two.

Find Possible Routes

It helps to understand the overall transformation before diving into the individual steps. We can get a good overview of the problem space by generalizing what has to happen: First an addition has occurred to connect the two reactants, then probably some proton transfers to set up the next step, an elimination of water. This allows us to draw a simplified problem space to guide our route decisions (Fig. 10.3).

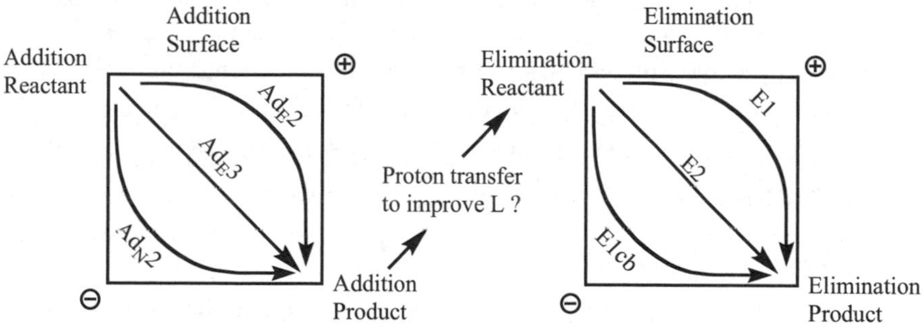

Addition Surface
Addition Reactant
Elimination Reactant
Elimination Surface
Proton transfer to improve L ?
Addition Product
Elimination Product

Figure 10.3 Simplified problem space for imine formation, example 10.5.2.

We can see from our simplified problem space that there are three possible addition routes that should be considered:
the hetero Ad_E2 addition (path p.t. followed by Ad_N),

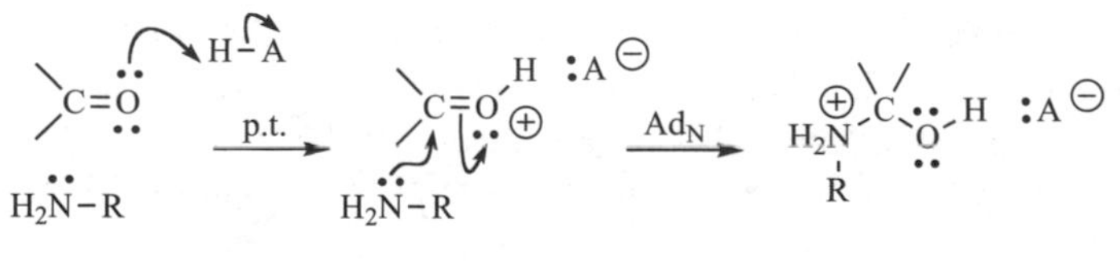

p.t.

the hetero Ad_E3 addition,

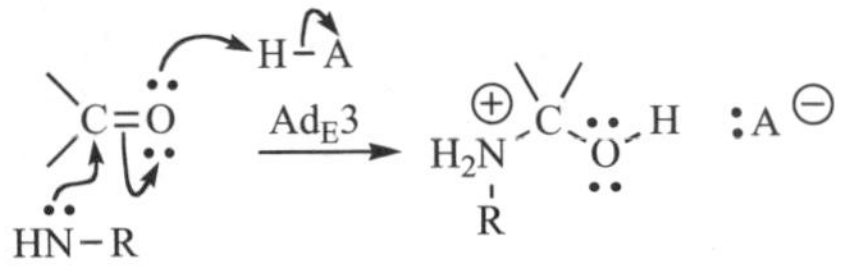

and the Ad_N2 addition (path Ad_N followed by p.t.),

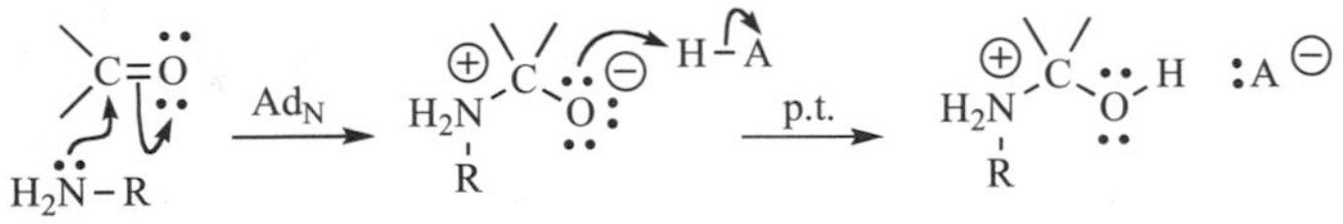

Evaluate and Cross-Check

The first step of the Ad_E2 is a proton transfer that creates a strong acid, $pK_a = -7$. Our best acid in acetate buffer is acetic acid, $pK_a = 4.8$, and so this proton transfer K_{eq} would be $10^{-11.8}$, which is slightly beyond our useful range.

The Ad_E3 has an incoming nucleophile whose conjugate base pK_a is about 10.6 and the addition is removing the proton from acetic acid, $pK_a = 4.8$, thereby creating a weaker base, and passing our crosscheck.

The Ad_N2 first step produces an ion pair that would be expected to have a pK_{abH} near 12. Our nearest model, choline $pK_a = 13.9$, is expected to be less acidic because it has an extra methylene $(CH_3)_3N^+CH_2CH_2OH$. The incoming nucleophile (conjugate base $pK_a = 10.6$) is only slightly less basic than the ion pair produced by the addition. The Ad_N2 second step also removes the proton from acetic acid, $pK_a = 4.8$, thereby creating a weaker base, so both steps pass our crosschecks.

We conclude either the Ad_E3 or Ad_N2 would be reasonable alternatives.

Understand the System

We need to set up the elimination. The nitrogen still has two protons on it, and the oxygen has only one. Hydroxide at $pK_{aHL} = 15.7$ is not a good leaving group but could be improved by protonation. The protons on the cationic nitrogen have a pK_a of 11, within range of acetate, $pK_{abH} = 4.8$. Even though oxygen's electronegativity makes its lone pair less basic than nitrogen's, at $pK_{abH} = -2.4$, the OH is still in range of being protonated by acetic acid at $pK_a = 4.8$.

Find Possible Routes

We need to deprotonate the nitrogen and protonate the oxygen, but which order is best? Draw both out and compare.

Oxygen first:

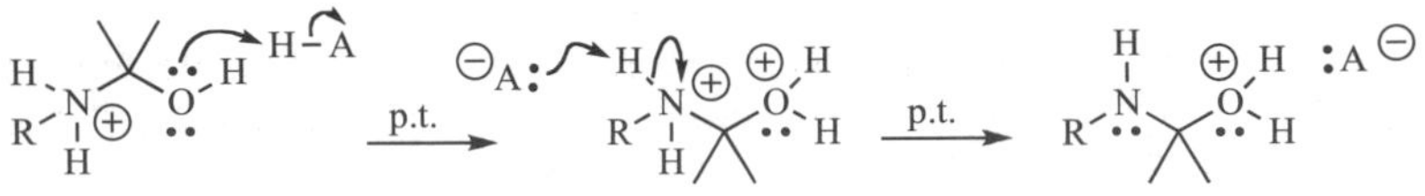

or nitrogen first:

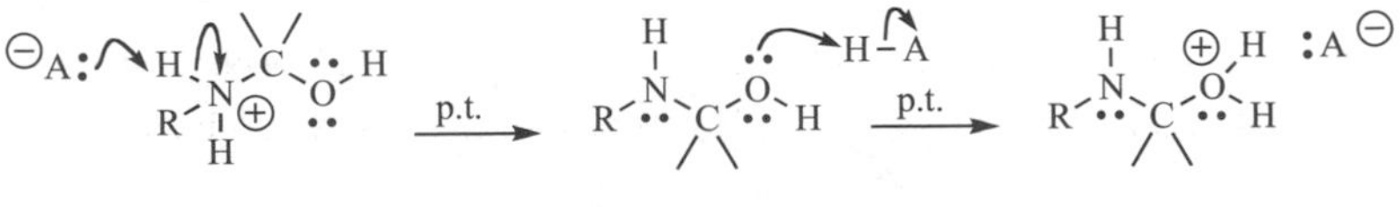

Evaluate and Cross-Check

If we protonated the oxygen first, there is a problem of the unfavorable electrostatics of having two like charges close to each other. Since we would like to avoid the electrostatic repulsion of like charges, it would be best to deprotonate the nitrogen before protonating the oxygen.

Understand the System

We are now ready to consider an elimination to remove the proton on nitrogen and kick out the water leaving group to get the final product. The proton on nitrogen is not very acidic, and our medium pH is just 5. Water, $pK_{aHL} = -1.7$, is a good leaving group.

Find Possible Routes

We can see from our problem space (Fig. 10.3) that there are three possible elimination routes that should be considered.

E1

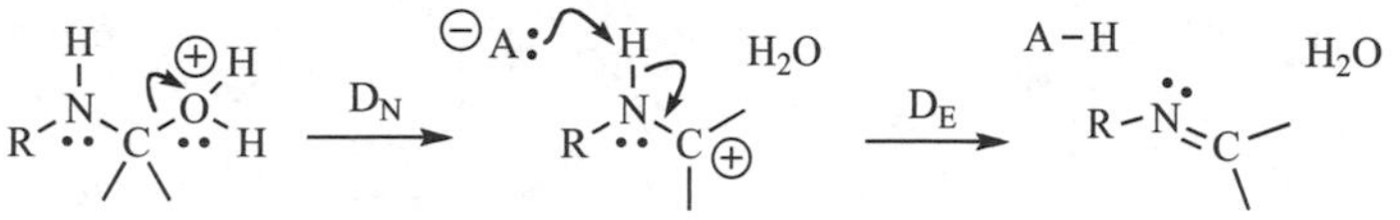
DN
DE
A-H
H2O

E2

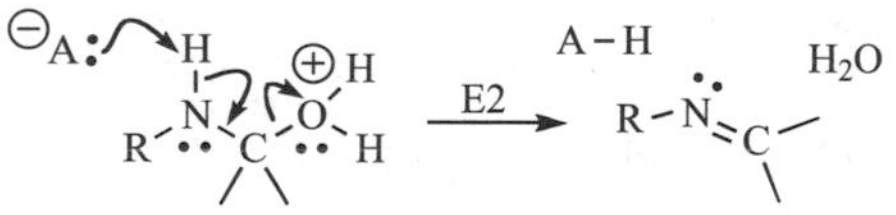
E2
A-H
H2O

E1cB

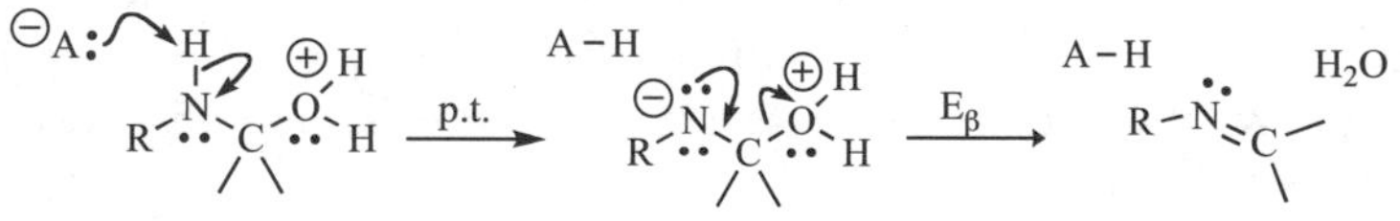
p.t.
Eβ
A-H
H2O

Evaluate and Cross-Check

The D_N step of the E1 requires three conditions to be fulfilled: a polar solvent, good leaving group of usually $pK_{aHL} < 0$, and a carbocation of secondary or better stability. The water buffer solvent is very polar. The water leaving group is $pK_{aHL} = -1.7$, and the carbocation formed has lone pair stabilization by resonance. The D_N step, and therefore the E1 route is reasonable.

The E2 route needs to pass a ΔpK_a test and an alignment test. The E2 passes the ΔpK_a test, for with the base as acetate, $pK_{abH} = 4.8$, and the pK_{aHL} of the leaving group is -1.7, the E2 reaction can create a much weaker base. Since the C–N single bond is free to rotate, the N–H and the C–L can assume the coplanar alignment that an E2 requires. Therefore the E2 route is also reasonable.

The proton transfer first step of the E1cB has problems. The conjugate base nitrogen anion has no resonance and only the ion pairing with the positive oxygen to stabilize it. Without the ion pairing, the conjugate base pK_{abH} would be 36. Field and inductive effects are less influential than resonance. For example, resonance stabilization of a nitrogen anion by a phenyl brings the pK_{abH} down to 27, still a long way from our strongest base in this reaction, acetate at $pK_{abH} = 4.8$. We conclude that the proton transfer, and thus also the E1cB, fails to be a reasonable hypothesis.

Overview

We are done, but let's take an overview of the process. Two addition and two elimination routes were reasonable. One acceptable reaction path is shown vertically in the center of Figure 10.4, and the alternate routes are shown to the side.

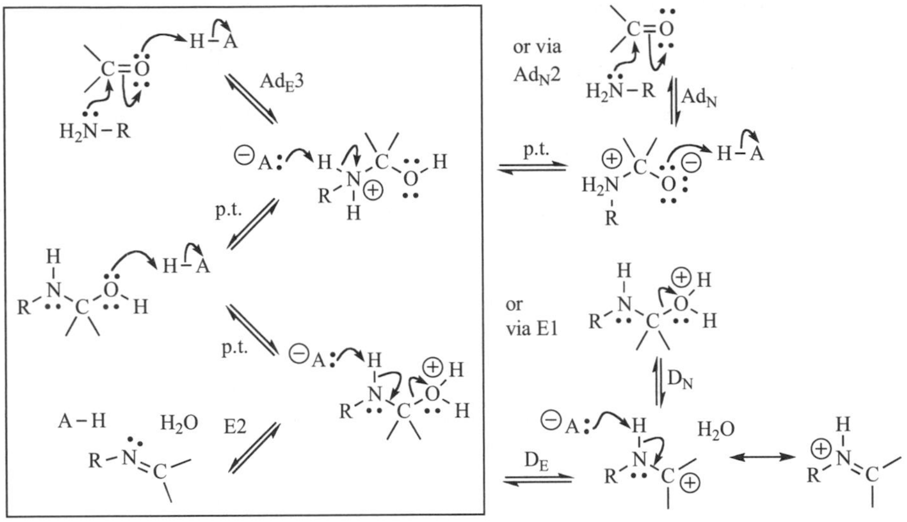
or via
Ad_N2
Ad_N
Ad_E3
p.t.
p.t.
p.t.
or
via E1
D_N
E2
D_E
H_2N-R
A−H
H_2O

Figure 10.4 Imine (Schiff base) formation overview.

Biochemical Example Problem

Imine formation is a very important biochemical process. It has an interesting behavior that shows a maximum in the pH rate profile (Fig. 10.5). Using the previous mechanism postulated for imine formation, examine extremes in pH to understand the figure. What would you expect to happen in strong acid to slow the reaction down? How about in strong base?

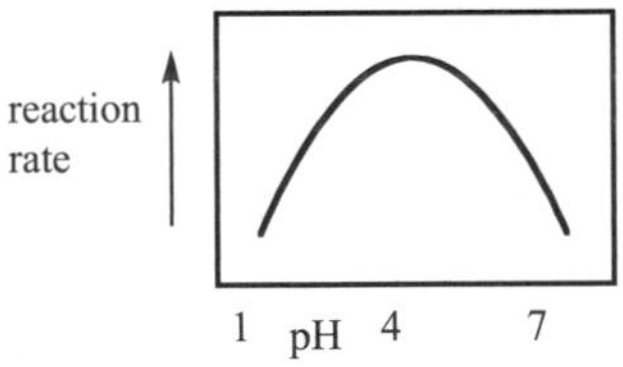
reaction
rate
1
pH
4
7

Figure 10.5 A pH rate profile for imine formation.

Answer: In strong acid, the amine nucleophile in the first step would be protonated and thus no longer nucleophilic. Without a good nucleophile the reaction would slow down greatly. In strong base, there would be no acid good enough to protonate the leaving group for the elimination step. With a poorer leaving group, the reaction would also slow greatly. Close to neutrality, there is enough free amine to serve as a nucleophile, and enough weak acid to protonate the leaving group, so the reaction proceeds nicely.

10.5.3 Aldol Condensation

Give a reasonable mechanism for the following aldol condensation.

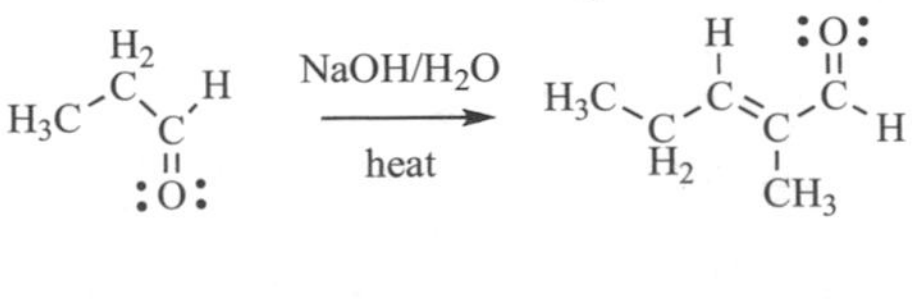
NaOH/H_2O
heat

Understand the System

Balanced? No, the left-hand side requires 2 mol of aldehyde, and the right-hand side needs a water molecule to balance. Find the pieces of the two reactants within the product. Note that the reaction has occurred between the aldehyde carbonyl (carbon 1) of one reactant and the carbon next to the aldehyde (carbon 2) on the other.

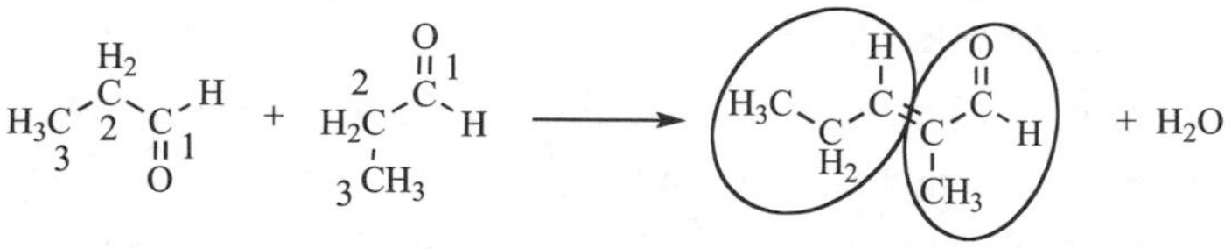

Generic process? An addition and an elimination have occurred. Medium? Definitely basic, predominant anion is hydroxide, pK_{abH} 15.7, whose pK_a would give a useful proton transfer K_{eq} up to about pK_a 26. Sources? The carbonyl lone pair, water lone pair, and hydroxide anion. Best source? Hydroxide anion, a lone pair source can behave as a nucleophile or as a base. Sinks? Polarized multiple bond, the aldehyde carbonyl. Acidic Hs? Water and the CH_2 next to the aldehyde, pK_a 16.7, are within range of hydroxide. Leaving groups? None. Resonance forms?

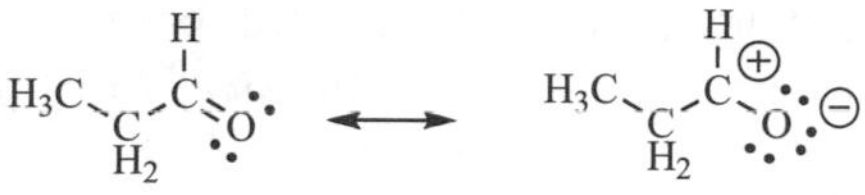

Find Possible Routes

We need to connect the two partners before we can make the double bond, so the addition must have occurred before the elimination. Since bonds are made by combination of nucleophile and electrophile, we can see that the addition is not ready to proceed. While the aldehyde carbonyl is an electrophilic electron sink, carbon 2 of the other partner is a mere methylene and not a nucleophile. Therefore we need to generate a nucleophile on carbon 2 before the reaction can proceed. The problem space for the overall process might look like Figure 10.6.

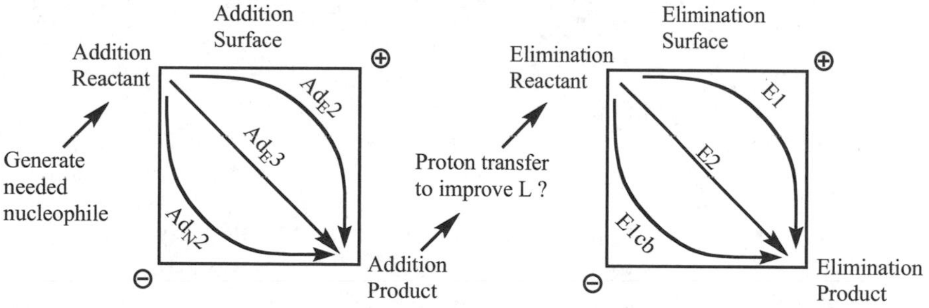
Addition Surface
Addition Reactant
Generate needed nucleophile
Ad_E2
Ad_E3
Ad_N2
Addition Product
Proton transfer to improve L ?
Elimination Surface
Elimination Reactant
E1
E2
E1cb
Elimination Product

Figure 10.6 Simplified problem space for aldol condensation, example 10.5.3.

As a first step, there are just two paths that fit the basic medium, sources and sinks: deprotonation to form an anion (path p.t.),

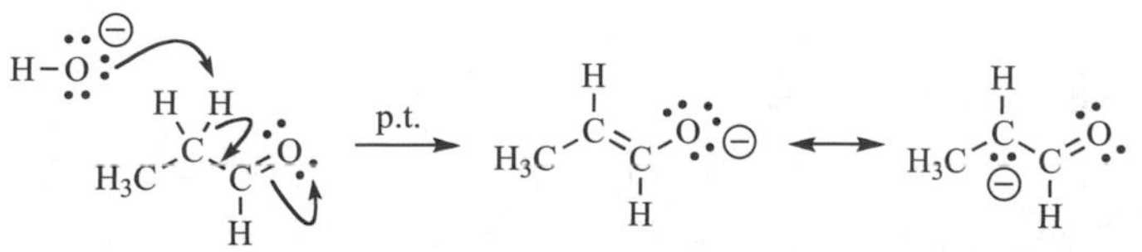

or addition to a polarized multiple bond (path Ad_N).

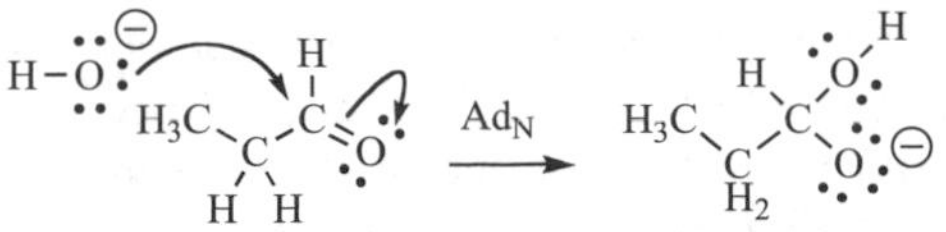

Evaluate and Cross-Check

Since the pK_a for both products can only be estimated, and they both come out to be around pK_{abH} 16 to 17 versus 14 to 15, they are too close in energy to distinguish with our crude ΔpK_a rule. We should explore both routes. A use of "product hindsight" shows that no additional hydroxyl groups are attached, thus making this nucleophilic attack path less probable. If this were a product prediction problem and not a mechanism problem, we would have to explore both routes.

Understand the System

We need to record only that which has changed from our original observations. We now have a new source, the delocalized anion just formed. This enolate anion is nucleophilic on carbon 2, just what is needed for our addition process. Resonance forms indicate the ambident nature of the enolate, an allylic source:

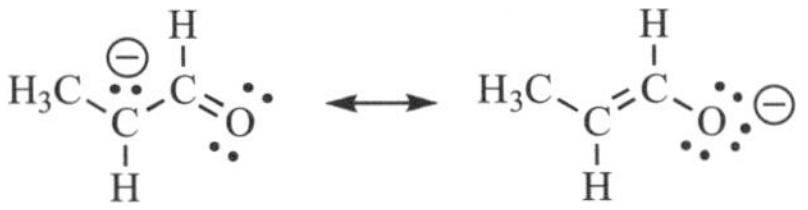

Find Possible Routes

Now we have a good nucleophile and a good electrophile for our addition process to occur. The three addition routes are:
the hetero Ad_E2 addition (path p.t. followed by Ad_N),

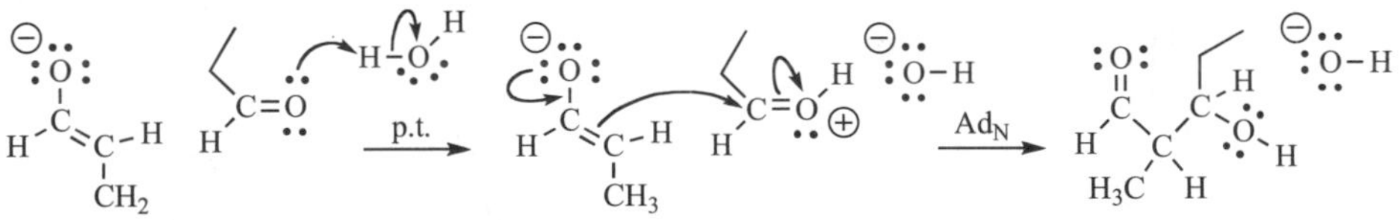

the hetero Ad_E3 addition,

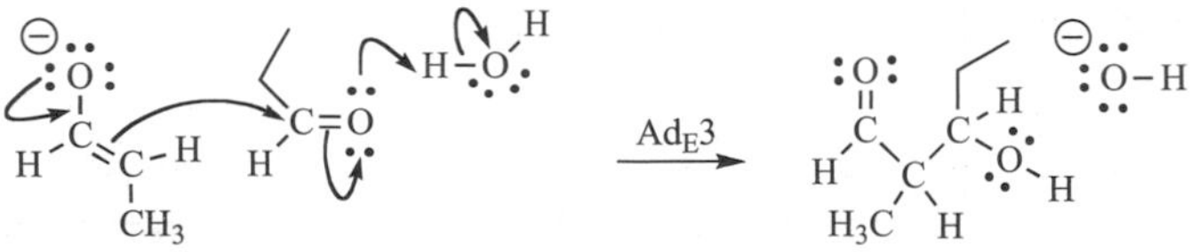

and the Ad_N2 addition (path Ad_N followed by p.t.).

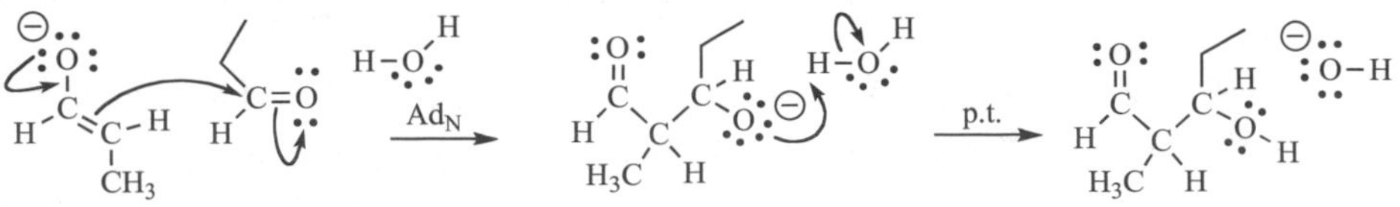

Evaluate and Cross-Check

For the Ad_E2, the best acid in basic water is water itself with a pK_a of 15.7, which is very weak. The product protonated aldehyde has a pK_a of –8, giving the proton transfer

K_{eq} a value of $10^{-23.7}$, very much beyond our minimum useful threshold of 10^{-10}. Another tip that something is wrong is that we have both a strong base and a strong acid present in an intermediate step, a violation of our pK_a span limit of no more than 15 pK_a units. The Ad_E3 addition is reasonable by the ΔpK_a rule, the incoming aldehyde enolate nucleophile is pK_{abH} 16.7, and the product of the nucleophilic attack is a hydroxide, pK_{abH} 15.7. The Ad_N2 addition is also reasonable by the ΔpK_a rule, the incoming nucleophile is pK_{abH} 16.7, and the product of the nucleophilic attack is a secondary alkoxide, having a pK_{abH} of about 17. Either Ad_E3 or Ad_N2 addition is energetically reasonable by the ΔpK_a rule.

Other Things That Can Happen

There are other paths that fit the medium, sources and sinks: protonation of an anion, p.t., and addition to a polarized multiple bond by the oxygen of the enolate by Ad_E3 or Ad_N2. There are two sites that can serve as a base or as a nucleophile on this ambident allylic source. We have three choices to evaluate: proton transfer to oxygen,

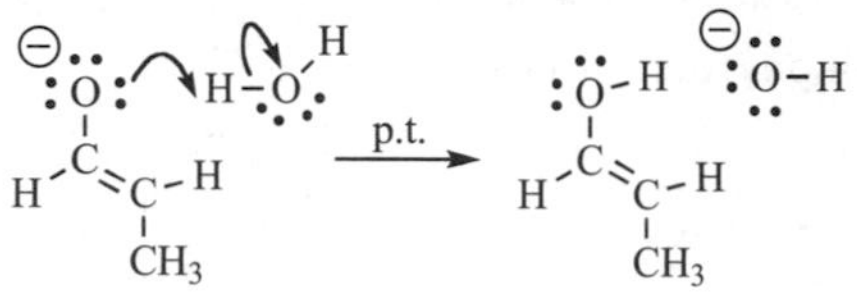

or to carbon,

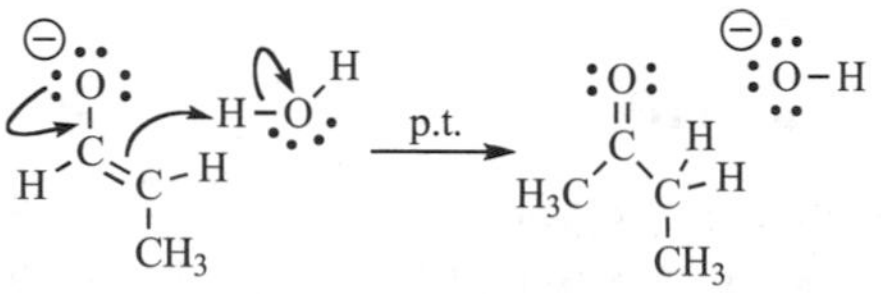

or addition to a polarized multiple bond by oxygen by either Ad_E3 or Ad_N2.

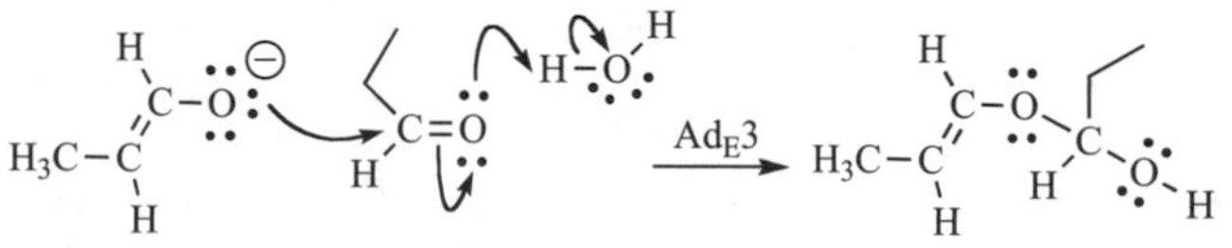

Evaluate and Cross-Check

Protonation of the anion takes us either to the enol or back to starting material, both reasonable alternatives. A ΔH calculation gives the enol as uphill in energy from the carbonyl. We know from our product that carbon, not oxygen, has been the nucleophile (as expected, Section 9.4). The oxygen attack on the carbonyl must have been reversible, and a ΔH calculation confirms the hemiacetal product is expected to be uphill in energy.

Understand the System

An elimination of water gets us to the product. We now are close enough to product to restrict paths to elimination pathways. From our simplified problem space in Figure 10.6, we need to determine whether the medium can improve our leaving group and then move to the reactant corner of the elimination surface (Section 4.3.1). Resonance forms:

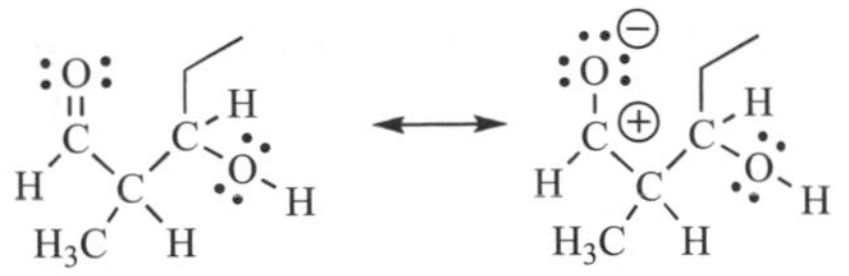

Find Possible Routes

The basic medium does not contain an acid strong enough (water is our best acid) to improve the leaving group by protonation; creating a protonated alcohol of pK_a −2.4 would have a K_{eq} of $10^{-18.1}$ with water as the acid. Therefore, because the hydroxide leaving group is poor, pK_{aHL} 15.7, and the E1 elimination path is not viable. There then are just two elimination paths that fit the medium, sources and sinks: the E2 elimination,

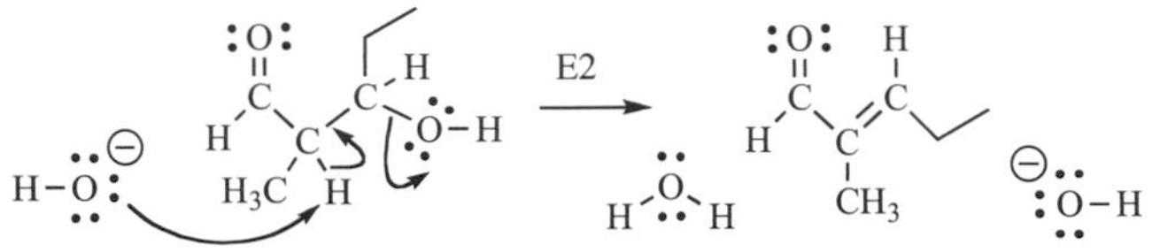

or the E1cB (path p.t. followed by path E_β).

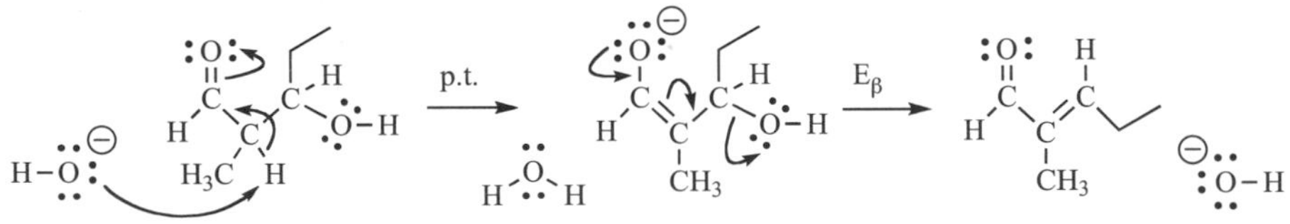

Evaluate and Cross-Check

Both elimination processes are reasonable, but the E1cB path is more probable because the H is acidic by virtue of being next to the aldehyde, and the leaving group is poor, pK_{aHL} 15.7, so an adjacent anion can help it leave. The pK_a span on all steps has been within reason.

Overview

We are done, but let's take an overview of the reasonable routes (Fig. 10.7). The reaction path is shown vertically in the center, and the alternate routes are shown to the side. The path abbreviation for the forward reaction only is shown. To summarize the route: Proton transfer generates the nucleophile, which reacts by Ad_N2, and then the E1cB process gives us the product.

10.5.4 Electrophilic Aromatic Substitution

Give a reasonable mechanism for the following reaction.

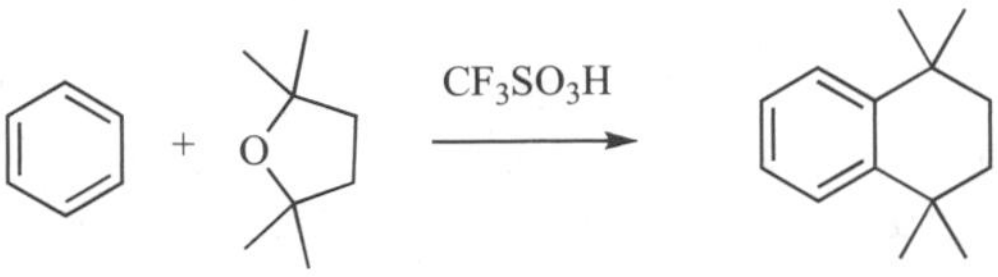
CF3SO3H

Understand the System

Balanced? No, water must be added to the right-hand side to balance the equation. Generic process? Benzene has substituted for the ether oxygen at both tertiary carbons.

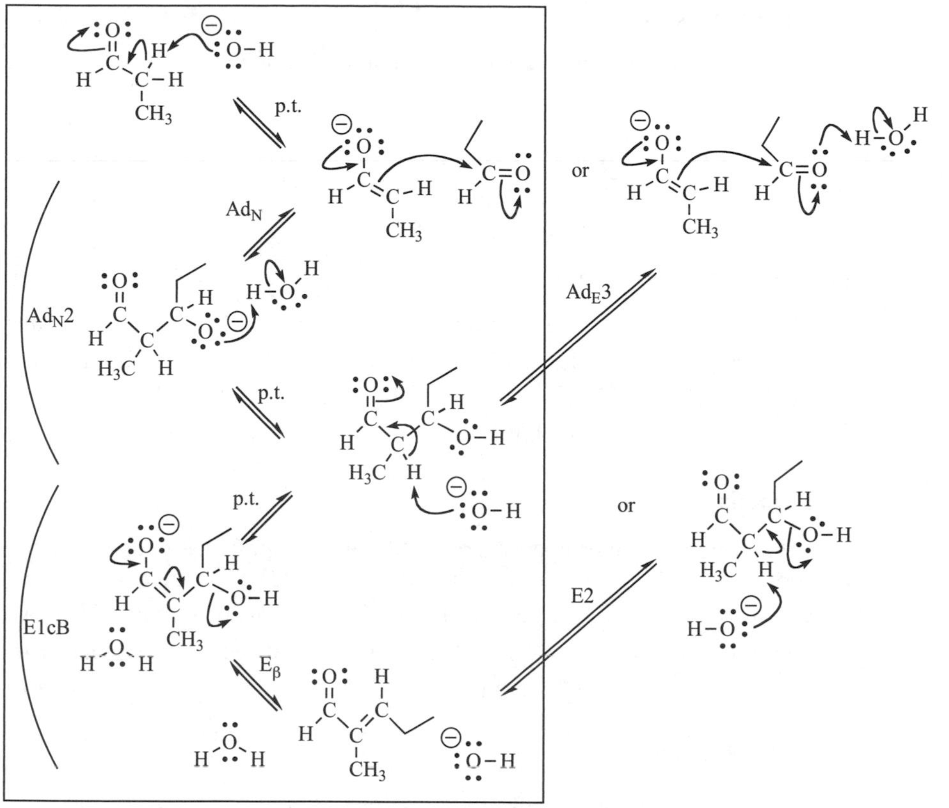

Figure 10.7 Aldol condensation overview. The most likely reaction path is shown boxed.

Medium? Highly acidic. Sources? The ether lone pairs are best, with the aromatic ring a poor second. Sinks? Trifluoromethanesulfonic acid (triflic acid) is a very strong acid with an estimated pK_a of −15. Acidic Hs? None. Leaving groups? The ether oxygen, but only if protonated. Resonance forms? Only benzene pi bond shift.

Every bond can be considered to be a combination of a nucleophile and an electrophile. Our best source is the ether lone pairs, and our best sink is the triflic acid. Since the reaction is in very acidic medium, try protonation as a first step. In base, try deprotonation as a first step.

To the ether electrophile it is a substitution process, but to the benzene nucleophile it is an electrophilic aromatic substitution that has two steps, electrophile addition and then rearomatization. The problem space for the overall process must incorporate both the substitution and rearomatization and also the fact that it happens twice (Fig. 10.8).

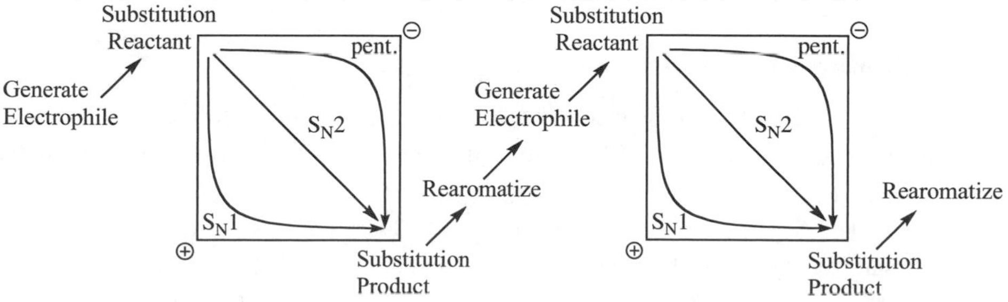
Substitution Reactant
Generate Electrophile
pent.
S_N2
S_N1
Rearomatize
Substitution Product
Substitution Reactant
Generate Electrophile
pent.
S_N2
S_N1
Rearomatize
Substitution Product

Figure 10.8 Problem space for electrophilic aromatic substitution example 10.5.4.

Find Possible Routes

There is only one reasonable path, proton transfer to the lone pair.

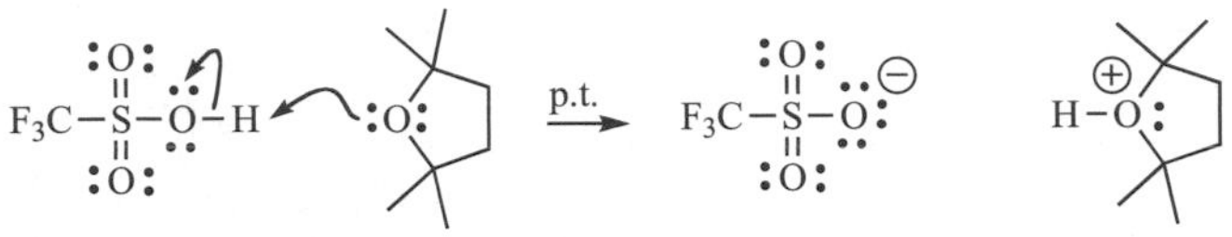
F3C–S–O–H
p.t.
F3C–S–O
H–O

Evaluate and Cross-Check

The reaction is downhill in energy if the acid catalyst has a pK_a of less than −3.5, the pK_a of the protonated ether. With the triflic acid pK_a of −15, this proton transfer is downhill by about 11.5 pK_a units. Another reaction that can happen is the reversible protonation of benzene, but all that will do is exchange the protons on the ring on return.

Understand the System

The protonated ether is now a good leaving group. A beginner might be tempted to overoptimize, to make a phenyl anion (pK_{abH} = 43) as a nucleophile and attack the electrophile, but anions that basic can't be formed in acidic media. Remember the pK_a span limit of 15 pK_a units. Powerful electrophiles and weak nucleophiles occur in acidic media, and the reverse is true in basic media. Our weak nucleophile is the aromatic ring.

Find Possible Routes

Of the three common routes on the substitution energy surface (Fig. 4.2), only two substitution routes are energetically possible because the site of attack is a carbon atom (the pentacoordinate path is out). The two alternatives are the S_N2,

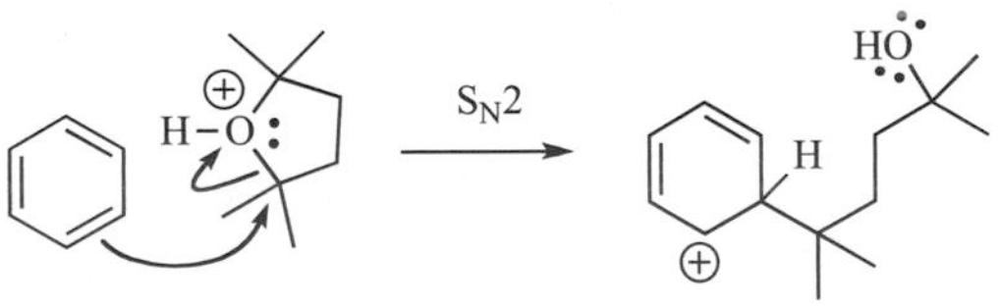
H–O
S_N2
HO
H

or the S_N1, path D_N followed by A_N (or from the point of the benzene, A_E).

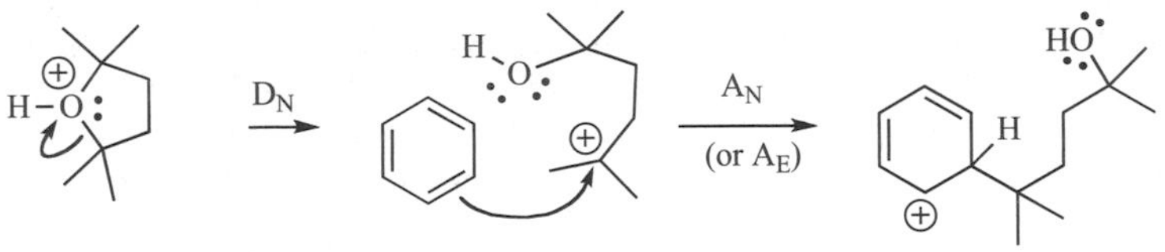
H–O
D_N
H
O
A_N
(or A_E)
HO
H

Evaluate and Cross-Check

The S_N2 has access problems, for the site of attack is very sterically hindered. The cutoff for a reasonable S_N2 is a secondary site, shown in Figure 4.13, so this tertiary site is just too hindered to proceed by S_N2.

The S_N1 must pass three criteria for the D_N step to proceed. The solvent needs to be polar, which this acidic medium is. The leaving group has to be very good, with a pK_{aHL} below zero; the neutral alcohol with a pK_{aHL} of −2.4 definitely qualifies. Finally, the carbocation must be reasonably stabilized, better than secondary, and this D_N would produce a tertiary carbocation, passing the last criterion. The S_N1 is very reasonable.

Understand the System

The aromatic ring is now a delocalized carbocation, but this sigma-complex is no longer aromatic because of the newly generated tetrahedral center at the point of attack. Usually the next step in an electrophilic aromatic substitution is to restore the stability of the aromatic system. A common beginner's mistake at this point is to remove the wrong H. This is especially true if line structure is used, which does not show the hydrogens. For this reason, always draw out the Lewis structure for any part of the molecule that interacts with an arrow. It is too easy to forget about hydrogens that are not shown.

Find Possible Routes

Restoring the aromaticity of the benzene ring requires a base to **remove the proton at the site of electrophilic attack.** The alcohol lone pair needs to get protonated to become a good leaving group if a second substitution is to occur. We can transfer this proton from the sigma complex to the alcohol either intermolecularly or intramolecularly. The intermolecular route is:

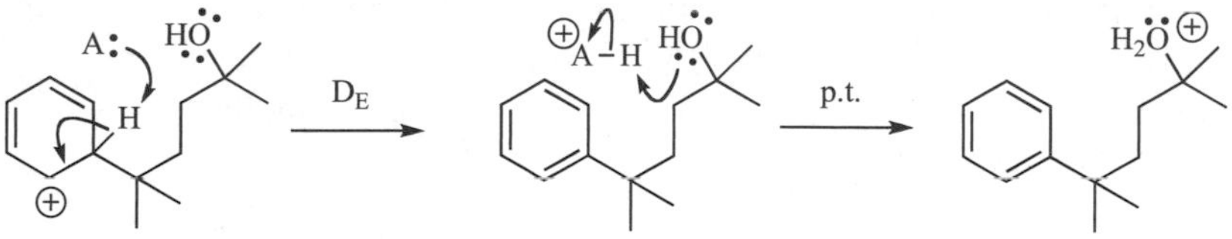
A:
HO
H
D_E
A–H
HO
p.t.
H_2O

or the intramolecular route is:

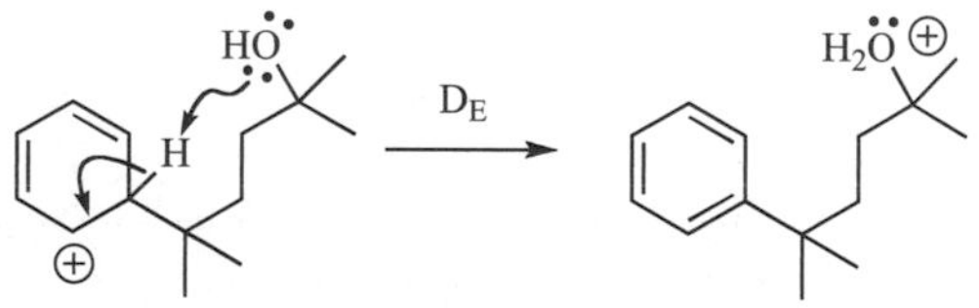
HO
H
D_E
H_2O

Evaluate and Cross-Check

Since the sigma-complex is also a strong acid, any neutral species in solution is basic enough to deprotonate it. Although both routes are reasonable, the alcohol lone pair is within reach (a 7-membered transition state) and reasonably basic for this medium, pK_{abH} of −2.4.

Find Possible Routes

Of the three common routes on the substitution energy surface (Fig. 4.2), only two substitution routes are energetically possible because the site of attack is a carbon atom (again, the pentacoordinate path is out). The two alternatives are the S_N2

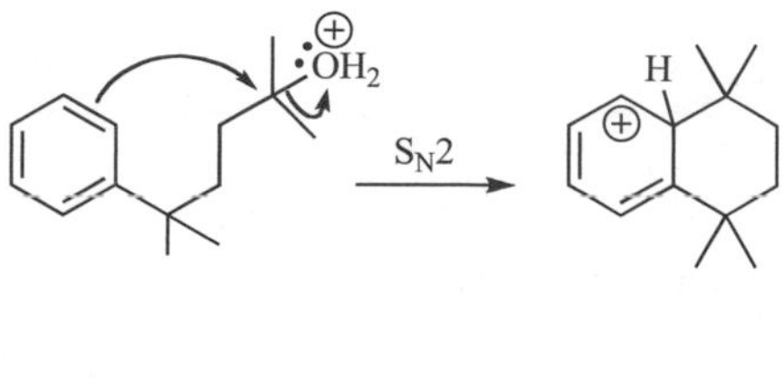
OH_2
S_N2
H

and the S_N1 (path D_N followed by A_N).

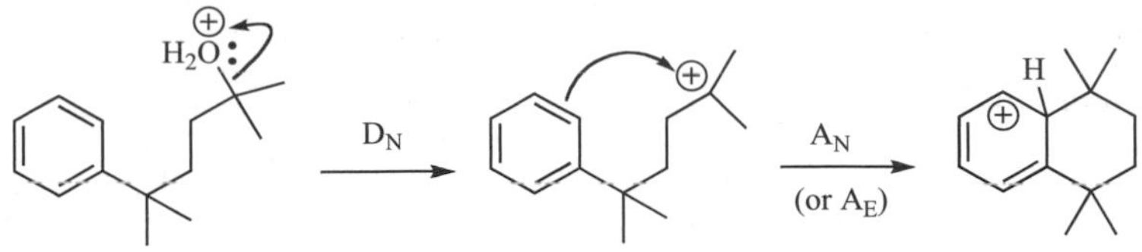
H2O
DN
AN
(or AE)
H

Evaluate and Cross-Check

Again, the S_N2 has access problems, for the site of attack is very sterically hindered, so this tertiary site is just too hindered to proceed by S_N2.

The S_N1 must pass three criteria for the D_N step to proceed. The solvent needs to be polar, which this acidic medium is. The leaving group has to be very good, with a pK_{aHL} below zero; the water leaving group with a pK_{aHL} of −1.7 definitely qualifies. Finally, the carbocation must be reasonably stabilized, better than secondary, and this D_N would produce a tertiary carbocation, passing the last criterion. The S_N1 is very reasonable.

Find Possible Routes

Restoring the aromaticity of the ring benzene ring gives the final product. The water we just created is basic enough, pK_{abH} of −1.7, to remove the proton at the site of electrophilic attack, thus finishing the last step of electrophilic aromatic substitution.

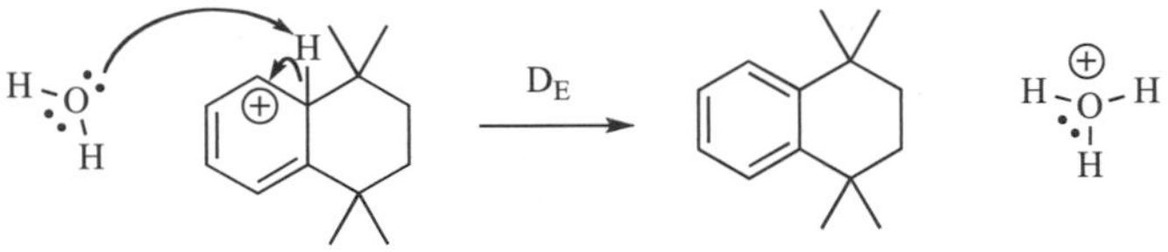
DE

Overview

Now let's back off from the problem far enough to see the entire process as a whole (Fig. 10.9). The mechanism for this transformation or "mechanistic sentence" can be made from our "mechanistic phrases" or path combinations: A proton transfer from the acid catalyst improves the leaving group, then an S_N1 substitution followed by a D_E rearomatization step that sets the alcohol up for a second S_N1 substitution, ending with a second D_E rearomatization step. The only alternative route is to use solvent to shuttle the proton from the sigma-complex to the alcohol.

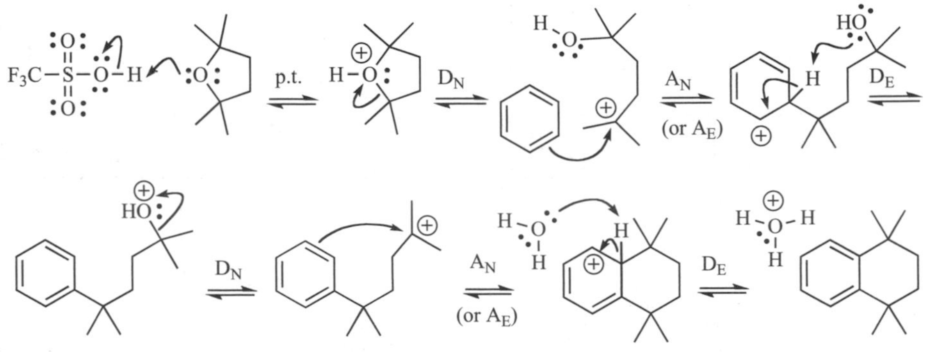
p.t.
DN
AN
(or AE)
DE
DN
AN
(or AE)
DE

Figure 10.9 Electrophilic aromatic substitution example overview.

10.5.5 Glucose to Fructose

Give a reasonable mechanism for the following biochemically important reaction.

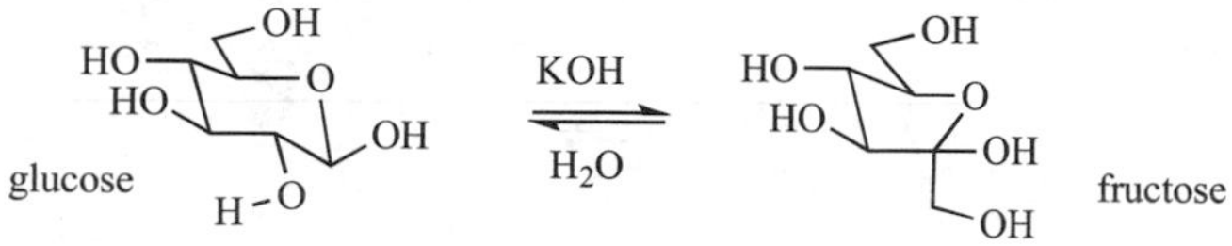
glucose
KOH
H_2O
fructose

Understand the System

Because the reactant and the product are structural isomers of one another, all pieces are accounted for in this balanced reaction. No overall addition of a nucleophile has occurred (there is nothing new attached), so we assume that hydroxide serves as a base. The carbons are now numbered to aid in the reaction analysis.

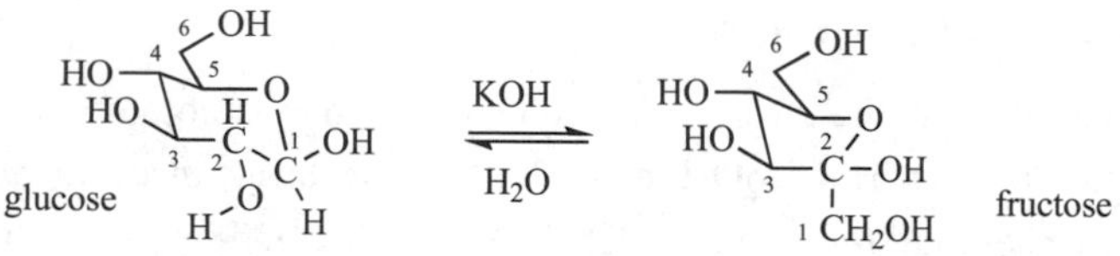
glucose
KOH
H_2O
fructose

It appears nothing has happened at carbons 3, 4, 5, and 6, so we leave those alone. The C–O bond to carbon 1 has been broken and a new C–O bond made to carbon 2. However, there is more; the line structure shorthand partially disguised that a C–H bond on carbon 2 has been broken and a new C–H bond made on carbon 1. The functional group at carbon 1 of the starting glucose is a hemiacetal, and the functional group at carbon 2 of the product fructose is a hemiketal.

The novice might be tempted to get this mechanism over with quickly by swapping C–O and C–H bonds between carbons 1 and 2. Even though it can be drawn with arrows, such a 1, 2 swap has never been observed in the real world. It is not one of the established electron flow pathways. We may not be able to visualize what a detailed problem space might look like, so we will have to trust where the known electron flow pathways might lead us.

Overall, we start with basic elimination reactions that will break our starting hemiacetal. We know at the end of the mechanism will be an addition reaction that forms the product hemiketal. What happens in between is uncertain at this point, but we have a rough idea of where to start and where to end on our problem space (Fig. 10.10).

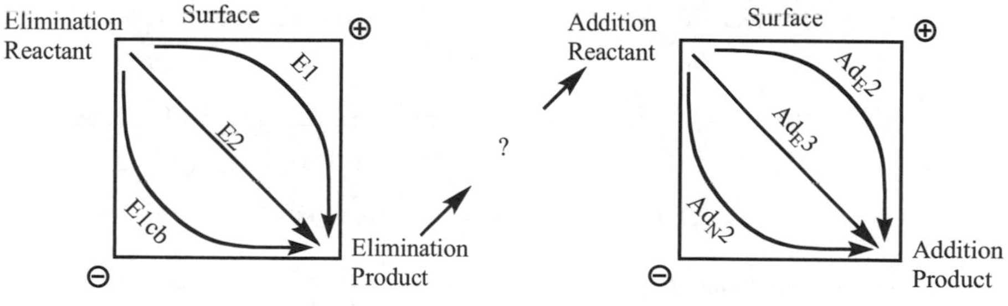
Elimination Reactant
Elimination Surface
E1
E2
E1cb
Elimination Product
?
Addition Reactant
Addition Surface
AdE2
AdE3
AdN2
Addition Product

Figure 10.10 Glucose to fructose partial problem space.

Find Possible Routes

First, try proton transfer. All the hydroxyls are within a few pK_a units of the hydroxide base, so all are available for proton transfer. Only carbon one bears both a hydroxyl and anything that can serve as a leaving group. All the other hydroxyls can only deprotonate, then reprotonate. Our two basic media elimination routes are the E1cB,

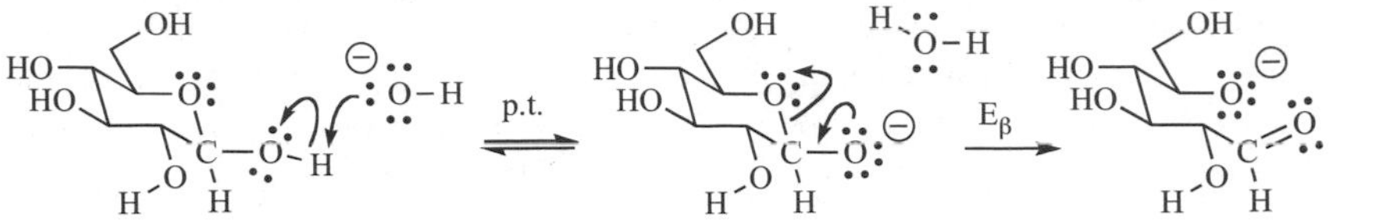
p.t.
Eβ

and the E2.

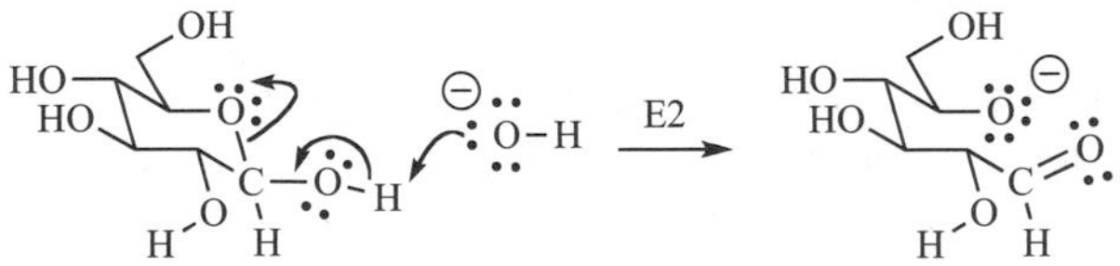
E2

Evaluate and Cross-Check

We ruled out the E1 because the leaving group is not good enough at a pK_{aHL} of 14 (the pK_a of $HOCH_2CH_2OH$ is 14.2), and the basic medium has no acids present to improve the L by protonation. The E1cB path combination (path p.t. followed by path E_β) accomplishes one goal of the transformation: It breaks the C–O bond to carbon 1. The product alkoxide is in equilibrium with the alcohol form in this medium. The literature value for the pK_a of glucose is 12.3. The alkoxide at this pK_{abH} has kicked out a leaving group at about pK_{aHL} 14, which checks with the ΔpK_a rule.

The E2 passes the same ΔpK_a rule test since the hydroxide base at pK_{abH} of 15.7 has kicked out a leaving group of pK_{abH} of 14, a weaker base. Since the OH can freely rotate, the E2 alignment restriction is not a problem. Both E1cb and E2 are reasonable.

Understand the System

The newly formed aldehyde is an electron-withdrawing group, making the proton on carbon 2 acidic; loss of this proton accomplishes the second goal of the transformation.

Find Possible Routes

Although hydrate formation by attack of hydroxide on the aldehyde is fast and reversible, it is also a dead end; a ΔH calculation shows us that the hydrate is uphill from the aldehyde. The alkoxide just kicked out could attack the aldehyde to return to starting material. The alkoxide is also a base and can remove the proton next to the aldehyde.

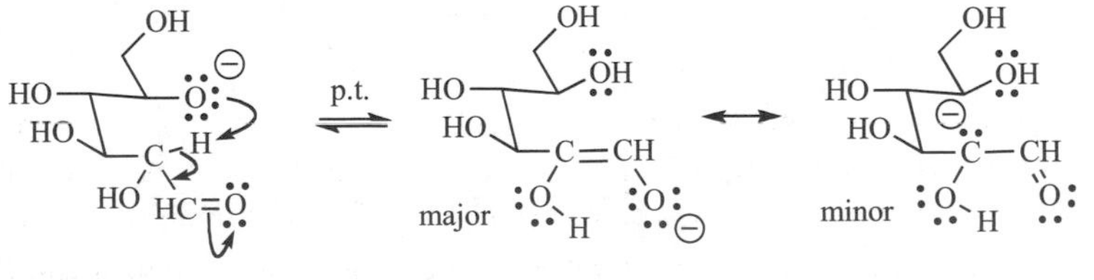
p.t.
major
minor

Evaluate and Cross-Check

The proton transfer step is the only unexplored new route. It is intramolecular and therefore expected to be fast. The alkoxide, whose pK_{abH} is 14, has removed a proton adjacent to an aldehyde at approximately the same pK_a, for a proton transfer K_{eq} of approximately 1. We could also have transferred this proton from carbon 2 to the alkoxide via the solvent, but this intramolecular proton transfer is expected to be faster.

Understand the System

The newly formed anionic intermediate is called an enediolate and has two resonance forms. It is internally hydrogen bonded. The enediolate is halfway along; two out of four goals have been achieved. Now another H is needed on carbon 1, and a C–O bond needs to be made to carbon 2. Because there is no electrophilic site at carbon 2, the C–O bond cannot yet be made.

Find Possible Routes

The internally hydrogen bonded enediolate intermediate undergoes rapid intramolecular proton transfer. This proton transfer could also be mediated by water.

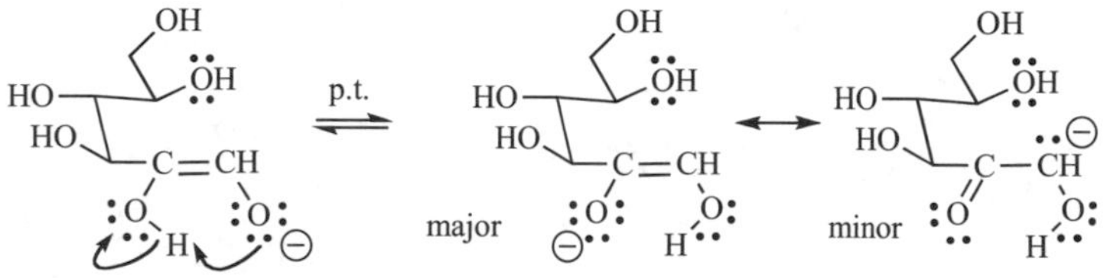

p.t.
major
minor

Evaluate and Cross-Check

Since the transition state is a five-membered ring, this intramolecular process is expected to be exceedingly fast. This proton transfer step is between two groups with the about same pK_a, for a proton-transfer K_{eq} close to 1.

Understand the System

The minor resonance form should provide a hint toward product. Proton transfer to this anion not only puts the H where it is needed on carbon 1 but also generates a site for nucleophilic attack on carbon 2.

Find Possible Routes

An intramolecular proton transfer can occur.

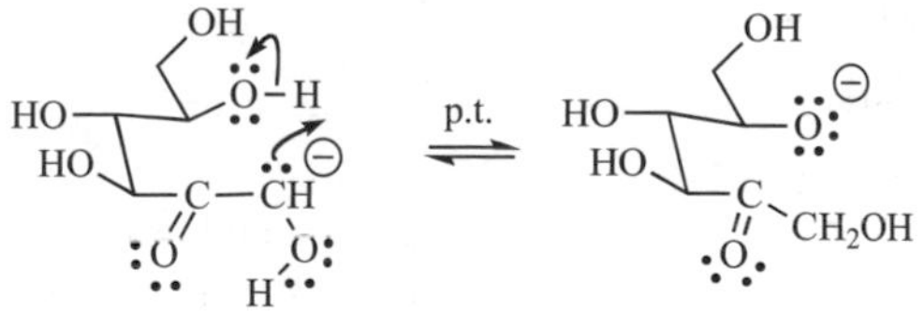

p.t.

Evaluate and Cross-Check

Since the transition state is a seven-membered ring, this intramolecular process is expected to be reasonably fast. This proton-transfer step is between two groups with about the same pK_a, for a proton-transfer K_{eq} of approximately 1. The water solvent can also mediate this proton-transfer step. The water that is hydrogen-bonded to the anion can be deprotonated, then in turn deprotonate the alcohol.

Understand the System

The carbonyl on carbon 2 is a new site for nucleophilic attack; the proton transfer has generated a good nucleophile.

Find Possible Routes

The C–O bond can now be made because we have a good nucleophile, a good sink, and a known route for the addition reaction, the Ad_N2,

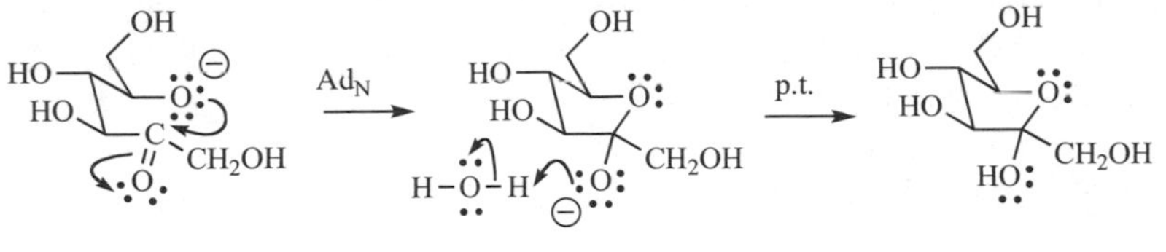

OH
HO
HO
C
CH2OH
AdN
H-O-H
p.t.

or the hetero AdE_3.

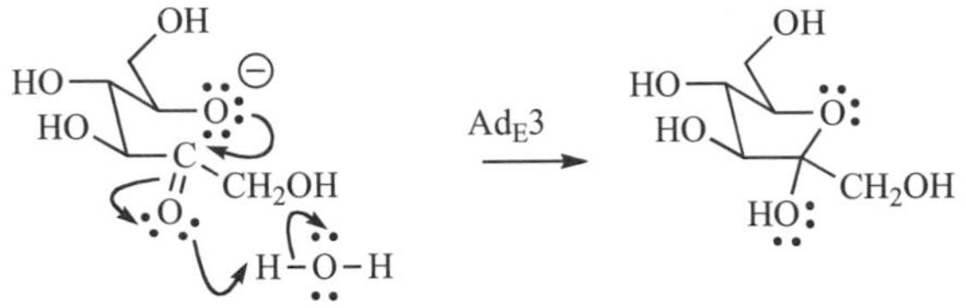

OH
HO
HO
CH2OH
H-O-H
AdE3

Evaluate and Cross-Check

The Ad_N2 addition of the alkoxide at pK_{abH} 14 has created a hemiketal alkoxide with a pK_{abH} at about the same value, which checks with the ΔpK_a rule. At this point we can see that protonation of the product yields fructose, with a reasonable K_{eq}. The alternate addition, the Ad_E3 path, passes the same ΔpK_a rule test, so it is also acceptable. The AdE2 is unlikely because there is not a strong enough acid present in base to protonate the carbonyl, pK_{abH} at –7.

Overview

Overall, the mechanism for this transformation was an E1cB (or E2) elimination, three proton transfers, and an Ad_N2 (or Ad_E3) addition (Fig. 10.11).

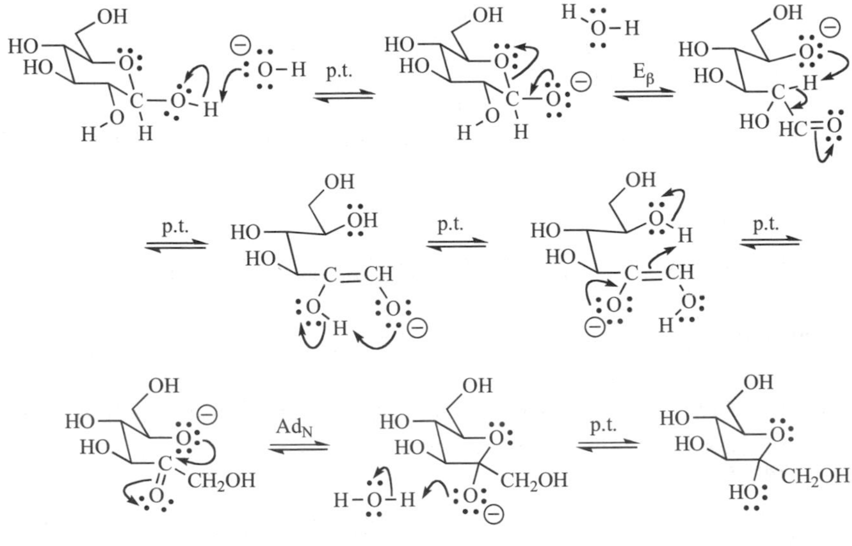

p.t.
Eβ
AdN

Figure 10.11 Glucose to fructose overview: E1cB top line, several p.t. steps, Ad_N2 bottom line.

10.6 PRODUCT PREDICTION STRATEGIES

Understand the System

Since structure determines reactivity, how are compounds in the mixture expected to behave? To understand this, we need to classify the reactants into generic sources and sinks, and then find the most reactive source and sink (Chapters 5 and 6). Being able to go from the specific to the general is an essential skill of a scientist. Ask yourself, "Have I seen anything like this?" Generally, how does a compound like this usually react? Often a first step is the reaction of an acid or base, so look for any reasonable proton transfer. Consider both the original reactants and the products of any proton transfer when you go to classify species into the generic classes of sources and sinks. Look for leaving groups, polarized single bonds, and polarized multiple bonds. Once we have the best source and sink, we can go looking for an electron flow pathway to combine them.

Find Possible Routes

Next, using the electron sink as a guide, find the few electron flow pathways that fit those sources and sinks (Chapter 7). Look at restrictions to limit the choice of pathways, such as acidic vs. basic media. Narrow down the choice of pathways further by using steric, electronic, and solvent limitations if available. Combine the best source with the best sink through an appropriate pathway to yield a preliminary product (Chapter 8). Only about half of all the possible sources and sinks are common, and those common combinations were outlined in the correlation matrix of Table 8.1 and discussed in the sections that followed. Be aware of alternatives like competing paths (Chapter 4) and decisions like substitution vs. elimination (Chapter 9).

Evaluate and Cross-Check

The difficult part is to decide whether or not the preliminary product is really the final product or just an intermediate. The ΔpK_a rule can be helpful to determine whether the process has climbed in energy or not. If you have formed an unstable anion or cation, of course it is just an intermediate. If there remains an easily lost leaving group or another reactive electron sink, look for a good source to react with it. You are following your energetics intuition downhill to a reasonable product. Basically look for any good process that can occur; if none are found, stop before you do anything rash.

Finally, have faith in the predictive process and allow it to carry you to the answer. Do not try to shorten the process to get to the answer by combining or skipping steps. Don't panic and force an answer because you can't see the answer from the beginning.

10.7 WORKED PRODUCT PREDICTION EXAMPLES

In these examples, if several routes go to the same compound, for simplicity only the more reasonable will be shown. Try to work each step before looking at the answer.

10.7.1 Predict the Product of Reaction of Ethyl Acetoacetate, Ethoxide, and 1-Bromobutane

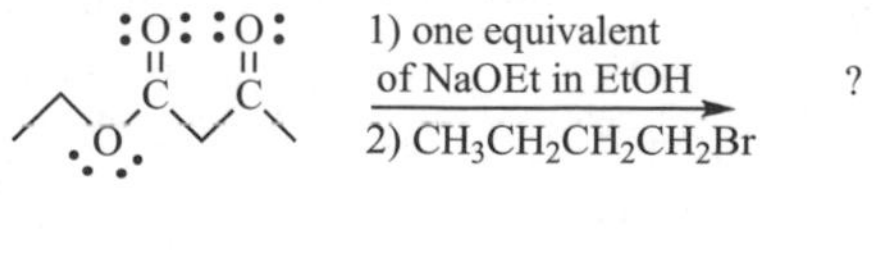
1) one equivalent of NaOEt in EtOH
2) CH3CH2CH2CH2Br
?

Understand the System

The first step is in basic medium; ethoxide has a pK_{abH} of 16. Sources? The ethoxide anion. Leaving groups? Only ethoxide on the ester. Sinks? The ketone is a polarized multiple bond, and the ester is a polarized multiple bond with a leaving group. Acidic Hs? The CH_2 between the two carbonyls the most acidic, pK_a 10.7; the methyl on the carbonyl is the next most acidic, pK_a 19. Resonance forms help us understand the polarization of our reactants.

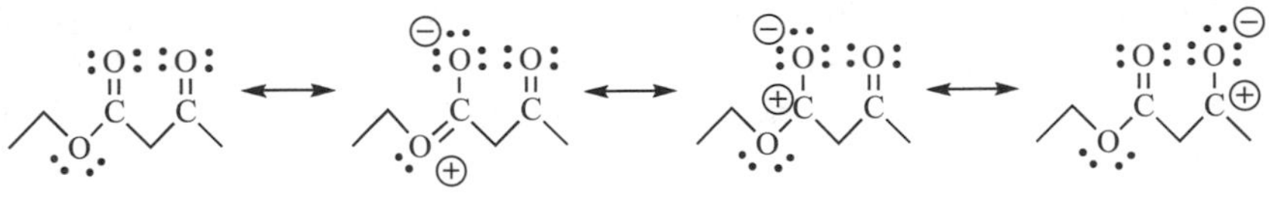

Find Possible Routes

There are two possible alternatives, ethoxide can serve as a nucleophile or as a base, and there are two possibilities with each, giving four routes to consider.
Ethoxide as nucleophile on the ketone, path Ad_N:

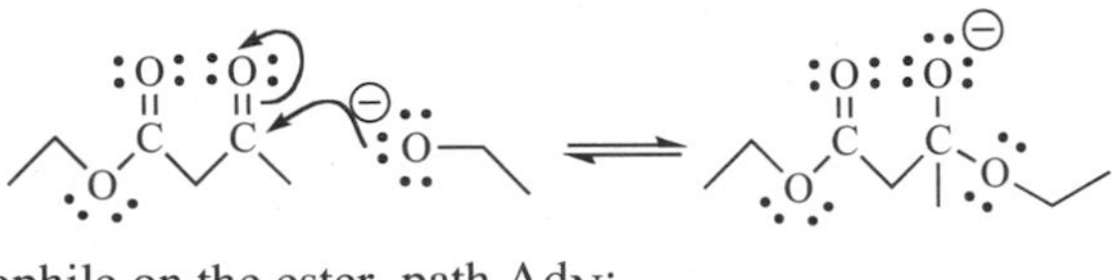

Ethoxide as nucleophile on the ester, path Ad_N:

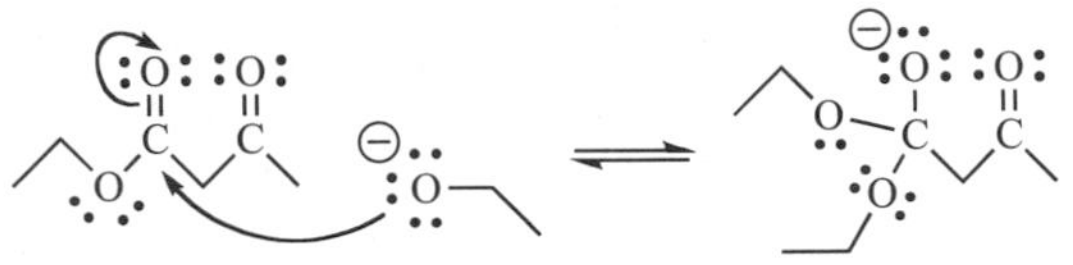

Ethoxide as base on the methyl, path p.t.:

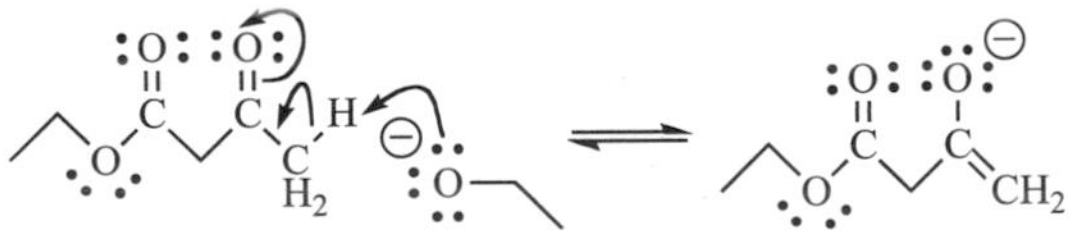
H
H2
CH2

Ethoxide as base on the methylene between the carbonyls, path p.t.:

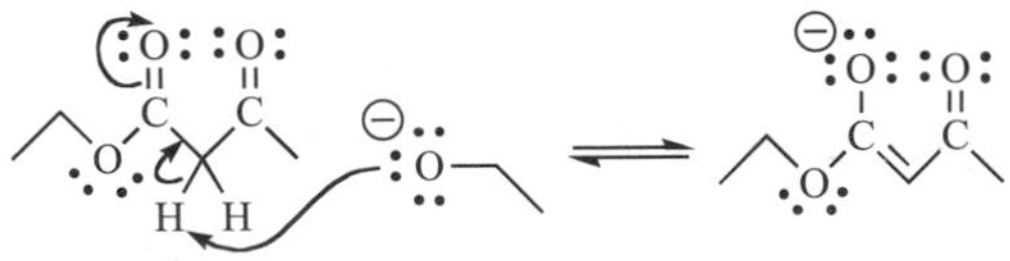
H H

Evaluate and Cross-Check

The reversible addition of the ethoxide to the ketone produces a hemiketal anion that is uphill in energy because a carbonyl is a stronger bond than two C–O single bonds. The reversible addition of the ethoxide to the ester produces a tetrahedral intermediate anion that, although uphill in energy, can kick out an ethoxide leaving group. If the ester were not an ethyl ester, this process would result in a noticeable exchange (transesterification). As the leaving group is the same as the nucleophile, the process proceeds without change.

Ethoxide as a base on the methyl reversibly produces a product anion of pK_{abH} 19.2, from a base of pK_{abH} 16, slightly uphill but within range. Ethoxide as a base on the methylene between the carbonyls produces a product anion of pK_{abH} 10.7, from a base of pK_{abH} 16, definitely energetically favorable (more favorable than deprotonation of the methyl by 8.5 pK_a units). Since this favorable proton transfer to make the ethyl acetoacetate enolate anion appears to be the best process, we carry this allylic source on as our first choice into step two.

Understand the System

Medium? Still basic. Sources? The ethyl acetoacetate enolate anion has a pK_{abH} of 10.7 and is an allylic source. Leaving groups? Bromide is a reactive sp^3 bound leaving group. Resonance forms help us understand the polarization of our reactants.

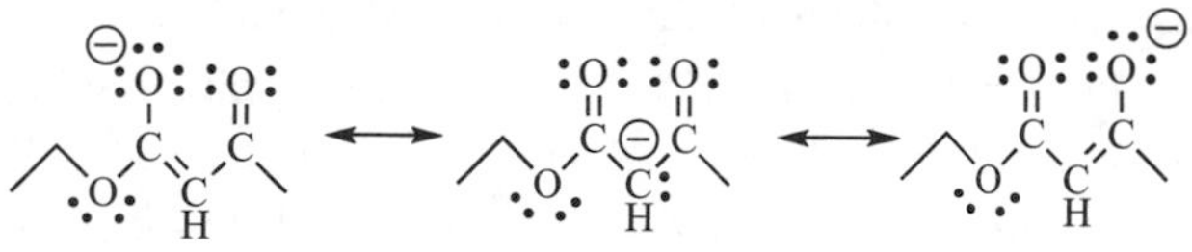

Find Possible Routes

In the interest of pruning the search tree, we can use what is known about the sources and sinks to limit the alternatives to the more reasonable choices.

Any of the three possible anionic atoms in the resonance forms above could serve as our source, but usually allylic sources attack on carbon because it is a softer, better nucleophile, and the product preserves the strong carbonyl double bond.

Our electron sink also presents two alternative routes, substitution or elimination. With a pK_{abH} of 10.7 and a soft carbon anion, this is a good nucleophile and a moderate base. Our 1-bromobutane presents a primary unhindered site so the decision is clearly in the substitution quadrant (Section 9.5). There are two possible substitutions, S_N1:

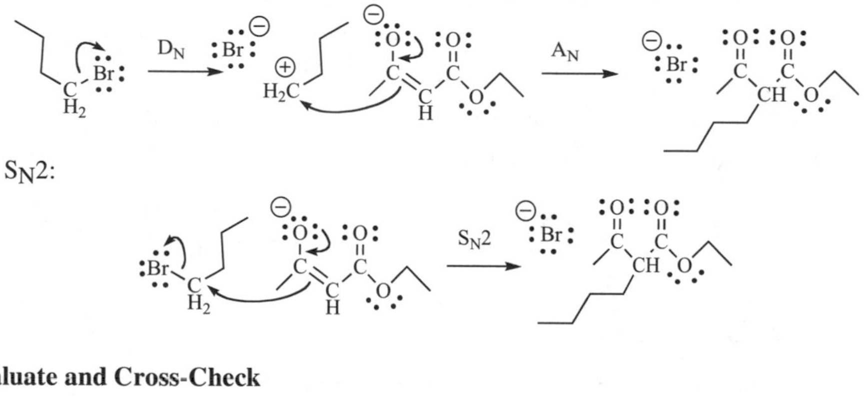

D_N
A_N

and S_N2:

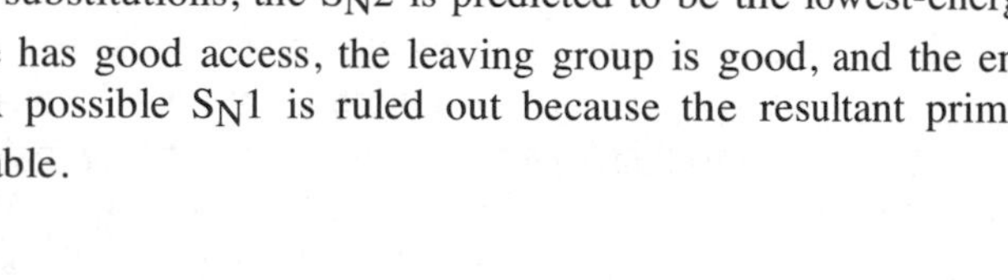

S_N2

Evaluate and Cross-Check

Of the two substitutions, the S_N2 is predicted to be the lowest-energy path because the primary site has good access, the leaving group is good, and the enolate is a good nucleophile. A possible S_N1 is ruled out because the resultant primary carbocation would not be stable.

Overview

We started with an anion of a pK_{abH} of 16 and now have a pK_{abH} of −9; the reaction is definitely downhill in energy and irreversible by the ΔpK_a rule. All the paths from this point are all uphill; it is a good time to stop.

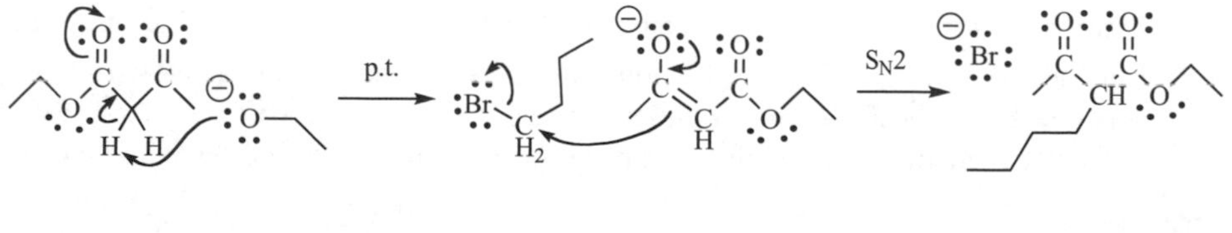

p.t.
S_N2

10.7.2 Predict the Product of Reaction of an Ester and Ethoxide

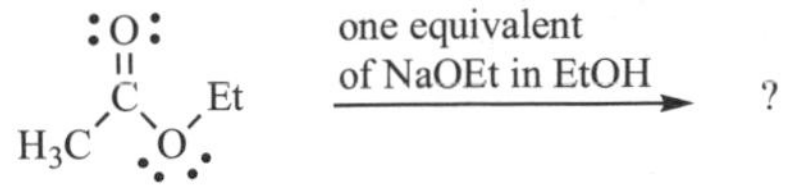
one equivalent
of NaOEt in EtOH
?

Understand the System

This reaction and its reverse are similar to important biochemical steps in the synthesis and metabolism of fatty acids. Medium? Basic, ethoxide's pK_{abH} is 16. Sources? Ethoxide. Sinks? The ester is a polarized multiple bond with a leaving group. Acidic Hs? The methyl next to the ester carbonyl has a pK_a of 25.6. Leaving groups? Ethoxide from the ester (poor leaving group). Resonance forms?

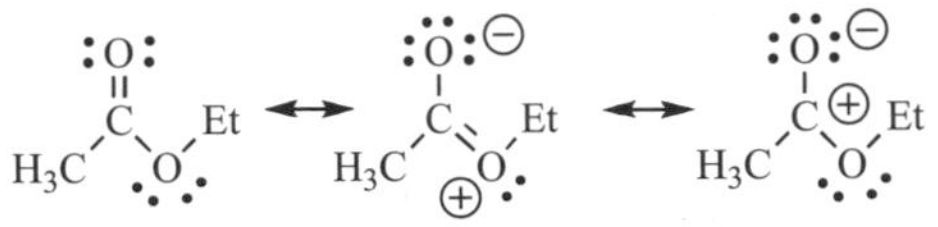

Find Possible Routes

Three routes are possible: Ad_N2 addition to a polarized multiple bond (Ad_N + p.t.),

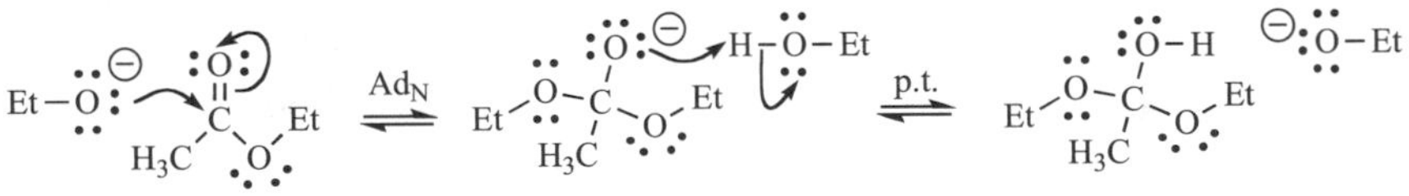

addition-elimination ($Ad_N + E_\beta$),

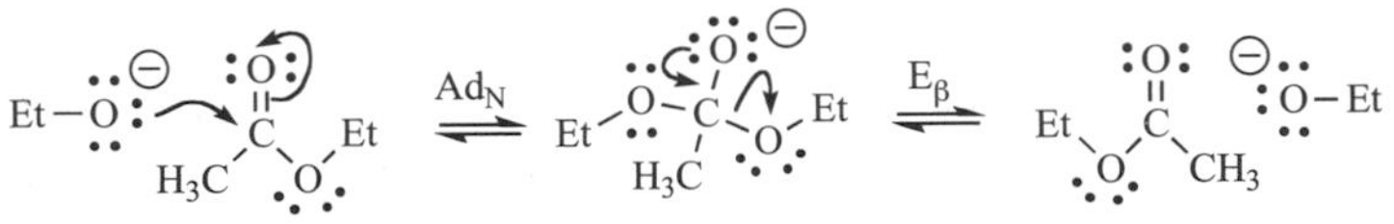

or deprotonation to form an anion.

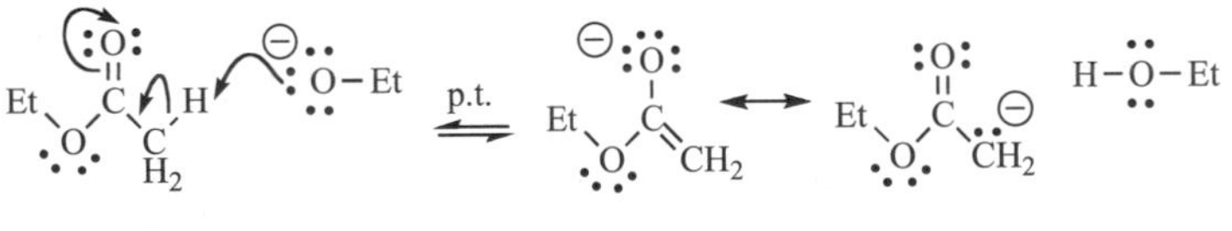

Evaluate and Cross-Check

Since the K_{eq} for proton transfer between the ester (pK_a = 25.6) and ethoxide (pK_{abH} = 16) is $10^{(16-25.6)} = 10^{-9.6}$ and therefore almost out of the useful range, it is reasonable to explore ethoxide as a nucleophile first.

The Ad_N2 addition creates a hemiorthoester that is just an ester version of a hemiketal. The tetrahedral intermediate is too hindered to serve as a nucleophile, and if it is protonated it yields a species that is uphill in energy from the starting ester. A ΔH calculation shows that a stronger carbonyl double bond (177 kcal/mol) was replaced with two C-O single bonds (2 × 86 = 172 kcal/mol), so this reaction is expected to be uphill by about 5 kcal/mol. Since this equilibrium will favor reactants, we should explore further.

Lone pair nucleophiles usually react with carboxyl derivatives by addition via path Ad_N, followed by beta elimination of the leaving group from the tetrahedral intermediate anion formed, path E_β. Because the leaving group is ethoxide, the same as the nucleophile, loss of either ethoxide returns us to the starting materials. This process is noticeable only if the nucleophile and leaving group are different (transesterification). The basic medium nucleophilic paths all return to the starting ester with this compound.

After exploring the more probable processes and ending up back at the start, we should now explore the energetically less favorable alternative, ethoxide as base, even though the K_{eq} is so unfavorable.

Understand the System

The ester enolate is an allylic source that can serve as a base or a nucleophile. The original ester has acidic hydrogens within range of the enolate. The ester carbonyl is an electrophilic sink (a polarized multiple bond with attached leaving group).

Find Possible Routes

Proton transfer easily occurs to the ester enolate on either carbon or oxygen. Protonation on carbon just exchanges hydrogens and returns us to reactants. The two new choices are that the enolate can act as a base to form the enol of the ester,

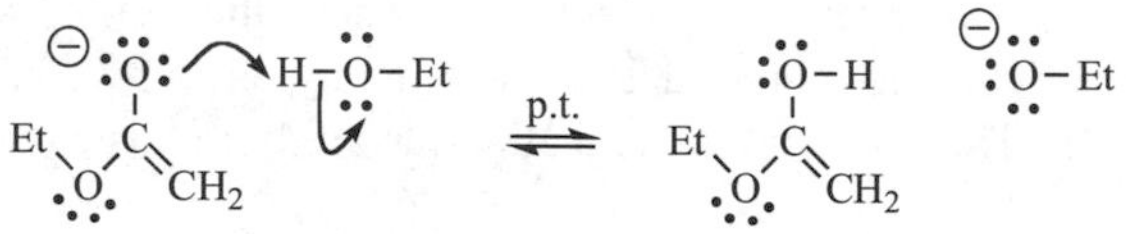

or the enolate can act as a nucleophile to form a new product by path Ad_N.

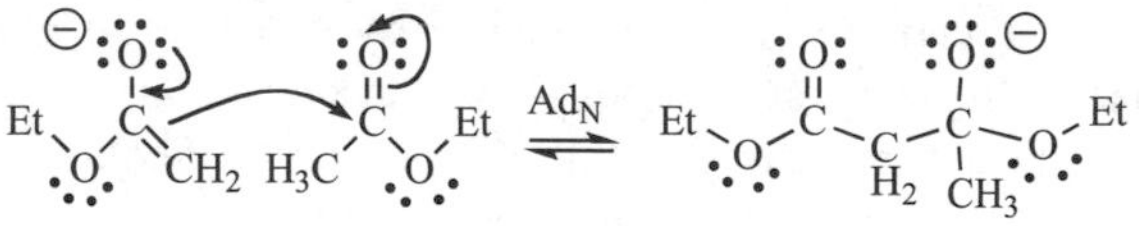

Evaluate and Cross-Check

The enol of the ester is uphill in energy from the ester since the C=C is much weaker than the C=O bond and is less basic and less nucleophilic than the enolate. The enolate Ad_N path did something new and is downhill because it forms a weaker base. The ester enolate pK_{abH} is 25.6; that of the tetrahedral intermediate is about 14.

Understand the System

The tetrahedral intermediate is basic (pK_{abH} = 14) and has two possible leaving groups, the enolate (pK_{aHL} = 25.6) and ethoxide (pK_{aHL} = 16), attached to the carbon adjacent to the oxyanion. Methyl (pK_{aHL} = 48) is not a leaving group.

Find Possible Routes

The tetrahedral intermediate can lose the enolate and return to reactants. The tetrahedral intermediate is too hindered to serve as a good nucleophile. Two unexplored paths are available. The tetrahedral intermediate can serve as a base (path p.t.),

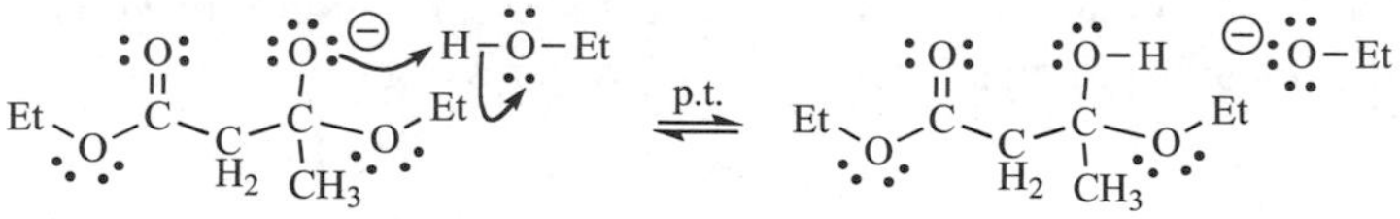

or it can lose the best leaving group it has, ethoxide (path E_β).

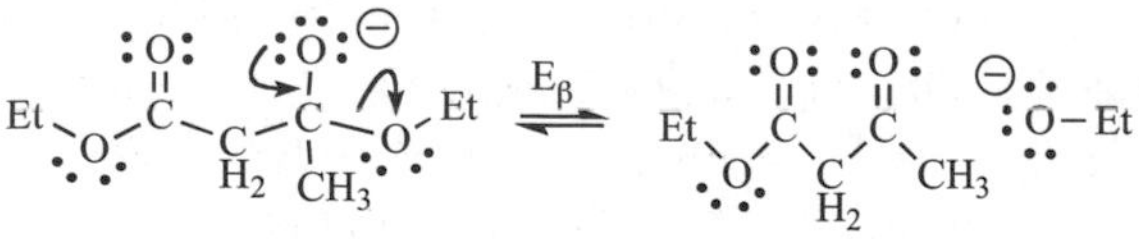

Evaluate and Cross-Check

Protonation gives the hemiketal, which by a ΔH calculation is energetically uphill. The hemiketal will simply return to reactants. The loss of the ester enolate would take us back to the previous structure. The leaving group trends tell us that ethoxide is the best leaving group and would be lost instead of the ester enolate. The loss of ethoxide has produced a slightly stronger base by about 2 units, well within the ΔpK_a rule.

A ΔH calculation is possible at this point since we now have a neutral species; two esters have formed the ketoester and ethanol. We have broken a C–H and a C–O and in return formed a C–C and an O–H for (99 + 86) – (83 + 111) = –9 kcal/mol (–38 kJ/mol). This tells us that the reaction has probably gone downhill in energy, but we need to explore further to see if there is anything farther downhill.

Understand the System

There are two sites for nucleophilic attack, the ketone and the ester carbonyl; the ketone carbonyl is the more reactive. The methylene between the two carbonyls is acidic ($pK_a = 10.7$). The ester enolate and ethoxide are both basic and nucleophilic.

Find Possible Routes

The ketone carbonyl can be attacked by ethoxide to give the previous tetrahedral intermediate. If the ketone carbonyl is attacked by the ester enolate, the intermediate formed is expected to reverse (multiple additions, Section 9.2). The ketoester can be deprotonated by either ethoxide or the ester enolate.

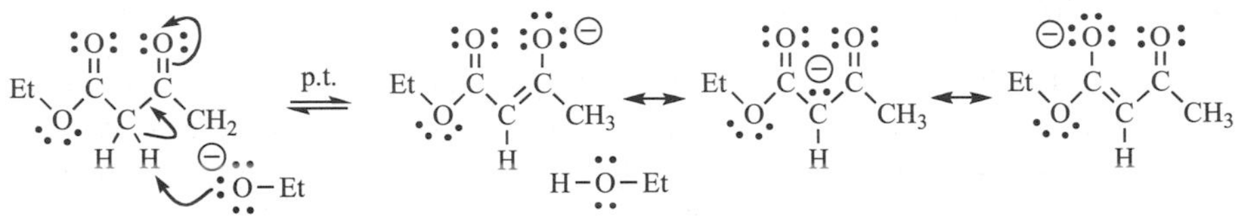
p.t.

Evaluate and Cross-Check

Deprotonation has a very favorable K_{eq} of $10^{+5.3}$ because it produces a stable anion (pK_{abH} is 10.7) stabilized by two electron-withdrawing groups. Because it is so stabilized, this anion is a weak base and is expected to be energetically downhill from the original starting reagents. If this highly delocalized anion acted as a nucleophile and attacked another ester, it would just preferentially fall off again because it is a better leaving group than ethoxide. Therefore an exploration of the surface has produced a stable anion downhill from the reagents that appears not to be able to do anything more under the reaction conditions. This stable anion is the product of the reaction. Mechanistic studies have shown that the deprotonation step to form this highly delocalized anion is the driving force for the reaction. In systems without an acidic hydrogen to deprotonate, this reaction easily reverses to reactants.

Overview

The overall reaction has been a deprotonation to produce the nucleophile for an addition–elimination reaction that is followed by a deprotonation to yield a resonance stabilized anion (Fig. 10.12). The reaction path is shown vertically on the left, and the side routes are shown to the right. This reaction is known as the Claisen condensation.

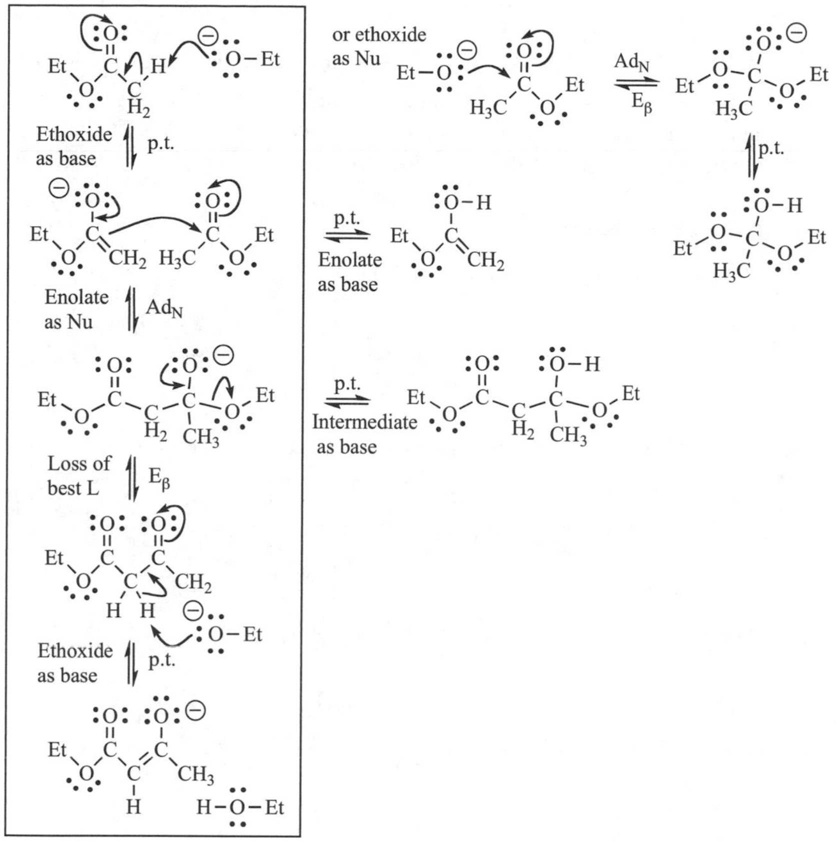
Ethoxide as base
p.t.
Enolate as Nu
AdN
Loss of best L
Eβ
Ethoxide as base
p.t.
or ethoxide as Nu
AdN
Eβ
p.t.
p.t.
Enolate as base
p.t.
Intermediate as base

Figure 10.12 Ester and ethoxide overview. The most likely reaction path is boxed.

10.7.3 Predict the Product of Reaction of a Ketal and Acidic Water

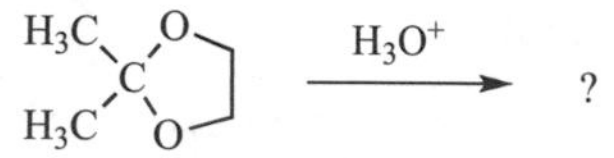
H_3O^+
?

Understand the system

In this example, a ketal is mixed with an excess of acidic water. Medium? Acidic water, pK_a −1.7. Sources? The lone pairs on the oxygens. Sinks? The acidic Hs of the acid. Leaving groups? No decent ones.

Find Possible Routes

The only path available at the start is proton transfer to the lone pair of the ketal.

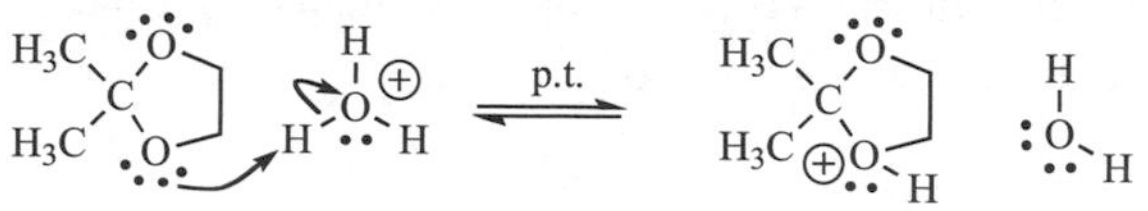
p.t.

Evaluate and Cross-Check

There is only one source and one sink and one path. The proton transfer has a K_{eq} of $10^{(-3.5 - (-1.7))} = 10^{-1.8}$, slightly uphill, although still a good process.

Understand the System

Protonation of the ketal oxygen makes it a good leaving group with a $pK_{aHL} = -2.4$. Since the reaction is in an acidic medium, the leaving group is good, and the delocalized tertiary carbocation formed upon loss of the leaving group would also be relatively stable, the reacting species can now enter the substitution/elimination surface (Fig. 9.2).

Find Possible Routes

The leaving group can ionize (path D_N).

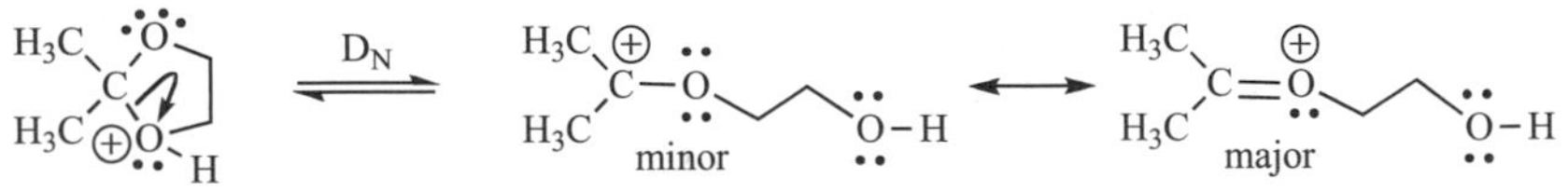

Either C–O bond to the protonated oxygen conceivably could break.

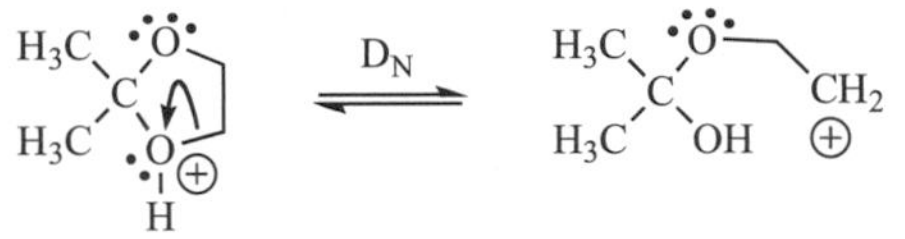

The same bonds could break by an S_N2 path with water as the nucleophile.

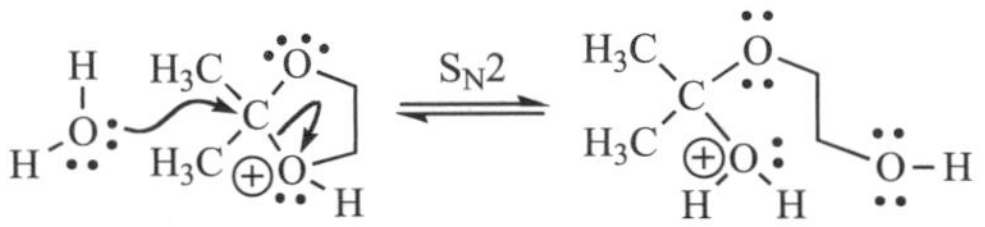

or

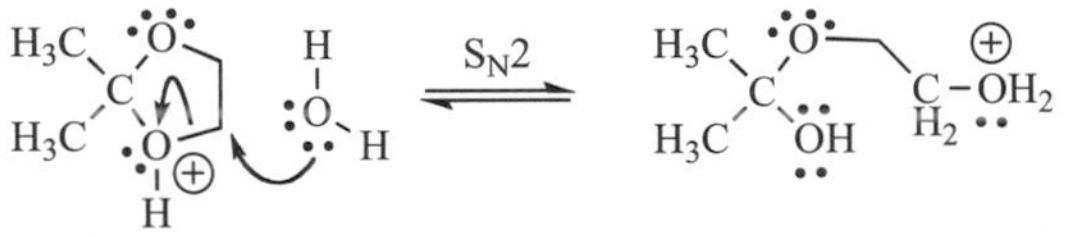

Evaluate and Cross-Check

The stability of both path D_N product carbocations must be checked: the lone-pair-stabilized tertiary cation is much more stable and therefore is favored over the unstabilized primary cation. Ionization of the leaving group (path D_N) creates a somewhat less stable cation than the protonated ketal, so ionization of the leaving group is uphill in energy.

The S_N2 path alternatives need to be explored next. An S_N2 attack on a tertiary center is too hindered to proceed and can be ruled out. The S_N2 attack on the CH_2 is reasonable and must be considered. In fact, it took a mechanistic study with ^{18}O-labeled water to settle this S_N1 vs. S_N2 question. The product with ^{18}O attached to the CH_2 group was not seen, ruling out the S_N2 mechanism. Therefore the best route is the D_N step to give the tertiary carbocation.

Understand the System

The lone-pair-stabilized tertiary cation is an excellent electron sink. The newly formed alcohol group and also the water solvent are both mildly basic and nucleophilic. However, nucleophilic attack on the carbocation by the alcohol group simply returns us to the previous intermediate, so let's go forward.

Find Possible Routes

This cation can lose an adjacent proton from the methyl (path D_E) to form the enol ether completing an E1 process,

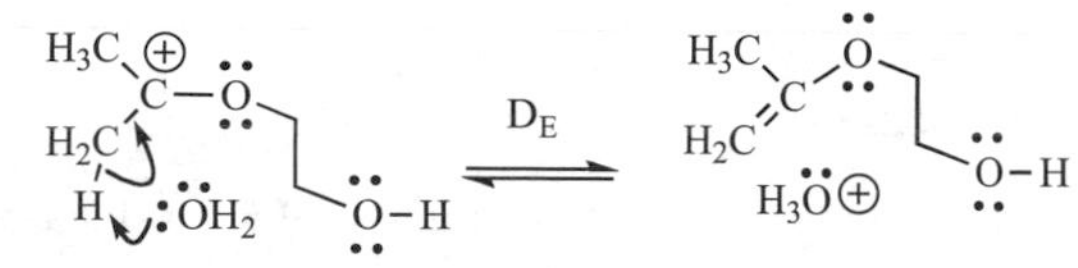

or the cation can get trapped by the most abundant nucleophile, water (path A_N), which completes an S_N1 process,

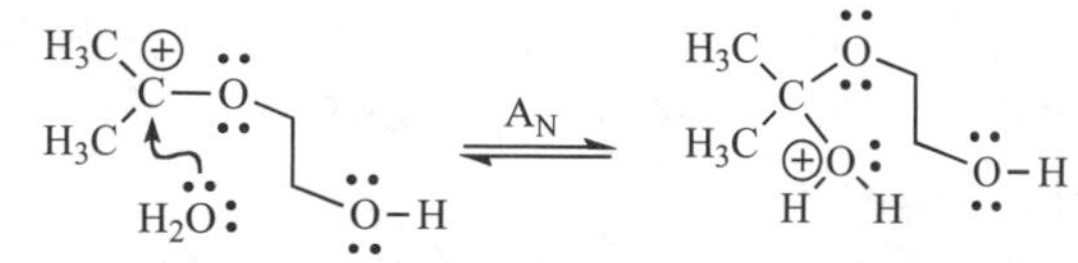

followed by proton loss to produce the hemiketal.

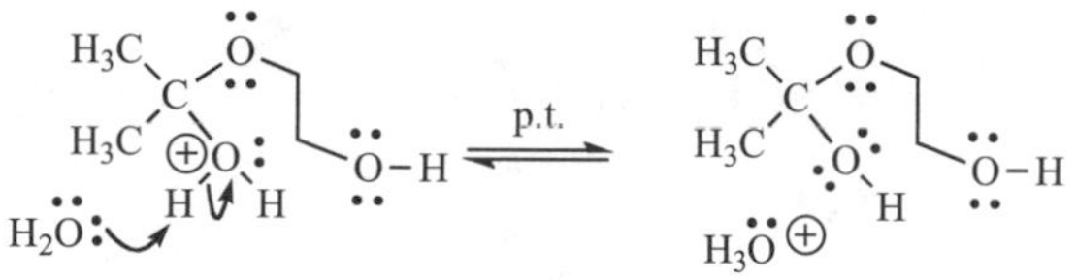

Evaluate and Cross-Check

Since both processes are reversible, the enol ether is uphill from the starting ketal [a ΔH calculation gives +11 kcal/mol (46 kJ/mol)], and there is an excess of nucleophile, S_N1 is favored. The proton transfer to water is favorable since each species has about the same pK_a.

Now we need to consider whether the hemiketal is downhill in energy from the ketal. A quick calculation of the ΔH of reaction yields zero, telling us that they are about the same in energy. It is best to keep exploring the energy surface to see if there is a lower-energy product.

Understand the System

The lone pairs on oxygen are basic, and there are no decent leaving groups. The media is still acidic.

Find Possible Routes

Protonation can occur on any lone pair of the three oxygen atoms. Protonation on the tertiary alcohol oxygen is fine, but it returns us to a previous structure. Protonation on the primary alcohol oxygen is also fine, but we know from the ^{18}O study that its methylene is not a site for attack. What is left is to protonate one of the ether oxygen's lone pairs.

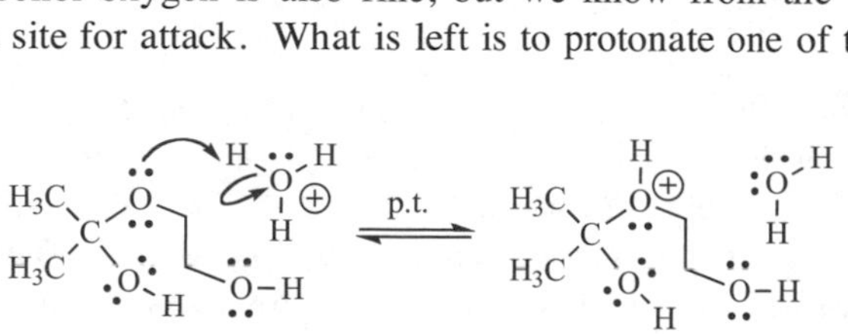

Evaluate and Cross-Check

Protonation on the slightly less basic ether lone pair converts it into a good leaving group. Similar to the first proton transfer step with the ketal, $K_{eq} = 10^{-1.8}$, the hemiketal is easily protonated in this acidic medium.

Understand the System

Protonation of the hemiketal oxygen makes it a good leaving group ($pK_{aHL} = -3.5$).

Find Possible Routes

Again, the reacting species enter the $S_N1/E1$ surface (the S_N2 was previously ruled out by the ^{18}O study). The protonated hemiketal can lose the leaving group in a process identical with that of the ketal (path D_N) to give the lone-pair-stabilized tertiary cation.

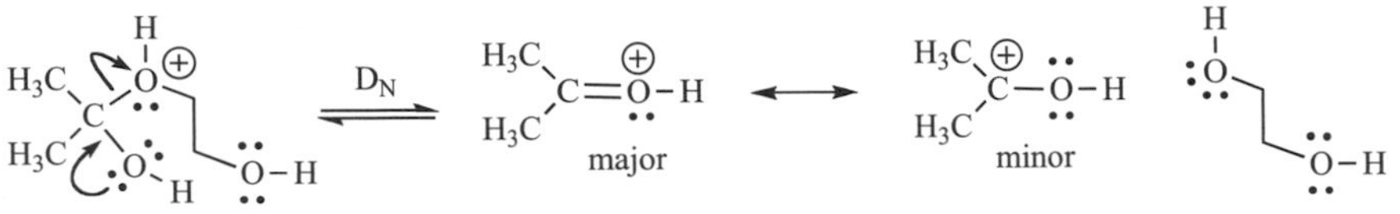

D_N
major
minor

The lone-pair-stabilized cation can lose a proton from carbon to form the enol (path D_E),

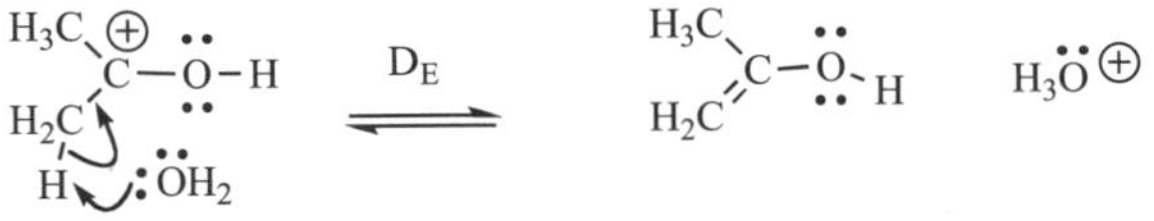

D_E

or can be trapped by water (path A_N) to yield, after deprotonation, the hydrate.

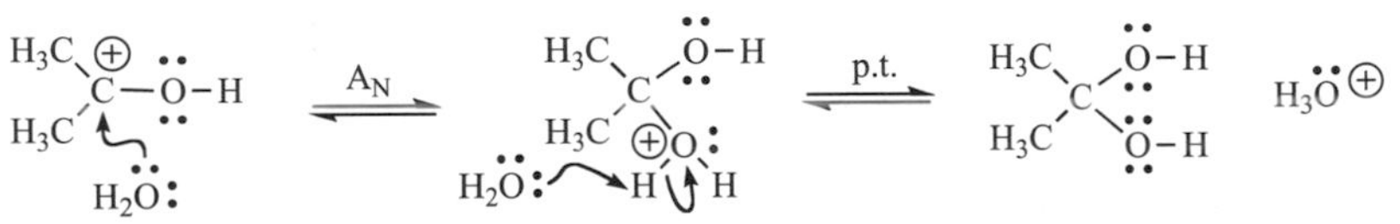

A_N
p.t.

But additionally it can lose a proton from the oxygen to form the ketone (path D_E).

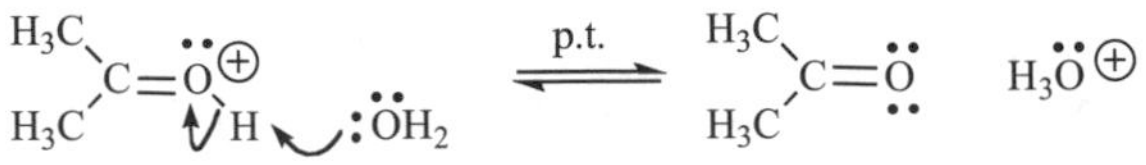

p.t.

Evaluate and Cross-Check

Since an equilibrium process favors the most stable product, we need to again calculate ΔH to find out which of the three products are preferred. The hydrate is at about the same energy as the ketal, and the enol is calculated to be 11 kcal/mol (46 kJ/mol) higher; the ketone is 5 kcal/mol (21 kJ/mol) lower in energy than the ketal and is therefore the product.

Overview

Occasionally a reaction will be driven to a product uphill in energy by displacement of the equilibrium, commonly an excess of one reagent combined with the removal of one product. Ketals are formed from ketones in this manner: An excess of alcohol is used, and the water formed is removed by distillation. The equilibria are the same as those discussed above, driven backward by mass balance effects. Watch reaction conditions for driven equilibria.

We have wandered over a rather complex energy surface (Fig. 10.13), exploring every possible side reaction until there was nothing more that was reasonable to do. Looking at the stability of all the possibilities on the surface, we arrived at the correct, most favored product. The reaction path is shown vertically on the left and the side routes are shown to the right side. The path name for the forward reaction only is shown.

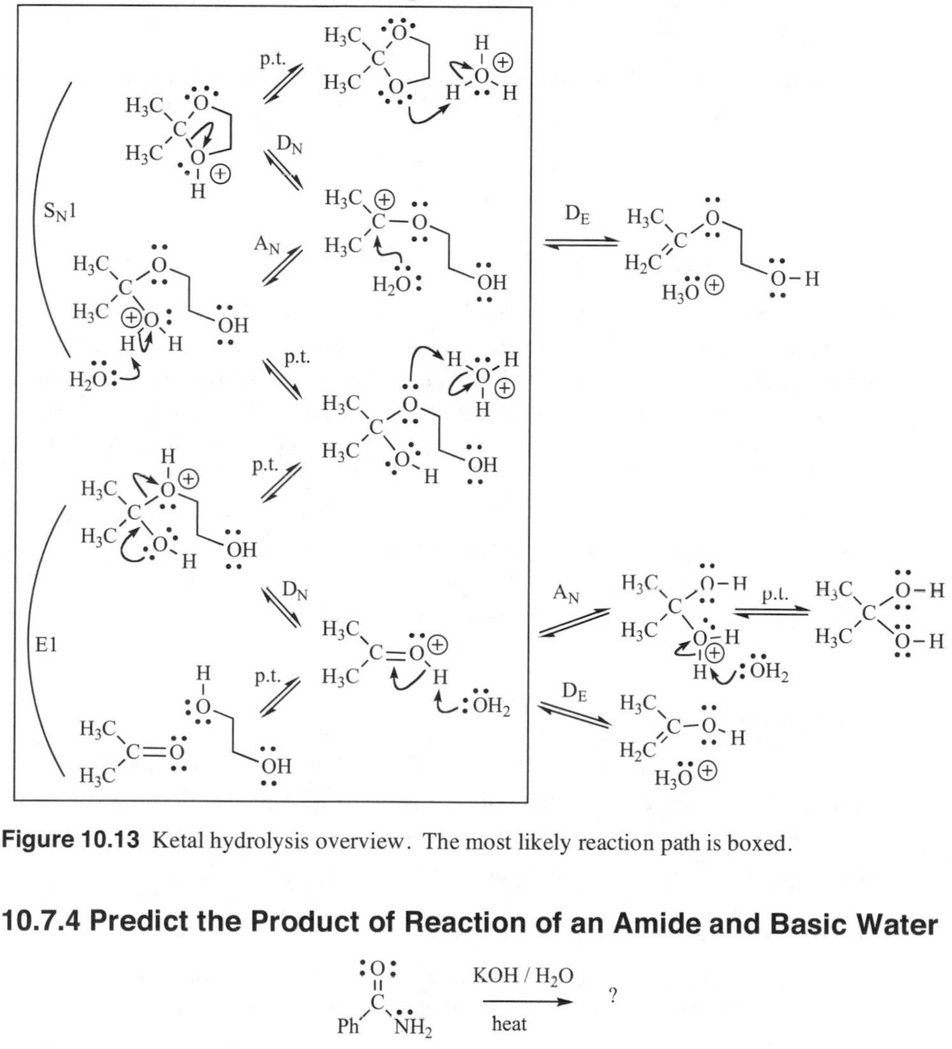
SN1
E1
p.t.
DN
AN
DE

Figure 10.13 Ketal hydrolysis overview. The most likely reaction path is boxed.

10.7.4 Predict the Product of Reaction of an Amide and Basic Water

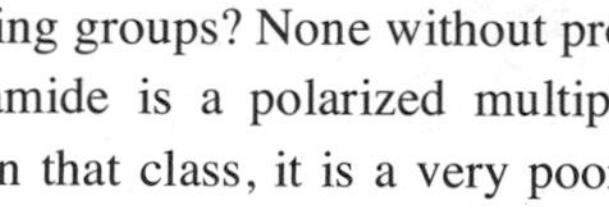
KOH / H2O
heat
?

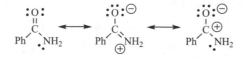

Understand the System

Biochemically similar reactions are responsible for the cleavage of the amide links of proteins into their constituent amino acids. Medium? Basic water. Sources? Hydroxide anion is the best source. Leaving groups? None without protonation to improve it (NH_2^- pK_{aHL} is 35). Sinks? The amide is a polarized multiple bond with leaving group. (Although the NH_2 places it in that class, it is a very poor leaving group.) Acidic Hs? The NH_2 of the amide is pK_a 17. Bases? Hydroxide, pK_{abH} = 15.7. Resonance forms?

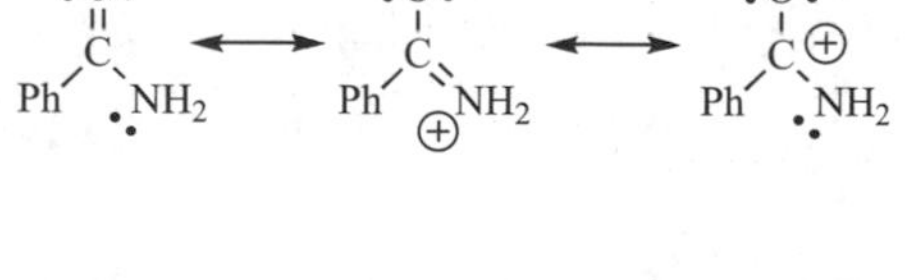

Find Possible Routes

There are two possible: We can use hydroxide as a base and remove an amide NH to form the amidate anion (path p.t.),

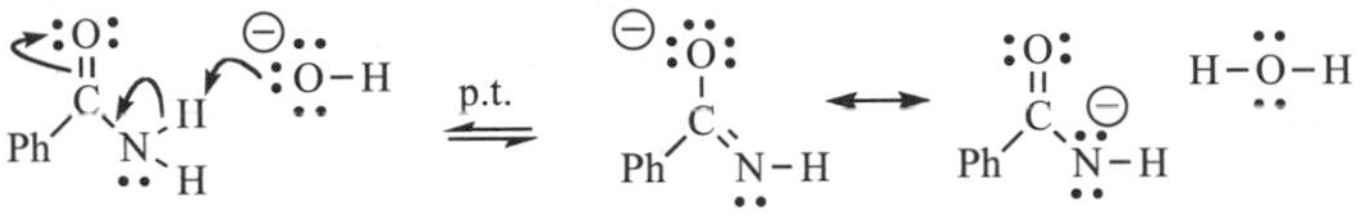

or use hydroxide as a nucleophile and attack the amide carbonyl (path Ad_N).

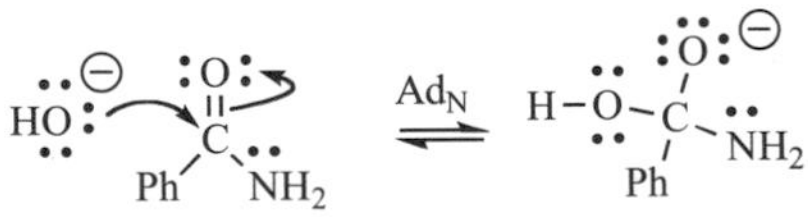

Evaluate and Cross-Check

The K_{eq} for proton transfer is $10^{-1.3}$, which slightly favors reactants. If the amidate anion were to act as a nucleophile, the only site for attack would be another amide, and then it would just be kicked right back out again since it would be the best leaving group. If the amidate acts as a base, it returns to the reactants. This short exploration down the amidate route reveals it as a dead end. Hydroxide as a nucleophile looks more promising.

The ΔpK_a rule is hard to use because we do not have a good value for the pK_a of the product of the addition of hydroxide. Steric hindrance from R raises the pK_{abH} of an alkoxide anion (poorer intermolecular solvation), whereas the inductive/field effect from the nitrogen should lower the pK_{abH}. If these effects balance out, the tetrahedral intermediate product of nucleophilic attack would have a pK_{abH} of about 13 ($HOCH_2OH$ is pK_a 13.3).

Understand the System

There is no leaving group other than the original nucleophile (kicking out NH_2^- with a $pK_{aHL} = 35$ from a tetrahedral intermediate 22 pK_a units less basic is **not** an option), so we should examine the possibilities for proton transfer. Hydroxide is a good base. In basic water, the most common proton source is neutral water.

Find Possible Routes

Proton transfer can occur from water to the alkoxide anion or to the nitrogen lone pair or to both. There are four common charge types of the tetrahedral intermediate.

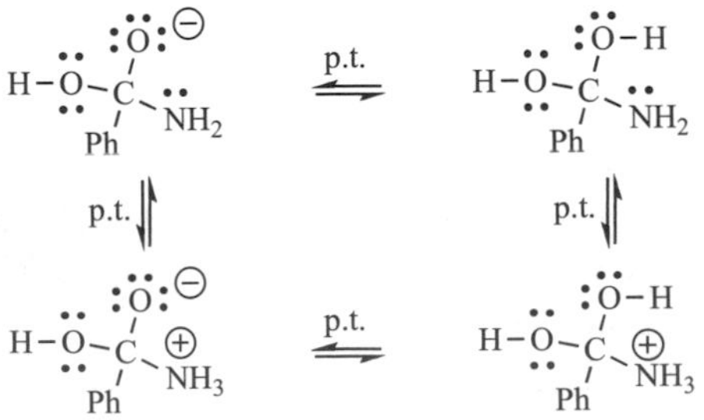

Evaluate and Cross-Check

We are again hampered by the need to estimate pK_as for these four charge types. The OH in the upper right structure was estimated at a pK_a of about 13. The NH_3^+ in the lower right structure should have a pK_a of about 7 owing to the inductive/field effect of two hydroxyls (a simple protonated amine has a pK_a of 11). Either OH in the lower right structure should also be acidic, pK_a estimated at 10, because of the inductive/field effect of NH_3^+. All these proton transfer reactions are reasonable.

If the medium were basic enough, we might expect the OH in the upper left structure to be deprotonated to give a dianion. It is difficult to gauge the reasonableness of the dianion since its pK_{abH} is difficult to estimate. Because of charge repulsion, it certainly has a pK_{abH} many units higher than 13, the estimated pK_{abH} of the upper left structure.

We seem to be at a proton transfer plateau. All four species are in equilibrium with each other. The top two species can do nothing except revert to starting materials, so in the absence of an obvious lower-energy route, we must explore paths from both of the bottom two species.

Understand the System

The NH_3^+ can serve as a fair leaving group (pK_{aHL} 9.2). **Note well: The unprotonated amine, pK_{aHL} 35, is not a leaving group at all.**

Find Possible Routes

Two elimination paths are possible: beta elimination from an anion (path E_β),

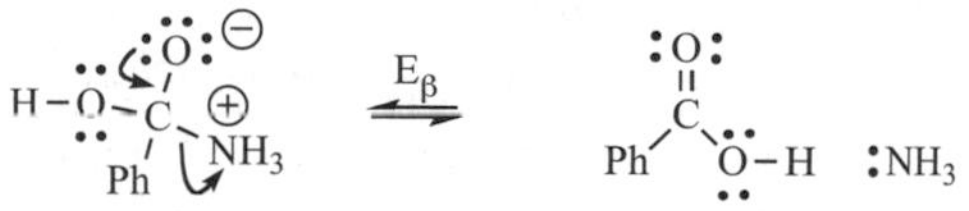

or E2 elimination.

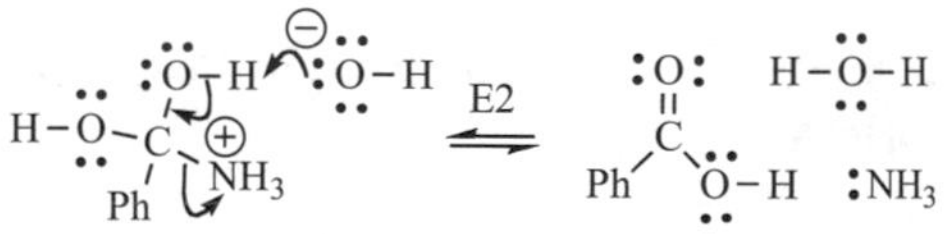

Evaluate and Cross-Check

Both are viable routes, and thankfully they both go to the same product. The E2 forms a weaker base of pK_{abH} 9.2 from a starting anion of pK_{abH} 15.7. The beta elimination forms a slightly weaker base and might be the preferred route for elimination of a fair leaving group.

Find Possible Routes

We can spot an irreversible proton transfer step, $K_{eq} = 10^{+10.9}$, that creates an unreactive carboxylate anion. With this proton transfer we are done.

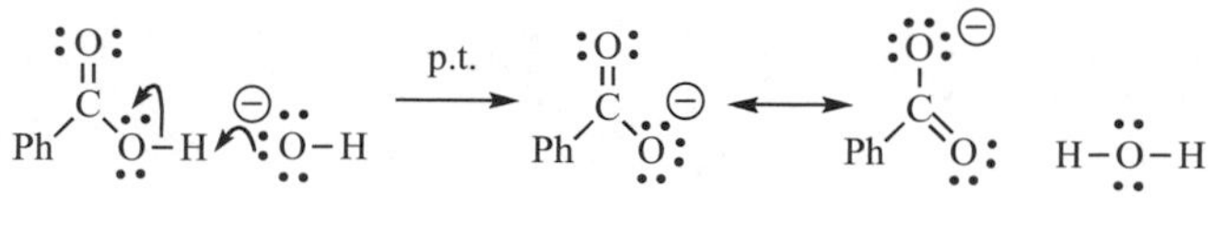

Overview

The overall reaction path is shown vertically, and the alternate paths are to the side (Fig. 10.14). The path name for the forward reaction only is shown.

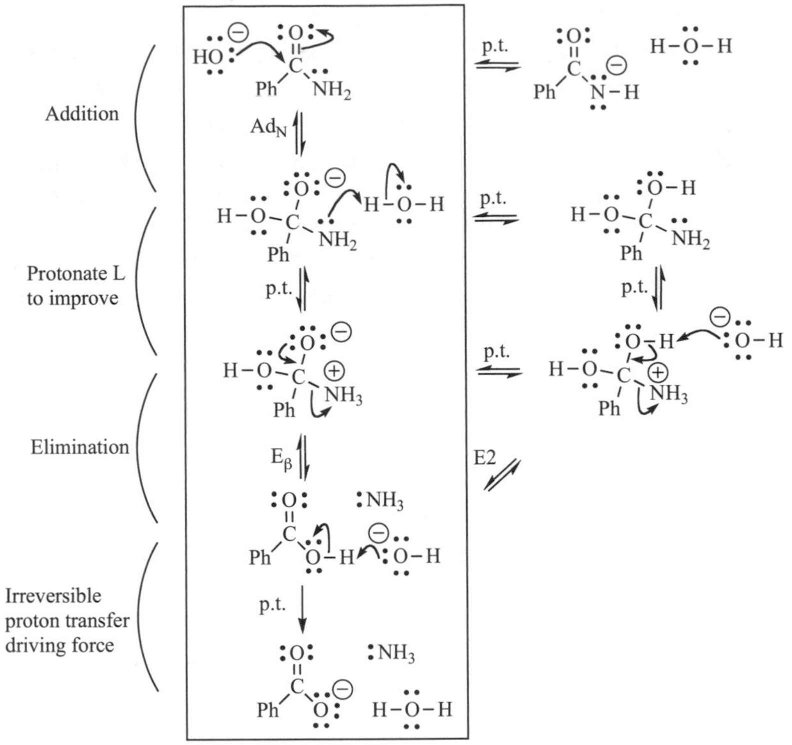
Addition
Protonate L
to improve
Elimination
Irreversible
proton transfer
driving force
p.t.
E2

Figure 10.14 Amide hydrolysis overview. The most likely reaction path is boxed. Additional paths that invoke extent of proton transfer variations like general acid and general base catalysis are also possible (Section 7.4.3).

10.7.5 Predict the Product of the Reaction of a Carboxylic Acid and Thionyl Chloride

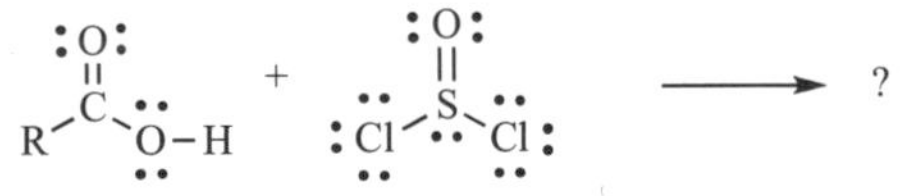

Understand the System

Medium? Acidic. Sources? The lone pairs on the carbonyl are the best source (much better than the lone pairs of the OH). Leaving groups? Chloride. Sinks? The best, $SOCl_2$, is a Y–L. The carboxylic acid is both an acid and a carboxyl derivative sink, but the OH is a poor leaving group. Acidic Hs? The carboxylic acid's OH. Bases? None. Resonance forms? By VSEPR $SOCl_2$ is tetrahedral, often drawn with an expanded octet resonance form containing a *d-p* pi bond.

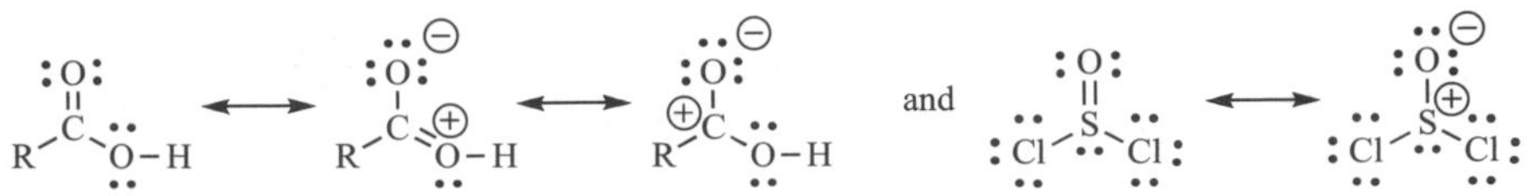

Find Possible Routes

We combine the best source (carbonyl lone pair) and the best sink (S–Cl weak bond) with the best path. The most probable path for a Y–L is the S_N2 substitution. It is currently debated whether sulfur goes through a pentacoordinate intermediate (path pent.); however, both routes yield the same product.

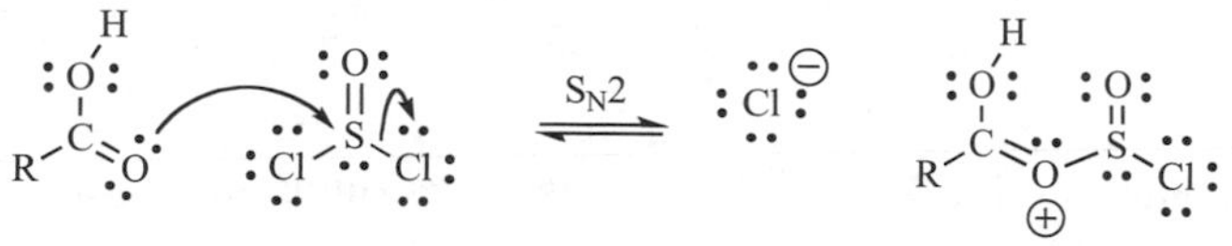

Evaluate and Cross-Check

If we had used the carboxylic acid's OH lone pair, the reaction would have produced a less stable cation, a positive oxygen between two electron-withdrawing groups, the carbonyl and the sulfoxide. The nucleophile's pK_{abH} is –6 and the leaving group's is –7. The attack has broken a weak S–Cl bond. The reaction is most likely downhill in energy.

Understand the System

The carbocation formed is tertiary and substituted by two lone pair donors. Resonance forms:

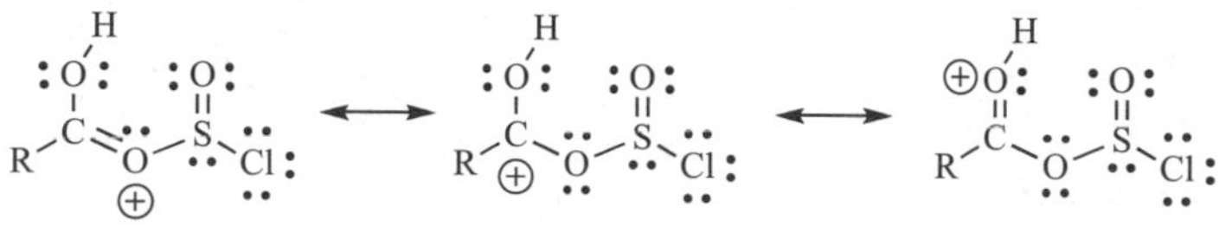

The carbocation is an excellent acid and a good site for nucleophilic attack.

Find Possible Routes

The two alternatives are deprotonation (path p.t.),

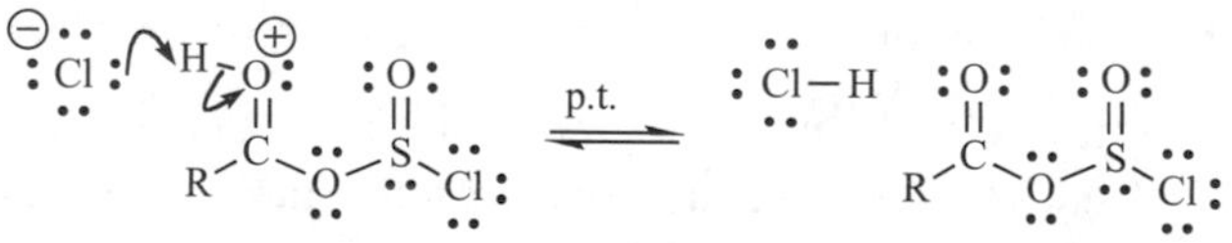

or nucleophilic attack (path Ad_N). The best nucleophile is now chloride, which can attack the protonated carbonyl to give a tetrahedral intermediate.

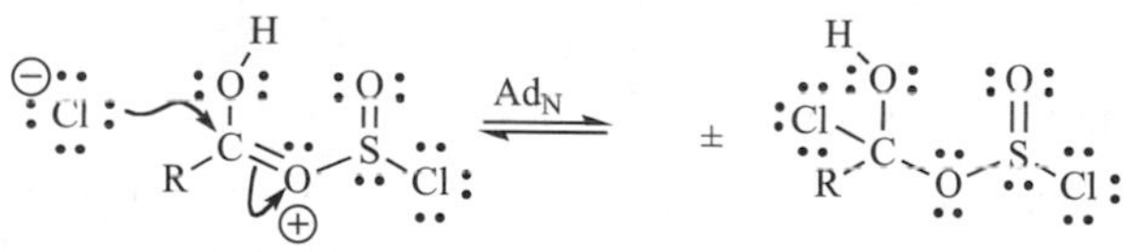

Evaluate and Cross-Check

Both processes should be favorable. The proton transfer occurs between two groups that have about the same pK_a to yield the acyl chlorosulfite. The nucleophilic attack to give the tetrahedral intermediate that has neutralized charge also should be favorable. Both processes are consistent with a very acidic medium. We should explore both routes.

Understand the System

Let's explore the product of the proton transfer route first. The reaction is generating HCl, pK_a –7, and the medium is becoming strongly acidic. The resonance forms for the acyl chlorosulfite are similar to those of the starting carboxylic acid.

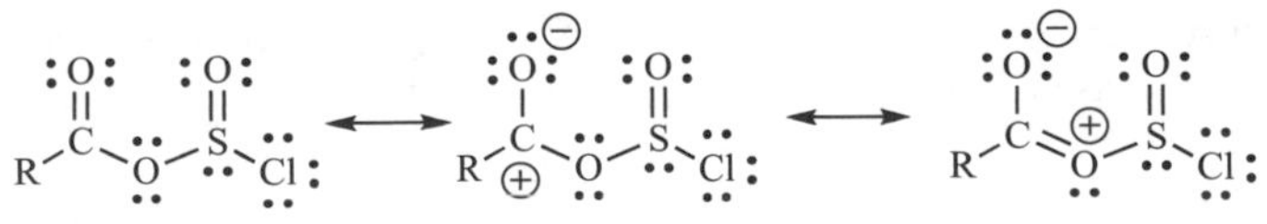

The acyl chlorosulfite carbonyl is a good site for nucleophilic attack. The best nucleophile is still chloride. Chloride attack on sulfur just returns to the starting point.

Find Possible Routes

The three alternatives are the Ad_N2 addition (path Ad_N then path p.t.),

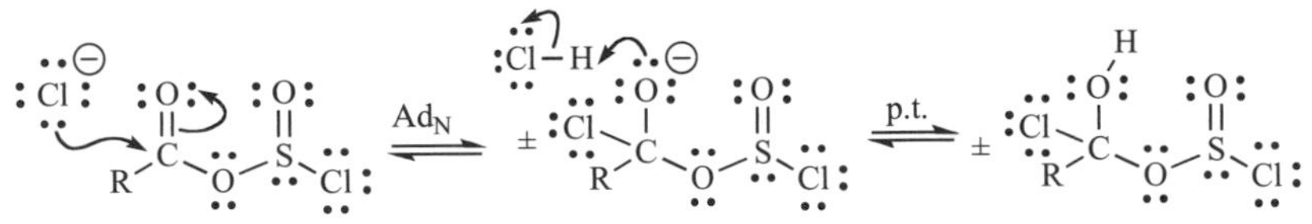

or the hetero Ad_E3 addition,

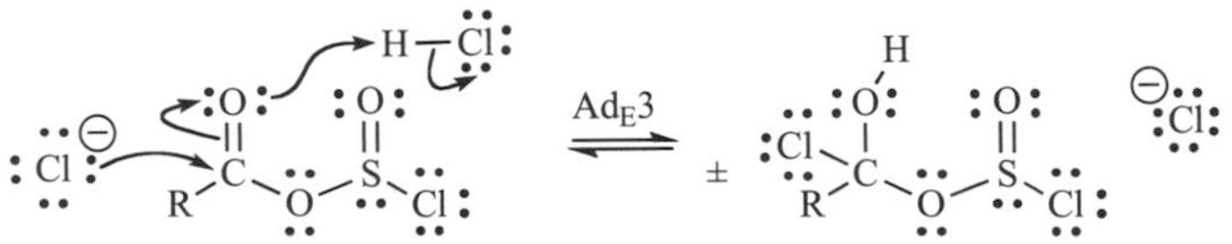

or the hetero Ad_E2 (path p.t. followed by path Ad_N).

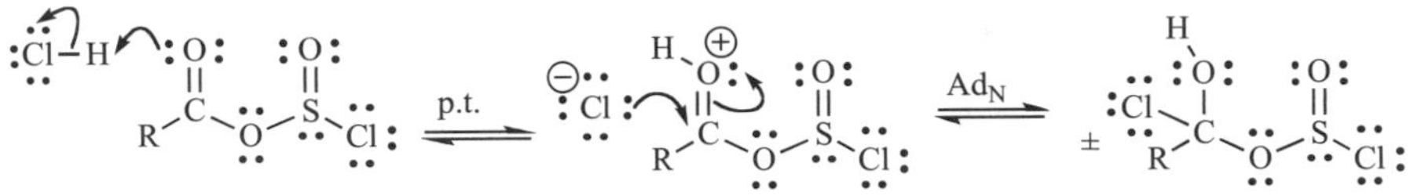

Evaluate and Cross-Check

The Ad_N2 is probably not a favored route because the nucleophilic attack of chloride, pK_{abH} of −7, to give the anionic tetrahedral intermediate, pK_{abH} about 10, goes against the ΔpK_a rule (a climb of 17 pK_a units) and forms a basic anion in a very acidic medium. The Ad_E3 and the hetero Ad_E2 are both reasonable routes by the ΔpK_a rule and are consistent with very acidic media.

Understand the System

We have come full circle back to the original tetrahedral intermediate. The tetrahedral intermediate has two attached leaving groups, the chloride and chlorosulfite. Loss of chloride reverts to reactants, but loss of chlorosulfite yields new compounds.

Find Possible Routes

There are four possible elimination routes that produce the unstable chlorosulfite anion, which immediately loses the good leaving group chloride to give sulfur dioxide gas. Chlorosulfite can be lost in an E2 elimination,

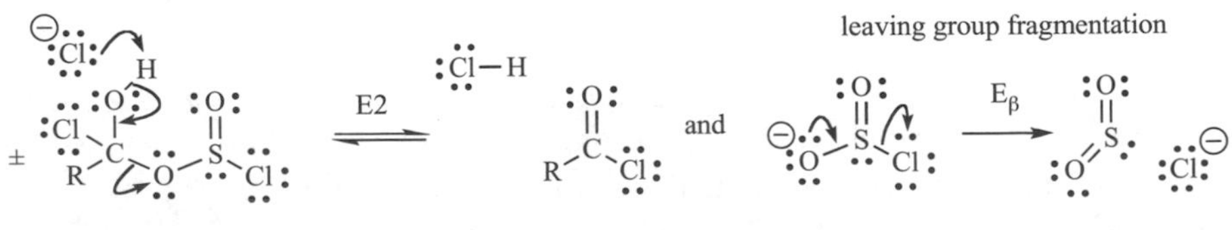
leaving group fragmentation
E2
and
Eβ

or in a lone-pair-assisted E1 (path E_β followed by path p.t.),

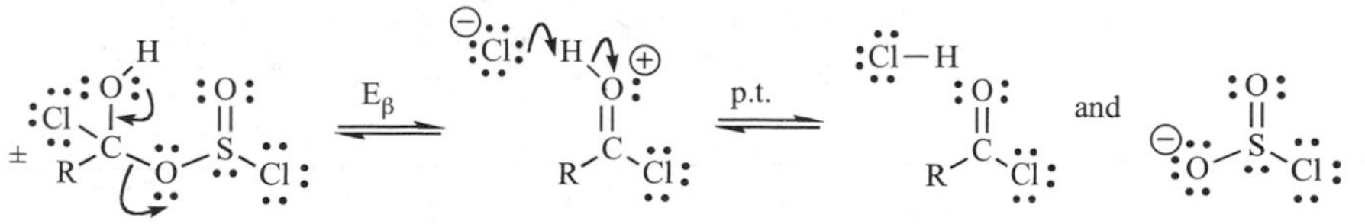

or by an E1cB (path p.t. followed by path E_β).

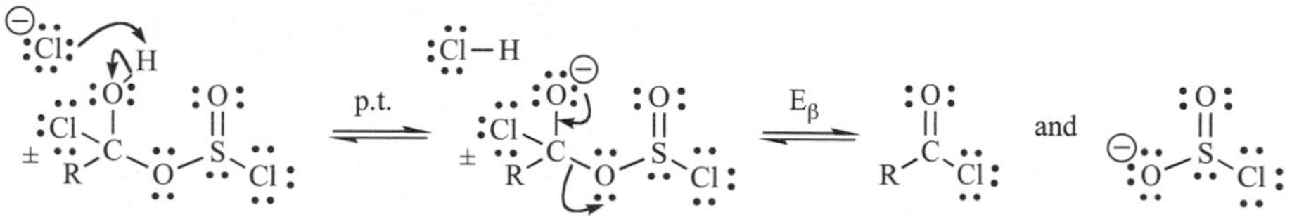

or by a *syn* internal elimination (path Ei),

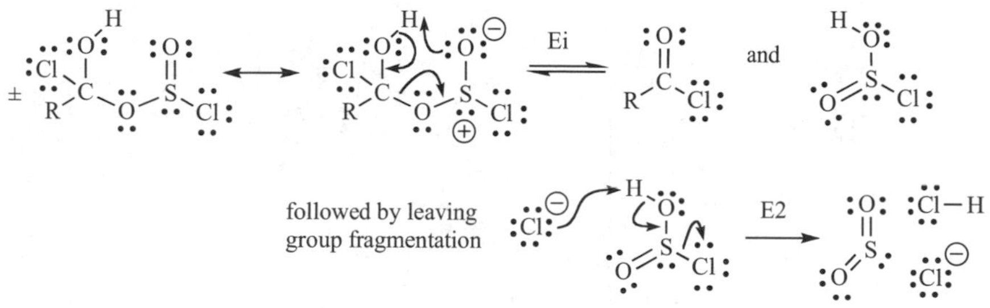

Evaluate and Cross-Check

The E1cB is probably not a good route because the deprotonation by chloride, pK_{abH} −7, to give the anionic tetrahedral intermediate, pK_{abH} about 10, goes against the ΔpK_a rule (a climb of 17 pK_a units) and forms a basic anion in a strongly acidic medium.

The other three eliminations are compatible with a strongly acidic medium. Because chlorosulfite is so unstable, no pK_a is available for its conjugate acid, and it is difficult to rank as a leaving group. The closest we can get to chlorosulfite on the pK_a chart is benzenesulfinic acid, $PhSO_2H$, whose pK_a is 2.1; the substitution of an electronegative chlorine for phenyl will make the acid much more acidic, dropping the pK_a at least several units; thus chlorosulfite is a good leaving group. It is difficult to rank which elimination route is lowest in energy. The Ei elimination has the advantage of being intramolecular and therefore may be the fastest.

Overview

Now two gases have been evolved and should escape the reaction mixture driving the transformation to completion (Fig. 10.15). The most probable route is in the center, with alternative routes drawn to either side. The path name for the forward reaction only is shown. The low-probability routes have been omitted for clarity.

10.8 METHODS FOR TESTING MECHANISMS

So far we have seen how to generate mechanistic hypotheses. The next step in the scientific method is the testing of them to rule out those incompatible with experimental

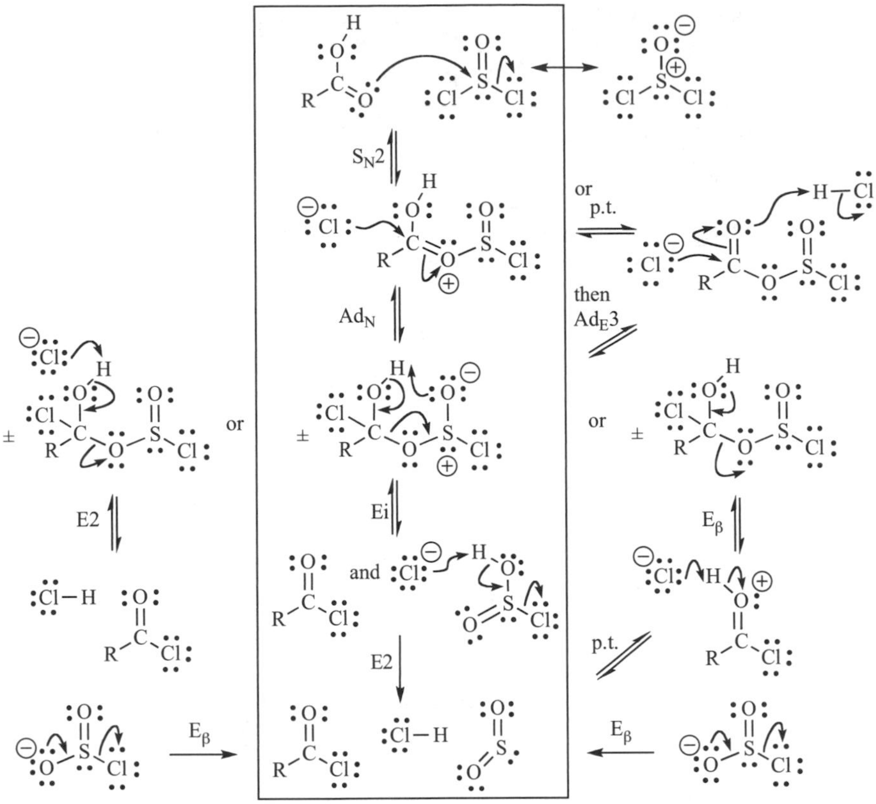

S_N2
or
p.t.
then
Ad_E3
Ad_N
or
or
E2
Ei
and
E2
$E_β$
p.t.
$E_β$
$E_β$

Figure 10.15 Acyl chloride formation overview. The most likely reaction path is boxed.

observations. We use the tools presented in this section to **disprove** these hypotheses, not to try to prove one correct. A reaction mechanism always remains a hypothesis and therefore can never be proven true. Experimental design is important; the experiment should be set up such that it will clearly exclude one of the mechanistic alternatives. The experiment must be carefully carried out so as to get a clean result. We repeat the process until only one hypothesis remains, and within the limits of our abilities to test, fails to be disproven.

Initial Studies

Identify the players. The first thing that needs to be done in studying a reaction is to know what substances enter into the reaction and what are produced. The exact stoichiometry of the reaction must be known. Any stereochemistry associated with the starting material and product needs to be determined. Side products should be identified.

Catalysis, Inhibition

What gets the reaction going? Sometimes nothing happens until a catalyst is added. Does the reaction require acidic or basic catalysis? Is a weak acid or base capable of catalyzing the reaction? Is a proton source, like trace water, required? Do Lewis acids catalyze the reaction? Do free radical inhibitors stop the reaction (see Chapter 11)? Does the reaction require a free radical initiator or light? For example, the addition of water to alkenes fails to go without an acid catalyst. Any postulated mechanism must explain the need for an acid catalyst.

Isolation, Detection, or Trapping of Intermediates

Who's on first? There are several spectroscopic methods that are capable of distinguishing reaction intermediates of a reaction in progress. For nuclear magnetic resonance (NMR), infrared (IR), and ultraviolet-visible (UV-VIS) spectra we can run the reaction in the instrument. Electron paramagnetic resonance (EPR) can detect free radicals (Chapter 11). There are even instruments that can record optical spectra at picosecond or less intervals. Also, it is occasionally possible to isolate intermediates from especially mild reaction conditions or low temperatures. Other reagents can be added to the reaction to trap the supposed reactive intermediate specifically and preferentially. Reactive intermediates of unimolecular reactions can be characterized spectroscopically by running the reaction in a glass such as frozen argon. Reactive intermediates can sometimes be trapped by an added reagent that reacts with them.

For example, an enzyme was suspected of reversibly forming an iminium ion with the substrate before cleaving it. Sodium borohydride was added, reducing the iminium ion to an amine, stopping the forward reaction, and preventing the reversal. The substrate was isolated attached to an amine of the enzyme. Not only did this confirm an iminium intermediate, but it also helped identify the amino acid responsible for the catalysis.

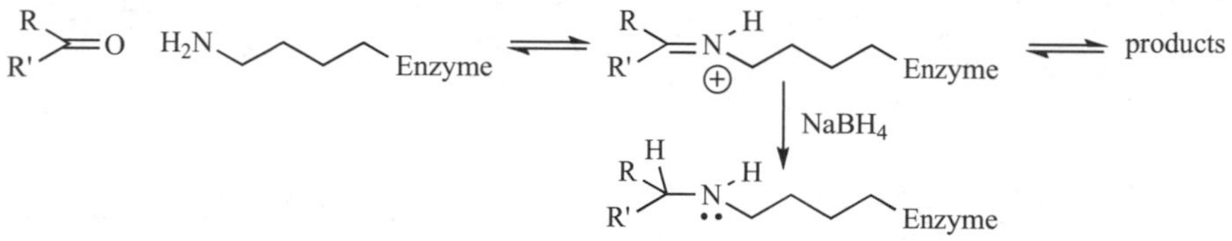

R
R'
O
H_2N
Enzyme
H
N
Enzyme
products
$NaBH_4$
H
H
N
Enzyme

Isotopic Labels

Find out where the atoms went. Labels are a convenient probe of a reaction mechanism, but care must be taken that whatever is chosen as a label does not alter the system under study. The label must be stable to the reaction conditions and not accidentally fall off into solution or scramble its position. Isotopic labels are a useful way of following an atom through a reaction. Common labels are ^2H (D), ^3H, ^{13}C, ^{14}C, ^{15}N, ^{18}O, and ^{32}P. The radioactive labels, ^3H, ^{14}C and ^{32}P, can be detected in very small quantities and so are useful in tracing products in biochemistry. However, extensive degradation schemes are required to determine the position of the radiolabel on the molecule. Mass spectra (MS) can be used to clarify the position of a label such as deuterium, ^{15}N, or ^{18}O from the fragmentation pattern of the molecule. With carbon NMR spectroscopy, finding the location of a ^{13}C label is especially easy to carry out. Labels can be very useful in determining the origins of a particular group in a rearrangement, a cleavage, or a biosynthesis.

Doubly labeled compounds provide an interesting test for an intermolecular versus intramolecular transfer of a group. The group transferred and another part of the same molecule are both labeled, and the reaction is run with an equal amount of unlabeled material. The product of an intramolecular transfer would be either doubly labeled or unlabeled. The intermolecular process would produce a statistical mixture of products: About half the product would be singly labeled.

Example problem

How would you use a label to disprove that a methyl ester underwent basic hydrolysis in KOH/water by an S_N2 mechanism?

Answer: Label the oxygen in the KOH/water with ^{18}O. An S_N2 on the methyl with labeled hydroxide will give the ^{18}O label in the methanol, whereas other mechanisms such as the addition–elimination path combination would not.

Regiochemistry, Stereochemistry, and Chirality

Probe orientation. Does the reaction have a preferred regiochemistry, like following Markovnikov's rule? Is stereochemistry lost or retained in the reaction? Is the process stereospecific? Does addition or elimination occur *anti* or *syn*? Is the chirality at the reacting center preserved, inverted, or lost?

For example, addition of bromine to *cis*-2-butene gives equal concentrations of (2*R*,3*R*)-2,3-dibromobutane and (2*S*,3*S*)-2,3-dibromobutane (a racemic mix), whereas addition of bromine to *trans*-2-butene gives the *meso* compound (2*S*,3*R*)-2,3-dibromobutane. Since the stereochemistry of the reactant is determining product stereochemistry, this reaction is stereospecific, and whatever mechanism is proposed must account for that.

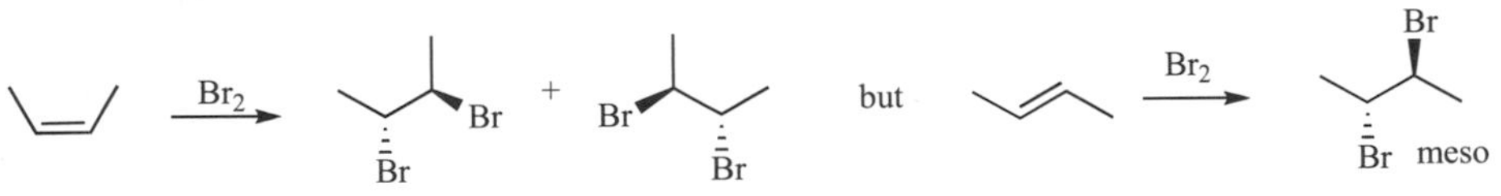
Br2
Br
+
Br
Br
Br
but
Br2
Br
Br
meso

Example problem

The addition of DBr to cyclohexene gives the *anti* addition product. Is this result more consistent with an Ad_E2 or an Ad_E3 mechanism?

Answer: The Ad_E3 is consistent with *anti* addition. The Ad_E2 would be expected to give a *syn*/*anti* mixture with *syn* addition preferred. The Ad_E2 intermediate ion pair collapses preferentially *syn*; the symmetrically solvated carbocation has no preference.

Kinetics

Who are rate-determining-step players? Kinetics can yield information about the rate-determining step of a reaction. At a constant temperature, the concentration of a particular reactant is changed and the effect on the rate of reaction is noted. For example, a reaction that is second order in compound A would quadruple its rate if the concentration of A were doubled. If the reaction were first order in A, the rate would double if the A concentration were doubled. In this way the kinetic order with respect to each reactant is determined. These kinetic orders can then be compared to that expected for the slow step of a postulated mechanism. Often, however the kinetic order can easily fit several possibilities.

Some examples of kinetic order have been presented already and are shown below. The magnitude of rate constant, *k*, is of course unique to each reaction and differing substrates.

Order	Example	Rate Expression
First	E1	Rate = k[R–L]
Second	S_N2	Rate = k[R–L][Nu]
Third	Ad_E3	Rate = k[alkene][HA][A^-]

Example problem

Propose a mechanism that is consistent with the fact that the rate of the acidic bromination of acetone is independent of the concentration of bromine.

Br2
+
O
Br
O
+ HBr

Answer: The independence of the rate on the bromine concentration means that bromine reacts after the slow step occurs. The rate is already determined before the bromine enters into the reaction. Acetone is a polarized multiple bond, and the methyl group is not nucleophilic. Under acidic conditions, the oxygen lone pair can be protonated, and this heteroatom-to-heteroatom proton transfer is usually fast. The protonated carbonyl is a great ewg, making the methyl hydrogens acidic. Loss of a methyl hydrogen produces the reactive enol, an allylic source that has a nucleophilic carbon that can attack bromine. The kinetic data tell us that the formation of the enol has to be slower than the enol attacking bromine. This is consistent with the enol tautomer being less stable than the keto form, so enol formation is uphill in energy. The final proton transfer is again expected to be fast.

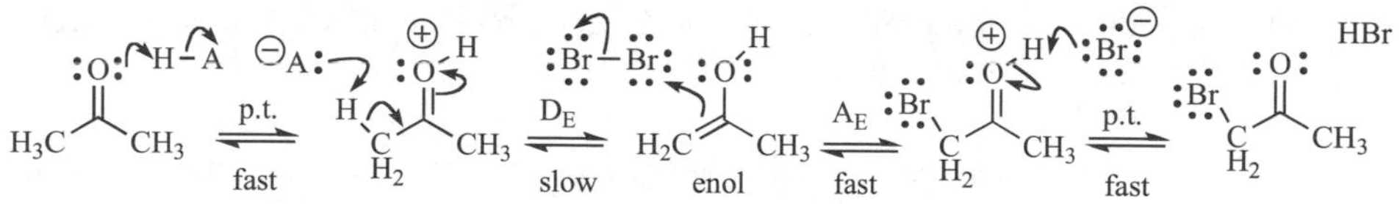
p.t.
fast
DE
slow
enol
AE
fast
p.t.
fast
HBr

Solvent Polarity Effects

Solvent effects probe charge concentration or dispersal. Solvent polarity can stabilize charged species, and can affect the rate of a reaction (Section 2.7). If the reaction forms a charged intermediate or product from neutral reactants, increased solvent polarity will speed up the reaction by lowering the energy of the charged species. However, if charge is neutralized or dispersed during the reaction, a more polar solvent will slow down the rate because reactant stabilization increases the reaction barrier.

S_N2 Charge Type	Charge at ‡	Products	Solvent Polarity Effect
Nu: and R–L	Increasing	$^+$Nu–R and L:$^-$	Polar increases rate
Nu:$^-$ and R–L$^+$	Decreasing	Nu–R and L:	Polar decreases rate
Nu:$^-$ and R–L	Dispersing	Nu–R and L:$^-$	Polar decreases rate
Nu: and R–L$^+$	Dispersing	$^+$Nu–R and L:	Polar decreases rate

Example problem

If the substitution of an alkyl chloride by an anionic nucleophile is greatly slowed as the reaction solvent becomes less polar, is this solvent effect more compatible with an S_N1 or S_N2 mechanism?

Answer: The S_N1 rate-determining step is the ionization of the leaving group to form a carbocation and a chloride ion. The solvent must stabilize these charged species, or the reaction will be slowed or stopped. The S_N2 reaction rate would be expected to increase slightly since the charge is dispersed in the transition state. The observed solvent effect is inconsistent with an S_N2 mechanism and consistent with the S_N1.

Substituent Effects

Substituent effects probe the electronic demand of the slow step. Electron-withdrawing or -releasing groups can greatly alter the rate of a reaction and sometimes even cause a change in mechanism. A common way to study substituent effects is to connect a benzene ring that has a *meta* or *para* substituent group to the site undergoing reaction. This substituent group is then varied, and the rate of the reaction is compared to the rate when the group is simply H. If electron-releasing groups accelerate the rate of reaction, the conclusion is drawn that either a positive charge is increased or a negative charge is decreased in the rate-determining step. If electron-withdrawing groups accelerate the rate, either a negative charge is increased or a positive charge is decreased in the rate-determining step.

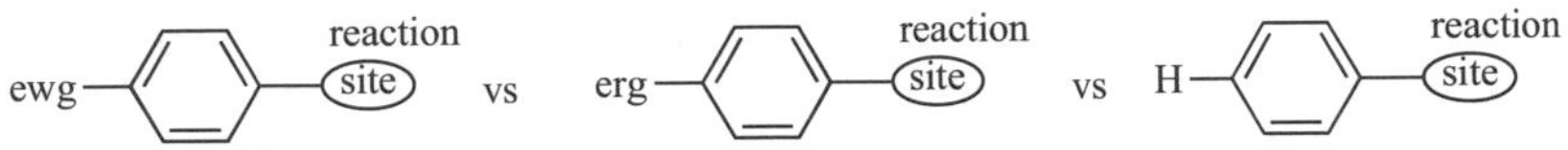
ewg
reaction
site
vs
erg
reaction
site
vs
H
reaction
site

Example problem

The rate of the substitution reaction of $Ar(CH_3)_2C{-}Cl$ with water is greatly favored by electron-releasing groups on the aryl (Ar) group. Is this substituent effect explained better by an S_N1 or S_N2 mechanism?

Answer: Since electron-releasing groups help, then positive charge is increased or a negative charge is decreased in the rate-determining step. The S_N1 rate-determining step is the formation of a carbocation that donors would stabilize. The neutral S_N2 transition state would not be expected to have a significant substituent effect.

Primary Deuterium Isotope Effects

Is C–H bond breaking in the slow step? The fact that a C–H bond is just slightly easier to break than a C–D bond can be used to probe reaction mechanisms that have the C–H bond breaking in their rate-determining step. In this case the rate of the deuterium-substituted reactant will be 2 to 7 times slower.

Example problem

The rate of the elimination reaction of deuterated 2-bromopropane with ethoxide is 6.7 times slower than the rate with undeuterated 2-bromopropane. Is this observation consistent with an E1 mechanism?

Answer: No, since the E1 rate-determining step is leaving group loss and not C–H bond breakage, it would not show a primary deuterium isotope effect; an E2 reaction would, because C–H (D) breaking occurs in the transition state of the concerted E2.

Barrier Data

Both $\Delta H^{\ddagger}$ and $\Delta S^{\ddagger}$ and therefore $\Delta G^{\ddagger}$ can be determined ($\Delta G^{\ddagger} = \Delta H^{\ddagger} - T\Delta S^{\ddagger}$) from how the rate constant changes as a function of temperature. Figure 10.16 shows an example of an Eyring plot in which the natural log of k/T (the rate constant divided by temperature in Kelvin) is plotted against $1/T$. We get $\Delta H^{\ddagger}$ from the slope and $\Delta S^{\ddagger}$ from the intercept.

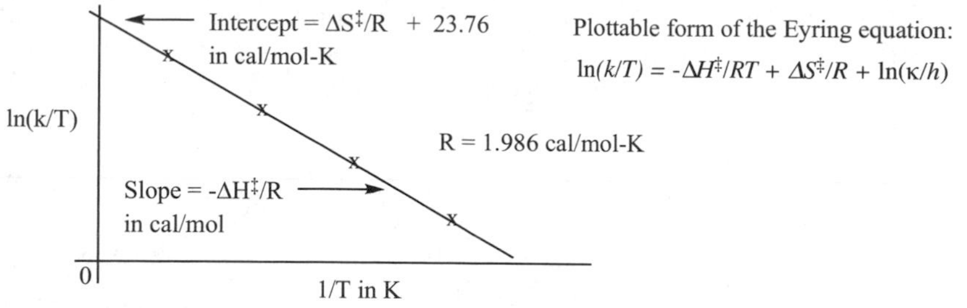
Intercept = ΔS‡/R + 23.76
in cal/mol-K
Plottable form of the Eyring equation:
ln(k/T) = -ΔH‡/RT + ΔS‡/R + ln(κ/h)
ln(k/T)
R = 1.986 cal/mol-K
Slope = -ΔH‡/R
in cal/mol
0
1/T in K

Figure 10.16 Example of an Eyring plot to determine $\Delta H^{\ddagger}$ and $\Delta S^{\ddagger}$ from plotting $\ln(k/T)$ vs. $1/T$.

Since bond breaking in the transition state increases the $\Delta H^{\ddagger}$, and bond formation diminishes it, processes that have simultaneous bond breakage and formation are expected to have a low $\Delta H^{\ddagger}$. Transition states that require a specific spatial arrangement of reactants will tend to have a negative $\Delta S^{\ddagger}$ because the degree of disorder decreases in going to the transition state. A low $\Delta H^{\ddagger}$ and negative $\Delta S^{\ddagger}$ for a gas phase reaction are indicative of a concerted process with an ordered transition state. However, a cautionary note must be sounded, for reactions in solution can have a negative $\Delta S^{\ddagger}$ because the degree of solvent disorder decreases in going to a highly solvated transition state, even as the actual reaction disorder increases.

10.9 LESSONS FROM BIOCHEMICAL MECHANISMS

Section 2.8 discussed two major contributors to enzymatic catalysis: Lowering the $\Delta G^{\ddagger}$ increases the number of molecules with sufficient energy to react, and confining the reacting partners within an enzyme's active site greatly increases the number of effective collisions. Binding in the substrate in the proper orientation for reaction significantly reduces the $T\Delta S^{\ddagger}$ contribution to the $\Delta G^{\ddagger}$, thus lowering the reaction barrier. The lowering of the $\Delta H^{\ddagger}$ component was addressed briefly with the statement that enzymes bind the transition state best. Section 2.8 ended with the promise that we would return for a general look at the chemical step of enzymatic catalysis, and how the $\Delta H^{\ddagger}$ for an enzyme catalyzed reaction can be significantly lowered.

Solution reactions have several disadvantages, but the greatest difficulty is the inherent disorder of a liquid. Since the probability of a three-molecule collision is rare, almost all collisions involve just two molecules, and paths like the Ad_E3 addition result from a reactant collision with weakly bound pi-complex or hydrogen bonded pair. On the other hand, an enzyme active site can prearrange all the catalytic groups so they are in just the correct position and orientation. The solvation within the active site is also preorganized to bind the transition state or reactive intermediate best; in this way solvation energy lowers the barrier. Since the amino acid side chains in the active site can vary from very polar to very nonpolar, the solvation provided to the substrate can vary in a similar way. The binding of the substrate in the active site can also destabilize, with binding raising the reactant's energy, thereby effectively lowering the reaction barrier.

Enzyme mechanisms can often avoid high-energy, unstable cationic or anionic intermediates that increase the reaction barrier. Triosephosphate isomerase catalyzes the tautomerization of the achiral dihydroxyacetone phosphate (DHAP) to *R*-glyceraldehyde-3-phosphate (G3P) by the mechanism shown below. Not only does the push–pull mechanism avoid forming a highly basic enolate, but the binding mode of DHAP determines which face of the enediol will be protonated in the second step, and therefore which enantiomer of G3P forms. Triosephosphate isomerase has reached catalytic perfection as it can produce just a single enantiomer of G3P as fast as DHAP can diffuse in. It is estimated that this enzyme is able to lower the barrier to reaction by about 12 kcal/mol.

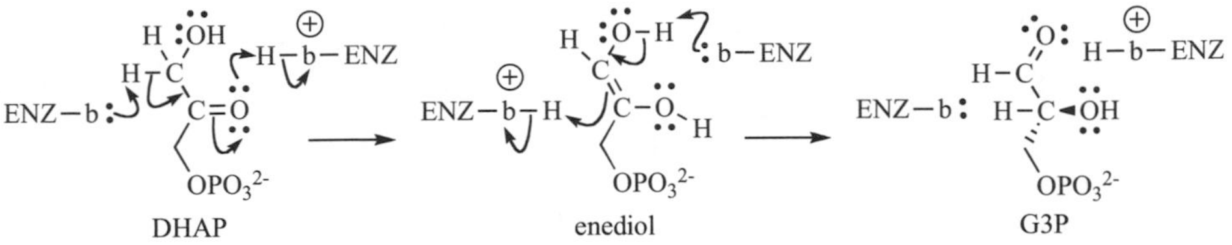
DHAP
enediol
G3P

This push–pull catalytic route is another type of general acid and general base catalysis discussed in Section 7.4.3, where the important bond making and breaking is occurring in the same step as proton transfer. Not only do enzyme cavities create their own solvent conditions, but in so doing they can adjust the pK_as of catalytic groups. A carboxylate anion placed in a nonpolar cavity is more basic than a free carboxylate. A cationic acidic group increases in acidity if placed in a nonpolar cavity, because poor solvation raises its energy. A good general acid will have a pK_a close to physiological pH of 7. If the acid is too acidic, it will be mostly deprotonated at physiological pH, and therefore very little of it will still have its acidic H to serve as an acid. Likewise, a good general base will have its pK_{abH} close to 7. Two groups that fit these criteria are the imidazole group of histidine (pK_{abH} about 6) and the carboxylic acid group of glutamic and aspartic acids (pK_a about 4), shown below.

common enzyme active site acids
common enzyme active site bases

Enzymes, in addition to providing hydrogen bonding to anions formed, use electrostatic catalysis to polarize the reactant and compensate for a charge formed on the substrate as it reacts. Although it may be difficult to protonate a carbonyl (pK_a = −7) at physiological pH, an enzyme can use Lewis acids such as Mg^{2+} to polarize the carbonyl instead. The Mg^{2+} can be viewed as a super proton at neutral pH for electrophilic catalysis (Section 6.2.3). In a similar manner, the Zn^{2+} ion in alcohol dehydrogenase (shown in Section 7.2.6) makes the acetaldehyde much more electrophilic, capable of reduction by NADH. Carbonic anhydrase, which converts carbon dioxide into bicarbonate ion in red blood cells, has a Zn^{2+} ion in the active site that complexes to water, making it easy to remove a proton and form hydroxide at neutral pH. It is estimated that the pK_a of the zinc bound water is about 7, so is easily deprotonated by an active site histidine. Carbonic anhydrase is one of the fastest of all enzymes, reacting with carbon dioxide as fast as it can diffuse in.

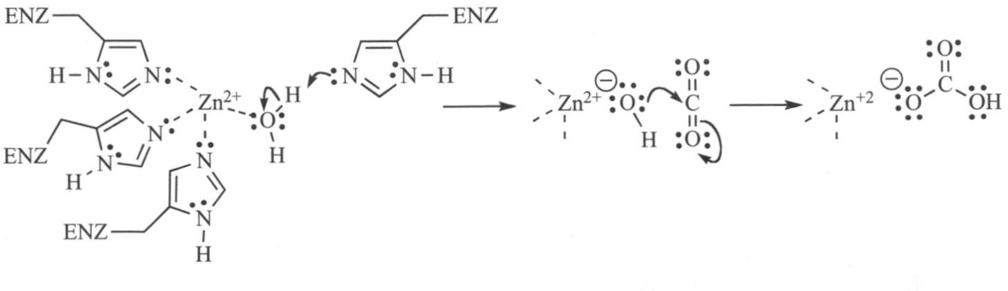

There are many other uses for metals in biochemistry: for binding of oxygen (hemoglobin), as templates bringing reactants together (aconitase), for electron transport (cytochrome *c*), for redox reactions (P450), for organometallics (vitamin B_{12}), and more.

Enzymes may use covalent catalysis, forming a covalent bond to the substrate that is later broken before product release. This catalysis can be nucleophilic or electrophilic. In Section 4.2.12, we saw an example of iodide nucleophilic catalysis of a substitution; iodide was a better nucleophile than the original nucleophile, water, and a better leaving group than the original leaving group, chloride. Enzymes often use specialized reagents for this covalent catalysis, coenzymes like thiamine, which in Figure 8.12 was used to decarboxylate pyruvate to acetaldehyde.

Aldolase is an example of an enzyme that uses electrophilic covalent catalysis. The amine of an active site lysine forms an imine (Section 10.5.2) with the carbonyl of fructose-1,6-bisphosphate. This more reactive imine electron sink allows a reverse aldol reaction to occur via the less basic enamine rather than the more basic enolate ion. Tautomerization of the resulting enamine to an imine, then hydrolysis, releases DHAP and returns the enzyme active site lysine to the free anime, ready for the next cycle.

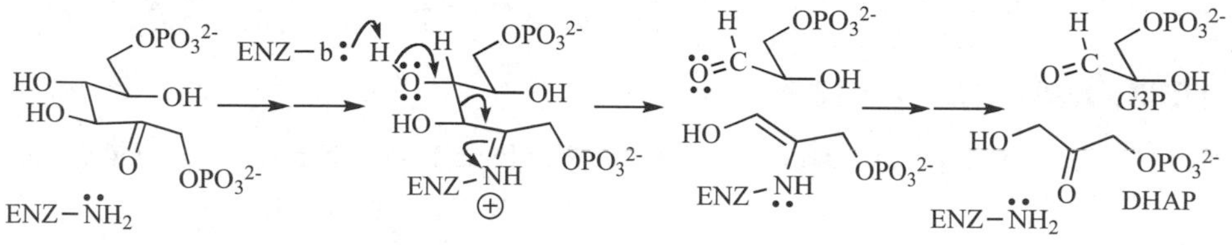
ENZ−b:
ENZ−NH2
ENZ−NH
G3P
DHAP

Enzymatic catalysis lowers $\Delta G^‡$ to accelerate the reaction rate, but does not affect the $\Delta G°$ that controls the position of equilibrium. However, cells also have an elegant ability to overcome an unfavorable equilibrium by coupling an endothermic reaction with exothermic ATP hydrolysis. For example, the reaction of acetate with coenzyme A (CoASH) giving acetyl CoA is uphill by about 7 kcal/mol. The hydrolysis of ATP to ADP and phosphate is downhill by about the same amount. By coupling these two reactions shown below for acyl-CoA synthetase, the cell achieves a total $\Delta G°$ of almost 0 for a K_{eq} of about 1. The phosphorylation of acetate by ATP improves the leaving group for the second reaction with CoASH. There are many enzymes whose catalytic process remains a mystery; enzymes still have much to teach us about reaction mechanisms.

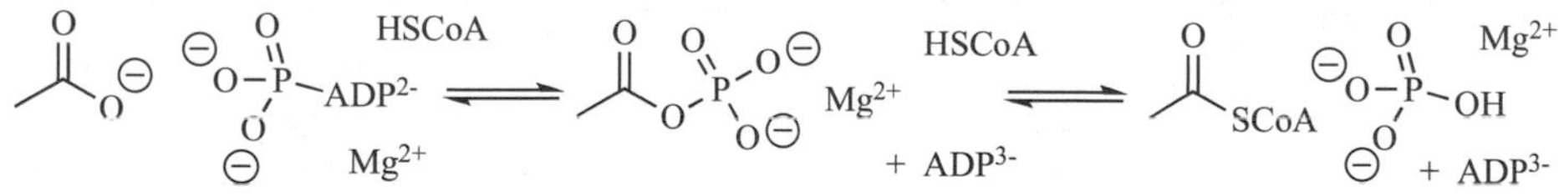
HSCoA
HSCoA
SCoA
+ ADP3-
+ ADP3-

10.10 SUMMARY

The entire emphasis of this chapter is to use the trends to guide a deliberate decision process that generates reasonable alternatives and then selects the best one. This decision process has a limited number of inputs (Fig. 10.17). Mechanism problems have the additional input of the structural information from the product. We can see what bonds were made and broken to guide our decisions. Although this "generate and select" process might seem slower than what you may be used to, it is the essence of the scientific method. As the problems you attempt get more difficult, a deliberate and organized search process for solutions not only is more efficient but also becomes absolutely necessary. If you can internalize this "best-first search" process, it will serve you well in many fields beyond organic chemistry.

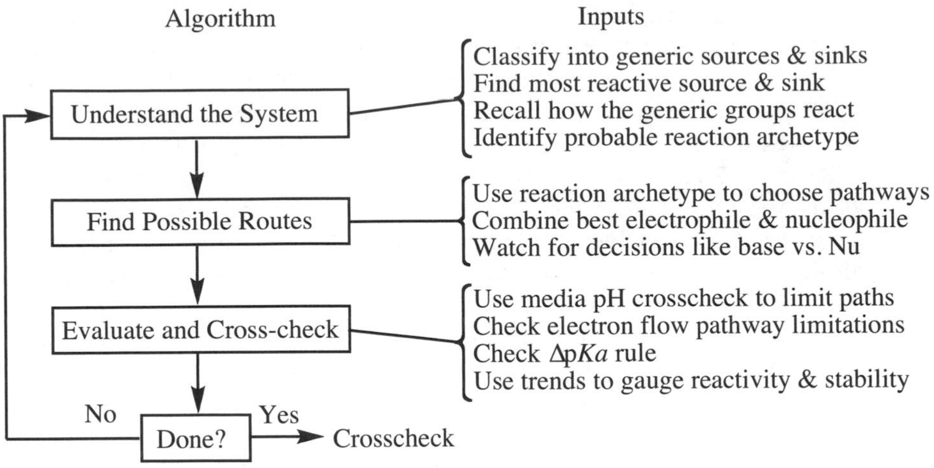
Algorithm
Inputs
Understand the System
Classify into generic sources & sinks
Find most reactive source & sink
Recall how the generic groups react
Identify probable reaction archetype
Find Possible Routes
Use reaction archetype to choose pathways
Combine best electrophile & nucleophile
Watch for decisions like base vs. Nu
Evaluate and Cross-check
Use media pH crosscheck to limit paths
Check electron flow pathway limitations
Check ΔpKa rule
Use trends to gauge reactivity & stability
No
Done?
Yes
Crosscheck

Figure 10.17 The organic problem solving algorithm and its common inputs.

ADDITIONAL EXERCISES

(See the Appendix: Hints to Selected Problems from Chapters 8, 9, and 10 if you need assistance.)

10.1 Provide a mechanism for the following transformations.

(a)

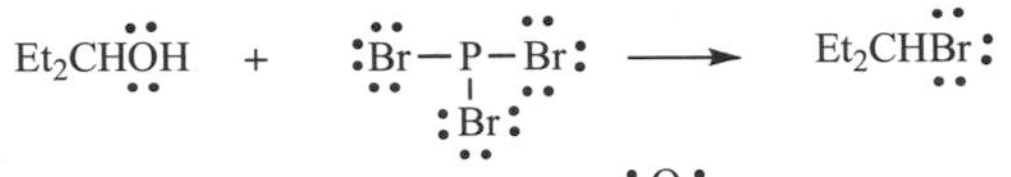

(b)

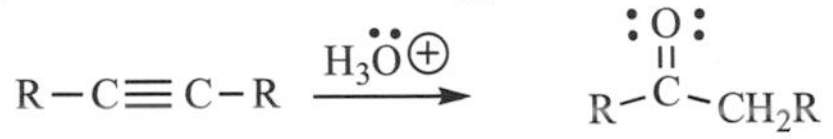

(c)

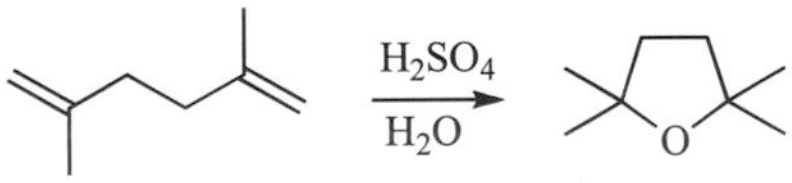

(d)

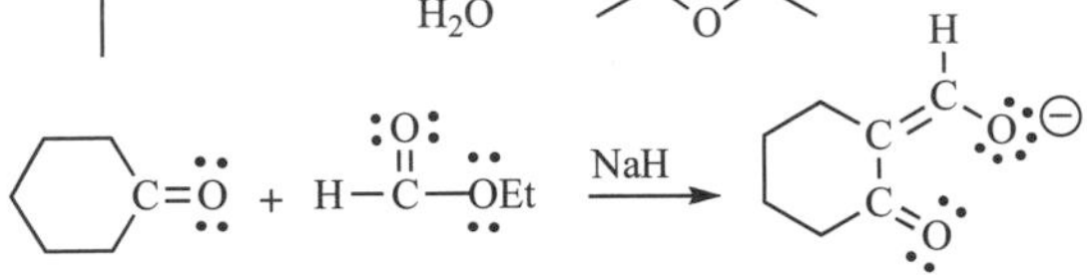

(e)

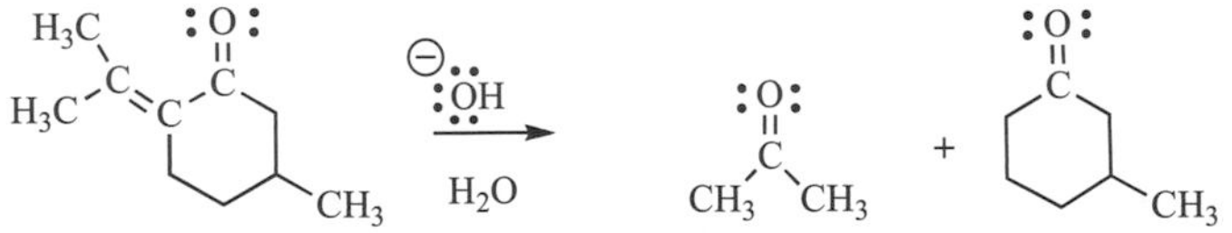

(f)

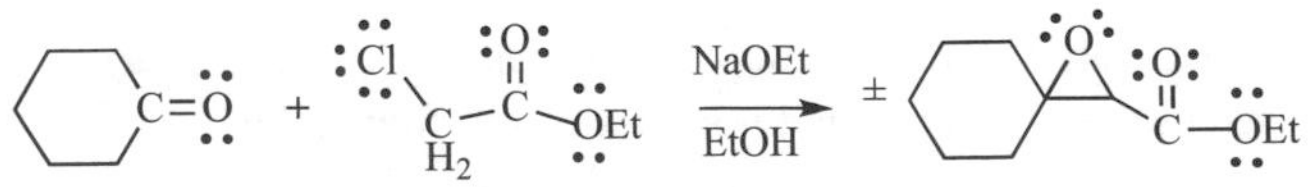

10.2 Provide a mechanism for the following transformations (medium difficulty).

(a)

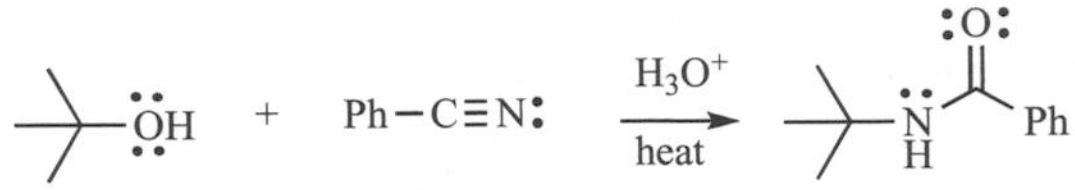

(b)

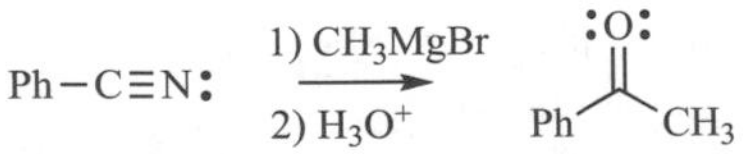

(c) 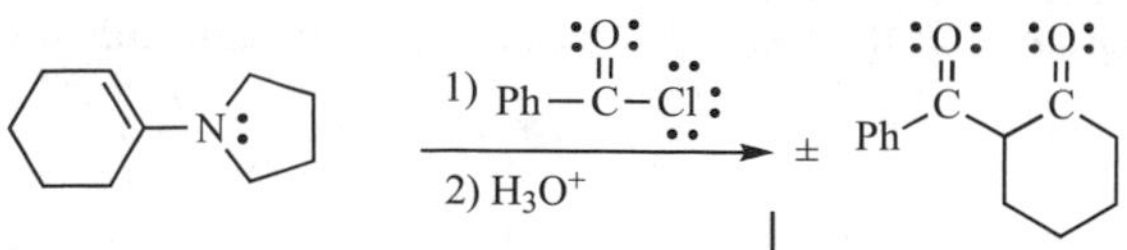

(d) 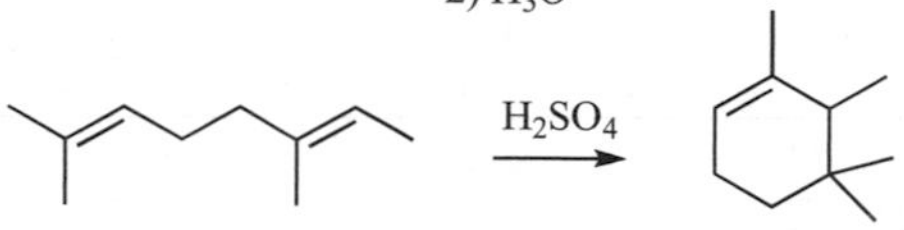

(e) 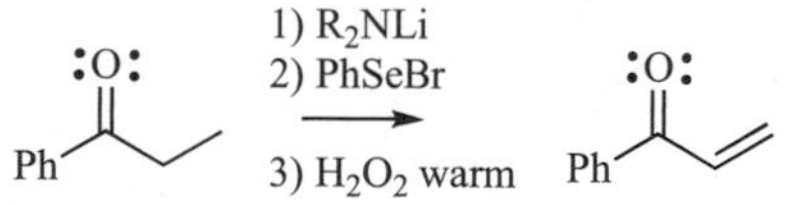

(f) 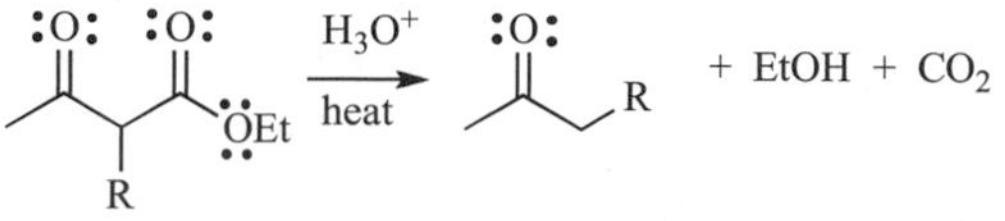

(g) 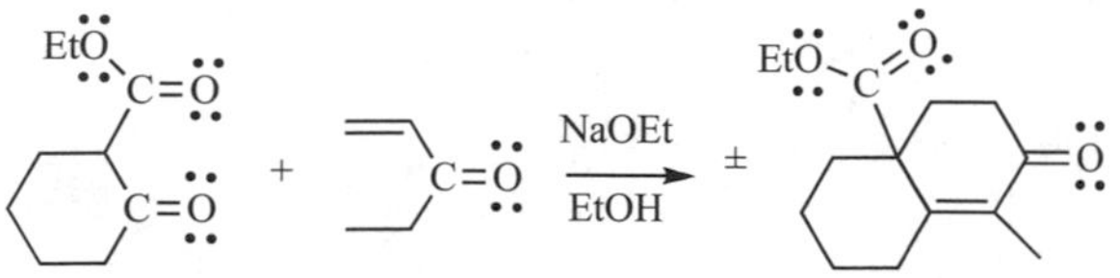

(h) 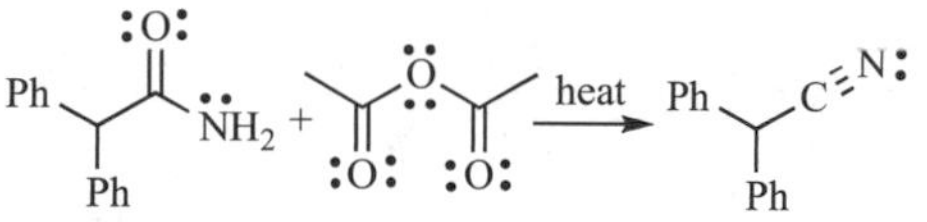

(i) 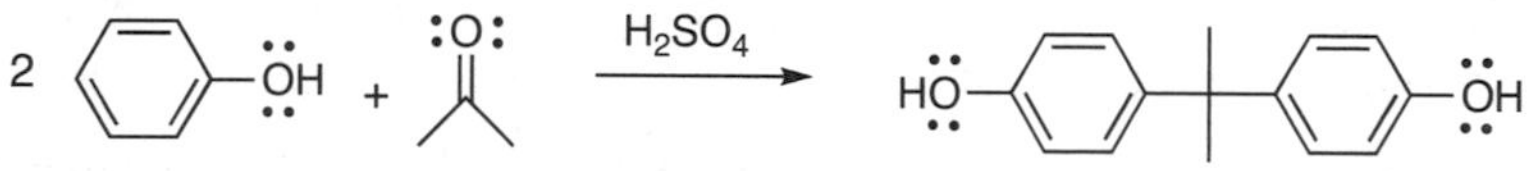

10.3 Give the product of the following reactions.

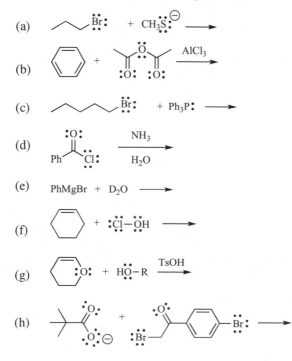

(a) 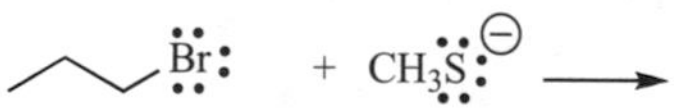

(b) 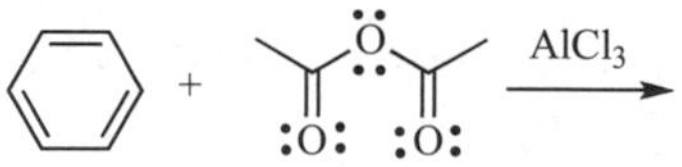

(c) 

(d) 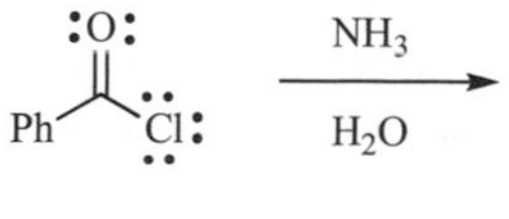

(e) $PhMgBr + D_2O \longrightarrow$

(f) 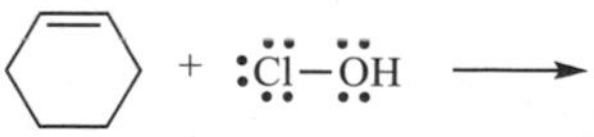

(g) 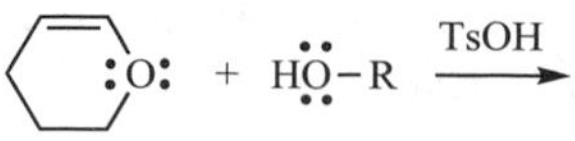

(h) 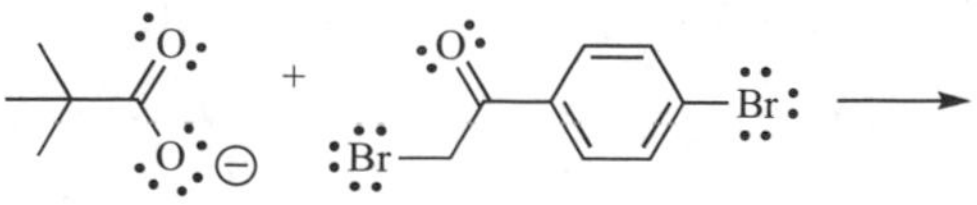

10.4 Give the product of the following reactions (more difficult).

(a) 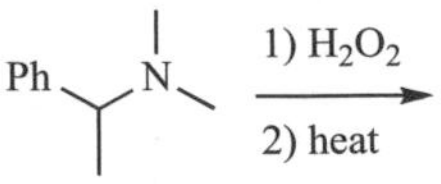
Ph
N
1) H_2O_2
2) heat

(b) 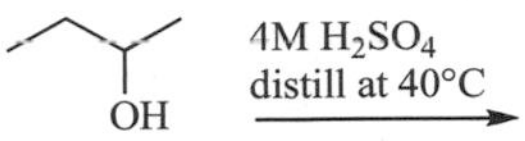
OH
4M H_2SO_4
distill at 40°C

(c) 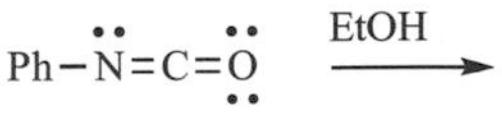
Ph−N=C=O
EtOH

(d) 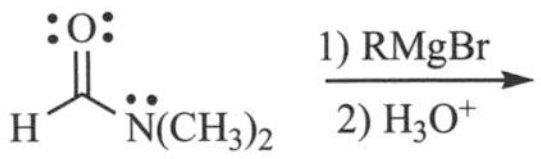
O
H
N(CH3)2
1) RMgBr
2) H_3O^+

(e) 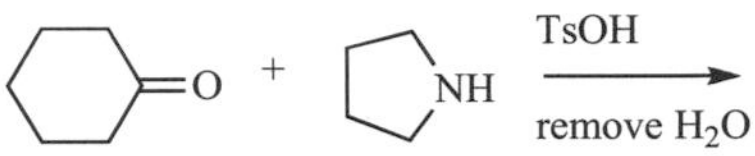
O
+
NH
TsOH
remove H_2O

(f) 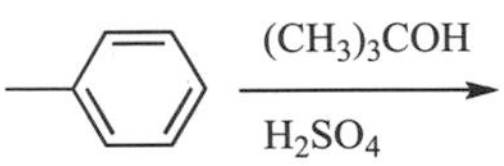
$(CH_3)_3COH$
H_2SO_4

(g) 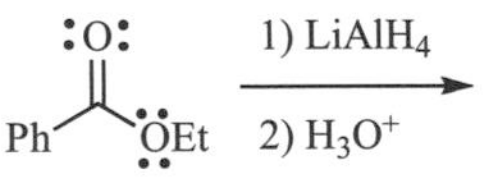
O
Ph
OEt
1) $LiAlH_4$
2) H_3O^+

10.5 Provide a mechanism for the following transformations that is consistent with the experimental results provided.

(a) Similar systems run in ^{18}O water show no ^{18}O in the alcohol product.

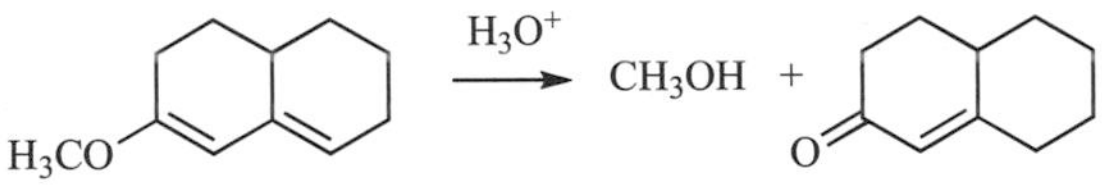
H_3CO
H_3O^+
CH_3OH +
O

(b) This reaction requires cyanide ion as a catalyst and fails to go if it is absent.

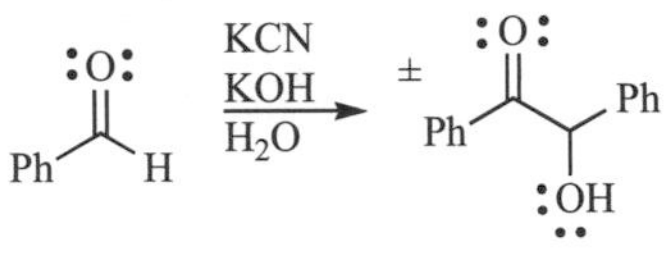
O
Ph
H
KCN
KOH
H_2O
±
O
Ph
Ph
OH

(c) Only this stereoisomer of product is formed in nitromethane solvent.

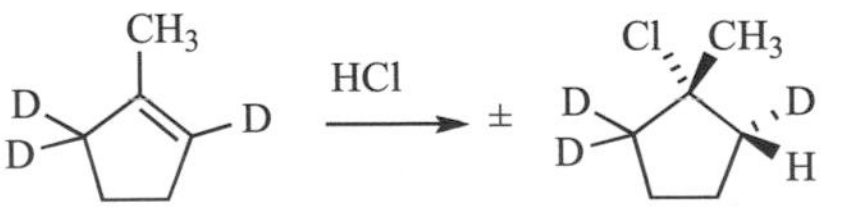
CH3
D
D
D
HCl
±
Cl
CH3
D
D
D
H

(d) The following reaction is unusual in that it goes with complete **retention** of configuration. Propose a mechanism that explains this.

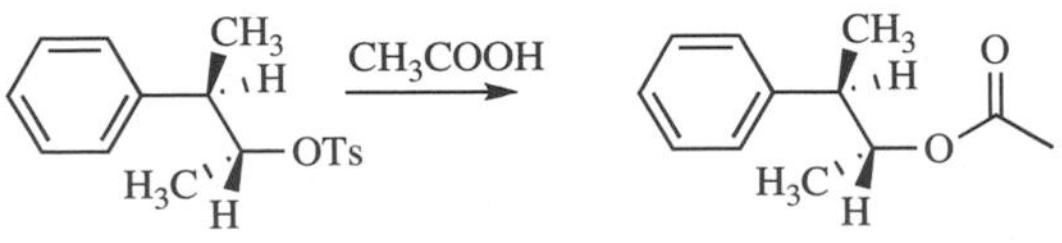
CH3
H
OTs
H3C
H
CH_3COOH
CH3
H
O
O
H3C
H

(e) Only this stereoisomer of product is formed at –72°C.

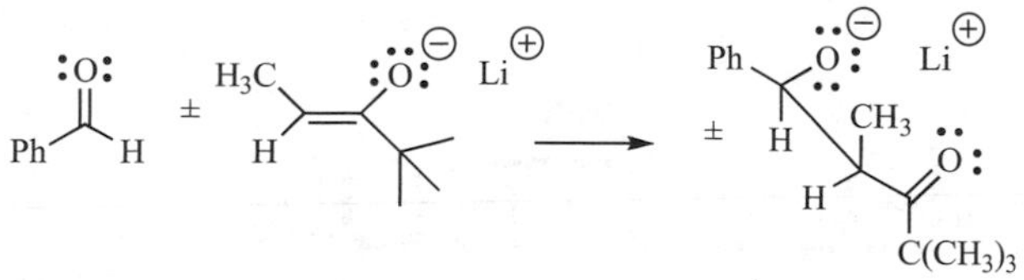

(f) Chiral optically active reactant gives a **racemic** product with overall **retention** of stereochemistry.

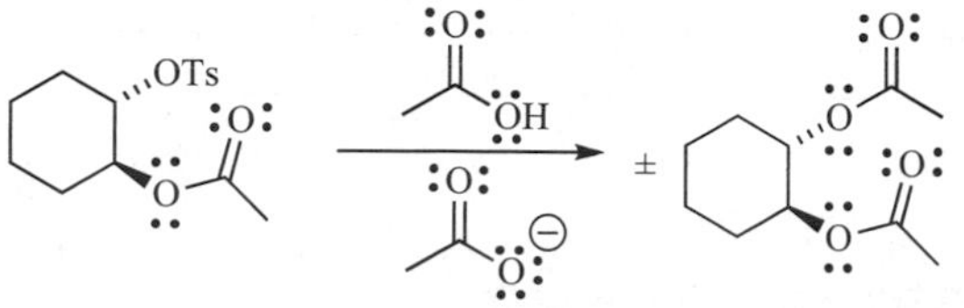

10.6 Provide a mechanism for the following transformations (more difficult).

(a)

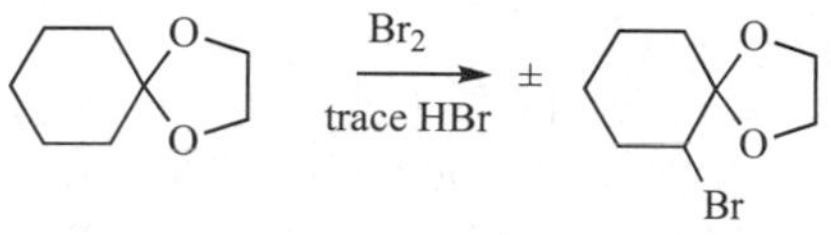

(b)

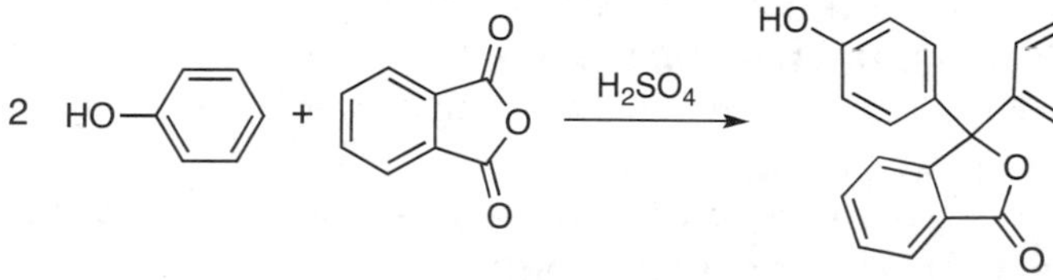

phenolphthalein, the pH indicator

(c)

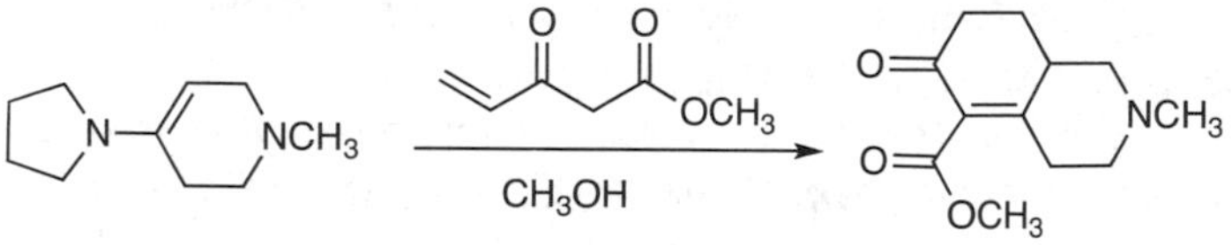

(d)

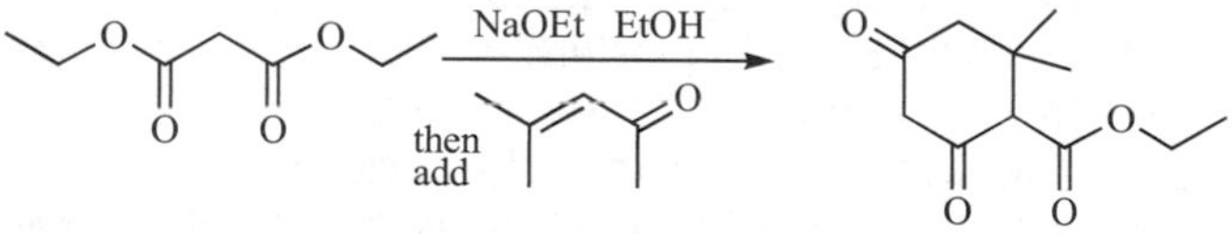

10.7 Part of the dark reactions of photosynthesis is interconversion of sugars with an enzyme called transketolase using thiamine pyrophosphate, TPP, as a catalyst (Section 8.12.8). Provide a reasonable mechanism for this enzymatic reaction. In addition to water, there are weak general acids and general bases present in the active site at pH 7.

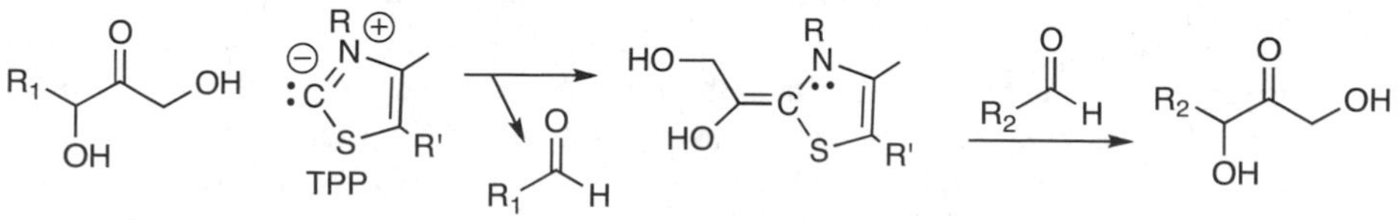

11

ONE-ELECTRON PROCESSES

11.1 RADICAL STRUCTURE AND STABILITY

One-electron processes belong to a conceptually different class of reactions and are best treated in this separate chapter. All one-electron processes require an initiation step, discussed in Section 11.2, which should serve as a flag to alert you to shift into using one-electron pathways. Because odd electron species, radicals, are usually reactive, their concentration in a reaction will be very low. Therefore, reactions of radicals with other

Electron Flow In Organic Chemistry: A Decision-Based Guide To Organic Mechanisms, Second Edition.
By Paul H. Scudder

radicals will be less common than reactions of radicals with the even electron species that make up the bulk of the reaction mixture. **The most common radical reactions involve the radical reacting with an even electron species;** the reaction usually gives an even electron product and a new radical. These paths are discussed in Section 11.3.

Radical formation is a concern for all aerobic organisms. Oxygen is a ground-state diradical and is very reactive and responsible for oxidative cell damage. Cells have radical-specific scavengers that minimize the damage that radicals can do.

Radical Structure

Simple alkyl radicals like the methyl radical are believed to be nearly planar, shallow pyramids at the radical center, rapidly inverting with a barrier of <2 kcal/mol (8 kJ/mol). The vinyl radical is planar and rapidly inverts. Thus all stereochemistry at the radical center is usually lost. In conjugated species, like the allyl radical, the radical is believed to be in a *p* orbital to maximize overlap with the rest of the pi system. Resonance forms allow us to predict the sites of radical reactivity in conjugated radicals.

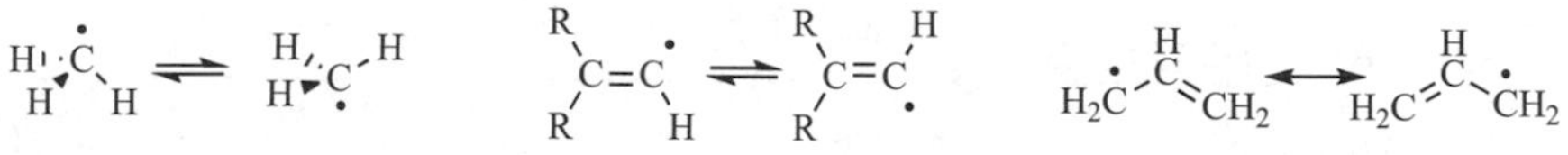

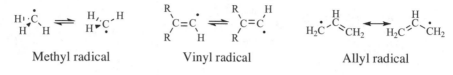

Methyl radical Vinyl radical Allyl radical

Bond Dissociation Energies

The energy required to break homolytically the R–H bond into R· and H· radicals is related to the stability of the R· radical. Common C–H bond dissociation energies in kcal/mol (kJ/mol) are given in Table 11.1.

Table 11.1 C–H Bond Dissociation Energies

Radical Name	C–H Bond Broken	Dissociation Energy, kcal/mol (kJ/mol)
Break sp^3 C–H bond		
Methyl	CH_3–H	105 (439)
Ethyl	CH_3CH_2–H	98 (410)
Isopropyl	$(CH_3)_2CH$–H	95 (397)
tert-Butyl	$(CH_3)_3C$–H	93 (389)
Break sp^2 C–H bond		
Vinyl	CH_2=CH–H	110 (460)
Phenyl	C_6H_5–H	110 (460)
Acyl	CH_3CO–H	86 (360)
Delocalized radicals		
Acylmethyl	CH_3COCH_2–H	96 (401)
Cyanomethyl	$NCCH_2$–H	86 (360)
Benzyl	$C_6H_5CH_2$–H	88 (368)
Allyl	CH_2=$CHCH_2$–H	86 (360)
Misc. radicals		
Hydroxymethyl	$HOCH_2$–H	94 (393)
Trifluoromethyl	F_3C–H	107 (448)
Trichloromethyl	Cl_3C–H	96 (402)

Radical Stabilities

Although steric relief in going to the nearly planar radical contributes to the dissociation energy, it is generally believed that the stability of simple alkyl radicals follows the trend: tertiary > secondary > primary > methyl. Both vinyl and phenyl radicals are destabilized relative to methyl. Whereas lone pair donors stabilize radicals somewhat, groups that increase conjugation have the greatest stabilizing effect. Since a radical is a neutral species, solvent polarity generally has little effect.

To conclude, radicals can be stabilized by either a pi donor or pi acceptor because both delocalize the odd electron. Radicals that have both a pi donor and a pi acceptor are especially stable. The following resonance forms demonstrate the delocalization of a radical by a pi donor and by a pi acceptor.

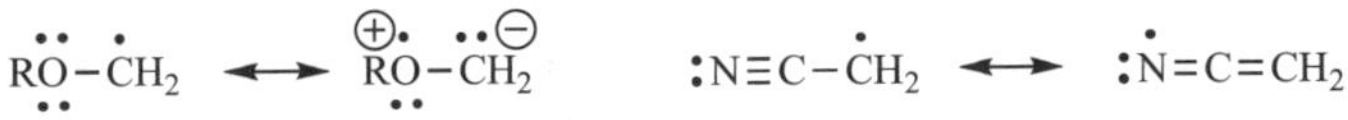
RO−CH2
RO−CH2
:N≡C−CH2
:N=C=CH2

pi donor stabilization pi acceptor stabilization

Example problem

Which of the following, $(CH_3)_3C\cdot$, $CH_3CH_2\cdot$, $PhCH_2\cdot$, is the most stable radical; which is the least stable?

Answer: The delocalized radical, $PhCH_2\cdot$, is the most stable. The least stable is the primary radical, $CH_3CH_2\cdot$, because it is less substituted than the tertiary radical, $(CH_3)_3C\cdot$. Bond strengths are $PhCH_2$–H, 88 kcal/mol (368 kJ/mol); $(CH_3)_3C$–H, 93kcal/mol (389 kJ/mol); CH_3CH_2–H, 98 kcal/mol (410 kJ/mol).

Interaction Diagrams for Radical Species

(A supplementary, more advanced explanation)

The stabilization of a radical by conjugation with a carbon–carbon pi bond can be understood if we consider the interaction of a singly occupied *p* orbital with a carbon–carbon double bond to form the allylic system. Although the singly occupied orbital is not changed in energy, the two electrons in the π orbital are stabilized by the interaction (Fig. 11.1).

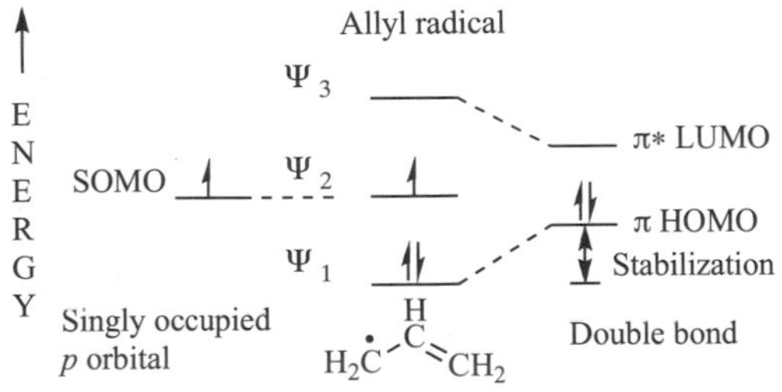
Allyl radical
ENERGY
Ψ3
Ψ2
Ψ1
SOMO
π* LUMO
π HOMO
Stabilization
Singly occupied
p orbital
Double bond
H
C
H2C
CH2

Figure 11.1 The interaction of a singly occupied *p* orbital with the π and π* orbitals of a double bond to form the three molecular orbitals of an allylic radical.

The odd electron resides in Ψ_2 of the allyl system and is called the **singly occupied molecular orbital (SOMO)**. The SOMO describes the distribution of the odd electron over the molecule. The resonance forms are an attempt to describe this SOMO in the language of lines and dots (Fig. 11.2).

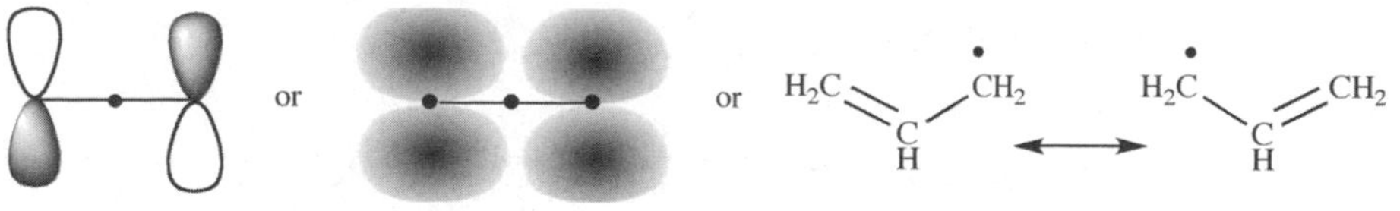
or
or
H2C
CH2
C
H
H2C
CH2
C
H

Figure 11.2 Alternate descriptions of Ψ_2 of the allyl radical.

Orbital interaction diagrams can easily explain how an adjacent radical center is stabilized by a pi acceptor (Fig. 11.3*a*). The pi acceptor has a low-lying empty orbital, LUMO, close in energy to the singly occupied molecular orbital, SOMO. The interaction of the SOMO with the empty LUMO forms a new pair of orbitals, bonding and antibonding. The new bonding orbital is lower in energy than either the original SOMO or LUMO. The single electron now resides in this new bonding orbital and is stabilized relative to a system that does not have this SOMO–LUMO interaction.

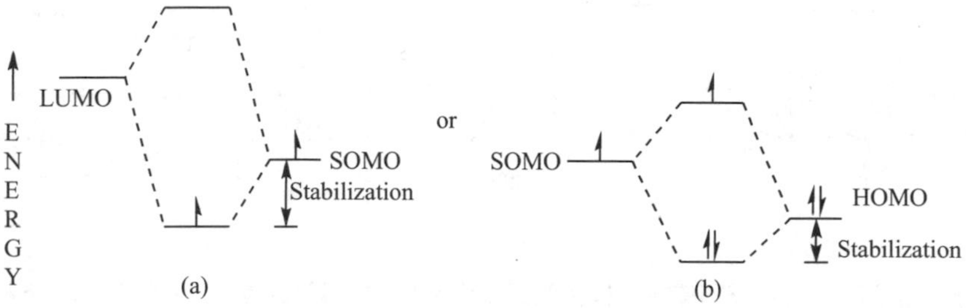
ENERGY
LUMO
SOMO
Stabilization
(a)
or
SOMO
HOMO
Stabilization
(b)

Figure 11.3 (*a*) The interaction of a singly occupied molecular orbital with the lowest unoccupied molecular orbital of a pi acceptor. (*b*) The interaction of a singly occupied molecular orbital with the highest occupied molecular orbital of a pi donor.

Similarly, orbital interaction diagrams can also explain the stabilization of a radical center by a pi donor (Fig. 11.3*b*). The pi donor has an accessible full orbital, HOMO, close in energy to the orbital bearing the single electron, SOMO. The interaction of the SOMO with a full HOMO destabilizes one electron and stabilizes two for a net stabilization of one electron overall. We have created a pi bond with one electron in the antibonding orbital; the pi bond order is thus half (our resonance structures were unable to indicate this partial pi bond with lines and dots).

11.2 RADICAL PATH INITIATION

Homolytic Cleavage by Heat or Light

Y–Z
Y
+
Z

Since almost all stable species are even electron and spin paired (molecular oxygen is the obvious exception), radical processes must start either by single electron transfer or by homolytic cleavage. It is important to recognize these initiation steps in order to know when to use the one-electron paths from this chapter rather than the two-electron paths described previously. Half-headed arrows are used to symbolize the movement of one electron. Homolytic cleavage is the simple extension of a bond-stretching vibration (Fig. 11.4). The process is **always** endothermic, for the barrier must be at least equal to the strength of the bond cleaved.

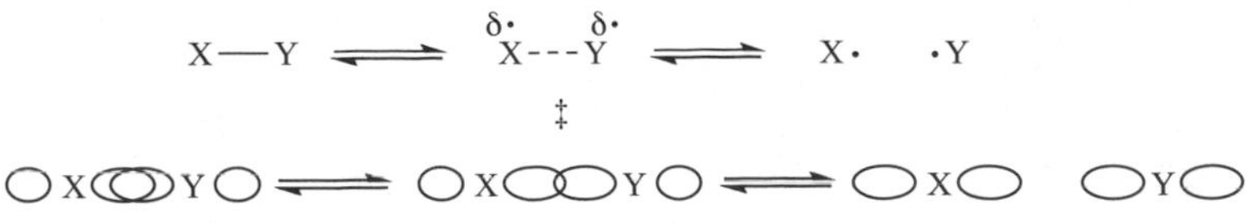

Figure 11.4 Homolytic cleavage of an X–Y bond with the transition state shown in the center.

The easiest bonds to break are weak sigma bonds between heteroatoms. Peroxide oxygen–oxygen single bonds, RO–OR, and the halogen–halogen single bonds, X–X, are easily broken by heat or light. Azo compounds, R–N=N–R, are also cleaved by heat or light into two R radicals and molecular nitrogen. A common radical initiator is azobisisobutyronitrile(AIBN), $(CH_3)_2C(CN)–N=N–C(CN)(CH_3)_2$.

The following are some common radical initiators and their cleavage into radicals.

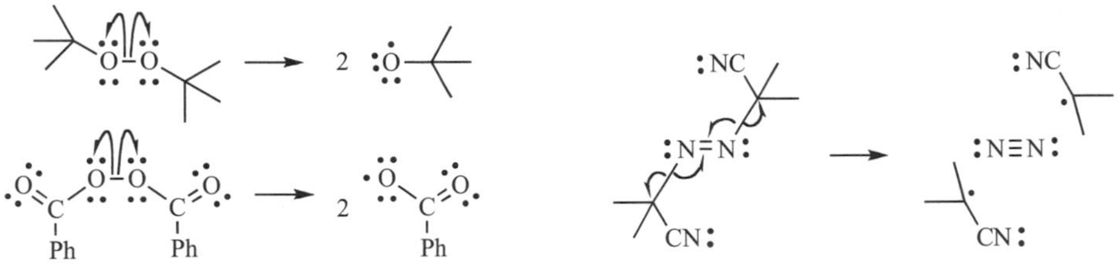

11.3 MAJOR PATHS FOR RADICALS REACTING WITH NEUTRALS

These two important reactions make up the core of most radical processes.

Abstraction

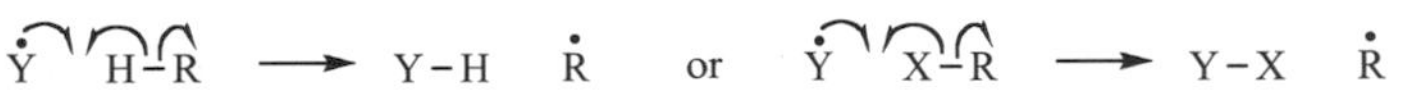

The initial radical abstracts an atom from a neutral molecule, giving a new radical (Fig. 11.5). The atom abstracted is commonly a hydrogen or a halogen atom.

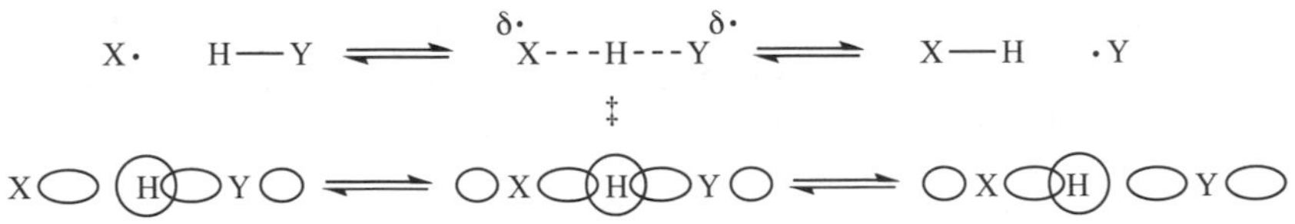

Figure 11.5 The abstraction of a hydrogen atom with the transition state shown in the center.

Overlap: No overlap restrictions other than the incoming radical center must collide with the atom to be abstracted.

Selectivity: There is a preference to abstract the hydrogen atom that would produce the most stable radical (breaking the weakest bond is energetically favored). Table 11.1 can usually be used to predict which C–H bond abstraction is preferred. Exceptions occur with abstractions by very electronegative radicals such as chlorine radical. With $CH_3CH_2CH_2$–ewg, chlorine radical preferentially abstracts the CH_2 away from the ewg since that CH_2 bears less of a partial plus.

Energetics: Bond strength tables allow the calculation of the ΔH for the process. Highly exothermic abstractions tend to have poor selectivity. For example, C–H abstractions by a bromine radical (endothermic) are much more selective than C–H abstractions by a chlorine radical (exothermic).

Hydrogen abstraction can occur intramolecularly and is useful for functionalizing rather remote unactivated sites. With freely rotating alkyl systems, there is a great preference for a five- or six-membered cyclic transition state. With rigid or conformationally restricted systems, molecular models are necessary to predict the site of intramolecular hydrogen abstraction.

Variations: Halogens are also commonly abstracted by radicals, and the reaction favors the weakest C–X bond, I > Br > Cl > F. Sulfur and selenium are also capable of being abstracted by radicals, with a transition state reminiscent of an S_N2 substitution.

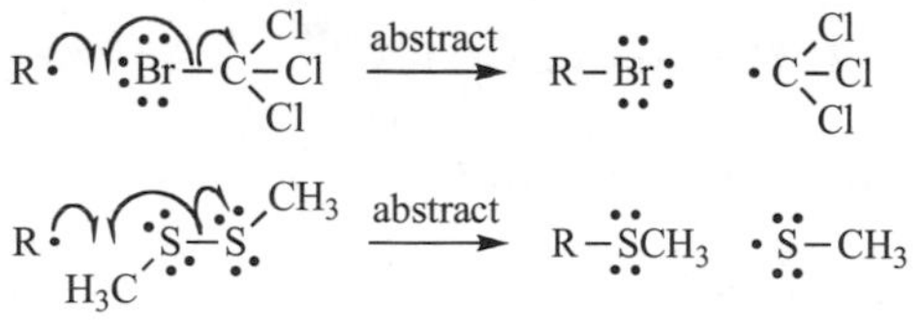
abstract
abstract

Example problem

Calculate the ΔH for abstraction of an H from methane by fluorine radical.

Answer: A methane C–H bond is broken, 105 kcal/mol (439 kJ/mol), and an H–F bond is formed, 135 kcal/mol (565 kJ/mol). The ΔH reaction is bonds broken – bonds made = 105 – 135 = –30 kcal/mol (126 kJ/mol). This abstraction is very exothermic.

Addition

The initial radical adds to a pi bond to form a new radical (Fig. 11.6).

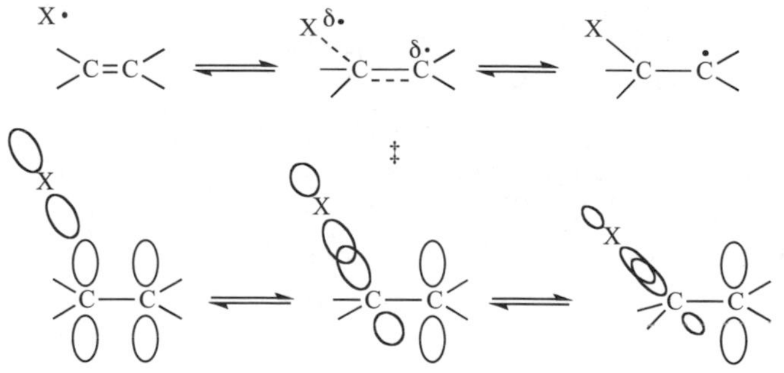

Figure 11.6 The addition of a radical to a pi bond with the transition state shown in the center.

Overlap: Calculations indicate that the preferred direction of attack of the radical on the end of the pi bond is at an angle of about 70° above the plane of the double bond (or approximately at 110° for the X–C–C angle in the figure).

Selectivity: Addition usually forms the more stable radical. Exceptions are relatively common since the radical stability difference can be less than a polar or steric effect.

Energetics: Since a pi bond is broken and usually a much stronger sigma bond is formed, this path is normally exothermic.

Variations: Cyclization is an intramolecular addition reaction. The preferred ring sizes for cyclization are usually 5 > 6 > 7 > 8; ring opening in these systems is rare. The closure to a four-membered ring has not been observed. Closure to a three-membered ring does occur, but the ring opening is preferred by a factor of 12,000.

Example problem

Calculate the ΔH of an ethyl radical adding to ethene.

Answer: A C=C bond is broken, 146 kcal/mol (611 kJ/mol), and two C–C single bonds are formed, 83 kcal/mol (347 kJ/mol) for each C–C. The ΔH for the reaction is 146 – 166 = –20 kcal/mol (83 kJ/mol). The addition is exothermic.

11.4 UNIMOLECULAR RADICAL PATHS

Elimination or Fragmentation

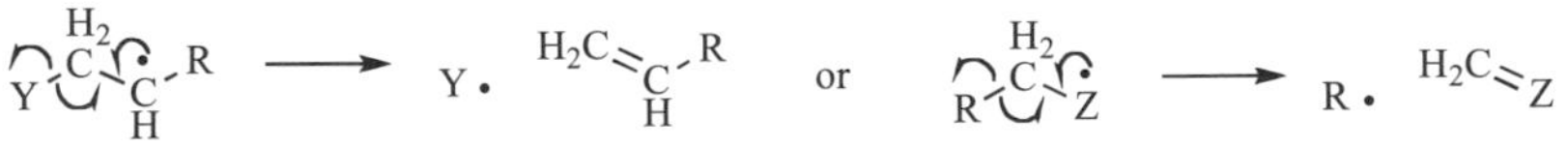

This path is the reverse of the addition path.

Overlap: By microscopic reversibility, the bond cleaved must be coplanar with the orbital bearing the odd electron (Fig. 11.6, but view the figure from right to left).

Energetics: This path is normally endothermic but can be favored by the formation of a strong pi bond and higher temperatures. When more than one group could be lost, the loss of the more stable radical is favored.

Radical decarboxylation is a very fast and common fragmentation reaction:

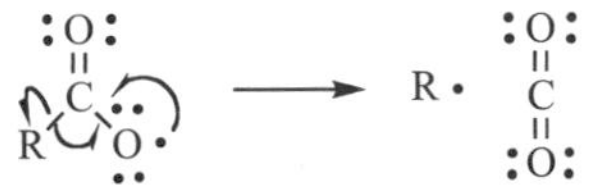

Decarbonylation

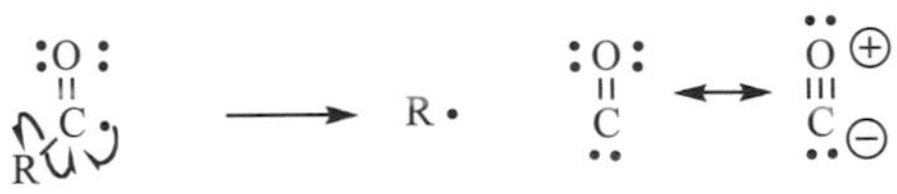

The loss of carbon monoxide is an unusual reaction. If the radical formed upon decarbonylation is not stabilized, addition or abstraction can easily compete.

Rearrangement

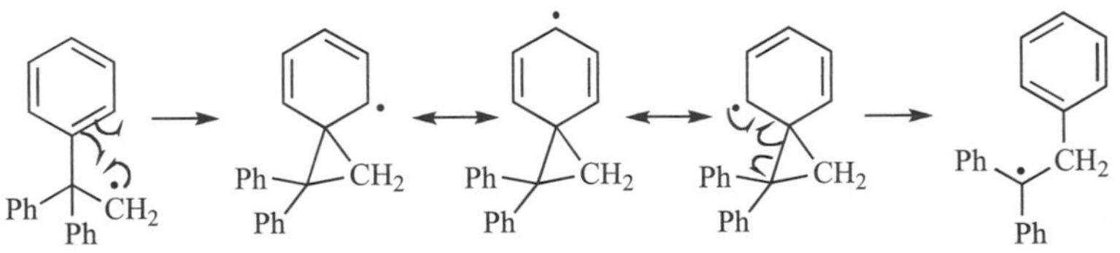

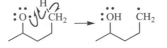

Rearrangements of radical species are rare. Migration occurs with unsaturated (aryl, vinyl, acyl) or halogen groups that can form a bridged intermediate. Rearrangement with unsaturated groups goes via intramolecular radical addition to form the three-membered ring followed by ring opening. The driving force for rearrangement is the formation of an equal or more stable radical or the relief of steric strain. Another way a radical can rearrange is by internal atom abstraction or fragmentation.

11.5 TERMINATION RADICAL PATHS

These two paths combine radicals to give nonradical products. These terminate the radical chain mechanisms discussed in Section 11.6, and so are often undesired but important reactions.

Radical-Radical Coupling

The coupling of two radicals to form a sigma bond is the reverse of the homolytic cleavage reaction and occurs without any activation barrier. Therefore, we might expect this reaction to be exceedingly fast. However, the rate of a reaction depends not only on the barrier height but also on the concentrations of the reacting species. Since the radical concentration in a usual reaction is low, radical–radical collisions are infrequent.

An obvious exception occurs when the reaction mixture is so inert that there are few species for radicals to react with except each other. Another exception is when both radicals are formed close together (for example, azo compound cleavage) and can react before they diffuse apart. Often collision with a solvent molecule will send one newly formed radical into the other, causing reaction; this is called the solvent cage effect. A third exception is that radicals react with molecular oxygen (which behaves as a diradical) to form R-O-O·, hydroperoxide radicals. Generally radical reactions are run in the absence of oxygen to prevent this reaction from occurring.

Radical-Radical Disproportionation

Alternative to coupling, two radicals on collision can undergo disproportionation. Disproportionation occurs when one radical abstracts a hydrogen atom adjacent to the other radical center.

11.6 RADICAL PATH COMBINATIONS

Chain reactions are characteristic of radicals. The radical chain is started with an *initiation* step. A chain *propagation* sequence occurs in which the radical species needed for the first propagation step is generated in the last step, and the sequence repeats many times. For a chain to be sustained, the overall process must be exothermic, and each individual step must have a low barrier. The chain undergoes *termination* whenever a coupling or disproportionation reaction removes the radical(s) needed to propagate the chain. If a propagation step is endothermic by more than 20 kcal/mol (84 kJ/mol) (the reaction barrier must be greater than this), this propagation step is slow enough that termination steps easily compete, stopping the chain. The average number of times a chain reaction repeats before termination, commonly in the thousands or more, is called the *chain length*.

The way to control a radical chain reaction is to control the initiation and termination steps. Radical chain reactions can be favored by adding radical initiators. Likewise, chain reactions can be greatly diminished by adding compounds called *inhibitors* that react with radicals to increase chain termination. The sensitivity of the radical reaction to radical initiators and inhibitors provides a convenient way to test for this mechanism.

S_H2, Substitution Chain (Substitution, Homolytic, Bimolecular)

abstract
abstract

The two propagation steps in the substitution chain are both radical abstractions. Radical chain halogenations illustrate different types of chain energetics. The CH_3–X bond strengths are 108 (452), 83.5 (349), 70 (293), and 56 (234) for F, Cl, Br, and I, respectively in kcal/mol (kJ/mol). The total reaction is the sum of the propagation steps with radical species that appear on both sides of the transformation arrow, canceling out. Initiation or termination steps are not included. The initiation step for halogenations is usually heat- or light-induced homolytic cleavage of the X–X bond into two X· radicals.

Fluorination radical chain (explosive)

Propagation step	CH_4 + F· → H–F + ·CH_3	ΔH = −30 (−126)
Propagation step	·CH_3 + F_2 → F–CH_3 + F·	ΔH = −71 (−297)
Total reaction	CH_4 + F_2 → F–CH_3 + H–F	ΔH = −101 (−423)

Chlorination radical chain (fast)

Propagation step	CH_4 + Cl· → H–Cl + ·CH_3	ΔH = +2.0 (+8)
Propagation step	·CH_3 + Cl_2 → Cl–CH_3 + Cl·	ΔH = −25.5 (−106)
Total reaction	CH_4 + Cl_2 → Cl–CH_3 + H–Cl	ΔH = −23.5 (−98)

Bromination radical chain (slower)

Propagation step	CH_4 + Br· → H–Br + ·CH_3	ΔH = +18 (+75)
Propagation step	·CH_3 + Br_2 → Br–CH_3 + Br·	ΔH = −24 (−100)
Total reaction	CH_4 + Br_2 → Br–CH_3 + H–Br	ΔH = −6 (−25)

Iodination radical chain (does not occur)

Propagation step	CH_4 + I· → H–I + ·CH_3	ΔH = +34 (+142)
Propagation step	·CH_3 + I_2 → I–CH_3 + I·	ΔH = −20 (−84)
Total reaction	CH_4 + I_2 → I–CH_3 + H–I	ΔH = +14 (+58)

The first propagation step of the bromination radical chain is significantly endothermic. A small change in product radical stability is reflected in the barrier for reaction and consequently in the rate of reaction. Thus the first propagation step (and the chain reaction) that goes the fastest forms the most stable product radical. Bromination will select for allylic and benzylic > tertiary > secondary > primary > vinyl and phenyl.

Allylic bromination is usually done with *N*–bromosuccinimide (NBS), which keeps the concentration of bromine low by reacting with the HBr formed in the first propagation step to produce the bromine needed for the second propagation step. This low bromine concentration suppresses the addition chain reaction, discussed next, by allowing time for the addition step to reverse before a bromine molecule is encountered.

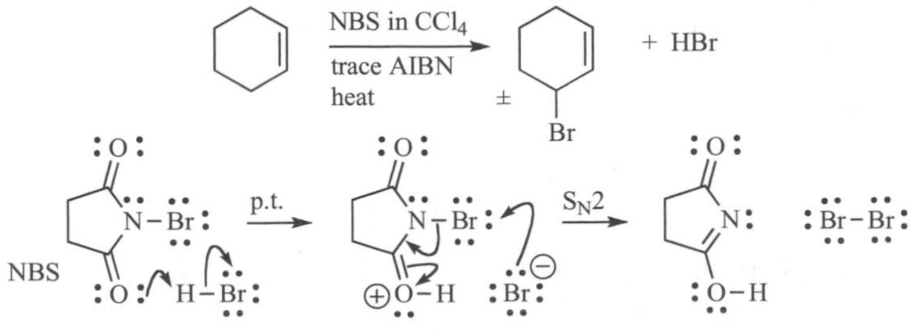

NBS in CCl_4
trace AIBN
heat
±
Br
+ HBr
NBS
p.t.
S_N2

The radical chain dehalogenation with $Bu_3Sn{-}H$ is a very useful reaction. The ease of dehalogenation follows the C–X bond strength: I > Br > Cl > F; the weakest C–X bond is preferred. The initiator for this reaction is commonly AIBN. An example mechanism involving this reaction is shown in Section 11.7, Approaches to Radical Mechanisms.

Dechlorination with Bu_3SnH

Propagation step	$R{-}Cl + Bu_3Sn\cdot \rightarrow Bu_3Sn{-}Cl + R\cdot$	$\Delta H = -39(-163)$
Propagation step	$R\cdot + Bu_3Sn{-}H \rightarrow R{-}H + Bu_3Sn\cdot$	$\Delta H = -19\ (-80)$
Total reaction	$R{-}Cl + Bu_3Sn{-}H \rightarrow Bu_3Sn{-}Cl + R{-}H$	$\Delta H = -58(-243)$

The oxidation radical chain (auto-oxidation) is very important in the spoiling of foodstuffs. The mechanism of auto-oxidation is complex, for hydroperoxide radicals easily add to multiple bonds, and those products also can cleave. The following is a much-simplified scheme. Free radical chain inhibitors are often added to the packaging of foods to retard this reaction.

Auto-oxidation

Propagation step	$O_2 + R\cdot \rightarrow R{-}O{-}O\cdot$
Propagation step	$R{-}O{-}O\cdot + R{-}H \rightarrow R{-}O{-}O{-}H + R\cdot$
Total reaction	$O_2 + R{-}H \rightarrow R{-}O{-}O{-}H$

Example problem

Predict the major radical substitution chain products of 1-phenylpropane heated with *t*-butyl hypochlorite.

Answer: The thermal homolysis of the weak oxygen–chlorine bond in *t*-butyl hypochlorite produces a *t*-butoxy radical that starts the chain. This radical will abstract a hydrogen atom from the benzylic methylene of 1-phenylpropane to give a resonance-delocalized benzylic radical, the most stable of all the possible alternatives. The propagation loop completes when the benzylic radical abstracts a chlorine atom from *t*-butyl hypochlorite and creates a *t*-butoxy radical to start the process over again. The products are 1-chloro-1-phenylpropane and *t*-butanol.

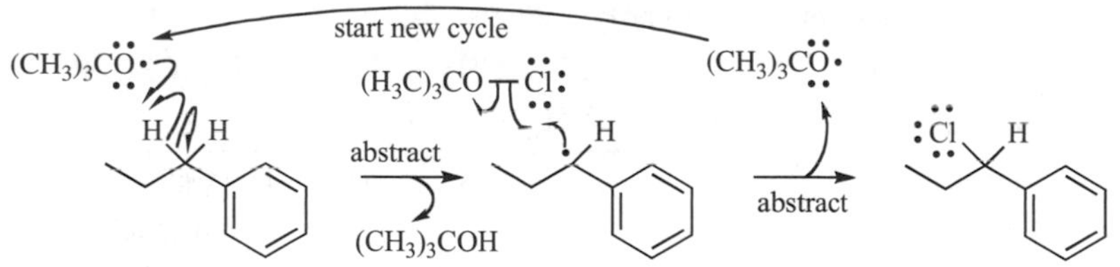
start new cycle
$(CH_3)_3CO\cdot$
$(H_3C)_3CO$–Cl
$(CH_3)_3CO\cdot$
H
H
H
Cl
H
abstract
$(CH_3)_3COH$
abstract

Ad_H2, Addition Chain (Addition, Homolytic, Bimolecular)

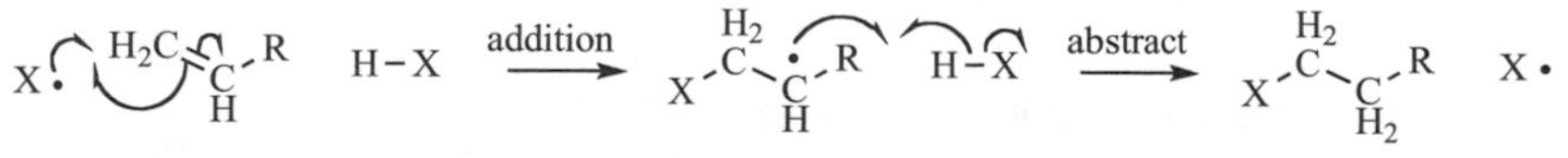
X
H_2C
C
R
H
H–X
addition
H_2
C
X
C
R
H
H–X
abstract
H_2
C
X
C
R
H_2
X•

The first propagation step in the addition chain is a radical addition. The second propagation step in the addition chain is a radical abstraction. Radical addition of HBr is a typical reaction. Bromine radical adds to the multiple bond to form the most stable of the possible radicals. The radical addition of HBr gives the opposite regiochemistry of addition as the polar addition of HBr.

Radical addition of HBr

Propagation	$CH_2{=}CH_2$ + Br· → ·CH_2–CH_2–Br	ΔH = –5(–21)
Propagation	·CH_2–CH_2–Br + H–Br → H–CH_2–CH_2–Br + Br·	ΔH = –12(–50)
Total	$CH_2{=}CH_2$ + H–Br → H–CH_2–CH_2–Br	ΔH = –17(–71)

The radical addition of HCl is not as useful since polymerization tends to compete with the addition chain. The radical addition of HI or HF does not occur because in each case a propagation step is too endothermic.

Example problem

Check the ΔH of the propagation steps for the radical chain addition of RSH to an alkene to see whether it is an energetically reasonable process.

Answer:

Step 1	$CH_2{=}CH_2$ + RS· → ·CH_2–CH_2–SR	ΔH = –2 (–8)
Step 2	·CH_2–CH_2–SR + H–SR → H–CH_2–CH_2–SR + RS·	ΔH = –16 (–67)
Total	$CH_2{=}CH_2$ + H–SR → H–CH_2–CH_2–SR	ΔH = –18 (–75)

The radical chain addition of RSH to an alkene is an energetically reasonable process because there are no steps more than 20 kcal/mol (84 kJ/mol) endothermic, and the entire process is exothermic.

Polymerization Chain

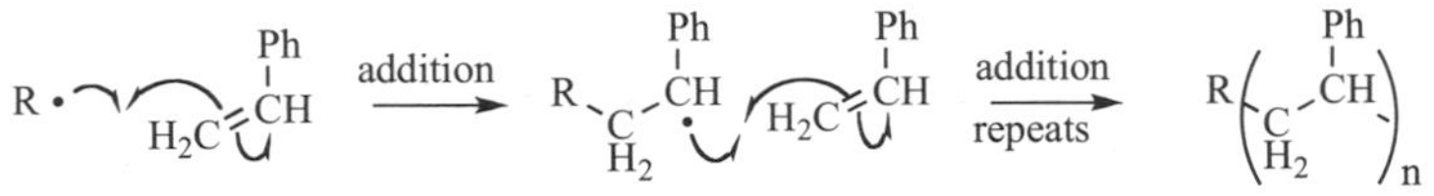
R
Ph
CH
H2C
addition
R
C
H2
CH
Ph
H2C
CH
Ph
addition
repeats
R
C
H2
CH
Ph
n

Polymerization is the repetition of an addition propagation step, like the polymerization of styrene to give polystyrene above. The initiator radical adds to a multiple bond to produce another radical that adds again and again until termination occurs. Polymerization is favored by a high concentration of the multiple bond reactant.

Polymerization of ethylene to polyethylene

Initiation step	$CH_2{=}CH_2$ + R· → ·CH_2–CH_2–R
Repeated propagation step	$CH_2{=}CH_2$ + ·CH_2–CH_2–R → ·$(CH_2–CH_2)_2$–R
Total reaction (after termination step)	n $CH_2{=}CH_2$ + R· → H–$(CH_2–CH_2)_n$–R (and other terminations)

11.7 APPROACHES TO RADICAL MECHANISMS

Common Radical Mechanism Errors

Because radicals are in minute concentration, the usual radical mechanisms involve a radical colliding with an even electron molecule in a chain process. A common error is to have a termination step instead of creating a regenerating loop. The three common radical path combinations, S_H2, Ad_H2, and radical polymerization, all have propagation steps in which radicals collide with an even electron species. creating a new radical.

Usually, the last propagation step creates the starting radical for the first propagation step, thus making a loop.

Initiators are added to the reaction in trace amounts and are just to get the radical chain loop going. The loop often goes for thousands of cycles before termination. Do not include this trace initiator in your propagation steps. Make your radical loop out of reactants present in good amounts.

Radical Mechanism Example

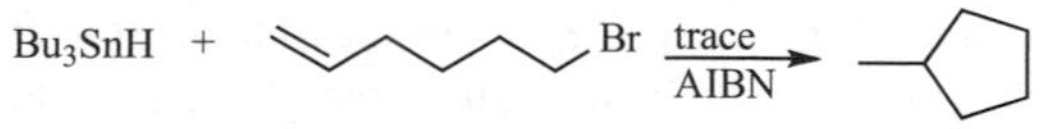
Bu3SnH +
Br
trace
AIBN

First, balancing the reaction and numbering the carbons of the reactant and product will give us more information about the reaction. Since carbons 1, 5, and 6 in the reactant were functionalized in the reactant, the most logical numbering of the product is:

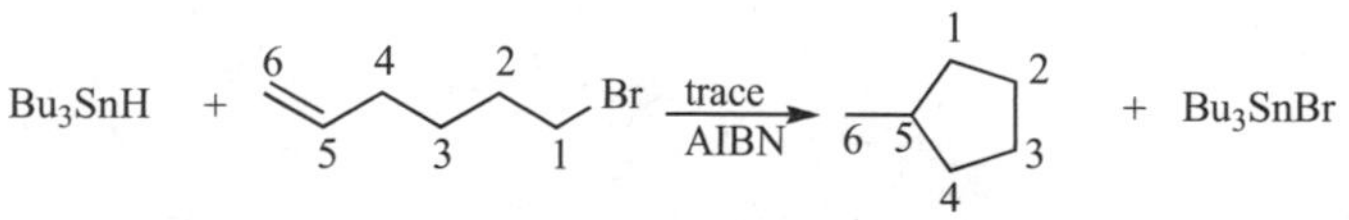
Bu3SnH +
6 4 2
5 3 1
Br
trace
AIBN
1 2 3 4 5 6
+ Bu3SnBr

In figuring out what bonds have changed, it may be helpful to draw in all the hydrogens. Changes are: carbon 6 has gone from a methylene to a methyl, so a new C–H bond was formed there; the bromine atom has been lost from carbon 1; finally, a new single bond between carbon 1 and carbon 5 has formed.

A radical initiator, AIBN, is present, and the product is suggestive of an intramolecular radical addition chain. To begin, homolytic cleavage of the AIBN initiator is used to create radicals that will begin the chain.

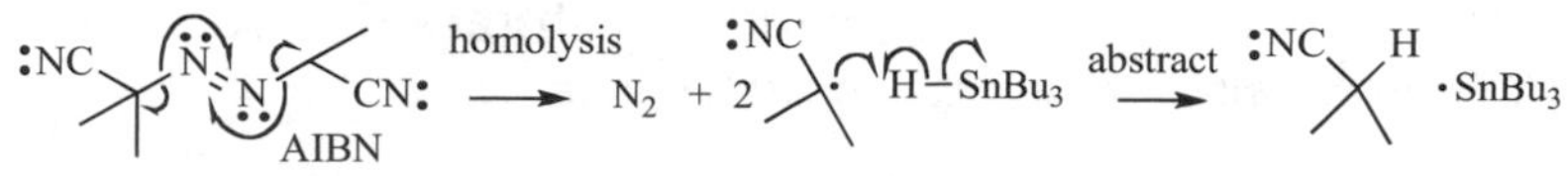
:NC
N
N
CN:
AIBN
homolysis
N2 + 2
:NC
H
SnBu3
abstract
:NC
H
·SnBu3

Now the propagation steps can start by having our tributyltin radical abstract a bromine atom from the alkyl bromide.

Br:
C
H2
·SnBu3
abstract
CH2
:BrSnBu3

This gives us the tributyltin bromide product and a primary carbon radical with which to do an intramolecular radical addition to a pi bond.

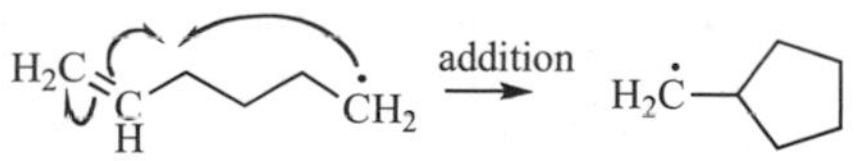
H2C
C
H
CH2
addition
H2C

Now it is time to close the loop and create the radical from the first propagation step, so that the loop can continue.

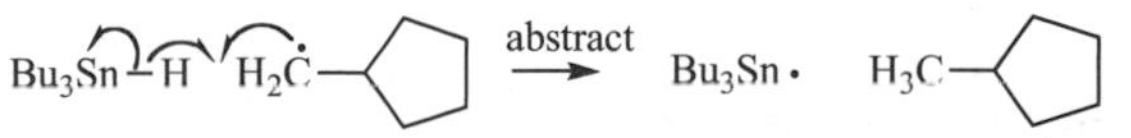
Bu3Sn
H
H2C
abstract
Bu3Sn·
H3C

Ideally, the propagation loop turns over many thousands of times before a termination step quenches the radical.

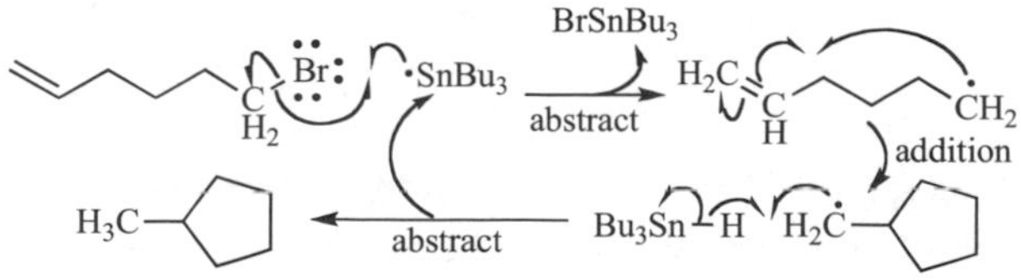
Br:
C
H2
·SnBu3
BrSnBu3
abstract
H2C
C
H
CH2
addition
H3C
abstract
Bu3Sn
H
H2C

11.8 SINGLE-ELECTRON TRANSFER, S.E.T., AND CHARGED RADICALS

$$A + e^- \longrightarrow A^{\overline{\bullet}} \qquad \text{or} \qquad A{:} - e^- \longrightarrow A^{\overset{+}{\bullet}}$$

Single-electron transfer is the one-electron oxidation or reduction of a species; no covalent bond formation occurs. Single-electron transfer to or from another chemical species or an electrode forms charged radicals. The chemical species must be picked using standard electrode potentials such that the electron transfer is energetically favorable. Single-electron transfers using an electrode are more versatile since the electrical potential of the electrode can be accurately set over a wide range.

Single-electron transfer from an alkali metal such as lithium, sodium, or potassium is a common method of producing a radical anion. A radical anion has its extra electron in an antibonding orbital, which weakens the corresponding bonding orbital, and makes the radical anion more reactive than the neutral species.

+ Na•
s.e.t.
Na

A radical cation can be chemically produced by single-electron transfer to commercially available stable amine radical cation salts. Again, the radical ion is more reactive than the neutral species.

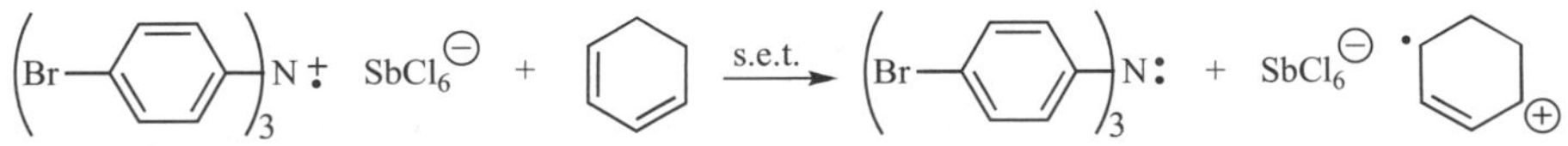
Br
N
SbCl6
s.e.t.
Br
N:
SbCl6

A Molecular Orbital Explanation of Single-Electron Transfer
(A Supplementary, More Advanced Explanation)

Single-electron transfer reduction adds one electron into the LUMO of the acceptor molecule. The lower in energy the acceptor LUMO is, the easier it is to reduce; therefore conjugated systems are easier to reduce than unconjugated systems. Single-electron transfer oxidation removes one electron from the HOMO of the donor molecule. The higher in energy the donor HOMO is, the easier it is to oxidize. For this reason conjugated systems are also easier to oxidize than unconjugated systems.

As electrons in the donor HOMO get higher in energy (as in nonbonded electron pairs on less electronegative elements), they can approach the energy of the LUMO of the acceptor. The single-electron transfer is then easily triggered by light or, if the gap is small enough, by thermal energy.

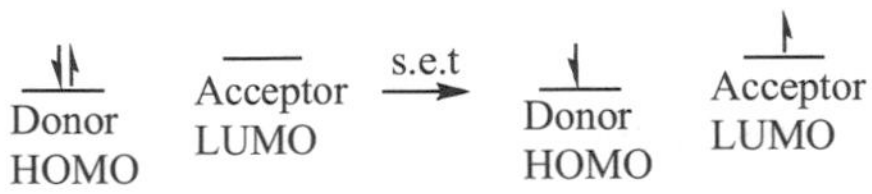
Donor
HOMO
Acceptor
LUMO
s.e.t
Donor
HOMO
Acceptor
LUMO

Organometallic Formation

The formation of organomagnesiums is believed to occur by single-electron transfer that triggers the loss of the halide followed by a second single-electron transfer.

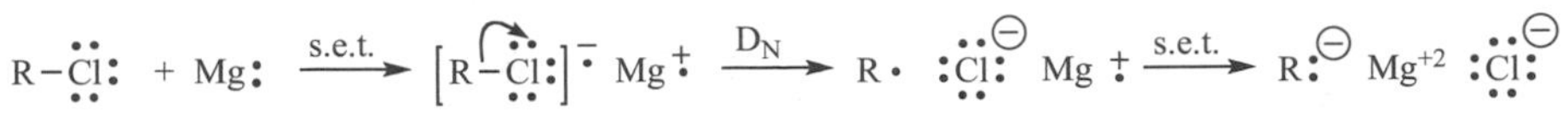
R–Cl: + Mg:
s.e.t.
Mg
DN
R•
Mg
s.e.t.
R:
Mg+2

The lowest unoccupied molecular orbital (LUMO) of R–L is usually the antibonding orbital of the C–L bond. When an electron is transferred into this LUMO, the result is two electrons in the bonding orbital and one in the antibonding for a net one-half bond order. This weakened C–L bond is now much easier to break.

11.9 DISSOLVING METAL REDUCTIONS

Alkali metals added to protic solvents in the presence of another reactant provide the opportunity for both single-electron transfer and two-electron processes to occur and are called dissolving metal reductions. Sodium metal dissolves in liquid ammonia to give a dark blue solution characteristic of a solvated electron. This highly reducing solution reacts with alkynes via single-electron transfer, forming a radical anion, protonation of this radical anion to give the neutral vinyl radical, and a second single-electron transfer to the radical, yielding a vinyl anion that then protonates. In this manner alkynes are reduced to *trans* alkenes. The second protonation gives the more stable *trans* alkene.

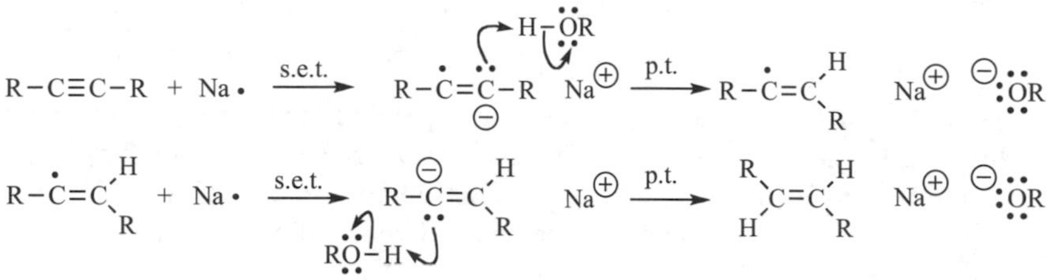

Aromatics and other conjugated systems can also be reduced by sodium in liquid ammonia with a trace of alcohol for a proton source.

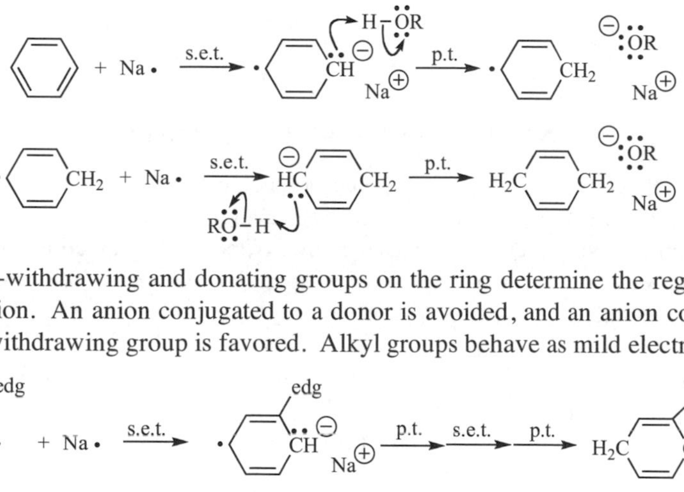

Electron-withdrawing and donating groups on the ring determine the regiochemistry of the reduction. An anion conjugated to a donor is avoided, and an anion conjugated to an electron-withdrawing group is favored. Alkyl groups behave as mild electron donors.

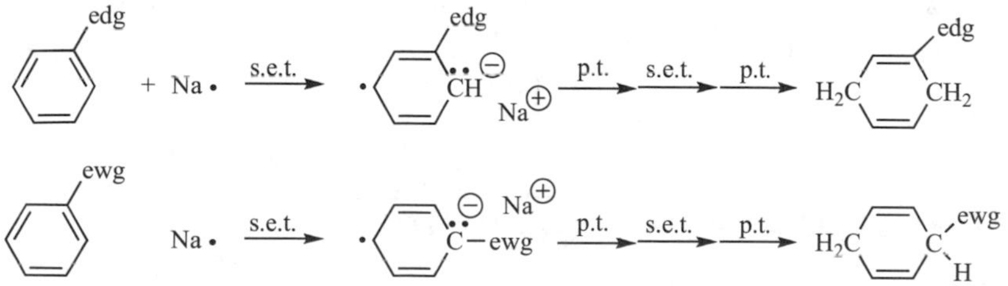

Another mixed process is a way to remove halogens or other leaving groups. Single-electron transfer forms a radical anion, which loses the leaving group anion to give the radical; a second single-electron transfer to the radical yields the anion, which protonates.

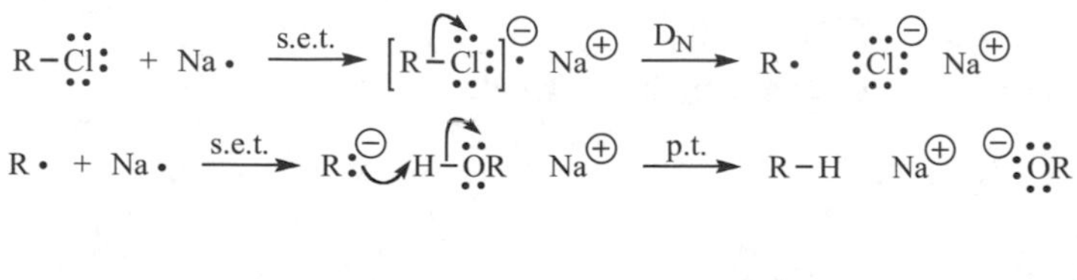

11.10 ELECTRON TRANSFER-INITIATED PROCESSES

$S_{RN}1$ Process (Substitution, Radical–Nucleophile, Unimolecular)

The $S_{RN}1$ process is a chain reaction that has both single-electron transfers and two-electron processes. A single-electron transfer forms a radical anion, which then loses the leaving group, forming the neutral radical that is attacked by a nucleophile to form a new radical anion. This new radical anion serves as the electron source for the initial single electron transfer. Aryl diazonium ions, $ArN{\equiv}N^+$, may react with nucleophiles this way.

Initiation step	$R{-}L + e^{\ominus} \rightarrow R{-}L^{\overset{\ominus}{\bullet}}$	then
Propagation step	$R{-}L^{\overset{\ominus}{\bullet}} \rightarrow R\bullet + :L^{\ominus}$	
Propagation step	$R\bullet + :Nu^{\ominus} \rightarrow R{-}Nu^{\overset{\ominus}{\bullet}}$	
Propagation step	$R{-}Nu^{\overset{\ominus}{\bullet}} + R{-}L \rightarrow R{-}L^{\overset{\ominus}{\bullet}} + R{-}Nu$	
Total reaction	$R{-}L + :Nu^{\ominus} \rightarrow R{-}Nu + :L^{\ominus}$	

If R is alkyl, commonly the nucleophile and/or the R group will bear a nitro group so that the single-electron transfer is favorable. If R is aryl, initiation can be by single-electron transfer from an electrode or an alkali metal. Initiation by light, hν, is also possible. The $S_{RN}1$ process can be stopped with radical inhibitors. Systems that would not have reacted by an S_N1 or S_N2 process may react by an $S_{RN}1$ mechanism since the R–L bond is considerably weakened in the radical anion compared to the neutral.

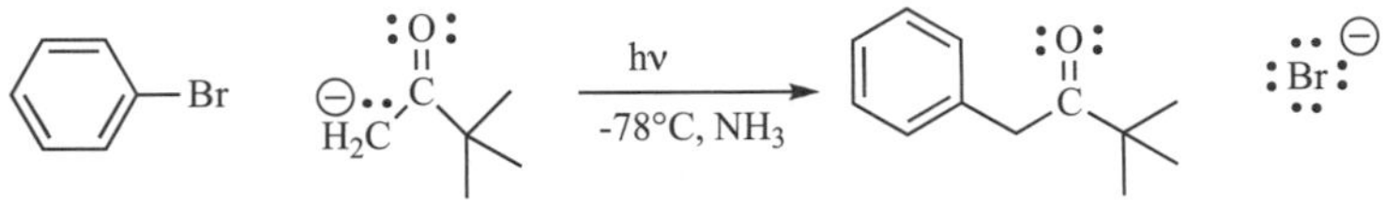
Br
hν
-78°C, NH3

Other Electron Transfer-Initiated Processes

The conjugate addition of some organometallics to carbonyl derivatives may occur by single-electron transfer. Enones are relatively easier to reduce than simple ketones, making the single-electron transfer more favorable. Organocopper reagents appear to use single-electron transfer followed by radical coupling in reacting with enones. The intermediates do not escape the solvent cage.

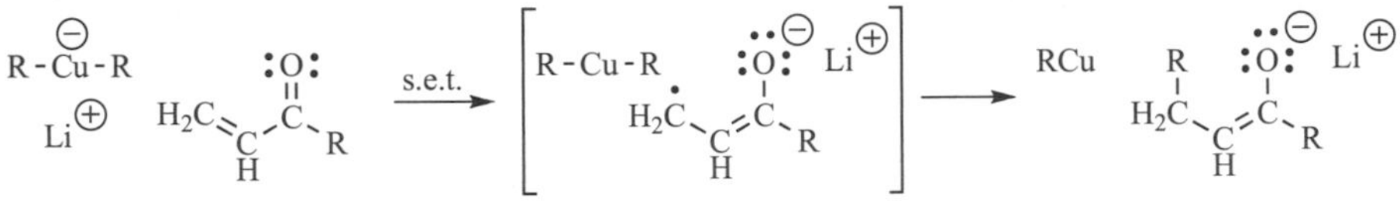
s.e.t.
RCu

11.11 ONE-ELECTRON PATH SUMMARY

Initiation Is Required

Single-electron transfer (s.e.t.)

$$A + e^- \longrightarrow A^{\overset{-}{\bullet}} \quad \text{or} \quad A\!: - e^- \longrightarrow A^{\overset{+}{\bullet}}$$

Homolytic cleavage by heat or light

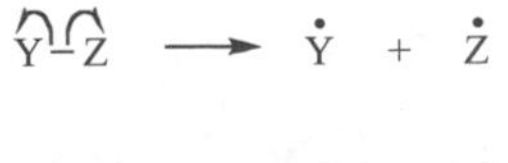

Radical Paths

Abstraction

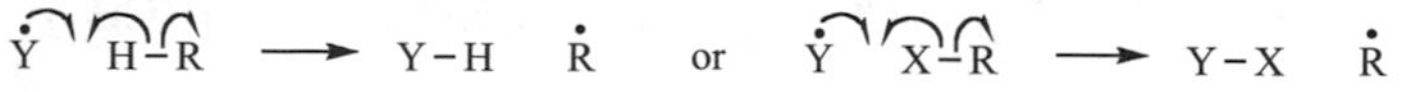

Addition

Elimination

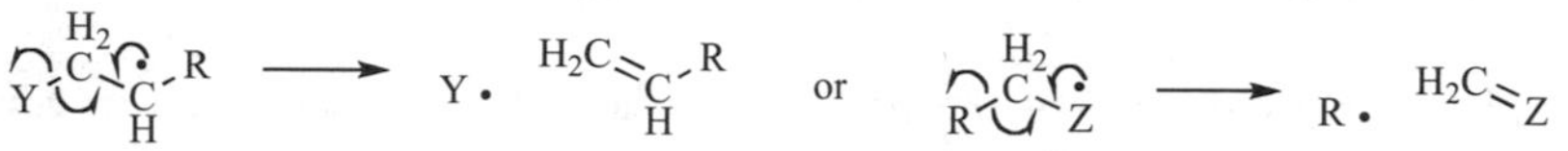

Decarbonylation

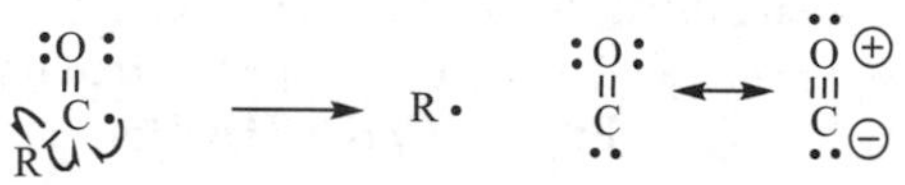

Rearrangement

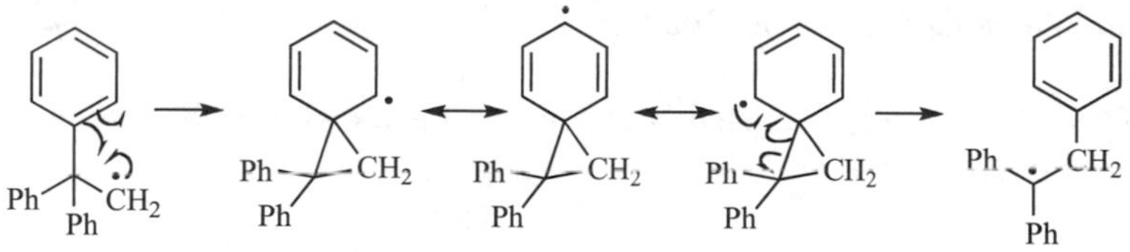

Radical Termination Paths

Coupling

Disproportionation

Common Radical Chain Reactions

Substitution Chain, S_H2

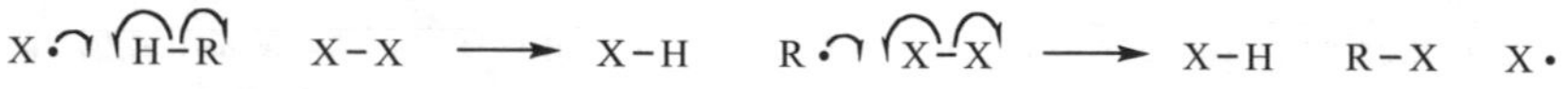

Addition Chain, Ad_H2

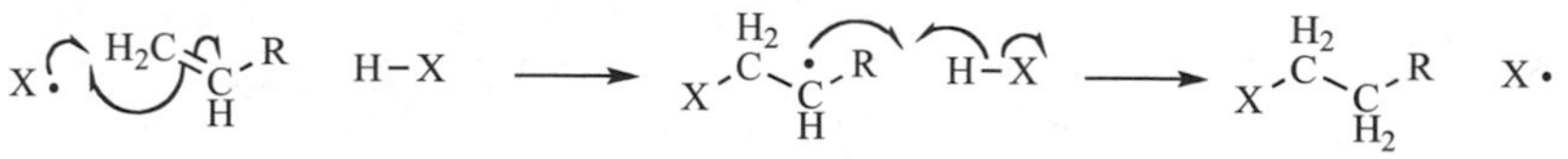

Polymerization Chain

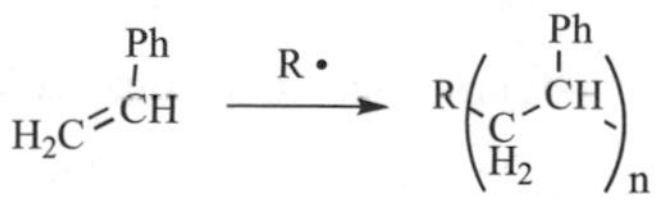

ADDITIONAL EXERCISES

11.1 Rank all species, using the numeral 1 to designate the most stable radical.

$\cdot CH{=}CH_2$ $\cdot CH_2CH{=}CH_2$ $\cdot CH_2CH_3$ $\cdot CH(CH_3)_2$ $\cdot CH_3$

11.2 Use radical stabilities to predict the regiochemistry of the radical addition chain reaction of HBr to propene (initiated by AIBN).

11.3 Calculate the ΔH for the abstraction of an H from ethane by a $\cdot CCl_3$ radical.

11.4 Calculate the ΔH of an iodine radical adding to a carbon–carbon double bond.

11.5 Check the ΔH of the propagation steps for the radical chain addition of HF to ethene to see whether it is an energetically reasonable process.

11.6 Use radical stabilities to predict the selectivity of the following radical chain substitution reaction (initiated by light). Stop after one bromine is attached.

$$Br_2 + C_6H_5CH_2CH_2CH_3 \rightarrow$$

11.7 Predict which you would expect to be more selective: radical chain chlorination or bromination of an alkane. Explain.

11.8 BHT is a radical chain terminator. Explain why H abstraction from oxygen yields a relatively unreactive radical.

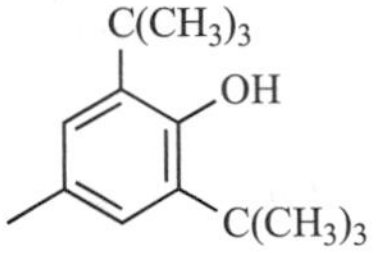
C(CH3)3
OH
C(CH3)3

11.9 Predict the product of the following radical reaction (initiated by light).

$$Br_2 + (CH_3)_2CHCH_2CH_2CH_3 \rightarrow$$

11.10 Give a mechanism for this radical chain reaction (initiated by light).

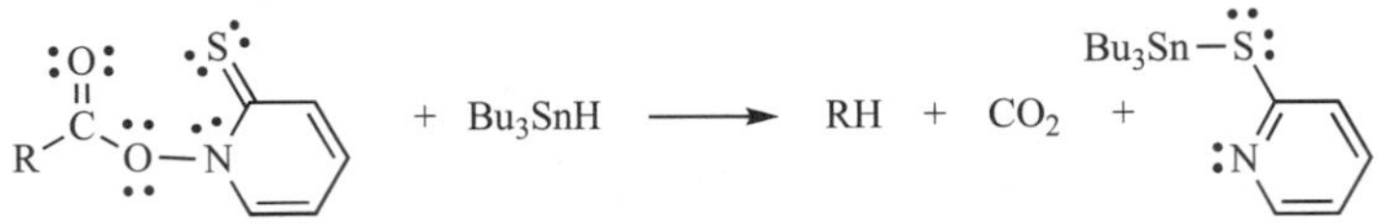
Bu3SnH
RH
CO2
Bu3Sn–S

11.11 Write out the $S_{RN}1$ process for phenyl diazonium ion reacting with iodide anion to produce nitrogen gas and phenyl iodide. Use s.e.t. from iodide anion to start the process.

11.12 Give a mechanism for this radical chain reaction (initiated by heat or light).

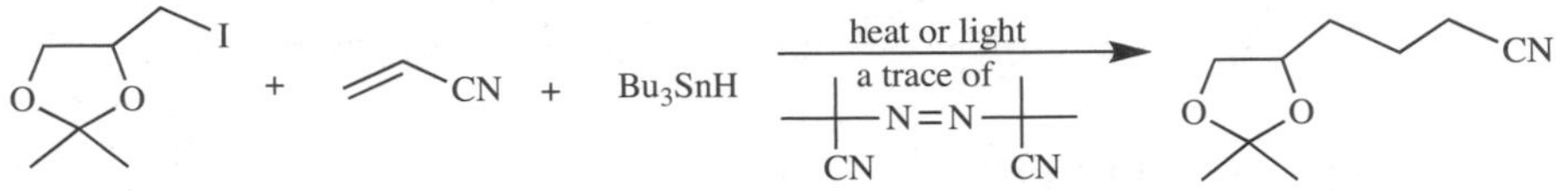
I
CN
Bu3SnH
heat or light
a trace of
N=N
CN
CN
CN

11.13 Give the product and mechanism of this radical chain addition initiated by light.

$$BrCCl_3 + H_2C{=}CEt_2 \longrightarrow$$

12

QUALITATIVE MOLECULAR ORBITAL THEORY AND PERICYCLIC REACTIONS

Electron Flow In Organic Chemistry: A Decision-Based Guide To Organic Mechanisms, Second Edition.
By Paul H. Scudder

This chapter is an introduction to qualitative molecular orbital theory and pericyclic reactions. Pericyclic reactions have cyclic transition states and electron flow paths that appear to go around in a loop. The regiochemistry and stereochemistry of these reactions are usually predictable by HOMO-LUMO interactions, so to understand them we need to understand molecular orbital theory, at least on a qualitative basis.

12.1 REVIEW OF ORBITALS AS STANDING WAVES

In Section 1.6 we saw that an electron can behave as a wave, and that standing waves are the result of a wave confined in space. These standing waves are described by mathematical equations called wave functions that describe the specific quantized orbitals that an electron may occupy. Figure 12.1 summarizes the one-dimensional set of standing waves. The lowest-frequency, longest wavelength that will fit within the limits is the first harmonic, or fundamental. The fundamental has one half-wavelength between the ends. The next one to fit is two half-wavelengths, the second harmonic. Three half-wavelengths gives the third harmonic, and so on. As we fit more half-wavelengths into this limited space, the actual wavelength, λ, gets shorter, and the frequency, ν, gets higher. The frequency is inversely proportional to the wavelength, and frequency is also directly proportional to the energy. The fundamental is of lowest energy, and each higher harmonic gets higher in energy. The nodes, shown by vertical dotted lines in the figure, are the regions where the wave goes through zero, the horizontal dotted line. When there is an odd number of half-wavelengths, as in the first and third harmonics, then the wave will have mirror symmetry about the center. These linear standing waves are very similar to molecular orbitals for linear conjugated pi systems in Section 12.2. As the number of *p* orbitals in the conjugated pi system increases, the standing wave is confined to a bigger space, so the wavelength of the fundamental gets longer and its energy drops.

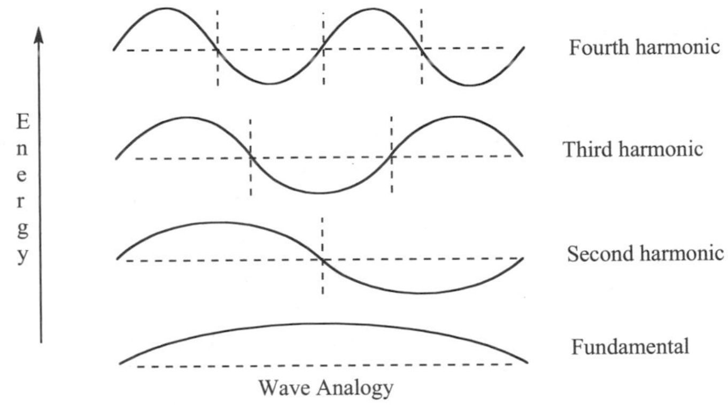
Energy
Fourth harmonic
Third harmonic
Second harmonic
Fundamental
Wave Analogy

Figure 12.1 The standing waves of a vibrating string. The fundamental (first harmonic) has no nodes; the second harmonic is half the wavelength and twice the frequency of the fundamental and has one node, shown by a dotted vertical line. Nodes increase as frequency and energy increase.

12.2 MOLECULAR ORBITAL THEORY FOR LINEAR PI SYSTEMS

12.2.1 Two *p* Orbitals, a Simple Pi Bond

Since the pi bonding portion of a molecule is perpendicular to the sigma bonded framework, it can generally be treated independently. In a double bond, **two *p* orbitals**

produce two molecular orbitals, π (bonding) and π* (antibonding), shown in Figure 12.2. The node in the wave corresponds to the mathematical sign change. The lowest-energy molecular orbital always has no nodes between the nuclei (we ignore the node in the plane of the atoms common to all pi systems). The electron pair in the bonding MO is shared between the nuclei. The antibonding MO has a node between the nuclei.

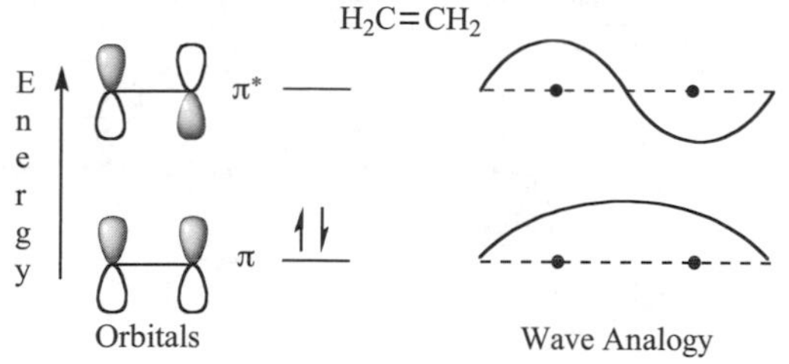
$H_2C{=}CH_2$
Energy
π*
π
Orbitals
Wave Analogy

Figure 12.2 Pi molecular orbitals for ethene, $CH_2{=}CH_2$. The dots in the wave analogy are approximate nuclei position. The standing wave extends beyond the nuclei because the orbitals do.

12.2.2 Three *p* Orbitals, the Allyl Unit

Three *p* orbitals, all aligned parallel to one another, make up the allyl unit and **produce three molecular orbitals.** The MOs for the following systems differ from the completely carbon allylic system only in the previously discussed electronegativity effects (the bonding MOs will have a larger contribution from the heteroatom *p* orbital and the antibonding MOs will have a lesser contribution, Section 2.3).

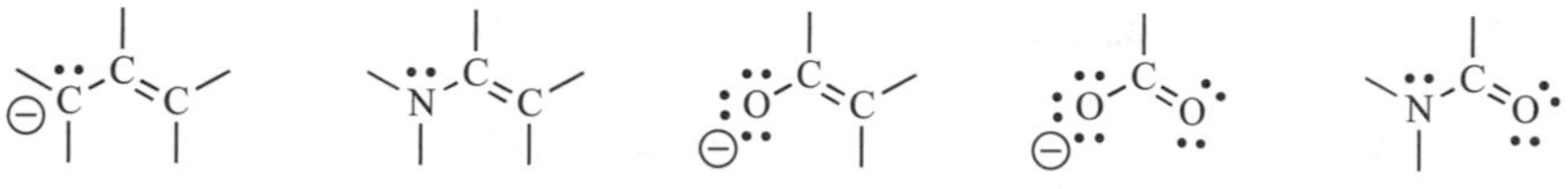

To determine how many electrons occupy any pi system, start by counting each double bond as contributing two electrons, but do not count any pi bonds that are perpendicular to the pi system of interest: For example, the second pi bond of a triple bond does not count. Because the maximum that a single atom can contribute is a single filled *p* orbital containing two electrons, if a heteroatom has several lone pairs it can contribute only two electrons. Only one *p* orbital on an atom can align itself with the pi system. If an atom is doubly bonded and contains a lone pair, the lone pair cannot be counted since it must be in an orbital that is perpendicular to the double bond. All the examples above have four electrons in a three-*p* orbital pi system.

Shown in Figure 12.3 are several different ways to describe the MOs for C=C–C:⁻, the allyl anion. The MOs are labeled from the lowest energy to the highest and are drawn with the lowest-energy MO at the bottom. The lowest-energy MO, ψ_1, has no nodes and will always be bonding. The highest-energy molecular orbital, ψ_3, has a node between each nucleus and will always be antibonding. In pi systems containing an odd number of *p* orbitals there is a *nonbonding* MO, in this case ψ_2, in which the wave puts a node through the central atom. If an atom lies on a node in an MO, it does not contribute an atomic orbital to that MO. Nodes are always symmetrical about the center; in waves with an odd number of nodes, one of the nodes must go through the center.

For C=C–C⁺, the allyl cation, there would be two electrons in the pi system, filling just ψ_1, leaving the others empty. The shape and energy of the MOs are independent of the occupancy. The allyl anion has four electrons in the pi system, filling both ψ_1 and ψ_2. The valence bond resonance forms, C=C–C:⁻ ↔ ⁻:C–C=C, describe the electron distribution in ψ_2 of the anion. The two electrons of the anion can be found on the ends of the pi system and not in the middle.

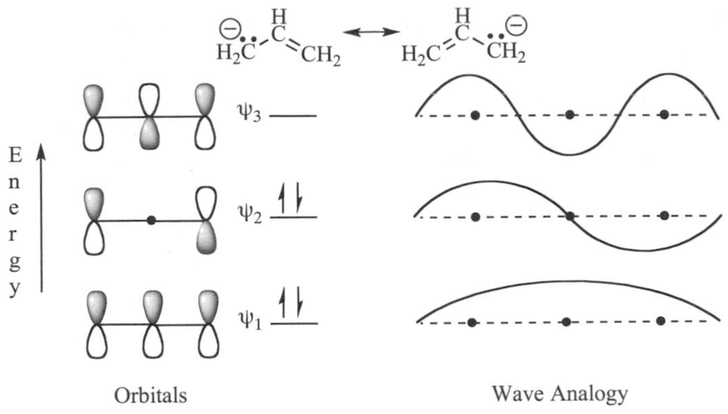
H
H₂C
CH₂
H
H₂C
CH₂
E
n
e
r
g
y
ψ3
ψ2
ψ1
Orbitals
Wave Analogy

Figure 12.3 The pi molecular orbitals of the allyl anion. The bonding MO is ψ_1, and ψ_2 is a nonbonding MO. The antibonding and highest energy MO is ψ_3 with a node between each atom.

12.2.3 Four *p* Orbitals, the 1,3-Diene Unit

Figure 12.4 gives the MOs for 1,3-butadiene, $H_2C{=}CH{-}HC{=}CH_2$, which has **four *p* orbitals** in the pi system and thus **four MOs**. The four electrons are placed in the two bonding MOs, leaving the two antibonding MOs empty. Again note the similarity to a vibrating string. The *highest occupied molecular orbital* (HOMO) is ψ_2, and the *lowest unoccupied molecular orbital* (LUMO) is ψ_3.

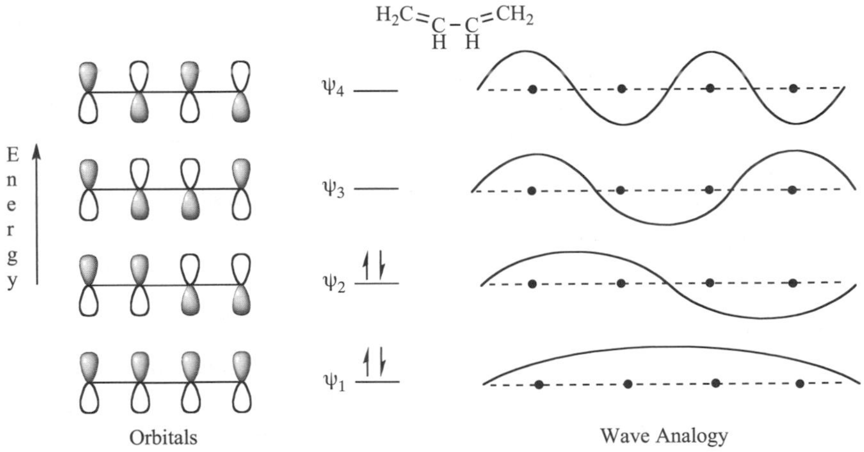
H₂C=C-C=CH₂
H H
ψ4
ψ3
ψ2
ψ1
E
n
e
r
g
y
Orbitals
Wave Analogy

Figure 12.4 The pi molecular orbitals of the 1,3-butadiene, $H_2C{=}CH{-}HC{=}CH_2$. The highest occupied MO (HOMO) is ψ_2, and the lowest unoccupied MO (LUMO) is ψ_3.

An alternative way to order the MOs is to count the bonding and antibonding overlap within each MO. ψ_1 is bonding between all the *p* orbitals, so is the lowest in energy. ψ_2 is bonding between the first *p* orbital and the second, antibonding between the second and the third, and bonding between the third and the fourth, giving two bonding and one antibonding interaction. ψ_3 is antibonding between the first *p* orbital and the second, bonding between the second and the third, and antibonding between the third and the fourth. Finally, ψ_4 is antibonding between all atoms, so is the highest-energy MO of all.

As we saw in Section 2.3, if we combine two atomic orbitals, we must get two molecular orbitals. The final number of MOs equals the starting number of AOs. The lower-energy in-phase bonding combination is stabilized slightly less than the out-of-phase antibonding combination is destabilized. The better the overlap between two orbitals, the greater their interaction. The closer in energy the two interacting orbitals are, the greater their interaction. When two orbitals of differing energy interact, the molecular orbital contains a greater percentage of the atomic orbital closest to it in energy. Along a row of the periodic table, increased electronegativity lowers the energy of an atomic orbital. An oxygen $2p$ orbital is lower in energy than a carbon $2p$ orbital.

Figure 12.5 shows how the interaction diagram of two ethylenes can produce the four molecular orbitals of 1,3-butadiene. Each π bonding orbital can be combined in phase or out of phase, and so can each antibonding π* orbital. This same set of orbitals for 1,3-butadiene can be generated by just looking at Figure 12.1 and orienting the *p* orbitals to create a similar set of harmonics.

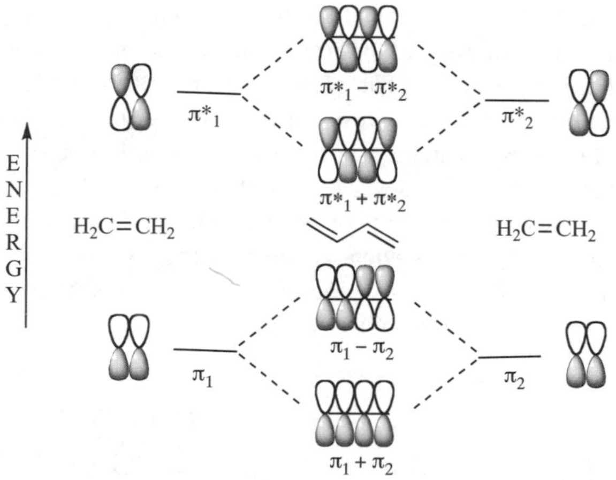
ENERGY
π*1
π*1 − π*2
π*2
π*1 + π*2
H2C=CH2
H2C=CH2
π1
π1 − π2
π2
π1 + π2

Figure 12.5 The interaction of two ethylenes to give the molecular orbitals for 1,3-butadiene.

12.2.4 The Progression of Orbitals and Energies for Linear Conjugated Systems

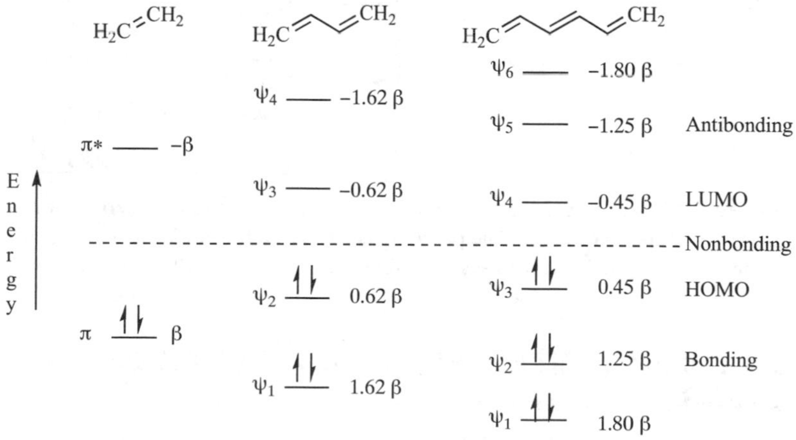
H2C=CH2
H2C=CH–CH=CH2
H2C=CH–CH=CH–CH=CH2
Energy
π* −β
π β
ψ4 −1.62 β
ψ3 −0.62 β
ψ2 0.62 β
ψ1 1.62 β
ψ6 −1.80 β
ψ5 −1.25 β Antibonding
ψ4 −0.45 β LUMO
Nonbonding
ψ3 0.45 β HOMO
ψ2 1.25 β Bonding
ψ1 1.80 β

Figure 12.6 A comparison of the relative orbital energies of ethene (left), 1,3-butadiene (middle), and 1,3,5-hexatriene (right). The stabilization of an electron in the π of ethene is defined as β .

Figure 12.6 shows how the relative energy of the pi orbitals changes as a second pi bond and then a third pi bond is added to form a longer conjugated system. **Additional conjugation raises the HOMO in energy and lowers the LUMO in energy**, as can be seen from comparing the energy of the MOs of ethene, $H_2C{=}CH_2$, 1,3-butadiene, $H_2C{=}CH{-}HC{=}CH_2$, and 1,3,5-hexatriene, $H_2C{=}CH{-}HC{=}CH{-}HC{=}CH_2$. Also, the energy gap between the HOMO and LUMO decreases with additional conjugation. The molecule can absorb light by promoting an electron from the HOMO to LUMO. Even 1,3,5-hexatriene absorbs in the high-energy ultraviolet, but with enough conjugation the HOMO-LUMO gap becomes small enough that the molecule can absorb longer-wavelength visible light and appear colored.

A summary of the progression of orbitals and energies for linear conjugated systems is shown in Figure 12.7. Note that the orbitals are always symmetrically distributed above and below the nonbonding level. The nodes are always symmetric about the center of the pi system. It is the phase of the ends of the HOMO and LUMO that will be important in pericyclic reactions. The ends of the fundamental (first harmonic) MO in any system are always in phase (both end *p* orbitals have the same shading on top) because the fundamental has no nodes. The second harmonic (one node) has the ends out of phase because the phase switches across the node, by definition. The third harmonic with two nodes has two phase switches, so the ends are back in phase. To generalize, all the MOs with an even number of nodes also have their ends in phase, but all the MOs with an odd number of nodes have the ends of opposite phase. So since we gradually add nodes one at a time **as we go up in energy, the phasing of the ends of the pi system always alternates.** This alternation of the phase of the ends of the pi system will be used to predict the outcome of pericyclic reactions.

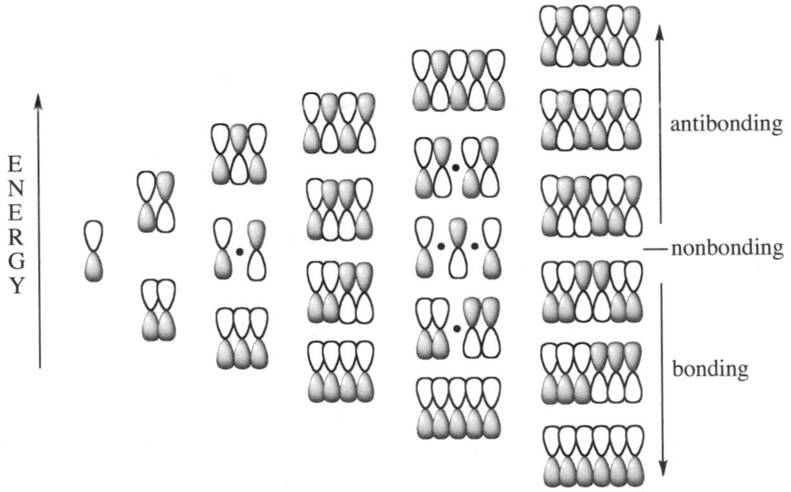
ENERGY
antibonding
nonbonding
bonding

Figure 12.7 A summary of the molecular orbitals and relative energies for linear conjugated pi systems of 1 to 6 *p* orbitals. The phase of the ends of a linear pi system molecular orbital alternates as the energy of the MO increases. The fundamental (lowest MO) is always symmetric.

12.3 MOLECULAR ORBITAL THEORY FOR CYCLIC CONJUGATED PI SYSTEMS

The molecular orbitals for benzene are shown in Figure 12.8. The ring is planar, and every atom in the ring bears a *p* orbital; we have a closed loop of overlapping *p* orbitals.

As with the linear MOs, the more nodes the MO has, the higher its energy. We have basically taken our linear system and brought the ends together so that now we have standing waves of a loop, similar to standing waves on a drumhead, like in Figure 1.4. Note that two pairs of MOs, called *degenerate pairs*, have the same energy, for they have the same number of nodes. When we place the six pi electrons in the MOs for benzene, a degenerate pair is completely filled; benzene has a filled shell.

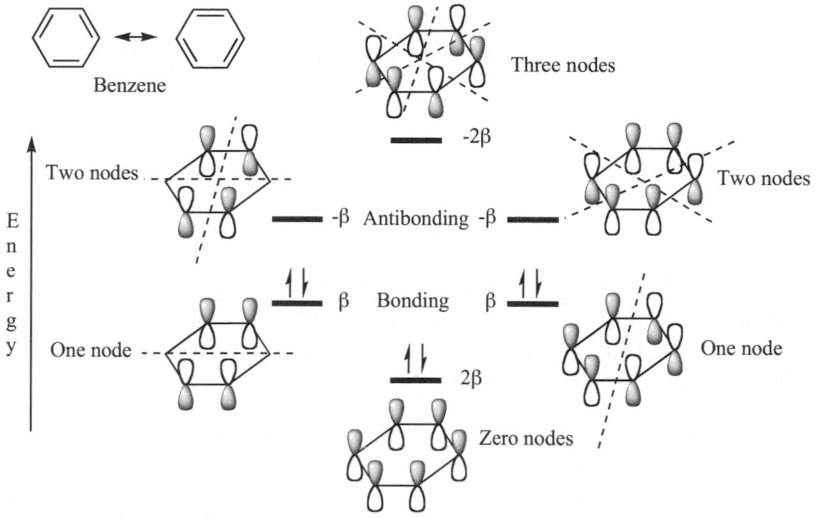
Benzene
Three nodes
-2β
Two nodes
Two nodes
-β Antibonding -β
Energy
β Bonding β
One node
One node
2β
Zero nodes

Figure 12.8 The pi molecular orbitals of benzene. The view is from the side. The nodes are planes that go through the center and cut both sides of the loop.

Cyclopentadienyl cation, on the other hand, is a very unstable molecule. Its MO energy levels are shown in Figure 12.9. When the four pi electrons are placed in the MOs for cyclopentadienyl cation, the degenerate pair is only half-filled; cyclopentadienyl cation has an incomplete shell. The instability of cyclopentadienyl cation comes as a direct result of these unpaired electrons in the half-filled shell.

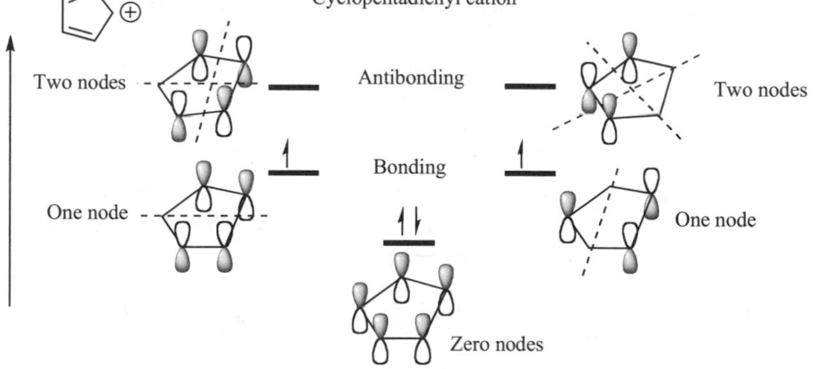
Cyclopentadienyl cation
Two nodes
Antibonding
Two nodes
Bonding
One node
One node
Zero nodes

Figure 12.9 The pi molecular orbitals of cyclopentadienyl cation.

The MO energy levels of cyclopentadienyl radical are shown in Figure 12.10. Cyclopentadienyl radical has one more electron than the cation, thus five electrons total. It has one unpaired electron and is a reactive radical. Radical reactions were covered in Chapter 11.

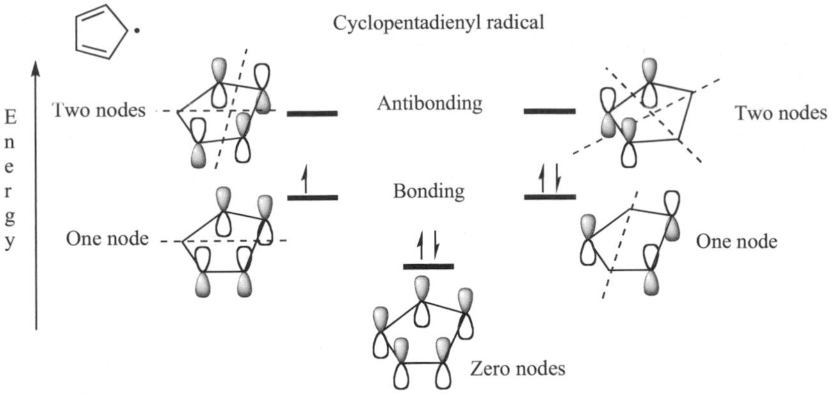
Cyclopentadienyl radical
Energy
Two nodes
Antibonding
Two nodes
Bonding
One node
One node
Zero nodes

Figure 12.10 The pi molecular orbitals of cyclopentadienyl radical.

The MO energy levels of cyclopentadienyl anion are shown in Figure 12.11. Cyclopentadienyl anion has two more electrons than the cation, thus six electrons total, and therefore has a filled degenerate pair and is predicted correctly to be aromatic and stable; it has enough electrons to fill the shell.

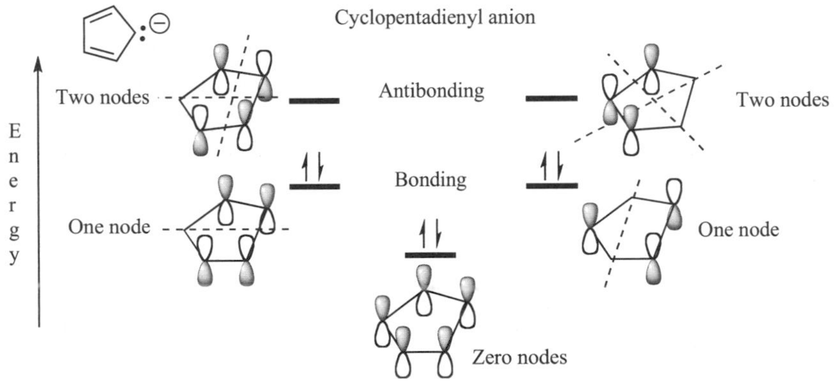
Cyclopentadienyl anion
Energy
Two nodes
Antibonding
Two nodes
Bonding
One node
One node
Zero nodes

Figure 12.11 The pi molecular orbitals of cyclopentadienyl anion.

The stabilization of benzene can be placed on a semiquantitative scale. If we define the stabilization of one electron in the pi orbital of ethene as β, then the stabilization of two electrons in a double bond is just 2β. The stabilization of six pi electrons in three double bonds is 6β. Benzene's MO energy levels are at 2β, 1β, 1β, and −1β, −1β, −2β; the first three are filled with two electrons each for a stabilization of 8β total (2 × 2β + 4 × 1β). The six electrons of benzene pi loop are then 2β more stabilized than the six electrons in three isolated double bonds.

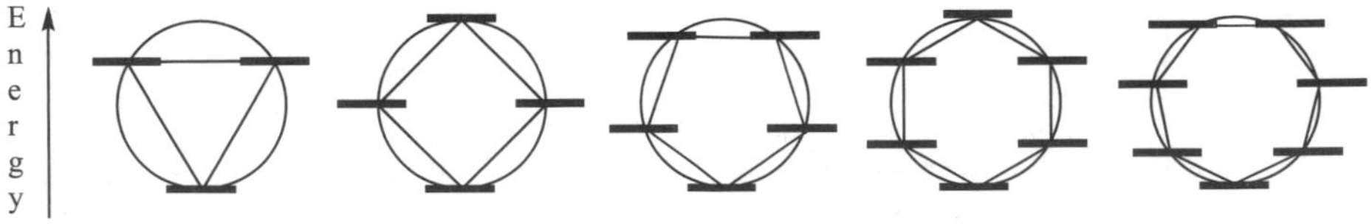
Energy

Figure 12.12 The pi molecular orbital energies of cyclic systems.

The MO energy levels for any simple cyclic conjugated polyene can be estimated by drawing the regular polygon of the ring point down inscribed in a circle of 2β radius (Fig.

12.12). The vertices of the polygon will correspond to the energies of the MOs on a vertical scale. The center of the circle corresponds to the nonbonding level (zero β) with the bonding levels below it and the antibonding above.

Since the degenerate pair must be completely filled or completely empty in order for the molecule to be aromatic, stable arrangements arise when 2, 6, 10, ... or $4n + 2$ pi electrons are in the cyclic conjugated pi system. Unstable half-filled degenerate shells will occur with 4, 8, 12, ... or $4n$ pi electrons. The $4n$ systems tend to distort from planarity to diminish pi overlap and this destabilizing antiaromatic conjugation.

12.4 PERTURBATION OF THE HOMO AND LUMO

The relative availability of the electrons in the HOMO determines the nucleophilicity of the pi system. The higher in energy the HOMO is, the more available that electron pair is for overlap with an electrophile's LUMO. Likewise, if the electrophile's LUMO is lower in energy, closer to the nucleophile's HOMO, the interaction of the two is greater, as in Figure 2.8. Recall that the closer in energy two interacting orbitals are, the greater their interaction. A favorable HOMO-LUMO interaction in a pericyclic reaction will predict a favorable reaction route. So it is important to understand what perturbs the energy of the HOMO and LUMO of a system.

One way to perturb a pi system is to swap out a carbon atom in the system for a more electronegative element such as oxygen, as shown in Figure 2.9. Replacement of carbon by **a more electronegative atom lowers both the HOMO and LUMO of the system.** In this way C=O bonds are more electrophilic than carbon–carbon double bonds.

When two groups are conjugated with each other, the new conjugated system has properties that are different from the original two groups. Figure 12.6 demonstrated that **simple conjugation with another pi bond raised the HOMO and lowered the LUMO.**

For example, we can look at the change that occurs in the molecular orbitals of a carbon–carbon pi bond as we attach a group that conjugates with it. One extreme would be to attach an atom with a fully occupied *p* orbital, an excellent pi donor. In general, any group that conjugates with a pi system and donates electron density to it is a pi donor. The other extreme would be to attach an atom with an empty *p* orbital, an excellent pi acceptor. Any group that conjugates with a pi system and withdraws electron density from it is a pi acceptor.

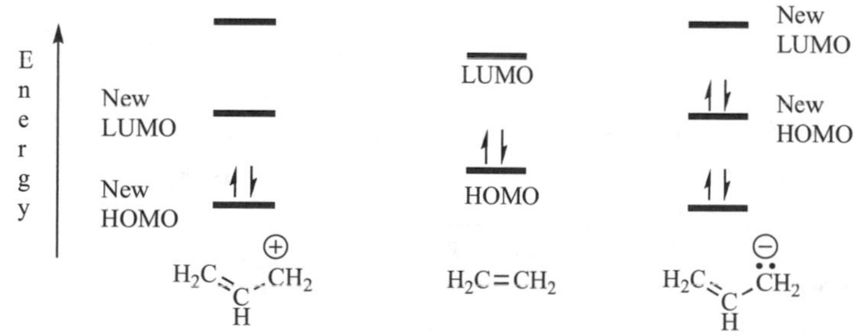
Energy
New LUMO
New HOMO
LUMO
HOMO
New LUMO
New HOMO
H2C=CH2

Figure 12.13 In the center is the HOMO and LUMO of an unperturbed pi bond. The left side shows that both the HOMO and LUMO drop in energy when a pi acceptor is attached. The right side shows that both the HOMO and LUMO are raised in energy when a pi donor is attached.

The right side of Figure 12.13 shows what happens if a good pi donor is attached to a double bond. The pi donor HOMO interacts with both the double-bond HOMO and LUMO to produce a new allylic system with a new HOMO and LUMO, which are raised from those of the double bond, displayed in the center of the figure. **Conjugation of a double bond with a pi donor raises the HOMO of the system, and makes the combination more nucleophilic.**

The left side of Figure 12.13 shows what happens if a good pi acceptor is attached to a double bond. The LUMO of the pi acceptor interacts with both the double bond HOMO and LUMO to produce a new allylic system with a new HOMO and LUMO, which are lowered from those of the double bond. **A pi acceptor bonded to a double bond makes the combination more electrophilic by lowering the LUMO of the system.** When the energy of the LUMO of the electrophile is lower, then it is softer, and the stabilization resulting from overlap with the average nucleophile's HOMO is much greater (recall Fig. 2.14).

A good covalent bond is a soft–soft interaction formed between an electrophile with a low-lying LUMO and a nucleophile with a high-lying HOMO. Optimally, the HOMO and LUMO are close in energy and overlap well, thus have a large interaction, releasing much energy, and forming a strong bond. It is possible to reduce the energy gap between the nucleophile HOMO and electrophile LUMO by raising the HOMO of the nucleophile and likewise lowering the LUMO of the electrophile, making both more reactive.

For example, consider the case of two neutral reactants, A and B, which initially have no difference in their HOMO and LUMO levels, as shown on the left in Figure 12.14. If the HOMO and LUMO levels in each are far enough apart, there is no reaction. The soft–soft interaction is poor, and since the reactants are not charged there is no hard–hard interaction. As an example, two molecules of ethylene do not normally react with each other. However, if a pi donor is added to reactant A, raising both HOMO and LUMO, and a pi acceptor is added to reactant B, lowering both HOMO and LUMO, then reaction occurs easily. The HOMO of A′ is now close in energy to the LUMO of B′, producing a good soft–soft interaction. Reactant A′ is now a good electron source, and reactant B′ is now a good electron sink. In essence, raising the source and lowering the sink has made bond formation, the flow of electrons from A′ to B′, a good process.

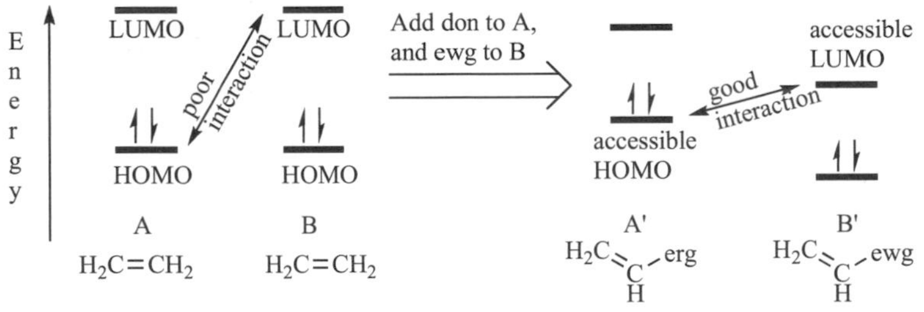
Energy
LUMO
LUMO
poor interaction
HOMO
HOMO
A
B
H2C=CH2
H2C=CH2
Add don to A, and ewg to B
accessible LUMO
good interaction
accessible HOMO
A'
B'
H2C=C(H)-erg
H2C=C(H)-ewg

Figure 12.14 The perturbation of two reactants by adding a pi donor (erg) to one and a pi acceptor (ewg) to the other, thus producing a more reactive pair.

The best electron sources have an accessible high-energy HOMO. When the energy of the HOMO of the nucleophile is high, the nucleophile is softer, and the stabilization resulting from overlap with the electrophile's LUMO is much greater. The best electron sinks have an accessible low-energy LUMO.

12.5 DELOCALIZATION OF SIGMA ELECTRONS (MORE ADVANCED)

Advanced molecular orbital calculations for the nitrogen molecule (:N≡N:) give a simple set of molecular orbitals (Fig. 12.15). Two perpendicular pairs of π and π^* molecular orbitals arise from the nitrogen–nitrogen triple bond.

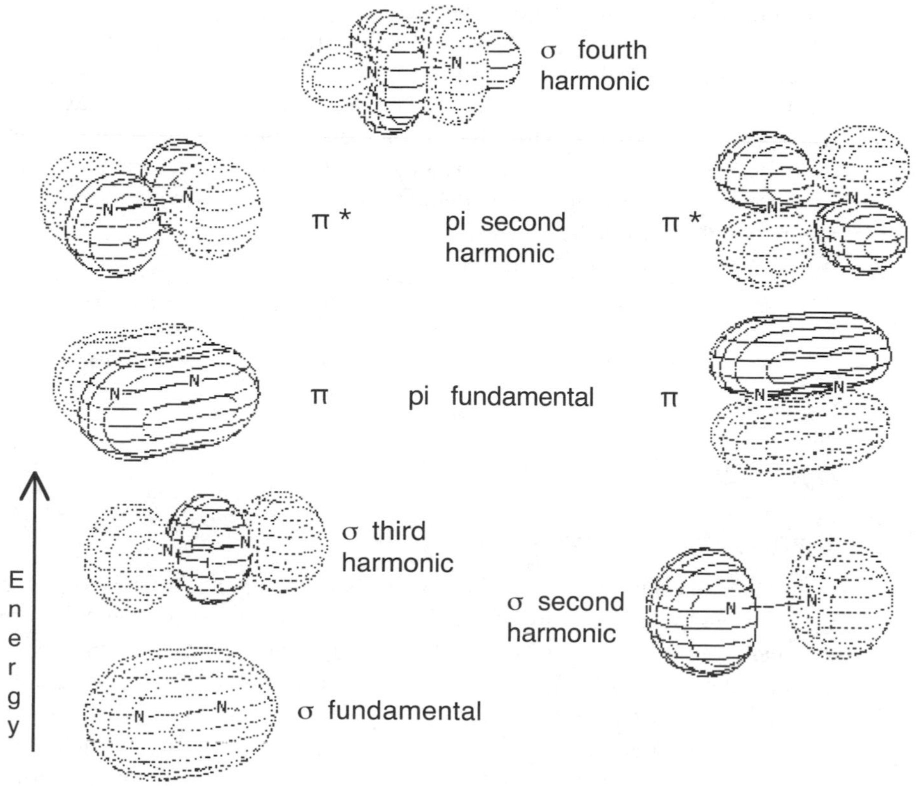
σ fourth
harmonic
π *
pi second
harmonic
π *
π
pi fundamental
π
σ third
harmonic
σ second
harmonic
σ fundamental
Energy
N

Figure 12.15 The molecular orbitals of nitrogen. The lowest five orbitals are filled. Modified with permission from W. L. Jorgensen and L. Salem, *The Organic Chemist's Book of Orbitals*, page 79, Copyright Elsevier 1973.

The sigma system from these calculations may be a bit of a surprise; the sigma bonding electrons are not constrained to individual bonds but are delocalized over the entire molecule. Instead of four localized orbitals, sigma bonding, sigma antibonding, and two lone pair orbitals, there are four sigma molecular orbitals: the fundamental, the second harmonic, the third harmonic, and the fourth harmonic.

Sigma molecular orbitals pose a difficult representational problem, for they usually are as large and as complex as the framework of the molecule itself. Commonly we treat the sigma framework as if it were completely localized, chopping up the representational problem into intellectually convenient little pieces, each piece corresponding to a particular bond or electron pair. Toward this end, the atomic orbitals are mixed to give hybridized orbitals that are more convenient for the discussion of these localized bonding orbitals. These localized bonding orbitals are easy to use and correspond to the lines and dots of our Lewis formula. There are very few penalties for using hybrid orbitals and treating the sigma framework as localized, when it actually is not.

12.6 CONCERTED PERICYCLIC CYCLOADDITION REACTIONS

Introduction to Pericyclic Reactions

Pericyclic reactions have cyclic transition states and electron flow paths that appear to go around in a loop. There are three main types of pericyclic reactions: cycloaddition reactions, electrocyclic reactions, and sigmatropic rearrangements. Figure 12.16 gives a representative example of each type. Cycloaddition reactions form a ring from two pi systems coming together. Electrocyclic reactions close the ends of a single pi system to form a ring or the reverse, open a ring to form a single pi system. Sigmatropic rearrangements move a sigma bond from one part of a pi system to another. The examples shown all have a transition state in which six electrons move around in a loop. The rules for prediction of the products will depend on many factors, including the number of electrons in the process.

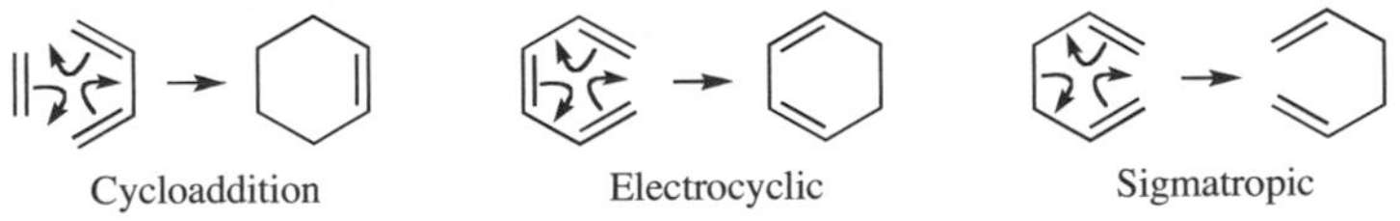
Cycloaddition
Electrocyclic
Sigmatropic

Figure 12.16 Examples of the three types of pericyclic reactions.

Cycloaddition Reactions

Four-Electron (4*n*) Cycloaddition Reactions

Concerted four electron cycloaddition reactions are not thermally allowed. All this really means is that the transition state for an allowed process is much lower than that of a nonallowed process. HOMO-LUMO theory can explain why the concerted 2 + 2 cycloaddition of two ethylenes is not favored. Bonding overlap between a HOMO and a LUMO lowers the transition state energy of the allowed process, making it favorable. With the concerted 2 + 2 cycloaddition of two ethylenes, no such transition state stabilization is achieved, as shown in Figure 12.17. Any bonding overlap is exactly cancelled out by an equivalent antibonding overlap, yielding no net stabilization.

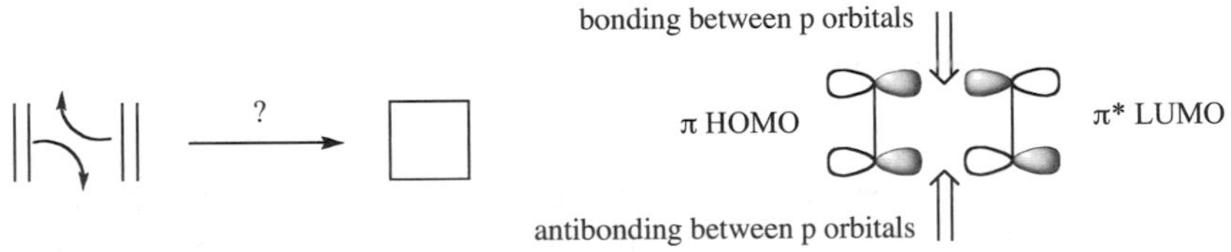
bonding between p orbitals
?
π HOMO
π* LUMO
antibonding between p orbitals

Figure 12.17 The thermal 2 + 2 concerted cycloaddition of two ethylenes is not favored because there is no net interaction possible between the HOMO of one partner and the LUMO of the other.

Six-Electron (4*n*+2) Cycloaddition Reactions

The Diels–Alder reaction is an allowed thermal six-electron (4 + 2) cycloaddition reaction between a diene (contributing four pi electrons to the transition state) and dienophile (contributing two pi electrons). It is an extremely useful cycloaddition because it goes in high yield and with predictable stereochemistry. The stereochemistry of both the starting pieces is preserved in the product; for example, if two groups on the dienophile are *cis*, they remain *cis* in the product.

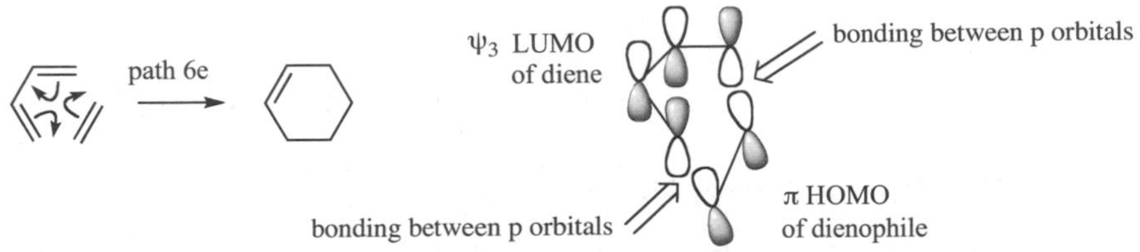
path 6e
ψ3 LUMO
of diene
bonding between p orbitals
π HOMO
of dienophile
bonding between p orbitals

Figure 12.18 Both ends of the HOMO and LUMO have bonding overlap in a 4+2 cycloaddition.

The net reaction breaks two pi bonds and makes two sigma bonds and therefore is about 40 kcal/mol (167 kJ/mol) exothermic. The six-membered transition state for the reaction resembles a folded cyclohexane (a boat conformation). Figure 12.18 shows the reaction with line structures and with orbitals. Notice how the *p* orbitals of the pi bonds rehybridize into the sigma bonds of the product. The Diels–Alder reaction has much higher yields if an electron-withdrawing group is conjugated with the dienophile pi bond. Surprisingly, the product is the one with the electron-withdrawing group underneath, *endo*, rather than *exo*, sticking out away. Favorable orbital interactions between the electron-withdrawing group and the diene are invoked to rationalize this endo stereochemistry, see Figure 12.20.

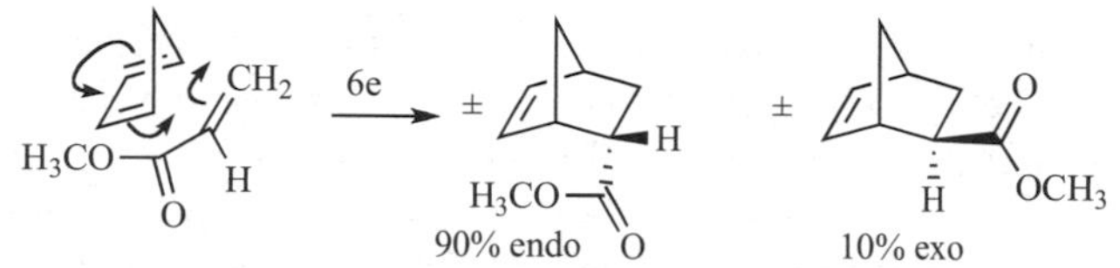
CH2
6e
±
H3CO
H
O
H
H3CO
90% endo
O
±
O
H
OCH3
10% exo

The five-membered transition state 4 + 2 cycloaddition is less common than the six. The four-pi electron piece is called a 1,3-dipole because it reacts as if it has a minus and a plus separated by one atom, $^{+}$A-B-C^{-}. The orbital arrangement for the 1,3-dipolar addition is similar to that for the Diels–Alder reaction:

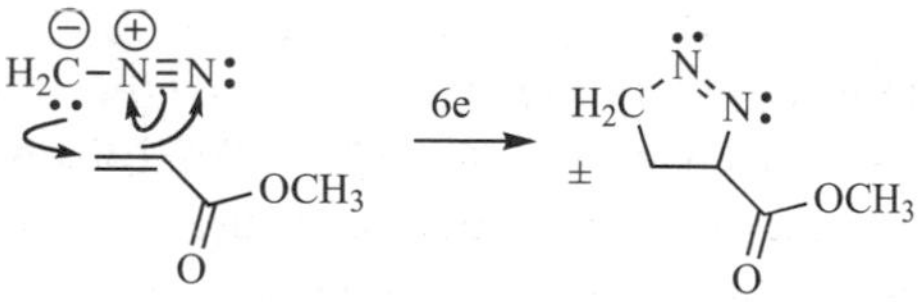
H2C–N≡N:
OCH3
O
6e
H2C
N
N:
±
OCH3
O

Cycloreversions are the reverse of cycloadditions and occur by the same mechanism. Because the cycloreversion breaks the molecule into two pieces, increasing degrees of freedom and therefore disorder, the ΔS of reaction is positive. The ΔG for the cycloreversion process becomes favorable (negative) with increased temperature because the $-T\Delta S$ term dominates over the ΔH term (recall $\Delta G = \Delta H - T\Delta S$). The reverse Diels–Alder reaction of dicyclopentadiene into two cyclopentadienes is a common example:

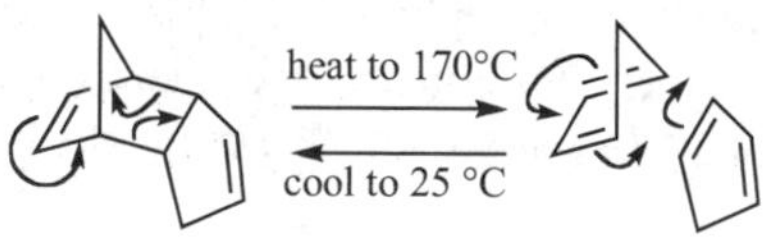
heat to 170°C
cool to 25 °C

4 + 2 Cycloadditions from a HOMO–LUMO Perspective

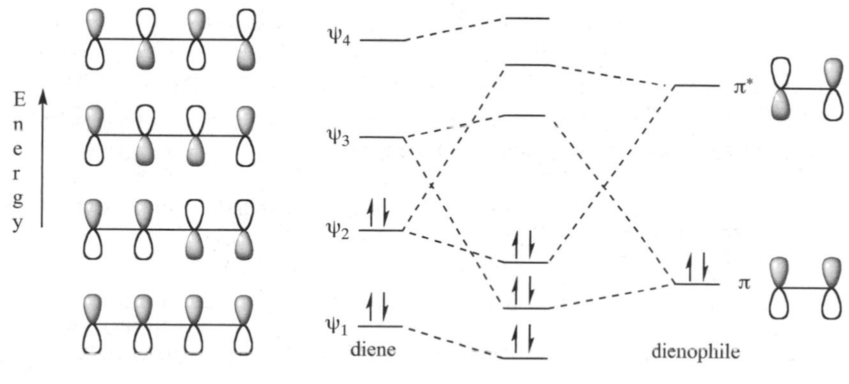
Energy
ψ4
ψ3
ψ2
ψ1
diene
π*
π
dienophile

Figure 12.19 The Diels–Alder (4 + 2 cycloaddition) interaction diagram.

The soft–soft interaction of filled with empty orbitals is the major interaction in the transition state because there is little hard–hard attraction. In an unsubstituted system two soft–soft interactions stabilize the transition state: The filled ψ_2 molecular orbital of the diene interacts with the unfilled π^* of the dienophile; also the empty ψ_3 of the diene interacts with the filled π of the dienophile. Figure 12.19 gives the interaction diagram.

However, for the simplest system, 1,3-butadiene reacting with ethene, the HOMO–LUMO interactions are average, for the interacting orbitals are reasonably far apart in energy. The interaction can be increased (Section 12.4), usually by placing an ewg on the dienophile (lowering the LUMO, π^*), and sometimes by placing a donor on the diene (raising the HOMO, ψ_2). In addition, these substituents polarize the pi bond they are attached to and introduce a hard–hard factor into the reaction. With substituents on the diene and dienophile, different constitutional isomers are possible. The regiochemistry is determined by the best overlap of the HOMO–LUMO pair closest in energy. The following structures show the regiochemistry for two common combinations (the controlling interaction is the unfilled π^* of the dienophile with the filled ψ_2 of the diene).

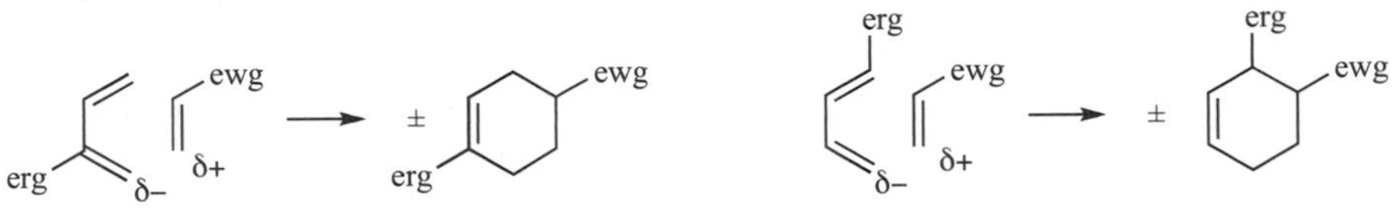
erg
ewg
δ+
δ−
±
erg
ewg
erg
ewg
δ+
δ−
±
erg
ewg

If the donor were on the dienophile and the ewg on the diene (inverse polarity), the controlling interaction would be the filled π of the dienophile with the empty ψ_3 of the diene. The regiochemistry for the inverse polarity Diels–Alder reaction is shown below.

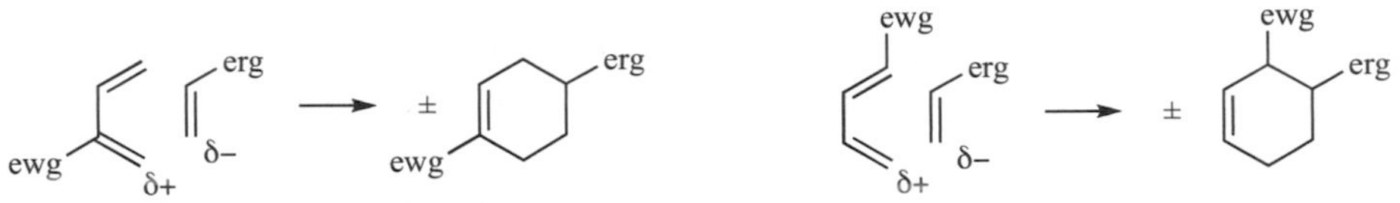
ewg
erg
δ−
δ+
±
ewg
erg
ewg
erg
δ−
δ+
±
ewg
erg

If the dienophile has a substituent that extends its pi system, the best transition state for the reaction maximizes overlap between the two pi systems. To achieve this greater degree of interaction the dienophile must place the substituent underneath the diene, endo. The Diels–Alder reaction of two dienes has such a transition state with an additional interaction between the unfilled diene ψ_3 MO and the filled diene ψ_2 MO (Fig. 12.20).

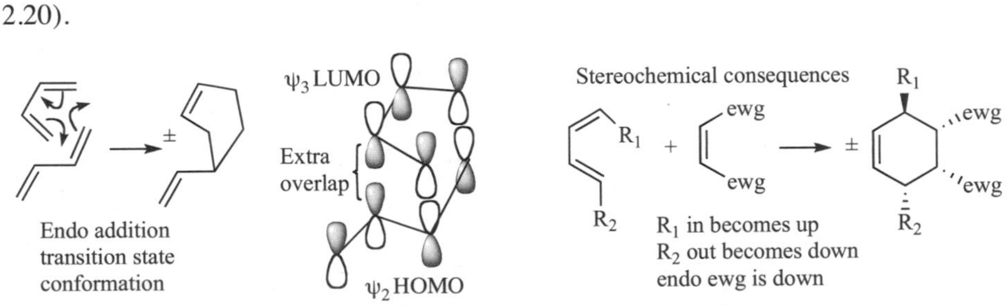
±
Endo addition
transition state
conformation
ψ3 LUMO
Extra
overlap
ψ2 HOMO
Stereochemical consequences
R1
ewg
+
ewg
±
R2
R1 in becomes up
R2 out becomes down
endo ewg is down
R1
ewg
ewg
R2

Figure 12.20 The Diels–Alder reaction of two dienes with the extra overlap of endo addition.

We saw that the ends of the two pi systems that become the new sigma bond must both overlap in a bonding manner. With the $4n$ cycloaddition they did not (Fig. 12.17), so the 2+2 reaction was not allowed. With the $4n$+2 cycloaddition (Fig. 12.18), the ends did match, so the 4+2 reaction was allowed. Figure 12.7 showed us that the phasing of the ends of each MO alternated as we increased in energy. In adding two more electrons to the $4n$ electron system, we moved up an MO so the ends now matched. In summary, face-to-face cycloadditions with $4n$ electrons are not thermally allowed, but the **corresponding $4n+2$ electron cycloadditions are thermally allowed.** Figure 12.21 shows graphically the interaction of just the ends of two pi systems in the transition state for a cycloaddition for an allowed process (bonding between both ends) and a disallowed one (bonding on one end but not the other). If it appears that the interaction on **both** ends is antibonding, reverse the shading on all the p orbitals of one MO, and this will give both ends bonding. For example, on the fundamental, it did not matter if we colored in the all the tops of the *p* orbitals or all the bottoms, as long as all were done in the same way.

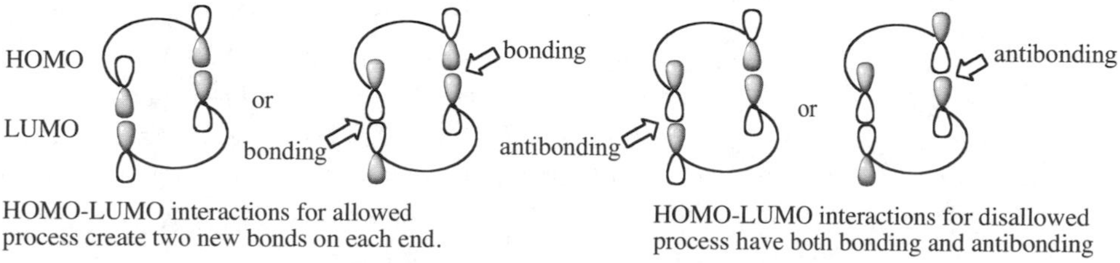
HOMO
LUMO
or
bonding
bonding
antibonding
or
antibonding
HOMO-LUMO interactions for allowed process create two new bonds on each end.
HOMO-LUMO interactions for disallowed process have both bonding and antibonding

Figure 12.21 The favorability of a cycloaddition between two pi systems is determined by how the phase of the ends of the HOMO and LUMO of two pi systems interact.

12.7 CONCERTED PERICYCLIC ELECTROCYCLIC REACTIONS

Four-Electron ($4n$) Electrocyclic Reactions

In an electrocyclic reaction both ends must rotate 90° to convert a sigma bond into a pi bond or the reverse. The direction of rotation can be identical, for example, both clockwise or both counterclockwise, which is called *conrotatory* (con). They can rotate in opposite directions, either both in or both out, which is *disrotatory* (dis). Dashed arrows will indicate orbital movement.

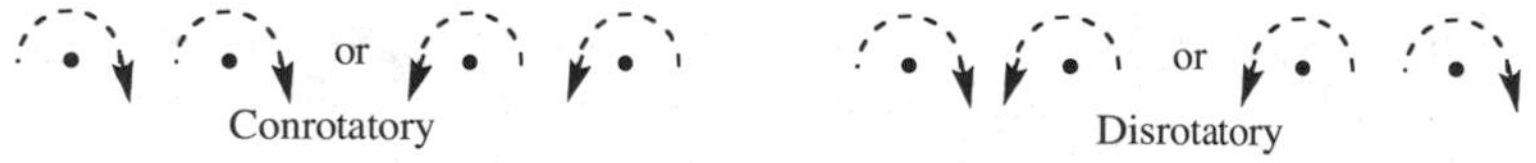
or
Conrotatory
or
Disrotatory

HOMO–LUMO principles allow us to predict whether the ends rotate conrotatory or disrotatory. If the process is favorable, as it is in Figure 12.22 when the rotation is in the same direction, then the σ-HOMO will overlap with the π-LUMO in a bonding manner to give a bonding molecular orbital of the product.

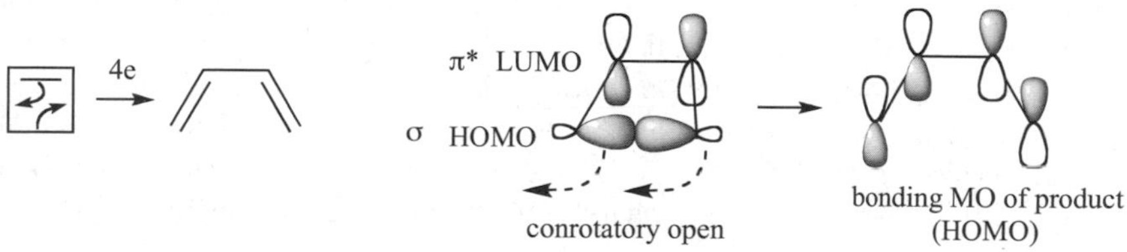
4e
π* LUMO
σ HOMO
conrotatory open
bonding MO of product
(HOMO)

Figure 12.22 The thermal conrotatory ring opening of cyclobutene.

Stereochemical labels on the cyclobutene confirm that this 4-electron electrocyclic ring opening is conrotatory. Another way to view the reaction is from the closure of the diene. The ends of the HOMO of the diene must rotate con to create a bonding overlap for the new sigma bond. Because of microscopic reversibility, opening and closure follow the same route; if the opening is conrotatory, then the closure will likewise be con.

Six-Electron (4*n*+2) Electrocyclic Reactions

As was seen for the four-electron electrocyclic ring opening (path 4e), the direction that the ends rotate when the sigma bond twists open is important. With two more electrons in the pi system, the six-electron electrocyclic ring opening (path 6e) of 1,3-cyclohexadiene prefers to open in a disrotatory manner as shown in Figure 12.23. The ring twists open so that the σ-LUMO overlaps with the π-HOMO in a bonding manner to produce a bonding molecular orbital of the product. Another way to view the reaction is from the closure of the triene. The ends of the HOMO of the triene would have to rotate dis to create a bonding overlap for the new sigma bond.

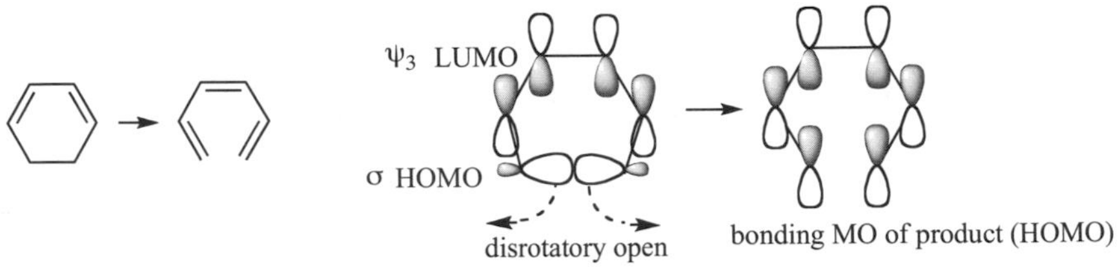
ψ_3 LUMO
σ HOMO
disrotatory open
bonding MO of product (HOMO)

Figure 12.23 The thermal disrotatory ring opening of 1,3-cyclohexadiene.

Once we decide con or dis, there is still another decision to be made, since each rotational mode has two ways it can happen. With alkyl groups, steric effects dominate; the conrotation that does not bump the alkyl groups is preferred.

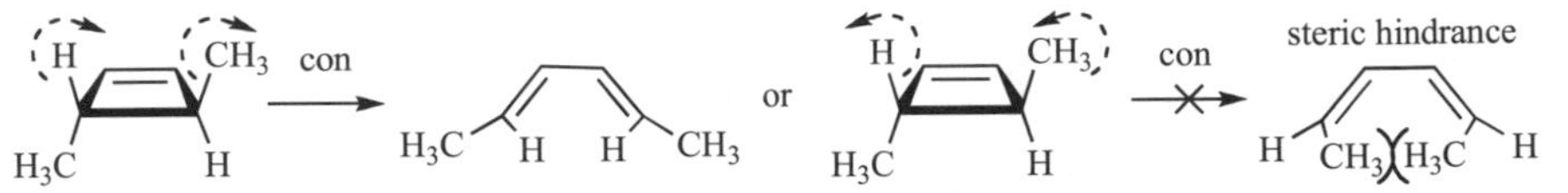
H
CH_3
con
H_3C
H
H_3C
H
H
CH_3
or
H
CH_3
con
H_3C
H
steric hindrance
H
CH_3
H_3C
H

With the conrotatory cyclobutene ring opening, secondary electronic effects contribute. Electron-releasing groups prefer to rotate outward, and electron-withdrawing groups prefer to rotate inward.

H
erg
con
H
H
H
H
H
erg
H
ewg
con
H
H
H
H ewg
H

We saw that the rules reverse from four-electron pericyclic reactions (more generally 4*n* electrons) to six-electron pericyclic reactions (more generally 4*n*+2 electrons). Looking at the closures of the HOMOs (Figs. 12.22 and 12.23), we can generalize the electrocyclic ring closure for any loop of *p* orbitals. If the ends of the HOMO are out of phase, then a con closure is required to get a sigma bonding orbital. If the ends of the HOMO are in phase, then a dis closure is now required (Fig. 12.24). Figure 12.7 showed the phase of the ends alternate as the MOs increase in energy, starting with the in-phase fundamental. If we add two more electrons we get the next MO up as our HOMO. Thus the ends of the HOMO of all 4*n* species will be out of phase, and the ends of the HOMO of all 4*n*+2 species will be in phase. Check Figure 12.7 to verify this yourself.

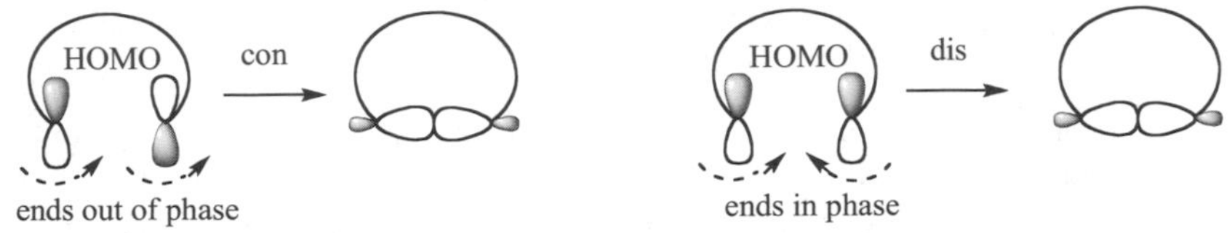
HOMO
con
ends out of phase
HOMO
dis
ends in phase

Figure 12.24 The thermal electrocyclic ring closure rotational mode of a loop of *p* orbitals is determined by the phase of the ends of the HOMO.

Let there be light! When an electron is promoted by light from the HOMO to the LUMO, the HOMO of the excited state is now the old LUMO. The HOMO has moved up one MO in energy, and the phase of the ends have reversed. We expect the rules to reverse, and they do. The $4n$ photochemical electrocyclic reactions go dis, and the $4n+2$ photochemical electrocyclic reactions go con, as predicted. This all boils down to an easy-to-remember rule: **$4n$ electrocyclics go thermal con, with the rules reversing for $4n+2$, and also reversing for photochemical reactions.**

12.8 CONCERTED PERICYCLIC SIGMATROPIC REARRANGEMENTS

Four-Electron (4*n*) Cyclic Rearrangements

The migrating group in a sigmatropic reaction can stay on the same face of the pi system it started from (suprafacial) or migrate to the opposite face (antarafacial) (Fig. 12.25). For consistency in this section, we will draw the sigma bond as the HOMO and the pi systems as the LUMO, even though it can be done reversed. We draw the sigma HOMO and pi LUMO initially in phase so that the newly formed pi bond will be bonding. Then we migrate the hydrogen to the same phase lobe of the other end of the pi system to form a new bonding sigma bond. The migrating group must maintain overlap with both ends at the transition state. This transition state is easy for suprafacial migrations once the ends are close. Antara migrations of hydrogen best occur if the pi system is long enough to spiral around on itself to pass the hydrogen between top and bottom faces.

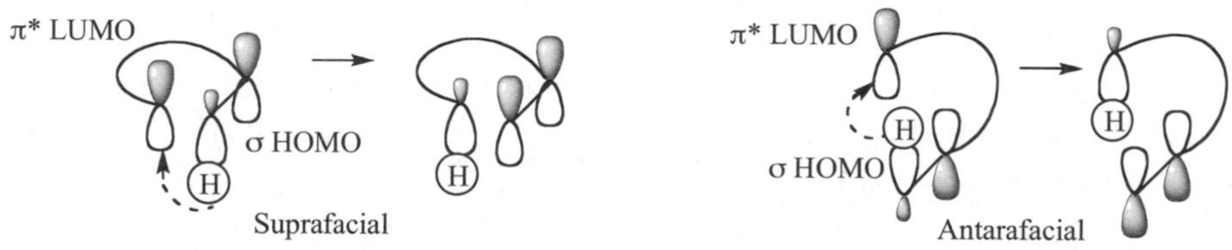

π* LUMO
σ HOMO
H
Suprafacial
π* LUMO
σ HOMO
H
Antarafacial

Figure 12.25 The suprafacial (same face) and antarafacial (spiral to migrate between opposite faces) sigmatropic rearrangement of a hydrogen atom across the ends of two different pi systems (pi LUMO ends in-phase on right, out-of-phase on left).

A common error in drawing a mechanism for tautomerization is passing the hydrogen internally, which is actually a concerted four-electron (two arrow) thermal 1,3 hydrogen sigmatropic rearrangement. The hydrogen in Figure 12.26 cannot migrate suprafacially because it would form an antibonding orbital. The ends of the π* LUMO are out of phase, so the allowed process is antarafacial. The hydrogen has to go from bottom to top, but the pi system is too short for the H to maintain continuous overlap on the allowed route. There is no low-energy allowed concerted process, so the reaction occurs stepwise.

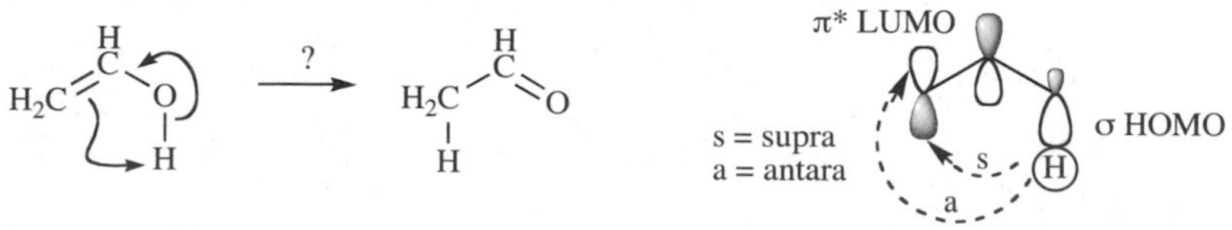

H
C
H2C
O
H
?
H
C
H2C
O
H
π* LUMO
s = supra
a = antara
s
a
H
σ HOMO

Figure 12.26 Concerted tautomerization—the thermal 1,3 hydrogen shift has a high barrier. HOMO-LUMO requires antara, but the migrating H atom can't maintain overlap with both ends.

Six-Electron (4*n*+2) Cyclic Rearrangements

Let's look at the similar six-electron (three arrow) thermal 1,5 hydrogen sigmatropic rearrangement. We expect the ends of the pi system to change phase as we add two more electrons. This means the rules will reverse and the allowed process will be suprafacial. Figure 12.27 shows the HOMO-LUMO prediction of suprafacial.

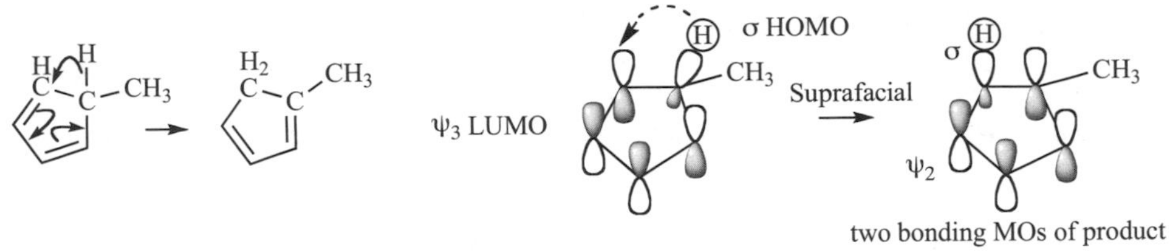

CH3
H2
CH3
ψ3 LUMO
σ HOMO
CH3
Suprafacial
σ
CH3
ψ2
two bonding MOs of product

Figure 12.27 The 1,5 hydrogen shift occurs easily. HOMO-LUMO requires suprafacial.

Sigmatropic Migration with Inversion

If the migrating group is carbon, not hydrogen, another variable is tossed in the mix. A carbon atom can migrate with retention (just like our H), or the carbon can undergo inversion (back side attack). The rules will reverse if the migrating center undergoes inversion (Fig. 12.28). The *p* orbital on the migrating center gives the option of a phase change. The transition state in the center of the figure shows that overlap can be maintained on this suprafacial inversion route. Therefore there are two types of four electron allowed paths: antara retention (Fig. 12.25 right) and supra inversion (Fig. 12.28 si). Likewise there are two types of six-electron allowed paths: supra retention (Fig. 12.25 left) and antara inversion (not shown).

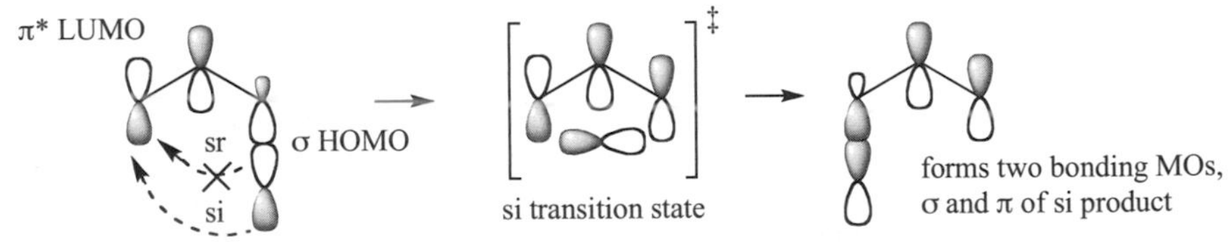

π* LUMO
sr
si
σ HOMO
‡
si transition state
forms two bonding MOs,
σ and π of si product

Figure 12.28 The 1,3 carbon shift. HOMO-LUMO requires suprafacial inversion (si) not supra retention (sr).

Again we remember (or derive) **one** sigmatropic allowed process, and then use the idea of phase reversals to generate all the other possibilities. **A 4*n* antara retention is thermally allowed, with the rules reversing for supra, reversing for 4*n*+2, reversing for inversion, and also reversing for photochemical reactions.** Two reversals are equivalent to no change at all.

Example problem

Is the following concerted sigmatropic rearrangement thermally allowed?

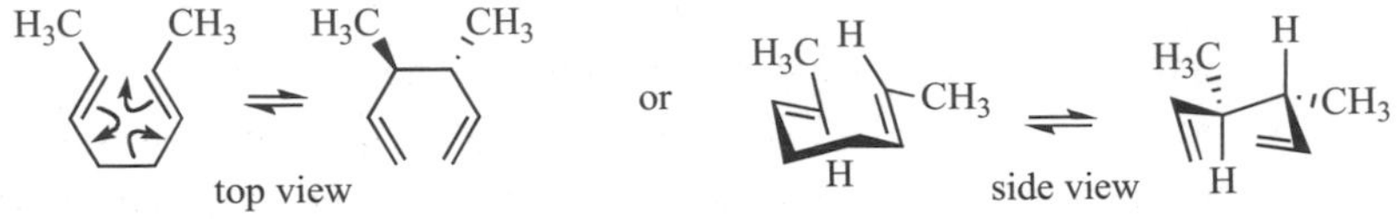

H3C
CH3
H3C
CH3
top view
or
H3C
H
CH3
H
side view
H3C
H
CH3
H

Answer: The process uses three arrows and so is a six-electron (4*n*+2) process. The side view shows that the rearrangement occurs on the same face and with retention. We know that $4n$ antara retention is thermally allowed and that this process is 4n+2 supra retention. There are two reversals: $4n$ to $4n$+2 and antara to supra; the reversals cancel, so we predict that this reaction is also thermally allowed. Experiment shows that this reaction easily proceeds thermally.

Y
Y

Two common six-membered rearrangements of 1,5-dienes have Y as either oxygen (Claisen rearrangement) or carbon (Cope rearrangement). Both are reversible and favor the more stable product. The preceding example problem demonstrated that they were thermally allowed $4n$+2 suprafacial retention processes. As in the Cope example problem above, the transition state for the reaction commonly resembles the chair conformation of cyclohexane. Figure 12.29 shows a biochemical example of the Claisen rearrangement.

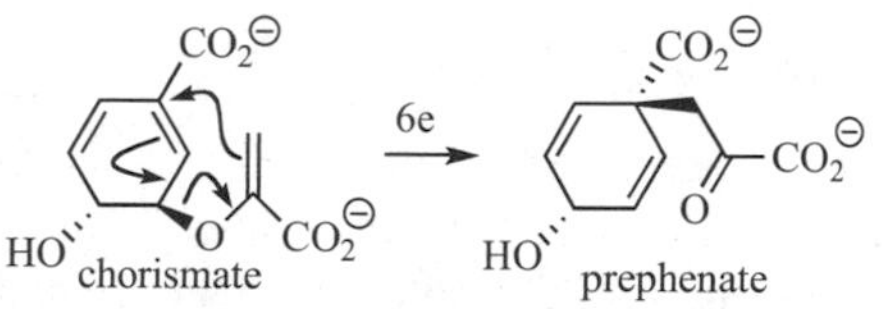
CO2⊖
6e
HO
O
CO2⊖
chorismate
CO2⊖
CO2⊖
O
HO
prephenate

Figure 12.29 The concerted thermal rearrangement of chorismate to prephenate is a biochemical example of a Claisen rearrangement that is part of the bacterial biosynthesis of the amino acids phenylalanine and tyrosine.

12.9 PERICYCLIC REACTIONS SUMMARY

The thermal favorability of pericyclic reactions can be predicted by looking for a bonding interaction between the HOMO of one reacting partner and the LUMO of the other partner. Pericyclic reactions usually involve the ends of the pi system(s). Therefore, the relative phase of the ends of the pi systems dictate whether a bonding or antibonding interaction results. The ends of the pi system are in phase in the fundamental MO and out of phase in the next higher one-node MO. This phase alternation continues as the energy increases.

Therefore, if we derive or remember one rule for a pericyclic reaction, then any time an MO phase change is added the rule will reverse. Two reversals cancel each other. For example, 4n face to face (supra-supra) cycloadditions are not thermally allowed. If we add two electrons, we fill the next highest MO, which has a phase reversal. This means 4n+2 cycloadditions are thermally favored. Thermal electrocyclic reactions of 4n species go conrotatory, whereas thermal 4n+2 electrocyclic reactions go disrotatory. Thermal sigmatropic reactions of 4n species go supra-inversion or antara-retention. Count arrows to tell whether the pericyclic reaction is 4n or 4n + 2. Phase reversals occur between retention/inversion at the migrating center, between antarafacial/suprafacial migration, with 4n vs. 4n+2 electrons, and between thermal and photochemically excited species.

To summarize for thermal $4n$ + 2 pericyclic reactions:

Thermally allowed $4n$ + 2 cycloadditions go supra-supra.

Thermally allowed $4n$ + 2 electrocyclic ring openings go disrotatory.

Thermally allowed $4n$ + 2 sigmatropic rearrangements go supra-retention.

The rules reverse for every MO phase reversal that occurs in going to $4n$ electrons, to inversion, to antarafacial, to conrotatory, or to photochemically excited systems.

ADDITIONAL EXERCISES

12.1 Draw the pi MOs of pentadienyl cation. What is the HOMO and LUMO?

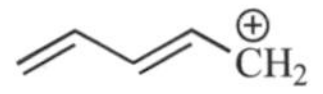

⊕
CH2

12.2 Which thermal electrocyclic ring closure mode, conrotatory or disrotatory, would the cation in problem 12.1 follow?

12.3 Would the thermal electrocyclic closure of 1,3,5-cycloheptatriene go con or dis?

12.4 Predict the regiochemistry of the Diels–Alder addition of these two reactants.

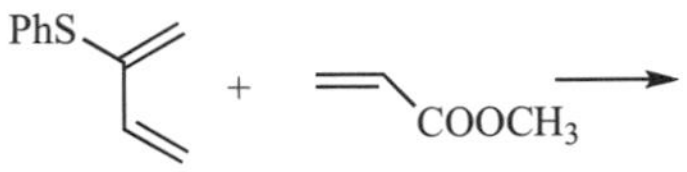

PhS
+
COOCH3

12.5 Is the following sigmatropic rearrangement thermally allowed?

12.6 Give an **allowed** thermal mechanism for the reaction below.

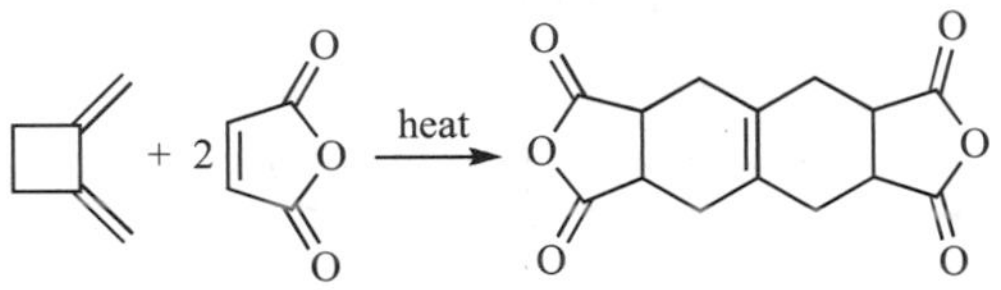

+ 2
O
O
O
heat
O
O
O
O
O
O

12.7 Give the product of this electrocyclic ring opening showing stereochemistry.

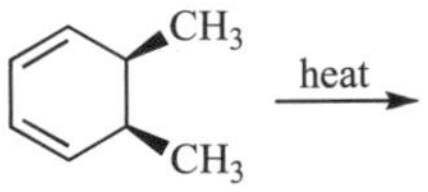

CH3
heat
CH3

12.8 Draw the photochemical ring closure product of the product from problem 12.7.

12.9 The compound below does not react with Diels–Alder dienophiles. Explain.

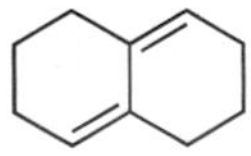

12.10 Give the product of this electrocyclic ring opening showing stereochemistry.

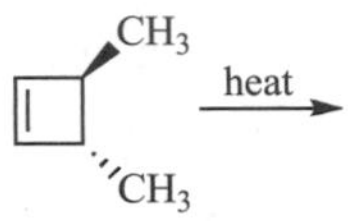

CH3
heat
CH3

12.11 Draw the photochemical ring closure product of the product from problem 12.10.

12.12 Give the regiochemistry of the following inverse electronic demand Diels–Alder.

12.13 Ozone is a 1,3 dipole and adds to alkenes to give a molozonide by the 6e path, then breaks by the reverse process, and adds together to form the ozonide by the 6e path. The ozonide is cleaved by dimethyl sulfide to give two carbonyls. Show the mechanism.

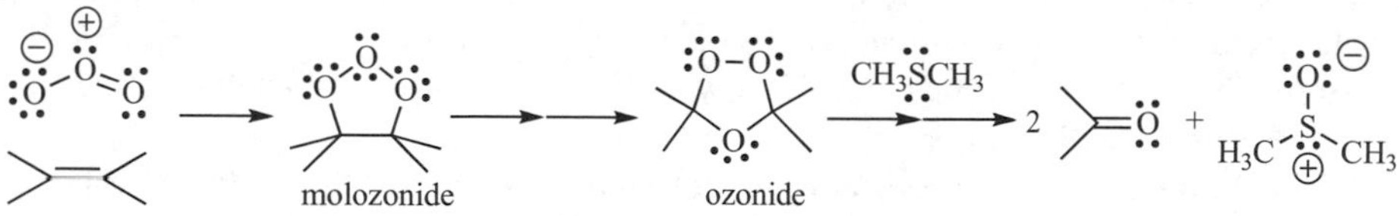
molozonide
ozonide
CH_3SCH_3
2
H_3C
CH_3

12.14 Give an **allowed** thermal mechanism for the reaction below.

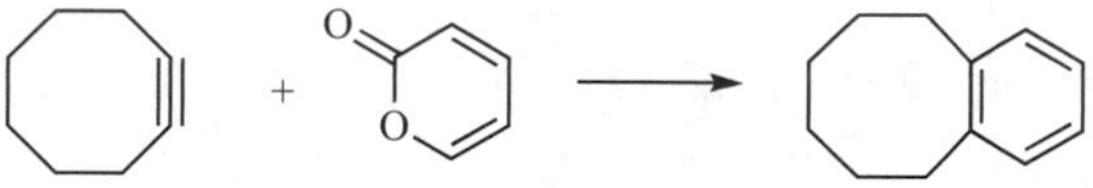

12.15 Predict the stereochemistry of the Diels–Alder addition of these two reactants.

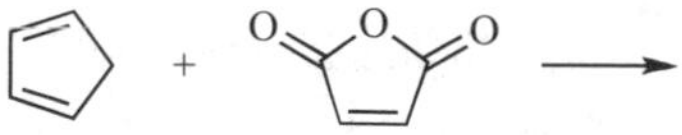

12.16 Predict the regiochemistry and stereochemistry of the Diels–Alder addition below.

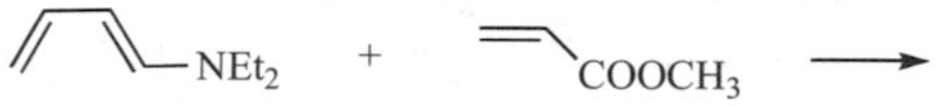
NEt_2
$COOCH_3$

12.17 Give the product of this thermally allowed reaction.

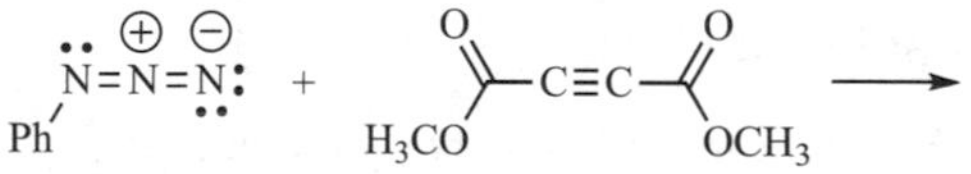
N=N=N:
Ph
C≡C
H_3CO
OCH_3

12.18 Provide a reasonable mechanism for the following reaction that has a pericyclic rearrangement as its last step.

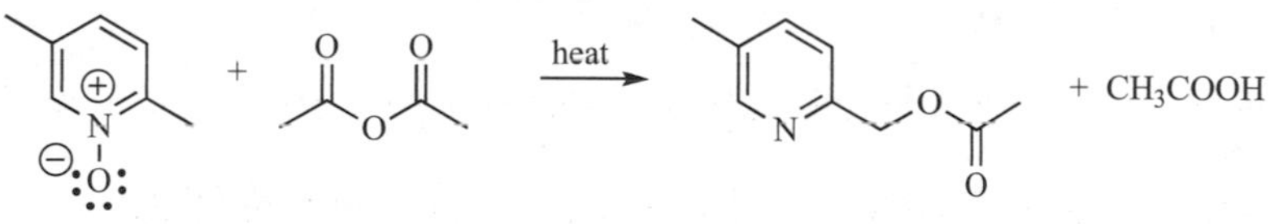
heat
+ CH_3COOH

12.19 Give the product of this thermally allowed rearrangement.

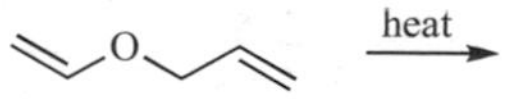
heat

12.20 Provide a reasonable mechanism for the following reaction that has a pericyclic rearrangement as its last step.

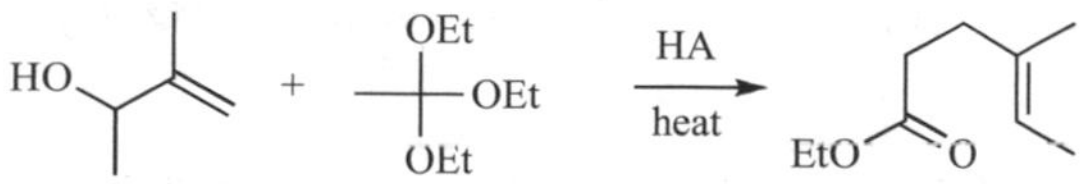
HO
OEt
OEt
OEt
HA
heat
EtO

APPENDIX

(A COLLECTION OF IMPORTANT TOOLS)

BIBLIOGRAPHY

Undergraduate Texts

Grossman, R. *The Art of Writing Reasonable Organic Reaction Mechanisms*; Springer, 2003.

Hornback, J. *Organic Chemistry*, 2nd ed.; Thomson, 2006.

Streitwieser, A.; Heathcock, C.; Kosower, E; *Introduction to Organic Chemistry*, 4th ed.; Pearson, 1998.

Sykes, P. *A Guidebook to Mechanism in Organic Chemistry*, 6th ed.; Prentice-Hall, 1986.

Weeks, D. *Pushing Electrons*, 3rd ed.; Thomson, 1998

Graduate Texts

Anslyn E. & Dougherty D. *Modern Physical Organic Chemistry*, University Science Books, 2006

Carey, F.& Sundberg, R. *Advanced Organic Chemistry*, 5th ed.; Springer, 2007.

Carroll, F. *Perspectives on Structure and Mechanism in Organic Chemistry*, 2nd ed.; Wiley, 2010.

Smith, M. & March, J. *March's Advanced Organic Chemistry: Reactions, Mechanisms, and Structure*, 6th ed.; Wiley, 2007.

Molecular Orbital Theory

Fleming, I. *Molecular Orbitals and Organic Chemical Reactions, Reference Edition*; Wiley, 2010.

Jorgensen, W. & Salem, L. *The Organic Chemist's Book of Orbitals*; Academic, 1973.

Biochemical Mechanisms

Fersht, A. *Enzyme Structure and Mechanism*; Freeman, 1985.

McMurry J. & Begley T. *The Organic Chemistry of Biological Pathways*, Roberts, 2005.

Walsh, C. *Enzymatic Reaction Mechanisms*; Freeman, 1979.

Inorganic Chemistry

Cotton, F., Wilkinson, G., Murillo, C., Bochmann, M. *Advanced Inorganic Chemistry*, 6th ed.; Wiley Interscience, 1999.

Crabtree R. *The Organometallic Chemistry of the Transition Metals*, Wiley, 2009

Bond Strengths

Benson, S. W. *J. Chem. Educ.* **1965**, *42*, 502.

Cottrell, T. *The Strengths of Chemical Bonds*, 2nd ed.; Butterworths, 1958.

pK_a Values

Perrin, D.; Dempsey, B.; Serjeant, E. *pK_a Prediction for Organic Acids and Bases*; Chapman and Hall, 1981.

Smith, M. & March, J. *March's Advanced Organic Chemistry: Reactions, Mechanisms, and Structure*, 6th ed.; Wiley, 2007 (primary source).

Speight, J., ed., *Lange's Handbook of Chemistry*, 16th ed.; McGraw-Hill, 2005.

Hard–Soft Acid–Base Theory

Pearson, R. G. *J. Chem. Educ.* **1968**, *45*, 581.

Nucleophilicity

Pearson, R. G. *J. Am. Chem. Soc.* **1968**, *90*, 319.

The ΔpK_a Rule and Computer Prediction of Organic Reactions

Salatin, T. D.; Jorgensen, W. L., *J. Org. Chem.* **1980**, *45*, 2043.

ABBREVIATIONS USED IN THIS TEXT

Ac	Acetyl $CH_3C{=}O$
Ar	Any aryl (aromatic) group
b	Brønsted base, proton acceptor
δ+	A partial positive charge
δ−	A partial negative charge
E	Electrophile, Lewis acid
Et	Ethyl CH_3CH_2
erg	Electron-releasing group
ewg	Electron-withdrawing group
G	Unspecified group
HA	Brønsted acid, proton donor
HOMO	Highest occupied molecular orbital
i-Pr	Isopropyl $(CH_3)_2CH$
L	Leaving group
LUMO	Lowest unoccupied molecular orbital
M	Metal atom
Me	Methyl CH_3
MO	Molecular orbital
n-Bu	Normal-butyl group $CH_3CH_2CH_2CH_2$
Nu	Nucleophile
Ph	Phenyl group, C_6H_5, a monosubstituted benzene
R	Any alkyl chain
t-Bu	Tertiary butyl $(CH_3)_3C$
Ts	Toluenesulfonyl, $CH_3C_6H_4SO_2$
X	Chlorine, bromine, or iodine
Y, Z	Heteroatoms, commonly oxygen, nitrogen, or sulfur
±	Racemic mixture
→→	Multistep process
‡	Transition state
......	Partially broken bond (or weak complexation)

For the path acronyms, please see the path summary in this Appendix.

Functional Group Glossary

Name	Functional Group	Example
Acetal	$H-C(\ddot{O}C)_2$	$H_3C-CH(\ddot{O}CH_3)_2$
Acid anhydride	$-C(=O)-O-C(=O)-$	$H_3C-C(=O)-O-C(=O)-CH_3$
Acyl halide	$-C(=O)-X$	$H_3C-C(=O)-Cl$
Acyloin	$-C(=O)-C(OH)<$	$H_3C-C(=O)-CHCH_3$ (with OH on CH)
Alcohol	$\geq C-O-H$	H_3C-O-H
Aldehyde	$-C(=O)-H$	$H_3C-C(=O)-H$
Alkane	$\geq C-C \leq$	H_3C-CH_3
Alkene	$>C=C<$	$H_2C=CH_2$
Alkoxide	$\geq C-O:^{\ominus}$	$H_3C-O:^{\ominus}$
Alkyl halide	$\geq C-X:$	$H_3C-Br:$
Alkyne	$-C \equiv C-$	$HC \equiv CH$
Allene (a cumulene)	$>C=C=C<$	$(H_3C)(H)C=C=CH_2$
Amide	$-C(=O)-N<$	$H_3C-C(=O)-NH_2$
Amidate	$-C(=O)-N^{\ominus}-$	$H_3C-C(=O)-NH^{\ominus}$
Amine	$\geq C-N<$	H_3C-NH_2
Amine oxide	$\geq N^{\oplus}-O:^{\ominus}$	$(H_3CH_2C)_3N^{\oplus}-O:^{\ominus}$
Aryl halide	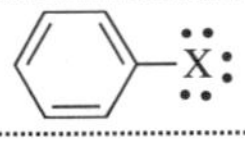	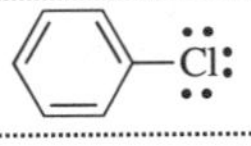
Azo compound	$-N=N-$	$H_3C-N=N-CH_3$
Borane	$\geq B$	BH_3
Borate ester	$CO-B(OC)_2$	$H_3CO-B(OCH_3)_2$
Carbamate (urethane)	$-O-C(=O)-N<$	$H_3C-O-C(=O)-N(CH_3)_2$
Carbene	$-\ddot{C}-$	$Cl-\ddot{C}-Cl$
Carbodiimide	$-N=C=N-$	$(H_3C)_2HC-N=C=N-CH(CH_3)_2$
Carbonate	$-O-C(=O)-O-$	$H_3C-O-C(=O)-O-CH_3$
Carboxylate	$-C(=O)-O:^{\ominus}$	$H_3C-C(=O)-O:^{\ominus}$

Carboxylic acid	
Diazonium	
Diene	
Disulfide	
Enamine	
Enediol	
Enol	
Enol ether	
Enolate	
Enone	
Epoxide	
Ester	
Ether	
Halohydrin	
Hemiacetal	
Hemiketal	
Hydrate	
Hypochlorite	
Imine	
Isocyanate	
Ketal	
Ketene	
Ketone	

Nitrile	>C−C≡N:	$H_3C-C\equiv N$:
Nitro compound	C−N⁺(=O)−O⁻	$H_3C-N^{\oplus}(=O)-O^{\ominus}$
Organometallic	R−M	H_3C-Li
Orthoester	C−C(OC)(OC)(OC)	$H_3C-C(OCH_3)_3$
Oxime	>C=N−OH	$(H_3C)_2C=N-OH$
Peroxide	C−O−O−C	$(H_3C)_3C-O-O-C(CH_3)_3$
Phosphate ester	O=P(OC)(OC)(OC)	$O=P(OCH_3)_3$
Phosphonate ester	O=P(C)(OC)(OC)	$O=P(H_3C)(OCH_3)_2$
Selenoxide	C−Se(=O)−C	$C_6H_5-Se(=O)-C_6H_5$
Sulfate ester	CO−S(=O)(=O)−OC	$H_3CO-S(=O)(=O)-OCH_3$
Sulfide	C−S−C	$H_3C-S-CH_3$
Sulfinic acid	C−S(=O)−OH	$C_6H_5-S(=O)-OH$
Sulfonate ester	C−S(=O)(=O)−OC	$H_3C-S(=O)(=O)-OCH_3$
Sulfone	C−S(=O)(=O)−C	$H_3C-S(=O)(=O)-CH_3$
Sulfonic acid	C−S(=O)(=O)−OH	$H_3C-S(=O)(=O)-OH$
Sulfonyl halide	C−S(=O)(=O)−X	$H_3C-S(=O)(=O)-Cl$
Sulfoxide	C−S(=O)−C	$H_3C-S(=O)-CH_3$
Thioester	−C(=O)−S−C	$H_3C-C(=O)-S-CH_3$
Thiol	C−S−H	H_3C-S-H
Urea	>N−C(=O)−N<	$H_2N-C(=O)-NH_2$
Vinyl halide	>C=C<X	$H_2C=CH-Cl$
Xanthate	−O−C(=S)−S−	$H_3C-O-C(=S)-SCH_3$

Composite pK_a Chart[1]

The Acidic H Is in Boldface on the Acid
The Basic Lone Pair Is Shown on the Base

Oxygen Acids

pK_a	Acid	Base
−15	$F_3C{-}S(=O)_2{-}OH$	$F_3C{-}S(=O)_2{-}O^-$
	$F{-}S(=O)_2{-}OH$	$F{-}S(=O)_2{-}O^-$
−12	$R{-}N^+(=O){-}OH$	$R{-}N^+(=O){-}O^-$
−10	$O_3Cl{-}OH$	$O_3Cl{-}O^-$
−10	$R{-}C(=O^+H){-}H$	$R{-}C(=O){-}H$
−9	$HO{-}S(=O)_2{-}OH$	$HO{-}S(=O)_2{-}O^-$
−9	$R{-}C(=O^+H){-}Cl$	$R{-}C(=O){-}Cl$
−7	$R{-}C(=O^+H){-}R$	$R{-}C(=O){-}R$
−6.5	$Ar{-}S(=O)_2{-}OH$	$Ar{-}S(=O)_2{-}O^-$
−6.5	$R{-}C(=O^+H){-}OR$	$R{-}C(=O){-}OR$
−6.4	$ArOH_2^+$	$ArOH$
−6	$R{-}C(=O^+H){-}OH$	$R{-}C(=O){-}OH$
−6	$Ar{-}O^+(H){-}R$	$Ar{-}O{-}R$
−3.5	$R{-}O^+(H){-}R$	$R{-}O{-}R$
−2.4	$CH_3CH_2OH_2^+$	CH_3CH_2OH
−1.7	H_3O^+	H_2O
−1.5	$Ar{-}C(=O^+H){-}NH_2$	$Ar{-}C(=O){-}NH_2$
−1.5	$(CH_3)_2S{=}OH^+$	$(CH_3)_2S{=}O$
−1.4	$^-O{-}N^+(=O){-}OH$	$^-O{-}N^+(=O){-}O^-$
−0.5	$R{-}C(=O^+H){-}NH_2$	$R{-}C(=O){-}NH_2$
0.5	$F_3C{-}C(=O){-}OH$	$F_3C{-}C(=O){-}O^-$

pK_a	Acid	Base
0.7	$C_5H_5N^+{-}OH$ (pyridine)	$C_5H_5N^+{-}O^-$
1.7	$O_2N{-}CH_2{-}C(=O){-}OH$	$O_2N{-}CH_2{-}C(=O){-}O^-$
1.8	$(H_3C)_3N^+{-}CH_2{-}C(=O){-}OH$	$(H_3C)_3N^+{-}CH_2{-}C(=O){-}O^-$
1.8	$HC{\equiv}C{-}C(=O){-}OH$	$HC{\equiv}C{-}C(=O){-}O^-$
2.0	$^-O{-}S(=O)_2{-}OH$	$^-O{-}S(=O)_2{-}O^-$
2.1	$Ph{-}S(=O){-}OH$	$Ph{-}S(=O){-}O^-$
2.2	$(HO)_2P(=O){-}OH$	$(HO)_2P(=O){-}O^-$
2.4	$H_3N^+{-}CH_2{-}C(=O){-}OH$	$H_3N^+{-}CH_2{-}C(=O){-}O^-$
2.5	$N{\equiv}C{-}CH_2{-}C(=O){-}OH$	$N{\equiv}C{-}CH_2{-}C(=O){-}O^-$
2.5	$CH_3{-}C(=O){-}C(=O){-}OH$	$CH_3{-}C(=O){-}C(=O){-}O^-$
2.9	$Cl{-}CH_2{-}C(=O){-}OH$	$Cl{-}CH_2{-}C(=O){-}O^-$
3.1	$F_3C{-}CH_2{-}C(=O){-}OH$	$F_3C{-}CH_2{-}C(=O){-}O^-$
3.3	HO-N=O	$^-O{-}N{=}O$
3.4	$O_2N{-}C_6H_4{-}C(=O){-}OH$	$O_2N{-}C_6H_4{-}C(=O){-}O^-$
3.6	$H_3C{-}C(=O){-}CH_2{-}C(=O){-}OH$	$H_3C{-}C(=O){-}CH_2{-}C(=O){-}O^-$
3.6	$H_3CO{-}CH_2{-}C(=O){-}OH$	$H_3CO{-}CH_2{-}C(=O){-}O^-$
4.2	$C_6H_5{-}C(=O){-}OH$	$C_6H_5{-}C(=O){-}O^-$
4.5	$H_3CO{-}C_6H_4{-}C(=O){-}OH$	$H_3CO{-}C_6H_4{-}C(=O){-}O^-$
4.6	$(CH_3)_3N^+{-}OH$	$(CH_3)_3N^+{-}O^-$
4.8	$H_3C{-}C(=O){-}OH$	$H_3C{-}C(=O){-}O^-$
6.4	$HO{-}C(=O){-}OH$	$HO{-}C(=O){-}O^-$
7.2	$^-O{-}P(=O)(OH){-}OH$	$^-O{-}P(=O)(OH){-}O^-$
7.2	$NO_2{-}C_6H_4{-}OH$	$NO_2{-}C_6H_4{-}O^-$

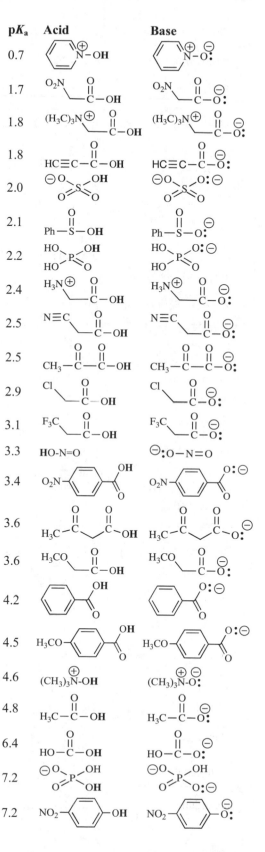

[1]Modified with permission from Smith, M.. March, J. *March's Advanced Organic Chemistry*, 6th ed; pp.359–364. Copyright ©2007 by John Wiley & Sons Inc.

Oxygen acids (continued)

pK_a	Acid	Base
7.5	$Cl-\mathbf{OH}$	$Cl-O:^{\ominus}$
10.0	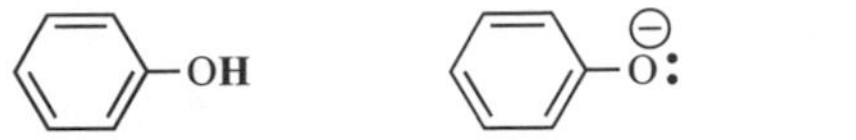	
10.2	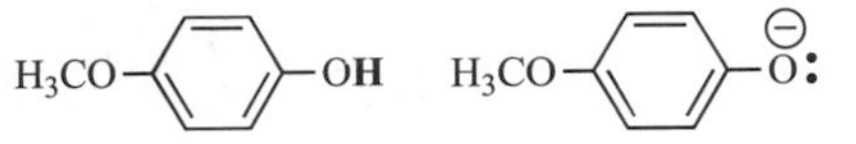	
10.3	$\mathbf{HO}-C(=O)-O^{\ominus}$	$^{\ominus}O-C(=O)-O:^{\ominus}$
11.6	HO-**OH**	$HO-O:^{\ominus}$
12.2	$(CH_3)_2C=N-\mathbf{OH}$	$(H_3C)_2C=N-O:^{\ominus}$
12.4		
12.4	$CF_3CH_2\mathbf{OH}$	$CF_3CH_2O:^{\ominus}$
13.3	$HOCH_2\mathbf{OH}$	$HOCH_2O:^{\ominus}$
13.9	$(CH_3)_3\overset{\oplus}{N}CH_2CH_2\mathbf{OH}$	$(CH_3)_3\overset{\oplus}{N}CH_2CH_2O:^{\ominus}$
14.2	$HOCH_2CH_2\mathbf{OH}$	$HOCH_2CH_2O:^{\ominus}$
15.5	$CH_3\mathbf{OH}$	$CH_3O:^{\ominus}$
15.7	$\mathbf{H}_2O$	$HO:^{\ominus}$
16	$CH_3CH_2\mathbf{OH}$	$CH_3CH_2O:^{\ominus}$
18	$(CH_3)_2CH\mathbf{OH}$	$(CH_3)_2CHO:^{\ominus}$
19	$(CH_3)_3C\mathbf{OH}$	$(CH_3)_3CO:^{\ominus}$

Nitrogen Acids

pK_a	Acid	Base
−10	$R-C\equiv\mathbf{NH}^{\oplus}$	$R-C\equiv N:$
−5	$Ar_3\mathbf{NH}^{\oplus}$	$Ar_3N:$
1	$Ar_2\mathbf{NH}_2^{\oplus}$	$Ar_2\ddot{N}H$
1.0	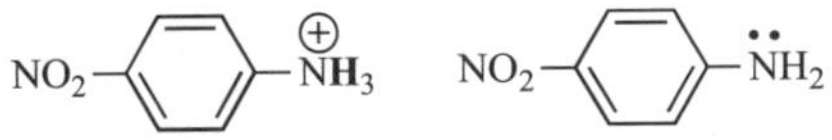	
4.6		
5.2		
5.4	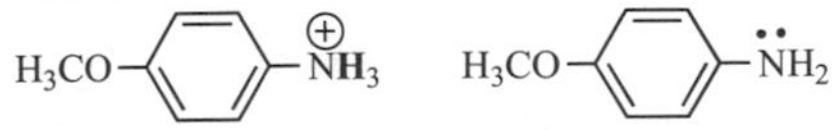	
5.8	$HO-\mathbf{NH}_3^{\oplus}$	$HO-\ddot{N}H_2$
7.0		
7.9	$H_2N-\mathbf{NH}_3^{\oplus}$	$H_2N-\ddot{N}H_2$
9.2	$\mathbf{NH}_4^{\oplus}$	$:NH_3$
9.3		
9.5		

Nitrogen acids (continued)

pK_a	Acid	Base
9.6		
9.8	$^{\ominus}O-C(=O)-CH_2\overset{\oplus}{N}\mathbf{H}_3$	$^{\ominus}O-C(=O)-CH_2\ddot{N}H_2$
10.1		
10.7	$Et_3\mathbf{NH}^{\oplus}$	$Et_3N:$
13.6	$(H_2N)_2C=\mathbf{NH}_2^{\oplus}$	$(H_2N)_2C=\ddot{N}H$
17	$R-C(=O)-\mathbf{NH}_2$	$R-C(=O)-\ddot{N}H^{\ominus}$
25.8	$((CH_3)_3Si)_2\mathbf{NH}$	$((CH_3)_3Si)_2N:^{\ominus}$
27		
35	$\mathbf{NH}_3$	$^{\ominus}\ddot{N}H_2$
36	$R_2\mathbf{NH}$	$R_2\ddot{N}^{\ominus}$

Carbon acids

pK_a	Acid	Base
−5	$\mathbf{H}C(CN)_3$	$^{\ominus}:C(CN)_3$
3.6	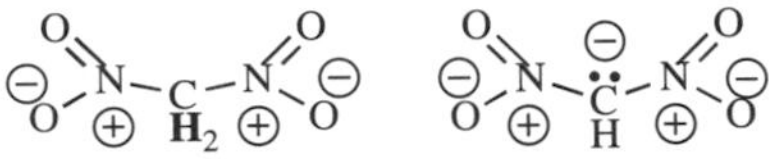	
5	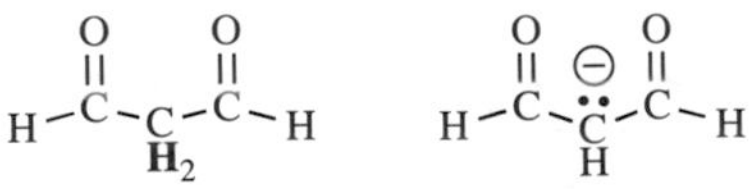	
9	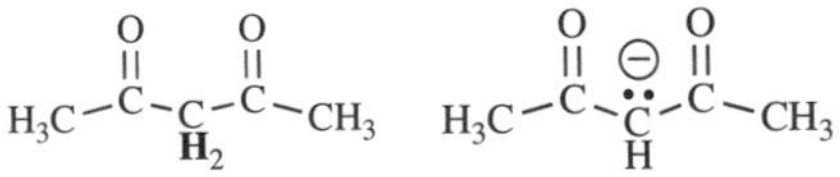	
9.2	$\mathbf{H}C\equiv N$	$^{\ominus}:C\equiv N$
10.2		
10.7	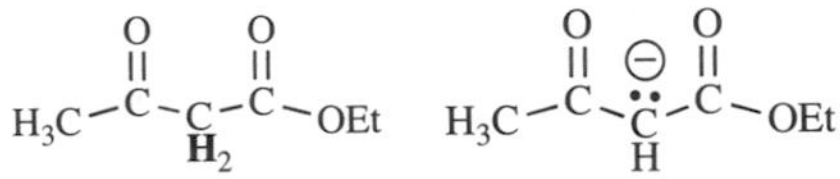	
11.2	$C\mathbf{H}_2(CN)_2$	$^{\ominus}:CH(CN)_2$
12.5	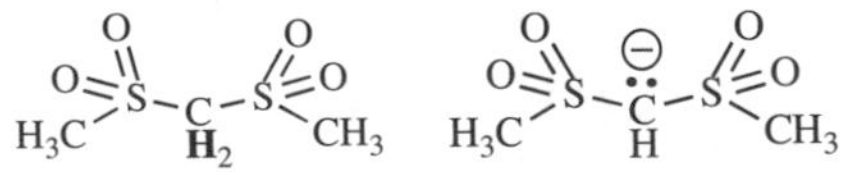	
13		
13.5	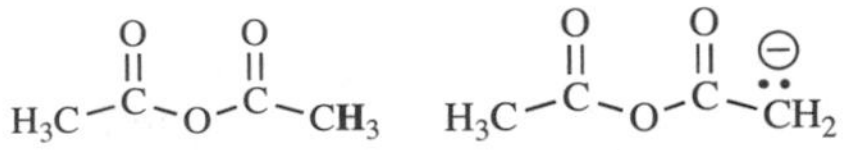	
15.9	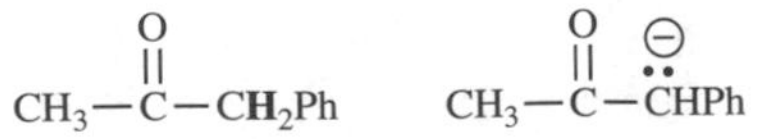	

Carbon acids (continued)

pK_a	Acid	Base
16		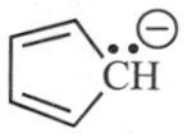
16	$H_3C-C(=O)-CH_2Cl$	$H_3C-C(=O)-\ddot{C}HCl^{\ominus}$
16.7	$H_3C-C(=O)-H$	$^{\ominus}\ddot{H_2C}-C(=O)-H$
18	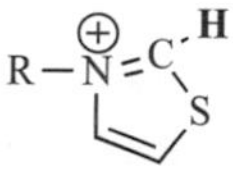	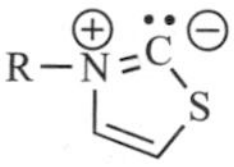
19.2	$H_3C-C(=O)-CH_3$	$^{\ominus}\ddot{H_2C}-C(=O)-CH_3$
21	$CH_3-C(=O)-SR$	$^{\ominus}\ddot{C}H_2-C(=O)-SR$
24	$HCCl_3$	$\ominus{:}CCl_3$
25	$H_3C-C\equiv N$	$^{\ominus}H_2\ddot{C}-C\equiv N$
25	$HC\equiv CH$	$\ominus{:}C\equiv CH$
25.6	$H_3C-C(=O)-OR$	$^{\ominus}\ddot{H_2C}-C(=O)-OR$
28	$H_3C-C(=O)-NR_2$	$^{\ominus}\ddot{H_2C}-C(=O)-NR_2$
≈30	$H_3C-\overset{\oplus}{P}Ph_3$	$^{\ominus}\ddot{H_2C}-\overset{\oplus}{P}Ph_3$
31	$H_3C-S(=O)_2-CH_3$	$^{\ominus}\ddot{H_2C}-S(=O)_2-CH_3$
31.1	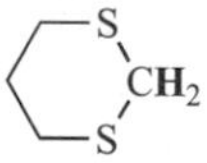	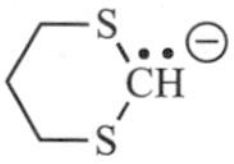
31.5	$HCPh_3$	$\ominus{:}CPh_3$
33.5	H_2CPh_2	$\ominus{:}CHPh_2$
35	$H_3C-S(=O)-CH_3$	$^{\ominus}\ddot{H_2C}-S(=O)-CH_3$
40	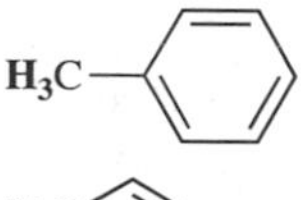	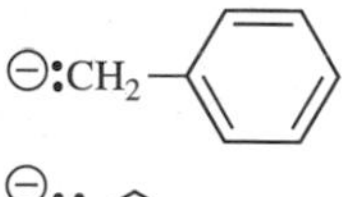
43	$H_3C-CH=CH_2$	$^{\ominus}\ddot{H_2C}-CH=CH_2$
43		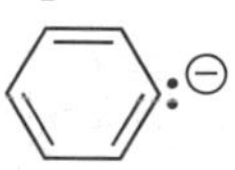
44	$H_2C=CH_2$	$^{\ominus}\ddot{H}C=CH_2$
48	H_4C	$\ominus{:}CH_3$
50	H_3C-CH_3	$^{\ominus}\ddot{H_2C}-CH_3$
51	$H_3C-CH_2-CH_3$	$H_3C-\ddot{C}H^{\ominus}-CH_3$
>52	$HC(CH_3)_3$	$\ominus{:}C(CH_3)_3$

Miscellaneous acids

pK_a	Acid	Base
−10	HI	$:I^{\ominus}$
−9	HBr	$:Br^{\ominus}$
−7	HCl	$:Cl^{\ominus}$
−7	$RSH_2^{\oplus}$	$R\ddot{S}H$
−5.3	$R_2SH^{\oplus}$	$R_2S:$
2.7	$Ph_3PH^{\oplus}$	$Ph_3P:$
3.2	HF	$:F^{\ominus}$
3.3	$H_3C-C(=O)-SH$	$H_3C-C(=O)-\ddot{S}:^{\ominus}$
3.9	H_2Se	$HSe:^{\ominus}$
6.5	PhSH	$PhS:^{\ominus}$
7.0	H_2S	$HS:^{\ominus}$
8.7	$Et_3PH^{\oplus}$	$Et_3P:$
10.6	EtSH	$EtS:^{\ominus}$
35	H_2	$H:^{\ominus}$

A strong acid has a low pK_a

A strong base has a high pK_{abH}

Proton Transfer Equilibria

Proton transfers go toward the formation of the weaker base.

$$\log K_{eq} = pK_{abH} - pK_{aHA}$$

$$K_{eq} = 10^{\{pK_{abH} - pK_{aHA}\}}$$

Subtract the pK_a of the acid from the pK_{abH} of the base to get the exponent of K_{eq}. If the proton transfer K_{eq} is equal to or greater than 10^{-10}, the proton transfer is within the useful range.

The ΔpK_a Rule:

Avoid intermediates that are more than 10 pK_a units *uphill* from the reactants (either 10 pK_a units more basic or 10 pK_a units more acidic). The energy drops if the pK_{abH} drops significantly.

Hard and Soft Acid–Base Principle:

Hard bases favor binding with hard acids; soft bases favor binding with soft acids.

BOND STRENGTH TABLE[2]

Average Single Bond Energies in kcal/mol (kJ/mol)

	I	Br	Cl	S	P	Si	F	O	N	C	H
H	71 (297)	87 (364)	103 (431)	83 (347)	77 (322)	76 (318)	135 (565)	111 (464)	93 (389)	99 (414)	104 (435)
C	51 (213)	68 (285)	81 (339)	65 (272)	63 (264)	72 (301)	116 (485)	86 (360)	73 (305)	83 (347)	
N			46 (192)				65 (272)	53 (222)	39 (163)		
O	48 (201)	48 (201)	52 (218)		141 (590)	108 (452)	45 (188)	47 (197)			
F	58 (243)	60 (251)	61 (255)	68 (285)	117 (490)	135 (565)	37 (155)				
Si	56 (234)	74 (310)	91 (381)			53 (222)					
P	44 (184)	63 (264)	78 (326)		48 (201)						
S		52 (218)	61 (255)	60 (251)							
Cl	50 (209)	52 (218)	58 (243)								
Br	42 (176)	46 (192)									
I	36 (151)										

Average Multiple Bond Energies in kcal/mol (kJ/mol)

C=C	146 (611)	C≡C	200 (837)
C=N	147 (615)	C≡N	213 (891)
C=O	177 (741)		
C=S	128 (536)		
N=N	100 (418)	N≡N	226 (946)
N=O	145 (607)		
O=O	119 (498)		

Note: this table contains average values that should be considered very approximate (± several kcal/mol); only the values that correspond to bond strengths of simple diatomic molecules (like HCl) have little error.

[2] Streitwieser, Andrew; Heathcock,: Kosower, *Introduction to Organic Chemistry*, Revised Printing, 4th edtition, ©1992. Adapted by permission of Pearson Education, Inc., Upper Saddle River, NJ. Additional values from Cottrell, T., *The Strength of Chemical Bonds*, 2nd ed.; Butterworths: London, 1958, and from Benson, S. W., *J. Chem. Educ.* **1965**, *42*, 502.

Generic Classification Guide

A big problem in organic chemistry is the sheer magnitude of information, so efficient ways of organizing the material are very valuable. **There are only 12 generic classes from which a set of arrows, an electron flow, can start. Likewise, there are only 18 generic classes of electron sinks in which an electron flow ends. It is much easier to handle multiple examples of familiar classes than hundreds of special cases.**

The best electron sources are electron rich, like anions and lone pairs, such that no bonds need be broken to use them as electron sources. Other excellent electron sources are highly ionic sigma bonds and pi bonds polarized by excellent electron-releasing groups. Table 5.2 repeated below shows the common electron sources, and Figure A.1 shows a flowchart of the classification process.

Table 5.2 Summary Reactivity Table for Common Electron Sources

Symbol & Name	Common Examples in Order of Decreasing Reactivity	Reactivity Trend Principle	Common Reactions Decision?	Related Generic Groups
Z: Heteroatom lone pairs as nucleophiles	RS^- I^- CN^- RO^-	Softer anions are more nucleophilic. If same atom, more basic, more Nu	Substitutions & Additions. Nucleophile vs. base decision	See allylic sources
Z: Heteroatom lone pairs as bases	R_2N^- t-BuO^- R_3N CH_3COO^-	The higher pK_{abH} is stronger	Proton transfer $K_{eq} > 10^{-10}$	NaH and KH as MH bases
R–M Organo-metallics	CH_3Li CH_3MgI $(CH_3)_2Cu^-Li^+$	The more ionic RM bond is more reactive	Substitutions & Additions. Deprotonates acidic Hs	See enolates (allylic sources)
MH_4^- Complex metal hydrides	$LiAlH_4$ $LiAlH(OR)_3$ $NaBH_4$ $NaBH_3CN$	The more ionic MH bond is more reactive	Substitutions & Additions. Deprotonates acidic Hs on heteroatoms	NaH and KH act as MH bases
C=C–Z: Allylic sources	Enolates $C{=}C{-}O^-$ Enamines $C{=}C{-}NR_2$ Enol ethers C=C–OR	The better donor on the pi bond is more reactive	Substitutions & Additions. Carbon vs. heteroatom decision	Extended enolates. Allylic alkyne sources
C=C Simple pi bonds	$R_2C{=}CR_2$ RHC=CHR $H_2C{=}CH_2$	The more stable the resultant carbocation, the more reactive the pi system.	Electrophilic Additions. Markovnikov's rule used for regiochemistry	Other pi systems: C=C–C=C C≡C C=C=C
ArH Aromatic rings	PhOR PhH PhCl PhCOR	A donor on the ring makes it more reactive; ewg makes ring less reactive.	Electrophilic aromatic substitution. Regiochemistry of addition	Hetero-aromatics, condensed aromatics

Electron sinks have a full, partial, or inducible plus charge on the atom that gets attacked by the electron source. The larger the partial plus is, the better the electron sink can attract a negatively charged electron source. Table 6.2 repeated below gives the common electron sinks, and Figure A.1 shows a flowchart of the classification process.

Table 6.2 Summary Reactivity Table for Common Electron Sinks

Symbol & Name	Common Examples in Order of Decreasing Reactivity	Reactivity Trend Principle	Common Reactions. Decision?	Related Generic Groups
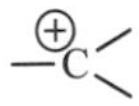 Carbocation	$(CH_3)_2CH^+$ $(CH_3)_3C^+$ Ph_3C^+ $(H_2N)_3C^+$	More stabilized, less reactive. Donors stabilize Lone pairs > pi bonds > R > H	C^+ trap Nu. Rearrange to equal or more stable C^+? Drop off H^+?	Inorganic Lewis acids
H–L or H–A Acids	H_2SO_4 HBr HCl	The lower pK_a is stronger	Proton transfer $K_{eq} \geq 10^{-10}$	Carbon acids H_3C-ewg
Y–L Leaving group on a heteroatom	Iodine I_2 Bromine Br_2 Chlorine Cl_2 Peroxides ROOR	A weaker bond is more reactive	Often Substitution Nu attacks δ+ end of Y–L if different	Leaving groups on phosphorus or silicon
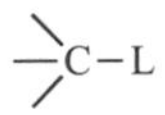 Leaving group on sp^3 carbon	Sulfonates $R–OSO_2R$ Alkyl iodides R–I Protonated alcohols ROH_2^+ Alkyl bromides R–Br Alkyl chlorides R–Cl	The better L is more reactive	Substitution versus Elimination Decision Section 9.5	2 Ls on C R_2CL_2 3 Ls on C RCL_3
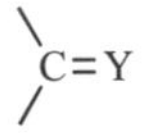 Polarized multiple bond without L	Protonated carbonyls $R_2C{=}OH^+$ Iminium ions $R_2C{=}NR_2^+$ Aldehydes RHC=O Ketones $R_2C{=}O$ Imines $R_2C{=}NR$	The larger partial plus carbon is more reactive	Addition. Following Substitution or Elimination? (8.5.1)	Polarized triple bonds R–C≡Y without leaving groups
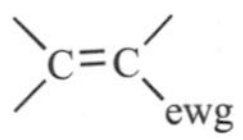 Conjugate acceptor	Conj. nitros $C{=}C{-}NO_2$ Enones C=C-COR Conj. esters C=C-COOR Conj. nitriles C=C-C≡N	The better the ewg, the more reactive	Addition 1,2 vs. 1,4 ? Conjugate addition or direct? (9.6)	Triply bonded conjugate acceptors C≡C–ewg
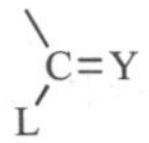 Polarized multiple bond with L	Acyl halides RCOCl Anhydrides RCOOCOR Thioesters RCOSR Esters RCOOR Amides $RCONR_2$	The poorer donor L is, the more reactive	Addition–Elimination Add a second Nu after add-elim? (9.2)	Carbonate derivatives $L_2C{=}Y$ L on triple bonds L-C≡Y

Flowcharts For the Classification of Electron Sources and Sinks

For some individuals, a graphical presentation of a process in the form of a flowchart greatly helps them understand and visualize the overall process (Fig. A.1).

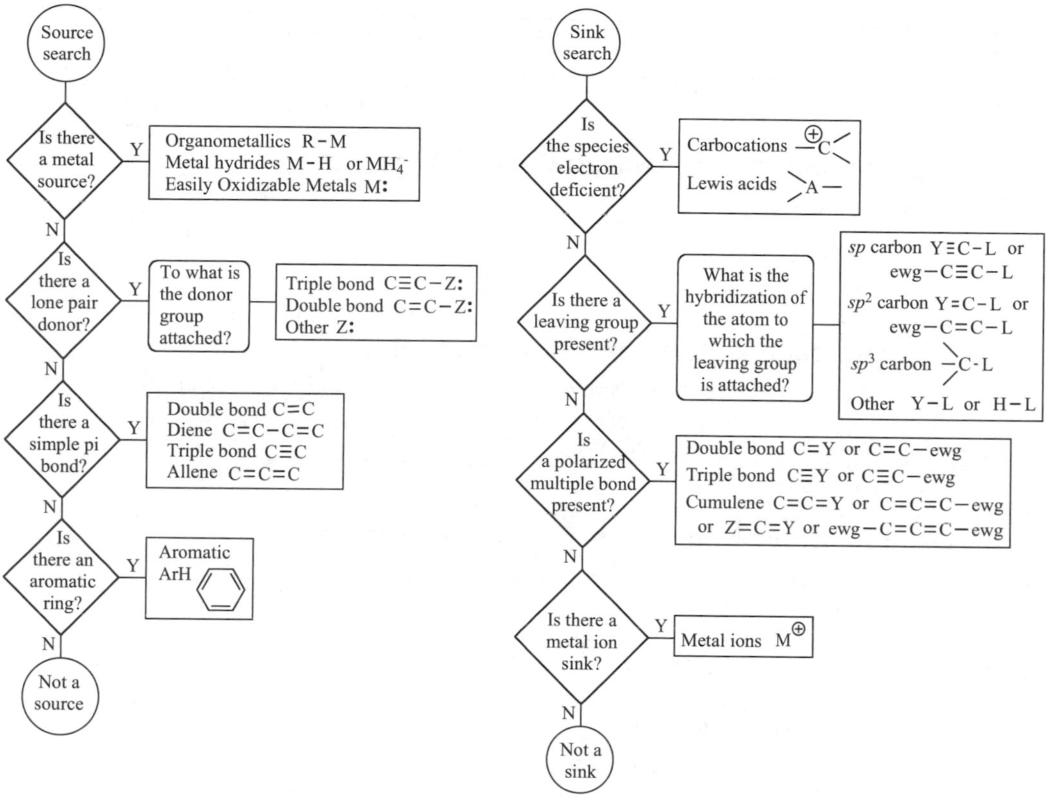
Source search
Is there a metal source?
Organometallics R−M
Metal hydrides M−H or MH4−
Easily Oxidizable Metals M:
Is there a lone pair donor?
To what is the donor group attached?
Triple bond C≡C−Z:
Double bond C=C−Z:
Other Z:
Is there a simple pi bond?
Double bond C=C
Diene C=C−C=C
Triple bond C≡C
Allene C=C=C
Is there an aromatic ring?
Aromatic ArH
Not a source
Sink search
Is the species electron deficient?
Carbocations
Lewis acids
Is there a leaving group present?
What is the hybridization of the atom to which the leaving group is attached?
sp carbon Y≡C−L or ewg−C≡C−L
sp2 carbon Y=C−L or ewg−C=C−L
sp3 carbon
Other Y−L or H−L
Is a polarized multiple bond present?
Double bond C=Y or C=C−ewg
Triple bond C≡Y or C≡C−ewg
Cumulene C=C=Y or C=C=C−ewg
or Z=C=Y or ewg−C=C=C−ewg
Is there a metal ion sink?
Metal ions M
Not a sink

Figure A.1 Flowcharts for the classification of electron sources and sinks.

Pathway Summary

Major Paths

p.t., Proton Transfer to and from an Anion or Lone Pair

Crosscheck: K_{eq} or ΔpK_a rule

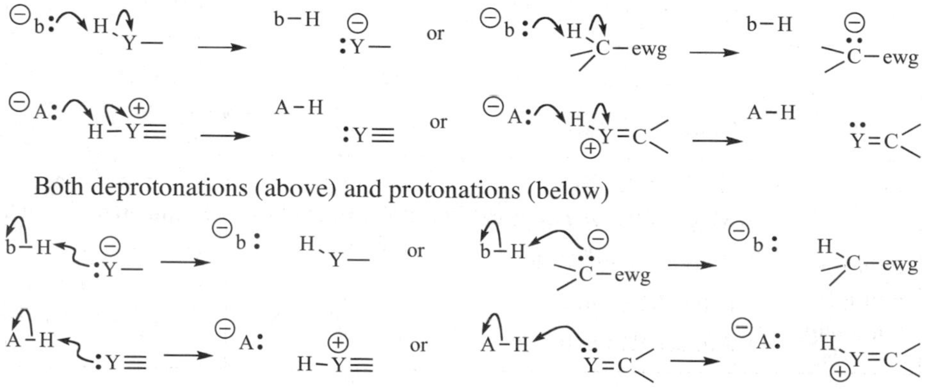
or
Both deprotonations (above) and protonations (below)

D_N, Ionization of a Leaving Group

Crosscheck: ΔpK_{aHL} less than zero, usually more stable than 2° cation, polar solvent

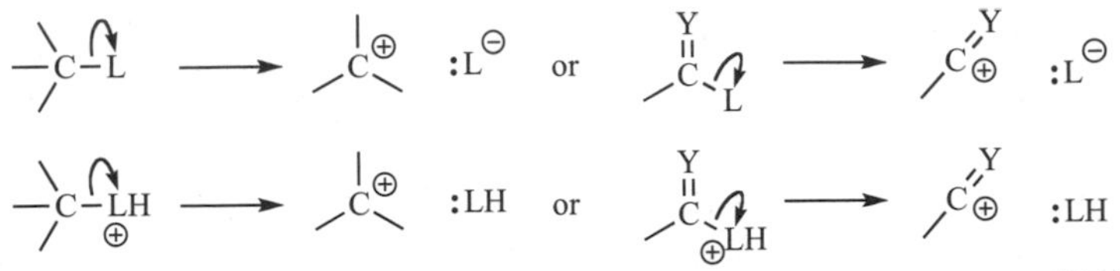

A_N, Trapping of an Electron-Deficient Species

Crosscheck: none, usually OK

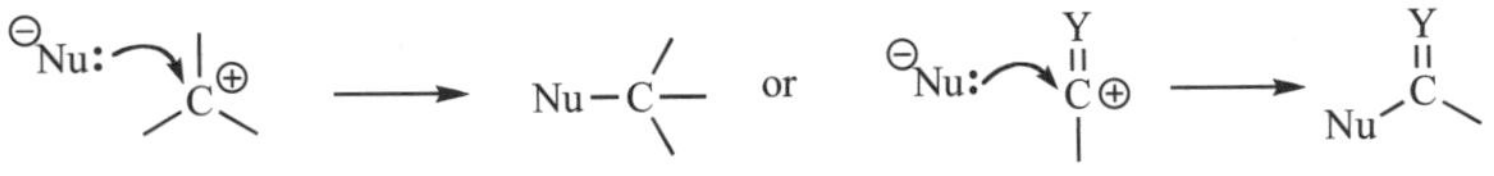

A_E, Electrophile Addition to a Multiple Bond

Crosscheck: Markovnikov's rule—Make the most stable carbocation

D_E, Electrofuge Loss from a Cation to Form a Pi Bond

Crosscheck: Electrofuge lost should be reasonably stable

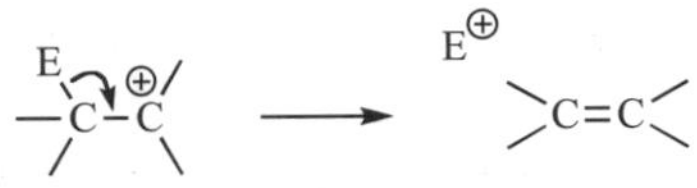

S_N2, The S_N2 Substitution

Crosscheck: ΔpK_a rule on Nu and L, $\Delta pK_{aHL} \leq 10$, good access to C attacked

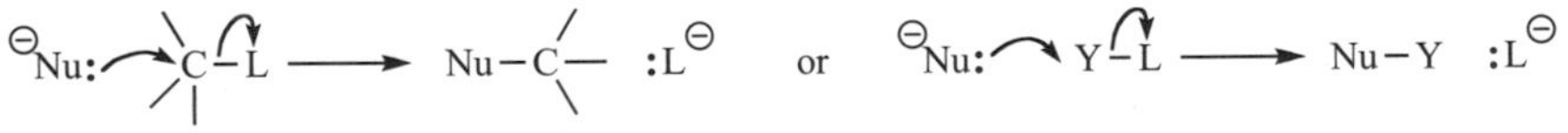

E2, The E2 Elimination

Crosscheck: ΔpK_a rule on base and L, C–H bond and C–L nearly coplanar

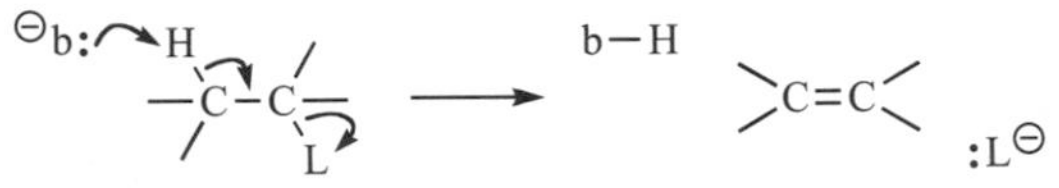

Ad_E3, The Ad_E3 Addition

Crosscheck: Markovnikov orientation

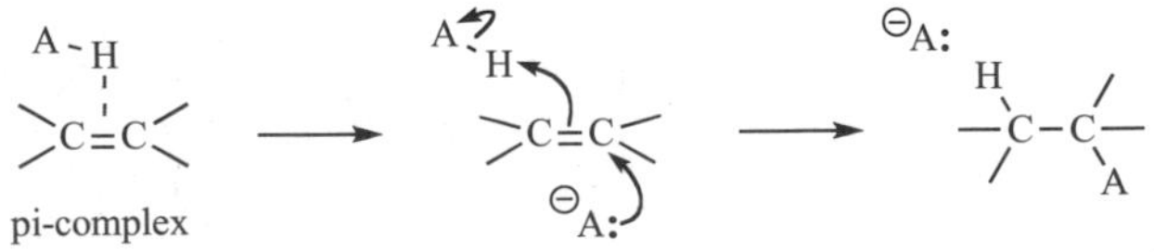
pi-complex

Ad_N, Nucleophilic Addition to a Polarized Multiple Bond

Crosscheck: ΔpK_a rule on Nu and anion formed

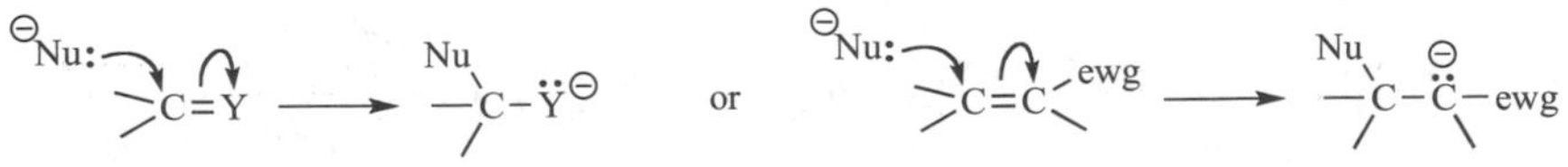

Eβ, Beta Elimination from an Anion or Lone Pair

Crosscheck: ΔpK_a rule on L and anion reactant

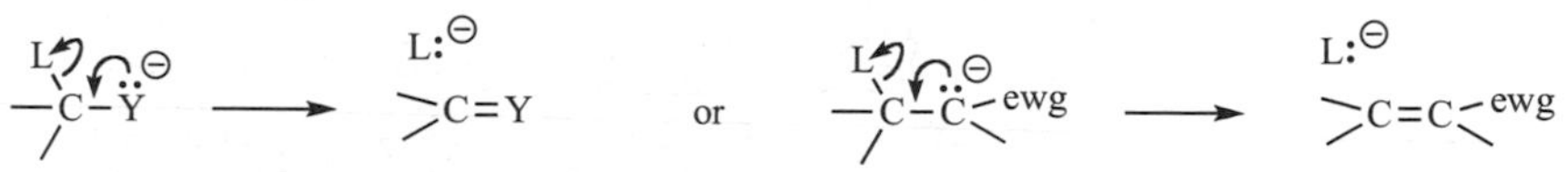

1,2R, Rearrangement of a Carbocation

Crosscheck: forms carbocation of equal or greater stability

1,2RL, 1,2 Rearrangement with Loss of Leaving Group

Crosscheck (if no pushing Y): good L, usually forms better than 2° cation, polar solvent

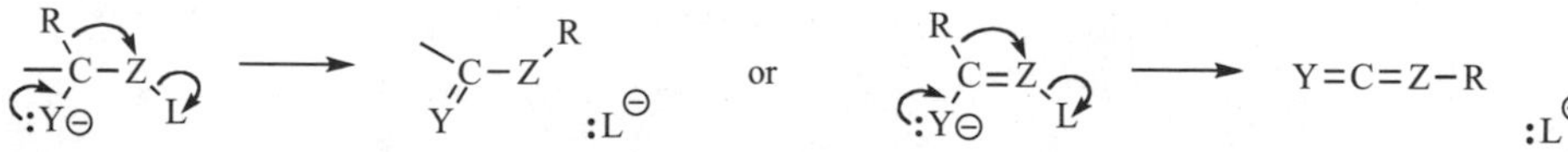

Minor Paths

Path pent., Substitution via a Pentacovalent Intermediate

(Not for carbon, only for third-row and higher elements)

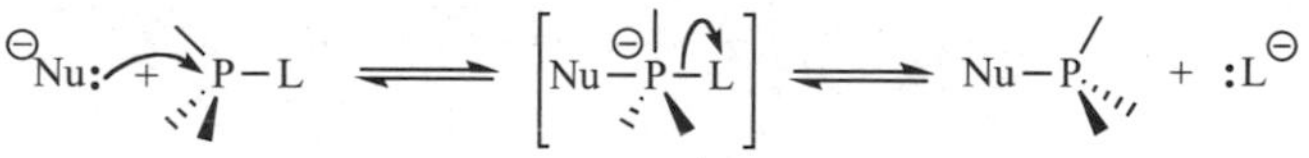

Path 6e, Concerted Six-Electron Pericyclic Reactions

Thermal rearrangements:

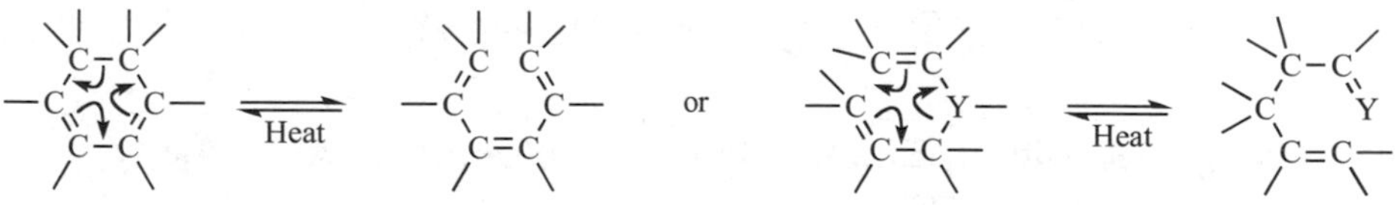

Thermal cycloadditions or cycloreversions (the reverse reaction):

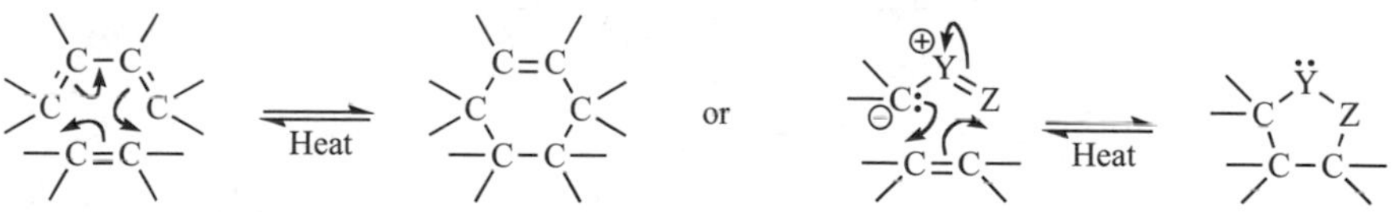

Metal-chelate-catalyzed additions:

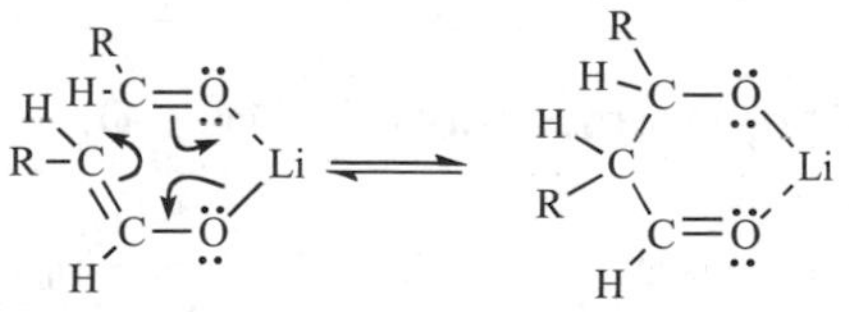

Path Ei, Thermal Internal *Syn* Elimination

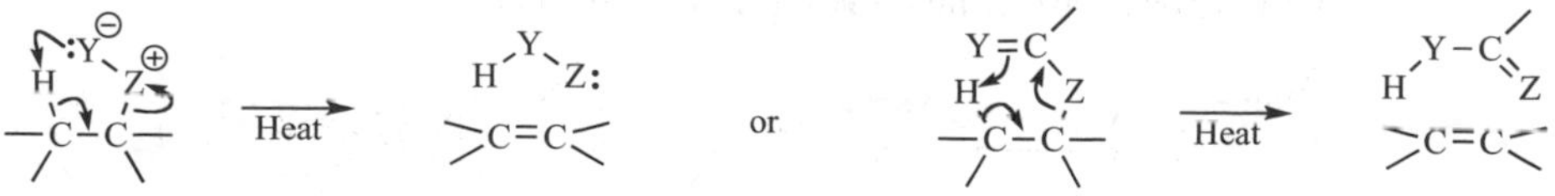

Path NuL, Nu–L Additions (three-membered ring formation)

Path 4e, Four-Center, Four-Electron (three cases only)

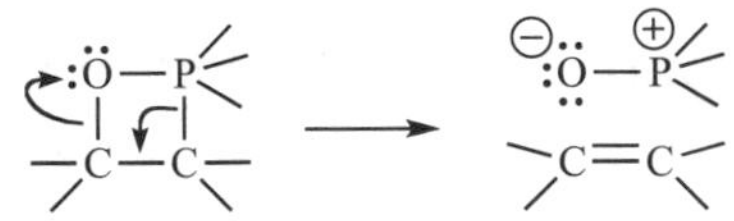

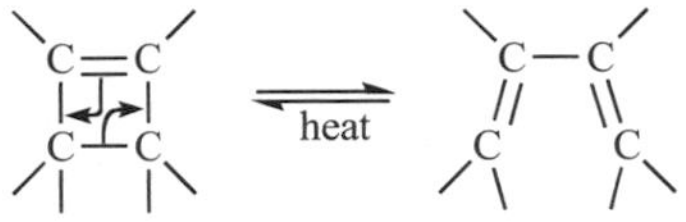

Path H^- t., Hydride Transfer to a Cationic Center

Common Path Combinations

S_N1 (Substitution, Nucleophilic, Unimolecular), $D_N + A_N$

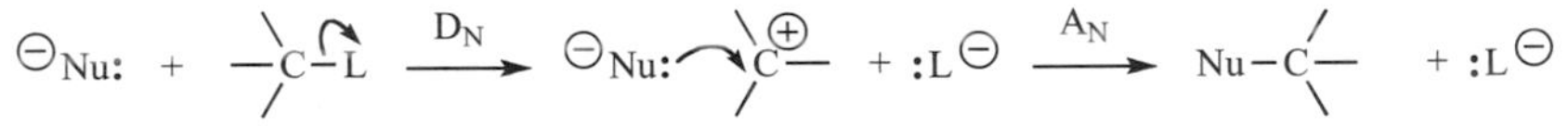

Ad$_E$2 (Addition, Electrophilic, Bimolecular), $A_E + A_N$

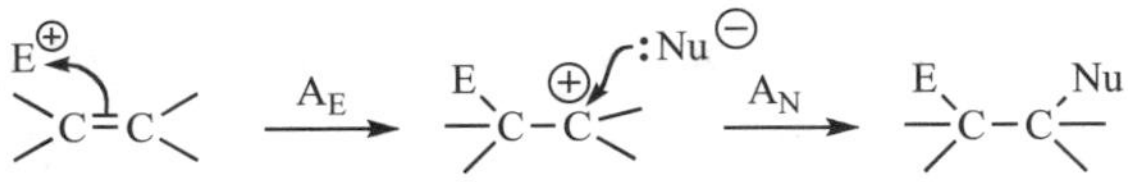

Hetero Ad$_E$2, p.t. + Ad$_N$

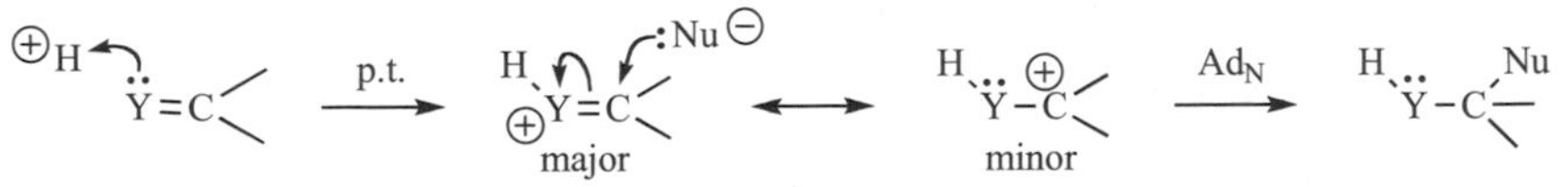

E1 (Elimination, Unimolecular), $D_N + D_E$

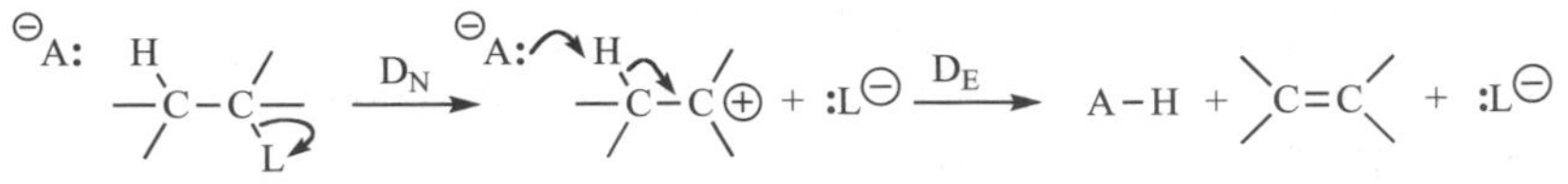

Lone-Pair-Assisted E1, E_β + p.t.

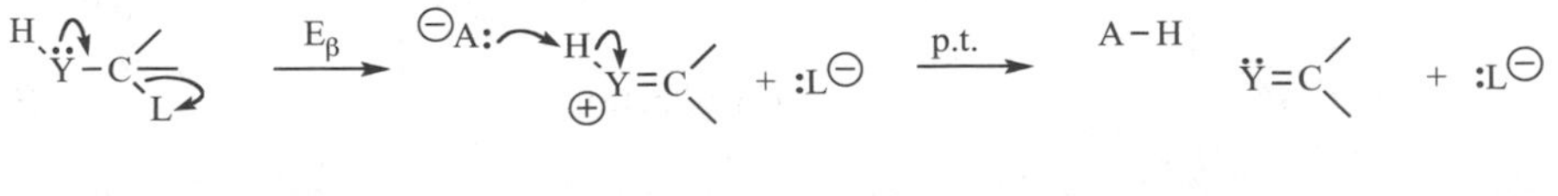

S_E2Ar Electrophilic Aromatic Substitution, $A_E + D_E$

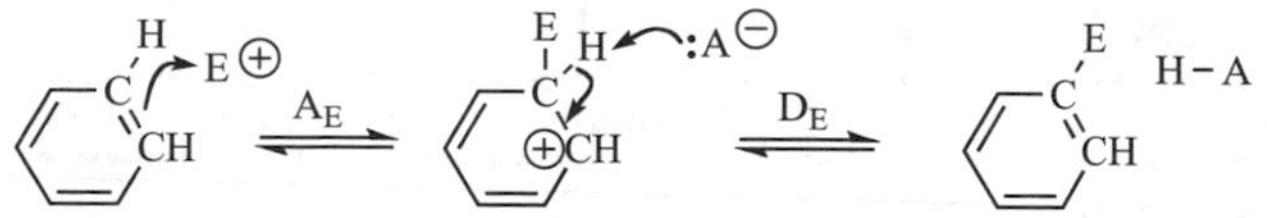

E1cB (Elimination, Unimolecular, Conjugate Base), p.t. + E_β

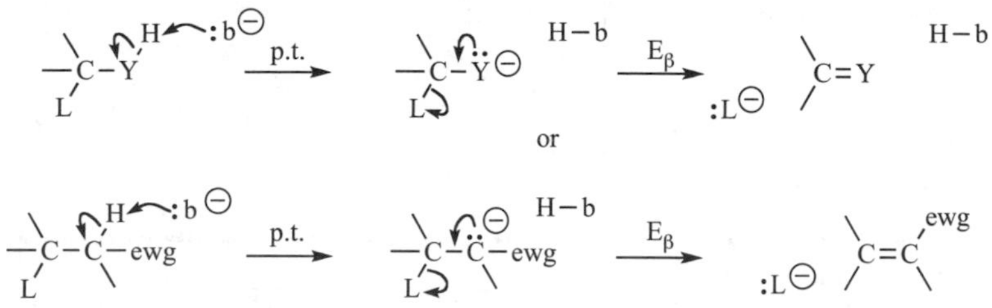

Ad_N2 (Addition, Nucleophilic, Bimolecular), Ad_N + p.t.

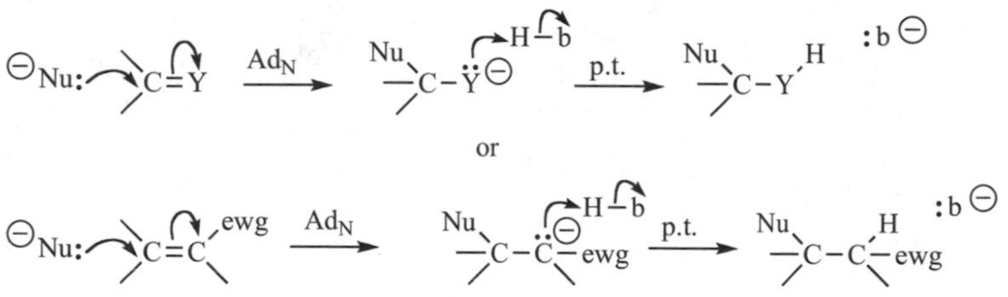

Addition–Elimination, $Ad_N + E_\beta$

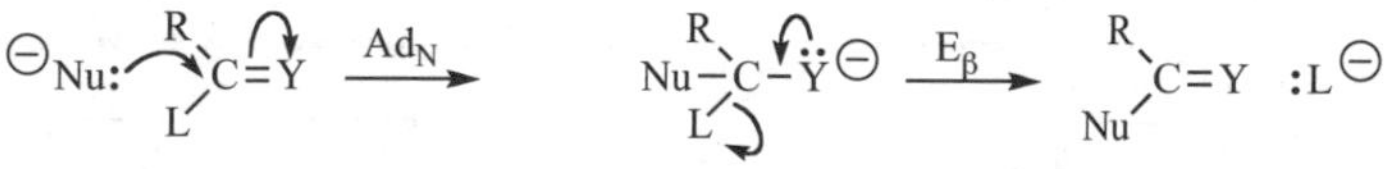

Tautomerization, taut.

$$\text{H-C-C=Z} \rightleftharpoons \text{C=C-Z-H}$$

Tautomerization is the shift of an H from a carbon adjacent to a carbon-heteroatom double bond to the heteroatom itself (and the reverse). It is an acid- or base-catalyzed equilibrium. Two examples are the keto/enol pair (Z = oxygen) and the imine/enamine pair (Z = nitrogen).

Base catalysis goes via the enolate anion.

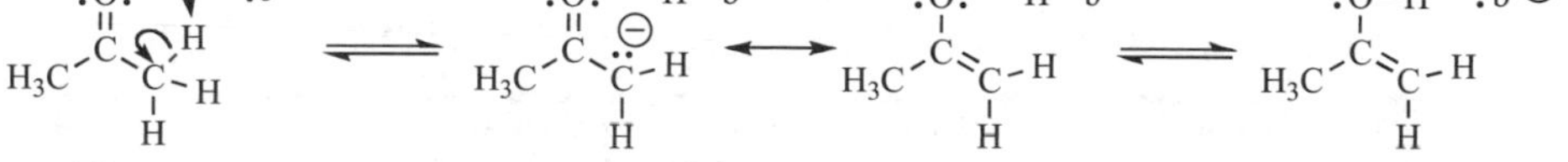

Keto Enolate Enol

Acid catalysis goes via the lone-pair-stabilized carbocation.

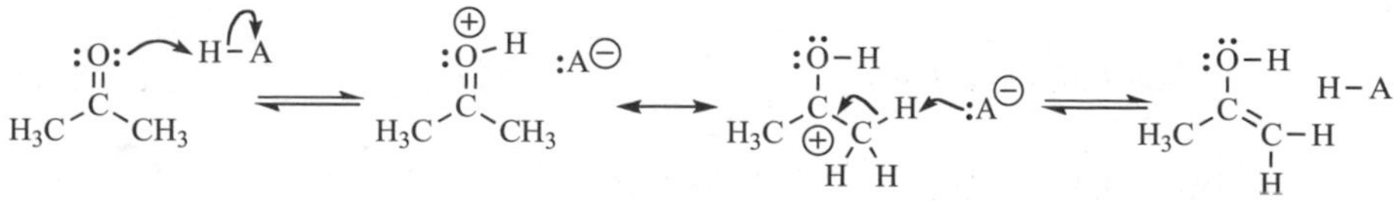

Keto Carbocation Enol

Trends Guide

Carbocation Stability Ranking (Table 4.1)

Carbocation Name	Structure	Comments on Stabilization
Stable carbocations		
Guanidinium cation	$(H_2\ddot{N})_3C^{\oplus}$	Three N lone pairs delocalize plus
Tropylium cation	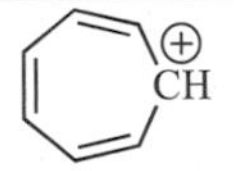	Stable by aromaticity
Protonated amide	$H_3C-C^{\oplus}(\ddot{O}H)(\ddot{N}H_2)$	Delocalized by N and O lone pairs
Moderately stable		
Protonated carboxylic acid	$H_3C-C^{\oplus}(\ddot{O}H)(\ddot{O}H)$	Delocalized by two O lone pairs
Triphenylmethyl cation	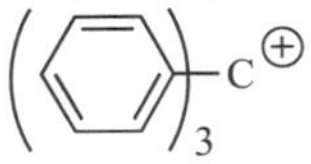	Delocalized by three phenyl groups
Protonated ketone	$H_3C-C^{\oplus}(\ddot{O}H)(CH_3)$	Delocalized by one O lone pair
Diphenylmethyl cation	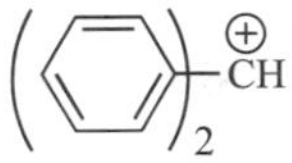	Delocalized by two phenyl groups
Average stability		
Tertiary alkyl cation (3°)	$(CH_3)_3C^{\oplus}$	Stabilized by three alkyl groups
Benzyl cation	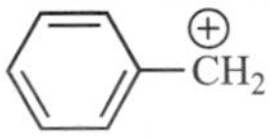	Delocalized by one phenyl group
Primary allyl cation	$H_2C=HC-\overset{\oplus}{C}H_2$	Delocalized by one pi bond
Acylium cation	$H_3C-\overset{\oplus}{C}=\ddot{O}$	Lone-pair-delocalized vinyl cation
Moderately unstable		**No resonance delocalization**
Secondary alkyl cation (2°)	$(CH_3)_2\overset{\oplus}{C}H$	Stabilized by two alkyl groups
Secondary vinyl cation	$H_3C-\overset{\oplus}{C}=CH_2$	Plus on more electronegative *sp* C
Unstable		**(Rarely formed)**
Primary alkyl cation (1°)	$CH_3\overset{\oplus}{C}H_2$	Stabilized by one alkyl group
Primary vinyl cation	$H-\overset{\oplus}{C}=CH_2$	Plus on more electronegative *sp* C
Phenyl cation	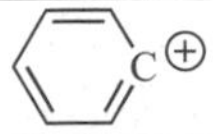	Bent vinyl cation and not delocalized
Methyl cation	$^{\oplus}CH_3$	No stabilization

Carbocations stabilized by resonance with a lone pair are more stable than those stabilized only by resonance with a carbon–carbon double bond, which in turn are more stable than those stabilized only by alkyl group substitution. The effects are additive; three of a lesser type of stabilization are at least as good as one of the better type.

Electron-Releasing Group Ranking (Table 4.2)

Excellent donors	Features of Group
$-\ddot{C}H_2^{\ominus}$	Anionic and least electronegative
$-\ddot{N}H^{\ominus}$	Anionic and more electronegative than C
$-\ddot{O}:^{\ominus}$	Anionic and more electronegative than N
Good donors	
$-\ddot{N}(CH_3)_2$	Neutral and less electronegative than O
$-\ddot{N}H_2$	Neutral and less electronegative than O
$-\ddot{O}H$	Neutral and more electronegative than N
$-\ddot{O}CH_3$	Neutral and more electronegative than N
$-\ddot{N}H-C(=\ddot{O})-CH_3$	Amide carbonyl decreases N lone pair availability
$-\ddot{S}CH_3$	Sulfur 3*p* orbital has poor overlap with a 2*p* orbital
Poor donors	
—Ph	Delocalization through resonance with phenyl
—R	Hyperconjugation only, no lone pair or pi resonance
—H	No substituent at all
Very poor donors	
$-\ddot{O}-C(=\ddot{O})-CH_3$	Electronegative, carbonyl decreases O lone pair availability
$-\ddot{C}l:$	Electronegative and poor overlap with a 2*p* orbital

Allylic Electron Source Reactivity Trends

The better the donor is, the better the allylic electron source is.

Alkene Stability Trends

The more substituted, the more stable.

Carbanion Stability Trends

The most stable carbanion will be the weakest base and have the lowest pK_{abH}.

Electron-Withdrawing Group Trends

CH_3–ewg has a low pK_a for a good ewg. The acidity of CH_3–ewg increases as groups better at withdrawing electrons more effectively stabilize $^-CH_2$–ewg, the anionic conjugate base. If the pK_as of CH_3–ewgs are compared, the more acidic (lower pK_a) the CH_3–ewg is, the better the ewg group is at withdrawing electrons. The following is a list of electron-withdrawing groups ranked by the pK_a of CH_3–ewg.

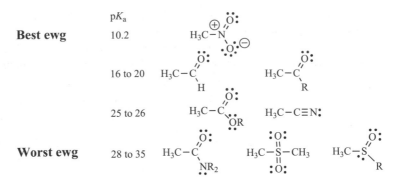

	pK_a
Best ewg	10.2
	16 to 20
	25 to 26
Worst ewg	28 to 35

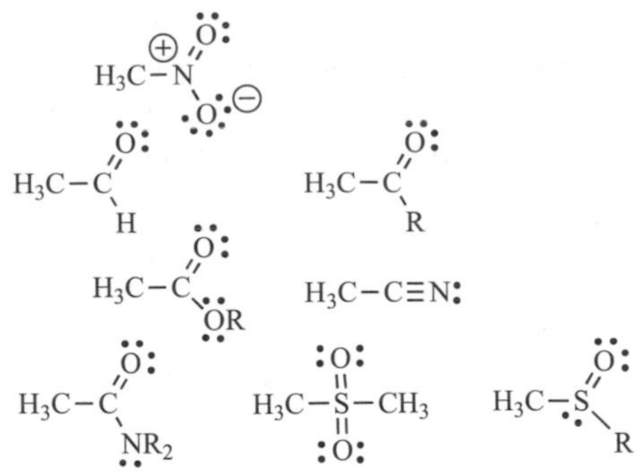

H_3C-N O O
H_3C-C O H
H_3C-C O R
H_3C-C O OR
$H_3C-C{\equiv}N$:
H_3C-C O NR_2
$H_3C-S-CH_3$ O O
H_3C-S O R

Leaving Group Trends

A good leaving group, L, has a negative pK_{aHL}. When using a pK_a chart to rank leaving groups, always look up the conjugate acid of the leaving group on the chart. Note that H^-, NH_2^-, and CH_3^-, each with a pK_{aHL} above 30, are not leaving groups.

Excellent	pK_{aHL}
N_2	<–10
$CF_3SO_3^-$	–15
$ArSO_3^-$	–6.5
$CH_3SO_3^-$	–6

Good	pK_{aHL}
I^-	–10
Br^-	–9
Cl^-	–7
EtOH	–2.4
H_2O	–1.7
CF_3COO^-	+0.5

Fair	pK_{aHL}
$O_2NC_6H_4COO^-$	+3.4
$RCOO^-$	+4.8
$O_2NC_6H_4O^-$	+7.2
NH_3	+9.2
RS^-	+10.6
NR_3	+10.7

Poor	pK_{aHL}
HO^-	+15.7
EtO^-	+16
$RCOCH_2^-$	+19
$ROCOCH_2^-$	+26

Aromatic Ring Reactivity Trends

The better the donor, the more reactive the ring is to electrophilic attack. Donors direct electrophiles *ortho–para*. The better the electron-withdrawing group, the less reactive the ring is to electrophilic attack. Electron-withdrawing groups direct electrophiles *meta*.

Carboxyl Derivative Trends

Donor groups deactivate carboxyl derivatives toward nucleophilic attack.

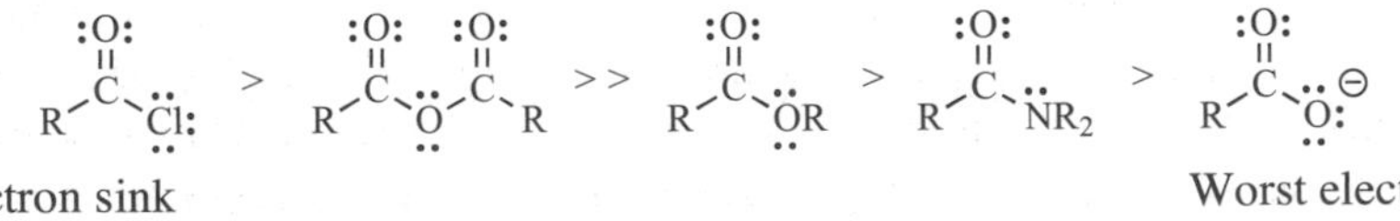

>
>>
>
>

Best electron sink — Worst electron sink

Lone Pair Nucleophilicity Trends

A pK_a chart can be used as a reference for nucleophilicity only if the difference in softness is considered. A partially plus carbon atom is a much softer electrophile than a proton. Soft ions are more nucleophilic in protic solvents because tighter solvation greatly decreases the nucleophilicity of the hard ions. Also, steric hindrance decreases nucleophilicity. To rank lone pair nucleophilicity in protic solvents with soft electrophiles such as R–X, first rank by softness, then by basicity (within the same attacking atom). However, for lone pair nucleophiles reacting with harder electrophiles like a proton or carbonyl, rank by basicity. Very reactive electrophiles like carbocations are not selective and react with the most abundant nucleophile (commonly the solvent).

Relative nucleophilicity toward CH_3I	
CH_3OH (solvent)	1
F^-	500
CH_3COO^-	20,000
Cl^-	23,000
Et_2S	220,000
NH_3	320,000
PhO^-	560,000
Br^-	620,000
CH_3O^-	1,900,000
Et_3N	4,600,000
CN^-	5,000,000
I^-	26,000,000
Et_3P	520,000,000
PhS^-	8,300,000,000
$PhSe^-$	50,000,000,000

Organometallic Reactivity Trends

The greater the electronegativity difference between R and M, the more reactive.

Electronegativity difference	**Most reactive**
1.62	R–Na
1.57	R–Li
1.24	R–MgX
0.90	R_2Zn
0.86	R_2Cd
0.65	$R_2Cu^-Li^+$
0.55	R_2Hg
0.22	R_4Pb
	Least reactive

Metal Hydride Reactivity Trends

The electronegativity difference also governs the reactivity of the metal hydrides. The metal hydride will become less reactive as the metal becomes more electronegative by attaching an inductive withdrawing group. As the hydride source becomes less reactive, it also becomes more selective. The complex metal hydride reactivity trend is:

$$\mathbf{AlH_4^- > HAl(OR)_3^- > BH_4^- > H_3BCN^-.}$$

Conjugate Acceptor Reactivity Trends

Trading C–ewg for Y is simply exchanging an electronegative carbon atom for an electronegative heteroatom. The better the electron-withdrawing group is, the better the electron sink will be. More than one good electron-withdrawing group on a double bond, $C{=}C(ewg)_2$, gives a very reactive electron sink because the resulting anion is stabilized to a much greater extent.

MAJOR ROUTES SUMMARY

Substitution Summary

- S_N2 must pass three tests: good access (primary or secondary), a leaving group with a pK_{aHL} of 10 or less, and a decent nucleophile.
- S_N2 does not occur on leaving groups attached to a trigonal planar center or on a highly hindered site like a tertiary or neopentyl center.
- Substitution version of the ΔpK_a rule: never kick out a leaving group more than 10 pK_a units more basic than the incoming nucleophile.
- S_N1 must pass three tests: usually a carbocation usually better than secondary, a polar solvent, and a leaving group with a pK_{aHL} of zero or less.
- The pentacovalent path is for third-row and higher elements and must pass ΔpK_a rule.

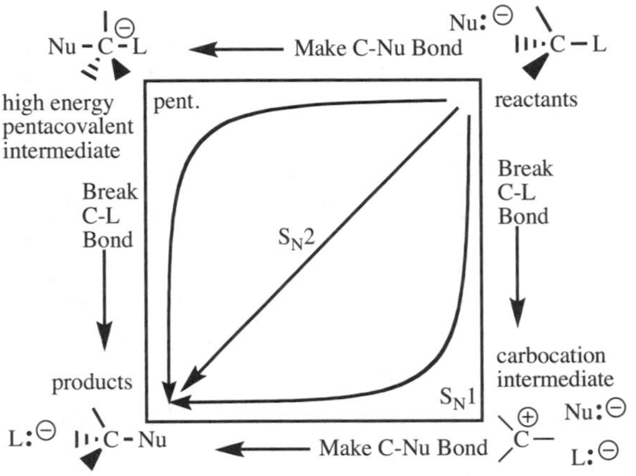
Make C-Nu Bond
reactants
high energy pentacovalent intermediate
pent.
Break C-L Bond
Break C-L Bond
S_N2
carbocation intermediate
products
S_N1
Make C-Nu Bond

Figure 4.2 Top view of a simplified energy surface for substitution at a tetrahedral center.

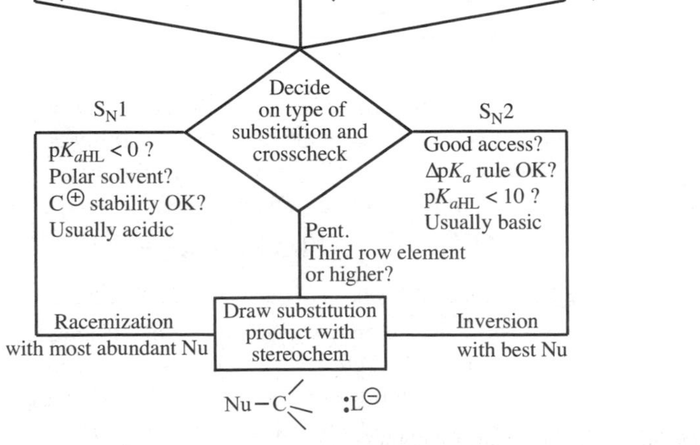
Substitution at a Tetrahedral Center
Decide on any proton transfer and check K_eq
Protonate L?
Acidic media
Deprotonate HNu?
Basic media
HNu
HNu
Decide on type of substitution and crosscheck
S_N1
pK_aHL < 0 ?
Polar solvent?
C⊕ stability OK?
Usually acidic
S_N2
Good access?
ΔpK_a rule OK?
pK_aHL < 10 ?
Usually basic
Pent.
Third row element or higher?
Racemization
with most abundant Nu
Draw substitution product with stereochem
Inversion
with best Nu

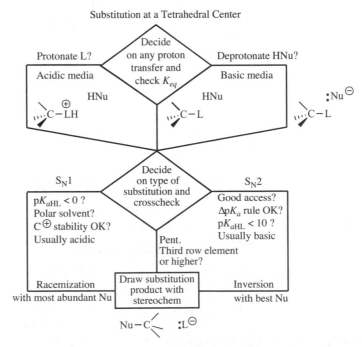

Figure 4.17 The flowchart for substitution at a tetrahedral center.

Elimination Summary

• Elimination version of the ΔpK_a rule: never kick out a leaving group more than 10 pK_a units more basic than the incoming base.

• E1 must pass three tests: a carbocation with a stability usually better than secondary, a polar solvent, and a leaving group with a pK_{aHL} of zero or less.

• E2 occurs when the carbanion and carbocation are not very stable: The energy surface folds down the middle. The C–H and C–L eliminated must be close to coplanar.

• E1cB requires an acidic hydrogen and a good base. Check that the proton transfer K_{eq} is at least 10^{-10}. E1cB can tolerate poorer leaving groups.

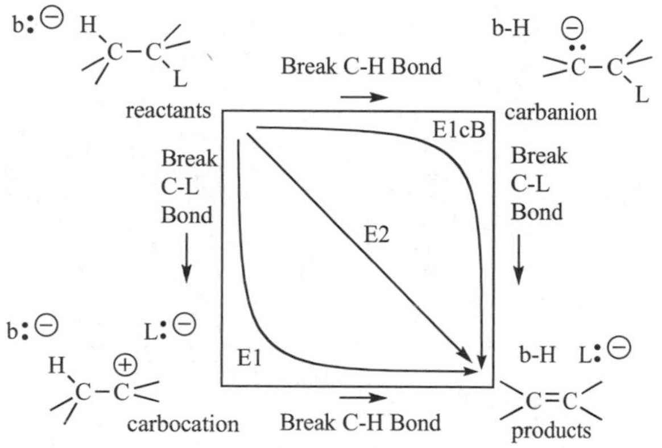
reactants
Break C-H Bond
carbanion
E1cB
Break C-L Bond
Break C-L Bond
E2
E1
carbocation
Break C-H Bond
products

Figure 4.18 Top view of the simplified energy surface for elimination.

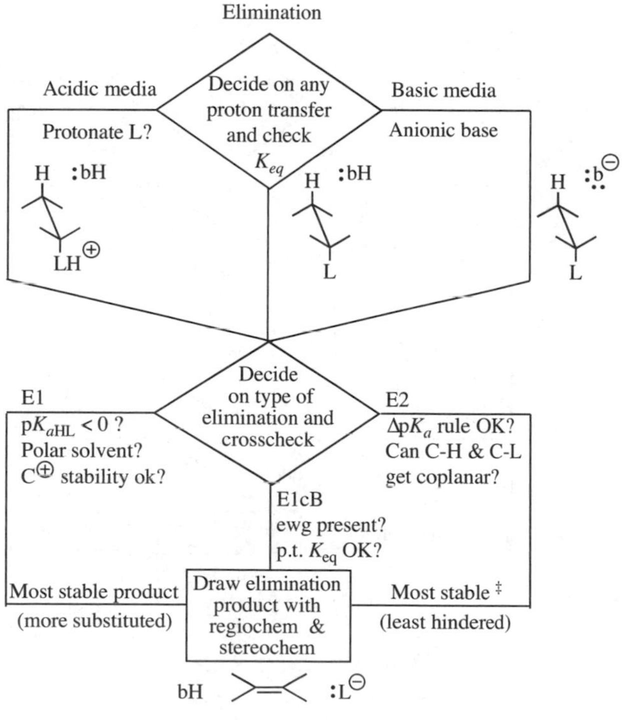
Elimination
Acidic media
Decide on any proton transfer and check K_eq
Basic media
Protonate L?
Anionic base
Decide on type of elimination and crosscheck
E1
pK_aHL < 0 ?
Polar solvent?
C⊕ stability ok?
E2
ΔpK_a rule OK?
Can C-H & C-L get coplanar?
E1cB
ewg present?
p.t. K_eq OK?
Most stable product
(more substituted)
Draw elimination product with regiochem & stereochem
Most stable ‡
(least hindered)

Figure 4.32 The flowchart for elimination.

Addition Summary

• Ad_E2 must form a reasonably stable carbocation usually better than secondary, and tends to produce a mix of *syn* and *anti* addition. Markovnikov's rule is followed: The electrophile adds to form the most stable carbocation.

• Ad_E3 occurs when the carbanion and carbocation are not very stable: The energy surface folds down the middle. The electrophile and nucleophile end up on opposite faces of the pi bond (*anti* addition). Markovnikov's rule is followed.

• Ad_N2 follows the ΔpK_a rule—Never form an anion more than 10 pK_a units more basic than the incoming nucleophile. For this to happen, the anion formed must usually be on an electronegative atom or stabilized by an electron-withdrawing group.

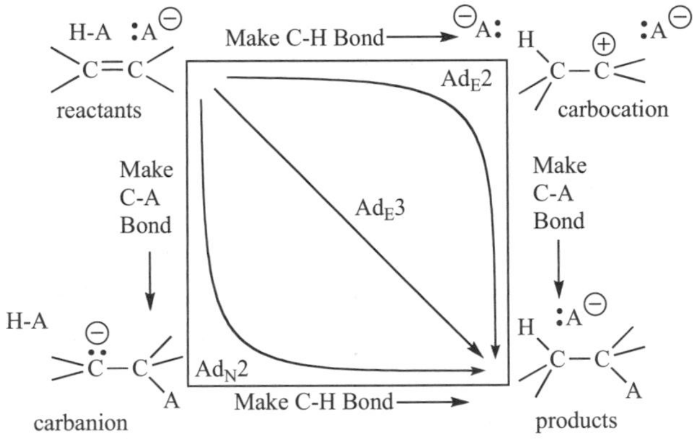

H-A :A⊖
reactants
Make C-H Bond
⊖A:
H
carbocation
AdE2
AdE3
AdN2
Make C-A Bond
Make C-A Bond
H-A
carbanion
A
Make C-H Bond
H :A⊖
A
products

Figure 4.34 Top view of the simplified energy surface for addition to a carbon–carbon pi bond.

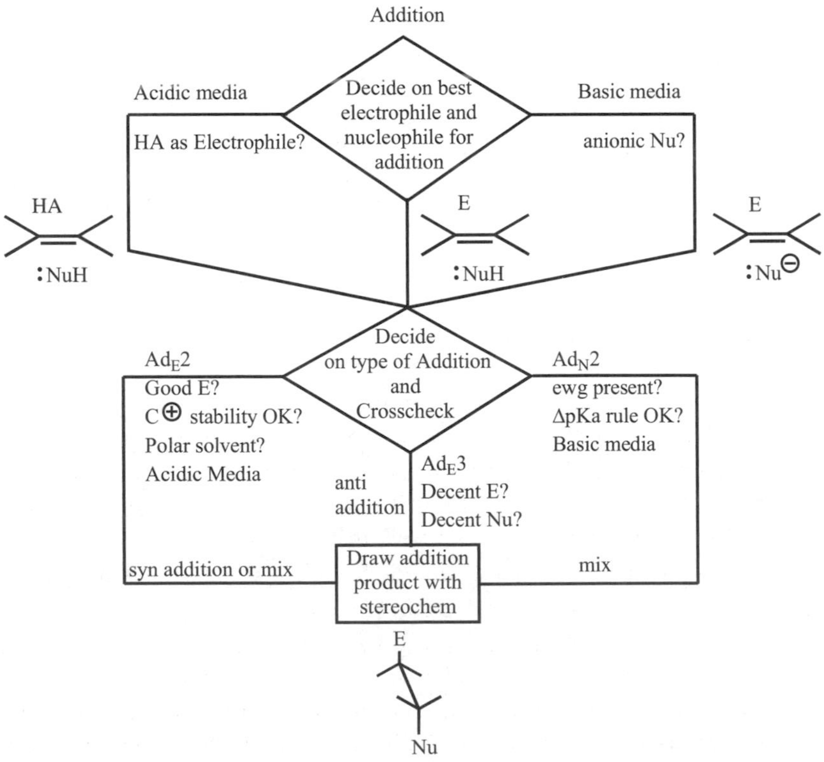

Addition
Acidic media
HA as Electrophile?
Decide on best electrophile and nucleophile for addition
Basic media
anionic Nu?
HA
:NuH
E
:NuH
E
:Nu⊖
Decide on type of Addition and Crosscheck
AdE2
Good E?
C⊕ stability OK?
Polar solvent?
Acidic Media
AdN2
ewg present?
ΔpKa rule OK?
Basic media
anti addition
AdE3
Decent E?
Decent Nu?
syn addition or mix
Draw addition product with stereochem
mix
E
Nu

Figure 4.45 The general flowchart for addition. A proton transfer step may follow.

Addition–Elimination of Carboxylic Acid Derivatives Summary

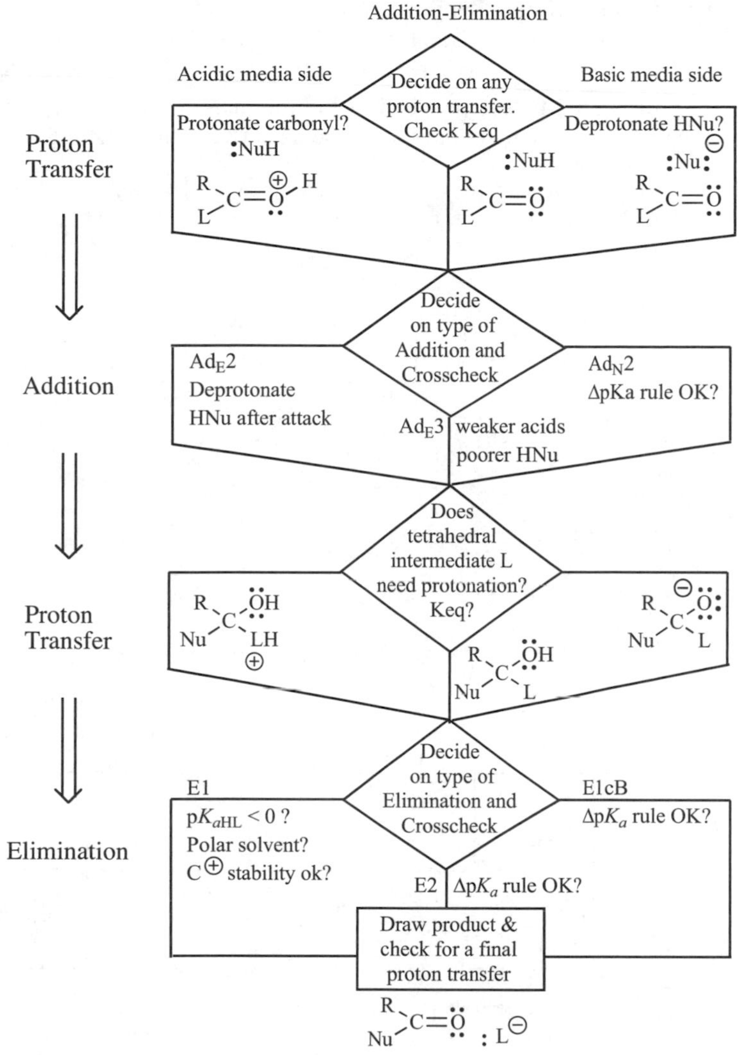
Addition-Elimination
Acidic media side
Decide on any proton transfer. Check Keq
Basic media side
Proton Transfer
Protonate carbonyl?
Deprotonate HNu?
Decide on type of Addition and Crosscheck
Addition
AdE2
Deprotonate HNu after attack
AdN2
ΔpKa rule OK?
AdE3
weaker acids poorer HNu
Does tetrahedral intermediate L need protonation? Keq?
Proton Transfer
Decide on type of Elimination and Crosscheck
Elimination
E1
pKaHL < 0 ?
Polar solvent?
C⊕ stability ok?
E1cB
ΔpKa rule OK?
E2
ΔpKa rule OK?
Draw product & check for a final proton transfer

Figure 4.50 The flowchart for addition–elimination on carboxylic acid derivatives.

- Ad_E2 usually occurs in acid because it forms a very acidic protonated carbonyl.
- Ad_E3 occurs in buffered media, where the medium is not acidic enough to protonate the carbonyl and not basic enough to support strong nucleophiles.
- Ad_N2 occurs in base with good nucleophiles and follows the ΔpK_a rule—never form an anion more than 10 pK_a units more basic than the incoming nucleophile.
- Elimination follows the ΔpK_a rule: never kick out a leaving group more than 10 pK_a units more basic than the tetrahedral intermediate.
- E1 tends to occur in acid and must pass three tests: a reasonably stable carbocation (better than secondary), a polar solvent, and a leaving group with a pK_{aHL} of zero or less.
- E2 usually occurs buffered media when the energy surface folds down the middle.
- E1cB usually requires basic media and an acidic hydrogen. Check that the proton transfer K_{eq} is more than 10^{-10}. It can tolerate poorer leaving groups.

Major Decisions Guide

Substitution vs. Elimination Decision Matrices.

The substitution/elimination decision becomes a function of three major variables: nucleophilicity, basicity, and steric hindrance. In acidic media the S_N1 competes with the E1 process; the first step of both is to lose the leaving group to form the carbocation. There normally is competition in basic media between the S_N2 and E2.

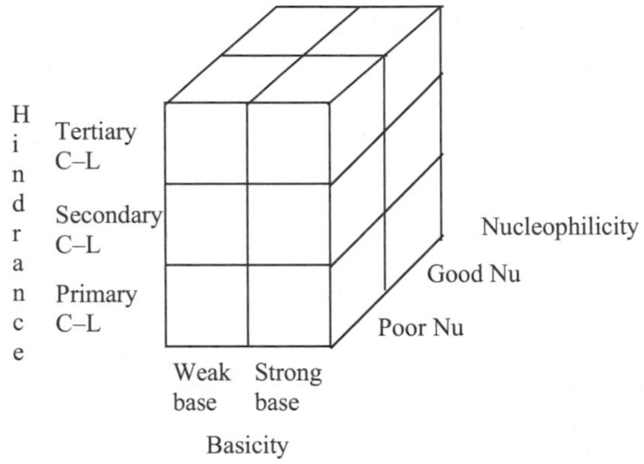
Hindrance
Tertiary C–L
Secondary C–L
Primary C–L
Weak base
Strong base
Basicity
Nucleophilicity
Good Nu
Poor Nu

Figure 9.5 A three-dimensional correlation matrix for substitution versus elimination.

Substitution vs. Elimination Decision

Example Reagents

	Weak base	Strong base
Good nucleophile	$I^{\ominus}$, $RS^{\ominus}$	$HO^{\ominus}$, $EtO^{\ominus}$
Poor nucleophile	H_2O, EtOH	$(CH_3)_3CO^{\ominus}$, $R_2N^{\ominus}$

Unhindered Primary

	Weak base	Strong base
Good nucleophile	Subst. by S_N2	Subst. by fast S_N2
Poor nucleophile	No rxn. or very slow S_N2	Mixture (depends on L)

Secondary

	Weak base	Strong base
Good nucleophile	Subst. by S_N2	Mixture (depends on temp.)
Poor nucleophile	No rxn. or slow S_N2 or S_N1	Elimin. by fast E2

Tertiary

	Weak base	Strong base
Good nucleophile	Mixture (depends on basic.)	Elimin. by fast E2 or E1
Poor nucleophile	Subst. by S_N1 (polar solvent)	Elimin. by fast E2 or E1

Figure 9.6 Substitution/elimination correlation matrix layers from Figure 9.5 sorted by hindrance. The matrix in the upper left gives some common examples of each type of reagents.

The unhindered primary layer has the strong base, poor nucleophile box as its mixture quadrant with all others as substitution. Notice how the mixture quadrant in the primary matrix moves around to the strong base, good nucleophile quadrant in the secondary matrix, then again to the weak base, good nucleophile quadrant in the tertiary matrix, each time leaving behind an elimination quadrant.

Multiple Addition Decision to Carboxylic Acid Derivatives

1. **Unfavorable Equilibrium:** The Nu is such a weak base that the second addition easily reverses and the equilibrium favors the more stable carbonyl. Heteroatom lone pair sources almost never add twice.

2. **Selective Nucleophile:** The initial electron sink must be more reactive than the product of the addition, and the electron source, although reactive enough to add once, is not reactive enough to make a second addition to the less reactive compound. Organometallics in the reactivity range of organocadmiums, organozincs, and organocoppers (metal electronegativities are 1.69, 1.65, 1.90, respectively) add quickly to the more reactive acyl chlorides but slowly to the ketone products.

3. **Poor Leaving Group:** The leaving group is so poor (in amides, $pK_{aHL} = 36$, for example) that it cannot be ejected easily from the tetrahedral intermediate. Since the carbonyl does not reform, there is no site for nucleophilic attack. The carbonyl forms in the acidic workup (no Nu left then) by protonating the nitrogen, making it a better L.

4. **Low Temperature:** One equivalent or less of the electron source was added to the sink at a low enough temperature (usually −78°C) so that the loss of the leaving group from the tetrahedral intermediate by path E_β does not occur to a significant extent before the electron source is used up.

Ambident Sink: 1, 2 versus 1, 4 Addition Decision to Conjugate Acceptors

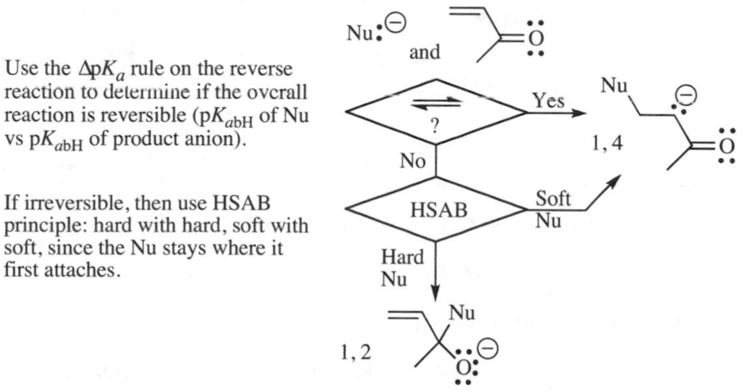
Use the ΔpK_a rule on the reverse reaction to determine if the overall reaction is reversible (pK_{abH} of Nu vs pK_{abH} of product anion).
If irreversible, then use HSAB principle: hard with hard, soft with soft, since the Nu stays where it first attaches.
Nu
and
?
Yes
1, 4
No
HSAB
Soft Nu
Hard Nu
1, 2

Figure 9.7 Flowchart for the ambident electrophile decision of an enone.

Ambident Source: C versus O Attack Decision of Allylic Sources

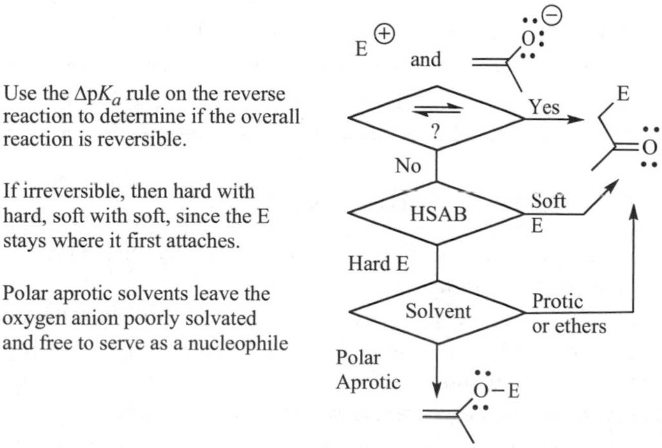
Use the ΔpK_a rule on the reverse reaction to determine if the overall reaction is reversible.
If irreversible, then hard with hard, soft with soft, since the E stays where it first attaches.
Polar aprotic solvents leave the oxygen anion poorly solvated and free to serve as a nucleophile
E
and
?
Yes
No
HSAB
Soft E
Hard E
Solvent
Protic or ethers
Polar Aprotic
O−E

Figure 9.1 Flowchart for the ambident nucleophile decision of an enolate.

Thermodynamics and Kinetics

Free Energy, $\Delta G°$, and K_{eq}

$$\Delta G° = -RT \ln K_{eq}$$

$\Delta G°$ is the free energy difference between the reactants and the products at standard conditions. R is 1.99×10^{-3} kcal/mol-K (8.33×10^{-3} kJ/mol-K), and T is temperature in K. At room temperature, every 1.36 kcal/mol (5.70 kJ/mol) changes the equilibrium constant by a factor of ten.

Table 2.1 $\Delta G°$ and K_{eq} Values for 25°C (room temperature)

$\Delta G°$ kcal/mol	K_{eq}	Reactant	Product	$\Delta G°$ (kJ/mol)
+5.44	0.0001	99.99	0.01	+22.79
+4.08	0.001	99.9	0.1	+17.09
+2.72	0.01	99	1	+11.39
+1.36	0.1	91	9	+5.70
+1.0	0.18	85	15	+4.18
+0.5	0.43	70	30	+2.09
0	1	50	50	0
−0.5	2.33	30	70	−2.09
−1.0	5.41	15	85	−4.18
−1.36	10	9	91	−5.70
−2.72	100	1	99	−11.39
−4.08	1,000	0.1	99.9	−17.09
−5.44	10,000	0.01	99.99	−22.79
−9.52	10^7	Essentially complete		−39.90

The Relationship of Free Energy to Enthalpy and Entropy

$$\Delta G = \Delta H - T\Delta S$$

Enthalpy, Heat of Reaction, ΔH

$$\Delta H = \Delta H(\text{bonds broken}) - \Delta H(\text{bonds made})$$

Kinetics, $\Delta G^‡$

The $\Delta G^‡$ of a reaction is the height of the free energy barrier to reaction measured from the reactants. The $\Delta G^‡$ of a reaction that proceeds at a reasonable rate at room temperature is 20 kcal/mol (84 kJ/mol). The diffusion-controlled limit of 10^{10} liters/mol-s corresponds to a reaction upon every collision. At room temperature, dropping the $\Delta G^‡$ by 1.36 kcal/mol (5.73 kJ/mol) increases the rate tenfold. The more reactive a species is, the less selective it is. The more stable a compound is, the less reactive it is.

Generation of Alternate Paths, Reaction Cubes

One of the more difficult tasks in working mechanism or product prediction problems is deciding whether the set of alternate reaction paths is complete or not. Keeping track of those paths and putting them in some logical order is also very important. Quite often reactants have four charge types, but rarely do more than three occur in the same medium; one is usually an acidic medium charge type, one a basic medium type, and two can often occur in either medium. Products likewise can have four charge types, again with the same media preferences. Frequently, these charge types are interrelated by proton transfer (Fig. A.2).

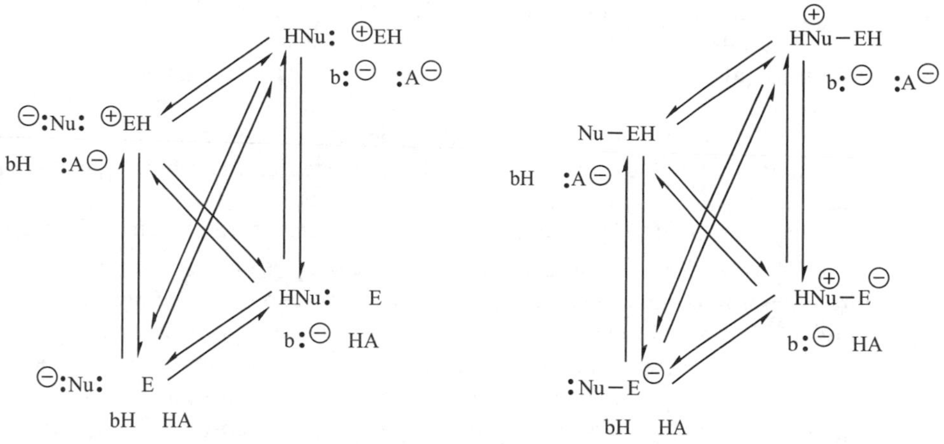

HNu: ⊕EH
b:⊖ :A⊖
⊖:Nu: ⊕EH
bH :A⊖
HNu: E
b:⊖ HA
⊖:Nu: E
bH HA
⊕ HNu−EH
b:⊖ :A⊖
Nu−EH
bH :A⊖
⊕ ⊖ HNu−E
b:⊖ HA
:Nu−E⊖
bH HA

Figure A.2 Conversions of charge types of reactants (left) and products (right) by proton transfer.

A data structure for describing the interrelation of eight variables—these charge types of both reactants and products—is a simple cube (Fig. A.3). Each charge type can then be related to any of the other seven by an edge, face diagonal, or body diagonal. The horizontal axis is the reaction coordinate for bond formation between the nucleophile and the electrophile. The base of the cube (and likewise the top, front, and back) forms the planar projection onto a horizontal plane of the simple energy surface relating two charge types of the reactants with two charge types of the products.

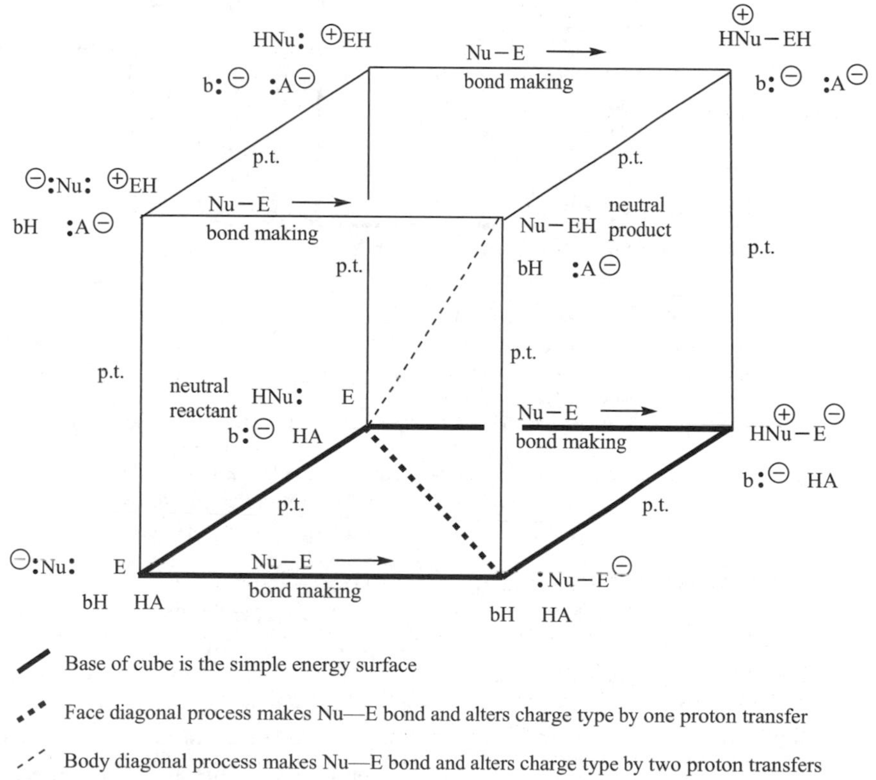

HNu: ⊕EH
b:⊖ :A⊖
Nu−E →
bond making
⊕ HNu−EH
b:⊖ :A⊖
p.t.
⊖:Nu: ⊕EH
bH :A⊖
Nu−E →
bond making
neutral product
Nu−EH
bH :A⊖
neutral reactant
HNu: E
b:⊖ HA
Nu−E →
bond making
⊕ ⊖ HNu−E
b:⊖ HA
⊖:Nu: E
bH HA
Nu−E →
bond making
:Nu−E⊖
bH HA
Base of cube is the simple energy surface
Face diagonal process makes Nu—E bond and alters charge type by one proton transfer
Body diagonal process makes Nu—E bond and alters charge type by two proton transfers

Figure A.3 A generic reaction cube illustrating the interaction of various charge types of reactants and products. p.t. = proton transfer.

Figure A.4 shows the usefulness of the reaction cube as a data structure. Additions to carbonyls often occur between different charge types, and frequently three-dimensional energy surfaces are used to clarify the various equilibria. We have seen two faces of this cube before as individual energy surfaces. The bottom face of the cube is Figure 7.16, polarized multiple bond addition/elimination mechanisms in basic media. The back face of the cube is Figure 7.17, polarized multiple bond addition/elimination mechanisms in acidic media.

One or more corners of this cube can be of higher energy because of a wide pK_a span (corner 3). When we add the dimension of energy to the reaction cube, in essence, it becomes a four-dimensional energy surface. Each corner can be ranked according to its probability of occurrence under the particular reaction conditions, and the most probable lowest-energy route would then be expected to avoid those high-energy corners. Another way to view a reaction cube is that each face is the planar projection of a three-dimensional energy surface. Conditions that raise the energy of a particular corner in the three-dimensional energy surface likewise affect that corner of the reaction cube.

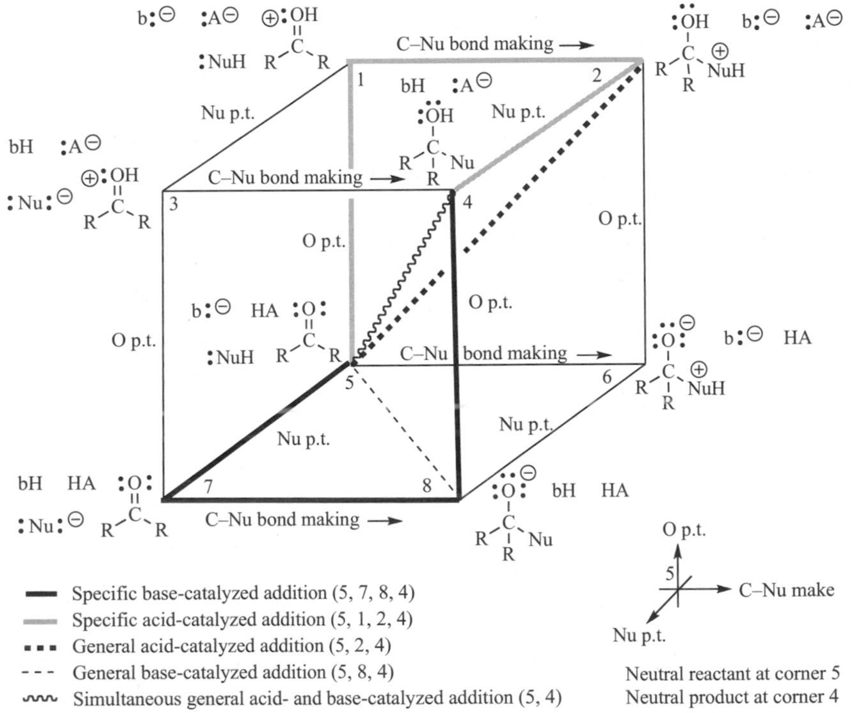
C–Nu bond making →
Nu p.t.
O p.t.
1
2
3
4
5
6
7
8
Specific base-catalyzed addition (5, 7, 8, 4)
Specific acid-catalyzed addition (5, 1, 2, 4)
General acid-catalyzed addition (5, 2, 4)
General base-catalyzed addition (5, 8, 4)
Simultaneous general acid- and base-catalyzed addition (5, 4)
C–Nu make
Neutral reactant at corner 5
Neutral product at corner 4

Figure A.4 The reaction cube for the addition of a nucleophile to a carbonyl in protic media (for hemiketal/hemiacetal, NuH = ROH) where O p.t. is proton transfer to oxygen and Nu p.t. is proton transfer to the nucleophile. (Modified with permission from P. H. Scudder, *J. Org. Chem.*, **1990**, *55*, 4238–4240. Copyright ©1990 by the American Chemical Society.)

There is a need for interrelating the various mechanisms in such a way that all alternatives are easily visible, so that none are left unexplored. With 12 edges, 12 face diagonals, and 4 body diagonals, there are 28 possible interactions that require checking. Specific acid or base processes lie along the edges of the cube. The general acid-catalyzed, Ad_E3, or general base-catalyzed processes constitute the face diagonals. The body diagonal is the simultaneous general acid, general base catalysis (push–pull catalysis). Reactions proceeding through an intermediate can be represented by cubes sharing a common face, edge, or vertex corresponding to the intermediate charge type(s).

Organic Structure Elucidation Strategies

Most chemists need to be able to solve structure problems: What compound will fit the spectral data? A proper attitude toward this type of problem solving is needed, and nobody has said it better than Sherlock Holmes:[1] "It is a capital mistake to theorize before one has data. Insensibly one begins to twist facts to suit theories, instead of theories to suit facts." Strategies are definitely needed. Generate possibilities, then select the best one that fits the data. If none fit, generate more possibilities and examine your assumptions.

Spectra problems are much like jigsaw puzzles. The first step is to find all the pieces, spread them out, and look at them. The next step is usually to find all the pieces that make up the border of the puzzle. Then pieces adjacent to the border are filled in, moving toward the center until the puzzle is complete. Don't "hammer home" any pieces. The most common error is to try to put the puzzle together before identifying all the pieces. Make sure you understand all the parts of the puzzle before trying to put it together. You will have to maintain flexibility so that you don't "lock yourself out" of the correct answer. Structure elucidation also is like a medical diagnosis; if it does not fit all the facts and make sense, it's probably incorrect.

The first piece of information we try to get is the molecular formula from the mass spectral formula weight and the percent composition from a combustion analysis if available. If we don't have a molecular formula, we can still get a minimum count on the number of carbons and hydrogens from the NMR spectra, which will be discussed later.

Mass Spectral Formula Weight

The mass spectrometer functions like an atomic bathroom scale. We get an exact weight of our molecule and also the weights of some of the fragments when it breaks apart. The molecular ion is the most important because we can get the molecular formula from an exact weight of the compound. We know the exact weights of every isotope of the common elements to four or more decimal places. Molecules that differ in elemental composition differ in exact weight. For example, N_2 and CO have the same nominal weight of 28 amu, but the exact weights differ: 28.0062 and 27.9949, respectively. Therefore, if we have a high-resolution mass spectrometer capable of getting the molecular ion weight to three or more decimal places, we could easily narrow down our possibilities for a molecular formula.

Even with a lower-resolution mass spectrometer we can get hints about the elemental composition. Molecular ions (M) have higher weight peaks at M+1, M+2 due to minor isotopes, which can tip you off to their presence. Common examples are shown below (with the intensity of the taller peak set at 100%). Hydrogen, nitrogen, oxygen, fluorine, and iodine have no significant isotope peaks (<0.4%). The presence of chlorine or bromine is important to identify in the mass spectrum.

Element	M (%)	M+1 (%)	M+2 (%)
Carbon	12 (100%)	13 (1.1%)	
Sulfur	32 (100%)	33 (0.8%)	34 (4.4%)
Chlorine	35 (100%)		37 (32.5%)
Bromine	79 (100%)		81 (98%)

Whether the molecular weight is odd or even is an important clue. The molecular ion will have an even molecular weight unless it contains an odd number of nitrogen atoms. The presence of one nitrogen atom allows an additional hydrogen atom in the molecular formula, making the weight odd.

---

1 Arthur Conan Doyle, "A Scandal in Bohemia," *The Adventures of Sherlock Holmes*, 1891

Combustion Analysis

If we burn a weighed compound in oxygen and measure the weight of CO_2 and H_2O formed, we can calculate the percentage of carbon and hydrogen in the compound. Other methods allow for calculation of the percentage of halogens, nitrogen, and other elements in the compound. The one element that cannot be found this way is oxygen, so it is assumed that the remaining percentage is oxygen. For example, if the combustion analysis is 54.5% carbon and 9.2% hydrogen, then the oxygen percentage is 100% − (54.5% + 9.2%) = 36.3%. Assume no other elements if the data are not given.

Molecular Formula

If we multiply the mass spectral molecular ion weight by an atom's percentage and divide by the atom's weight, we get the number of those atoms in the molecular formula. For example, ethane's MS formula weight is 30 and it is 80% carbon by combustion. The number of carbons in ethane is (30)(0.80)/(12) = 24/12 = 2. This is much easier than going via an empirical formula. A molecular formula not only tells you how many of each type of atom is present but also gives the number of rings and/or pi bonds present. The number of Hs actually present is subtracted from the number expected from the Hs expected if there were no rings or pi bonds; this gives the total number of Hs missing. A compound has two less Hs for every ring or pi bond. For propane below, every C is paired with two Hs; each end is capped with an H, giving $2n$+2 Hs for n Cs for any molecule without rings or pi bonds. The next two structures show that to make a ring, we lose the 2 capping Hs; pi bonds also lose 2 Hs. An oxygen does not change the H count. For every N atom add one H. Count the halogens as if they were an H.

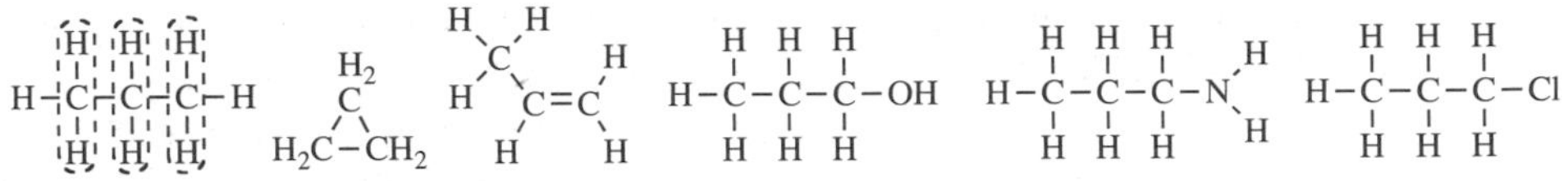

Infrared Spectra—IR

These vibrational spectra give information about functional groups. They can be good evidence that a group is absent if an expected strong spectral band is missing, but since IR spectra often have many bands in a particular region, they give much weaker evidence that a group may be present. The intensity of an IR band is dependent on the vibration changing the dipole moment, so symmetrical species can have weak IR bands.

First look at the molecular formula to find out what functional groups are possible. There is no sense in looking for nitrogen-containing functional groups if the compound has no nitrogen. There is also no sense in looking for functional groups containing a double or triple bond if the molecular formula indicates the compound has no rings or pi bonds. Using the many available IR band tables, try to rule out possible functional groups. Note that negative information is often more important than positive: We can rule out possibilities with negative information. If there is no band between 3600 and 3300 cm^{-1} then the structure does not have an OH or NH (if a small band appears in that region it may just be a wet sample). Triple bond stretches of unsymmetrical alkynes and nitriles occur between 2250 and 2100 cm^{-1}. The C=O stretch is a strong peak near 1700 cm^{-1} whose exact position gives useful information. The different C=O stretch bands of ketones (1740–1710 cm^{-1}), esters (1750–1735 cm^{-1}), anhydrides (1820–1750 cm^{-1}), acyl chlorides (1800–1770 cm^{-1}), amides (1680–1640 cm^{-1}), and carboxylic acids (1760–1710 cm^{-1}) are useful in characterizing each compound. The weaker C=C stretch of unsymmetrical double bonds is near 1640 cm^{-1} and of aromatic rings near 1610 cm^{-1}. The C–O stretch of ethers and esters is around 1300–1000 cm^{-1}. Below 1000 cm^{-1} is the fingerprint region where peaks are much harder to assign. Extract what functional group information you can from the IR, but do not try to assign more than a few major peaks.

Carbon NMR Spectra—CMR

Carbon-13 NMR gives information on the different types of carbons present in the molecule. Since the carbon NMR peaks are spread out over a relatively large range (0–220 ppm), small differences in the types of carbons can be distinguished (less than 0.01 ppm). The tetramethylsilane (TMS) internal standard is set at 0 ppm. The CMR general ranges are simple alkyl 10–50 ppm, C–O 50–90 ppm, alkene and aryl 100–170 ppm, carboxylic acid derivatives 150–185 ppm, and aldehydes and ketones 180–220 ppm. A technique known as DEPT will distinguish between carbon peaks coming from a CH_3, CH_2, CH, or C. Carbon environments in a molecule can be made identical by symmetry. For example, an internal mirror plane in $HOCH_2CH_2CH_2CH_2OH$ means that there are only two unique carbon environments. Check the number of carbons in the molecular formula against the number of CMR peaks to determine whether some symmetry is present. Even if you do not have a molecular formula, the structure must have at least as many carbons as there are CMR peaks. It is best to start with the CMR to get the initial pieces, then use the PMR discussed next to help assemble these carbon NMR pieces.

Example CMR Problem

Mass Spectra: molecular ion at m/e 72.1 gives the molecular weight. Since it is even, then zero or an even number of nitrogen atoms are present.

Combustion analysis: 66.63% carbon and 11.18% hydrogen does not sum to 100%; therefore the remaining 22.19% must be oxygen. Also, no other elements are present.

Molecular formula: (72.1)(0.6663)/12 = 4 carbons, (72.1)(0.1118)/1 = 8 hydrogens, and (72.1)(0.2219)/16 = 1 oxygen for a formula of C_4H_8O. Since no rings or pi bonds would be 10 Hs but we have 8 Hs, we are missing 2 Hs, so there is one ring or pi bond.

CMR (DEPT): 25.4 (CH_2) and 67.6 ppm (CH_2).

One of the first things to notice is that we have half the number of CMR peaks as carbons in the molecular formula; there must be symmetry. There are no pi bonds from the CMR shifts, so we must have a ring. There are only CH_2s and no CHs or CH_3s in the DEPT. Our jigsaw puzzle has no edges, since CH_2s are middle pieces. Draw a line for the symmetry, then put all unique pieces on that line and identical pairs on both sides of the line. The 67.7 ppm CMR shift is correct for a carbon deshielded by an oxygen.

CH2
CH2
CH2
CH2
O
H2C−CH2
H2C
CH2
O

Proton NMR Spectra—PMR

Proton NMR spectra can give **three pieces of information** about the hydrogen atoms in a molecule: The chemical shift: (position) gives details about the magnetic environment that each set of protons is in, the integration (peak area) tells the relative number of protons in each environment, and the coupling pattern (peak complexity) of a proton resonance shows what sort of neighbors it has.

Chemical shift reflects local electron density. Electronegative groups attached to a proton remove the shielding electron density, and the peak for that proton is more deshielded (higher ppm). Resonance effects from erg or ewg groups can respectively add electron density (shield) or remove electron density (deshield). To tell if a resonance is deshielded by an electronegative group we need to know where it would come if undisturbed, necessitating a few numbers to be memorized: CH_3, CH_2, and CH come between 0.9 and 1.5 ppm; vinyl C=CH comes around 5.5 ppm; and aromatics near 7.2 ppm. The magnitude of the effect is also helpful to know: A carbonyl will pull a methyl group from 0.9 down to 2.1 ppm, while an ether oxygen will pull it all the way down to

3.3 ppm. Some groups have characteristic proton shifts; the COOH is at 12 ppm and CHO is at 9 ppm. The deshielding effects of groups is roughly additive, so two weakly deshielding groups may deshield an H to the same position as one, more deshielding group. For this reason, keep your options open when looking at chemical shift tables.

The integration: looks like a staircase and gives the relative numbers of protons in each environment. The total integral step height divided by the number of hydrogens in the molecular formula gives the step height per proton. Dividing the integral step height for a particular resonance by the step height per proton gives the number of protons in the resonance. The integral must sum to the same number of Hs in the formula. Correlate the DEPT data with the PMR integration. Remember that both the CMR pieces and PMR pieces must be consistent; don't assign a CMR methyl to a two-hydrogen PMR peak.

The coupling of protons to their neighboring protons gives very valuable information on how to fit the pieces back together. For the simpler spectra, *n* number of adjacent equivalent protons will split the resonance into *n*+1 lines. Common patterns and peak intensities are shown below. A proton whose peak is a quartet has three adjacent protons; a triplet has two adjacent protons; a doublet has one adjacent proton; and a singlet none.

1 : 1
doublet
1 neighbor
1 : 2 : 1
triplet
2 neighbors
1 : 3 : 3 :1
quartet
3 neighbors
1 : 4 : 6 : 4 :1
quintet
4 neighbors
1 : 5 : 10 :10 : 5 : 1
hexet
5 neighbors

But keep an open mind; the smaller peaks in a larger pattern may get lost in the noise. As two proton resonances get closer together, the pattern distorts; inner lines of each pattern get taller and outer lines get smaller, until finally the two patterns merge into one. Protons at the same shift do not show coupling. Therefore, symmetry in the CMR can show up as a lack of coupling in the PMR. The PMR of 1,2-dichloroethane, $ClCH_2CH_2Cl$, is only one singlet, not two triplets, and it has just one CMR peak.

One further precaution: Acidic hydrogen atoms on heteroatoms like OH and NH groups tend to intermolecularly exchange faster than the NMR can take a measurement; the NMR sees just an average of those hydrogen shifts and no coupling. Ethanol's OH is a singlet under normal conditions and does not couple to the adjacent methylene.

Example PMR Problem

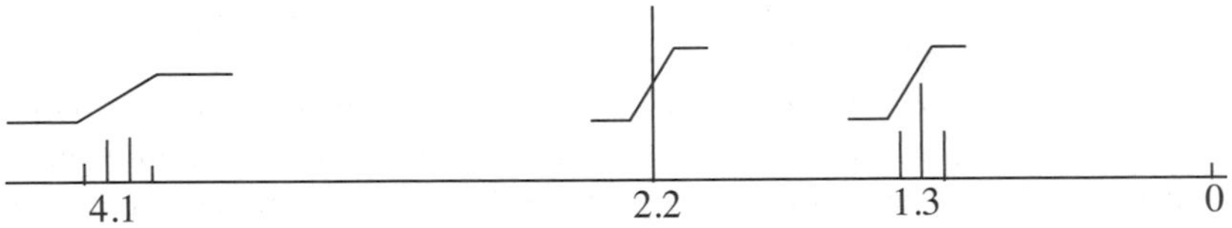
4.1
2.2
1.3
0

Mass Spectra: molecular ion at m/e 88.1 gives the molecular weight. Since it is even, then zero or an even number of nitrogen atoms are present.

Combustion analysis: 54.53% carbon and 9.15% hydrogen does not sum to 100%; therefore the remaining 36.32% must be oxygen. Also, no other atoms are present.

Molecular formula: (88.1)(0.5453)/12 = 4 carbons, (88.1)(0.0915)/1 = 8 hydrogens, and (88.1)(0.3632)/16 = 2 oxygens for a formula of $C_4H_8O_2$. Since no rings or pi bonds would be 10 Hs but we have 8 Hs, we are missing 2 Hs, so there is one ring or pi bond.

IR spectra: no band 3400 to 3300 cm^{-1} rules out OH and a strong band at 1740 cm^{-1} suggests a C=O.

CMR (DEPT): 14.2 (CH_3), 21.0 (CH_3), 60.5 (CH_2), and 171.4 ppm (C).

PMR: 1.3 ppm 3H triplet, 2.2 ppm 3H singlet, and 4.2 ppm 2H quartet.

Piecing a Spectral Jigsaw All Together

A table is a useful way of showing NMR data. Under possibilities the number in parenthesis is the number of adjacent protons, and the arrow signifies that the resonance is deshielded. Write down all possibilities so that you do not artificially restrict the search for pieces. Check that your pieces are consistent with those from the CMR and DEPT spectra. Assemble those CMR pieces using the PMR!

Shift ppm	Integration # of Hs	Coupling	Adjacent H's	Possibilities (use CMR DEPT pieces)
1.3	3H	triplet	2	$-CH_3$ (2H)
2.2	3H	singlet	0	$\leftarrow CH_3$
4.1	2H	quartet	3	$\leftarrow\leftarrow CH_2$ (3H)

First find the "borders" of the molecule, those groups that are bound to only one other thing. In this case it is a pair of methyls, so we write them down with plenty of space in the middle for the rest of the pieces:

$CH_3\rightarrow$ $\quad\quad$ $-CH_3$ (2H)

Like solving a jigsaw puzzle, we work in from these edges, fitting in the next pieces carefully, using the adjacent H's information from the coupling data. The $-CH_2-$ adjacent to three Hs will fit next to the $-CH_3$ adjacent to two.

$CH_3\rightarrow$ $\quad\quad$ $\leftarrow\leftarrow CH_2\text{-}CH_3$

These two pieces must now be connected to electronegative groups that bear no protons to couple with. We have used up all our Hs and have one carbon atom and two oxygen atoms left (plus a ring or pi bond). Since the IR spectra suggest a carbonyl (verified by the CMR), that would take care of the pi bond in the molecule. We have two electronegative groups to add and know that an oxygen atom is more electronegative than a carbonyl. We can now attach them correctly. We have used all the pieces, and can plug the two halves together. When you end up with more than two pieces, remember that the middle pieces may be joined to each other and to the ends in several ways. Use chemical shift data to help decide alternatives if the coupling information is not there.

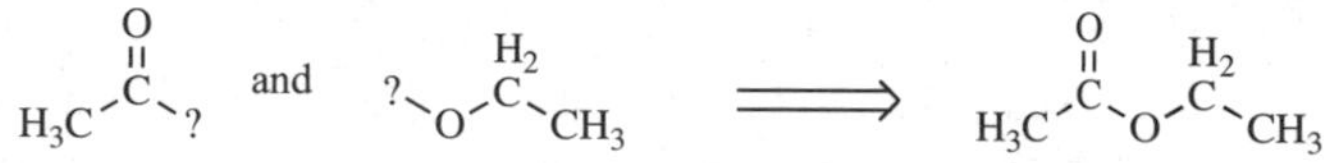

Now comes the hard part: not stopping and quitting, but going back and looking for alternative structures that also fit. Do not "fall in love" with the first answer you find. The order in which you find answers does not correlate with their correctness. You might have not noticed that the chemical shift of the quartet required that the oxygen be attached to the methylene and so might have had the central ester piece reversed. Taking the first answer that comes to you without checking other possibilities is not thinking. Generate all possibilities, then select the one that best fits the data. Spectra puzzles are fun in that when you finally hit upon the right answer it will "click." You will see hidden clues in the spectra that are obvious now that you have the final structure.

Example Problem with Symmetry

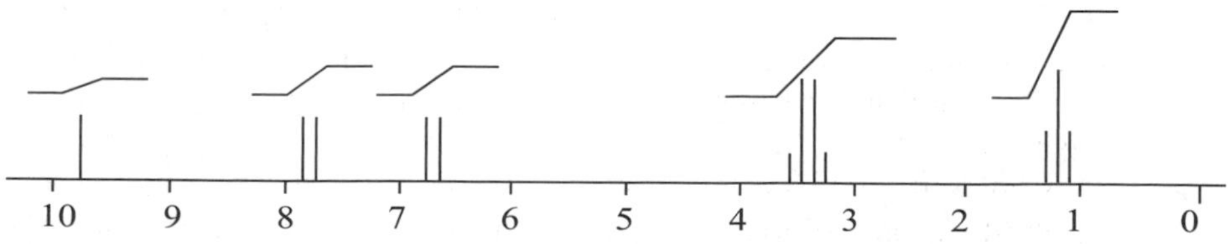
10
9
8
7
6
5
4
3
2
1
0

The molecular formula $C_{11}H_{15}NO$ has 5 rings/pi bonds. IR spectra: nothing 3800–3100 cm^{-1} means no OH or NH, the major band at 1714 cm^{-1} is C=O stretch. The PMR is shown above. CMR (DEPT): 13 (CH_3), 45 (CH_2), 111 (CH), 125 (C), 132 (CH), 152 (C), 190 ppm (CH). Take a moment to think about this problem before continuing on.

Eleven carbons give seven CMR peaks. If the CMR has fewer peaks than there are carbons in the molecular formula, then the molecule has some symmetry. A most common symmetry element is a mirror plane, and we can use the integration of the PMR to figure out which CMR peaks are doubled. For a preliminary assignment, we can assume anything that deshields the H will also deshield the C it's bonded to. For example the methyl at 13 ppm in the CMR best fits with the PMR shift at 1.2, but that integrates for six hydrogens, not three, so it must be doubled. Likewise, the CH_2 at 45 in the CMR fits the PMR shift at 3.4 ppm, and is also doubled. That CH_2 won't fit the PMR shift at 6.7 or 7.7 since those shifts are in the *sp*^{2} H region but the carbon shift is not. The deshielded PMR peaks at 6.7 and 7.7 both integrate for 2H but there are only two *sp*^{2} CH peaks left in the CMR, at 111 and 132, so both of these must be also doubled. Below is an NMR table with a new symmetry column indicating how many times the CMR piece must be present to get the PMR integration.

NMR Table

Shift ppm	Integral # of Hs	Coupling	Adjacent Hs	Possibilities	Possible CMR assignment	Symmetry to achieve 11C
1.2	6H	Triplet	2	$-CH_3$ (2H)	13 CH_3	Two identical
3.4	4H	Quartet	3	$\leftarrow CH_2$ (3H)	45 CH_2	Two identical
6.7	2H	Doublet	1	=CH (1H)	111 =CH	Two identical
7.7	2H	Doublet	1	=CH (1H)	132 =CH	Two identical
9.7	1H	Singlet	0	HC=O (no H)	190 HC=O	One
	sum 15H				125 C & 152 C	One each

Always identify pieces before attempting to assemble the molecule. Two identical isolated HC=CH pieces satisfy the coupling. The $-CH_3$ (2H) is coupled to the $\leftarrow CH_2$ (3H) to give an identical pair of $\leftarrow CH_2CH_3$ (no H). The ends of the molecule are the two identical $\leftarrow CH_2CH_3$ (no H) pair and the aldehyde –CHO (no H). We need to assemble the molecule with symmetry. Again, draw a line for the symmetry and then put all unique pieces on that line and identical pairs on both sides of the line. The only electronegative group to deshield the two identical ethyls is the nitrogen atom, and the shift fits. The two identical HC=CH shifts of 6.7 and 7.7 are in the aromatic region, and the coupling could be explained with a 1,4-disubstituted benzene ring. The molecule assembles in three steps. Now the most important step— Look for another correct solution and cross-check the one you have. This structure passes all checks, with no alternative structures found.

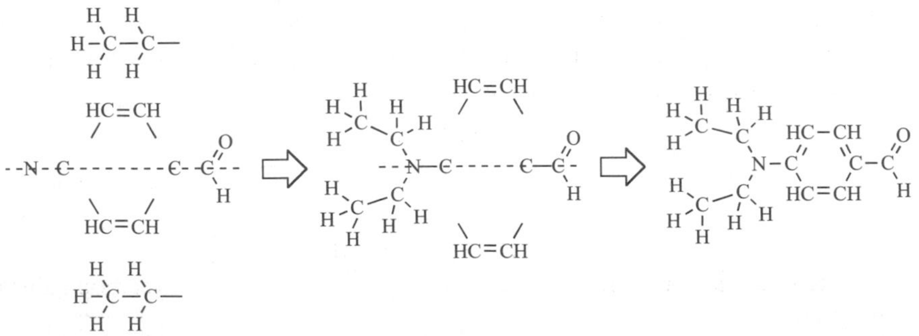

NOTES ON NOMENCLATURE

This section is meant to be a brief compilation of the important aspects of organic nomenclature and is by no means complete. The most important organic nomenclature to know is the names of functional groups, since they are sites of reactivity. Since structure determines reactivity, knowing (for instance) that the structure contains a ketone, and therefore is in the polarized multiple bond class, is more important than being able to name all the grease hanging off a complex structure. The most common functional groups are in Table 1.3; a much larger set is in the functional group glossary in this appendix.

Common Names

Some frequently used compounds do not have a systematic name or have an infrequently used systematic name. Many compounds were named before their structure was known, so the common name has little or no structural information. Some important examples are shown below.

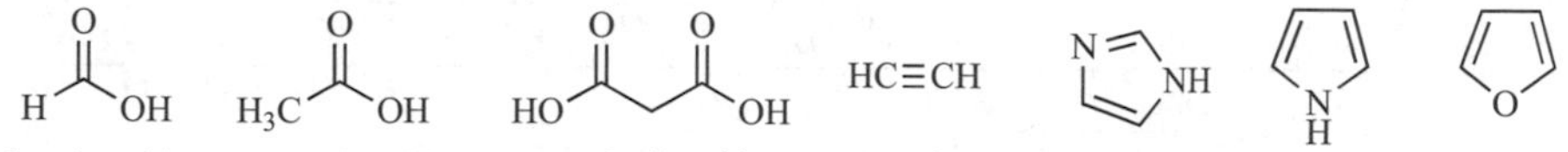
O
H
OH
O
H_3C
OH
O
O
HO
OH
HC≡CH
N
NH
N
H
O

formic acid, acetic acid, malonic acid, acetylene, imidazole, pyrrole, furan

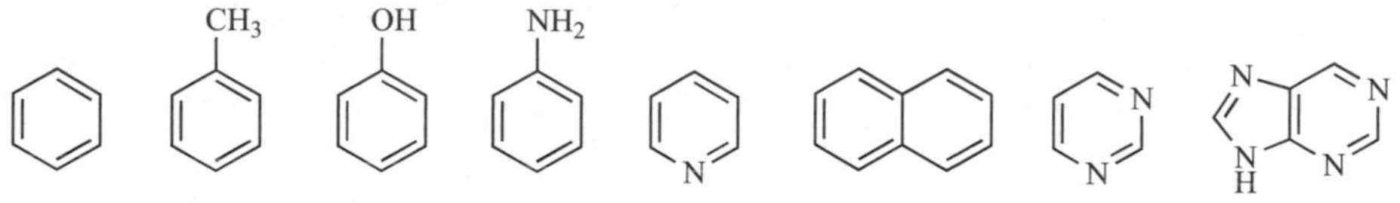
CH_3
OH
NH_2
N
N
N
N
N
N
N
H

benzene, toluene, phenol, aniline, pyridine, naphthalene, pyrimidine, purine

Group Names

Sometimes a group attached to a compound has a special name that has to be learned separately; listed below are the most common groups.

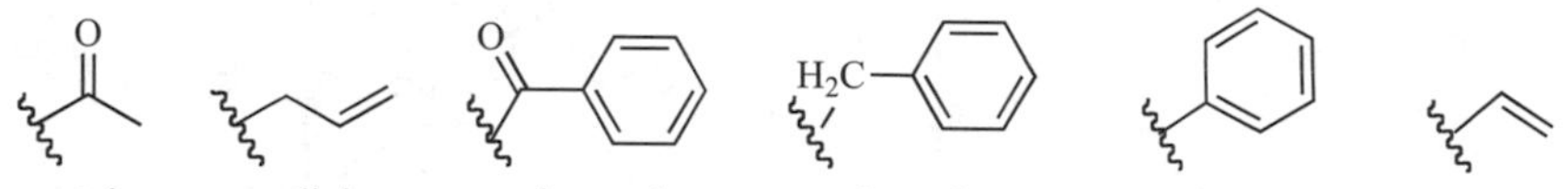
O
O
H_2C

acetyl, allyl, benzoyl, benzyl, phenyl, vinyl

Oxidation State of Carbon

To determine oxidation state, we assign the shared bonding electrons to the more electronegative atom. For organic molecules this means: Carbon-carbon bonds don't count; each carbon to hydrogen bond gives the carbon a −1, and each carbon to heteroatom bond gives the carbon a +1, so the total for any carbon atom can vary from −4 in methane to +4 in carbon dioxide. Therefore, in comparing two organic compounds, the more hydrogen atoms bonded to a carbon atom, the more reduced it is. The more heteroatoms bonded to the carbon, the more oxidized it is. Usually, we just want to know if a redox reaction has occurred and not the actual oxidation states, because oxidation and reduction use special redox reagents.

Oxidation example: 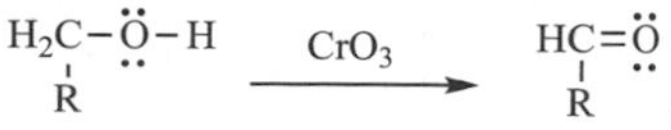
$H_2C-\ddot{O}-H$
R
CrO_3
$HC=\ddot{O}$
R
(carbon loses hydrogens)

Reduction example: 
$HC=\ddot{O}$
R
$LiAlH_4$
$H_2C-\ddot{O}-H$
R
(carbon gains hydrogens)

Prefix-Root-Suffix Naming Convention

This is section is just the basics of naming to allow us to understand simple structural names. The longest chain that includes the primary functional group determines the length of the root chain. Knowing the names of the first ten alkanes, shown below, is usually sufficient to name the root for small molecules.

List of the First Ten Alkanes

# of C	Alkane name	Group name	Alkane
1	Methane	Methyl	CH_4
2	Ethane	Ethyl	CH_3CH_3
3	Propane	Propyl	$CH_3CH_2CH_3$
4	Butane	Butyl	$CH_3CH_2CH_2CH_3$
5	Pentane	Pentyl	$CH_3CH_2CH_2CH_2CH_3$
6	Hexane	Hexyl	$CH_3CH_2CH_2CH_2CH_2CH_3$
7	Heptane	Heptyl	$CH_3CH_2CH_2CH_2CH_2CH_2CH_3$
8	Octane	Octyl	$CH_3CH_2CH_2CH_2CH_2CH_2CH_2CH_3$
9	Nonane	Nonyl	$CH_3CH_2CH_2CH_2CH_2CH_2CH_2CH_2CH_3$
10	Decane	Decyl	$CH_3CH_2CH_2CH_2CH_2CH_2CH_2CH_2CH_2CH_3$

The naming precedence for the functional groups is based on the most oxidized, as shown in the table below. The prefix identifies the position and groups attached to the root chain. The suffix identifies the functional group. Numbering is from the end of the root chain that gives the functional group the lowest number.

Precedence of Functional Groups and Examples

	Functional Group	Suffix	Example	Example Name
1	Carboxylic acid	-oic acid	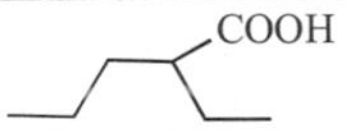 COOH	2-Ethylpentanoic acid
2	Esters	-oate	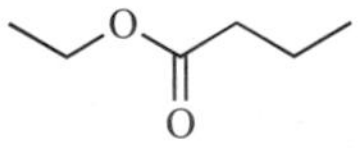 O O	Ethyl butanoate
3	Amides	-amide	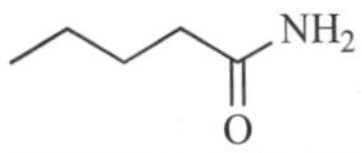 NH2 O	Pentanamide
4	Nitriles	-nitrile	 CN	Hexanenitrile
5	Aldehydes	-al	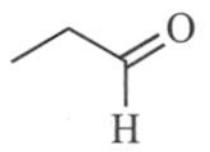 O H	Propanal
6	Ketones	-one	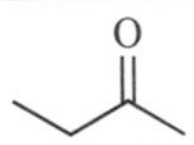 O	Butan-2-one
7	Alcohols	-ol	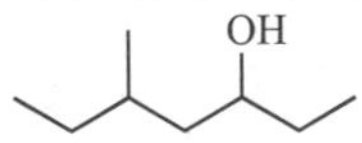 OH	5-Methylheptan-3-ol
8	Amines	-amine	 NH2	2-Methylpropan-2-amine
9	Halogen	As prefix	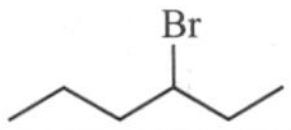 Br	3-Bromohexane

Compounds with two functional groups are named using the higher-precedence group for determining the root numbering and for the group suffix.

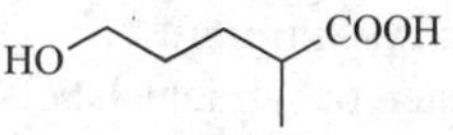

5-hydroxy-2-methylpentanoic acid

Primary, 1°, Secondary, 2°, Tertiary, 3°, and Quaternary, 4°, Notation

For carbocations, the 1°, 2°, 3° designation is based on how many R groups (not counting hydrogen atoms) are attached to the positive carbon atom, as shown below. For alcohols, it is the number of R groups attached to the carbon bearing the OH group. For amines, it is the number of R groups attached to the nitrogen atom.

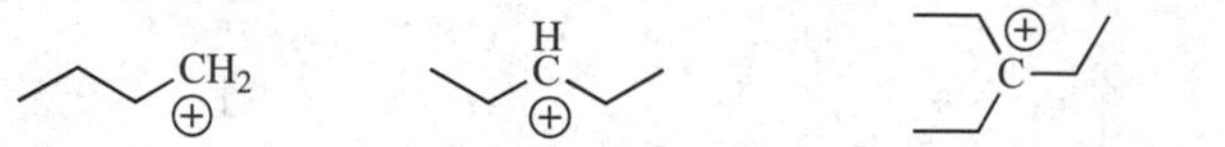

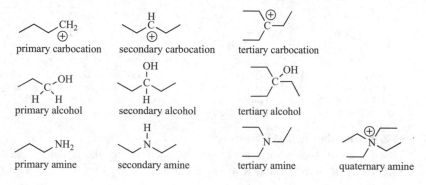

primary carbocation | secondary carbocation | tertiary carbocation

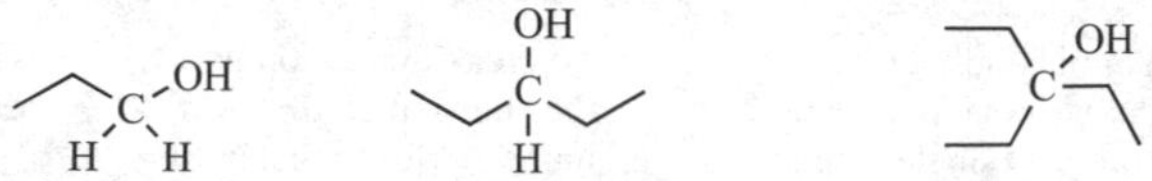

primary alcohol | secondary alcohol | tertiary alcohol

primary amine | secondary amine | tertiary amine | quaternary amine

Projections for Stereochemistry

We also need to be able to describe stereochemistry, the different arrangements of atoms in space, since reaction mechanisms often have stereochemical consequences. We need to have a unambiguous convention for drawing a three-dimensional molecule flat on a page. Shown below are several projections of the same molecule, D-glyceraldehyde. A wedged bond comes out of the page and a dashed bond goes into the page. Molecular models help in visualizing these projections as three-dimensional.

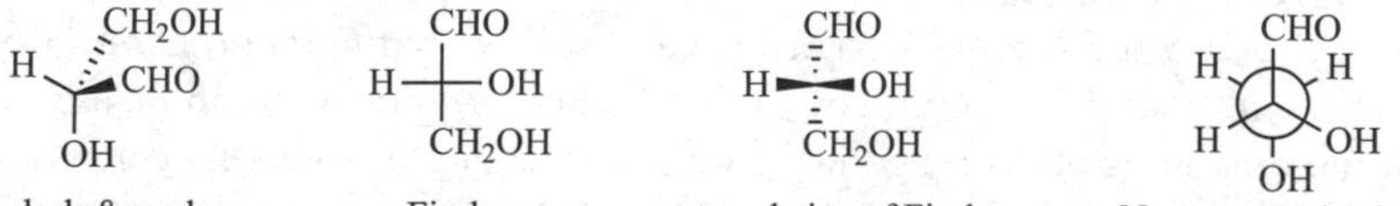

dash & wedge (back) & (out) | Fischer projection | translation of Fischer into dash & wedge | Newman projection down the C2-C3 bond

***Cis* and *Trans* on Alkenes**

Double bonds with two attached groups can be designated as *trans*, *cis*, or *geminal*, shown below. Double bonds are named by the lower number of the two double bond carbons; so if it is called a 2-alkene, then the double bond is between carbons 2 and 3.

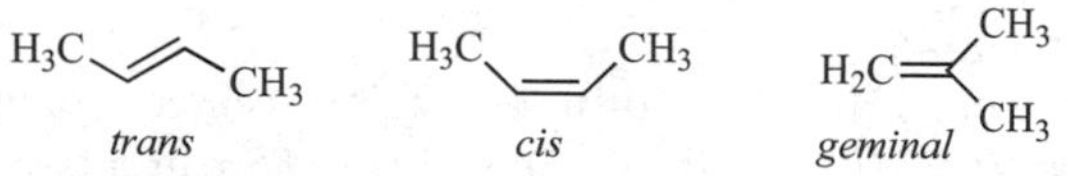

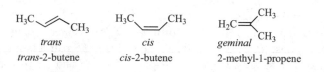

trans | *cis* | *geminal*

trans-2-butene | *cis*-2-butene | 2-methyl-1-propene

***Cis* and *Trans* on Cycloalkanes**

Cis and *trans* can be used to describe the relationships between groups on a cycloalkane. Groups on the same side of a ring (viewed from the top) are also called *cis*. If the one group is below the ring and the other above they are called *trans* (see below).

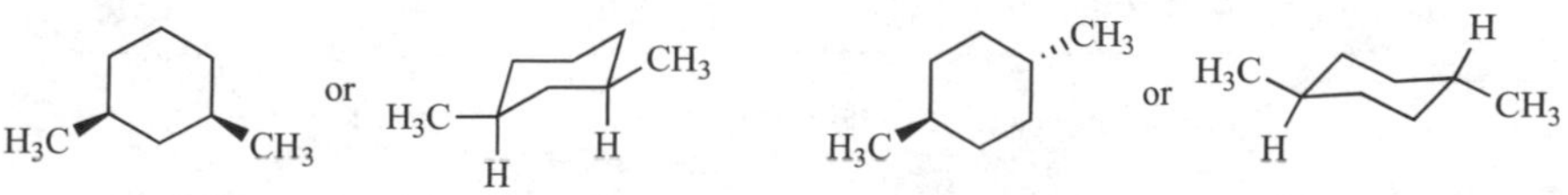

cis-1,3-dimethylcyclohexane | *trans*-1,4--dimethylcyclohexane

Cahn-Ingold-Prelog Priority Assignment

For the next set of stereochemical labels, we need to be able to assign priorities to groups. The Cahn-Ingold-Prelog sequence rules give highest atomic number the top priority. For example, in assigning priorities to the groups on the carbon of bromofluoroiodomethane, CHBrFI, top priority goes to iodine, second to bromine, third to fluorine, and lowest to hydrogen. When there is a tie, the atoms next out from the tying atoms are considered; the highest-priority atoms directly bonded to the tied atoms are compared until a difference is found. See examples below. Double bonds are counted as if they were two single bonds.

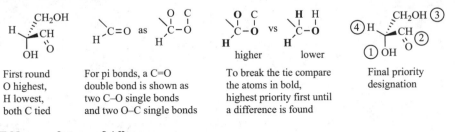

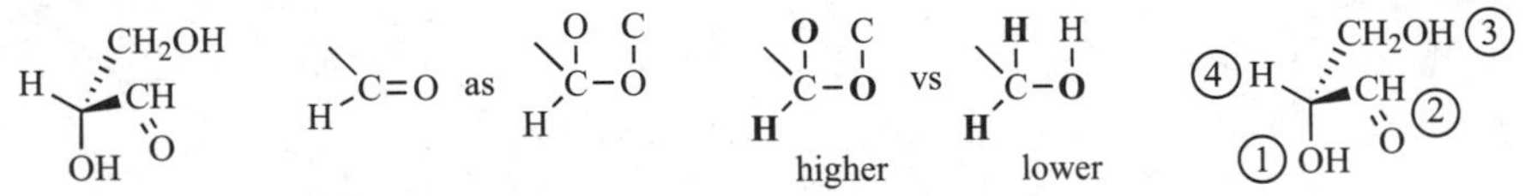
as
vs
higher
lower

First round O highest, H lowest, both C tied

For pi bonds, a C=O double bond is shown as two C–O single bonds and two O–C single bonds

To break the tie compare the atoms in bold, highest priority first until a difference is found

Final priority designation

E and Z Nomenclature of Alkenes

If the two highest-priority groups on each end of the double bond are adjacent to each other, then the stereochemistry is called *Z*; if they are across the pi bond, then the stereochemistry is called *E*.

Z-2-Bromo-2-butene

E-2-Bromo-2-butene

R and S Absolute Configuration

R and *S* are used to designate the absolute configuration about a chiral center, an atom with four different groups. With the lowest-priority group facing away to the back, if the remaining priorities go in a clockwise manner it is *R*; if they go counterclockwise it is *S*. See example below. Molecular models help in assigning *R* and *S*.

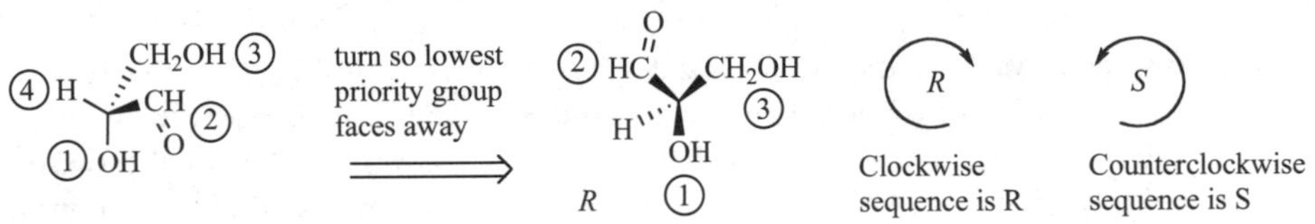
turn so lowest priority group faces away
R
R
S
Clockwise sequence is R
Counterclockwise sequence is S

Multiple Chiral Centers

If a compound has two different chiral centers, then there are four possible combinations, *RR*, *RS*, *SR*, and *SS*. The *RR* and *SS* pair are mirror images (enantiomers), and likewise so are the *RS* and *SR* pair. The *SS* and *SR* pair are stereoisomers but not mirror images, and are called diastereomers.

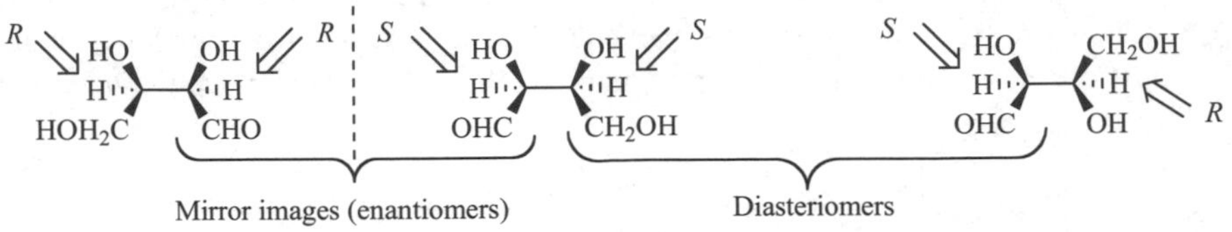
Mirror images (enantiomers)
Diasteriomers

If a compound has two identical chiral centers, then the *RS* and *SR* compounds can be superimposed; they are identical and are called a *meso* compound. *Meso* compounds can usually be recognized by having an internal mirror plane, and are not chiral.

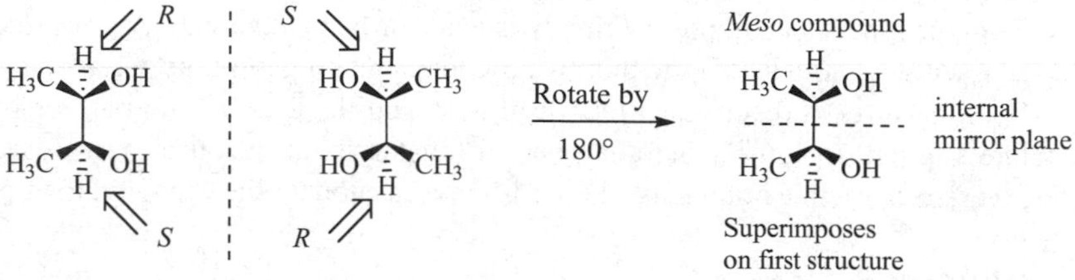
R
S
S
R
Rotate by
180°
Meso compound
internal
mirror plane
Superimposes
on first structure

D and L Relative Configuration

Relative configuration was developed before the age of crystallography and is still in use today for sugars and amino acids. If the highest numbered chiral center in a sugar has the same stereochemistry as D-glyceraldehyde, then it is a D sugar. With amino acids, if the chiral carbon bearing the amine has the same stereochemistry as L-glyceraldehyde then it is an L amino acid. If the Fischer projection is drawn with the most oxidized end to the top, then the D sugars have the OH of the highest numbered chiral center (arrow below) drawn on the right and L amino acids have the alpha carbon amine drawn to the left (arrow).

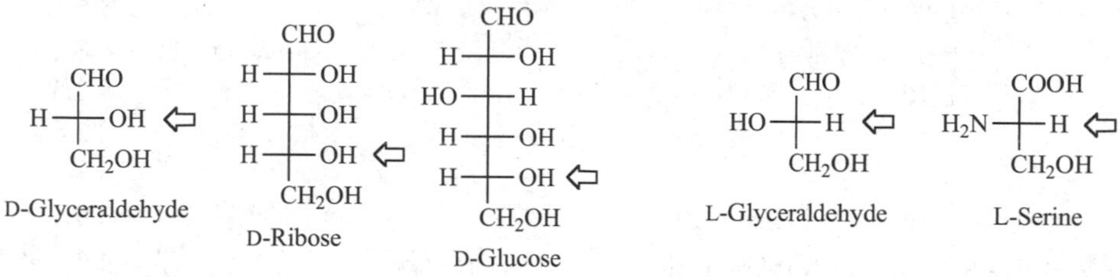
D-Glyceraldehyde
D-Ribose
D-Glucose
L-Glyceraldehyde
L-Serine

Prochiral Relationships

When, for example, an enzyme creates a chiral compound from an achiral reactant, we need a way to distinguish between the two identical groups in the reactant that will become different in the chiral product. To do this we "promote" one of the two identical groups slightly and then assign *R* or *S* to the "new" chiral center. If the group we promoted would create an *R* chiral center, then we call that promoted group pro-*R*; if it would create an *S* center, then that promoted group is pro-*S*.

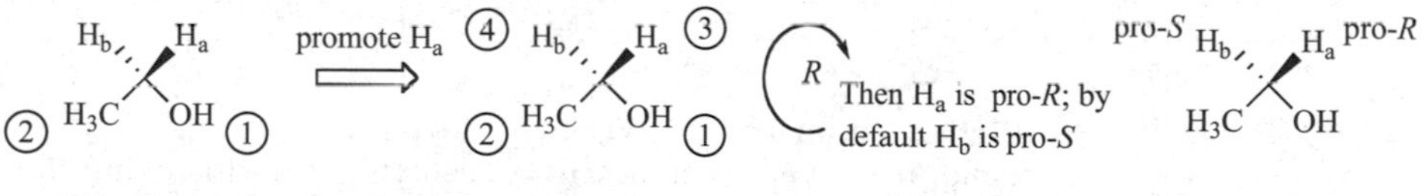
promote Ha
R
Then Ha is pro-R; by
default Hb is pro-S
pro-S
pro-R

When an enzyme shows selectivity for a face of a reactant, a carbonyl addition, for example, we need a way to label each face of the reactant. If the priorities go in a clockwise manner, then the face showing is the *re* face; if they go in a counterclockwise manner, that face is the *si* face.

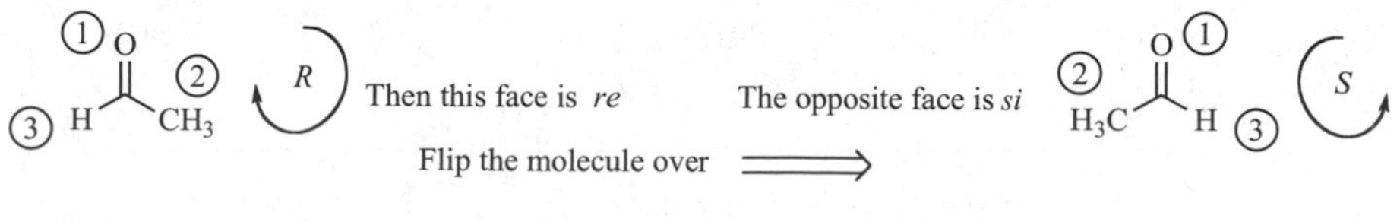
R
Then this face is re
The opposite face is si
Flip the molecule over
S

HINTS TO SELECTED PROBLEMS FROM CHAPTERS 8, 9, AND 10

The most important point to remember in classifying into sources and sinks is to **stay flexible**. Almost all electron sinks have lone pairs that can serve as sources. Lone pairs on sinks can complex with Lewis acids or be protonated. A polarized multiple bond can serve as an electron-withdrawing group, making protons on adjacent atoms acidic. Looking at both reactive partners can help you decide. If one partner is clearly a sink, like sulfuric acid, then the other partner must serve as a source, even though it may be a polarized multiple bond sink like a carbonyl; protonate the carbonyl lone pair as a source.

Chapter 8

Problem	Source Class (first step)	Sink Class (first step)	Chapter 8 section
8.1	R–M	Y=C–L	8.8.4
8.2	C=C	H–A	8.2.6
8.3	R–M	C=C–ewg	8.7.4
8.4	Z (O lone pair)	sp^3 C–L (and adjacent CH)	8.4.2
8.5	MH_4	C=Y	8.5.3
8.6	C=C	Y–L	8.3.6
8.7	R–M	sp^3 C–L	8.4.4
8.8	Z ($^-$OH lone pair)	C≡Y	8.6.1
8.9	M–H (as base)	Y=C–L (and adjacent CH)	8.8.2
8.10	Z–C=C	Y=C–L	8.8.5
8.11	C=C	H–A	8.2.6
8.12	R–M	Y–L	8.3.4
8.13	Z (Br lone pair)	A (Lewis acid)	8.9.1
8.14	Z (O lone pair)	sp^3 C–L	8.4.1 and 8.4.2
8.15	Z (carbonyl lone pair)	H–A	8.2.1
8.16	Z (N lone pair)	H–A	8.2.1
8.17	Z (Cl lone pair)	A (Lewis acid)	8.9.1

Chapter 9

Problem	Site	Base	Nucleophile	Minor Variable
9.1a	Primary	Weak	Good	
9.1b	Primary	Strong	Poor	
9.1c	Primary	Strong	Poor	
9.1d	Hindered primary	Weak	Good	
9.1e	Secondary	Weak	Poor	
9.1f	Secondary	Strong	Good	Heated
9.1g	Primary	Strong	Poor	
9.1h	Secondary	Weak	Good	
9.1i	Secondary	Weak	Good	
9.1j	Hindered secondary	Weak	Good	Heated
9.1k	Tertiary	Weak	Poor	
9.1l	Tertiary	Strong	Good	
9.1m	Tertiary	Weak	Good	

Problem	Source Class (first step)	Sink Class (first step)	Decision Type
9.2a	R–M	Y=C–L	Multiple addition
9.2b	Z–C=C	C=C–ewg	Ambident electrophile and nucleophile
9.2c	Z–C=C	Y=C–L	Ambident nucleophile
9.2d	Z (O lone pair)	*sp*^{3} C–L	Intermolecular versus intramolecular
9.2e	Z (N lone pair)	C=Y	Thermodynamic versus kinetic enolate
9.2f	Z–C=C	*sp*^{3} C–L	Ambident nucleophile
9.2g	R–M	C=C–ewg	Ambident electrophile
9.2h	Z (O lone pair)	H–A	To migrate or not
9.4a	R–M	Y=C–L	Multiple addition
9.4b	R–M	C=C–ewg	Ambident electrophile
9.4c	R–M	C=C–ewg	Ambident electrophile
9.4d	Z (O lone pair)	*sp*^{3} C–L	Substitution/elimination
9.4e	R–M	Y=C–L	Multiple addition
9.4f	Z–C=C	Y–L	Ambident nucleophile
9.9a	Z (O lone pair)	*sp*^{3} C–L	Intermolecular versus intramolecular
9.9b	R–M	C=C–ewg	Ambident electrophile
9.9c	Z–C=C	C=C–ewg	Ambident electrophile and nucleophile

Chapter 10

Problem	Source Class (for first step)	Sink Class (for first step)	First Step Pathway
10.1a	Z (O lone pair)	Y–L	pent. or S_N2
10.1b	C≡C	H–A	Ad_E3
10.1c	C=C	H–A	A_E
10.1d	M–H	C=Y (and adjacent CH)	p.t.
10.1e	Z ($^-$OH lone pair)	C=C–ewg	Ad_N
10.1f	Z ($^-$OEt lone pair)	Y=C–L (and adjacent CH)	p.t.
10.2a	Z (O lone pair)	H–A	p.t.
10.2b	R–M	C≡Y	Ad_N
10.2c	Z–C=C	Y=C–L	Ad_N
10.2d	C=C	H–A	A_E
10.2e	Z (N lone pair)	C=Y (and adjacent CH)	p.t.
10.2f	Z (carbonyl lone pair)	H–A	p.t.
10.2g	Z ($^-$OEt lone pair)	Y=C–L (and adjacent CH)	p.t.
10.2h	Z (O lone pair)	Y=C–L	Ad_N
10.2i	Z (carbonyl lone pair)	H–A	p.t.
10.3a	Z (S lone pair)	sp^3 C–L	S_N2
10.3b	Z (carbonyl lone pair)	A (Lewis acid)	A_N
10.3c	Z (P lone pair)	sp^3 C–L	S_N2
10.3d	Z (N lone pair)	Y=C–L	Ad_N

Problem	Source Class (for first step)	Sink Class (for first step)	First Step Pathway
10.3e	R–M	H–A	p.t.
10.3f	C=C	Nu–L	NuL
10.3g	Z–C=C	H–A	A_E
10.3h	Z (O lone pair)	sp^3 C–L	S_N2
10.4a	Z (N lone pair)	Y–L	S_N2
10.4b	Z (O lone pair)	H–A	p.t.
10.4c	Z (N lone pair)	H–A	p.t.
10.4d	R–M	Y=C–L	Ad_N
10.4e	Z (N lone pair)	C=Y and H–A	Ad_E3
10.4f	Z (O lone pair)	H–A	p.t.
10.4g	MH_4	Y=C–L	Ad_N
10.5a	Z–C=C–C=C	H–A	A_E
10.5b	R–M	C=Y and H–A	Ad_E3
10.5c	C=C	H–A	Ad_E3
10.5d	ArH	sp^3 C–L	S_N2
10.5e	C=C–Z	C=Y	6e
10.5f	Z (carbonyl lone pair)	sp^3 C–L	S_N2
10.6a	Z (O lone pair)	H–A	p.t.
10.6b	Z (carbonyl lone pair)	H–A	p.t.
10.6c	C=C–Z	C=C–ewg	Ad_N
10.6d	Z ($^-$OEt lone pair)	Y=C–L (and adjacent CH)	p.t.
10.7	Z (C lone pair)	C=Y and H–A	Ad_E3

A

C

D

E

F

G

H

I

J

K

L

M

P

Q

R

T